Lippincott's Illustrated Reviews: Physiology

Lippincott's Illustrated Reviews: Physiology

Robin R. Preston, Ph.D.
Formerly Associate Professor
Department of Pharmacology and Physiology
Drexel University College of Medicine
Philadelphia, Pennsylvania

Thad E. Wilson, Ph.D.
Associate Professor of Physiology and Medicine
Departments of Biomedical Sciences and Specialty Medicine
Ohio University Heritage College of Osteopathic Medicine
Athens, Ohio

Wolters Kluwer Health | Lippincott Williams & Wilkins
Philadelphia • Baltimore • New York • London
Buenos Aires • Hong Kong • Sydney • Tokyo

Acquisitions Editor: Crystal Taylor
Product Manager: Jennifer Verbiar
Development Editor: Kelly Horvath
Marketing Manager: Joy Fisher-Williams
Designer: Holly McLaughlin
Compositor: Absolute Service, Inc.

351 West Camden Street
Baltimore, MD 21201

Two Commerce Square
2001 Market Street
Philadelphia, PA 19103

Printed in China

9 8 7 6 5 4 3 2

Library of Congress Cataloging-in-Publication Data

Preston, Robin R.
Physiology / Robin R. Preston, Thad E. Wilson.
p. ; cm. — (Lippincott's illustrated reviews)
ISBN 978-1-60913-241-5 (pbk.)
I. Wilson, Thad E. II. Title. III. Series: Lippincott's illustrated reviews.
[DNLM: 1. Physiological Phenomena—Examination Questions. 2. Physiological Phenomena—Outlines. QT 18.2]

612.0076—dc23

2012016111

To purchase additional copies of this book, call our customer service department at **(800) 638-3030** or fax orders to **(301) 223-2320**. International customers should call **(301) 223-2300.**

Visit Lippincott Williams & Wilkins on the Internet: http://www.lww.com. Lippincott Williams & Wilkins customer service representatives are available from 8:30 am to 6:00 pm, EST.

CCS1015

Dedication

To Barbara and Kristen,
whose unwavering support
and encouragement have
made this book possible.

Acknowledgments

Many entertain the notion of writing a book without fully comprehending what such an endeavor entails, the authors included. We owe a debt of gratitude to a number of individuals who have helped us bring this lengthy project to fruition.

First and foremost, we thank Kelly Horvath (Development Editor) and Matt Chansky (Artist). Kelly has been a constant voice of enthusiasm since the project's inception and has patiently guided us through the various drafting stages. Kelly's lightness of heart allowed us to retain a sense of humor even when deadlines were looming. This book and its potential success will owe much to her insights, suggestions, and literary skills. The "look" of *LIR Physiology* is thanks to the artistic prowess of Matt Chansky. Matt ran with our ideas for art and made them a reality, working with us closely to find ways to animate ions, make transporters spin, and add a spark of excitation to membranes. We are also grateful to the diligence and compositing skills of Harold Medina and his team at Absolute Service, Inc. Harold cheerfully implemented multiple 11th-hour changes and thereby allowed us to make significant improvements to the content.

The original outline for *LIR Physiology* was compiled by Pamela Champe, Ph.D. (in memorium) and Richard Harvey, Ph.D. We are grateful to Richard for his vision for the LIR series and continued endorsement of this book and its authors. We are also indebted to Crystal Taylor (Acquisitions Editor, LWW) and for her ongoing support and to Jenn Verbiar (Product Manager, LWW) for her help during the early development and production phases.

Our sincere thanks go to the numerous individuals who have read and made suggestions for improvements on drafts of one or more chapters. Chief among these are Kristen Metzler-Wilson, P.T., Ph.D. (Lebanon Valley College), who read and commented on all chapters during various stages of development, and Barbara Mroz, M.D. (The Southeast Permanente Medical Group) for her contributions and for editing most of the clinical material.

LWW solicited reviews from many faculty and students. Special thanks go to Sandra K. Leeper-Woodford, Ph.D. (Mercer University School of Medicine), who read and reviewed most chapters and whose insights greatly improved the text. R. Tyler Morris, Ph.D. (Vanderbilt University) also offered many helpful suggestions.

We also extend thanks to faculty colleagues Brian Clark, Ph.D. (Ohio University Heritage College of Osteopathic Medicine, OUHCOM); Leslie Consent, Ph.D. (OUHCOM); Scott Davis, Ph.D. (Southern Methodist University); John Howell, Ph.D. (OUHCOM); Richard Klabunde, Ph.D. (OUHCOM); Anne Loucks, Ph.D. (Ohio University); and to medical and physician assistant students Micah Boehr (OMS III), Jacqueline Fisher (OMS III), Derek Gross (OMS III), Aiwane Iboaya (OSM II), Andrew Jurovick (OMS I), Sarah Mann (PAS I), Christa Tomc (OMS IV), and Jeffrey Turner (OMS III).

Preface

Take a look in the mirror. The image that stares back is familiar, its distinctive features identifying you as "you" to others. However, a face is just a front for the more than 10 trillion cells that make up a human body.

Lean a little closer.

The contours of your face are sculpted by bone, padded with fat, and covered with a continuous sheet of cells called skin. Eyebrows and facial hairs are the product of specialized secretory glands (hair follicles). Your eye movements are coordinated by delicate muscles that contract in response to orders from your brain. The pulse at your temple reflects a wave of pressure generated by a heart beating within your thorax. Lower down, your stomach grinds your most recent meal while two kidneys strain your blood. Virtually all of this activity goes unnoticed until it goes wrong.

LIR Physiology is the story of who we are, how we live, and, ultimately, how we die. It follows the organization of the human body, each unit treating a different organ system and considering its role in the life of the individual. Physiology texts typically take a "macro-to-micro" approach, their descriptions of organs following the history of human physiologic discovery (gross anatomy, microanatomy, cellular biology, and, finally, molecular biology). We begin most units by identifying organ function and then showing how cells and tissues are designed to fulfill that function. Although physiologic design is shaped by natural selection not by purpose, this teleologic approach can help us understand why cells and organs are structured the way they are. Understanding the "why" aids retention and gives future health care providers a powerful tool for anticipating how and understanding why disease processes present clinically in the way that they do.

What does LIR Physiology cover? Physiology is a burgeoning discipline, which no single text can cover exhaustively. We used three principal guides to help us decide what material to include:

- Topics currently being tested by the U.S. Medical Licensing Examiners
- Learning objectives that many medical school physiology courses cover (American Physiological Society and the Association of Chairs of Departments of Physiology, 2006)
- Topics covered in *BRS Physiology*, LWW's popular board review book

Who should use this book? *LIR Physiology* is intended to help medical students preparing for their licensing examinations, but the material is presented with a clarity and level of detail that also suits it for a primary course text by any of the allied health disciplines as well as a reference for clinicians.

Format: *LIR Physiology* follows a lecture-note format, with minimal introductions, history, or discussions of ongoing research—the chapters quickly cut to the chase in a narrative form customized for fast assimilation. The subheadings break the presentation in easy-to-absorb paragraphs that are appropriate to skim for review, yet sufficiently detailed to instruct a student who may be new to or unsure of a topic. The writing style is engaging yet succinct, rendering complex topics accessible and memorable.

Art: The text is also heavily illustrated with step-by-step guides to help the visual learner and for ease of review by students preparing for examinations. Art and text combine seamlessly to tell the story of physiology in a completely new way. More than 600 original and energetic full-color line drawings are supplemented by abundant clinical images that

together illustrate physiology with a dynamism belying their two-dimensionality. Legends are deliberately minimal to allow the art itself to "speak." Step-by-step dialogue boxes *guide* the viewer through physiologic processes.

Features: *LIR Physiology* incorporates multiple features to facilitate comprehension of material:

- **Real-world examples:** Physiologic concepts are notoriously difficult to grasp, so we have used real-world examples wherever possible to aid comprehension.
- **Clinical Applications:** All chapters include Clinical Applications—many with accompanying clinical images—that show how physiology gone awry can present clinically.
- **Equations boxes:** Real numbers are run through tricky equations for equilibrium potential, alveolar—arterial oxygen difference, and renal clearance, and featured in yellow margin boxes to show students examples they might encounter in practice.
- **Consistency:** Cellular physiology can be overwhelming in its details, especially where transport physiology is concerned. We have kept details in artwork to a bare minimum and consistently use the same colors to denote different ion species throughout the text:
  - Sodium = red
  - Calcium = indigo
  - Potassium = purple
  - Anions (chloride and bicarbonate) = green
  - Acid = orange

  Readers will quickly become familiar with visual cues and spend less time reading labels. Easy-to-follow-and-recall flow charts and concept maps are also widely used.
- **Infolinks:** These cross references among the LIR series provide resources for students to delve more deeply into related topics across several disciplines, including biochemistry, pharmacology, microbiology, neuroscience, immunology, and cell and molecular biology.
- **Cross references:** Linking topics across chapters, intravolume cross references in an easy-to-locate format specify section number to the nearest heading level, for example, (see 25·III·B). Chapter number and section level are provided at the head of every page for ease of location.
- **Practice questions:** Each unit is accompanied by several pages of sample USMLE-style questions that students can use to self-assess their physiologic knowledge. These often integrative questions test for understanding of physiologic concepts and the ability to draw connections among multiple organ systems rather than mere recall of minor details. Choosing "best" answers are the goal insofar as "correct" answers suggest absolute states that rarely—if ever—exist in physiology. Additional questions with a simpler, more traditional textbook format can be found online at thePoint. These varying styles give readers graduated levels with which to increasingly challenge themselves.

Bonus material: A companion Web site on thePoint provides additional resources, including an interactive board-style (i.e., USMLE and COMLEX) question bank, which includes full answer explanations. Students using LWW's *BRS Physiology* can also download a map cross-referencing *BRS Physiology* to *LIR Physiology*, which allows them to quickly locate a more complete explanation of a difficult concept than that found in the review series.

Comments? Our current understanding of physiologic mechanisms is evolving constantly in the light of new research findings. Subsequent editions of *LIR Physiology* will be updated to take into account new findings and reader feedback. If you have any suggestions for improvement or other comments on content or the LIR approach, you are welcome to submit them to the publisher directly at http://www.lww.com or contact the authors by e-mail at LIRphysiology@gmail.com.

Contents

UNIT I

Principles of Physiologic Function

Cell and Membrane Physiology

1

I. OVERVIEW

The human body comprises several distinct organs, each of which has a unique role in supporting the life and well-being of the individual. Organs are, in turn, composed of tissues. Tissues are collections of cells specialized to perform specific tasks that are required of the organ. Although cells from any two organs may appear strikingly dissimilar at the microscopic level (compare the shape of a red blood cell with the branching structure of a nerve cell's dendritic tree, for example, as in Figure 1.1), morphology can be misleading because it masks a set of common principles in design and function that apply to all cells. All cells are enclosed within a membrane that separates the inside of the cell from the outside. This barrier allows the cells to create an internal environment that is optimized to support the biochemical reactions required for normal function. The composition of this internal environment varies little from cell to cell. Most cells also contain an identical set of membrane-bound organelles: **nuclei**, **endoplasmic reticulum** (**ER**), **lysosomes**, **Golgi apparatuses**, **mitochondria**. Specialization of cell and organ function is usually achieved by adding a novel organelle or structure, or by altering the mix of membrane proteins that provide pathways for ions and other solutes to move across the barrier. This chapter reviews some common principles of molecular and cellular function that will serve as a foundation for later discussions of how the various organs contribute to maintaining normal bodily function.

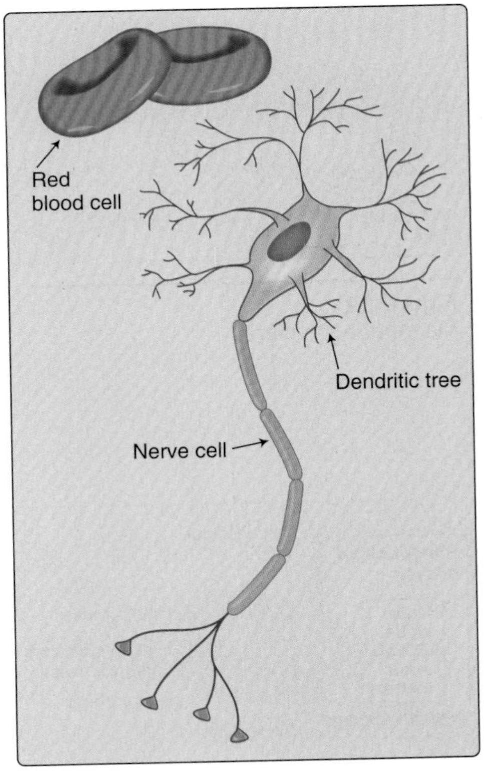

Figure 1.1
Differences in cell morphology.

II. CELLULAR ENVIRONMENT

Cells are bathed in an **extracellular fluid** (**ECF**) that contains ionized sodium (Na^+), potassium (K^+), magnesium (Mg^{2+}), chloride (Cl^-), phosphate (PO_4^{3-}), bicarbonate (HCO_3^-), glucose, and small amounts of protein (Table 1.1). It also contains around 2 mmol free calcium (Ca^{2+}). Ca^{2+} is essential to life, but many of the biochemical reactions required of cells can only occur if free Ca^{2+} concentrations are lowered ten-thousandfold, to around 10^{-7} mol. Thus, cells erect a barrier that is impermeable to ions

Table 1.1: Extracellular and Intracellular Fluid Composition

Solute	ECF	ICF
Na^+	145	12
K^+	4	120
Ca^{2+}	2.5	0.0001
Mg^{2+}	1	0.5
Cl^-	110	15
HCO_3^-	24	12
Phosphates	0.8	0.7
Glucose	5	<1
Proteins (g/dL)	1	30
pH	7.4	7.2

Values are approximate and represent free concentrations under normal metabolic conditions. All values (with the exception of protein concentration and pH) are given in mmol/L. ECF = extracellular fluid; ICF = intracellular fluid.

(the **plasma membrane**) to separate **intracellular fluid** ([**ICF**] or **cytosol**) from ECF and then selectively modify the composition of ICF to facilitate the biochemical reactions that sustain life. ICF is characterized by low Ca^{2+}, Na^+, and Cl^- concentrations compared with ECF, whereas the K^+ concentration is increased. Cells also contain more free protein than does the ECF, and the pH of ICF is slightly more acidic.

III. MEMBRANE COMPOSITION

Membranes comprise lipid and protein (Figure 1.2). Lipids form the core of all membranes. Lipids are ideally suited to a barrier function because they are **hydrophobic**: They repel water and anything dissolved in it (**hydrophilic** molecules). Proteins allow cells to interact with and communicate with each other, and they provide pathways that allow water and hydrophilic molecules to cross the lipid core.

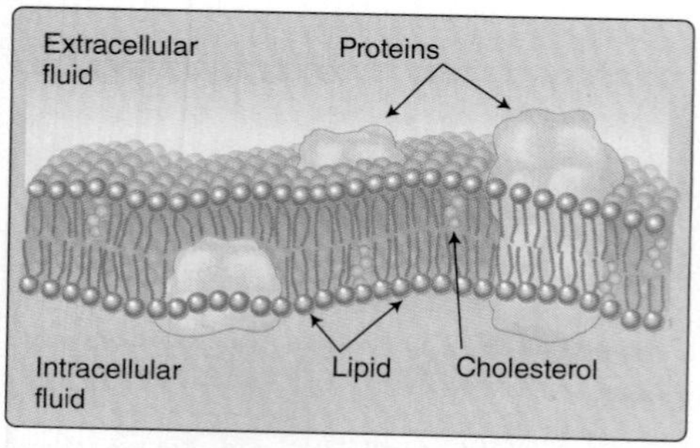

Figure 1.2
Membrane structure.

A. Lipids

Membranes contain three predominant types of lipid: **phospholipids**, **cholesterol**, and **glycolipids**. All are **amphipathic** in nature, meaning that they have a polar (hydrophilic) region and a nonpolar (hydrophobic) region. The polar region is referred to as the **head group**. The hydrophobic region is usually composed of fatty acid **"tails"** of variable length. When the membrane is assembled, the lipids naturally gather into a continuous bilayer (Figure 1.3). The polar head groups gather at the internal and external surfaces where the two layers interface with ICF and ECF, respectively. The hydrophobic tail groups dangle down from the head groups to form the fatty membrane core. Although the two halves of the bilayer are closely apposed, there is no significant lipid exchange between the two membrane leaflets.

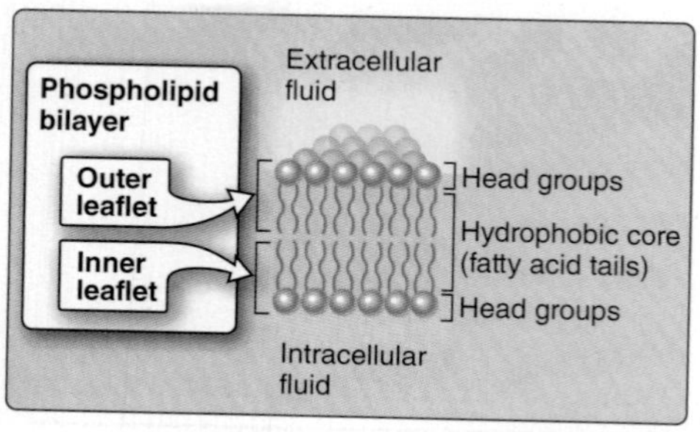

Figure 1.3
Membrane lipid bilayer.

1. **Phospholipids:** Phospholipids are the most common membrane lipid type. Phospholipids comprise a fatty acid tail coupled via glycerol to a head group that contains phosphate and an attached alcohol. Dominant phospholipids include **phosphatidylserine**, **phosphatidylethanolamine**, **phosphatidylcholine**, **phosphatidylinositol**, and **phosphatidylglycerol**. **Sphingomyelin** is a related phospholipid in which glycerol has been replaced by sphingosine. The alcohol group in sphingomyelin is choline.

2. **Cholesterol:** Cholesterol is the second most common membrane lipid. It is hydrophobic but contains a polar hydroxyl group that draws it to the bilayer's outer surface, where it nestles between adjacent phospholipids (Figure 1.4). Between the hydroxyl group and the hydrocarbon tail is a steroid nucleus. The four steroid carbon rings make it relatively inflexible, so adding cholesterol to a membrane reduces its fluidity and makes it stronger and more rigid.

3. **Glycolipids:** The outer leaflet of the bilayer contains glycolipids, a minor but physiologically significant lipid type comprising a fatty acid tail coupled via sphingosine to a carbohydrate head group. The glycolipids create a carbohydrate cell coat that is involved in cell-to-cell interactions and that conveys antigenicity.

B. Proteins

The membrane's lipid core seals the cell in an envelope across which only lipid-soluble materials, such as O_2, CO_2, and alcohol can cross. Cells exist in an aqueous world, however, and most of the molecules that they need to thrive are hydrophilic and cannot penetrate the lipid core. Thus, the surface (**plasma**) membrane also contains proteins whose function is to help ions and other charged molecules across the lipid barrier. Membrane proteins also allow for intercellular communication and provide cells with sensory information about the external environment. Proteins are grouped on the basis whether they localize to the membrane surface (**peripheral**) or are **integral** to the lipid bilayer (Figure 1.5).

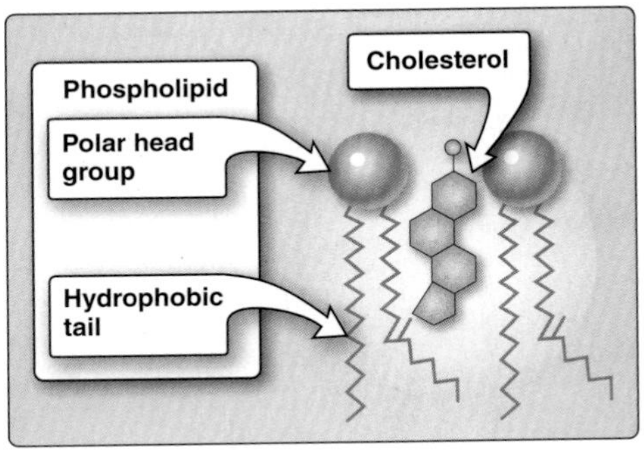

Figure 1.4
Cholesterol location with the membrane.

1. **Peripheral:** Peripheral proteins are found on the membrane surface. Their link to the membrane is relatively weak and, thus, they can easily be washed free using simple salt solutions. Peripheral proteins associate with both the intracellular and extracellular plasma membrane surfaces.

   a. **Intracellular:** Proteins that localize to the intracellular surface include many enzymes; regulatory subunits of ion channels, receptors, and transporters; and proteins involved in vesicle trafficking and membrane fusion as well as proteins that tether the membrane to a dense network of fibrils lying just beneath its inner surface. The network is composed of spectrin, actin, ankrin, and several other molecules that link together to form a **subcortical cytoskeleton** (see Figure 1.5).

   b. **Extracellular:** Proteins located on the extracellular surface include enzymes, antigens, and adhesion molecules. Many peripheral proteins are attached to the membrane via **glycophosphatidylinositol** ([**GPI**] a glycosylated phospholipid) and are known collectively as **GPI-anchored proteins**.

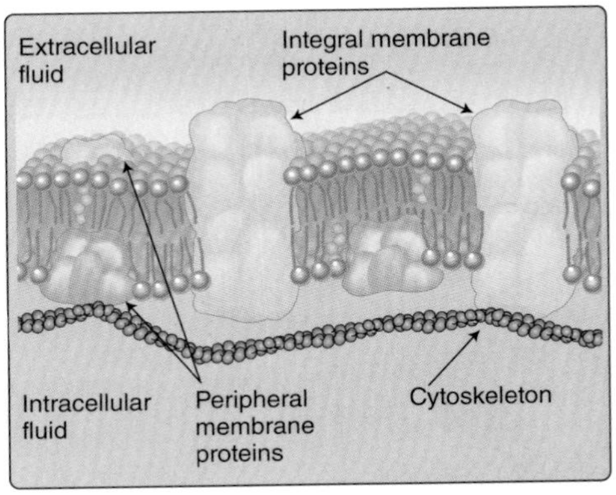

Figure 1.5
Membrane proteins.

2. **Integral:** Integral membrane proteins penetrate the lipid core. They are anchored by covalent bonds to surrounding structures

Clinical Application 1.1: Paroxysmal Nocturnal Hemoglobinuria

Paroxysmal nocturnal hemoglobinuria (PNH) is a rare, inherited disease caused by a defect in the gene that encodes phosphatidylinositol glycan A. This protein is required for synthesis of the glycophosphatidylinositol anchor used to tether peripheral proteins to the outside of the cell membrane. The gene defect prevents cells from expressing proteins that normally protect them from the immune system. The nighttime appearance of hemoglobin in urine (hemoglobinuria) reflects red blood cell lysis by immune complement. Patients typically manifest symptoms associated with anemia. PNH is associated with a significant risk of morbidity, in part, because patients are prone to thrombotic events. The reason for the increased incidence of thrombosis is not well delineated.

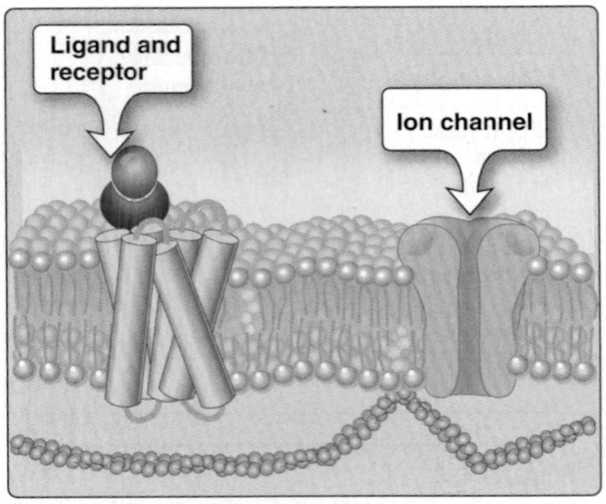

Figure 1.6
Membrane-spanning proteins.

and can only be removed by experimentally treating the membrane with a detergent. Some integral proteins may remain localized to one or the other of the two membrane leaflets without actually traversing its width. Others may weave across the membrane many times (**transmembrane proteins**) as shown in Figure 1.6. Examples include various classes of **ion channels**, **transporters**, and **receptors**.

IV. DIFFUSION

Movement across a membrane requires a motive force. Most substances cross the plasma membrane by **diffusion**, their movement driven by a transmembrane concentration gradient. When the concentration difference across a membrane is unfavorable, however, then the cell must expend energy to force movement "uphill" against the concentration gradient (**active transport**).

A. Simple diffusion

Consider a container filled with water and divided into two compartments by a pure lipid membrane (Figure 1.7). Blue dye is now dropped into the container at left. Initially, the dye remains concentrated and restricted to its small entry area, but molecules of gas, water, or anything dissolved in water are in constant thermal motion. These movements cause the dye molecules to distribute randomly throughout the entire chamber, and the water eventually becomes a uniform color, albeit lighter than the original drop. The example shown in Figure 1.7 assumes that the dye is unable to cross the membrane, so the chamber on the right remains clear, even though the difference in dye concentrations across the barrier is very high. However, if the dye is lipid soluble or is provided with a pathway (a protein) that allows it to cross the barrier, diffusion will carry the molecules into the second chamber, and the entire tank will turn blue (Figure 1.8).

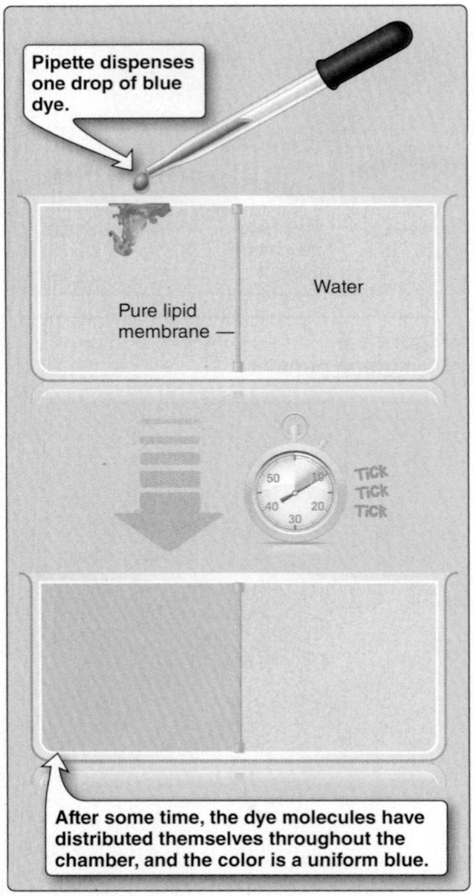

Figure 1.7
Simple diffusion in water.

B. Fick law

The rate at which molecules such as blue dye cross membranes can be determined using a simplified version of the **Fick law**:

$$J = P \times A\,(C_1 - C_2)$$

where J is diffusion rate (in mmol/s), P is a **permeability coefficient**, A is **membrane surface area** (cm^2), and C_1 and C_2 are dye concentrations (mmol/L) in compartments 1 and 2, respectively. The permeability coefficient takes into account a molecule's **diffusion coefficient**, **partition coefficient**, and the **thickness** of the barrier that it must traverse.

1. **Diffusion coefficient:** Diffusion rates increase when a molecule's velocity increases, which is, in turn, determined by its **diffusion coefficient**. The coefficient is proportional to temperature and inversely proportional to molecular radius and the viscosity of the medium through which it diffuses. In practice, small molecules diffuse quickly through warm water, whereas large molecules diffuse very slowly through cold, viscous solutions.

2. **Partition coefficient:** Lipid-soluble molecules, such as fats, alcohols, and some anesthetics can cross the membrane by dissolving in its lipid core, and they have a high partition coefficient. Conversely, ions such as Na^+ and Ca^{2+} are repelled by lipids and have a very low partition coefficient. A molecule's partition coefficient is determined by measuring its solubility in oil compared with water.

3. **Distance:** Net diffusion rate slows when molecules have to traverse thick membranes compared with thin ones. The practical consequences of this relationship can been seen in the lungs (see 22·II·C) and fetal placenta (see 37·III·B), organs designed to maximize diffusional rates by minimizing diffusional distance between two compartments.

4. **Surface area:** Increasing the surface area available for diffusion also increases the rate of diffusion. This relationship is used to practical advantage in several organs. The lungs comprise 300,000,000 small sacs (**alveoli**) that have a combined surface area of ~80 m^2 that allows for efficient O_2 and CO_2 exchange between blood and the atmosphere (see 22·II·C). The lining of the small intestine is folded into fingerlike **villi** (Figure 1.9), and the villi sprout **microvilli** that, together, create a combined surface area of ~200 m^2 (see 31·II). Surface area amplification allows for efficient absorption of water and nutrients from the gut lumen. Efficient exchange of nutrients and metabolic waste products between blood and tissues is ensured by a vast network of small vessels (**capillaries**) whose combined surface area exceeds 500 m^2 (see 19·II·C).

5. **Concentration gradient:** The rate at which molecules diffuse across a membrane is directly proportional to the concentration difference between the two sides of the membrane. In the example shown in Figure 1.8, the concentration gradient (and, thus, the dye diffusion rate) between the two compartments is high initially, but the rate slows and eventually ceases as the gradient dissipates and the two sides equilibrate. Note that thermal motion causes the dye molecules to continue moving back and forth between the two compartments at equilibrium, but net movement between the two is zero. If there were a way of removing dye continually from the chamber on the right, the concentration gradient and diffusion rate would remain high (we would also have to keep adding dye to the left chamber to compensate for movement across the barrier).

O_2 movement between the atmosphere and the pulmonary circulation occurs by simple diffusion, driven by an air–blood O_2 concentration gradient. Blood carries away O_2 as fast as it is absorbed, and breathing movements constantly renew the O_2 content of the lungs, thereby maintaining a favorable concentration gradient across the air–blood interface (see 23·V).

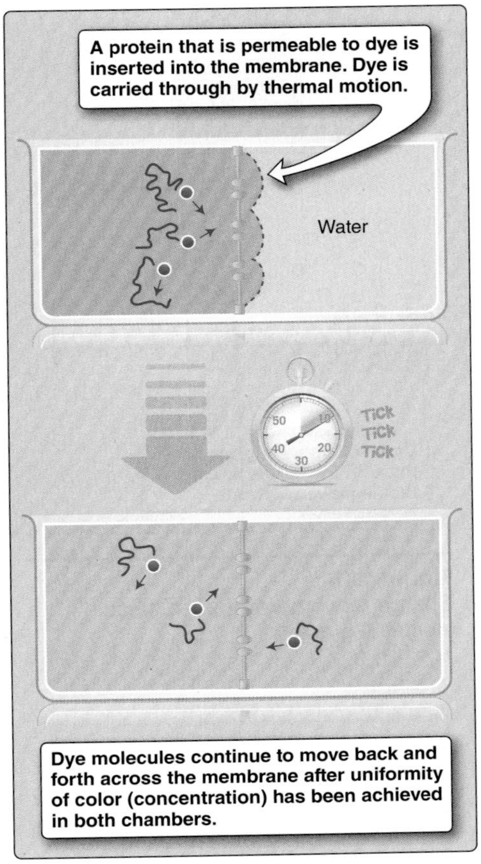

A protein that is permeable to dye is inserted into the membrane. Dye is carried through by thermal motion.
Water
50
10
40
20
30
Tick
Tick
Tick
Dye molecules continue to move back and forth across the membrane after uniformity of color (concentration) has been achieved in both chambers.

Figure 1.8
Diffusion through a lipid bilayer.

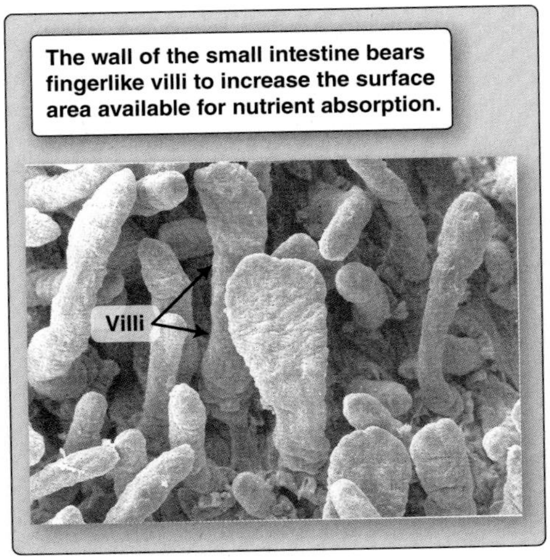

Figure 1.9
Intestinal villi.

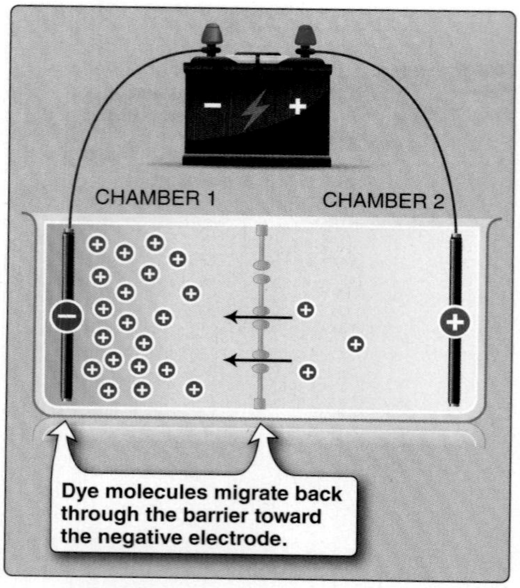

Figure 1.10
Charge movement induced by an electrical gradient.

C. Charged molecules

The prior discussion assumes that the dye is uncharged. The same basic principles apply to diffusion of a charged molecule (an **electrolyte**), but electrolytes are also influenced by electrical gradients. Positively charged ions such as Na^+, K^+, Ca^{2+}, and Mg^{2+} (**cations**) and negatively charged ions such as Cl^- and HCO_3^- (**anions**) are attracted to and will move toward their charge opposites. Thus, if the dye in Figure 1.8 carries a positive charge, and an electrical gradient is imposed across the container, dye molecules will move back through the membrane toward the negative electrode (Figure 1.10). The electrical gradient in Figure 1.10 was generated using a battery, but the same effect can be achieved by adding membrane-impermeant anions to Chamber 1. If Chamber 1 is filled with cations, the dye molecules will be repelled by the positive (like) charge and will accumulate in Chamber 2 (Figure 1.11). Note that the *electrical* gradient causes the dye *concentration* gradient between the two chambers to reform. Dye molecules will continue migrating from Chamber 2 back to Chamber 1 until the concentration gradient becomes so large that it equals and opposes the electrical gradient, at which point an **electrochemical equilibrium** has been established. As discussed in Chapter 2, most cells actively expel Na^+ ions to create an electrical gradient across their membranes. They then use the power of the combined electrical and chemical gradient (the **electrochemical gradient**) to move ions and other small molecules (e.g., glucose) across their membranes and for electrical signaling.

V. PORES, CHANNELS, AND CARRIERS

Small, nonpolar molecules (e.g., O_2 and CO_2) diffuse across membranes rapidly, and they require no specialized pathway. Most molecules common to the ICF and ECF are charged, however, meaning that they require assistance from a **pore**, **channel**, or **carrier** protein to pass through the membrane's lipid core.

A. Pores

Pores are integral membrane proteins containing unregulated, water-filled passages that allow ions and other small molecules to cross the membrane. Pores are relatively uncommon in higher organisms because they are always open and can support very high transit rates (Table 1.2). Unregulated holes in the lipid barrier potentially

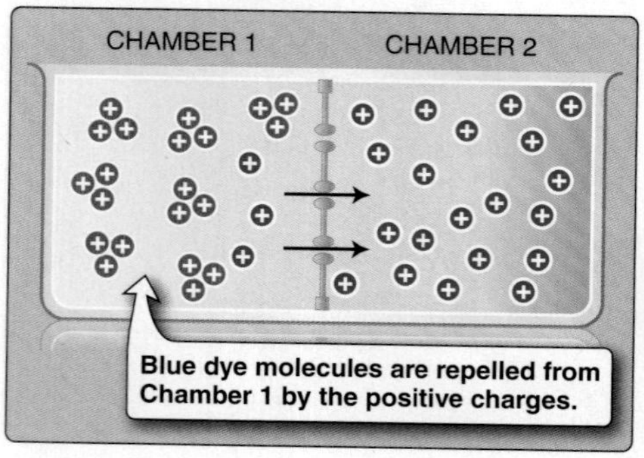

Figure 1.11
Repellent effects of like charges.

Table 1.2: Approximate Transit Rates for Pores, Channels, and Carriers

Pathway	Example	Molecule(s) Moved	Transit Rate (Number/s)
Pores	Aquaporin-1	H_2O	3×10^9
Channels	Na^+ ClC1	Na^+ Cl^-	10^8 10^6
Carriers	Na^+-K^+ ATPase	Na^+, K^+	3×10^2

Clinical Application 1.2: *Staphylococcus aureus*

Staphylococcus aureus is a skin-borne bacterium that is a leading cause of community- and hospital-acquired, bloodborne "staph" infections (bacteremia). The incidence of such infections has been rising steadily in recent decades and is of growing concern because of the high associated mortality rates and the increasing prevalence of antibiotic-resistant strains such as MRSA (methicillin-resistant *Staphylococcus aureus*).[1] The bacterium's lethality is due to its producing a toxin (α-hemolysin) that kills blood cells. The toxin targets the plasma membrane, where toxin monomers assemble into a multimeric protein with an unregulated pore at its center. The cell loses control of its internal environment as a result. The ion gradients dissipate, and the cell swells and then blisters and ruptures, releasing nutrients that are believed to nourish the bacterium as it proliferates.

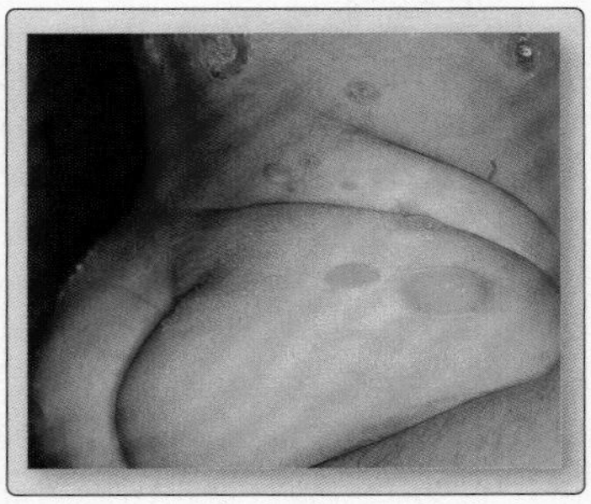

Skin lesions caused by *Staphylococcus aureus*.

can kill cells by allowing valuable cytoplasmic constituents to escape and Ca^{2+} to flood into the cell from the ECF. **Aquaporin** (**AQP**) is a ubiquitous water-selective pore. There are 13 known family members (AQP0–AQP12), three of which are expressed widely throughout the body (AQP1, AQP3, and AQP4). AQP is found wherever there is a need to move water across membranes. AQPs play a critical role in regulating water recovery from the renal tubule (see 27·V·C), for example, but they are also required for lens transparency in the eye (AQP0), keeping skin moist (AQP3), and mediating brain edema following insult (AQP4). Because AQP is always open, cells must regulate their water permeability by adding or removing AQP from the membrane.

B. Channels

Ion channels are transmembrane proteins that assemble so as to create one or more water-filled passages across the membrane. Channels differ from pores in that the permeability pathways are revealed transiently (**channel opening**) in response to a membrane-potential change, neurotransmitter binding, or other stimulus, thereby allowing small ions (e.g., Na^+, K^+, Ca^{2+}, and Cl^-) to enter and traverse the lipid core (Figure 1.12). Ion movement is driven by simple diffusion and powered by the transmembrane electrochemical gradient. Ions are forced to interact with the channel pore briefly so that their chemical nature and suitability for passage can be established (a **selectivity filter**), but the rate at which ions traverse the membrane via channels can be as high as 10^8 per second (see Table 1.2). All cells express ion channels, and there are numerous types, including the voltage-gated Na^+ channel that mediates nerve action potentials and voltage-gated Ca^{2+} channels that mediate muscle contraction. Ion channels are discussed in detail in Chapter 2.

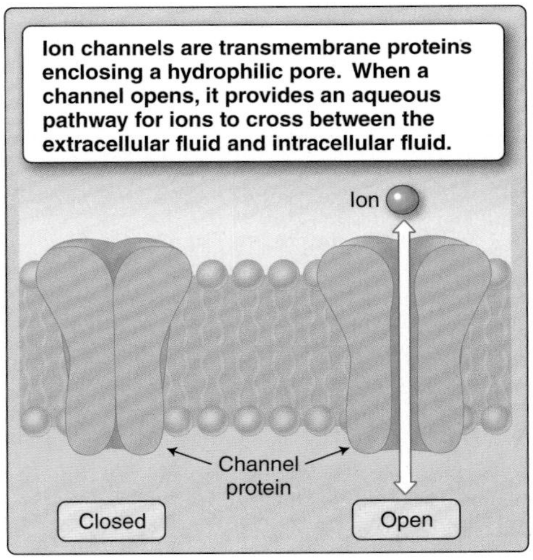
Ion channels are transmembrane proteins enclosing a hydrophilic pore. When a channel opens, it provides an aqueous pathway for ions to cross between the extracellular fluid and intracellular fluid.
Ion
Channel protein
Closed
Open

Figure 1.12
Ion channel opening.

INFO LINK [1]For more information on *Staphylococcus aureus* and MRSA, see *LIR Microbiology*, 3e, Chapter 8.

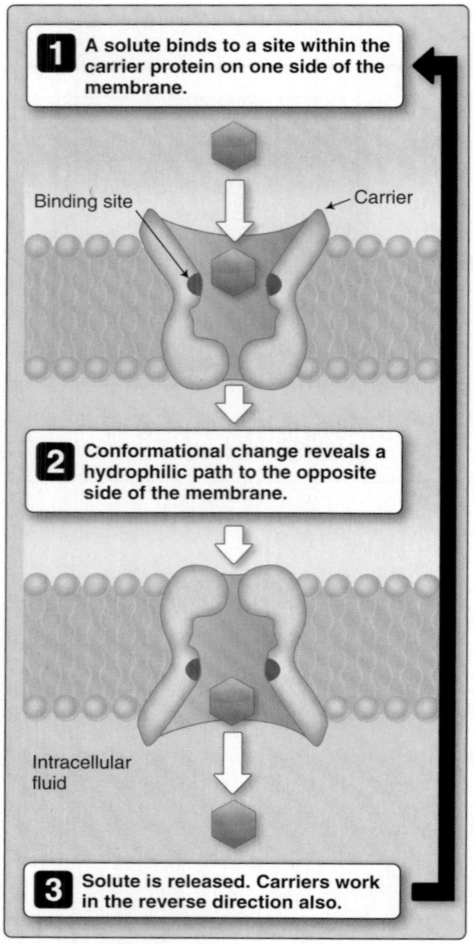

Figure 1.13
Model for transport by a carrier protein.

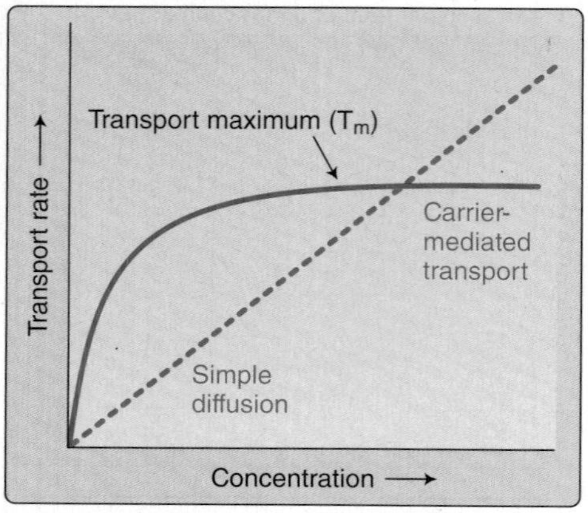

Figure 1.14
Carrier saturation kinetics.

C. Carriers

Larger solutes, such as sugars and amino acids, are typically assisted across the membrane by carriers. Carriers can be considered enzymes that catalyze movement rather than a biochemical reaction. Translocation involves a binding step, which slows transport rate considerably compared with pores and channels (see Table 1.2). There are three principal carrier modes: **facilitated diffusion**, **primary active transport**, and **secondary active transport**.

1. **Transport kinetics:** Carriers, like enzymes, show substrate specificity, saturation kinetics (Michaelis-Menten kinetics), and susceptibility to competition. A general scheme for carrier-mediated transport envisions a solute-binding step, a change in carrier conformation that reveals a conduit through which the solute may pass, and then release on the opposite side of the membrane (Figure 1.13). When solute concentrations are low, carrier-mediated transport is more efficient than simple diffusion, but a finite number of solute binding sites means that a carrier can saturate when substrate concentrations are high (Figure 1.14). The transport rate at which saturation occurs is known as the **transport maximum** (**T_m**) and is the functional equivalent of V_{max} that defines maximal reaction velocity catalyzed by an enzyme.[1]

2. **Facilitated diffusion:** The simplest carriers use electrochemical gradients as a motive force (facilitated diffusion) as shown in Figure 1.15A. They simply provide a selective pathway by which organic solutes, such as glucose, organic acids, and urea, can move across the membrane down their electrochemical gradients. The binding step ensures selectivity of passage. Common examples of such carriers includes the GLUT family of glucose transporters and the renal tubule urea transporter (see 27·V·D). The GLUT1 transporter is ubiquitous and provides a principal pathway by which all cells take up glucose. GLUT4 is an insulin-regulated glucose transporter expressed primarily in adipose tissue and muscle.

3. **Primary active transport:** Moving a solute uphill against its electrochemical gradient requires energy. **Primary active transporters** are **ATPases** that move or **"pump"** solutes across membranes by hydrolyzing adenosine triphosphate (ATP) as shown in Figure 1.15B. There are three main types of pump, all related P-type ATPase family members: a **Na^+-K^+ ATPase**, a group of **Ca^{2+} ATPases**, and a **H^+-K^+ ATPase**.

   a. **Na^+-K^+ ATPase**: The Na^+-K^+ ATPase (**Na^+-K^+ exchanger** or **Na^+-K^+ pump**) is common to all cells and uses the energy of a single ATP molecule to transport three Na^+ out of the cell, while simultaneously bringing two K^+ back from the ECF. Movement of both ions occurs uphill against their respective electrochemical gradients. The physiologic importance of the Na^+-K^+ ATPase cannot be overstated. The Na^+ and K^+ gradients it establishes permit electrical signaling in neurons and

[1]For further discussion of enzymatic maximal velocity, see *LIR Biochemistry*, 5e, p. 56.

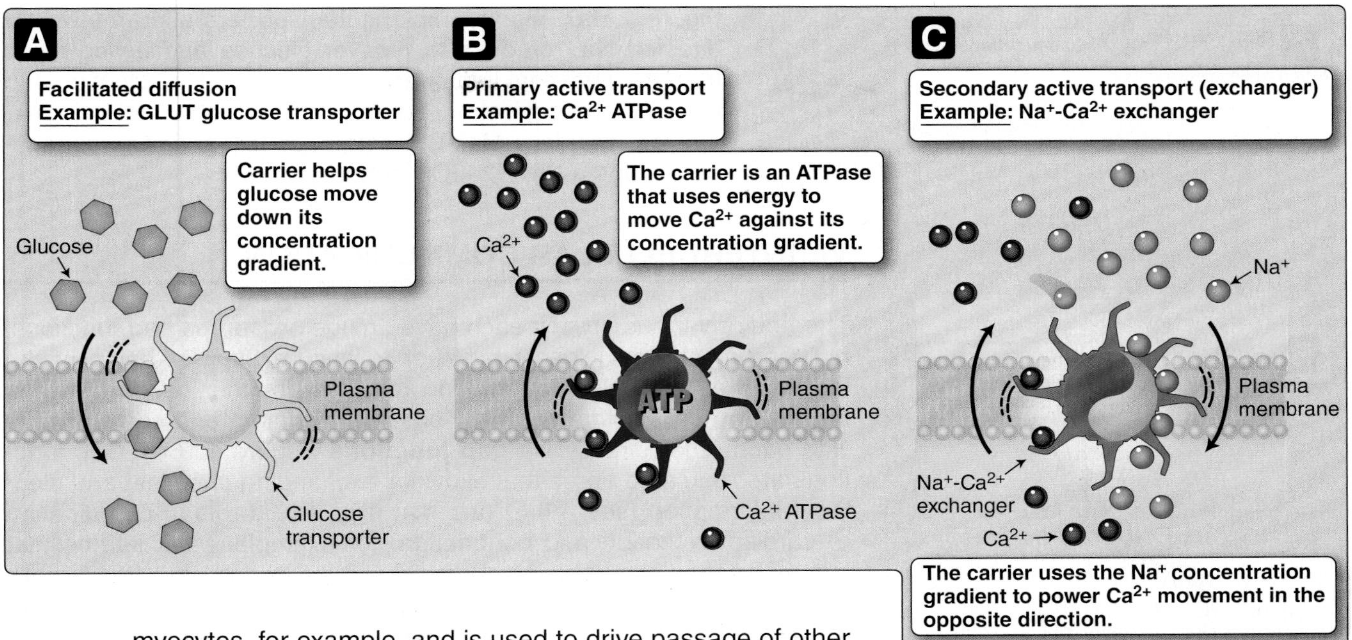
A
Facilitated diffusion
Example: GLUT glucose transporter
Carrier helps glucose move down its concentration gradient.
Glucose
Plasma membrane
Glucose transporter
B
Primary active transport
Example: Ca2+ ATPase
The carrier is an ATPase that uses energy to move Ca2+ against its concentration gradient.
Ca2+
ATP
Plasma membrane
Ca2+ ATPase
C
Secondary active transport (exchanger)
Example: Na+-Ca2+ exchanger
Na+
Plasma membrane
Na+-Ca2+ exchanger
Ca2+
The carrier uses the Na+ concentration gradient to power Ca2+ movement in the opposite direction.

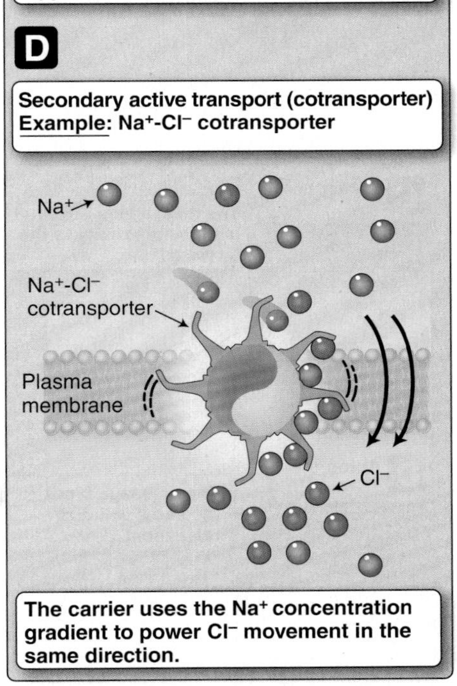
D
Secondary active transport (cotransporter)
Example: Na+-Cl− cotransporter
Na+
Na+-Cl− cotransporter
Plasma membrane
Cl−
The carrier uses the Na+ concentration gradient to power Cl− movement in the same direction.

Figure 1.15
Principal modes of membrane transport. ATP = adenosine triphosphate.

myocytes, for example, and is used to drive passage of other solutes into and out of virtually all cells by **secondary active transport** (see below).

b. **Ca^{2+} ATPases:** All cells express a **plasma membrane Ca^{2+} ATPase (PMCA)** that pumps Ca^{2+} out of the cytoplasm and is primarily responsible for maintaining intracellular Ca^{2+} concentrations at submicromolar levels. A related **sarco(endo) plasmic reticulum Ca^{2+} ATPase (SERCA)** is expressed in the sarcoplasmic reticulum of myocytes and the ER of other cells. SERCA sequesters Ca^{2+} in intracellular stores.

c. **H^+-K^+ ATPase:** The H^+-K^+ ATPase pumps acid and is responsible for lowering stomach pH, for example (see 30·IV·C). It is also found in the kidney, where it is involved in pH balance (see 27·IV·D).

4. **Secondary active transport:** A second class of active transporters use the energy inherent in the electrochemical gradient of one solute to drive uphill movement of a second solute (**secondary active transport**). Such carriers do not hydrolyze ATP directly, although ATP may have been used to create the gradient being harnessed by the secondary transporter. Two transport modes are possible: **countertransport** and **cotransport**.

a. **Countertransport: Exchangers** (**antiporters**) use the electrochemical gradient of one solute (e.g., Na^+) to drive flow of a second (e.g., Ca^{2+}) in the opposite direction to the first (see Figure 1.15C). The Na^+-Ca^{2+} exchanger helps maintain low intracellular Ca^{2+} concentrations by using the inwardly directed Na^+ gradient to pump Ca^{2+} out of the cell. Other important exchangers include a Na^+-H^+ exchanger and a Cl^--HCO_3^- exchanger.

b. **Cotransport:** Cotransporters (**symports**) use the electrochemical gradient of one solute to drive flow of a second or even a third solute in the same direction as the first (see

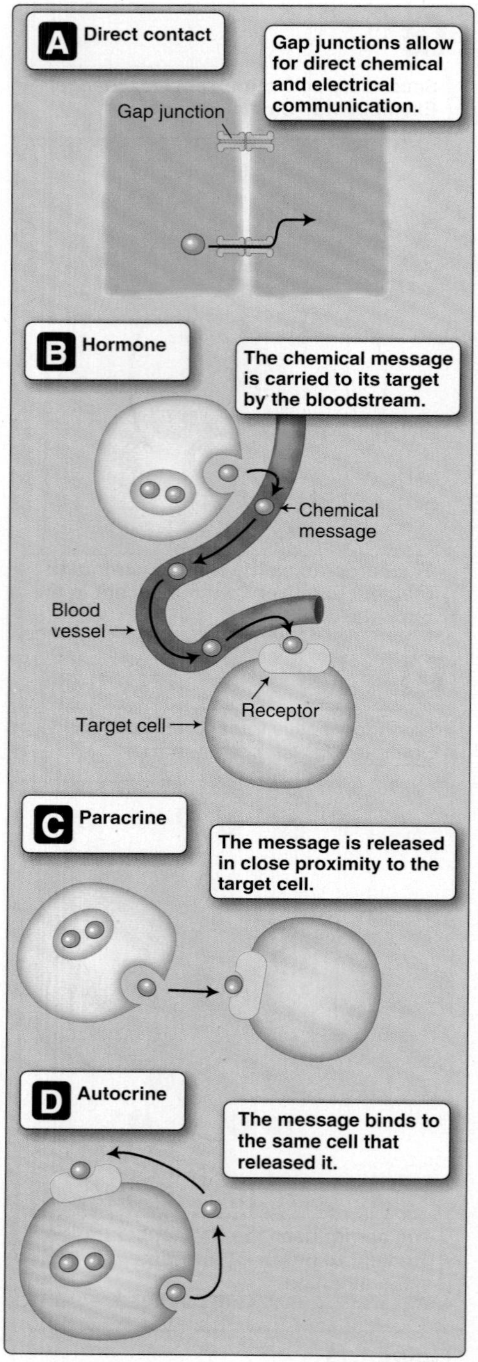

Figure 1.16
Chemical signaling pathways.

Figure 1.15D). For example, cotransporters use an inwardly directed Na^+ gradient to recover glucose and amino acids from the intestinal lumen and renal tubule (Na^+-glucose and Na^+–amino acid cotransporters, respectively), but other examples include a Na^+-Cl^- cotransporter, a K^+-Cl^- cotransporter, and a Na^+-K^+-$2Cl^-$ cotransporter.

VI. INTERCELLULAR COMMUNICATION

The body's various organs each have unique properties and functions, but they must work closely together to ensure the well-being of the individual as a whole. Cooperation requires communication between organs and cells within organs. Some cells contact and communicate with each other directly via **gap junctions** (Figure 1.16A). Gap junctions are regulated pores that allow for exchanging chemical and electrical information (see 4·II·F) and that play a vital role in coordination of cardiac excitation and contraction, for example. Most intercellular communication occurs using chemical signals, which have traditionally been classified according to the distance and route they have travel to exert a physiologic effect. **Hormones** are chemicals produced by endocrine glands and some nonendocrine tissues that are carried to distant targets by the vasculature (see Figure 1.16B). Insulin, for example, is released into the circulation by pancreatic islet cells for carriage to muscle, adipose tissue, and the liver. **Paracrines** are released from cells in very close proximity to their target (see Figure 1.16C). For example, the endothelial cells that line blood vessels release nitric oxide as a way of communicating with the smooth muscle cells that make up the vessel walls (see 20·II·E·1). Paracrines typically have a very limited signaling range because either they are degraded or are taken up rapidly by neighboring cells. **Autocrine** messengers bind to receptors on the same cell that released them, creating a negative feedback pathway that modulates autocrine release (see Figure 1.16D). Autocrines, like paracrines, have a very limited signaling range.

VII. INTRACELLULAR SIGNALING

Once a chemical message arrives at its destination, it must be recognized as such by the target cell and then transduced into a form that can modify cell function. Most chemical messengers are charged and cannot permeate the membrane, so recognition has to occur at the cell surface. Recognition is accomplished using receptors, which serve as cellular switches. Hormone or neurotransmitter binding trips the switch and elicits a preprogrammed instruction set that culminates in a cellular response. Receptors are typically integral membrane proteins such as **ligand-gated channels**, **G protein–coupled receptors** (**GPCRs**), or **enzyme-associated receptors**. Lipophilic messengers can cross the plasma membrane and are recognized by **intracellular receptors**.

A. Channels

Ligand-gated ion channels facilitate communication between neurons and their target cells, including other neurons (see 2·VI·B). For example, the nicotinic acetylcholine (ACh) receptor is a ligand-gated

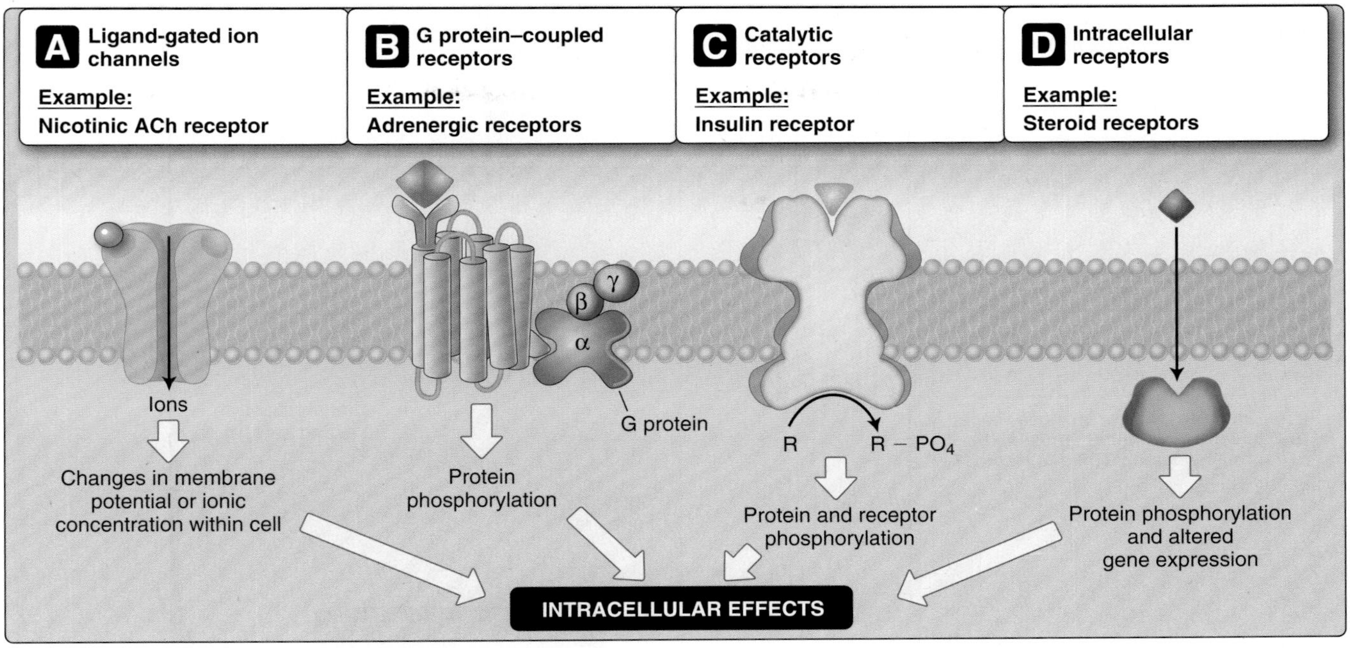

Figure 1.17
Neurotransmitter and hormone receptors. ACh = acetylcholine.

ion channel that allows skeletal muscle cells to respond to excitatory commands from α-motor neurons. Neurotransmitter binding to its receptor causes a conformational change that opens the channel and allows ions such as Na^+, K^+, Ca^{2+}, and Cl^- to flow across the membrane through the pore (Figure 1.17A). Charge movement across the membrane constitutes an electrical signal that influences target cell activity directly, but channel-mediated Ca^{2+} influx can have additional and potent effects on cell function by activating various Ca^{2+}-dependent signal transduction pathways (see below).

B. G protein–coupled receptors

GPCRs sense and transduce a majority of chemical signals, and the GPCR family is large and diverse (the human genome contains >900 GPCR genes). They are found in both neural and nonneural tissues. Common examples include the muscarinic ACh receptor, α- and β-adrenergic receptors, and odorant receptors. GPCRs all share a common structure that includes seven **membrane-spanning regions** that weave back and forth across the membrane (Figure 1.18). Receptor binding is transduced by a **G protein** (guanosine triphosphate [GTP]-binding protein), which then activates one or more **second messenger** pathways (see Figure 1.17B). Second messengers include **cyclic 3′5′-adenosine monophosphate** (**cAMP**), **cyclic 3′5′-guanosine monophosphate** (**cGMP**), and **inositol trisphosphate** (**IP_3**). Multistep signal relay pathways allow for profound amplification of receptor-binding events. Thus, one occupied receptor can activate several G proteins, each of which can yield multiple second messenger molecules that, in turn, can activate multiple effector pathways (Figure 1.19).

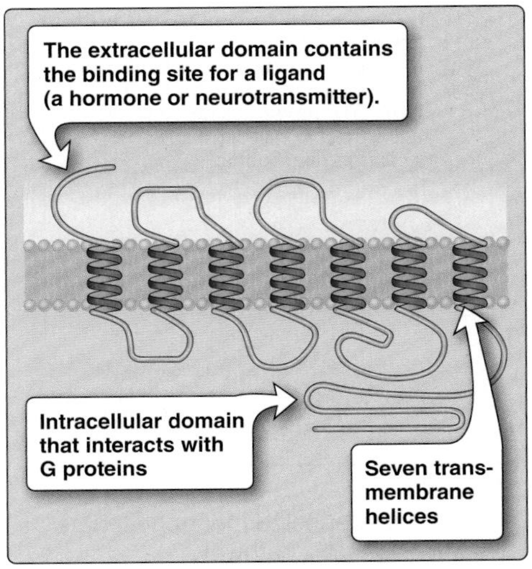

Figure 1.18
G protein–coupled receptor structure.

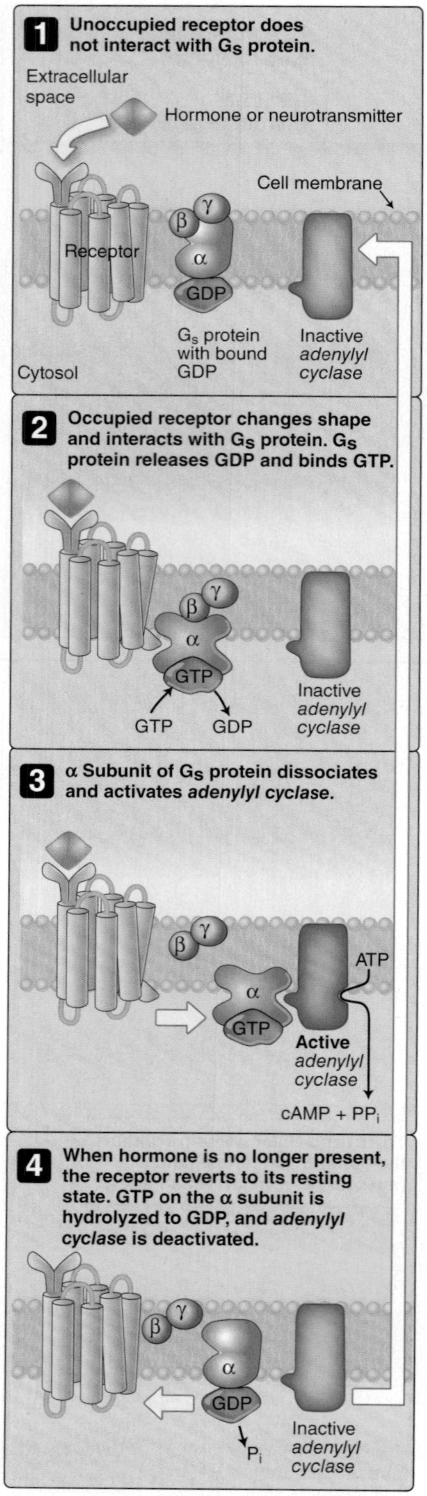

Figure 1.20
The cyclic adenosine monophosphate (cAMP) signaling pathway. ATP = adenosine triphosphate; G_s = stimulatory G protein; GDP = guanosine diphosphate; GTP = guanosine triphosphate; P_i = inorganic phosphate; PP_i = pyrophosphate.

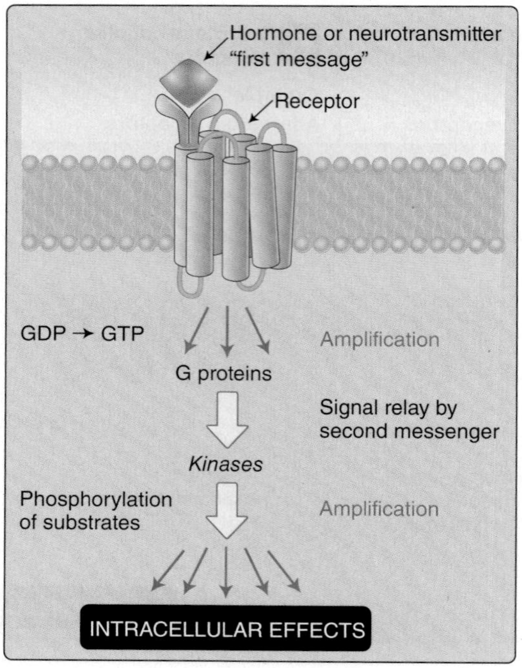

Figure 1.19
Signal amplification by second messengers. GDP = guanosine diphosphate; GTP = guanosine triphosphate.

1. **G proteins:** G proteins are small membrane-associated proteins with *GTPase* activity. Two types of G protein have been described. The class that associates with hormone and neurotransmitter receptors are assemblies of three subunits: α, β, and γ. The *GTPase* activity resides in the α-subunit (G_α), which is normally bound to GDP. Receptor binding causes a conformational change that allows it to interact with its G-protein partner. The α-subunit then releases GDP, binds GTP, and dissociates from the protein complex (Figure 1.20). An occupied receptor can activate many G proteins before the hormone or transmitter dissociates. Active G_α subunits can interact with a variety of second messenger cascades, the principal ones being the cAMP and IP_3 signaling pathways. The duration of G_α's effects are limited by the protein's intrinsic *GTPase* activity. The rate of hydrolysis is slow, but, once GTP has been converted to GDP (and inorganic phosphate), the subunit loses its ability to signal. It then redocks with the $G_{\beta\gamma}$ assembly in the surface membrane and awaits a further opportunity to bind to an occupied receptor.

At least 16 different G_α subunits have been described. They can be classed according to their effects on a target pathway. $G_{\alpha s}$ subunits are stimulatory. $G_{\alpha i}$ subunits are inhibitory, meaning that they suppress second messenger formation when active.

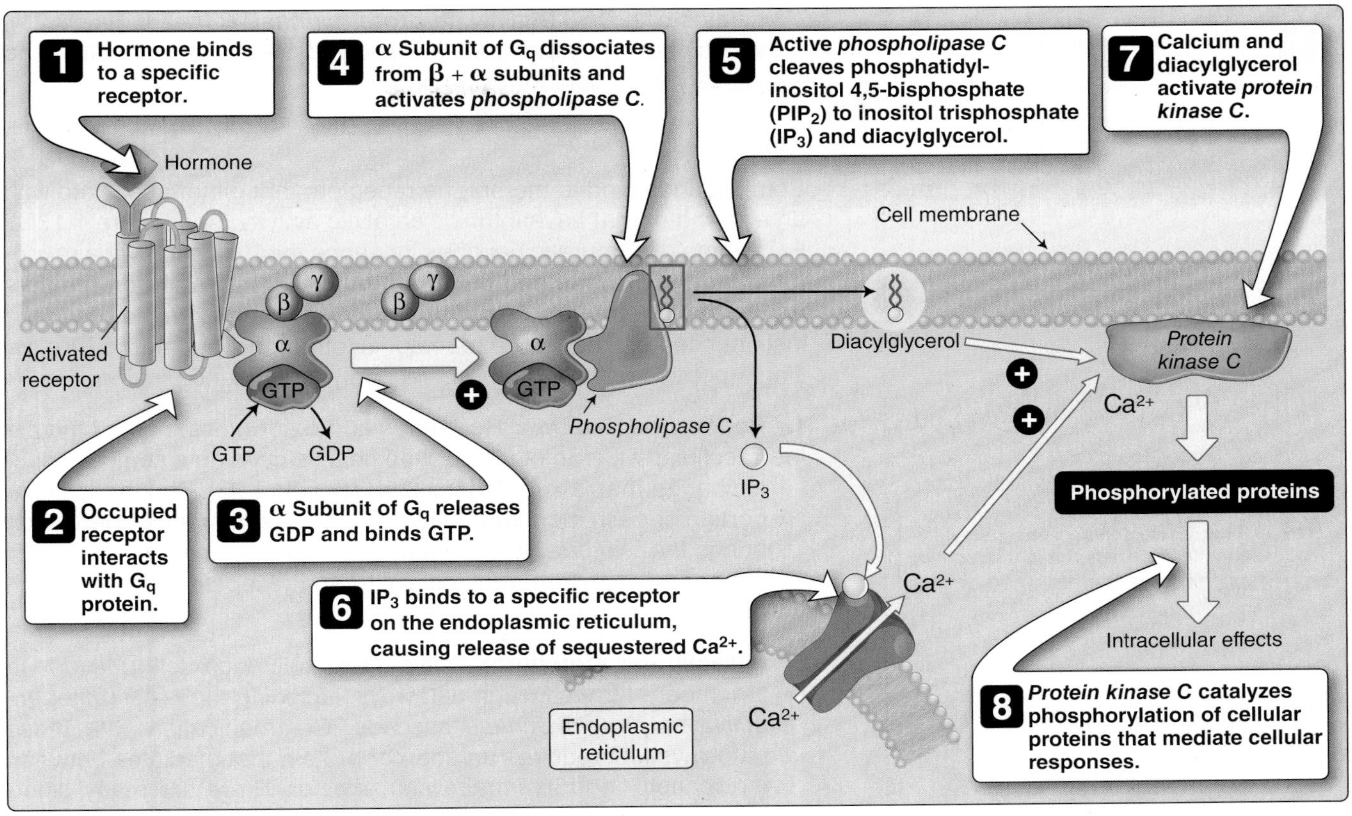

Figure 1.21
The inositol trisphosphate (IP_3) signaling pathway. G_q = stimulatory G protein; GDP = guanosine diphosphate; GTP = guanosine triphosphate.

2. **cAMP signaling pathway:** cAMP is a second messenger that is synthesized from ATP by ***adenylyl cyclase***. *Adenylyl cyclase* is regulated by G proteins. $G_{\alpha s}$ stimulates cAMP formation, whereas $G_{\alpha i}$ inhibits it. cAMP activates ***protein kinase A*** (***PKA***), which phosphorylates and modifies the function of a variety of intracellular proteins, including enzymes, ion channels, and pumps. The cAMP signaling pathway is capable of tremendous signal amplification, so two checks are in place to limit its effects. *Protein phosphatases* counter the effects of the *kinase* by dephosphorylating the target proteins. The effects of the *adenylyl cyclase* are countered by a ***phosphodiesterase*** that converts cAMP to 5′-AMP.

3. **IP_3 signaling pathway: $G_{\alpha q}$** is a G-protein subunit that liberates three different second messengers via activation of ***phospholipase C*** (***PLC***) as shown in Figure 1.21. The messengers are IP_3, **diacylglycerol** (**DAG**), and Ca^{2+}. *PLC* catalyzes the formation of IP_3 and DAG from **phosphatidylinositol 4,5-bisphosphate** (**PIP_2**), a plasma membrane lipid. DAG remains localized to the membrane, but IP_3 is released into the cytoplasm and binds to a Ca^{2+} release channel located in the ER. Ca^{2+} then floods out of the stores and into the cytosol, where it binds to **calmodulin** (**CaM**) as shown in Figure 1.22. CaM mediates Ca^{2+}-activation of many enzymes and other intracellular effectors. Ca^{2+} also coor-

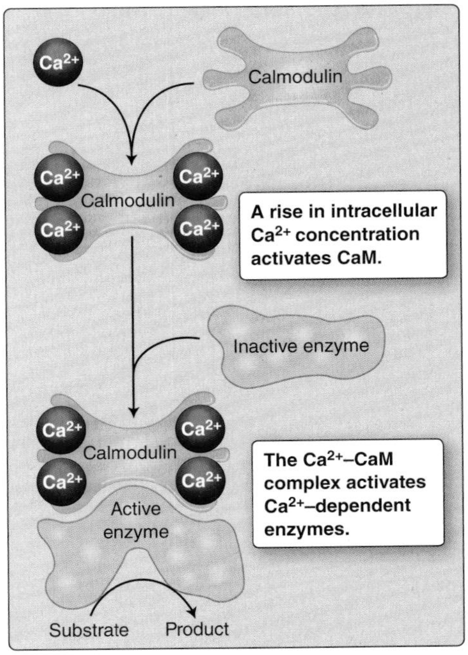

Figure 1.22
Ca^{2+}-calmodulin (CaM)-dependent enzyme activation.

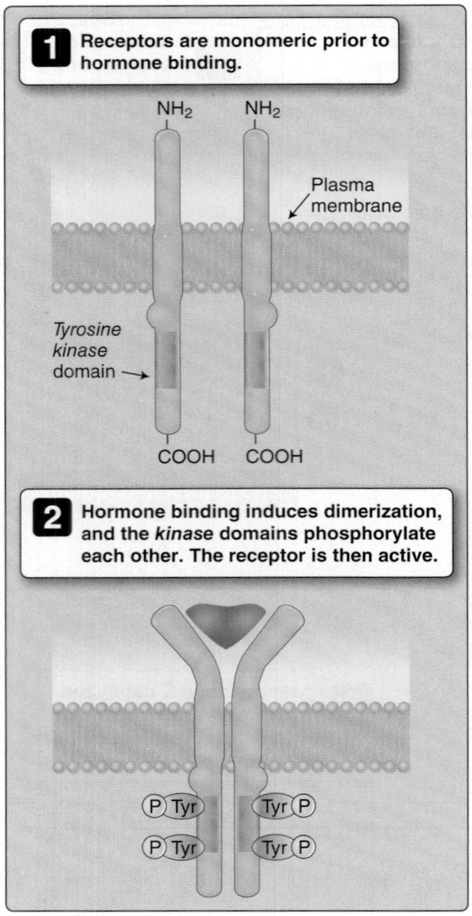

Figure 1.23
Tyrosine receptor kinase activation.

dinates with DAG to activate ***protein kinase C*** (see Figure 1.21), which phosphorylates proteins involved in muscle contraction and salivary secretion, for example.

C. Catalytic receptors

Some ligands bind to membrane receptors that either associate with an enzyme or that have intrinsic catalytic activity (see Figure 1.17C). For example, natriuretic peptides influence renal function via a receptor ***guanylyl cyclase*** and cGMP formation. Most catalytic receptors are ***tyrosine kinases*** (***TRKs***), the most common example being the insulin receptor. The insulin receptor is tetrameric, but most *TRKs* are single-peptide chains that associate only after ligand binding.

1. **Receptor activation:** Hormones and other messengers bind extracellularly to one of the peptide chains, causing a conformational change that favors dimerization (Figure 1.23). The intracellular portion of each monomer contains a *kinase* domain. Dimerization brings the two catalytic domains into contact, and they phosphorylate each other, thereby activating the receptor complex, which begins signaling.

2. **Intracellular signaling:** Active *TRKs* influence cell function via a number of transduction pathways, including the *MAP* (*mitogen-activated protein*) *kinase* cascade. Communication with these pathways first requires an adapter protein that mediates between the receptor and its intracellular effector. There are many different adapter proteins, but they all contain *Src* homology domains named SH2 and SH3. The SH2 domain recognizes the phosphorylated tyrosine domains on the activated *TRK* and allows the adapter protein to bind to the signaling complex.

D. Intracellular receptors

A fourth receptor class is located intracellularly and includes receptors for thyroid hormone and a majority of steroid hormones (see Figure 1.17D). All are transcription factors that influence cell function by binding to DNA and altering gene expression levels. Some of the receptors are cytoplasmic, whereas others are nuclear and may be associated with DNA. The cytoplasmic receptors are normally bound to a "heat shock" protein, which is displaced by the conformational change caused by steroid binding. The occupied receptor then translocates to the nucleus and binds to a **hormone response element** within the promoter region of the target gene. Nuclear receptors act in a similar way. Once bound, the receptor induces gene transcription, and the product alters cell function.

Chapter Summary

- All cells erect a lipid barrier (the **plasma membrane**) to separate the inside of the cell from the outside, and then selectively modify the ionic composition in the intracellular environment to facilitate the biochemical reactions that sustain life. **Intracellular fluid** contains very low concentrations of Ca^{2+} compared with **extracellular fluid**. Na^{+} concentrations are also lower inside, but K^{+} levels are higher.
- The plasma membrane contains three principal lipid types: **phospholipids**, **cholesterol**, and **glycolipids**. Phospholipids dominate the structure, cholesterol adds strength, and glycolipids mediate interactions with other cells.
- Movement across membranes occurs primarily by **diffusion**. Diffusion rate is dependent on the transmembrane **concentration difference**, molecular size, membrane thickness and surface area, temperature, viscosity of the solution through which the molecule must diffuse, and the molecule's solubility in lipid (**partition coefficient**).
- Integral membrane proteins such as **pores**, **channels**, and **carriers** provide pathways by which hydrophilic molecules may cross the lipid barrier.
- Pores are always open and are rare, the principal example being **aquaporin**, a ubiquitous water channel. Channels are regulated pores that open transiently to allow passage of small ions, such as Na^{+}, Ca^{2+}, K^{+}, and Cl^{-}. Movement through pores and channels occurs by simple diffusion down electrical and chemical concentration gradients (**electrochemical gradient**).
- Carriers selectively bind ions and small organic solutes, carry them across the membrane, and then release them on the opposite side. Carriers operate by two modes of transport: **facilitated diffusion** and **active transport**. Facilitated diffusion moves solutes "downhill" in the direction of the electrochemical gradients (e.g., glucose transport by the GLUT transporter family). Active transport uses energy to move solutes "uphill" from an area containing low solute concentration to an area of higher concentration.
- **Primary active transporters**, or **pumps**, use adenosine triphosphate to drive solutes uphill against their electrochemical gradient. Pumps include the Na^{+}-K^{+} ATPase that is present in all cells, Ca^{2+} ATPases, and the H^{+}-K^{+} ATPase.
- **Secondary active transporters** move solutes uphill by harnessing the energy inherent in electrochemical gradients for other ions. **Exchangers** move two solutes across the membrane in opposite directions (e.g., the Na^{+}-Ca^{2+} exchanger). **Cotransporters** (e.g., Na^{+}-K^{+}-$2Cl^{-}$ cotransporter and the Na^{+}-glucose cotransporter) move two or more solutes in the same direction.
- The plasma membrane also contains receptor proteins that allow cells to communicate with each other using chemical messages. Signaling can occur over long distances via the release of **hormones** (e.g., insulin) into the bloodstream. Cells that are in close proximity to each other communicate using **paracrines** (e.g., nitric oxide). **Autocrines** are chemical signals that target the same cell that released them.
- **Receptor binding** is transduced in a variety of ways. **Ligand-gated ion channels** transduce binding using changes in membrane potential. Other receptor classes release **G proteins** to activate or inhibit **second messenger** pathways. Many receptors possess intrinsic *kinase* activity and signal occupancy through protein phosphorylation. A fourth class of receptor is located inside the cell. Intracellular receptors affect levels of gene expression when a message binds.
- G proteins modulate two major second-messenger cascades. The first involves **cyclic adenosine monophosphate (cAMP)** formation by ***adenylyl cyclase***. cAMP acts primarily through regulation of ***protein kinase A*** and protein phosphorylation.
- Other G proteins activate ***phospholipase C*** and cause the release of **inositol trisphosphate (IP_3)** and **diacylglycerol (DAG)**. IP_3, in turn, initiates Ca^{2+} release from intracellular stores. Ca^{2+} then binds to calmodulin and activates Ca^{2+}-dependent transduction pathways. Ca^{2+} and DAG together activate *protein kinase C* and cause phosphorylation of target proteins.
- Receptors with intrinsic ***tyrosine kinase*** activity autophosphorylate when the message binds. This allows them to complex with adapter proteins that initiate signal cascades affecting cell growth and differentiation.
- **Intracellular receptors** translocate to the nucleus and bind to **hormone response elements** within the promoter region of target genes. Cell function is altered through increased levels of target gene expression.

2 Membrane Excitability

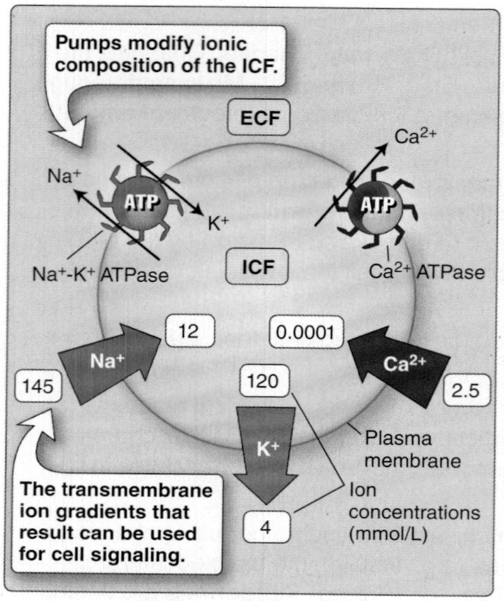

Figure 2.1
Intracellular fluid (ICF) modification by ion transporters. ATP = adenosine triphosphate; ECF = extracellular fluid.

I. OVERVIEW

All cells selectively modify the ionic composition of their internal environment to support the biochemistry of life (see 1·II) as shown in Figure 2.1. Moving ions into or out of a cell creates a charge imbalance between the intracellular fluid (ICF) and the extracellular fluid (ECF) and thereby allows a voltage difference to form across the surface membrane (a **membrane potential**, or $\mathbf{V_m}$). This process creates an electrochemical driving force for diffusion that can be used to move charged solutes across the membrane or that can be modified transiently to create an electrical signal for intercellular communication. For example, nerve cells use changes in V_m (**action potentials**) to signal to a muscle that it needs to contract. The muscle cell, in turn, uses a change in V_m to activate Ca^{2+} release from its internal stores. Ca^{2+} release then facilitates actin and myosin interactions and initiates muscle contraction. Neuronal and muscle action potentials both involve carefully coordinated sequences of **ion channel** events that allow selective transmembrane passage of ions (e.g., Na^+, Ca^{2+}, and K^+) between ICF and ECF.

II. MEMBRANE POTENTIALS

The term "membrane potential" refers to the voltage difference that exists across the plasma membrane. By convention, the ECF is considered to be at zero volts, or electrical **"ground."** Inserting a fine electrode across the surface membrane reveals that the cell interior is negative with respect to the ECF by several tens of millivolts. A typical nerve cell has a **resting potential** of −70 mV, for example (Figure 2.2). V_m is established by membrane-permeant ions traveling down their respective concentration gradients and generating **diffusion potentials**.

A. Diffusion potentials

Imagine a model cell in which the plasma membrane is composed of pure lipid, the ICF is rich in potassium chloride (KCl, which dissociates into K^+ and Cl^-), and ECF is pure water (Figure 2.3). Although there is a strong KCl concentration gradient for diffusion across the membrane, the lipid barrier prevents both K^+ and Cl^- from leaving the cell and thereby constrains both ions to the ICF. The charges carried by K^+ and Cl^- cancel each other out and, thus, there is no voltage difference between ECF and ICF. If a protein that permits passage of K^+ alone is inserted into the lipid barrier, K^+ is now free to diffuse down

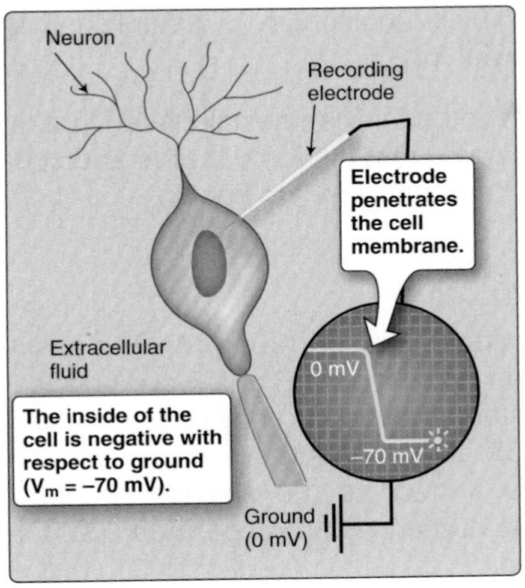

Figure 2.2
Membrane potential (V_m).

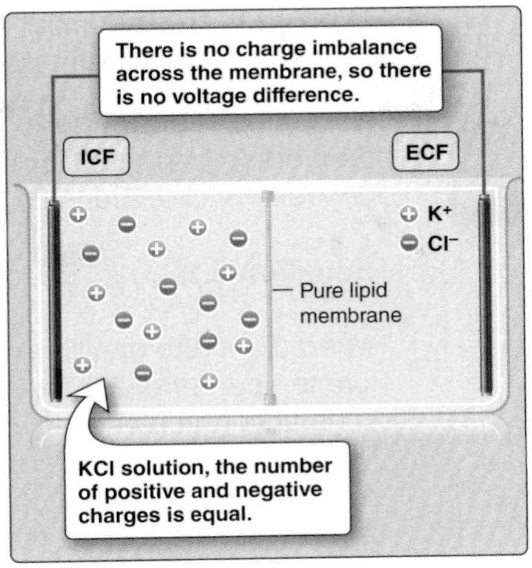

Figure 2.3
Charge distribution in a model cell with an impermeable membrane. ECF = extracellular fluid; ICF = intracellular fluid.

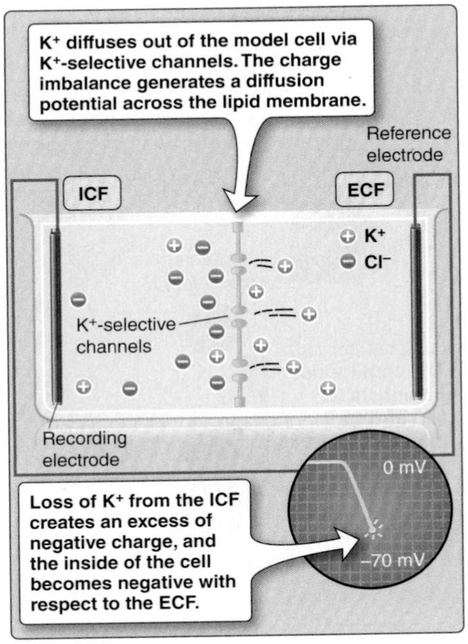

Figure 2.4
Origin of a diffusion potential. ECF = extracellular fluid; ICF = intracellular fluid.

its concentration gradient from ICF to ECF, and the membrane is said to be **semipermeable** (Figure 2.4). Because potassium ions carry charge, their movement causes a **diffusion potential** to form across the membrane in direct proportion to the magnitude of the concentration gradient. The potential may be significant (tens of millivolts) but involves relatively few ions.

> The **principle of bulk electroneutrality** notes that the number of positive charges in a given solution is always balanced by an equal number of negative charges. The ICF and ECF are also subject to this rule, even though all cells create a negative V_m by altering charge distribution between the two compartments. In practice, V_m is established by just a few charges moving in the immediate vicinity of the cell membrane and their *net effect* on overall charge distribution within the bulk of the ICF and ECF is negligible.

B. Equilibrium potentials

When K^+ crosses the membrane down its concentration gradient, it leaves a negative charge in the form of Cl^- behind. Net charge magnitude builds in direct proportion to the number of ions leaving the ICF (see Figure 2.4), but, because opposite charges attract, K^+ movement slows and eventually stops when the attraction of the negative charges inside the cell precisely counters the outward driving force created by the concentration gradient (**electrochemical**

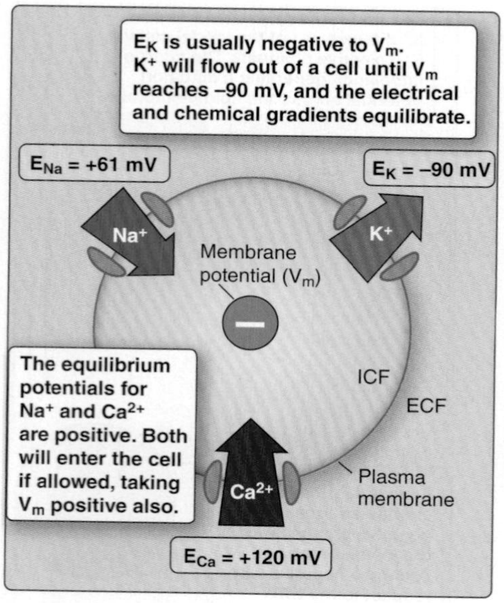

Figure 2.5
Equilibrium potentials for Na^+ (E_{Na}), Ca^{2+} (E_{Ca}), and K^+ (E_K). ECF = extracellular fluid; ICF = intracellular fluid.

equilibrium). The potential at which equilibrium is established is known as the **equilibrium potential** for K^+.

1. **Nernst equation:** Equilibrium potentials can be calculated for any membrane-permeant ion assuming that the ion's charge and concentrations on either side of the membrane are known:

Equation 2.1 $$E_X = \frac{RT}{zF}\, ln \frac{[X]_o}{[X]_i}$$

where E_X is the equilibrium potential for ion X (in mV), T is absolute temperature, z is the valence of the ion, R and F are physical constants (the ideal gas constant and the Faraday constant), and $[X]_o$ and $[X]_i$ are ECF and ICF concentrations of X (in mmol/L), respectively. Equation 2.1 is known as the **Nernst** equation. If T is assumed to be normal human body temperature (37°C), Equation 2.1 can be simplified:

Equation 2.2 $$E_X = \frac{60}{z} \log_{10} \frac{[X]_o}{[X]_i}$$

> Most of the common inorganic ions (Na^+, K^+, Cl^-, HCO_3^-) have an electrical valence of 1 (**monovalents**). Ca^{2+} and Mg^{2+} have a valence of 2 (**divalents**).

Example 2.1

A cell has an intracellular free Mg^{2+} concentration of 0.5 mmol/L and is bathed in a saline solution with a Mg^{2+} composition that approximates plasma (1.0 mmol/L). The saline is held 37°C. If the cell has a membrane potential (V_m) of −70 mV, and the membrane contains a gated channel that is Mg^{2+} permeable, will Mg^{2+} flow into or out of the cell when the channel opens?

We can use Equation 2.2 to calculate the Mg^{2+} equilibrium potential for (E_{Mg}):

$$E_{Mg} = \frac{60}{z} \log_{10} \frac{[Mg^{2+}]_o}{[Mg^{2+}]_i}$$
$$= \frac{60}{+2} \text{ mV} \log_{10} \frac{1.0 \text{ mmol/L}}{0.5 \text{ mmol/L}}$$
$$= 30 \text{ mV} \log_{10} 2.0$$
$$= 30 \text{ mV } (0.3)$$
$$= 9 \text{ mV}.$$

E_{Mg} tells us that Mg^{2+} will flow into the cell, its positive charges tending to drive V_m toward 9 mV.

2. **Equilibrium potentials:** The ICF and ECF are both strictly regulated, and their ionic composition is well known (see Figure 2.1, also see Table 1.1). Using known values for concentrations of the common ions, we can use the Nernst equation to predict that, for most cells in the body, $E_K = -90$ mV, $E_{Na} = +61$ mV, and $E_{Ca} = +120$ mV. Intracellular Cl^- concentrations can vary considerably, but E_{Cl} usually lies very close to V_m. If any of these ions are provided with a pathway that allows them to diffuse across the plasma membrane, they will drag V_m toward the equilibrium potential for that ion (Figure 2.5).

C. Resting potential

The plasma membranes of living cells are rich in ion channels that are permeable to one or more of the ions mentioned above, and some of these channels are open at rest. Resting V_m (**resting potential**) thus reflects the sum of the diffusion potentials generated by each of these ions flowing through open channels. V_m can be calculated mathematically as follows:

$$V_m = \frac{g_{Na}}{g_T} E_{Na} + \frac{g_K}{g_T} E_K + \frac{g_{Ca}}{g_T} E_{Ca} + \frac{g_{Cl}}{g_T} E_{Cl}$$

where g_T is total membrane conductance (membrane conductance is the reciprocal of membrane resistance, in $Ohms^{-1}$); g_{Na}, g_K, g_{Ca}, and g_{Cl} are individual conductances for each of the common ions (Na^+, K^+, Ca^{2+}, and Cl^-, respectively); and E_{Na}, E_K, E_{Ca}, and E_{Cl} are equilibrium potentials for these ions (in mV). V_m can also be calculated using the Goldman-Hodgkin-Katz (GHK) equation, which is similar in

form to the Nernst equation above (Equation 2.1). The GHK equation derives V_m using relative membrane permeabilities for each of the ions that contribute to membrane potential.

> In practice, most cells at rest have a negligible permeability to either Na^+ or Ca^{2+}. Cells do have a significant resting K^+ conductance, however. Thus, V_m typically rests close to the equilibrium potential for K^+ (Figure 2.6). The approximate value for resting potential in neurons is −70 mV, −90 mV in cardiac myocytes, −55 mV in smooth muscle cells, and −40 mV in hepatocytes, for example.

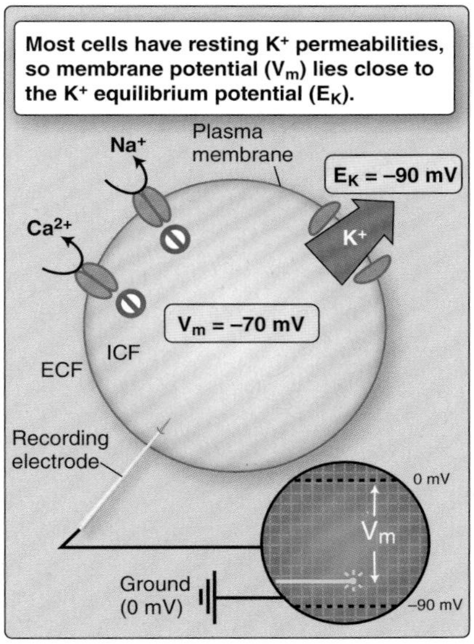

Figure 2.6
Resting potential origins. ECF = extracellular fluid; ICF = intracellular fluid.

D. Extracellular ion effects

The ionic composition of the ECF is regulated within a fairly narrow range, but significant disturbances can occur through inadequate or excessive ingestion of salts or water. Because resting membrane permeability to Na^+ and Ca^{2+} is low, V_m is relatively insensitive to changes in ECF concentration of either ion. V_m *is* sensitive to changes in extracellular K^+ concentration, however, because resting potential is closely tied to the equilibrium potential for K^+ (see Figure 2.6). Increasing extracellular K^+ concentration (**hyperkalemia**) reduces the electrochemical gradient that drives K^+ efflux, causing the membrane to depolarize (Figure 2.7). Conversely, lowering the extracellular K^+ concentration (**hypokalemia**) steepens the gradient, and V_m becomes more negative.

E. Transporter contribution

The Na^+-K^+ ATPase that resides in the plasma membrane of all cells drives three Na^+ out of the cell while simultaneously transferring two

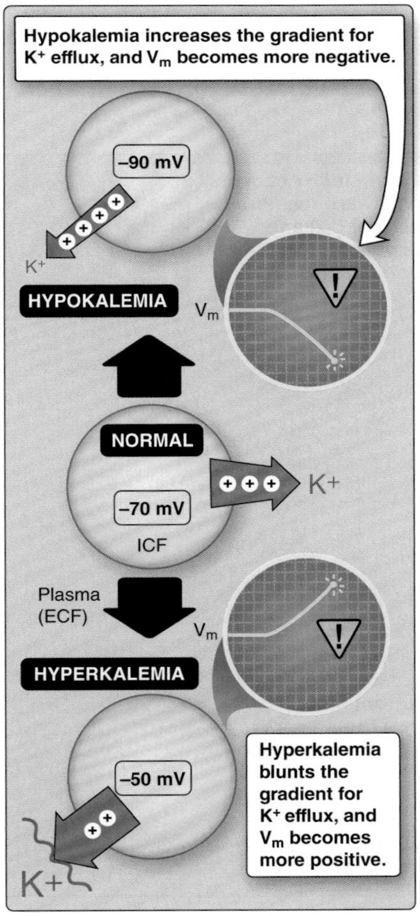

Figure 2.7
Membrane-potential (V_m) dependence on K^+ concentration in the extracellular fluid (ECF). ICF = intracellular fluid.

Clinical Application 2.1: Hypokalemia and Hyperkalemia

Excitable cell function critically depends on maintaining membrane potential within a narrow range, so plasma levels normally range between 3.5 and 5.0 mmol/L. Hypokalemia and hyperkalemia are both commonly encountered clinically, however. Hypokalemia is generally of less concern than hyperkalemia, although some individuals with a rare inherited disorder (hypokalemic periodic paralysis) can experience muscle weakness when plasma K^+ concentrations dip such as following a meal. Hyperkalemia is, potentially, a more serious condition. The slow depolarization caused by rising plasma K^+ levels inactivates Na^+ channels that are required for muscle excitation, resulting in skeletal muscle weakness or paralysis and cardiac arrhythmias and conduction abnormalities. Hyperkalemia usually results from kidney failure and impaired ability to excrete K^+. Treatment typically requires either diuresis or dialysis to remove excess K^+ from the body.

K^+ from ECF to ICF. The three-for-two exchange results in an excess of positive charges being removed from the cell. Because the transporter creates a charge imbalance across the membrane, it is said to be **electrogenic**. The direct contribution of this exchange to V_m is insignificant, however. The main role of the Na^+-K^+ ATPase is to maintain a K^+ concentration gradient across the membrane, because it is the K^+ gradient that ultimately determines V_m via the K^+ diffusion potential.

III. EXCITATION

Many cell types use changes in V_m and transmembrane ion fluxes as a means of signaling or initiating intracellular events. Sensory cells (e.g., mechanosensors, olfactory receptors, and photoreceptors) transduce sensory stimuli by generating a V_m change called a **receptor potential**. Neurons signal to each other and to effector tissues using action potentials. Myocytes and secretory cells also use changes in V_m to increase intracellular Ca^{2+} concentration, thereby facilitating contraction and secretion, respectively. All such cells are said to have **excitable membranes**.

A. Terminology

The electrical changes caused by increased membrane permeability to ions do not consider the permeant ion's species (e.g., Na^+ *versus* K^+ or Cl^-), only the charge that it carries.

1. **Membrane potential changes:** The inside of a cell at rest is always negative with respect to the ECF. When a positively charged ion (**cation**) flows into a cell, negative charges (**anions**) are neutralized, and the membrane loses polarization. The influx is said to have **depolarized** the cell, or caused **membrane depolarization**. By convention, depolarization is shown as an upward pen deflection on a voltage record (Figure 2.8). Conversely, if a cation leaves the cell, V_m becomes more negative: The efflux **hyperpolarizes** the cell (**membrane hyperpolarization**) and yields a downward pen deflection on a recording device.

2. **Currents:** When positive charges flow into a cell, they generate an **inward current**. By convention, recording devices, such as oscilloscopes and chart recorders, are configured so that inward currents cause a downward deflection (see Figure 2.8). Positive charges leaving the cell cause an **outward current** and an upward deflection on a recording device.

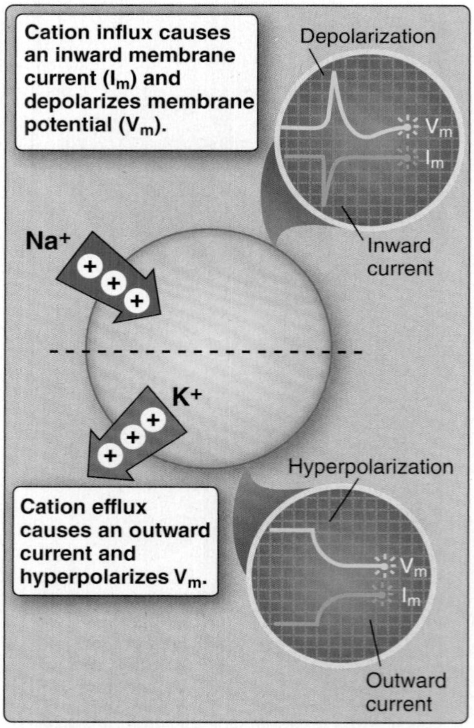
Cation influx causes an inward membrane current (I_m) and depolarizes membrane potential (V_m).
Depolarization
V_m
I_m
Inward current
Na^+
K^+
Hyperpolarization
Cation efflux causes an outward current and hyperpolarizes V_m.
V_m
I_m
Outward current

Figure 2.8
Membrane potential changes and ion currents.

> Anions and cations are equally effective in changing V_m, but, because anions carry negative charges, their effects are opposite to those of cations. When an anion enters the cell from the ECF, it hyperpolarizes the membrane and yields an outward current. Conversely, anions leaving a cell create an inward current, and the cell depolarizes.

B. Action potentials

Action potential size, shape, and timing may vary widely between the different cell types, but there are several common characteristics, including the existence of a **threshold** for action potential formation, **all-or-nothing** behavior, **overshoots**, and **afterpotentials** (Figure 2.9). The discussion below focuses on a nerve action potential whose **upstroke** is mediated by voltage-dependent Na^+ channels, but voltage-dependent Ca^{2+} channels can support action potentials also (e.g., see 17·IV·B·3).

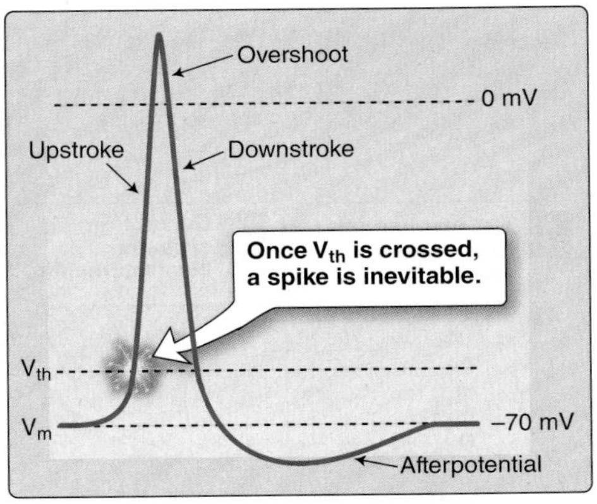

Figure 2.9
An action potential. V_m = membrane potential; V_{th} = threshold potential.

1. **Threshold potential:** Because action potentials are explosive membrane events that have consequences (e.g., initiating muscle contraction), they must be triggered with care. V_m normally fluctuates over a range of a few millivolts with changes in extracellular K^+ concentration and other variables, even at rest, but such changes do not trigger spikes. Neurons only fire action potentials when V_m depolarizes sufficiently to cross the voltage threshold for action potential formation (V_{th}), which, in a neuron, usually resides at around −60 mV. V_{th} corresponds to the voltage needed to open the number of voltage-dependent Na^+ channels required to trigger an action potential.

2. **All or nothing:** Voltage-dependent Na^+ channels that mediate action potentials are typically present in the membrane in high numbers. When V_m crosses threshold, they open to allow a massive inward current, and the membrane depolarizes in a self-perpetuating (**regenerative**) fashion toward E_{Na} (+61 mV). This "all-or-nothing" behavior can be likened to breaching a dam wall. Once depolarization begins, it does not stop until the ionic flood is complete.

3. **Overshoot:** The action-potential peak typically does not reach E_{Na}, but it often "overshoots" the zero-potential line, and the inside of the cell becomes positively charged with respect to the ECF.

4. **Afterpotentials:** Action potentials are transient events. The **downstroke** is caused in part by voltage-dependent K^+ channels that open to allow K^+ efflux, causing V_m to repolarize. In some cells, the action potential may be followed by an afterpotential of varying size and polarity. A hyperpolarizing afterpotential takes the membrane negative to V_m for a period before eventually settling at the normal resting potential.

C. Action potential propagation

When neurons fire an action potential, the electrical event does not instantaneously involve the entire cell, but rather the spike begins at one end of the cell and then **propagates** at speeds of up to 120 m/s to the far end (Figure 2.10). Muscle cells behave similarly, although **conduction velocities** are typically lower in muscle than in nerve (~1 m/s). The advantage to propagation is that it allows a message to be carried unlimited distances. By way of analogy, the travel distance of a written message within a hollow baton thrown to a recipient is limited by the strength of the throw. Pass the message to a team of relay runners, however, and travel distance is limited only by the number of runners available. In practice, signal propagation allows spinal neurons to communicate with the feet, which are typically a meter away!

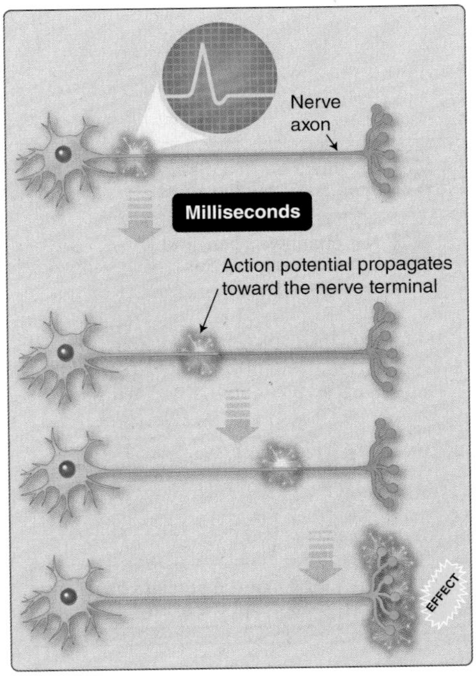

Figure 2.10
Action potential propagation.

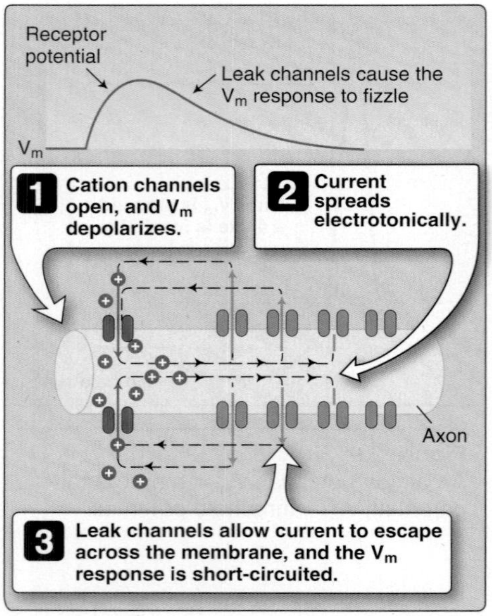

Figure 2.11
Passive current spread and degradation in a neuron. V_m = membrane potential.

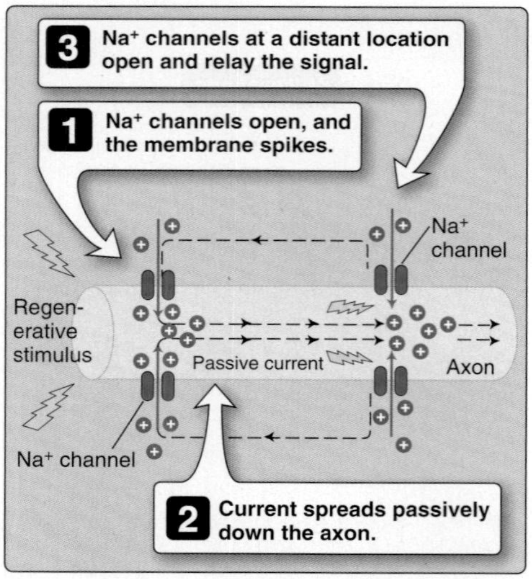

Figure 2.12
Regenerative signal propagation in a neuron.

Neuronal signaling involves a number of sequential steps, including membrane excitation, action potential initiation, signal propagation, and recovery.

1. **Excitation:** Action potentials are typically initiated by sensory receptor potentials or dendritic postsynaptic potentials, for example. These are minor membrane events whose amplitude is graded with input intensity. Their reach is limited, much as throwing a baton is limited by arm muscle strength. The potential spreads passively and instantaneously (**electrotonically**), using the same physical principles by which electricity travels in a wire. Its reach is limited because the local currents created by the receptor potential are short-circuited by **leak channels**, which are found in all excitable membranes (Figure 2.11). Leak channels (typically K^+ channels) are open at rest, allowing voltage changes to fizzle before they can travel far by "leaking" current across the membrane.

> Electrical impulses travel through conductive materials like shock waves. A Newton cradle (i.e., classic desk toy comprising five silver balls suspended side by side within a frame) provides a good visual analogy. When a ball at one end is lifted and released, it impacts its neighbor and imparts its kinetic energy via a shock wave to the ball at the opposing end without disturbing the three intervening balls. The ball rises on its nylon line with little apparent energy loss. Electricity similarly creates shock waves between adjacent metal atoms within a copper wire that are transmitted at close to the speed of light. Electrical currents cause electrons to move also, but they travel at speeds closer to that of cold molasses.

2. **Initiation:** If the receptor potential is sufficiently large to cause V_m in a region of the membrane that contains voltage-dependent Na^+ channels to cross threshold, it will trigger a spike.
3. **Propagation:** Na^+-channel opening allows Na^+ to flow into the cell, driven by the electrochemical gradient for Na^+ and generating an **active current** (Figure 2.12). The current then spreads electrotonically and causes a V_m depolarization that extends some distance down the axon. If the distant region contains Na^+ channels and the change in V_m is sufficiently large to cross threshold, the channels in this region open and **regenerate** the signal, much as a relay runner picks up a baton. The cycle of Na^+ influx, electronic spread, and regenerative Na^+-channel opening is repeated (propagates) down the length of the cell.
4. **Recovery:** Na^+ channels are inactivated by depolarization within a few milliseconds, which temporarily inhibits further excitation and prevents action potentials endlessly boomeranging back and forth along an axon. Excitation is followed by a period of recovery, during which time ion gradients are renormalized by ion pumps and channels recover from excitation.

All cells have a membrane potential but not all are excitable. By definition, nonexcitable cells do not generate action potentials, but many do show functional changes in V_m. For example, glucose causes the membrane potential of pancreatic β cells to oscillate in a sustained, rhythmic fashion. The electrical events correlate with insulin release.

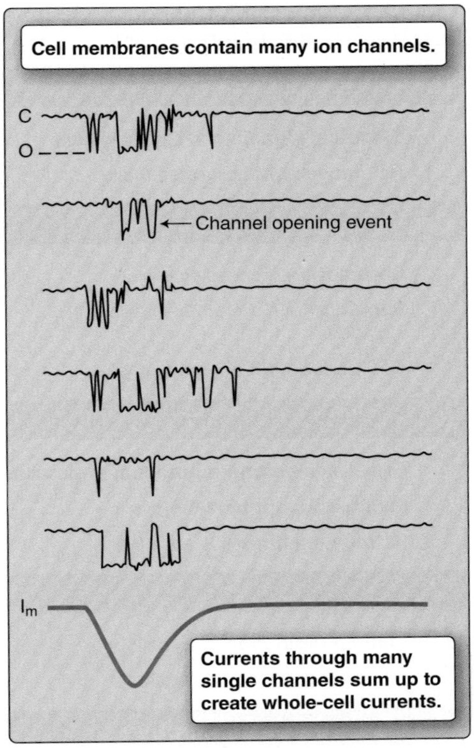

Figure 2.13
Single-channel and whole-cell ion currents. C = closed; I_m = membrane current; O = open.

D. Currents

Action potentials are gross membrane events reflecting net charge movements through many thousands of individual ion channels. Each channel-opening event generates a **unitary current**, the size of which is directly proportional to the number of charges moving through its pore (Figure 2.13). The sum of individual Na^+ channel-opening events yields a whole-cell Na^+ current. Similarly, the sum of individual K^+ channel events yields a whole-cell K^+ current. Because there are many different classes of ion channel with selective permeabilities for all of the common inorganic ions, a whole-cell K^+ current (for example) may represent K^+ efflux through two or more discrete K^+-channel types. Such currents can be dissected into their individual components based on their physical properties using voltage-clamp (whole-cell recordings) and patch-clamp (recordings made from small membrane patches) techniques.

IV. ION CHANNELS

Ion channels are integral membrane proteins containing one or more hydrophilic pores that open transiently to allow ions to cross to the membrane. Channels have several distinguishing features that identify them as such, including an **activation** mechanism, a **selectivity filter**, and a finite **conductance**. Many channels also inactivate with time during prolonged stimulation.

A. Activation

Ion channels create holes in the lipid barrier separating the ICF from ECF. If they were unregulated, ions would continue to flow across the membrane and collapse their respective concentration gradients, along with V_m. Thus, most channels have activation gates that regulate passage through the pore (Figure 2.14). When a channel is in its **closed state**, the gate seals the pore, and ions cannot pass. Channel activation (e.g., in response to a change in V_m; see below) initiates a change in protein conformation that opens the gate and allows ions to pass (i.e., the **open state**). Some channels transition between the open and closed states thousands of times per second, with the net open time (or, **open probability**) increasing in direct proportion to the strength of the activating stimulus.

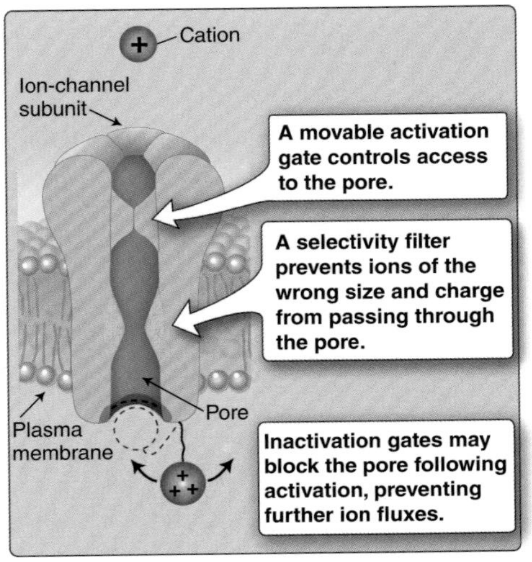

Cation
Ion-channel subunit
A movable activation gate controls access to the pore.
A selectivity filter prevents ions of the wrong size and charge from passing through the pore.
Pore
Plasma membrane
Inactivation gates may block the pore following activation, preventing further ion fluxes.

Figure 2.14
Ion-channel structure.

Table 2.1: Typical Ion-Channel Functions Based on Permeant Ion Species

Na^+	
	Electrical signaling in excitable cells
	Na^+ **movement** across epithelia
Ca^{2+}	
	Contraction of muscle cells
	Secretion from nerve terminals and glands
	Ca^{2+} **movement** across epithelia
K^+	
	Establish **membrane potential** in all cells
	Membrane repolarization
	K^+ **movement** across epithelia
Cl^-	
	Volume regulation in all cells
	Cl^- **movement** across epithelia
Na^+, Ca^{2+}, K^+ nonselective	
	Sensory transduction

B. Selectivity

Before an ion can traverse the membrane, it must pass through a selectivity filter that determines its suitability for passage (see Figure 2.14). Selectivity filters reside within the pore and comprise regions where the permeant ion is forced to interact with one or more charged groups that limit passage based on molecular size and charge density. Thus, **Na^+ channels** only pass Na^+, **K^+ channels** are selective for K^+, **Ca^{2+} channels** are Ca^{2+} selective, and **Cl^- channels** pass only Cl^- (Table 2.1). Other **nonselective** channels may allow passage of two or more different ions. Note that there are many different classes of each type of channel, differentiated by their mode of activation, kinetics, conductance, regulatory mechanisms, tissue specificity, and pharmacology.

> The human genome encodes genes for over 400 ion channels. Almost half of these are K^+ channels. Although they all have selectivity for K^+ over other ions, the individual members of this large family all have unique properties (mode of activation, activation kinetics, conductance, regulatory mechanism, for example) and roles to play in membrane physiology (membrane repolarization, absorption, and secretion, for example).

C. Conductance

When a channel opens, membrane resistance falls because the channel allows current to flow across the lipid barrier. The extent to which resistance falls is dependent on the number of ions flowing through its pore per unit time, or its **conductance**. A channel's maximum conductance is one of its distinguishing hallmarks and is usually expressed in picoSiemens (pS).

D. Inactivation

Some channels possess an "inactivation gate" that is tripped upon activation, causing it to seal the channel pore and prevent further passage of ions (see Figure 2.14). The timescale of inactivation can vary from milliseconds to many tens of seconds, depending on channel class. Regardless, once a channel has been inactivated, it remains unresponsive to new stimuli, no matter how large the activating stimulus might be. Reactivation can only occur once the inactivation gate has been reset, a process that also has a variable timescale depending on channel type.

V. CHANNEL STRUCTURE

There are many ways in which a protein can be configured so as to create a transmembrane channel. Most mammalian channels follow a similar design principle, however, in which up to four to six subunits

assemble around a central, water-filled pore. Tetrameric channels are the most common form (see below), as shown in Figure 2.15, but many ligand-gated channels are pentameric, and connexin channels are hexameric (see 4·II·F). Tetrameric channel subunits typically comprise six membrane-spanning domains (S1–S6). The S5 and S6 domains include charged residues that fashion a pore and selectivity filter when the subunits are assembled. Voltage-dependent Na^+ and Ca^{2+} channels are products of a single gene incorporating four subunit-like domains, but most channels are assembled from independent proteins. The advantage to a modular approach to channel design is that it allows for infinite channel diversity. Changes in a single subunit can cause a voltage-gated channel to become a second messenger–gated channel, for example, or change its selectivity or its regulatory mechanism.

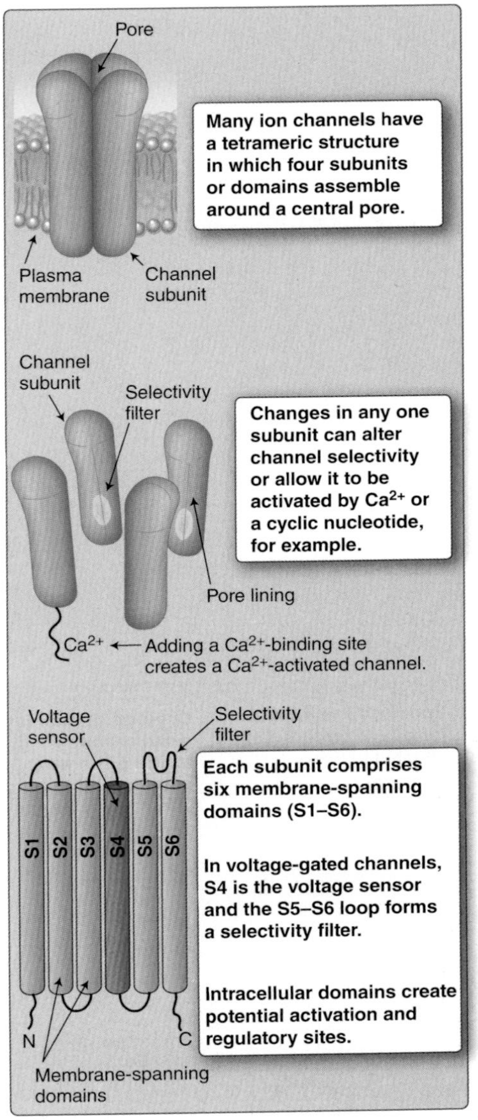

Figure 2.15
Voltage-gated ion-channel structure.

VI. CHANNEL TYPES

Channels are usually identified on the basis of their ion selectivity and their **gating** (activation) mechanism. Thus, a "voltage-dependent Na^+ channel" is activated by membrane depolarization and is Na^+ selective. Several different gating mechanisms are known.

A. Voltage gated

Voltage-gated Na^+ channels, Ca^{2+} channels, and K^+ channels all belong to the same channel superfamily related by a common tetrameric structure. The voltage-gated Na^+ channel responsible for the action-potential upstroke in nerve and muscle cells comprises a single pore-forming α subunit that associates with several smaller regulatory β subunits. The α subunit contains four related subunit-like domains that assemble around a central pore, as discussed in section V above. Each domain includes a highly charged peptide sequence (the S4 region) that functions as a voltage sensor. Membrane depolarization alters the charge distribution between the inner and outer membrane surfaces, and the voltage sensor shifts within the membrane, initiating a conformational change that opens the gate and reveals the channel pore.

B. Ligand gated

Ligand-gated channels transduce chemical signals and are the principal means by which neurons communicate with their targets. The diversity of the ligand-gated channel family is discussed in more detail in Chapter 6, but there are six principal classes that can be placed in three groups: ***cys*-loop receptors**, **ionotropic glutamate receptors**, and **adenosine triphosphate** (**ATP**) **receptors**.

1. ***Cys*-loop superfamily:** The *cys*-loop family includes the **nicotinic acetylcholine receptor** (**nAChR**), the **5-hydroxytryptamine** (**5-HT**) **receptor**, the **γ-aminobutyric acid** (**GABA**) receptor, and the **glycine** receptor. All family members share a short, highly conserved amino acid sequence that gives the family its name, and all comprise five subunits arranged around a central

Table 2.2: Second Messenger–Gated Channels

Second Messenger	Permeant Ion	Functions	Typical Locations	Notes
Ca^{2+}	K^+	Membrane repolarization	All cells	Three classes based on conductance, eight known members of the group
	Cl^-	Membrane repolarization	All cells (?)	Not well understood
	Ca^{2+}	Contraction	Sarcoplasmic reticulum	
G protein	K^+	Heart rate control	Heart	Four subunits assembled in heteromeric complexes
		Membrane repolarization	Nervous system	
	Ca^{2+}	Signaling; regulation	Nervous system	
cAMP, cGMP	Na^+, K^+, Ca^{2+}	Signal transduction	Visual and olfactory systems	Closely related to voltage-gated channels
		Pacemaker	Heart	
		Na^+ absorption	Kidney	
IP_3	Ca^{2+}	Contraction; secretion; transcription; others	Endoplasmic reticulum of all cells	Three related genes; four assembled subunits

cAMP = cyclic adenosine monophosphate; cGMP = cyclic guanosine monophosphate; IP_3 = inositol trisphosphate.

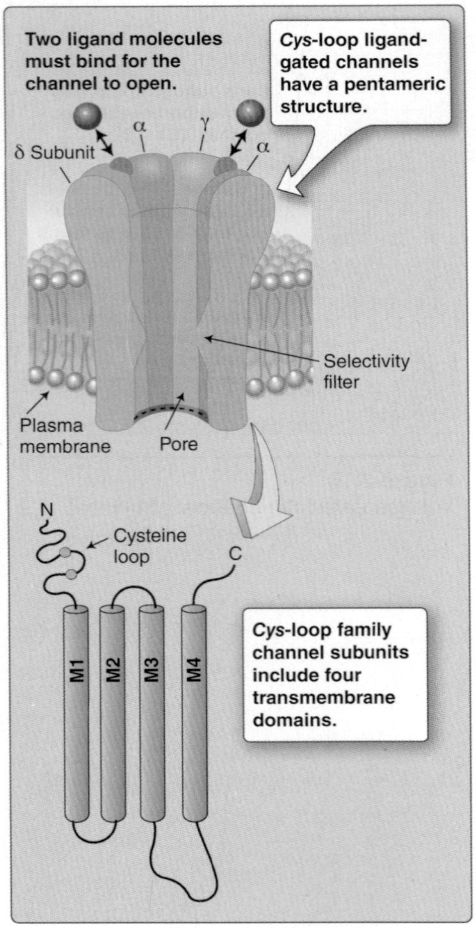
Two ligand molecules must bind for the channel to open.
Cys-loop ligand-gated channels have a pentameric structure.
α
γ
α
δ Subunit
Selectivity filter
Plasma membrane
Pore
N
Cysteine loop
C
M1
M2
M3
M4
Cys-loop family channel subunits include four transmembrane domains.

Figure 2.16
Cys-loop ligand-gated channel structure.

pore (Figure 2.16). The nAChR and serotonin receptors are relatively nonspecific cation channels that support a mixed Na^+, K^+, and Ca^{2+} influx upon ligand binding. The resultant membrane depolarization is excitatory. GABA and glycine receptors are anion channels that mediate Cl^- fluxes. These fluxes tend to stabilize V_m around resting potential and thereby inhibit membrane excitation. The nAChR and other family members have two ligand-binding sites that must be occupied simultaneously before the channel opens.

2. **Ionotropic glutamate receptors:** Ionotropic glutamate receptors are common in the central nervous system, where they play a critical role in learning and memory. All are tetrameric structures that support relatively nonselective Na^+ and K^+ fluxes when active. There are three principal groups that are differentiated pharmacologically (see Table 5.2): **AMPA** (α-amino-3-hydroxy-5-methyl-4-isoxazolepropionic acid receptor) receptors, **kainate** receptors, and **NMDA** (N-methyl-D-aspartate) receptors.

3. **Adenosine triphosphate receptors:** ATP-gated channels are P2X-family purinoreceptors that are activated by ATP and support a nonspecific Na^+, K^+, and Ca^{2+} flux when open. They are believed to form trimeric channels *in vivo*. ATP receptors are involved in taste transduction (see 10·II).

C. Second messenger gated

A third class of channel opens or closes in response to changes in intracellular messenger concentration (Table 2.2). Ca^{2+}-gated channels are ubiquitous, opening any time intracellular Ca^{2+} levels rise, regardless of whether the source of Ca^{2+} is an intracellular store or the ECF via a voltage-gated Ca^{2+} channel. Other channels are activated by G proteins, cyclic nucleotides, IP_3, and a number of additional messengers.

D. Sensory channels

Transient receptor-potential channels (**TRPs**) form a large and diverse group of channels that function as cellular sensors transducing temperature, taste, pain, and mechanical stress (cell swelling and shear stress), for example. TRPs are also required for Ca^{2+} and Mg^{2+} reabsorption from the renal tubule (see 27·III). TRPs are currently the subject of intense study, and many aspects of their behavior *in vivo* have yet to be delineated, but they are known to be tetrameric assemblies similar to the voltage-gated channels described above. Most family members are weakly cation selective, passing Na^+, K^+, and Ca^{2+}, the net result being membrane depolarization. The TRP family comprises six structurally distinct groups whose functions are summarized in Table 2.3.

Table 2.3: Sensory Channels

Channel	Function
TRPC (Canonical [TRPC1-7])	Ubiquitous
TRPV (Vanilloid [TRPV1-6])	Ca^{2+}-selective; transduce noxious heat and "hot" chemicals (e.g., capsaicin [TRPV1]); osmosensory (TRPV4); Ca^{2+} recovery from the renal tubule (TRPV5)
TRPM (Melastatin [TRPM1-8])	Transduce taste (TRPM5), cold, and "cold" chemicals (e.g., menthol [TRPM8]); recover Mg^{2+} from renal tubule (TRPM6)
TRPP (Polycystin [TRPP2, 3, 5])	Mutations cause polycystic kidney disease
TRPML (Mucolipin [TRPML1-3])	Lysosomal (?)
TRPA (Ankyrin [TRPA1])	Mechanosensor

Chapter Summary

- All cells modify their internal ionic environment using ion pumps (ATPases), which causes chemical concentration gradients to form across their surface membrane. Ions diffusing back down these concentration gradients create **diffusion potentials**. **Membrane potential** represents the sum of the diffusion potentials for all permeant ions (Na^+, Ca^{2+}, K^+, Mg^{2+}, and Cl^-).
- Ions are also influenced by electrical gradients, so their tendency to cross a membrane is governed by the net **electrochemical gradient**. The potential at which the chemical and electrical gradients balance precisely (the **equilibrium potential**) can be calculated using the **Nernst equation**.
- Most cells are impermeable to Na^+ and Ca^{2+} at rest, but the presence of a significant resting K^+ conductance causes **resting potential** to settle at close to the K^+ equilibrium potential. The resting K^+ conductance makes resting potential highly susceptible to changes in extracellular K^+ concentration (**hypokalemia** and **hyperkalemia**).
- Excitable cells use changes in membrane potential (**action potentials**, or **spikes**) to communicate with each other and to trigger cellular events, such as muscle contraction and secretion. Action potentials are effected by the sequential opening and closing of **ion channels**. **Voltage-dependent Na^+-channel** opening facilitates an **inward Na^+ current** to cause **membrane depolarization**. Membrane **repolarization** is effected (in part) by an **outward K^+ current** through voltage-dependent K^+ channels.
- Action potentials are initiated locally at the site of stimulation and then **propagate** in a self-sustaining, **regenerative** fashion along the length of a cell.
- Most cells express many different ion-channel classes in their surface membrane, which can be distinguished on the basis of their mode of activation (**gating**), ion **selectivity**, **activation** and **inactivation kinetics**, **conductance**, and pharmacology.
- Voltage-dependent Na^+ channels, K^+ channels, Ca^{2+} channels, and Cl^- channels are activated by changes in membrane potential. **Ligand-gated channels** are activated by neurotransmitters, including acetylcholine, γ-aminobutyric acid, and glutamate. **Second messenger–gated channels** are sensitive to intracellular Ca^{2+}, G proteins, cyclic nucleotides, and inositol trisphosphate. **Transient receptor-potential channels** are cellular sensors, mediating responses to chemicals, hot and cold temperatures, and mechanical stress.

3 Osmosis and Body Fluids

I. OVERVIEW

One of the more memorable quotes from the popular television series *Star Trek: The Next Generation* came from a silicon-based alien life form that referred to the intrepid Captain Picard as an "ugly bag of mostly water." The average human body comprises 50%–60% water by weight, depending on body composition, gender, and age of the individual. The proportion of water in cells is even greater (~80%) as shown in Figure 3.1, the remainder largely comprising proteins. Water is the universal solvent, facilitating molecular interactions, biochemical reactions, and providing a medium that supports molecular movement between different cellular and subcellular compartments. The biochemistry of life is highly sensitive to solute concentration, which, in turn, is determined by how much water is contained within a cell. Thus, the **autonomic nervous system** (**ANS**) closely monitors **total body water** (**TBW**) and adjusts intake and output pathways (drinking and urine formation, respectively) to maintain water balance (see 28·II). Although TBW is tightly regulated, water moves freely across cell membranes and between the body's different fluid compartments. Loss of water from the cell raises intracellular solute concentrations and, thereby, interferes with normal cell function. The body does not contain a transporter capable of redistributing water between compartments, so its approach to water management at the cellular and tissue level is to manipulate solute concentrations within intracellular fluid (ICF), extracellular fluid (ECF), and plasma. This approach is effective because water is enslaved to solute concentration by **osmosis**.

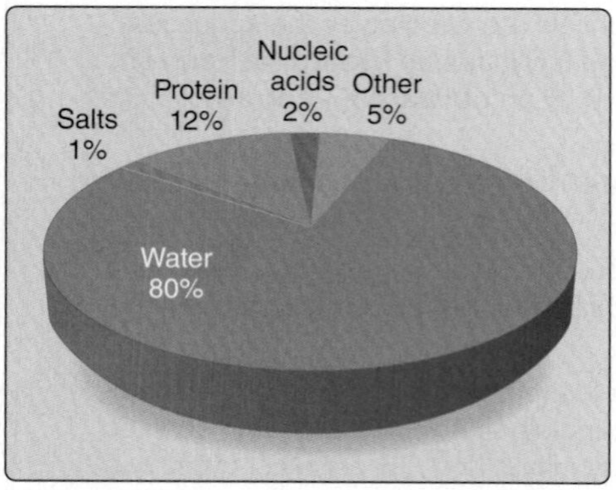

Figure 3.1
Cellular composition.

II. OSMOSIS

Osmosis describes a process by which water moves passively across a semipermeable membrane, driven by a difference in water concentration between the two sides of the membrane. Pure water has a molarity of >55 moles/L. Although cells do not contain pure water, it is nevertheless a superabundant chemical. The concentration difference required to generate physiologically significant water flow across membranes is very small, so, in practice, it is far more convenient to discuss osmosis in terms of the amount of *pressure* that water is capable of generating as it moves down its concentration gradient. Thus, a chemical concentration gradient becomes an **osmotic pressure gradient**.

A. Osmotic pressure

Osmotic pressure gradients are created when solute molecules displace water, thereby decreasing water concentration. An apparent peculiarity of the process is that pressure is determined entirely by solute particle number and is largely independent of the size, mass, chemical nature of the solute, or even its electrical valence. Therefore, two small ions such as Na^+ generate a higher osmotic pressure than a single complex glucose polymer such as starch (MW >40,000) as shown in Figure 3.2. The osmotic pressure of a solution (π; measured in mm Hg) can be calculated from:

Equation 3.1 $$\pi = nCRT$$

where n is the number of particles that a given solute dissociates into when in solution, C is solute concentration (in mmol/L), and R and T are the universal gas constant and absolute temperature, respectively. Osmotic pressure can be measured physically as the amount of pressure required to precisely counter water movement between two solutions with dissimilar solute concentrations (Figure 3.3).

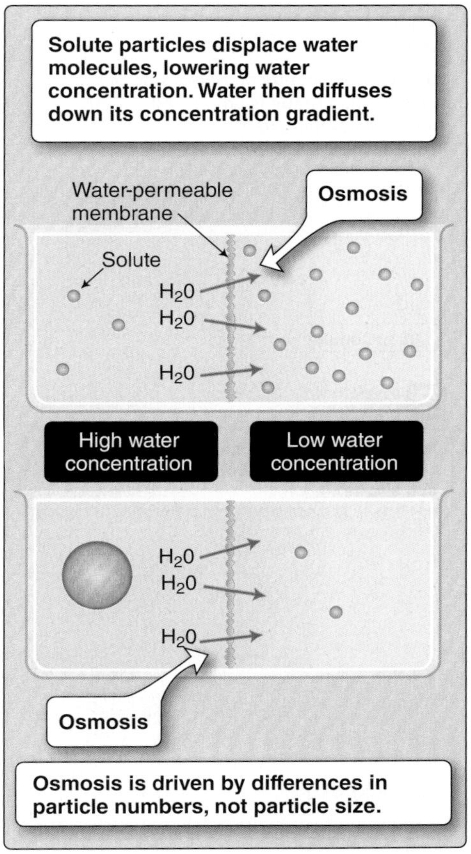

Figure 3.2
Osmosis.

B. Osmolarity and osmolality

Osmolarity is a measure of a solute's ability to generate osmotic pressure that takes into account how many particles a solute dissociates into when dissolved in water. Glucose does not dissociate in solution, so a 1-mmol/L glucose solution has an osmolarity of 1 milliOsmole (mOsm). NaCl dissociates into two osmotically active particles in solution (Na^+ and Cl^-) and, thus, a 1 mmol/L–NaCl solution has an osmolarity of ~2 mOsm. $MgCl_2$ dissociates into three particles ($Mg^{2+} + 2Cl^-$) and, thus, a 1 mmol/L–$MgCl_2$ solution has an osmolarity of 3 mOsm.

Osmolality is an almost identical measure to osmolarity but uses water mass in place of volume (i.e., Osm/kg H_2O). A liter of water has a mass of 1 kg at 4°C, but water volume increases with temperature, which causes osmolarity to fall slightly. Because mass is invariant, Osm/kg H_2O is the preferred unit for use in discussions of human physiology.

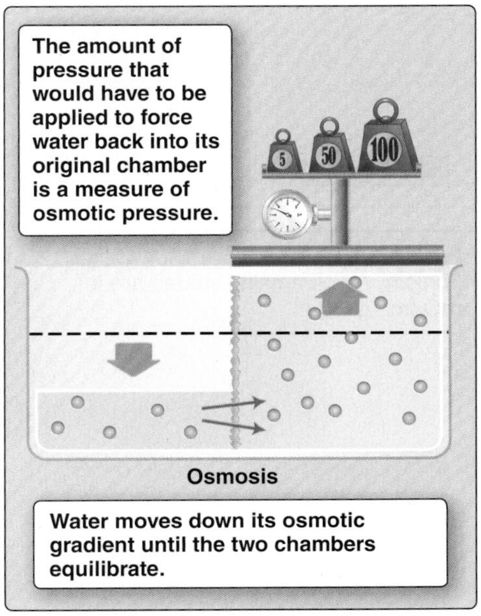

The amount of pressure that would have to be applied to force water back into its original chamber is a measure of osmotic pressure.
5
50
100
Osmosis
Water moves down its osmotic gradient until the two chambers equilibrate.

Figure 3.3
Osmotic pressure.

C. Tonicity

Tonicity measures a solute's effect on *cell volume*, the term recognizing that membrane-permeant solutes cause cells to shrink or swell through effects on ICF osmolality.

1. **Nonpermeant solutes:** Sucrose cannot cross the plasma membrane of most cells. Therefore, if a cell is placed in a sucrose solution whose osmolality matches that of the ICF (300 mOsm/kg H_2O), cell volume will remain unchanged because the solution is **isotonic** (Figure 3.4A). Volume changes only occur when there is an osmotic gradient across the plasma membrane that forces water to enter or leave the cell.

> Note that ICF typically has an osmolality of 290 mOsm/kg H_2O *in vivo*. The value of 300 mOsm/kg H_2O used in this and the following examples is for ease of illustration only.

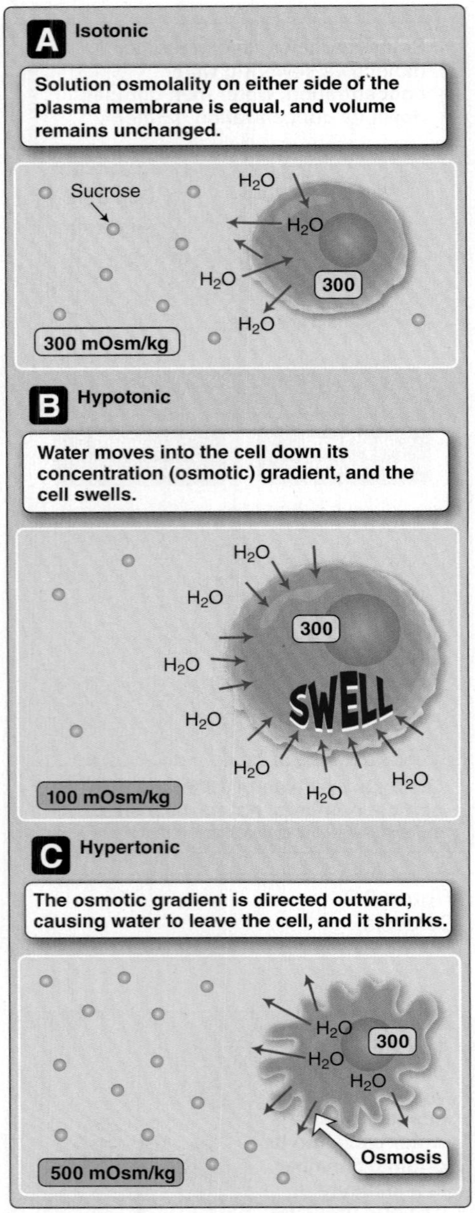

Figure 3.4
Tonicity. All osmolality values are in mOsm/kg H_2O.

A 100–mOsm/kg H_2O sucrose solution is **hypotonic** compared with the ICF. Water molecules will migrate across the membrane from ECF to ICF following the osmotic gradient, and the cell will swell (see Figure 3.4B). Conversely, a 500–mOsm/kg H_2O sucrose solution is **hypertonic**: Water will be drawn out of the cell by osmosis, causing the cell to shrink (see Figure 3.4C).

2. **Permeant solutes:** Urea is a small (60 MW) organic molecule that, unlike sucrose, readily permeates the membranes of most cells via a urea transporter (UT). Thus, although 300–mOsm/kg H_2O urea and 300–mOsm/kg H_2O sucrose have identical osmolalities (i.e., they are **isosmotic**), they are *not* isotonic. When a cell is placed in a 300–mOsm/kg H_2O urea solution, urea crosses the membrane via UT and raises ICF osmolality. Water then follows urea by osmosis, and the cell swells. A 300–mOsm/kg H_2O urea solution is, thus, considered to be **hypotonic**.

3. **Mixed solutions:** A solution containing 300 mOsm/kg H_2O urea *plus* 300 mOsm/kg H_2O sucrose has an osmolarity of 600 mOsm/kg H_2O and is, thus, **hyperosmotic** relative to the ICF. It is also functionally isotonic, however, because urea rapidly crosses the membrane until the intracellular and extracellular urea concentrations equilibrate at 150 mOsm/kg H_2O. With solution osmolality on both sides of the membrane now standing at 450 mOsm/kg H_2O, the driving force for osmosis is zero, and cell volume remains unchanged.

4. **Reflection coefficient:** When calculating the osmotic potential of a solution that bathes a cell, it is necessary to add a reflection coefficient (σ) to Equation 3.1 above.

$$\pi = \sigma nCRT$$

The reflection coefficient is a measure of the ease with which a solute can traverse the plasma membrane. For highly permeant solutes such as urea, σ approaches 0. The reflection coefficient for nonpermeant solutes (such as sucrose and plasma proteins) approaches 1.0.

D. Water movement between intracellular and extracellular fluids

The plasma membrane's lipid core is hydrophobic, but water enters and exits the cell with relative ease. Some water molecules slip between adjacent membrane phospholipid molecules, whereas others are swept along with solutes in ion channels and transporters. Most cells also express **aquaporins** (**AQPs**) in their surface membrane, large tetrameric proteins that form water-specific channels across the lipid bilayer. AQPs, unlike most ion channels, are always open and water permeable (see 1·V·A).

E. Cell volume regulation

ECF solute composition is maintained within fairly narrow limits by the pathways involved in TBW homeostasis (see 28·II), but ICF osmolality changes constantly with changing activity levels. When cell metabolism increases, for example, nutrients are absorbed, metabolic waste products accumulate, and water moves into the cell by

osmosis, causing it to swell. Cells that exist on the boundary between the internal and external environment (e.g., intestinal and renal epithelial cells) are also subject to acute changes in extracellular osmolality, causing frequent changes in cell volume. The mechanisms by which cells sense and transduce volume changes are still not well defined, but they respond to osmotic shrinkage and swelling by enacting a **regulatory volume increase** (**RVI**) or a **regulatory volume decrease** (**RVD**), respectively.

1. **Regulatory volume increase:** When ECF osmolality rises, water is drawn out of the cell by osmosis, and it shrinks. The cell responds with an RVI, which, in the short term, involves accumulation of Na^+ and Cl^- through increased Na^+-H^+ exchanger and Na^+-K^+-$2Cl^-$ cotransporter activity (Figure 3.5). Na^+ and Cl^- uptake raises ICF osmolality and restores cell volume by osmosis. In the longer term, cells may accumulate small organic molecules, such as betaine (an amino acid), sorbitol, and inositol (polyalcohols) to maintain increased ICF osmolality and retain volume.

2. **Regulatory volume decrease:** Cell swelling initiates an RVD, which principally involves K^+ and Cl^- efflux via swelling-activated K^+ channels and Cl^- channels. The resulting fall in ICF osmolality causes water loss by osmosis, and cell volume renormalizes. Cells may also release amino acids (principally glutamate, glutamine, and taurine) as a way of reducing their osmolality and volume.

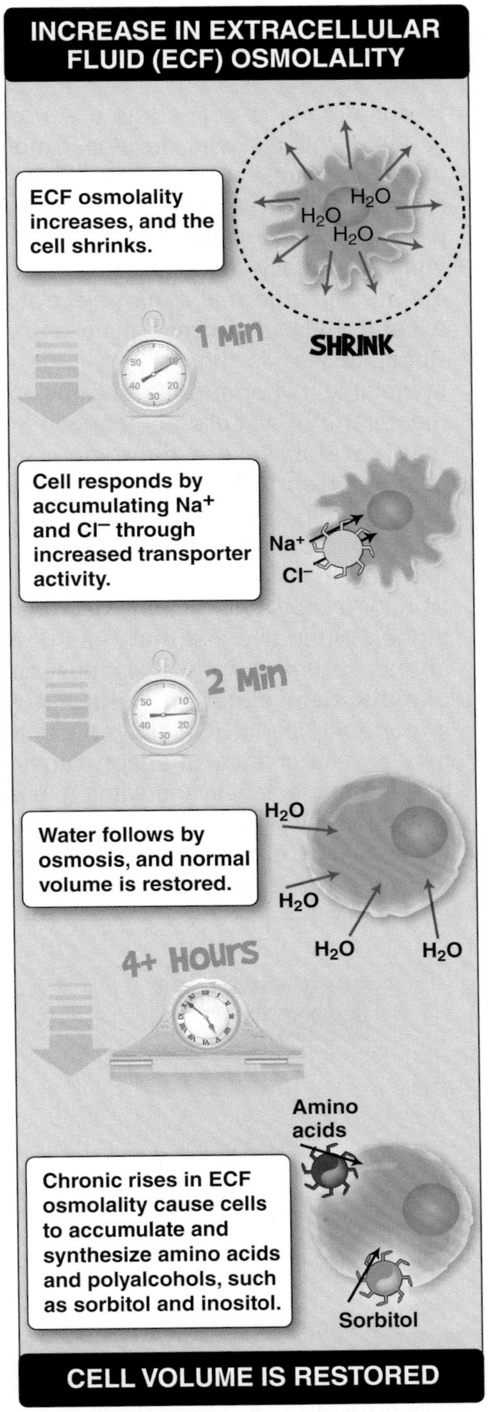
INCREASE IN EXTRACELLULAR FLUID (ECF) OSMOLALITY
ECF osmolality increases, and the cell shrinks.
H_2O
H_2O
H_2O
1 Min
SHRINK
Cell responds by accumulating Na^+ and Cl^- through increased transporter activity.
Na^+
Cl^-
2 Min
Water follows by osmosis, and normal volume is restored.
H_2O
H_2O
H_2O
H_2O
4+ Hours
Amino acids
Chronic rises in ECF osmolality cause cells to accumulate and synthesize amino acids and polyalcohols, such as sorbitol and inositol.
Sorbitol
CELL VOLUME IS RESTORED

Figure 3.5
Regulatory volume increase.

III. BODY FLUID COMPARTMENTS

A 70-kg male contains 42 L of water, or around 60% of total body weight. Females generally have less muscle and more adipose tissue as a percentage of total body mass than do males. Because fat contains less water than muscle, their total water content is correspondingly lower (55%). TBW usually decreases with age in both sexes due to loss of muscle mass (**sarcopenia**) associated with aging.

A. Distribution

Two thirds of TBW is contained within cells (ICF = ~28 L of the 42 L cited above). The remainder (14 L) is divided between the interstitium and blood plasma (Figure 3.6).

1. **Plasma:** The cardiovascular system comprises the heart and an extensive network of blood vessels that together hold ~5 L of blood, a fluid composed of cells and protein-rich plasma. Approximately 1.5 L of total blood volume is contained within blood cells and is included in the value given for ICF above. Plasma accounts for 3.5 L of ECF volume.

2. **Interstitium:** The remaining 10.5 L of water resides outside the vasculature and occupies spaces between cells (the **interstitium**). Interstitial fluid and plasma have very similar solute compositions because water and small molecules move freely between the two compartments. The main difference between plasma and interstitial fluid is that plasma contains large amounts of proteins, whereas interstitial fluid is relatively protein free.

Clinical Application 3.1: Hyponatremia and Osmotic Demyelination Syndrome

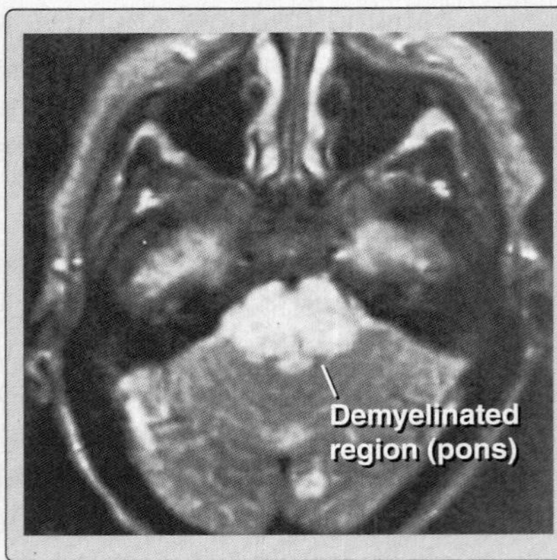

Osmotic demyelination in the pons region of the brain.

Hyponatremia is defined as a serum Na^+ concentration of 135 mmol/L or less. Patients who develop hyponatremia usually have an impaired ability to excrete water, often due to an inability to suppress antidiuretic hormone (ADH) secretion. Hyponatremia with appropriate ADH suppression is also seen with advanced renal failure and low dietary sodium intake. Normally, the kidneys can excrete 10–15 L of dilute urine per day and maintain normal serum electrolyte levels, but higher flow rates may exceed their solute resorptive capabilities, and hyponatremia ensues. Because Na^+ is the primary determinant of extracellular fluid (ECF) osmolality, hyponatremia creates an osmotic shift across the plasma membrane of all cells and causes them to swell. Hyponatremic patients may develop severe neurologic symptoms (i.e., lethargy, seizures, coma), which typically only occur with acute and severe hyponatremia (serum sodium concentration $<$120 mmol/L), and rapid correction with hypertonic saline is necessary in this clinical scenario. Hyponatremia that develops slowly and chronically (more commonly the case) allows time for a regulatory volume decrease, and severe symptoms may be delayed until serum Na^+ levels fall even further. When hyponatremia has developed slowly, and a patient has no neurologic symptoms, correction to normal serum sodium levels must also be undertaken slowly to avoid a treatment complication known as the **osmotic demyelination syndrome** ([**ODS**] formerly called **central pontine myelinolysis**). ODS occurs when a too-rapid rise in ECF Na^+ concentration creates an osmotic gradient that draws water from neurons before they have a chance to adapt, causing cell shrinkage and demyelination (myelin is a lipid-rich layered membrane that electrically insulates axons to enhance their conduction velocity; see 5·V·A). ODS may manifest as confusion, behavioral changes, quadriplegia, difficulties with speech or swallowing (dysarthria and dysphagia, respectively), or coma. Because these devastating changes may not be reversible, the maximum rate of correction in stable patients with chronic hyponatremia should not exceed ~10 mmol/L in the first 24 hours.

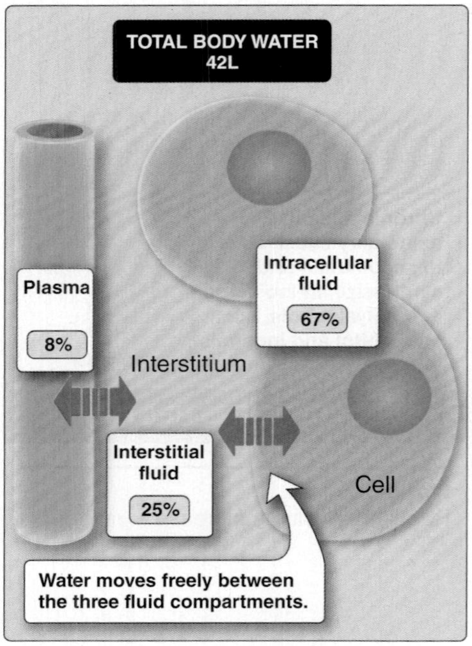

Figure 3.6
Total body water distribution.

A variable amount of fluid is held behind cellular barriers that separate it from plasma and interstitial fluid (**transcellular fluid**). This includes cerebrospinal fluid, fluid within the eye (aqueous humor), joints (synovial fluid), bladder (urine) and intestine. Transcellular fluid volume averages between 1–2 L and is not considered in calculations of TBW.

B. Restricting water movement

Water moves freely and rapidly across membranes and capillary walls, which creates the possibility of one fluid compartment (the ICF, for example) becoming hypohydrated or hyperhydrated relative to the other compartments to the detriment of body function (Figure 3.7). Thus, the body puts mechanisms in place that independently control the water content and that limit net water movement between the ICF, ECF, and plasma.

1. **Intracellular fluid:** ICF osmolality typically averages ~275–295 mOsm/kg H_2O, due primarily to K^+ and its associated anions (Cl^-,

phosphates, and proteins). The ICF's K^+-rich composition is due to the plasma membrane Na^+-K^+ ATPase, which concentrates K^+ within the ICF and expels Na^+. Net water loss or accumulation from the interstitium is prevented by regulatory volume increases and decreases, respectively, as discussed above.

2. **Extracellular fluid:** Plasma and interstitial fluid also have an osmolality of ~275–295 mOsm/kg H_2O, but principal solutes here are Na^+ and its associated anions (Cl^- and HCO_3^-). ECF water content is tightly controlled by centrally located osmoreceptors acting through antidiuretic hormone (ADH). When TBW falls as a result of excessive sweating, for example (see Figure 3.7, panel 1), ECF osmolality rises because its solutes have concentrated. The rise in osmolality draws water from ICF by osmosis (see Figure 3.7, panel 2) and triggers a RVI in all cells, but not before the central osmoreceptors have initiated ADH release from the posterior pituitary as shown in Figure 3.7, panel 3 (also see 28·II·B). ADH stimulates thirst and enhances AQP expression by the renal tubule epithelium, permitting increased water recovery from urine. TBW and ECF osmolality are restored to normal as a result (see Figure 3.7, panel 4). When TBW is too high, AQP expression is suppressed, and the excess water is expelled from the body.

3. **Plasma:** Plasma is the smallest but also the most vital of the three internal fluid compartments. The heart absolutely depends on blood volume to generate pressure and flow through the vasculature (see 18·III). Plasma volume must be preserved even if ECF volume is falling due to prolonged sweating or reduced water ingestion, for example. The body cannot regulate plasma volume directly because most small blood vessels (capillaries and venules) are inherently leaky and, thus, plasma and interstitial fluid (the two ECF components) are always in equilibrium with each other. The solution to maintaining adequate plasma volume lies with plasma proteins, such as albumin, which are synthesized by the liver and remain trapped in the vasculature by virtue of their large size. Here, they exert an osmotic potential (**plasma colloid osmotic pressure**) that draws fluid from the interstitium, regardless of changes in bulk ECF osmolality or ECF volume depletion as shown in Figure 3.7, panel 2 (also see 19·VII·A).

IV. BODY FLUID pH

H^+ is a common inorganic cation that is similar in many ways to Na^+ and K^+. It is attracted to and binds to anions, and it depolarizes cells when it crosses the plasma membrane. H^+ deserves special consideration and cellular handling because its small atomic size allows it to form strong bonds with proteins. Such interactions alter a protein's internal charge distribution, weaken interactions between adjacent polypeptide chains, and cause conformational changes that may inhibit function such as hormone binding (Figure 3.8). High H^+ concentrations denature proteins and cause cell degradation. Thus, the pH of fluid in which cells are bathed must be tightly controlled at all times.

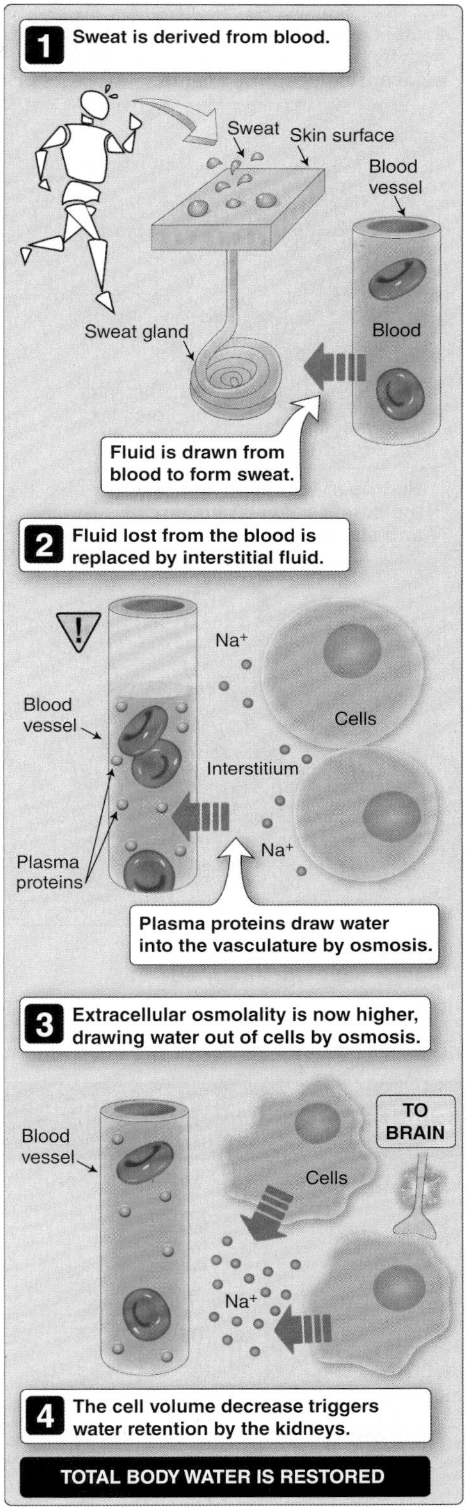

1 Sweat is derived from blood.
Sweat
Skin surface
Blood vessel
Sweat gland
Blood
Fluid is drawn from blood to form sweat.
2 Fluid lost from the blood is replaced by interstitial fluid.
Na+
Blood vessel
Cells
Interstitium
Plasma proteins
Na+
Plasma proteins draw water into the vasculature by osmosis.
3 Extracellular osmolality is now higher, drawing water out of cells by osmosis.
Blood vessel
TO BRAIN
Cells
Na+
4 The cell volume decrease triggers water retention by the kidneys.
TOTAL BODY WATER IS RESTORED

Figure 3.7
Movement between fluid compartments during dehydration.

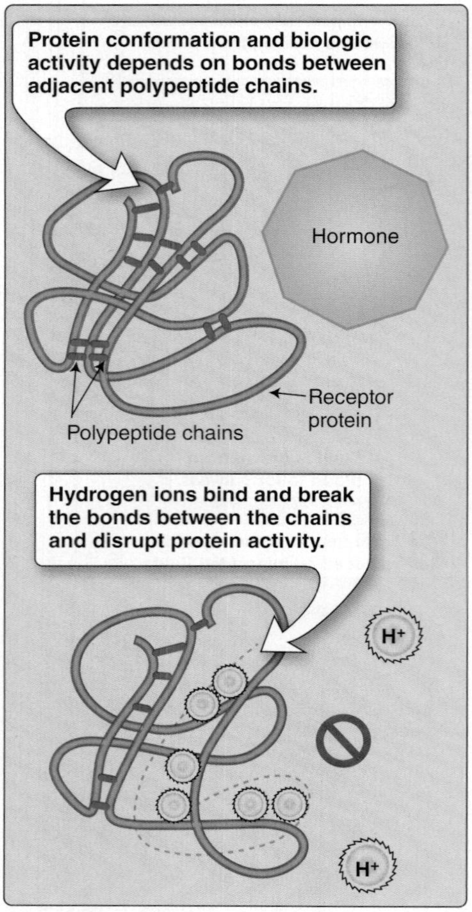

Protein conformation and biologic activity depends on bonds between adjacent polypeptide chains.
Hormone
Receptor protein
Polypeptide chains
Hydrogen ions bind and break the bonds between the chains and disrupt protein activity.
H+
H+

Figure 3.8
Protein denaturation by acid.

Table 3.1: Serum Electrolytes

Electrolyte	Reference Range (mmol/L)
Na^+	136–145
K^+	3.5–5.0
Ca^{2+}	2.1–2.8
Mg^{2+}	0.75–1.00
Cl^-	95–105
HCO_3^-	22–28
Phosphorus (inorganic)	1.0–1.5

Clinical Application 3.2: Electrolytes

Extracellular fluid (ECF) is Na^+ rich, but it also contains a number of other charged solutes, or **electrolytes**, the bulk comprising common inorganic ions (K^+, Ca^{2+}, Mg^{2+}, Cl^-, phosphates, and HCO_3^-). All cells are bathed in ECF. Because changes in the concentrations of any of these electrolytes can have significant effects on cell function, serum levels are maintained within a fairly narrow range, principally through modulation of kidney function (see Chapter 28). Blood tests typically include a standard electrolyte panel that measures serum Na^+, K^+, and Cl^- (Table 3.1). Serum Na^+ and Cl^- levels are measured, in part, to assess kidney function, but also because they determine ECF osmolality and total body water. K^+ is measured because normal cardiac function depends on stable serum K^+ levels.

A. Acids

Blood has a pH of 7.4 and seldom varies by more than 0.05 pH units. This corresponds to a H^+ concentration range of 35–45 nmol/L, which is impressive given that metabolism of carbohydrates, fats, and proteins pours ~22 moles of acid into the vasculature every day! Acid generated by metabolism comes in two forms: **volatile** and **nonvolatile** (Figure 3.9).

1. **Volatile:** The vast majority of daily acid output comes in the form of carbonic acid (H_2CO_3), which is created when CO_2 dissolves in water. CO_2 is generated from carbohydrates (such as glucose) during aerobic respiration ($C_6H_{12}O_6 + 6O_2 \rightarrow 6CO_2 + 6H_2O$). Carbonic acid is known as a volatile acid because it is converted back to CO_2 and water in the lungs and then liberated to the atmosphere (see Figure 3.9).

2. **Nonvolatile:** Metabolism also generates smaller amounts (~70–100 mmol per day) of **nonvolatile** or **fixed acid** that cannot be disposed of via the lungs. Nonvolatile acids include sulphuric, nitric, and phosphoric acids, which are formed during catabolism of amino acids (e.g., cysteine and methionine) and phosphate compounds. Nonvolatile acids are excreted in urine (see Figure 3.9).

3. **Range:** Life can only exist within a relatively narrow pH range (pH 6.8–7.8, corresponding to a H^+ concentration of 16–160 nmol/L), so excreting H^+ in a timely manner is critical for survival. A decrease in plasma pH below 7.35 is called **acidemia**. **Alkalemia** is an increase in plasma pH above 7.45. **Acidosis** and **alkalosis** are more general terms referring to processes that result in acidemia and alkalemia, respectively.

B. Buffer systems

Cells produce acid continually. Their intracellular structures are protected from the deleterious effects of this acid by buffer systems, which immobilize H^+ temporarily and limit its destructive effects until it can be disposed of. The body contains three primary buffer

systems: the **bicarbonate buffer system**, **phosphate buffer system**, and proteins.

1. **Bicarbonate:** HCO_3^- is the body's primary defense against acid. HCO_3^- is a base that combines with H^+ to form carbonic acid, H_2CO_3:

$$HCO_3^- + H^+ \leftrightarrows H_2CO_3 \leftrightarrows CO_2 + H_2O$$
carbonic anhydrase

H_2CO_3 can then be broken down to form CO_2 and water, both of which are readily expelled from the body via the lungs and kidneys, respectively. Spontaneous conversion to CO_2 and H_2O occurs too slowly for the HCO_3^- buffer system to be of any practical use, but the reaction becomes essentially instantaneous when catalyzed by *carbonic anhydrase* (*CA*). *CA* is a ubiquitous enzyme expressed by all tissues, reflecting the central importance of the HCO_3^- buffer system.

> There are at least 12 different functional *CA* isoforms, many of which are expressed in virtually all tissues. *CA-II* is a ubiquitous cytosolic isoform. *CA-I* is expressed at high levels in red blood cells, whereas *CA-III* is found primarily in muscle. *CA-IV* is a membrane-bound isoform that is expressed on the surface of pulmonary and renal epithelia, where it facilitates acid excretion.

2. **Phosphate:** The phosphate buffer system employs hydrogen phosphate to buffer acid, the end-product being dihydrogen phosphate:

$$H^+ + HPO_4^{2-} \leftrightarrows H_2PO_4^-$$

HPO_4^{2-} is used to buffer acid in the renal tubule during urinary excretion of nonvolatile acids.

3. **Proteins:** Proteins contain numerous H^+-binding sites and, therefore, make a major contribution to net intracellular and extracellular buffering capacity. One of the most important of these is hemoglobin (Hb), a protein found in red blood cells (RBCs), that buffers acid during transit to the lungs and kidney.

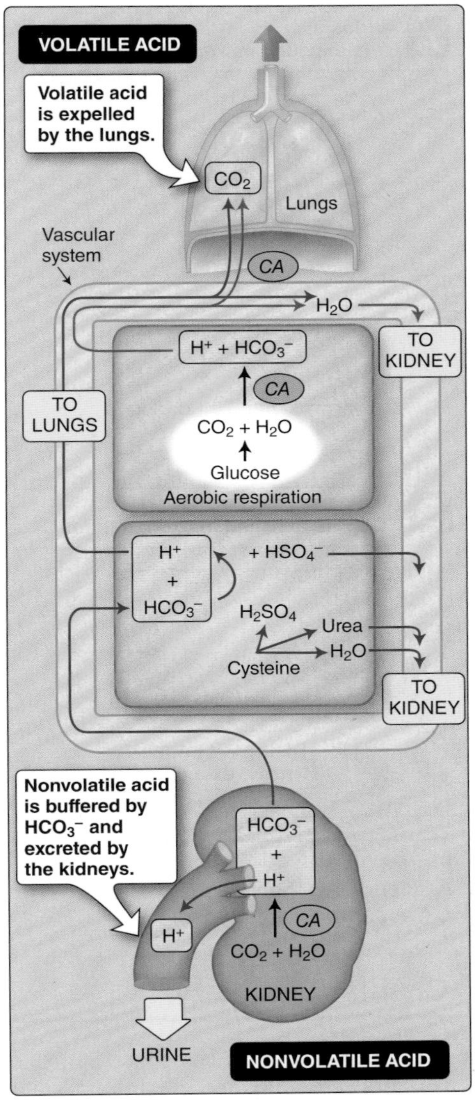

Figure 3.9
Excretion of volatile and nonvolatile acids. *CA* = *carbonic anhydrase.*

C. Acid handling

Most acid is generated intracellularly at sites of active metabolism and then is transported in the vasculature to the lungs and kidneys for disposal. pH is carefully controlled by buffers and pumps at all stages of handling.

1. **Cells:** Intracellular structures are shielded from locally produced acid by buffers, the most important being intracellular proteins and HCO_3^-. Cells also actively control their internal pH using transporters, although the pathways involved in cellular pH control have not been well delineated.

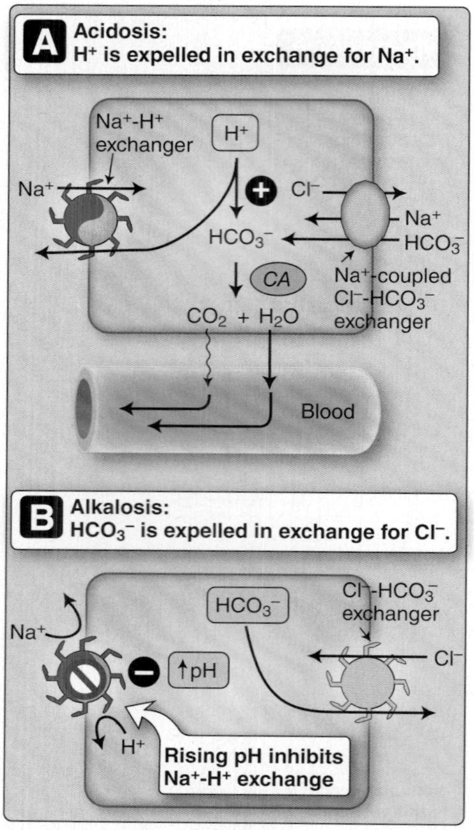

Figure 3.10
Acid and base handling by cells.
CA = carbonic anhydrase.

a. **Acid:** Most cells express a Na^+-H^+ exchanger to expel acid and can also take up HCO_3^- from the ECF if the need arises, via Na^+-coupled Cl^--HCO_3^- exchange (Figure 3.10A).

b. **Base:** Most cells also express a Cl^--HCO_3^- exchanger to expel excess base. Alkalosis simultaneously suppresses Na^+-H^+ exchange to help lower intracellular pH (see Figure 3.10B).

2. **Lungs:** CO_2 produced by cells during aerobic respiration rapidly diffuses across the cell membrane and crosses through the interstitium to the vasculature. RBCs express high levels of *CA-II*, which facilitates conversion of CO_2 and H_2O to HCO_3^- and H^+ (see 23·VII). H^+ then binds to Hb for transit to the lungs. Pulmonary epithelia also contain high levels of *CA*, which facilitates conversion back to CO_2 for transfer to the atmosphere (see Figure 3.9).

3. **Kidneys:** H^+ that is formed from protein metabolism (nonvolatile acid) is pumped into the renal tubule lumen and excreted in urine as shown in Figure 3.9 (also see 28·V). The urinary epithelia are protected during excretion by buffers, primarily phosphate and ammonium, which the renal tubule secretes specifically for this purpose. Nonvolatile acid is generated at distant sites, however, and the cells responsible must be protected from this acid until transport to the kidney can be arranged. Thus, the renal epithelium also expresses high levels of *CA-IV*, which generates HCO_3^- and releases it to the vasculature for transport to the sites of acid generation (see Figure 3.9). H^+ that is formed during HCO_3^- synthesis is pumped into the tubule lumen and excreted.

Chapter Summary

- The human body is composed largely of water that distributes between three principal compartments: **intracellular fluid**, **interstitial fluid**, and **plasma**. The latter two compartments together comprise **extracellular fluid**. Water movement between these compartments occurs principally by **osmosis**.
- Osmosis is driven by **osmotic pressure gradients** that are created by local differences in solute particle number. Water moves from regions containing low particle number toward regions with high particle numbers, generating osmotic pressure.
- **Osmolarity** and **osmolality** measure a solute's ability to generate osmotic pressure, whereas **tonicity** is governed by a solution's effect on cell volume.
- Most cells contain channels (**aquaporins**) that allow water to move easily between intracellular fluid and extracellular fluid (ECF) in response to transmembrane osmolality gradients. Increases in ECF osmolality cause water to leave the cell and its volume decreases. Cells respond by accumulating solutes (Na^+, Cl^-, and amino acids) to recruit water from the ECF by osmosis (a **regulatory volume increase**). Cell volume increases elicit a **regulatory volume decrease**, involving volume-activated K^+ and Cl^- channel opening and secretion of small organic solutes (amino acids and polyalcohols).
- Regulatory volume changes allow cells to control intracellular water content. Kidney function is modulated to control total body Na^+ content, which, in turn, determines how much water is retained by extracellular fluid (ECF). Plasma proteins determine how much of this ECF is retained by the vasculature.
- All cells rely on **buffer systems** to maintain the pH of intracellular and extracellular fluids within a narrow range. Acid is produced continually as a result of carbohydrate metabolism and amino acid catabolism. Carbohydrate metabolism yields CO_2, which dissolves in water to form carbonic acid (a **volatile** acid). Amino acid breakdown yields sulphuric and phosphoric acids (**nonvolatile** acids).
- The **bicarbonate buffer system** represents the body's primary defense against acid. The buffer system relies on the ubiquitous enzyme ***carbonic anhydrase*** to facilitate bicarbonate formation from CO_2 and water. Volatile acid is expelled as CO_2 from the lungs, whereas nonvolatile acid is excreted in urine by the kidneys.

Epithelial and Connective Tissue

4

I. OVERVIEW

The human body comprises a diverse assemblage of cells that can be placed in one of four groups based on structural and functional similarities. These groups are known as **tissues**: **epithelial tissue**, **nervous tissue**, **muscle tissue**, and **connective tissue**. The four tissue types associate with and work in close cooperation with each other. Epithelial tissue comprises sheets of cells that provide barriers between the internal and external environment. Skin (the **epidermis**) is the most visible example (see Chapter 16), but there are many unseen internal interfaces that are lined with epithelia also (e.g., lungs, gastrointestinal [GI] tract, kidneys, and reproductive organs). Nervous tissue comprises **neurons** and their support cells (**glia**) that provide pathways for communication and coordinate tissue function, as will be discussed in greater detail in Unit II. Muscle tissue is specialized for contraction. There are three types of muscle: **skeletal muscle** (see Chapter 12), **cardiac muscle** (see Chapter 13), and **smooth muscle** (see Chapter 14). Connective tissue is a mix of cells, structural fibers, and **ground substance** that connects and fills the spaces between adjacent cells and gives tissues their strength and form. Bone is a specialized connective tissue that is mineralized to provide strength and to resist compression (see Chapter 15). This chapter considers the structure and varied functions of epithelial tissue (Table 4.1) and connective tissue.

Table 4.1: Epithelial Functions

Function	Examples
Protection	Epidermis; mouth; esophagus; larynx; vagina; anal canal
Excretion	Kidney
Secretion	Intestine; kidney; most glands
Absorption	Intestine; kidney
Lubrication	Intestine; airways; reproductive tracts
Cleansing	Trachea; auditory canal
Sensory	Gustatory, olfactory, and vestibular epithelia
Reproduction	Germinal, uterine, and ovarian epithelia

II. EPITHELIA

Epithelia are continuous sheets of cells that line all body surfaces and create barriers that separate the internal and external environments. They help protect us from invasion by microorganisms, and they limit fluid loss from the internal environment: They "keep our insides in." Epithelia are much more than just barriers, however. Most epithelia have additional specialized secretory or absorptive functions that include sweat formation, food digestion and absorption, and excretion of waste products.

A. Structure

The simplest epithelia comprise a single layer of cells adhered to each other by a variety of **junctional complexes** that impart mechanical strength and create pathways for communication between adjacent

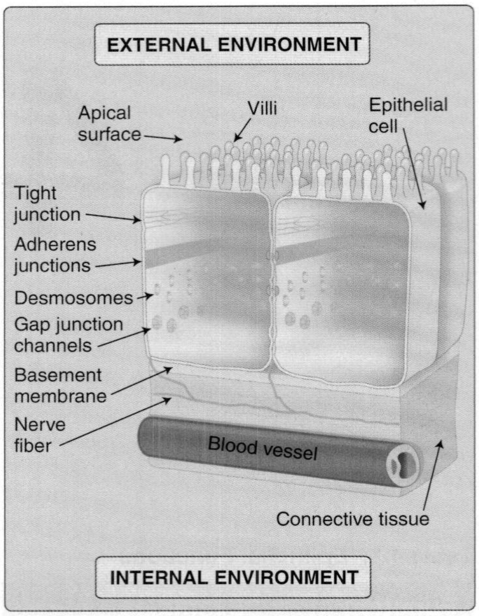

Figure 4.1
Epithelial cell structure.

cells (Figure 4.1). The **apical surface** interfaces with the external environment (or an internal body cavity), whereas the **basal surface** rests on a **basement membrane** that provides structural support. The basement membrane comprises two fused layers. The **basal lamina** is synthesized by the epithelial cells that it supports and is composed of collagen and associated proteins. The inner layer (**lamina reticularis**, or **reticular lamina**) is formed by underlying connective tissue. Epithelia are avascular, relying on blood vessels lying close to the basement membrane to deliver O_2 and nutrients, but they are innervated.

B. Types

Epithelia are classified based on their morphology, which is usually a reflection of their function. There are three types: **simple epithelia**, **stratified epithelia**, and **glandular epithelia**.

1. **Simple:** Many epithelia are specialized to facilitate exchange of materials between their apical surface and the vasculature. For example, the pulmonary epithelium facilitates gas exchange between the atmosphere and the pulmonary circulation (see 22·II·C). The renal tubule epithelium transfers fluid between the tubule lumen and blood (see 26·II·C), whereas the epithelium that lines the small intestine transfers materials between the intestinal lumen and the circulation (see 31·II). Exchange and transport functions require that the barrier separating the two compartments be minimal, so all of the above structures are lined with "simple" epithelia. Simple epithelia comprise a single cell layer and can be further subdivided into three groups according to epithelial cell shape. Pulmonary alveoli and blood vessels are lined with **simple squamous epithelium**.

Clinical Application 4.1: Squamous Cell Carcinoma

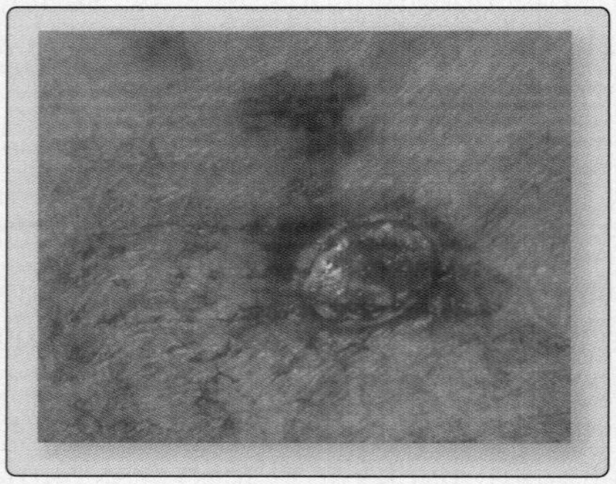

Squamous cell carcinoma.

Squamous cell carcinomas are one of the most common forms of cancer that arise from most epithelia, including the skin, lips, buccal lining, esophagus, lungs, prostate, vagina, cervix, and urinary bladder. Cutaneous squamous cell carcinoma is a prevalent skin cancer that typically occurs in sun-exposed skin areas. Squamous cell malignancies are believed to arise from uncontrolled division of epithelial stem cells rather than squamous epithelial cells. Squamous cell cancers usually remain localized and can be treated by Mohs surgery (a specialized dermatologic surgery for skin malignancies), cryotherapy, or surgical excision.

Squamous epithelial cells are extremely thin to maximize diffusional exchange of gases. Many segments of renal tubules and glandular ducts are lined with cube-shaped (**simple cuboidal**) epithelial cells. Their shape reflects the fact that they actively transport materials and, thus, they must accommodate mitochondria to produce the adenosine triphosphate (ATP) needed to support primary active transporter function. **Simple columnar epithelia** comprise sheets of cells that are long and narrow to accommodate large numbers of mitochondria and are found in the distal regions of the renal tubule (see 27·IV·A) and in the intestines, for example.

2. **Stratified:** Epithelia that are subject to mechanical abrasion are composed of multiple cell layers (Figure 4.2). The layers are designed to be sacrificed to prevent exposure of the basement membrane and deeper structures. The inner epithelial cell layers are renewed continually and the damaged outer layers sloughed off. Examples include the skin and the lining of the mouth, esophagus, and vagina. The skin suffers constant exposure to mechanical stress associated with contact with and manipulation of external objects, so the outer layers are reinforced with keratin, a resilient structural protein (see 16·III·A). **Transitional epithelium** (also known as **urothelium**) is a specialized stratified epithelium that lines the urinary bladder, ureters, and urethra (see 25·VI). A transitional epithelium comprises cells that readily stretch and change shape (from cuboidal to squamous) without tearing to accommodate volume changes within the structures that they line.

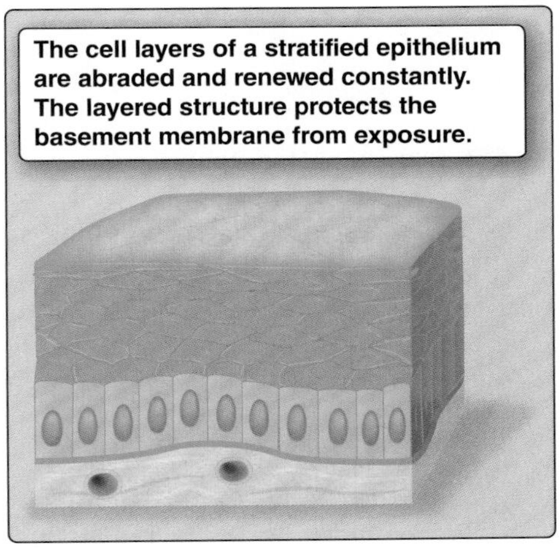

The cell layers of a stratified epithelium are abraded and renewed constantly. The layered structure protects the basement membrane from exposure.

Figure 4.2
Stratified epithelial structure.

3. **Glandular:** Glandular epithelia produce specialized proteinaceous secretions (Figure 4.3). Glands are formed from columns or tubes of surface epithelial cells that invade the underlying structures to form invaginations. Glandular secretions are then released either via a duct or ductal system onto the epithelial surface (**exocrine glands**), or across the basement membrane into the bloodstream (**endocrine glands**). Endocrine glands include the adrenal glands (which secrete epinephrine), the endocrine pancreas (which secretes insulin and glucagon), and reproductive glands, which are considered in Unit VIII. Sweat glands, salivary glands, and mammary glands are all examples of exocrine glands. Exocrine glands are typically composed of two epithelial cell types: **serous** and **mucous**.

   a. **Serous:** Serous cells produce a watery secretion containing proteins, typically enzymes. Salivary serous cells produce salivary *amylase,* gastric chief cells produce pepsinogen (a *pepsin* precursor) and pancreatic exocrine serous cells produce trypsinogen, chymotrypsinogen, pancreatic *lipase*, and pancreatic *amylase.*

   b. **Mucous:** Mucous cells secrete **mucus**, a slippery glycoprotein (**mucin**)-rich secretion that lubricates the surface of mucous membranes. Many glands contain a mix of serous and mucous cells that together create an epithelial barrier layer

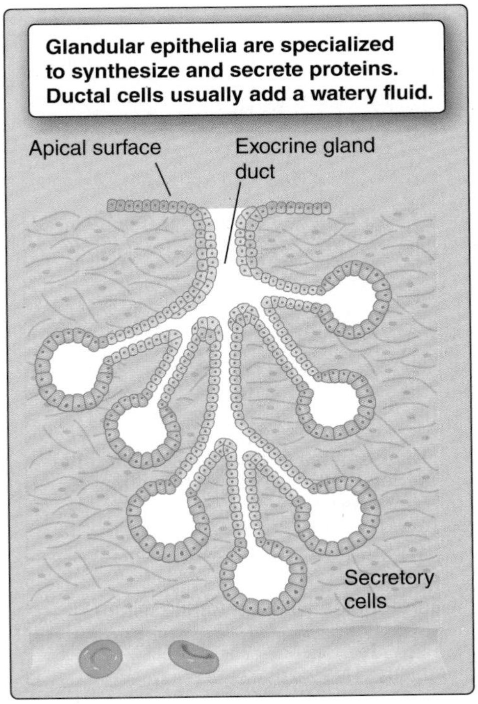

Figure 4.3
Glandular epithelial structure.

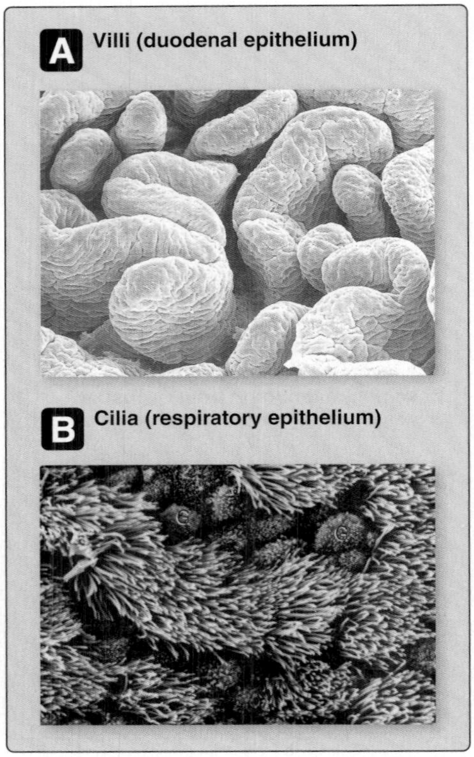
A Villi (duodenal epithelium)
B Cilia (respiratory epithelium)

Figure 4.4
Apical surface specializations.

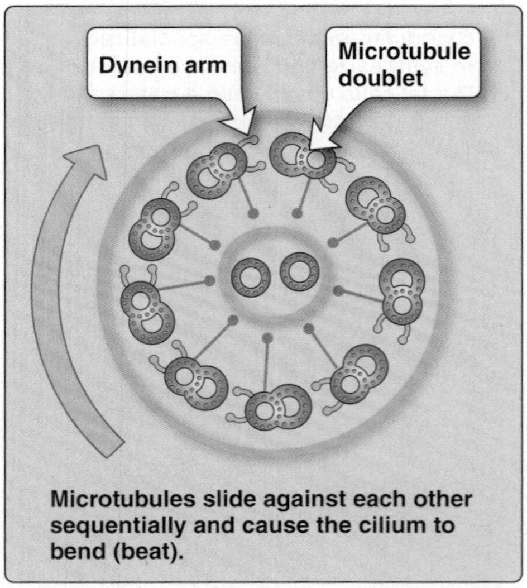
Dynein arm
Microtubule doublet
Microtubules slide against each other sequentially and cause the cilium to bend (beat).

Figure 4.5
Microtubules within a motile cilium.

enriched with antibacterial agents such as lactoferrin to help ward off infection (e.g., pancreatic glands) or enriched with HCO_3^- to neutralize acid (e.g., gastric epithelium).

C. Apical specializations

Several epithelia support apical modifications that amplify surface area or serve motile or sensory functions, including **villi**, **cilia** (**motile** and **sensory**), and **stereocilia** (Figure 4.4).

1. **Villi:** Epithelia that are specialized for high-volume fluid uptake or secretion (e.g., epithelia lining the renal proximal tubule and small intestine) are folded extensively to create fingerlike projections (**villi**) that serve to amplify the surface area available for diffusion and transport (see Figure 4.4A). The epithelial cells that cover villi may also support **microvilli**, plasma-membrane projections that enhance surface area even further. Villi and microvilli are nonmotile.

2. **Motile cilia:** The epithelia that line the upper airways, brain ventricles, and fallopian tubes are covered with motile cilia. Cilia are hairlike organelles containing a 9 + 2 arrangement of microtubules that run the length of the organelle (Figure 4.5). Two microtubules are located centrally, and nine microtubule doublets run around the ciliary circumference. Adjacent microtubule doublets are associated with **dynein** (dynein arms are shown in Figure 4.5), which is a molecular motor (an ATPase). When activated, dynein causes adjacent microtubule doublets to slide against each other sequentially around the ciliary circumference, causing the cilium to bend or "**beat**." The synchronized beating of many thousands of cilia cause the mucus (e.g., in the airways; see 22·II·A) or cerebrospinal fluid ([CSF] see 6·VII·D) in which they are immersed to move over the epithelial surface (see Figure 4.4B). Respiratory cilia propel mucus and trapped dust, bacteria, and other inhaled particles upward and away from the blood–gas interface. In the brain, ciliary beating helps circulate CSF.

3. **Sensory cilia:** Epithelial cells lining the renal tubule each sprout a single central cilium that is nonmotile and is believed to monitor flow rates through the tubule. The olfactory epithelium also bears nonmotile cilia whose membranes are dense with odorant receptors (see 10·III·B).

4. **Stereocilia:** The sensory epithelium that forms the lining of the inner ear expresses mechanosensory **stereocilia** that transduce sound waves (organ of Corti; see 9·IV·A) and detect head motion (vestibular apparatus; see 9·V·A). Stereocilia are nonmotile epithelial projections more closely related to villi than to true cilia.

D. Basolateral membrane

The membranes of two adjacent epithelial cells come into close apposition just below the apical surface to form **tight junctions** (**zona occludens**) as shown in Figure 4.1. Tight junctions comprise continuous structural bands that link adjacent cells together, much as

beverage cans are held together by plastic six-pack rings (Figure 4.6). Tight junctions effectively seal the apical surface of an epithelium and create a barrier, which, in some epithelia (e.g., distal segments of the renal tubule), is impermeant to water and solutes. Tight junctions also divide the epithelial plasma membrane into two distinct regions (apical and basal) by preventing lateral movement and mixing of membrane proteins. The membrane located on the basal side of the tight junction includes the lateral and basal membranes, which are contiguous and together form a functional unit known as the **basolateral membrane**. The basolateral membrane usually contains a different complement of ion channels and transporters from the apical side (e.g., the Na^+-K^+ ATPase is usually restricted to the basolateral membrane) and may be folded to increase the surface area available for transporter proteins (e.g., some portions of the nephron). The basolateral membrane faces the vasculature across an interstitial space.

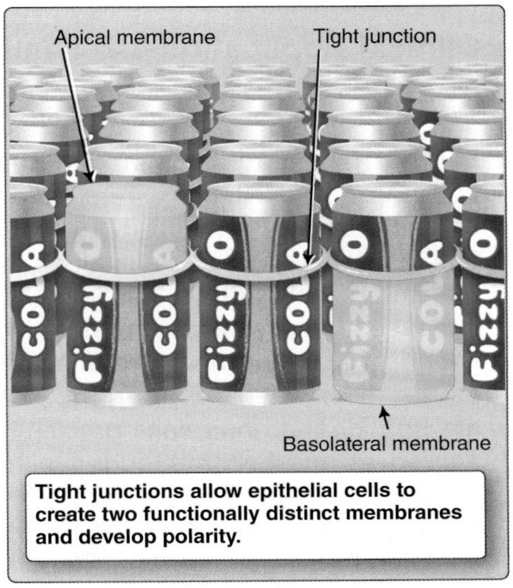
Apical membrane
Tight junction
Basolateral membrane
COLA
Fizzy
Tight junctions allow epithelial cells to create two functionally distinct membranes and develop polarity.

Figure 4.6
Model for an epithelium.

E. Tight junctions

Tight junctions contain numerous different proteins, the principal ones being **occludin** and **claudin**. Tight junctions serve several important functions: They form molecular "fences," they determine tight-junction "leakiness," and they regulate water and solute flow across epithelia.

1. **Fences:** Tight junctions prevent apical and basolateral membrane proteins from mixing (a fence function) and, thereby, allow epithelial cells to develop **functional polarity** (Figure 4.7). The apical membrane becomes specialized for moving material between the external environment and the cell interior, whereas the basolateral membrane moves material between the inside of the cell and the bloodstream (see below).

2. **Leakiness:** Adjacent cells within an epithelium are separated by a narrow space that creates a physical pathway for transepithelial fluid flow (the **paracellular pathway**). Tight junctions act as gates that limit paracellular fluid movement and, in so doing, define epithelial leakiness.

   a. **Leaky epithelia:** The tight junctions in a **"leaky" epithelium** (e.g., renal proximal tubule; see 26·II) are highly permeable and allow solutes and water to pass with relative ease (see Figure 4.7). Leakiness prevents an epithelium from being able to create strong solute concentration gradients between external and internal surfaces, but leaky epithelia are capable of taking up large fluid volumes by paracellular flow.

   b. **Tight epithelia:** Tight junctions in **"tight" epithelia** effectively bar paracellular flow of water and solutes and allows an epithelium to become highly selective in what it absorbs or secretes (e.g., nephron distal segments; see 27·IV) as shown in Figure 4.8. An epithelium's leakiness is defined by its electrical resistance. Because the tight junctions in tight epithelia restrict passage of ions, they have a high resistance to current flow (>50,000 Ohm), whereas leaky epithelia have low resistance (<10 Ohm).

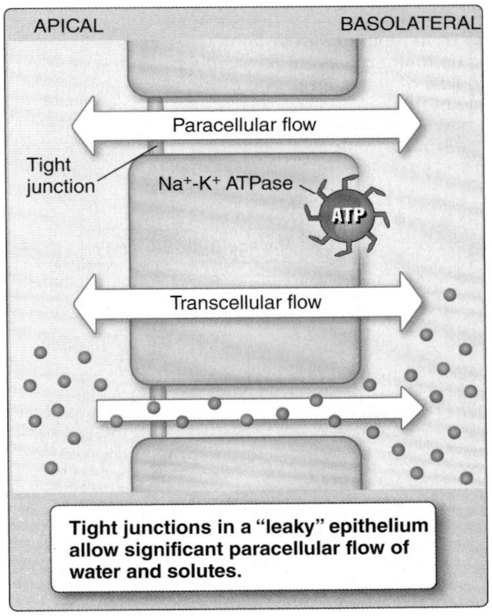
APICAL
BASOLATERAL
Paracellular flow
Tight junction
Na+-K+ ATPase
ATP
Transcellular flow
Tight junctions in a "leaky" epithelium allow significant paracellular flow of water and solutes.

Figure 4.7
Flow across a leaky epithelium.
ATP = adenosine triphosphate.

Clinical Application 4.2: Familial Hypomagnesemia with Hypercalciuria and Nephrocalcinosis

Familial hypomagnesemia with hypercalciuria and nephrocalcinosis (FHHNC) is a rare autosomal recessive disorder characterized by an inability to reabsorb Mg^{2+} from the renal tubule. Plasma Mg^{2+} levels fall as a consequence (hypomagnesemia). The mutation also impairs Ca^{2+} reabsorption, which increases urinary excretion rates (hypercalciuria) and the likelihood of kidney stone formation (nephrocalcinosis). Kidney stones form when urinary Ca^{2+} and Mg^{2+} concentrations are so high that their salts precipitate as crystals, which then aggregate and become lodged within the renal tubule (intrarenal calculi), ureters (ureteral calculi), or bladder. FHHNC is caused by claudin-16 gene mutations (the human genome contains 24 claudin genes). Claudin-16 forms a divalent cation-specific pathway (**paracellin-1**) for Mg^{2+} and Ca^{2+} reabsorption from the thick ascending limb of the loop of Henle. Mutations in claudin-19 can similarly produce renal Mg^{2+} wasting. Affected individuals typically require magnesium supplements and frequent lithotripsy to mechanically fragment kidney stones and allow them to pass out of the body.

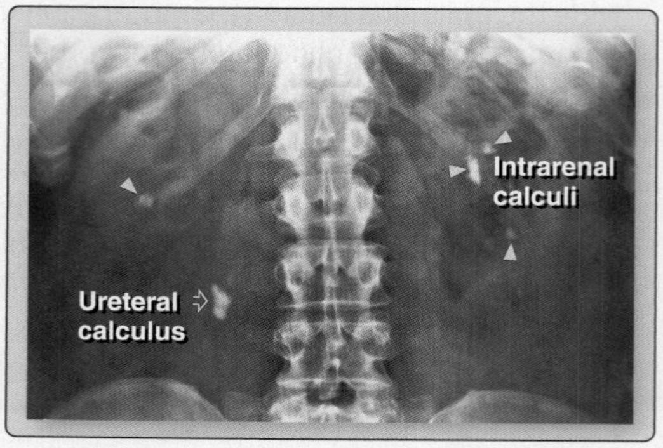

Intrarenal calculi.

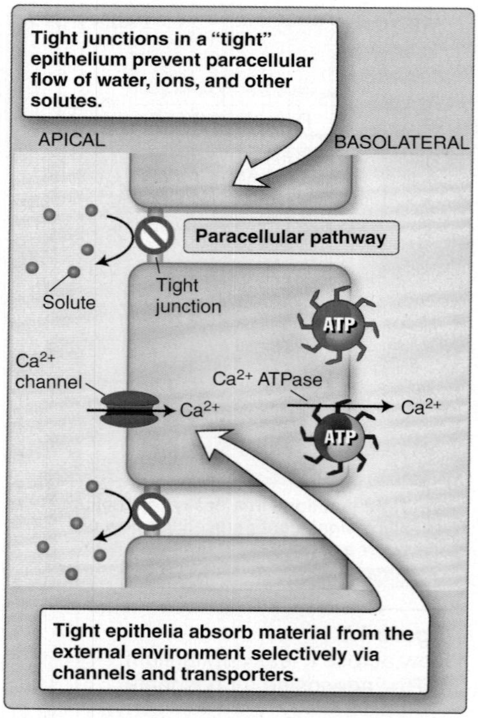

Figure 4.8
Tight epithelia. ATP = adenosine triphosphate.

3. **Regulation:** Although tight junctions in leaky epithelia are highly permeable to water and solutes, they are selective about what they give passage to. Tight-junction permeability can also be regulated to increase or decrease net uptake of water and solutes via the paracellular route. For example, transcellular Na^+-glucose transport by intestinal epithelia increases paracellular Na^+-glucose transport through changes in tight-junction permeability. The mechanisms involved are not well delineated, but claudins clearly have a central role in determining the size and charge of permeant solutes, and *myosin light-chain kinase* is involved in regulating junctional permeability.

F. Gap junctions

Gap junctions are the location of gap-junction channels that provide pathways for communication between adjacent cells. They are found in many areas (including muscle and nervous tissue), but are so abundant in some epithelia (e.g., intestinal epithelia) that they are packed into dense crystalline arrays, each containing thousands of individual channels. Gap-junction channels are formed by the association of two **connexin hemichannels** (**connexons**) as shown in Figure 4.9.

1. **Connexins:** Gap-junction channels are formed by six connexin subunits that assemble around a central pore, and each subunit contains four membrane-spanning domains (see Figure 4.9). The human genome contains 21 connexin isoforms that yield channels with distinct gating properties, selectivities, and regulatory mechanisms when expressed. All 21 isoforms are associated with hereditary diseases, which underscores the significance of the gap-junction communication pathway.

Connexin gene mutations produce disorders ranging from idiopathic atrial fibrillation, congenital cataracts, hearing loss, and oculodentodigital dysplasia, to an X-linked form of Charcot-Marie-Tooth disease (CMT). CMT includes a diverse group of demyelinating disorders that primarily affect peripheral nerves, resulting in sensory loss, muscle wasting, and paralysis.

2. **Connexons:** A gap-junction channel forms when **hemichannels** from two adjacent cells contact each other end to end, align, and form a tight association (see Figure 4.9). Gap junctions, have a significant intercellular adhesion function also.
3. **Gating:** Gap-junction channels are gated by numerous factors, including the potential difference across the junction, membrane-potential changes, Ca^{2+}, pH changes, and by phosphorylation. At rest, when the transjunctional potential is 0 mV, gap-junction channels are usually open.
4. **Permeability:** The gap-junction channel pore is sufficiently large to allow passage of ions, water, metabolites, second messengers, and even small proteins of up to around 1,000 MW. Gap junctions allow all cells within an epithelium to communicate with each other both electrically and chemically.

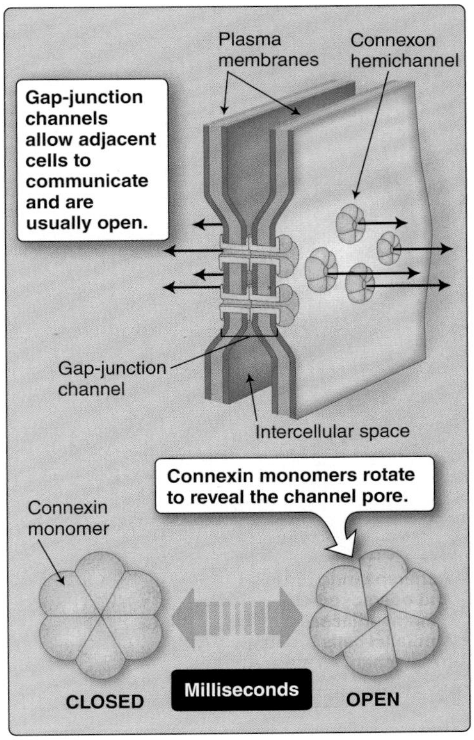

Figure 4.9
Gap-junction channels.

Clinical Application 4.3: Pemphigus Foliaceus

Pemphigus foliaceus is a rare autoimmune disorder that presents as scaly, crusting skin blisters located on the face and scalp, primarily, although the chest and back may become involved in later stages. Affected individuals express antibodies to desmoglein 1, an integral membrane protein that forms a part of the desmosomal complex. Symptoms typically are triggered by drugs (e.g., penicillin) and are caused by desmoglein 1 being targeted and degraded by the immune system. Adjacent skin epithelial cells become detached from one another, and the skin blisters. The blisters ultimately slough off and leave sores. Treatment includes immunosuppressive therapy.

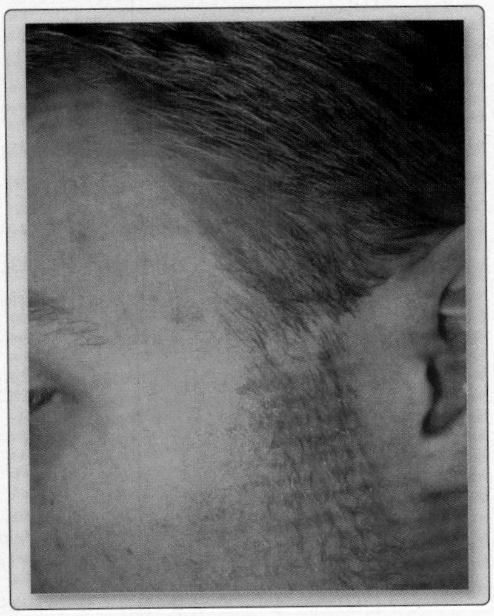

Pemphigus foliaceus.

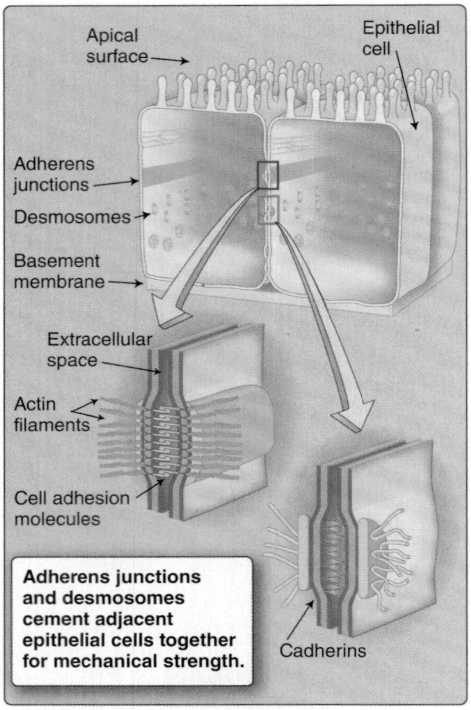

Figure 4.10
Adherens junction and desmosome structures.

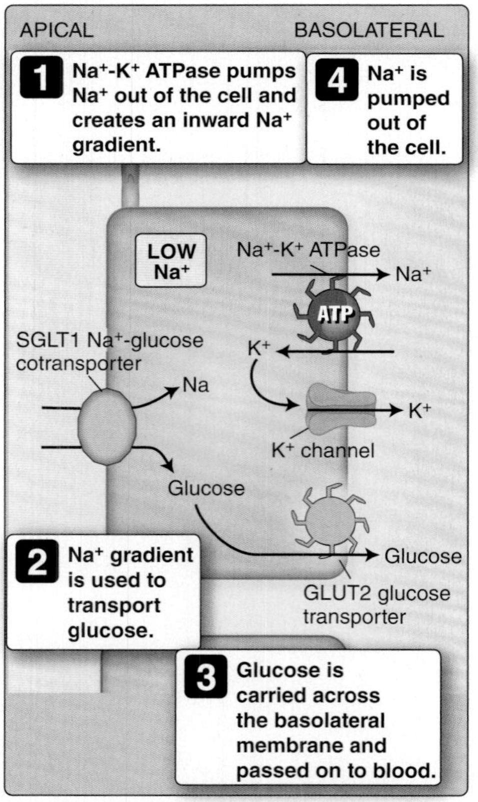

Figure 4.11
Epithelial transport principles. ATP = adenosine triphosphate.

G. Other junctional structures

Two additional structures provide support to the epithelial sheet: **adherens junctions** and **desmosomes** (Figure 4.10).

1. **Adherens junctions:** All cells in an epithelial sheet are tethered together by bands of protein complexes that lie just below the tight junction known as adherens junctions (**zonula adherens**; see Figure 4.10). The complexes straddle two adjacent cells and then link to the cell cytoskeleton.

2. **Desmosomes:** Adjacent cells within an epithelium are also tightly adhered by **desmosomes** (**macula adherens**) as shown in Figure 4.10. Desmosomes are small, rounded, membrane specializations that function much like the spot welds used to join metal body panels to an automobile chassis. Protein complexes link the membrane to the cytoskeleton on the intracellular side, whereas adhesion proteins (**cadherins**) bridge the gap between cells and fuse the two surfaces together. Desmosomes are particularly important for maintaining the integrity of epithelia that are stressed mechanically (e.g., the urinary bladder transitional epithelium).

III. MOVEMENT ACROSS EPITHELIA

Transepithelial flow of water and solutes occurs via regulated pathways (channels and transporters) and is driven by the same physical forces discussed previously in relation to flow across membranes (i.e., diffusion and carrier-mediated transport; see 1·IV). The principal difference is in the availability of a paracellular route for transepithelial transport.

A. Transcellular transport

Transport epithelia (e.g., intestinal epithelia) are specialized to move large volumes of water and solutes between the outside of the body (e.g., GI tract or renal tubule) and the vasculature by way of the interstitium. **Secretory epithelia** transfer water and solutes to the outside of the body, whereas **absorptive epithelia** take up water and solutes from the outside and transfer them to the vasculature. The example below considers the steps involved in glucose uptake by the small intestine (or the renal proximal tubule) as an example, but secretory and absorptive epithelia use the same basic transport principles regardless of body location. The first step involves establishing a Na^+-concentration gradient across the surface membrane (Figure 4.11). The steps below correspond to the steps in the figure.

1. **Step 1—Create a sodium gradient:** Transepithelial transport involves work, the energy for which is supplied by ATP. ATP is used to power primary active transport, which, in virtually all instances, involves the ubiquitous Na^+-K^+ ATPase located in the basolateral membrane. The Na^+-K^+ ATPase takes up K^+ and expels Na^+, thereby creating an inwardly directed Na^+-concentration gradient and an outwardly directed K^+-concentration gradient.

2. **Step 2—Glucose uptake:** Gut luminal glucose concentration is usually lower than that in intracellular fluid, meaning that sugar must be transported "uphill" against a concentration gradient. Uptake is powered by the inward Na^+-concentration gradient using a SGLT1 Na^+-glucose cotransporter (secondary active transport).

3. **Step 3—Glucose absorption:** Na^+-glucose cotransport raises intracellular glucose concentration and creates an outwardly directed concentration gradient that favors glucose movement from the cell to the interstitium. Glucose diffuses across the cell and exits via a GLUT2 transporter in the basolateral membrane. Glucose subsequently diffuses through the interstitium, enters a capillary, and is carried away in the bloodstream.

4. **Step 4—Sodium removal:** Na^+ that crossed the apical membrane during glucose transport is removed from the cell by the basolateral Na^+-K^+ ATPase.

5. **Step 5—Potassium removal:** The Na^+-K^+ exchange during Step 4 above raises intracellular K^+ concentrations, but the gradient favoring K^+ efflux is already very strong, and excess K^+ exits the cell passively via K^+ channels. K^+ channels are usually present in the basolateral membrane but may be located apically also.

B. Water movement

Water cannot be actively transported across epithelia, but a tried-and-true maxim of transport physiology notes that **"water follows solutes"** (by osmosis). The steps outlined in section (A) caused Na^+ and glucose to be translocated from the intestinal lumen to the interstitium, an act that created a transepithelial osmotic gradient that is then used to absorb water. There are two potential routes for water absorption: transcellular and paracellular (Figure 4.12).

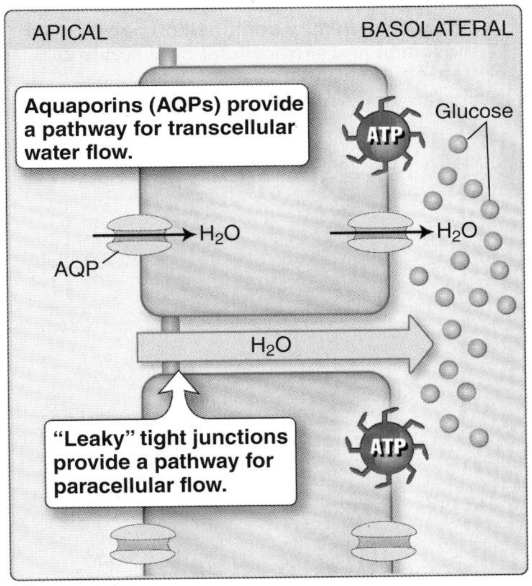

Figure 4.12
Transepithelial water movement. ATP = adenosine triphosphate.

1. **Transcellular flow:** Transcellular water movement only occurs if water is provided with a clear passage through the epithelial cell. In practice, this requires that water channels (aquaporins [AQPs]) be present in both the apical and basolateral membranes. Transport epithelia typically express high AQP levels, which support high volumes of transcellular water uptake (or secretion). The epithelium that lines renal collecting ducts actively regulates its water permeability by modulating apical AQP expression levels (see 27·V·C). When there is a need to reabsorb water from the tubule lumen, AQPs are recruited to the apical membrane from stores located intracellularly in vesicles. When the body contains water in excess of homeostatic requirements, AQPs are removed from the apical membrane, and the epithelium becomes water impermeant.

2. **Paracellular flow:** Paracellular water flow is also driven by osmotic pressure gradients created by solute transport. The availability of the paracellular route is determined by epithelial leakiness, which is, in turn, determined by tight junctions.

Clinical Application 4.4: Oral Rehydration Therapy

Intestinal epithelia are capable of transporting high volumes of watery fluids. In a healthy person, virtually all of the ~10 L of fluids secreted by intestinal epithelia during the digestive phase are subsequently reabsorbed, so that <200 mL/d is lost from the body in stool. The bacterium *Vibrio cholerae* secretes a toxin that increases intestinal epithelial Cl^- permeability and raises intestinal luminal osmolality.[1] Copious amounts of fluid are drawn osmotically across the epithelium as a result. Almost all of the secreted fluid is lost to the external environment, either as a result of vomiting or frequent, watery stools. Death usually occurs as a result of hyponatremia, hypovolemia, and loss of blood pressure. Cholera can be treated and death prevented fairly simply using oral rehydration therapy (ORT). ORT takes advantage of the fact that Na^+ and glucose are rapidly absorbed by the intestinal epithelia (via SGLT1), creating an inwardly directed osmotic gradient that drives water reabsorption. Typical home remedies involve giving patients a solution containing 6 tsp sugar and 1/2 tsp salt (NaCl) per liter of water. The advantage of ORT is that it is highly effective, easy to administer, and cheap, which is of particular advantage in developing countries where cholera is endemic and resources are typically limited.

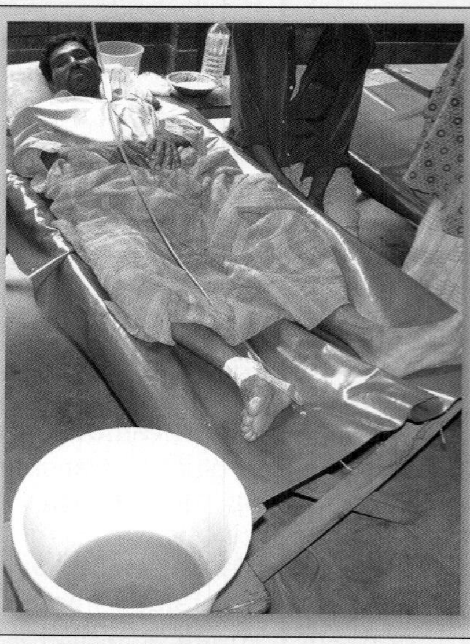

Cholera patients excrete large volumes of watery stool.

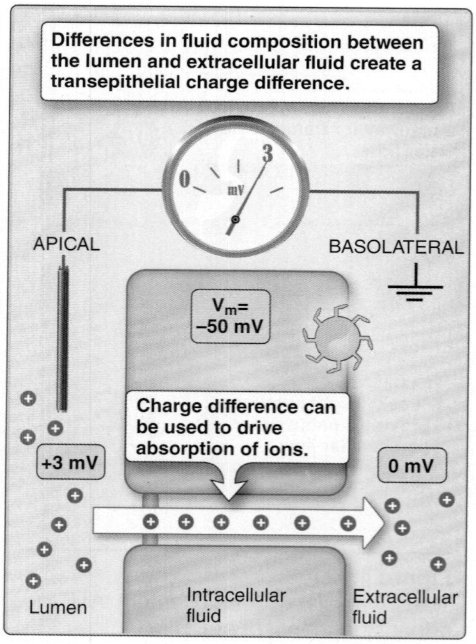

Figure 4.13
Transepithelial voltage difference. V_m = membrane potential.

C. Solvent drag

Intestinal epithelia secrete and absorb ~10 L of water per day, whereas the renal tubule reabsorbs almost as much on an hourly basis. These secretory and absorptive functions generate high water flow rates, both transcellularly and paracellularly. The resulting water streams carry ions and other small solutes with them, much as a fast-flowing river sweeps along sand and other fine particles. This phenomenon is known as **solvent drag** and can contribute significantly to transepithelial solute movement. The net result of all this solute and water flow is that the secreted or absorbed fluid usually has a composition that is isosmotic relative to the source (**isosmotic flow**).

D. Transepithelial voltage effects

Epithelial cells are located at the interface between two compartments that may have very different chemical compositions. The basolateral membrane faces the inside of the body and is bathed in extracellular fluid (ECF) whose chemical composition is well controlled. The apical membrane is bathed in external fluid whose composition may be indeterminate and variable. Charge differences between the two fluids create a transepithelial voltage difference that influences transport (Figure 4.13).

[1]For more information on the pathogenesis and treatment of cholera, see *LIR Microbiology*, 3e, pp. 122–123.

1. **Transport:** The distal segment of the renal proximal tubule, for example, is positive charged (~3 mV) with respect to ECF. Although the voltage difference is small, it provides a motive force that drives significant amounts of Na^+ out of the tubule and toward the interstitium (see 26·X·B).
2. **Local membrane potential effects:** Epithelial cells, like all cells in the body, establish a membrane potential (V_m) across their surface membrane, inside negative. V_m is measured with respect to ECF and is uniform throughout the cell. However, the fact that the apical surface is bathed in a medium of different ionic composition can create local differences in V_m. Thus, if V_m is −50 mV and the tubule lumen is +3 mV with respect to ECF, the potential across the apical membrane will be −53 mV relative to the lumen.

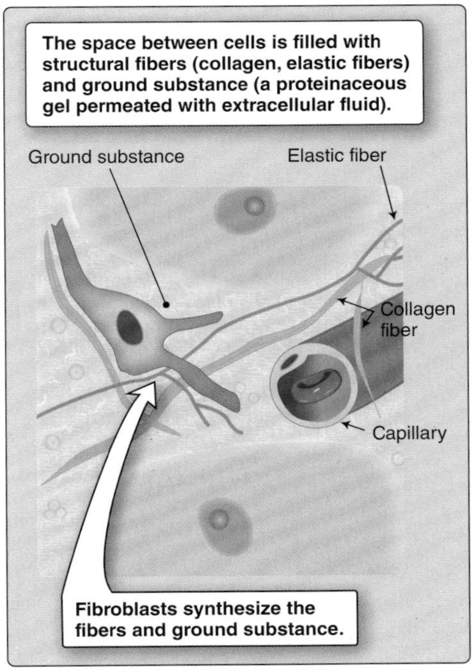

The space between cells is filled with structural fibers (collagen, elastic fibers) and ground substance (a proteinaceous gel permeated with extracellular fluid).
Ground substance
Elastic fiber
Collagen fiber
Capillary
Fibroblasts synthesize the fibers and ground substance.

Figure 4.14
Connective tissue.

IV. CONNECTIVE TISSUE

Connective tissue is the most abundant tissue class that can be found in all areas of the body. There are several different connective tissue types, but they all follow a common organizational principle (Figure 4.14). Connective tissues are composed of specialized cells, structural proteins, and a fluid-permeated ground substance.

A. Types

There are three main types of connective tissue: **embryonic** (not considered further here), **specialized connective tissue**, and **connective tissue proper**.

1. **Specialized:** Specialized connective tissue includes cartilage, bone (see Chapter 15), hematopoietic tissue and blood, lymphatic tissue, and adipose tissue. Cartilage is a flexible connective tissue that cushions bones at sites of articulation and that gives shape to the nose and ears, for example. Lymphatic tissue comprises a system of vessels that drain fluid from the extracellular space (see 19·VII·C). Adipose tissue is composed largely of adipocytes whose primary function is to store energy in the form of triglycerides. Fat conducts heat poorly, so it is layered beneath the skin (subcutaneous fat) to help insulate the body. Adipose tissue deposits can also be associated with internal organs (visceral fat) and in yellow bone marrow.
2. **Proper:** Connective tissue proper forms the **extracellular matrix** (**ECM**) that occupies the interstitial space. Connective tissue proper can be further subdivided into **loose connective tissue**, a highly pliable form that occupies the space between most cells, **dense connective tissue** (tendons, ligaments, and fibrous fascia and capsules that enclose muscles and organs), and **reticular connective tissue** that forms the scaffolding upon which blood vessels, muscle, and the liver is built, for example.

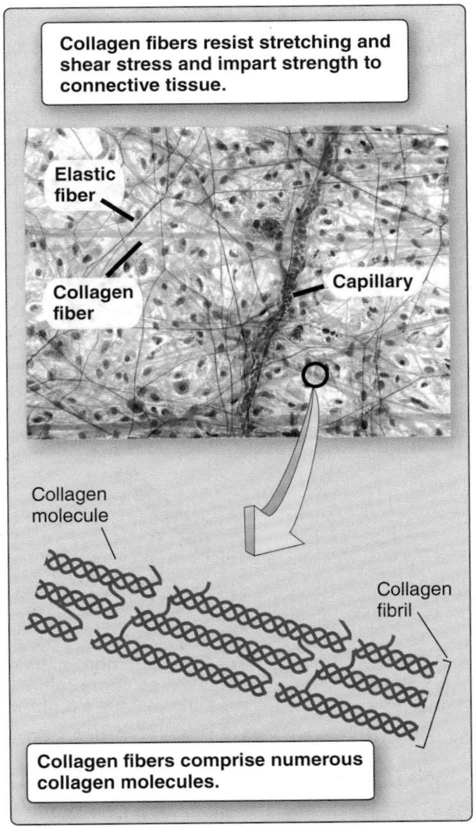

Collagen fibers resist stretching and shear stress and impart strength to connective tissue.
Elastic fiber
Collagen fiber
Capillary
Collagen molecule
Collagen fibril
Collagen fibers comprise numerous collagen molecules.

Figure 4.15
Collagen structure.

B. Extracellular matrix

The ECM is a mix of cells (fibroblasts), structural proteins (collagen and elastic fibers), ground substance, and ECF. The ECM imparts

form and strength to tissues and provides pathways for chemical diffusion and for immune system cell migration (e.g., macrophages).

1. **Fibroblasts:** Fibroblasts are motile cells that continually synthesize and secrete structural protein precursors and ground substance. They are essential for ECM maintenance and for wound healing.

2. **Structural proteins:** The ECM is filled with an interconnected structural matrix comprising collagen and elastic fibers.

   a. **Collagen:** Collagen is a tough, fibrous protein that possesses high tensile strength and resistance to shear stress. The body contains 28 different collagen types, but four forms (types I, II, III, and IV) predominate. Type I collagen is abundant in skin and vascular walls and is bundled to form ligaments, tendons, and bone. Type IV organizes into meshlike networks that make up the basal lamina of epithelia, for example. Collagen molecules are composed of three polypeptide chains braided into a triple helix, and then cross-linked extensively for enhanced stress resistance[1] (Figure 4.15).

   b. **Elastic fibers:** Elastic fibers are composed of elastin and glycoprotein microfibrils (e.g., **fibrillin** and **fibullin**). Elastic fibers stretch like rubber bands when stressed and then recoil and assume their original shape when allowed to relax. Elastic fibers are found in the walls of arteries and veins, which allows them to stretch when intraluminal pressures increase. Elastic fibers also allow lungs to expand during inspiration as well as help reduce stress on teeth during chewing (periodontal fibers). Elastic fibers are synthesized by fibroblasts, which first lay down structural scaffolding made of fibrillin and then deposit tropoelastin monomers onto the scaffolding. Four adjacent elastin monomers are then cross-linked to form an irregular network that comprises the mature elastin molecule (Figure 4.16).

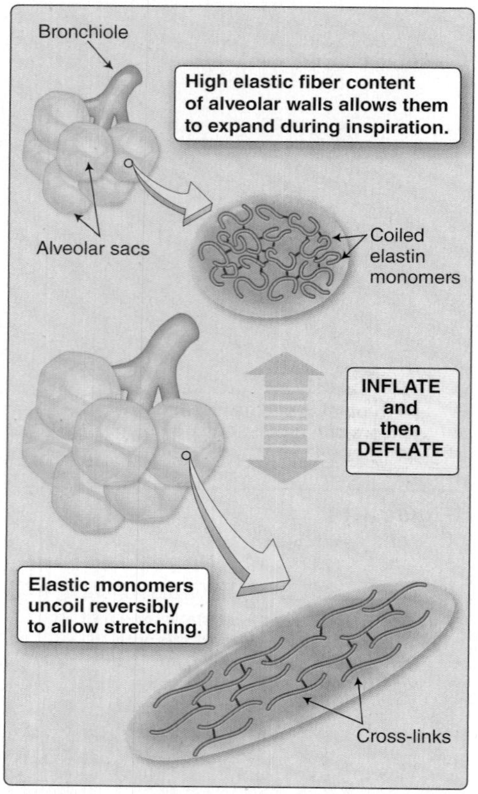

Figure 4.16
Elastin properties.

3. **Ground substance:** Ground substance is a mix of various proteins (principally proteoglycans) and ECF that creates an amorphous gel filling the spaces between cells and structural fibers. The high water content of the gel facilitates chemical diffusion between cells and the vasculature, yet the structural fibers simultaneously impede movement of invading pathogens. Proteoglycans are formed by attaching numerous glycosaminoglycan (GAG) molecules to a core protein, the final structure resembling a bottlebrush or pipe cleaner (Figure 4.17). GAGs possess a high negative-charge density, which allows them to attract and loosely trap water molecules within the gel. The interstitium contains >10 L of ECF in an average person, representing a substantial buffer volume that helps minimize the impact of changes in total body water on cell and cardiovascular function (see 3·III and 19·VIII·C).

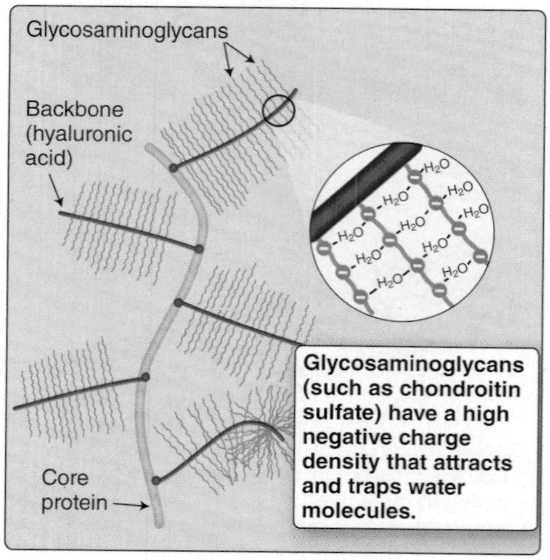

Figure 4.17
Proteoglycan structure.

[1]For more information on collagen synthesis and assembly, see *LIR Biochemistry*, 5e, p. 43.

Chapter Summary

- The human body is composed of four tissue types: **epithelial tissue**, **nervous tissue**, **muscle tissue**, and **connective tissue**. Epithelial tissue comprises sheets of tightly packed cells that line all external and internal body surfaces (e.g., skin and the pulmonary and gastrointestinal linings). Epithelia form **barriers** that protect the body from invasion by microbes, but many also have specialized transport functions.
- Epithelia are classed morphologically. **Simple epithelia** are composed of a single cell layer (e.g., pulmonary epithelium). **Stratified epithelia** (e.g., skin) comprise multiple layers that are sloughed off and renewed. **Glandular epithelia** (endocrine and exocrine glands) are specialized for secretion.
- Epithelia are **polarized** with functionally distinct **apical** and **basolateral surfaces**. The apical surface faces the external environment, the lumen of a hollow organ, or a body cavity. Apical surfaces may be specialized to include **villi**, **cilia**, and **stereocilia** that amplify surface area, propel mucus layers, or serve a sensory role, respectively.
- The basolateral membrane communicates with the body interior via the **interstitium** and the vasculature. It rests on a **basement membrane** that anchors the epithelium to underlying connective tissues.
- Polarization of epithelia is made possible by **tight junctions**, which are structural bands encircling all cells in an epithelium close to their apical surface. The junctions form tight seals with an important barrier function. The junctions also segregate apical and basolateral membrane proteins, thereby allowing for specialization of membrane function.
- Tight-junction permeability is regulated by **claudins**, which determine how much water and solutes cross an epithelium via the space between adjacent cells (**paracellular flow**). The junctions effectively block passage of all water and solutes across a **"tight" epithelium**. In contrast, **"leaky" epithelia** secrete and absorb significant amounts of fluid.
- **Gap junctions** are hexameric channels comprising **connexin** monomers that connect adjacent cells and allow all cells in an epithelium to communicate chemically and electrically. **Adherens junctions** and **desmosomes** are junctional structures that provide strength to an epithelium and help prevent it from tearing when stressed mechanically.
- Many epithelia have **secretory** and **absorptive functions**. Transepithelial transport usually occurs **transcellularly** and **paracellularly**, although both routes are regulated. The electrochemical and osmotic driving forces for movement of solutes and ions are established by primary and secondary active transporters (e.g., Na^+-K^+ ATPase and Na^+-coupled transport).
- **Connective tissue** comprises cells; **structural fibers**; and an amorphous, fluid-permeated **ground substance**. **Specialized connective tissues** include cartilage, bone, and adipose tissue. **Connective tissue proper** forms the **extracellular matrix**.
- The extracellular matrix fills the space between all cells, providing mechanical strength and support as well as a loose, gel-like medium that facilitates chemical diffusion and cell migration.
- The extracellular matrix (ECM) is synthesized and maintained by **fibroblasts**. ECM structural proteins include **collagen fibers** for strength and **elastic fibers** to allow stretching. Elastic fibers are composed primarily of **elastin**. Ground substance is a proteinaceous matrix containing large quantities of **proteoglycans**. Proteoglycans have a high negative-charge density that allows them to attract and immobilize ~10 L of **extracellular fluid** in an average person.

Study Questions

Choose the ONE best answer.

I.1 A fluid that is composed of 120 mmol/L K^+, 12 mmol/L Na^+, and 15 mmol/L Cl^- but is virtually Ca^{2+} free (<1 μmol/L) would best approximate which body fluid compartment?

A. Transcellular
B. Plasma
C. Interstitial
D. Intracellular
E. Extracellular

Best answer = D. Intracellular fluid should be recognized by its relatively high K^+ concentration, which is due to the Na^+-K^+ ATPase found in the membrane of virtually all cells (1·II). Extracellular fluid (ECF) has a lower K^+ and higher Na^+, Cl^-, and Ca^{2+} concentration compared with ECF and can be further subdivided into plasma (fluid within the vascular space) and interstitial fluid (fluid outside the vascular space; 3·III·A). Because the barrier between these two ECF compartments does not prevent ion movement, their ionic composition is similar. Transcellular fluid (including cerebrospinal fluid, synovial fluid, and urine) composition is variable depending on the location and, therefore, not the best choice.

I.2 Dye indicators are important physiologic tools used to calculate unknown volumes or concentrations within the body. If a dye is membrane permeable, which of the following changes will most likely increase dye diffusion rate?

A. Lowering dye concentration
B. Increasing membrane surface area
C. Increasing membrane thickness
D. Decreasing fluid temperature
E. Lowering the dye partition coefficient

Best answer = B. Increasing membrane surface area increases the opportunity for dye to cross the membrane, which increases its diffusion rate (1·IV·B). It takes longer for molecules to diffuse across thick membranes than thin ones, and decreasing a fluid's temperature increases its viscosity, which also slows diffusion rate. Concentration gradients provide the driving force for diffusion, so lowering dye concentration flattens the gradient and reduces diffusion rate. A partition coefficient is a measure of dye lipid solubility. Molecules with high lipid solubility diffuse through membranes faster than poorly soluble molecules, so lowering partition coefficient would decrease diffusion rate.

I.3 A 66-year-old male is treated with the loop diuretic furosemide (Na^+-K^+-$2Cl^-$ cotransport inhibitor) to reduce symptoms associated with congestive heart failure. Which of the following best describes this cotransporter's mode of action?

A. It is a primary active transporter.
B. It is electrogenic.
C. A rise in intracellular K^+ would decrease transport rate.
D. It transports Na^+ and K^+ into the cell and 2 Cl^- out of the cell.
E. It transports Na^+ against its electrochemical gradient.

Best answer = C. Cotransporters, by definition, move two or more ions in the same direction (1·V·C). The Na^+-K^+-$2Cl^-$ cotransporter simultaneously carries two anions and two cations across the plasma membrane and, thus, is not electrogenic. Primary active transporters use adenosine triphosphate to pump ions against their electrochemical gradients. Transporters that move Na^+ in one direction while simultaneously bringing other ions back in the opposite direction are exchangers, not cotransporters. The Na^+ gradient established by the basolateral Na^+ pump (Na^+-K^+ ATPase) provides the electrochemical driving force for K^+ and Cl^- uptake, but transport rate is sensitive to transmembrane K^+ and Cl^- gradients. Increasing intracellular concentrations of either ion will slow net uptake.

I.4 Serum electrolytes levels are ordered on a 12-year-old boy with a gastrointestinal infection, which induced prolonged and severe vomiting episodes. Plasma K^+ concentrations were found to be abnormally low (2 mmol/L). Which of the following might be expected to result from mild hypokalemia?

A. Resting potentials would shift positive.
B. K^+ equilibrium potential would shift negative.
C. Neuronal action potentials would be inhibited.
D. Na^+ channels would inactivate.
E. K^+-channel activation would yield K^+ influx.

Best answer = B. Hypokalemia, or reduced extracellular K^+ concentrations, enhances the electrochemical gradient favoring K^+ efflux from cells and causes the K^+ equilibrium potential to shift negative (see 2·II·B). Because membrane potential is determined largely by the transmembrane K^+ gradient, resting membrane potential (V_m) would shift negative also. A negative shift in V_m means that a stronger depolarization would be necessary to take V_m to the threshold for voltage-gated Na^+ channel activation (2·III·B), but, once reached, an action potential would be initiated. K^+-channel activation always causes K^+ efflux, except in rare instances (e.g., in the inner ear; 9·IV·C).

I.5 A 35-year-old man carries an epilepsy gene. The gene mutation affects the neuronal voltage-dependent Na^+ channel, causing it to inactivate more slowly (~50%). How might expression of this epilepsy gene affect nerve function?

A. Resting potential would settle close to 0 mV.
B. Action potentials would no longer overshoot 0 mV.
C. Action potentials would be prolonged.
D. Action potentials would rise very slowly.
E. There would be no action potentials.

Best answer = C. The voltage-dependent Na^+ channel is opened by membrane depolarization to yield the upstroke of the neuronal action potential (2·VI·A). An inactivation gate closes shortly after activation, blocking passage of Na^+ and allowing membrane potential to return to resting levels. If inactivation were slowed, membrane recovery would be delayed, and the action potential would be prolonged. Resting potential should not be affected by an inactivation defect unless it prevented the channel from closing, causing a sustained Na^+ influx. Activation and inactivation are separate processes, and, therefore, the rate at which the action potential rises should be normal.

I.6 An agricultural worker is packing hot chili peppers for transport. He removes his protective mask and becomes incapacitated with a sensation of nasal burning caused by capsaicin from the peppers. What receptor type is capsaicin stimulating?

A. Transient receptor-potential channels
B. Purinergic receptors
C. Ionotropic glutamate receptors
D. *Cys*-loop family receptors
E. Voltage-gated Na^+ channels

Best answer = A. The transient receptor potential channel (TRP) family transduces a variety of sensory stimuli, including heat, cold, and osmolality (2·VI·D). TRPV1 channels are stimulated by capsaicin. Purinergic, glutamate, and *cys*-loop receptors are activated by specific ligands (i.e., adenosine triphosphate, L-glutamate, and acetylcholine, respectively). Capsaicin is not an agonist for these receptor classes. Voltage-gated Na^+ channels are activated primarily by membrane depolarization.

I.7 During a serological analysis, red blood cells (RBCs) were transferred from blood to a solution containing 100 mmol/L $CaCl_2$ and 100 mmol/L urea and then monitored using light microscopy. How would you expect this transfer to affect RBC volume?

A. The solution is isosmotic, so no long-term effect.
B. The solution is isotonic, so no long-term effect.
C. Transient swelling would occur.
D. Swelling to the point of lysis would occur.
E. The cell would shrink by ~50%.

Best answer = B. $CaCl_2$ dissociates into three particles (1 Ca^{2+} and 2 Cl^-) in water. A 100-mmol/L $CaCl_2$ solution has an osmolality of 300 mOsm/kg H_2O, which approximates that of RBC intracellular fluid (ICF). 100 mmol/L urea brings total osmolality to 400 mOsm/kg H_2O (the solution is hyperosmotic), but urea would rapidly enter the cell until intracellular and extracellular fluids equilibrated at ~350 mOsm/kg H_2O (3·II·C). The solution is, thus, isotonic. Cell shrinkage would occur if urea was impermeant, but most cells are highly permeable to urea. Cell swelling in this example would only occur if ICF osmolality rose due to active accumulation of one or more of the three solutes.

I.8 Liver damage may result in decreased synthesis of plasma proteins such as albumin. What is the most significant effect of low plasma albumin on osmosis or fluid transport?

A. Interstitial fluid volume increases.
B. Vascular fluid volume increases.
C. Plasma colloid osmotic pressure increases.
D. Plasma osmolality increases.
E. Plasma osmolality decreases.

Best answer = A. Blood contains large amounts of albumin (3.5–5 g/dL) that is trapped in the vascular compartment by virtue of its large size (3·III·B). Its function is to help create an osmotic potential (known as plasma colloid osmotic pressure) that draws extracellular fluid (ECF) into the vasculature. A decrease in plasma albumin concentration would, therefore, allow fluid to leave the vasculature and enter the interstitium. Plasma osmolality does not change significantly with changes in protein concentration. The main determinants of ECF osmolality are ions (e.g., Na^+ and Cl^-) and other solutes (e.g., glucose and urea, measured as blood urea nitrogen, or "BUN").

I.9 A 95-year-old man with widely metastatic cancer is receiving morphine to help alleviate pain. Brainstem respiratory center function has been depressed as a result, causing hypoventilation. Which of the following might be expected to result from reducing ventilation?

A. Alkalemia
B. Decreased plasma HCO_3^- levels
C. Decreased renal HCO_3^- reabsorption
D. Increased interstitial pH
E. Increased urinary H^+ excretion

Best answer = E. Metabolism generates large amounts of volatile acid (H_2CO_3) that is excreted via the lungs (3·IV·A). Decreasing ventilation allows this acid to accumulate, producing a (respiratory) acidemia. The kidneys help compensate by increasing H^+ excretion. Decreasing renal HCO_3^- reabsorption would exacerbate the acidemia through loss of buffer to urine. Accumulation of volatile acid raises plasma HCO_3^- levels because H_2CO_3 dissociates in solution to form HCO_3^- and H^+. H^+, along with other small molecules, moves freely between the blood and interstitium. Thus, if blood is acidic, the interstitium will also have a low pH.

I.10 A researcher investigating the properties of intestinal epithelium from a patient with inflammatory bowel disease noted that the diseased areas have a low electrical resistance, whereas the healthy areas have a high resistance. What might be inferred about the properties of the healthy epithelium?

A. It forms weak transepithelial ionic gradients.
B. It is specialized for isosmotic transport.
C. It has a thick basement membrane.
D. The tight junctions are highly impermeable.
E. It lacks a basolateral Na^+-K^+ ATPase.

Best answer = D. High electrical resistance is characteristic of a "tight" epithelium, a property conferred in part by the impenetrability of the tight junctions between cells to ions and water (4·II·E). Tight epithelia are notable for their ability to establish strong osmotic and ionic concentration gradients. The areas of inflammation have a low electrical resistance, which makes them "leakier" to ions. Leaky epithelia are usually specialized for high-volume isosmotic transport. Basement membranes do not directly contribute to epithelial electrical resistance. All intestinal epithelia express a basolateral Na^+-K^+ ATPase.

I.11 A 52-year-old woman presents with heart palpitations and lightheadedness. An electrocardiogram shows her to be in atrial fibrillation, which has been linked to increased connexin 43 expression. Which of the following best describes connexins normally?

A. They open during membrane depolarization.
B. They are highly ion selective.
C. They mediate Ca^{2+} influx from the cell exterior.
D. They allow electrical propagation through tissues.
E. They are found only in the heart.

Best answer = D. Connexins form hexameric assemblies (connexons) with a pore at their center (4·II·F). Connexons from two adjacent cells fuse to create a gap-junction channel that provides a pathway for electrical and chemical communication between cells. They are widely distributed. In the heart, they allow waves of contraction to spread across the myocardium (17·III). Gap-junction channels are characterized by their wide, nonselective pores that can allow passage of small peptides. They are usually open at rest and may close upon depolarization. Ca^{2+} channels mediate Ca^{2+} influx across surface membranes, not gap-junction channels.

I.12 Hypokalemia is relatively rare in healthy individuals, but which of the following would favor increased K^+ uptake by a transport epithelium for transfer to the circulation?

A. Lumen-negative potential difference
B. Increased paracellular water uptake
C. Increased Na^+-K^+ ATPase activity
D. High interstitial K^+ concentrations
E. Apical glucose cotransport

Best answer = B. The paracellular route is a significant pathway for solute and water movement across many epithelia (4·III·B). High paracellular water flow rates generate solvent drag, whereby inorganic ions and other solutes are swept along with water. Because K^+ is a cation, a renal or gastrointestinal lumen (for example) that is negatively charged with respect to blood decreases net uptake. The Na^+-K^+ ATPase increases intracellular K^+ concentrations, which decreases the driving force for apical K^+ uptake. High interstitial K^+ concentrations also decrease the electrochemical gradient favoring net K^+ uptake. Glucose cotransporters generally couple glucose movement with Na^+, not K^+.

UNIT II

Sensory and Motor Systems

Nervous System Organization

5

I. OVERVIEW

To a casual observer, a microscopic pond organism such as *Paramecium* behaves with apparent intent and coordination that suggests the involvement of a sophisticated nervous system. If it bumps into an object, it stops, swims backward, and then moves off in a new direction (Figure 5.1). This simple behavior minimally requires a sensory system to detect touch, an integrator to process information from the sensor, and a motor pathway to effect a response. *Paramecium* does not possess a nervous system, however. It is a single cell. The human brain contains over a trillion neurons. It has evolved sophisticated structures and networks that allow for self-awareness, creativity, and memory. Yet the human nervous system's basic organizational principles share many similarities with our unicellular cousins. Unicells and neurons both use changes in membrane potential (V_m) to integrate and respond to divergent and, sometimes, conflicting inputs. On an organismal level, humans, like unicells, have sensory systems to inform them about their immediate environment, integrators to process sensory data, and motor systems to effect an appropriate response.

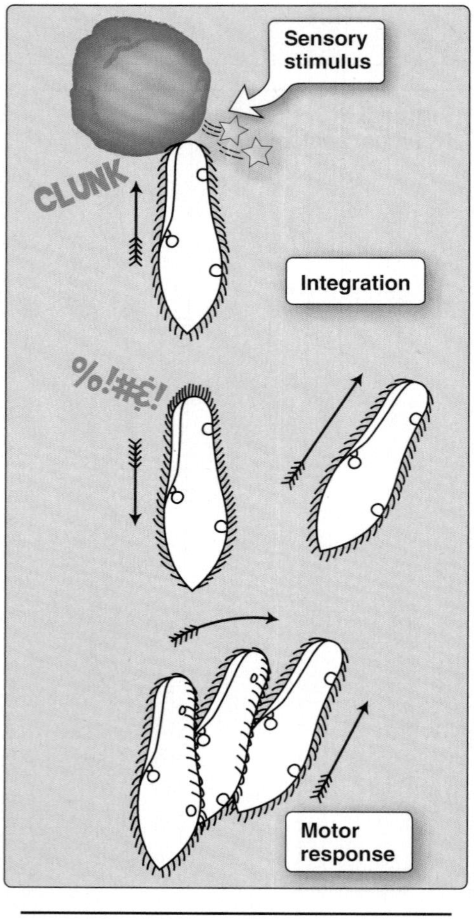

Figure 5.1
Sensory response in *Paramecium*.

II. NERVOUS SYSTEM

In discussing how the nervous system works, it is useful to define three partially overlapping subdivisions.

- The **central nervous system** (**CNS**) includes the neurons of the brain and spinal cord. The CNS is the nervous system's integrative and decision-making arm.
- The **peripheral nervous system** (**PNS**) collects sensory information and conveys it to the CNS for processing. It then directs motor commands from the CNS to the appropriate targets. The PNS includes neurons that originate in the cranium and spinal cord and extend beyond the CNS.
- The **autonomic nervous system** (**ANS**) is central to many discussions of human physiology because it regulates and coordinates **visceral** organ function, including the gastrointestinal system, lungs, heart, and

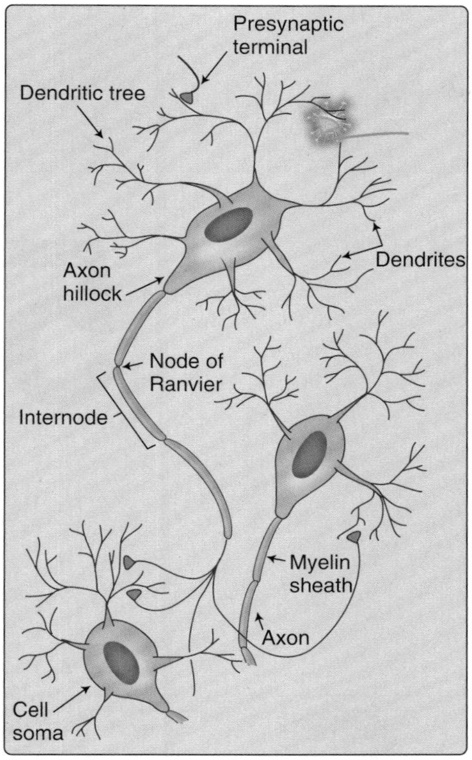

Figure 5.2
Neuronal anatomy.

vasculature. The distinction between the ANS and the other two divisions is functional rather than anatomic. The ANS can be further subdivided into the **sympathetic nervous system** and **parasympathetic nervous system**. Both divisions function largely independently of voluntary control.

III. NEURONS

The nervous system comprises a network of **neurons**. Although their shape may vary according to function and location within the body, the basic principles of neuronal design and operation are universal. Their role is to transmit information as rapidly as possible from one area of the body to the next. In the brain, the distance involved may be a few micrometers, but, in the periphery, it can exceed a meter. Because speed is achieved using electrical signals, a neuron can be thought of as a biologic wire. Unlike a wire, however, a neuron has the ability to integrate incoming signals before transmitting information to a recipient.

A. Anatomy

A neuron can be divided into four anatomically distinct regions: the **cell body**, **dendrites**, an **axon**, and one or more **nerve terminals** (Figure 5.2).

1. **Cell body:** The cell body (**soma**) houses the nucleus and components required for protein synthesis and other normal cellular housekeeping functions.

2. **Dendrites:** Dendrites are branched projections of the cell body that radiate in multiple directions ("dendrite" is derived from *dendros*, the Greek word for tree). Some neurons have dense and elaborate dendritic trees, whereas others may be very simple. Dendrites are cellular antennae waiting to receive information from the neural net. Many tens of thousands of nerve terminals may synapse with a single neuron via its dendrites.

3. **Axon:** An axon is designed to relay information at high speed from one end of the neuron to the other. It arises from a swelling of the soma called an **axon hillock**. An axon is long and thin like a wire. It is often wrapped with an insulating material (**myelin**) that enhances signal-transmission rate (see below). Myelination begins some distance distal to the axon hillock, leaving a short **initial segment** that is unmyelinated. The **axoplasm** (axonal cytoplasm) is filled with parallel arrays of microtubules and microfilaments. They are partly structural, but they also act like railway tracks in a mineshaft. "Ore carts" (vesicles) filled with neurotransmitters and other materials attach to the tracks and then motor along at relatively high speed ($\sim$2 μm/s) from one end of the cell to the other. Movement away from the cell body toward the nerve terminal (**anterograde transport**) is powered by **kinesin**. The return trip (**retrograde transport**) relies on a different molecular motor (**dynein**).

4. **Nerve terminal:** The nerve terminal is specialized to convert an electrical signal (an action potential) into a chemical signal for dispatch to one or more recipients. The junction between the

Clinical Application 5.1: Polio

Retrograde transport is believed to be the mechanism by which polio and many other viruses enter the central nervous system from the periphery.[1] Poliovirus is an enterovirus distributed by fecal–oral contact that causes **paralytic poliomyelitis**. After infecting the host, the virus enters and spreads through the nervous system via nerve terminals. After fusing with the surface membrane and entering the axoplasm, the viral capsid (a protein shell) attaches to the retrograde transport machinery and motors to the cell body. Here, it proliferates and, ultimately, destroys the neuron. The result is a flaccid paralysis of the musculature, classically affecting the lower limbs, but it can also cause fatal paralysis of the respiratory musculature. Polio has been largely eradicated in North America and Europe but is endemic in many other regions of the world. Defects in axonal transport are also believed to have a role in precipitating the neuronal death that accompanies Alzheimer disease, Huntington disease, Parkinson disease, and several other adult-onset neurodegenerative diseases.

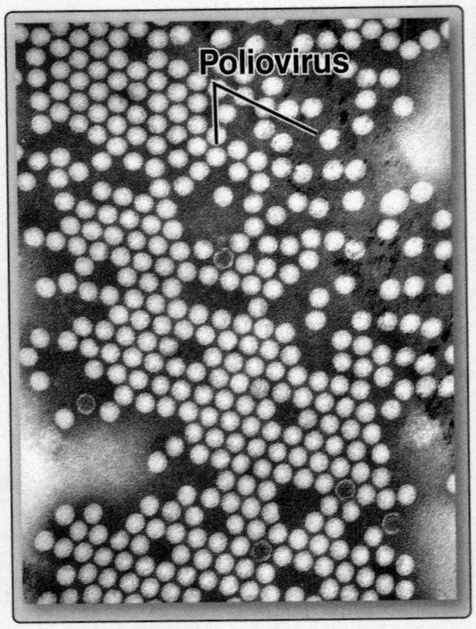

Poliovirus.

terminal and its target is called a **synapse**. The **presynaptic** and **postsynaptic** cell membranes are separated by a ~30–50 nm **synaptic cleft**. Facing the terminal across the cleft may be any of a number of different postsynaptic effector cells, including myocytes, secretory cells, or even a **dendrite** extending from the cell body of another neuron.

B. Excitability

The speed with which neural nets process and output data is limited by the rate at which signals are transmitted from one component to the next. Extracting maximal speed from a neuron is achieved by using a fast action potential, by optimizing axonal geometry, and by insulating the axons.

1. **Action potentials:** Axons that convey signals over long distances typically display action potentials that have a very simple form and function as binary digits on the neural information net. Neuronal action potentials are mediated primarily by voltage-gated Na^+ channels, which are very fast activating (Figure 5.3). When Na^+ channels open, Na^+ flows into the neuron down its electrochemical gradient, and V_m depolarizes rapidly toward the equilibrium potential for Na^+ (see 2·II·B). It is the rapidity of Na^+-channel opening ("gating kinetics") that allows electrical signals to propagate at high speeds down an axon's length. Membrane repolarization occurs largely as a result of Na^+-channel inactivation. Voltage-dependent K^+ channels activate during a spike also, but their numbers are small and, thus, their contribution to membrane repolarization is limited.

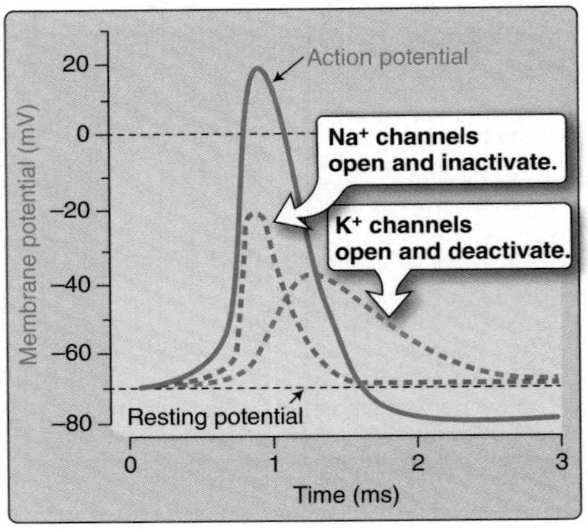

Figure 5.3
Time course of ion channel events during a neuronal action potential.

[1]For a discussion of enteroviruses, see *LIR Microbiology*, 2e, pp. 283–286.

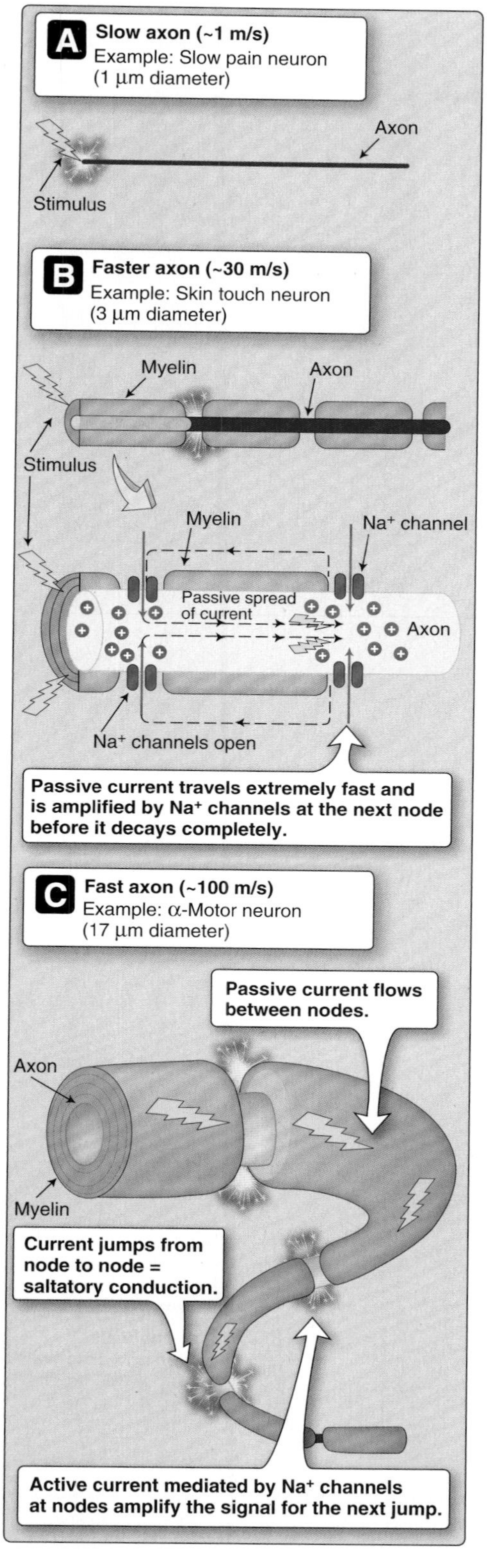

Figure 5.4
Myelin and diameter effects on axonal conduction velocity.

2. **Axon diameter:** The rate at which electrical signals travel down an axon increases with axonal diameter (Figure 5.4). This is because internal resistance, which is inversely proportional to diameter, determines how far passive current can reach down the axon's length before the signal decays and needs amplifying by an active current (i.e., a Na^+-channel–mediated current). The amplification step is slow compared with transmission of passive current, so wide axons transmit information over long distances much faster than do thin ones.

3. **Insulation:** The passive currents that flow during excitation dissipate with distance because the membrane contains K^+ "leak" channels that lose current to the extracellular medium (see Figure 2.11). Leak channels are always open. Conduction velocity is improved significantly by insulating the axon with myelin to prevent such leak (see Figure 5.4B). Myelin is formed by **glial** cells and comprises concentric layers of **sphingomyelin-rich membrane** (see Section V below). Insulation increases conduction velocity up to 250-fold.

4. **Saltatory conduction:** An axon's **myelin sheath** is not continuous. Every 1–2 mm is a 2–3-μm segment of exposed axonal membrane known as a **node of Ranvier**. Nodes are tightly packed with Na^+ channels, whereas the **internodal regions** (the areas lying hidden beneath the myelin sheath) have virtually no channels. In practice, this means that an action potential leapfrogs from one node to the next down the length of the axon, a behavior known as **nodal**, or **saltatory**, **conduction** (see Figure 5.4C).

C. Classification

CNS neurons are a diverse group of cells, and there are many ways of classifying them. Morphologically, they can be grouped on the basis of the number of **neurites** (processes, such as axons and dendrites) extending from the cell body.

1. **Pseudounipolar: Pseudounipolar neurons** are usually sensory. The cell body gives rise to a single process (the axon) that then splits into two branches. One branch returns sensory information from the periphery (the **peripheral branch**), whereas the other branch projects and conveys this information to the CNS (**central branch**).

2. **Bipolar: Bipolar neurons** are usually specialized sensory neurons. Bipolar neurons can be found in the retina (see 8·VII·A) and olfactory epithelium (see 10·III·B), for example. Their cell body gives rise to two processes. One conveys sensory information from the periphery, and the other (the axon) travels to the CNS.

3. **Multipolar: Multipolar neurons** have a cell body that gives rise to a single axon and numerous dendritic branches. Most CNS neurons are multipolar. They can be further subcharacterized based on the size and complexity of their dendritic tree.

D. Neurons as integrators

The unicell mentioned in the introduction is capable of integrating multiple sensory signals (e.g., mechanical, chemical, thermal) through changes in V_m. For example, a noxious signal that depolarizes V_m

and increases the tendency to turn might be ignored if an attractant signal indicating nearby food hyperpolarizes the membrane and negates or overrides noxious signal input. A paramecium is not capable of conscious thought, yet it makes a decision that affects behavior based on the summed effect of multiple stimuli on V_m. The dendritic trees of higher cortical neurons receive tens of thousands of competing inputs. The likelihood that neuronal output (spiking) will be modified on the basis of these signals is similarly determined by their net effect on V_m.

1. **Incoming signals:** Neurons hand off information to each other via dendrites. When a presynaptic neuron fires, it releases transmitter into the synaptic cleft. If the neuron is excitatory, transmitter binding to the postsynaptic dendritic membrane causes a transient depolarization known as an **excitatory postsynaptic potential** (**EPSP**) as shown in Figure 5.5A. Inhibitory neurons release transmitters that cause transient hyperpolarizations known as **inhibitory postsynaptic potentials** (**IPSPs**). EPSP and IPSP amplitudes are graded with incoming signal(s) strength.

2. **Filtering:** Much of the information being received by neurons at their dendrites represents sensory "noise." Isolating the strongest and most relevant signals is accomplished using a noise filter that takes advantage of a dendrite's natural electrical properties. Postsynaptic potentials (PSPs) are passive responses that degrade rapidly as they travel toward the cell body (see Figure 2.11). Degradation is enhanced by a dendrite's inherent electrical leakiness and its lack of myelin. In practice, this means that a small PSP may never reach the cell body. PSPs generated by strong presynaptic activity activate voltage-gated ion currents along the length of the dendrite (see Figure 2.12). These enhance the signals and, thereby, increase their likelihood of reaching the cell body.

3. **Integration:** Signal integration also begins at the dendritic level. PSPs may meet and combine with PSPs arriving from other synapses as they travel toward the soma. This phenomenon is known as **summation** and is reminiscent of the way in which waves (e.g., sound waves and ripples spreading across a pond surface) interfere constructively and destructively. There are two types of summation: **spatial** and **temporal**.

    a. **Spatial summation:** If EPSPs from two different dendrites collide, they combine to create a larger EPSP (see Figure 5.5B). This is known as **spatial summation** and applies to IPSPs also. EPSPs and IPSPs can also summate to yield an attenuated membrane response (see Figure 5.5C).

    b. **Temporal summation:** Two EPSPs (or IPSPs) traveling along a dendrite in rapid succession can also combine to produce a single, larger event. This is known as **temporal summation** (see Figure 5.5D).

4. **Output:** The net effect of multiple PSPs on V_m determines the likelihood and intensity of neuronal output. If a depolarization is sufficiently strong, it may elicit a train of spikes. Spikes arise from the initial segment (also known as the **spike initiation zone**) and travel down the length of the axon toward the presynaptic terminal.

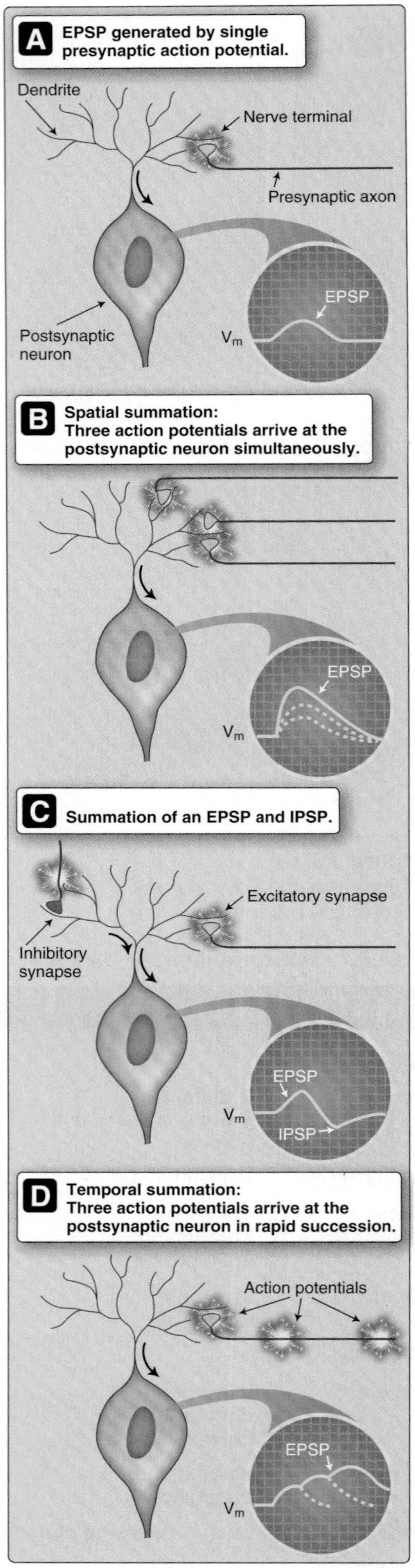

Figure 5.5
Summation. EPSP = excitatory postsynaptic potential; IPSP = inhibitory postsynaptic potential; V_m = membrane potential.

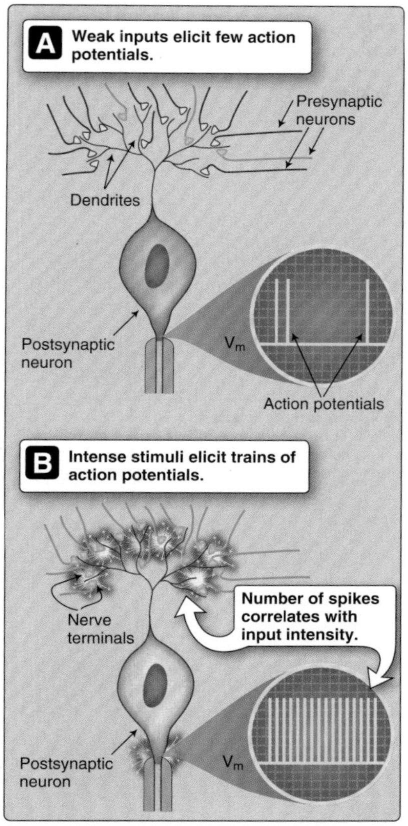

Figure 5.6
Digital encoding by neurons. V_m = membrane potential.

Table 5.1: Neurotransmitter Classes

Class	Name
Small Molecule	
Amino acid	**Glutamate** **γ-Aminobutyric acid** **Glycine**
Cholinergic	**Acetylcholine**
Cate-cholamine	**Dopamine** **Norepinephrine** **Epinephrine**
Monoamine	**Serotonin** **Histamine**
Peptides	
Opioid	**Dynorphins** **Endorphins** **Enkephalins**
Tachykinin	**Neurokinins** **Substance P**
Enteric	**Gastrin-releasing peptide**
Others	
Gas	**Nitric oxide** **Carbon monoxide**
Purine	**Adenosine triphosphate**

5. **Encoding:** Action potentials are all-or-nothing events, so neurons must pass on information about signal strength using digital encoding. Weak stimuli may yield one or two spikes. Strong stimuli elicit spike trains (volleys) that travel in rapid succession down the axon's length. There is tremendous variability in the size, shape, and frequency of spikes generated by different neurons. As a general rule, the number of spikes in a volley reflects incoming stimulus strength (Figure 5.6).

IV. NEUROTRANSMISSION

Neurons communicate with each other at synapses, specialized regions where two cells come into close apposition with each other. Communication typically occurs chemically via neurotransmitter release and is unidirectional. Although an inherently slow form of communication, placing a neurotransmitter receptor in the signaling pathway allows for a great diversity of responsiveness and unlimited opportunities for regulation.

A. Neurotransmitters

There are two main classes of neurotransmitter: small-molecule transmitters and peptides. A third, lesser group includes gases and other unconventional transmitters such as adenosine triphosphate (Table 5.1). Many tens of neuroactive peptides have been described also, many of which are coreleased along with a small-molecule transmitter. Most neural interactions involve just a handful of molecules, whose synthetic pathways are summarized in Figure 5.7.

B. Synaptic vesicles

Neurotransmitters are released into the synaptic cleft from synaptic vesicles. The vesicles are synthesized in the presynaptic cell body and then shuttled by fast axonal transport to the nerve terminal. Here, they are filled with locally produced neurotransmitter for storage and eventual release. Mature vesicles then dock at specialized release sites on the presynaptic membrane and remain there awaiting the arrival of an action potential.

> Peptide transmitters are synthesized and prepackaged in vesicles within the cell body rather than in the nerve terminal.

C. Release

Neurotransmitter release occurs when an action potential arrives at the nerve terminal and opens voltage-dependent Ca^{2+} channels in the nerve terminal membrane (Figure 5.8). Ca^{2+} influx raises local Ca^{2+} concentrations and initiates a Ca^{2+}-dependent secretory event. The details are complex and not fully resolved. The Ca^{2+} signal is sensed by a vesicle-associated Ca^{2+}-binding protein called **synaptotagmin**, which activates a **SNARE**-protein complex that includes **synaptobrevin**, **syntaxin**, and **SNAP-25**. The vesicle then fuses with the presynaptic membrane at an **active zone**, and the contents are emptied into the synaptic cleft. Each vesicle releases a single **quantum** of neurotransmitter ("**quantal signaling**").

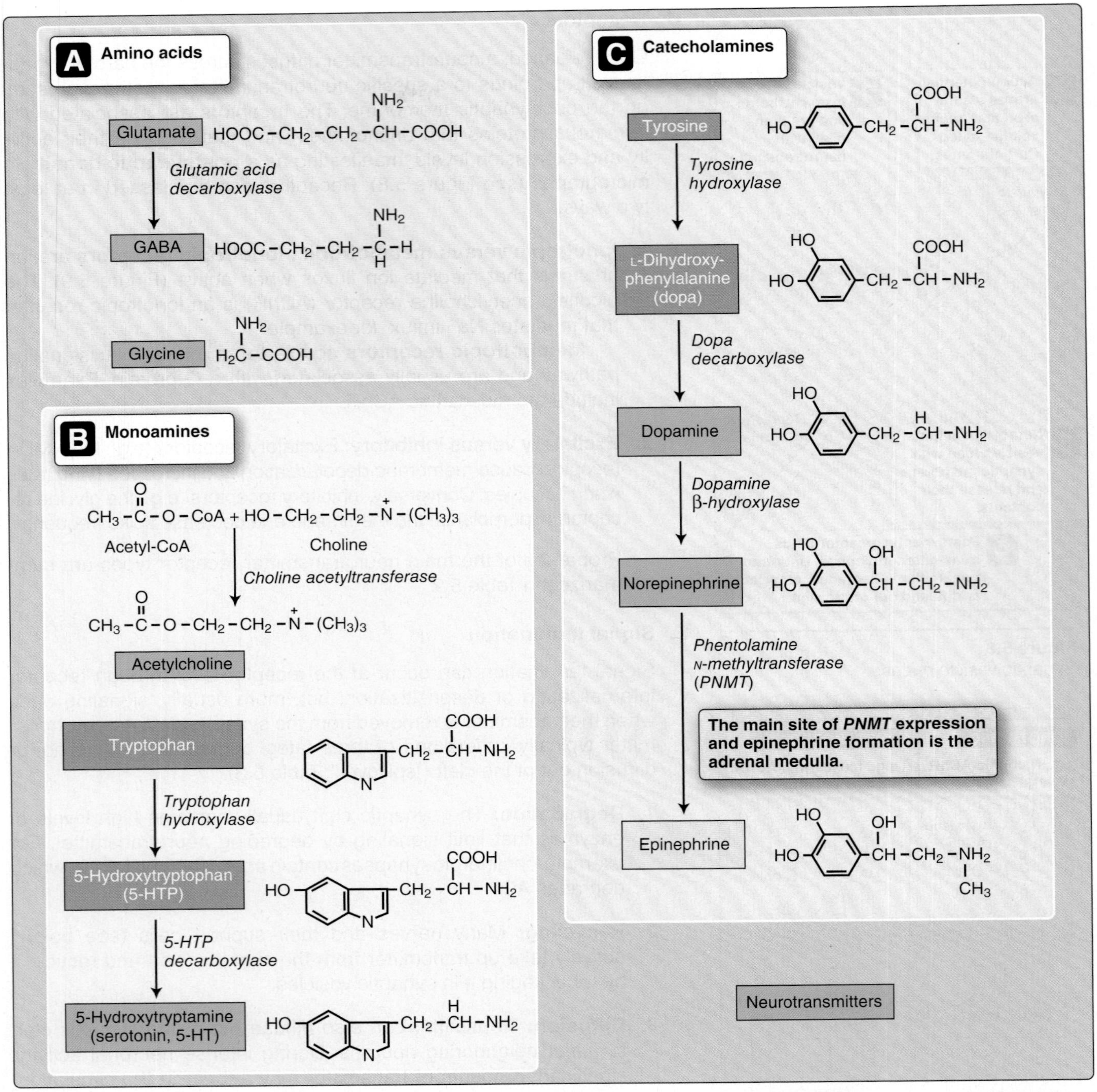

Figure 5.7
Common small-molecule neurotransmitters and their synthesis pathways. GABA = γ-aminobutyric acid.

> Botulinum toxin and tetanus toxin, two of the most lethal known neurotoxins, both paralyze their victims by targeting the synaptic terminal and disrupting neurotransmitter release. Both toxins degrade SNAPS and SNARES by virtue of their intrinsic *protease* activity.

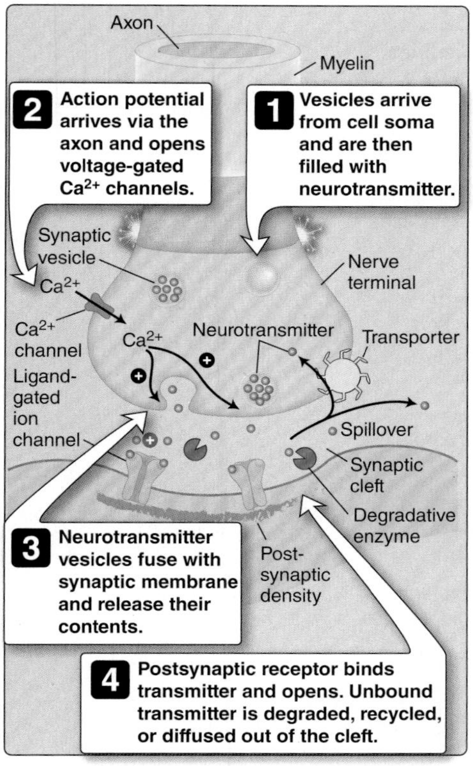

Figure 5.8
Synaptic vesicle release.

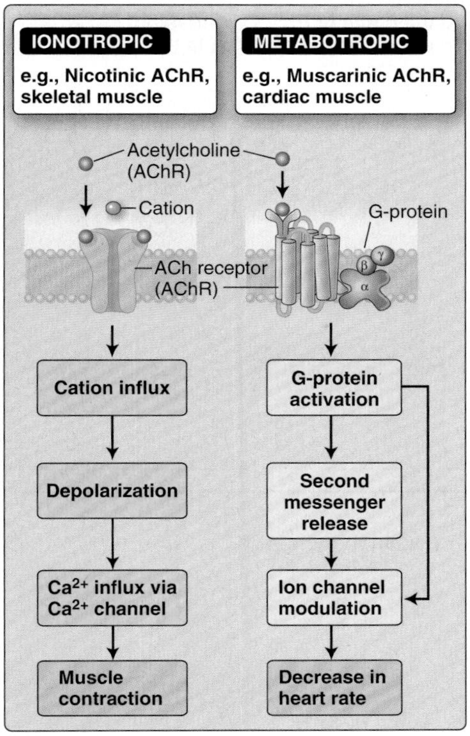

Figure 5.9
Ionotropic versus metabotropic receptors.

D. Receptors

Once released, a neurotransmitter diffuses across the narrow synaptic cleft and binds to a specific neurotransmitter receptor expressed on the postsynaptic membrane. The receptors are associated with numerous proteins that anchor them and that regulate their activity and expression levels, manifesting as a **postsynaptic density** in micrographs (see Figure 5.8). Receptors can be classed in at least two ways.

1. **Ionotropic versus metabotropic: Ionotropic receptors** are ion channels that mediate ion fluxes when active (Figure 5.9). The nicotinic acetylcholine receptor (AChR) is an ionotropic receptor that mediates Na^+ influx, for example.

   Metabotropic receptors couple to an intracellular signaling pathway and are usually associated with a G protein. Examples include the muscarinic AChR.

2. **Excitatory versus inhibitory:** Excitatory receptors (e.g., the NMDA receptor) cause membrane depolarization and increased firing rates when occupied. Conversely, inhibitory receptors (e.g., the glycine receptor) hyperpolarize the membrane and decrease spike frequency.

   Properties of the main neurotransmitter receptor types are summarized in Table 5.2.

E. Signal termination

Signal termination can occur at the receptor level through receptor internalization or desensitization, but, more usually, signaling ends when the transmitter is removed from the synaptic cleft. A neurotransmitter typically suffers one of three fates: degradation, recycling, or diffusion out of the cleft ("spillover"; Table 5.3).

1. **Degradation:** The synaptic cleft usually contains high levels of enzymes that limit signaling by degrading neurotransmitter. For example, cholinergic synapses contain *acetylcholinesterase*, which degrades ACh.
2. **Recycling:** Many nerves and their support cells (see below) actively take up transmitter from the synaptic cleft and recycle it by repackaging it in synaptic vesicles.
3. **Diffusion:** Transmitter can also diffuse out of the synaptic cleft to affect neighboring neurons. During intense neuronal activity, significant amounts of transmitter may appear in the circulation, ultimately being degraded by systemic enzymes or being excreted by the kidneys.

V. NEUROGLIA

Glia (or **neuroglia**) are inexcitable cells that support many aspects of neuronal function. In addition to forming and maintaining myelin, they control local ion concentrations, help recycle neurotransmitters, and provide neurons with nutrients. They are found throughout the PNS and CNS (Table 5.4), where neurons and glia are present in equal numbers.

Table 5.2: Neurotransmitter Receptors

Receptor	Type		Transduction	Agonists	Antagonists	Localization
GluN	I		$\uparrow I_{Na}$, I_{Ca}	Glutamate, NMDA	Phencyclidine	CNS
GluA	I		$\uparrow I_{Na}$	Glutamate, AMPA		CNS
GluK	I		$\uparrow I_{Na}$	Glutamate, kainate		CNS
mGluR	M	Group I	G_q, $\uparrow IP_3$	Glutamate		CNS
mGluR	M	Group II, III	G_i, $\downarrow$cAMP	Glutamate		CNS
$GABA_A$	I		$\uparrow I_{Cl}$	GABA, ibotenate	Bicuculline	CNS
$GABA_B$	M		G_i, $\uparrow I_K$	GABA, baclofen		ANS
nAChR (nicotinic)	I	Muscle	$\uparrow I_{Na}$	ACh, nicotine	Pancuronium	Skeletal muscle
	I	Ganglion	$\uparrow I_{Na}$	ACh, nicotine	Trimethaphan	ANS
mAChR (muscarinic)	M	M_1	G_q, $\downarrow I_K$	ACh, muscarine	Atropine, diphenhydramine, ipratropium	ANS ganglia
	M	M_2	G_i, $\uparrow I_K$ $\downarrow I_{Ca}$	ACh, muscarine		Heart
	M	M_3	G_q	ACh, pilocarpine		GI glands, eye
	M	M_4	G_i, $\uparrow I_K$ $\downarrow I_{Ca}$	ACh		CNS
Dopamine	M	D_1	G_s, $\uparrow$cAMP	Dopamine		CNS
	M	D_2	G_i, $\downarrow$cAMP	Dopamine	Clozapine	CNS
Adrenergic	M	α_1	G_q, $\uparrow IP_3$	NE, phenylephrine	Prazocin	Vasculature
Epinephrine (Epi), norepinephrine (NE)	M	α_2	G_i, $\downarrow$cAMP	Epi, clonidine	Phentolamine	Heart, vasculature
	M	β_1	G_s, $\uparrow$cAMP	Epi, dobutamide	Propanolol, sotalol	Heart
	M	β_2	G_s, $\uparrow$cAMP	Epi, isoproterenol	Propanolol	Vasculature
Serotonin	I	5-HT_3	$\uparrow I_{Na}$, I_{Ca}	5-HT	Granisetron	CNS, GI
(5-hydroxytryptamine, or 5-HT)	M	5-HT_1	G_i, $\downarrow$cAMP	5-HT, triptans		CNS, vasculature
	M	5-HT_2	G_q, $\uparrow IP_3$	5-HT, lysergic acid	Clozapine	CNS, GI, smooth muscle, vasculature
	M	5-HT_4	G_s, $\uparrow$cAMP	5-HT		CNS, GI
	M	5-HT_5	G_i, $\downarrow$cAMP	5-HT		CNS
	M	5-HT_6	G_s, $\uparrow$cAMP	5-HT		CNS
	M	5-HT_7	G_s, $\uparrow$cAMP	5-HT		CNS, ANS
Histamine	M	H_1	G_q, $\uparrow IP_3$	Histamine	Diphenhydramine	CNS, airways, vasculature
	M	H_2	G_s, $\uparrow$cAMP	Histamine	Ranitidine	Heart, stomach
	M	H_3	G_i, $\downarrow$cAMP	Histamine	Ciproxifan	CNS
	M	H_4	G_i, $\downarrow$cAMP	Histamine		Mast cells
Substance P		NK1		Substance P		CNS, pain fibers
Neuropeptide Y		$Y_{1\text{-}2}$, $Y_{4\text{-}5}$	G_i, $\downarrow$cAMP	Neuropeptide Y		CNS
Nitric Oxide		*GC*	$\uparrow$cGMP	Nitric oxide		CNS, ANS
Purinergic	I	P2X	$\uparrow I_{Na}$, I_{Ca}	ATP	Suramin	Widespread

ACh = acetylcholine; AMPA = α-amino-3-hydroxy-5-methyl-4-isoxazolepropionic acid; ANS = autonomic nervous system; ATP = adenosine triphosphate; cAMP = cyclic adenosine monophosphate; cGMP = cyclic guanosine monophosphate; CNS = central nervous system; GABA = γ-aminobutyric acid; *GC* = *guanylyl cyclase*; GI = gastrointestinal; I = Ionotropic; I_{Ca} = Ca^{2+} current; I_{Cl} = Cl^- current; I_K = K^+ current; I_{Na} = Na^+ current; IP_3 = inositol trisphosphate; M = metabotropic; NMDA = N-methyl-D-aspartatic acid.

Table 5.3: Mechanisms of Signal Termination

Neurotransmitter	Fate
Glutamate	Reuptake by the glutamate transporters in neurons and glia
γ-Aminobutyric acid (GABA)	Reuptake by neuronal GABA reuptake transporters and by glia
Acetylcholine	Degraded in synaptic cleft by *acetylcholinesterase*
Dopamine, norepinephrine, serotonin	Degraded in cleft by *catechol-O-methyltransferase,* reuptake by Na^+-Cl^--dependent transporters and recycled, or degraded by *monoamine oxidases*
Histamine	Degraded by *histamine methyltransferase* and *histaminases*
Substance P	Internalization of transmitter–receptor complex
Nitric oxide	Oxidized
Adenosine triphosphate	Degraded

Table 5.4: Glial Cells and Their Functions

Glial Cell Type	Localization	Function
Peripheral Nervous System		
Schwann cells	Axons	Myelination, phagocytosis
Satellite cells	Ganglia	Regulate chemical environment
Enteric Nervous System		
Enteric glial cells	Ganglia	Various
Central Nervous System (CNS: Astrocytes)		
Protoplasmic	Gray matter	Nutrient delivery, blood–brain barrier function
Fibrous	White matter	Repair
Müller cells	Retina	Repair
Bergmann glia	Cerebellum	Synaptic plasticity
Oligodendrocytes	White matter (some in gray)	Myelination
Microglia	Throughout CNS	Responses to trauma
Ependymal cells	Ventricles	Regulate exchange between cerebrospinal fluid and brain extracellular fluid

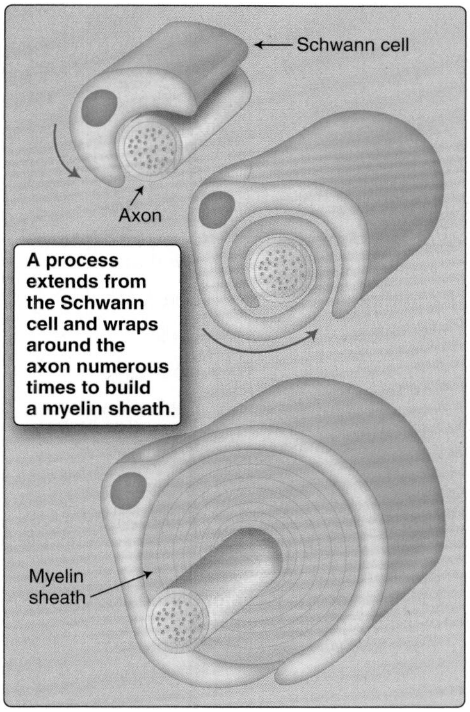

Schwann cell
Axon
A process extends from the Schwann cell and wraps around the axon numerous times to build a myelin sheath.
Myelin sheath

Figure 5.10
Myelin sheath formation.

A. Myelination

Myelin is formed by **Schwann cells** (in the PNS) and **oligodendrocytes** (in the CNS). Oligodendrocytes may simultaneously myelinate the axons of multiple neurons, but Schwann cells remain dedicated to a single axon. Glia form myelin by extending processes that rotate around an axon over 100 times (Figure 5.10). Cytoplasm is squeezed out from between the membrane layers as the myelin builds, so that the lipid layers become highly compacted. Glial cells remain viable after the layering is complete, with the nucleus and residual cytoplasm occupying the outermost layer.

B. Potassium homeostasis

Renormalization of V_m after neuronal excitation involves K^+ release (see Figure 5.3). During intense neuronal activity, extracellular K^+ concentration can rise significantly as a result of this release. Because V_m is dependent on the transmembrane K^+ gradient (see 2·II·C), K^+ buildup can be detrimental to neuronal function. **Astrocytes** (the predominant glial cell type in the CNS) are inexcitable, but they do possess K^+ channels and K^+ transporters that allow them to siphon K^+ away from active neurons and redistribute it over nonactive regions of the CNS (Figure 5.11). "**Spatial buffering**" takes advantage of the fact that adjacent astrocytes are tightly coupled via gap junctions (see 4·II·F) that provide pathways for K^+ to flow down their concentration gradient from the active zone to a remote site.

Clinical Application 5.2: Multiple Sclerosis

Myelin is essential for normal neural communication, so diseases that affect myelin or the cells that produce it have devastating physiologic effects. **Multiple sclerosis** (**MS**) is a demyelinating disease of central nervous system neurons, and the cause is unknown. Symptoms arise when autoreactive immune cells produce antibodies against one or more sheath components. The myelin swells and degrades, and axonal conduction is interrupted, producing pathologic changes that can be visualized using computed tomography (commonly known as a CT scan). Patients can present with a variety of neurologic symptoms, including tremors, visual disturbances, autonomic dysfunction, weakness, and fatigue. MS may be relapsing in nature, characterized by acute onset of clinical symptoms, followed by a period of remission with full or partial recovery of function. There is no known treatment for MS, however, and the disease usually progresses over a period of 10 to 20 years.

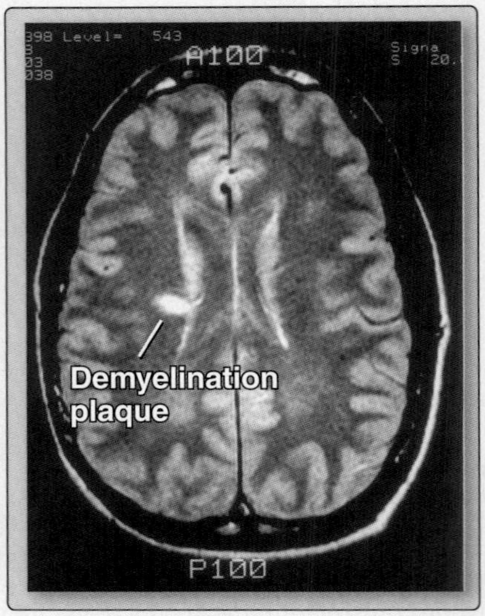

Periventricular frontal lobe demyelination.

C. Neurotransmitter uptake and recycling

Glutamate, γ-aminobutyric acid (GABA), and glycine are three of the most widely used neurotransmitters in the CNS. Many neurons employing these transmitters spike at such high frequencies that their ability to control synaptic transmitter levels and prevent spillover into adjacent regions is challenged. Intense activity also strains the transmitter synthesis pathways. Glial cells assist with both issues. They

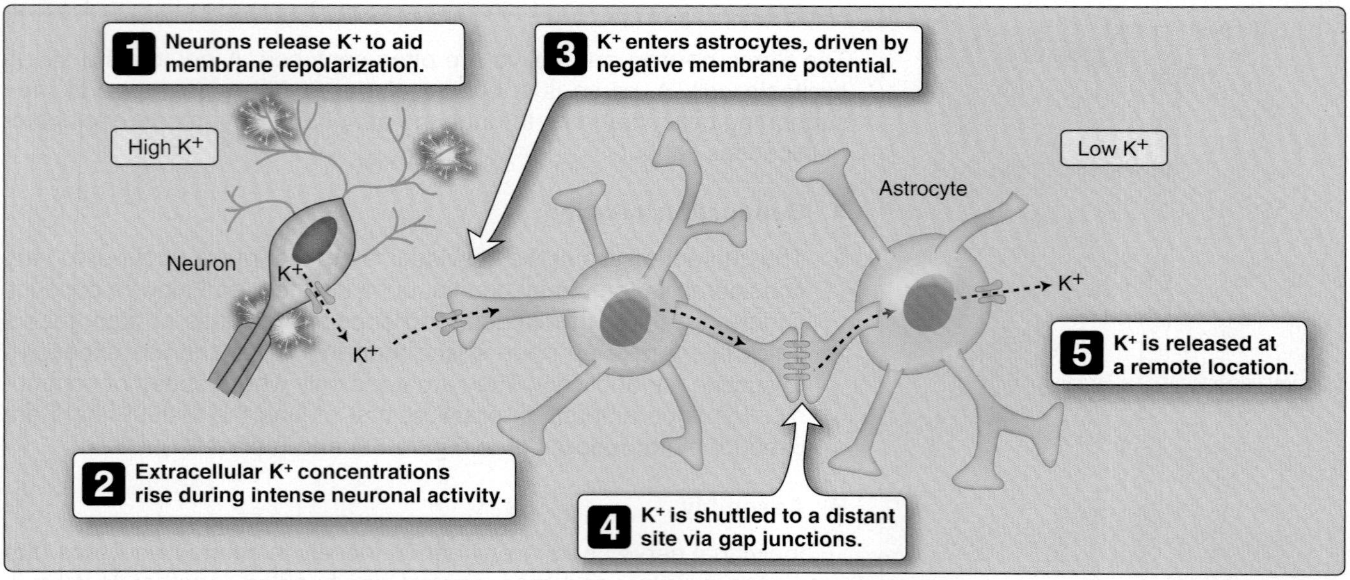

Figure 5.11
Spatial buffering.

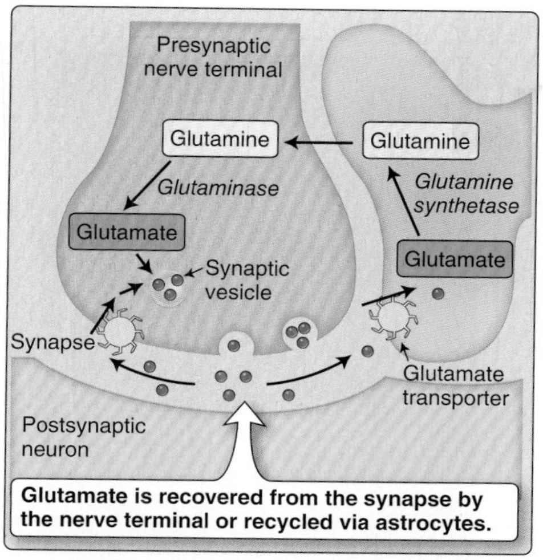

Figure 5.12
Neurotransmitter recycling by astrocytes.

extend foot processes that surround the synapse and rapidly take up transmitter from the cleft using high-affinity transport systems (Figure 5.12). Glutamate is subsequently converted to glutamine by *glutamine synthetase* and returned to the presynaptic terminal for conversion back to glutamate by *glutaminase*. Glutamine is also distributed to inhibitory neurons for GABA synthesis. GABAergic neurons have very limited glutamine reserves and are dependent on glia to supply the substrates necessary for continued signaling.

D. Nutrient supply

Neurons are highly dependent on O_2 and glucose for continued activity, the nervous system accounting for 20% of total body usage. Glia play a unique role in ensuring that these needs are met (see also 21·II for a discussion of the role of glial cells in maintaining the blood-brain barrier).

1. **Lactate shuttle:** Astrocytes ferry glucose from the vasculature to neurons. They extend foot processes that surround cerebral capillaries and absorb glucose from blood using transporters. Glucose then diffuses through the glial network via gap junctions. Some glucose is converted to glycogen, and the rest is metabolized to lactic acid. Lactate is then excreted into the extracellular fluid for uptake by surrounding neurons, a process known as the **lactate shuttle**.

2. **Storage:** Neurons have few energy reserves. They rely on astrocytes to maintain a steady lactate supply in the face of changes in neuronal activity or waning blood glucose levels. Astrocytes contain extensive glycogen stores and the necessary pathways to convert them to lactate when needed.

VI. NERVES

The terms **neuron** and **nerve** are often confused. A *neuron* is a single excitable cell. A *nerve* is a bundle of **nerve fibers** (axons and their supporting cells) that runs through the periphery like a modern telecommunications cable.

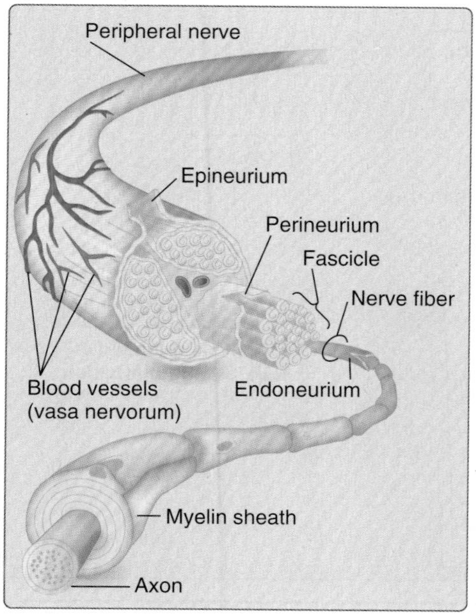

Figure 5.13
Peripheral nerve anatomy.

A. Conduction velocity

The characteristics of the individual fibers that make up a nerve vary considerably. Some may be thin, unmyelinated, and slow to conduct. Others are thick, myelinated, and conduct impulses at high speed. Thick fibers take up more space than thin ones and are expensive to sustain metabolically. They are used only where speed of communication is paramount. In practice, that means the fastest fibers are used for motor reflexes (see Figure 5.4 and Table 11.1).

B. Assembly

Individual nerve fibers are wrapped loosely in connective tissue (the **endoneurium**), and then several are bundled together to form a **fascicle** (Figure 5.13). A fascicle is wrapped in yet more connective tissue (the **perineurium**), and, finally, several fascicles are gathered together with blood vessels to form a **nerve**. The nerve is heavily

protected with a dense layer of connective tissue (the **epineurium**). Peripheral nerves are subjected to considerable mechanical stress associated with locomotion, so the multiple reinforcing layers are essential for protection.

C. Ganglia

The cell bodies of axons that make up a nerve are clustered in swellings called **ganglia**. Ganglia are the sites of information relay between neurons. Ganglia may also contain intrinsic circuits that allow for reflex arcs and signal processing.

D. Types

Nerves can be classified according to the type of information that they carry. **Afferent nerves** contain fibers relaying sensory information from the various body regions to the CNS. **Efferent nerves** originate in the CNS and contain **somatic** motor neurons that innervate skeletal muscles and ANS **visceral** motor neurons. Some peripheral nerves contain a mix of both sensory and motor fibers (**mixed nerves**). This can result in reflex arcs that travel within the same nerve. The vagus nerve, for example, informs the CNS when food is entering the stomach. The CNS responds with a motor command that makes the stomach relax, and it travels via an efferent fiber contained within the vagus nerve (see 30·IV·A). The arc is called a **vagovagal reflex**.

Chapter Summary

- The **nervous system** comprises the **central nervous system** (**CNS**) and the **peripheral nervous system** (**PNS**). The **autonomic nervous system** (**ANS**) is a functional subdivision of the CNS and PNS. The CNS includes **brain** and **spinal cord** neurons. The PNS includes nerves that carry **sensory** and **motor information** to the periphery. The ANS monitors and controls internal organ function.
- Neurons are **excitable cells** that communicate with each other and with target organs using **action potentials**. Electrical signals are relayed from one cell to the next at a **synapse** using **chemical neurotransmitters**.
- Neurotransmitters are a diverse group that includes gases and polypeptides. The central nervous system (CNS) uses small molecules such as **amino acids** (glutamate, aspartate, glycine, and γ-aminobutyric acid) as transmitters. **Acetylcholine** is the neurotransmitter used at the **neuromuscular junction** and by some neurons of the autonomic nervous system (ANS). The ANS and CNS also use **monoamines** (dopamine and norepinephrine) as transmitters.
- A transmitter may excite or inhibit the postsynaptic cell, depending on the nature of the receptor on the postsynaptic membrane. **Excitatory transmitters** cause postsynaptic cell **depolarization** (an **excitatory postsynaptic potential**), whereas **inhibitory transmitters** cause **hyperpolarization** and reduced excitability (**inhibitory postsynaptic potential**).
- Central nervous system neurons may receive thousands of synaptic inputs via their dendritic trees. The electrical properties of neuronal dendrites ensure that weak inputs do not propagate. Stronger inputs may **summate** to push membrane potential beyond the **threshold for excitation**, causing the neuron to fire an action potential.
- Neuronal function is supported by **glial cells**. Glia may regulate extracellular ion concentrations, aid in uptake and **recycling** of neurotransmitters, supply nutrients, and encase the axons in **myelin**.
- Myelin is an insulating material laid down by **Schwann cells** (peripheral nervous system) and **oligodendrocytes** (central nervous system). Myelin increases the speed at which electrical signals propagate down an axon. Myelin is formed from compacted layers of glial surface membrane. Myelinated neurons are used in **reflex arcs** where response timing is critical. **Motor neurons** also have axons that are wider than normal to further increase signal propagation rates.
- **Nerves** are bundles of axons and their supporting cells. Peripheral nerves may convey afferent signals from sensory cells to the central nervous system (CNS), efferent signals from the CNS to effector cells, or a signal mixture.

6 Central Nervous System

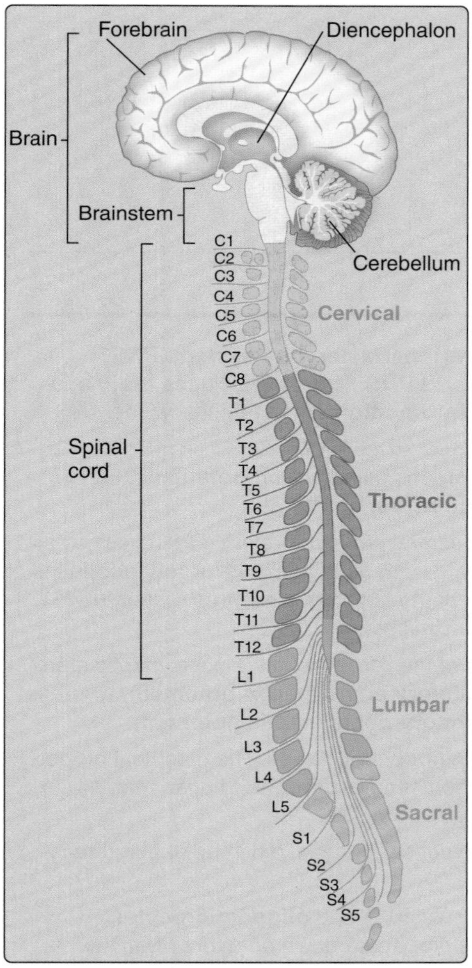

Figure 6.1
Central nervous system. C1–C8, T1–T12, L1–L5, and S1–S5 are spinal nerves.

I. OVERVIEW

The central nervous system (CNS) comprises the spinal cord and brain (Figure 6.1). The spinal cord is a thick communications tract that relays sensory and motor signals between the peripheral nervous system (PNS) and the brain. The cord also contains intrinsic circuits that support certain muscle reflexes. The brain is a highly sophisticated data processor containing neural circuits that analyze sensory data and then execute appropriate responses via the spinal cord and PNS efferents. Large portions of the brain are devoted to associative functions that integrate information from the various senses and allow us to assign meaning to sounds, associate smells with specific memories, and recognize objects and faces, for example. Associative regions also provide for abstract thinking, language skills, social interactions, and learning and memory. The human body has bilateral symmetry, and the structures of the spinal cord and brain are, for the most part, mirrored about a midline. Sensory and motor information generally crosses the midline at some point in its journey between the brain and the periphery. In practice, this means that the left side of the brain controls the right side of the body and *vice versa*. For the purposes of discussion, the brain can be divided into four principal areas: the **brainstem**, **cerebellum**, **diencephalon**, and the **cerebral hemispheres** (**triencephalon**). A full discussion of CNS function is beyond the scope of this book, which focuses on the sensory and motor aspects of CNS function. For more information on higher brain functions, see *LIR Neuroscience*.

II. SPINAL CORD

The spinal cord is housed within the vertebral canal. It extends from the foramen magnum at the base of the skull caudally to the second lumbar vertebra.

A. Segments

The vertebral column consists of a series of stacked vertebrae divided anatomically into five regions: **cervical**, **thoracic**, **lumbar**, **sacral**, and **coccygeal**. The cervical, thoracic, and lumbar vertebrae are separated by intervertebral disks that allow the bones to articulate, but the sacral and coccygeal vertebrae are fused to form the **sacrum** and **coccyx**, respectively. The spinal cord can be divided

into 31 named segments. Thirty-one pairs of spinal nerves (one on each side of the body) emerge from corresponding segments (see Figure 6.1). Although the spinal cord terminates before it reaches the sacrum, spinal nerves continue caudally within the vertebral canal until they reach an appropriate exit level.

> **Rostral** and **caudal** are anatomical terms meaning "beak" (or mouth) and "tail," respectively. They are commonly used to indicate direction of information flow in the CNS.

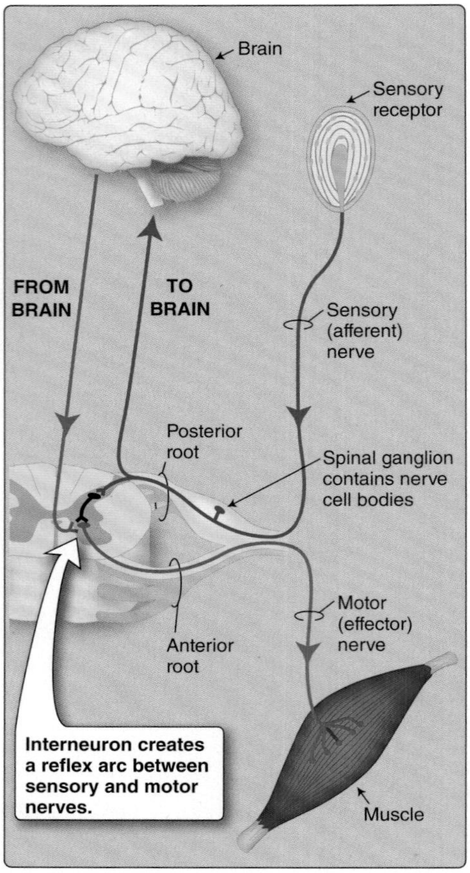

Figure 6.2
Sensory and motor pathways.

B. Nerves

Spinal nerves are a component of the PNS. The nerves contain **sensory afferent** and **motor efferent** fibers (spinal nerves are sometimes called **mixed spinal nerves** for this reason) that generally serve tissues on the same level as the nerves. Thus, nerves emerging from the cervical region (C2) control head and neck movements, whereas sacral nerves (S2 and S3) project to the bladder and large intestine.

1. **Sensory:** Somatic and autonomic sensory fibers travel to the spinal cord via peripheral nerves (Figure 6.2). They relay sensations of pain, temperature, and touch from the skin; proprioceptive signals from muscle and joint receptors; and sensory signals from numerous visceral receptors. Multiple peripheral nerves come together to form the **posterior root** of a spinal nerve and enter the vertebral canal via an **intervertebral foramen**. The cell bodies of these nerves cluster within a prominent **spinal ganglion** located within the foramen. The posterior root then divides into a number of **rootlets** and joins the spinal cord. Sensory nerves travel rostrally to synapse within nuclei *en route* to the brain. Branches of sensory afferents may also synapse directly with motor neurons or on interneurons that synapse with motor neurons, which makes local spinal cord–mediated reflexes possible (see 11·III).

2. **Motor:** Motor efferents from the brain travel caudally and synapse with peripheral motor nerves within the spinal cord. These nerves include both somatic and autonomic motor efferents. They leave the spinal cord via **anterior rootlets**, which join to form an **anterior root** and then travel out to the periphery alongside sensory fibers in spinal nerves.

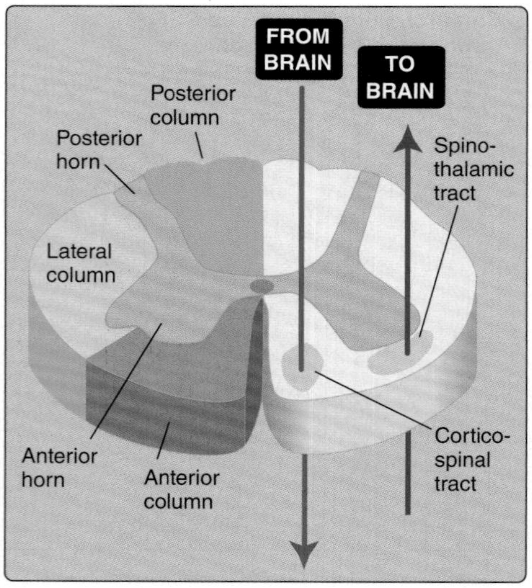

Figure 6.3
Spinal cord organization.

C. Tracts

The spinal cord's interior is roughly organized into a butterfly-shaped central area of gray matter surrounded by white matter (Figure 6.3). The white matter contains bundles of nerve fibers with common origins and destinations that relay information between the PNS and the brain. Sensory nerve fibers from the periphery travel rostrally to the brain in discrete **ascending tracts**. **Descending tracts** carry bundles of motor efferents from the CNS en route to the periphery.

The tracts are named according to their origin and destination. For example, the **spinothalamic tract** carries pain fibers from the spine upward to the thalamus. The **corticospinal tract** carries motor fibers from the cortex downward to the spine. The tracts (also known as **fasciculi**) are grouped in **posterior**, **lateral**, and **anterior columns** (also known as **funiculi**). The "wings" of the gray butterflies are divided into **posterior** and **anterior horns** and act as synaptic relay stations for information flow between neurons. They contain neuronal cell bodies, which may be clustered in functionally related groups, or **nuclei**. The gray matter on either side of the cord is connected by **commissures** containing bundles of fibers that allow for information flow across the midline.

CNS tissue typically appears white or gray in color. White matter is largely composed of myelinated nerve axons (it gets its white color from myelin). Gray matter is composed of cell bodies, dendrites, and unmyelinated axons.

III. BRAINSTEM

All sensory and motor information flowing to and from the brain passes through the brainstem (Figure 6.4). The brainstem contains several important nuclei that act as relay stations for information flow between brain and periphery. Many of the 12 **cranial nerves** (**CNs**) originate from nuclei within the brainstem also (Figure 6.5). The CNs provide sensory and motor innervation to the head and neck and include nerves that mediate vision, hearing, smell, and taste, among many other functions. Intrinsic circuits within the brainstem create control centers that allow for reflex responses to sensory data. The location and functions of these centers are discussed in more detail in Chapter 7. The brainstem can be subdivided anatomically into three areas:

- **Medulla:** The medulla contains autonomic nuclei involved in the control of respiration and blood pressure and in coordination of swallowing, vomiting, coughing, and sneezing reflexes.
- **Pons:** The pons helps control respiration.
- **Midbrain:** The midbrain contains areas involved in controlling eye movements.

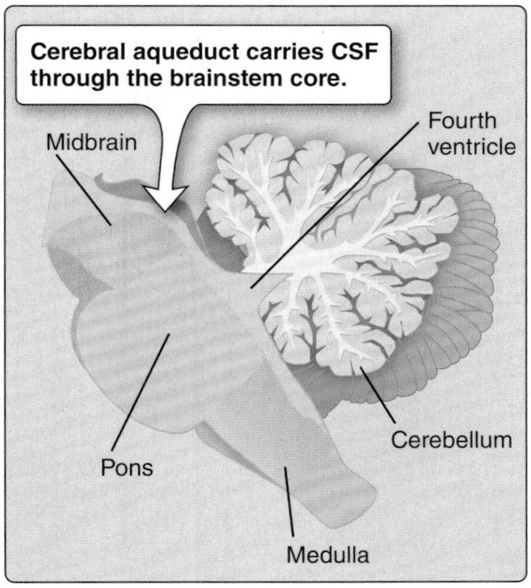

Figure 6.4
Brainstem organization.
CSF = cerebrospinal fluid.

CN I and CN II do not originate in the brainstem. CN I, the olfactory nerve, is a sensory nerve that relays information from the olfactory epithelium in the roof of the nasal cavities directly to the olfactory bulb. CN II, the optic nerve, enters the brain at the level of the diencephalon.

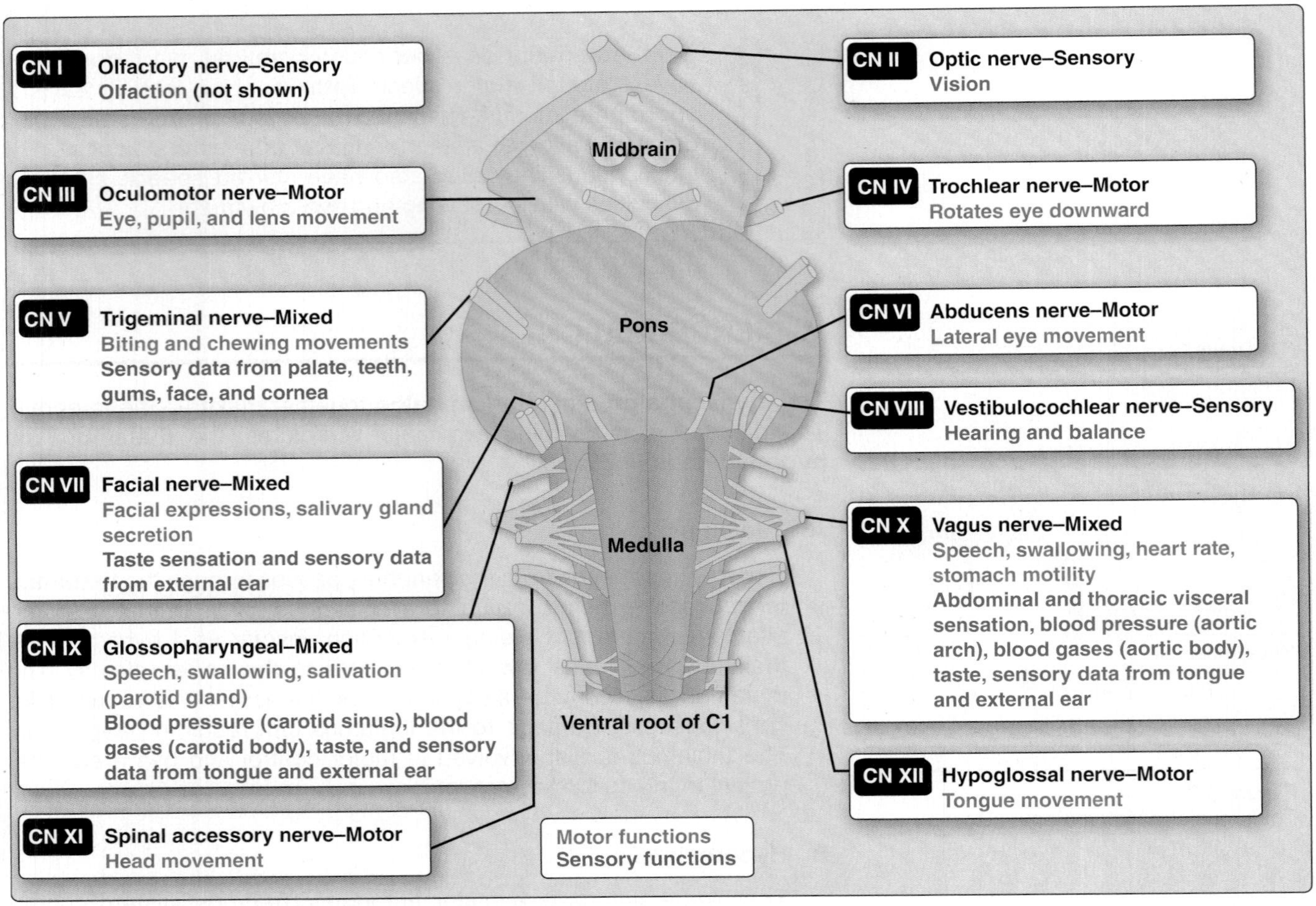

Figure 6.5
Cranial nerve (CN) functions. C1 = first cervical vertebra.

IV. CEREBELLUM

The cerebellum fine-tunes motor control and facilitates smooth execution of learned motor sequences (see 11·IV·C). Cerebellar function requires massive integrative and computational capabilities, which is why this small area contains more neurons than the rest of the brain combined, even though it accounts for only ~10% of total brain mass! The cerebellum is attached to the brainstem by three **peduncles** that contain thick afferent and efferent nerve fiber bundles. The cerebellum receives sensory data from muscles, tendons, joints, skin, and the visual and vestibular systems and inputs from all regions of the CNS involved in motor control. It also sends signals back to most of these areas and modifies their output (Figure 6.6). Integration of sensory data with motor commands is achieved using feedback and feedforward circuits that include the **Purkinje cell**, a neuronal type renowned for its immense dendritic tree. The dendrites are sites of information flow from hundreds of thousands of presynaptic neurons. The cerebellar circuits allow movements to be finessed with reference to incoming sensory data, even as they are being executed.

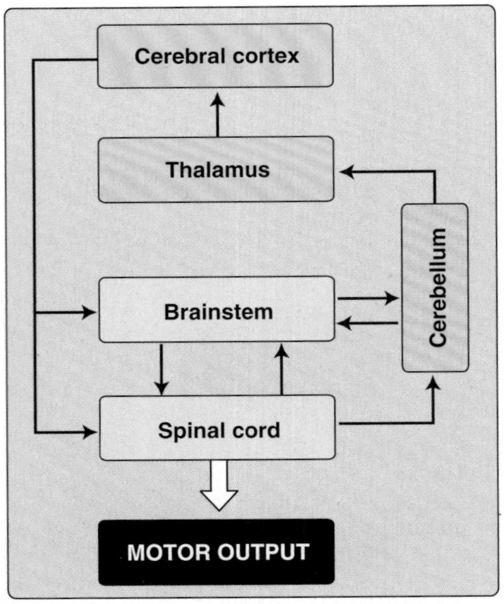

Figure 6.6
Functional relationships among central nervous system components.

> Cerebellar injury does not cause paralysis, but it does have profound motor effects (**ataxia**, or an inability to coordinate muscle activity). Patients with cerebellar damage walk with a staggering gait that mimics alcohol intoxication. They may also have slurred speech and difficulties with swallowing and eye movement.

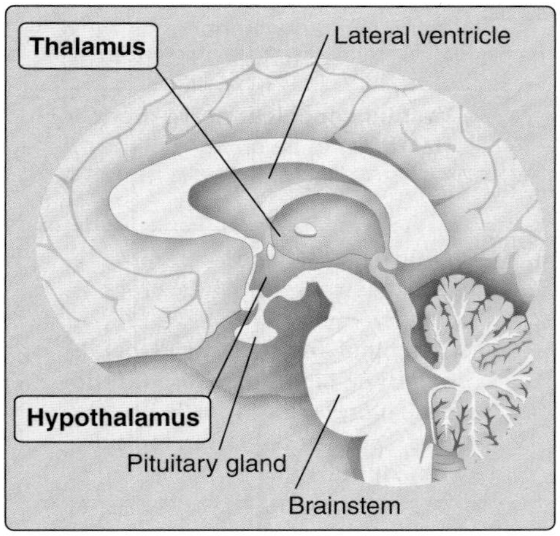

Figure 6.7
Thalamus and hypothalamus location.

V. DIENCEPHALON

The **diencephalon** and **telencephalon** together make up the forebrain. The diencephalon contains two major structures: the **thalamus** and **hypothalamus** (Figure 6.7).

A. Thalamus

Sensory information from the periphery passes through the **thalamus** for processing before reaching a conscious level. Output from the olfactory system is the single exception, insofar as it bypasses the thalamus and feeds raw olfactory data to the cortex directly. The thalamus also controls sleep and wakefulness and is required for consciousness. Damage to the thalamus can result in deep coma. The thalamus is also involved in motor control and has areas that project to the cortical motor regions.

B. Hypothalamus

The hypothalamus is a major autonomic nervous system control center that is discussed in detail in Chapter 7. Its functions include control of body temperature, food intake, thirst and water balance, and blood pressure, and it also controls aggression and rage. The hypothalamus exerts control through direct neural connections to autonomic centers in the brainstem, but it also controls the endocrine system. Endocrine control occurs directly through hormonal synthesis and release (oxytocin and antidiuretic hormone) and indirectly by secreting hormones that affect release of pituitary hormones.

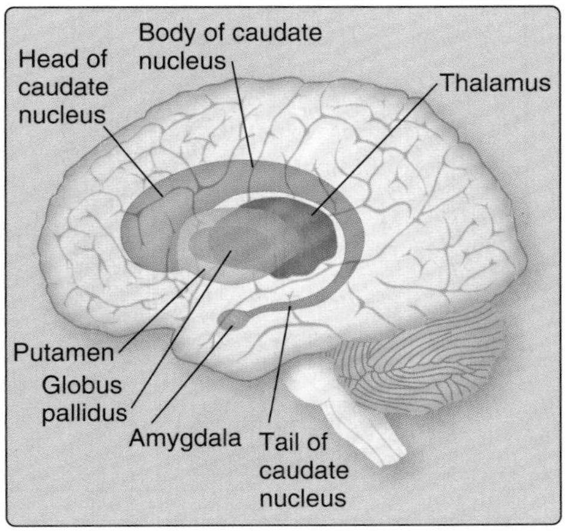

Figure 6.8
Basal ganglia.

VI. TELENCEPHALON

The **telencephalon**, or **cerebrum**, is the seat of human intellect. It is organized into two cerebral hemispheres comprising the **basal ganglia** and the **cerebral cortex**.

A. Basal ganglia

The basal ganglia are a group of functionally related nuclei (Figure 6.8) that work closely with the cerebral cortex and thalamus to effect motor control. Their function is discussed at length in Chapter 11. Major structures within the basal ganglia include the **caudate nucleus** and **putamen** (together forming the **striatum**) and the **globus pallidus**.

B. Cerebral cortex

The cerebral cortex is involved in conscious thought, awareness, language, and learning and memory.

1. **Anatomy:** The cortex comprises a sheet of neural tissue organized in six layers that is folded to accommodate the 15 to 20 billion (1.5–2.0 × 10^{10}) neurons contained within. The folds (**gyri**) are separated by **sulci** (grooves). Deep **fissures** separate the cortex into four lobes: frontal, parietal, occipital, and temporal (Figure 6.9). The lobes contain discrete areas that can be distinguished on a cytoarchitectural basis and that correlate with regions of specialized function.

2. **Function:** The cortex can be functionally divided into three general areas that stretch across both hemispheres: **sensory**, **motor**, and **associative**.

   a. **Sensory:** Sensory regions process information from the sensory organs (see Chapters 8–10). **Primary sensory regions** receive and process information directly from the thalamus. Spatial information is preserved as data flows from the senses to the sensory areas and then accurately maps onto the cortex (**topographic mapping**). Thus, the pattern of light falling on the retina is faithfully replicated in the pattern of excitation within the primary visual cortex.

   b. **Motor:** Motor areas are involved with planning and executing motor commands. **Primary motor areas** execute movements. Axons from these areas project to the spinal cord, where they synapse with and excite motor neurons. **Supplementary motor areas** are involved with planning and fine control of such movements (see Chapter 11).

   c. **Associative:** The majority of cortical neurons are involved in associative functions. Each cortical sensory region feeds information to a corresponding association area. Here, patterns of color, light, and shade are recognized as a human face, for example, or a series of notes can be recognized as coming from a songbird. Other associative areas integrate sensory information from other parts of the brain to allow for higher mental functions. These include abstract thinking, acquisition of language, musical and mathematical skills, and the ability to engage in social interactions.

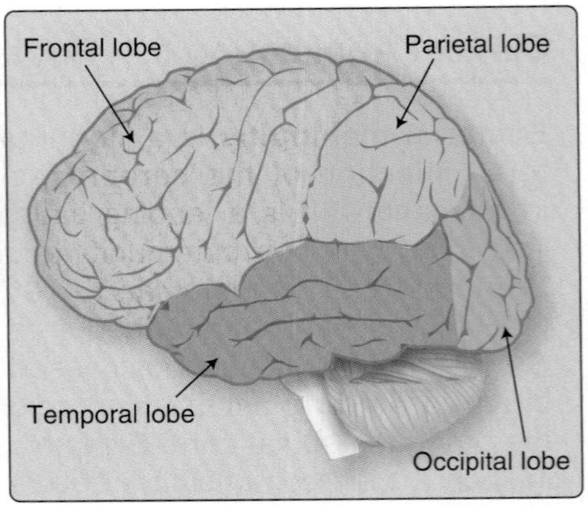

Figure 6.9
Cortical lobes.

VII. CEREBROSPINAL FLUID

Because the CNS has a central role in all aspects of life, its neurons are provided with multiple layers of protection and support.

A. Protective layers

The role of the blood–brain barrier in protecting CNS neurons against bloodborne chemicals is discussed in Chapter 21 (see 21·II·B). The CNS is also enclosed within five protective layers, including three membranes (the **meninges**), a layer of cerebrospinal fluid (CSF), and an outer layer of bone (Figure 6.10). The meninges comprise the **pia mater**, the **arachnoid mater**, and the **dura mater**.

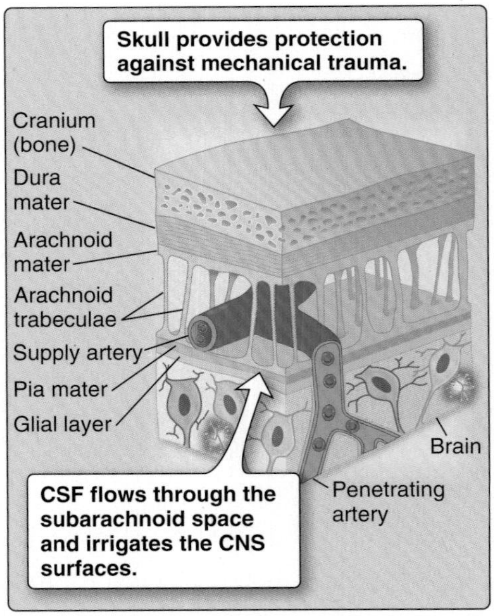

Figure 6.10
Layers that protect the brain and provide a pathway for cerebrospinal fluid (CSF) flow. CNS = central nervous system.

Clinical Application 6.1: Bacterial Meningitis

Bacterial meningitis is a life-threatening disease caused by bacterial infection of the cerebrospinal fluid (CSF) and meningeal inflammation.[1] It is a leading cause of infectious death worldwide. The most common community-acquired causes of meningitis are *Streptococcus pneumoniae* (~70%) and *Neisseria meningitiditis* (12%), whereas hospital-acquired cases are usually caused by *Staphylococcus*. Infection is caused by bacteria crossing the blood–brain barrier and establishing colonies in the CSF, affecting both the brain and spinal cord. Symptoms are usually rapid in onset and include a triad comprising a severe headache, nuchal (neck) rigidity, and a change in mental status. Most patients also present with a high fever. Nuchal rigidity is caused by pain and muscle spasm when attempting to flex or turn the head, reflecting meningeal inflammation in the cervical region. Immediate treatment to reduce swelling and address the infection usually leads to full recovery.

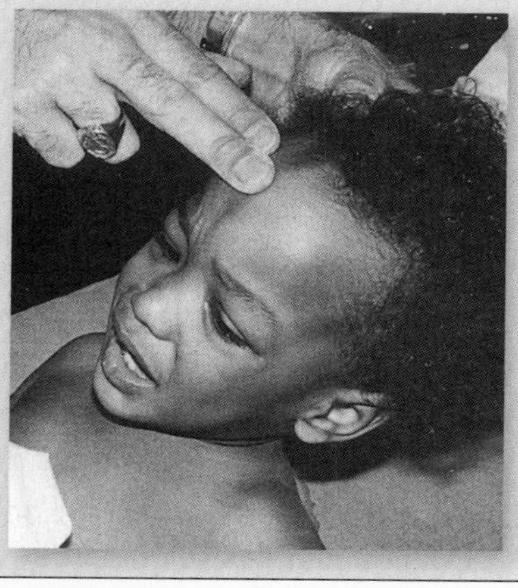

Patient lifts her shoulders rather than flexing her neck when her head is elevated (nuchal rigidity).

1. **Pia mater:** The entire surface of the brain and spinal cord is tightly adhered to a thin, fibrous membrane called the **pia mater** (Latin for "dutiful mother"). The pia's cerebral portion is held in place by a continuous layer of astrocyte foot processes.

2. **Arachnoid mater:** The arachnoid mater comprises a layered epithelial membrane that is loosely connected to the pia by **trabeculae**, small structural supports that give the arachnoid mater a cobwebbed appearance. The trabeculae create a **subarachnoid space** through which CSF flows unhindered over the brain's surface. The layer of CSF has multiple functions (see below), including cushioning the brain against trauma.

3. **Dura mater:** The "tough mother" is a thick, leathery membrane comprising two layers. An inner, **meningeal** layer is firmly attached to the arachnoid mater and covers the entire surface of both brain and spinal cord. The cranium is lined by a **periosteal** layer. The two layers separate in places to create an **intracranial venous sinus** that drains blood and CSF from the brain and channels it to the circulation.

4. **Bone:** The brain is protected by the cranium. The spinal cord lies within the vertebral canal, protected by the vertebral column.

B. Functions

CSF is a highly purified, sterile, colorless fluid nearly devoid of proteins that surrounds and bathes the tissues of the CNS. It has four main functions: providing buoyancy, absorbing shock, permitting limited intracranial volume changes, and maintaining homeostasis.

INFO LINK [1]For a more complete discussion of bacterial meningitis, see *LIR Microbiology*, 2e, p. 376.

1. **Buoyancy:** The brain's high lipid content gives it a relatively high specific gravity compared with CSF (1.036 *versus* 1.004). In practice, this means that the brain floats in CSF. The advantage is that flotation distributes brain mass evenly and helps prevent cerebral tissues from being compressed by gravity against the skull. Compression must be avoided because it impedes blood flow through the cerebral vasculature and causes ischemia.

2. **Shock absorption:** CSF surrounds the brain on all sides and envelops it in a liquid cushion. Cushioning reduces the chance of mechanical trauma to the brain when the skull is impacted or impacts an object at speed.

3. **Volume changes:** During periods of intense activity, neurons and glia tend to swell due to accumulation of metabolites and other osmotically active materials. CSF allows water to shift from CSF to cells without causing any gross change in CNS volume. Because the CNS is constrained by bone on all sides, volume changes can potentially compress the cerebral vasculature and cause ischemia (see 21·II·D).

4. **Homeostasis:** Cell membrane potential (V_m) and neuronal excitability is highly sensitive to changes in extracellular K^+ concentration. Plasma K^+ concentrations can rise by >40% even under normal conditions (normal plasma K^+ concentration = 3.5–5.0 mmol/L), changes that are unacceptable for a V_m-dependent organ such as the CNS. CSF K^+ concentrations are tightly maintained at a relatively low level (2.8–3.2 mmol/L), thereby insulating neurons from large swings in plasma concentration. CSF is also free of potentially neuroactive compounds (such as glutamate and glycine) that constantly circulate in blood. CSF thus provides the CNS with a stable, rarified extracellular environment that is renewed constantly to prevent buildup of neuronal waste products, transmitters, and ions.

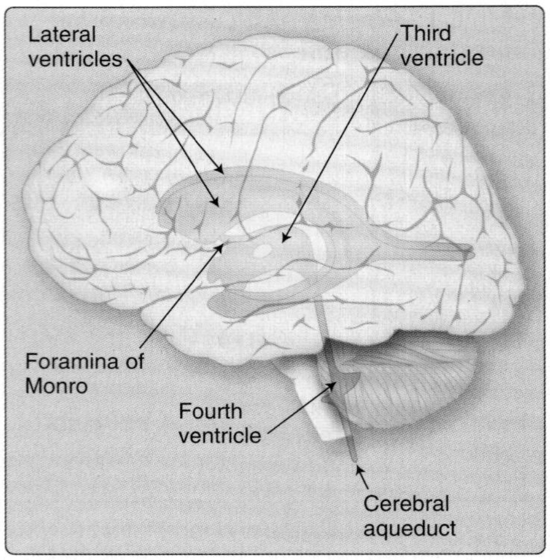

Figure 6.11
Location of the cerebrospinal fluid-filled ventricles and cerebral aqueduct.

C. Choroid plexus

CSF is formed by the **choroid plexus**, a specialized epithelium that lines four fluid-filled **ventricles** within the brain's core (Figure 6.11).

1. **Ventricles:** The brain contains four ventricles: two lateral ventricles and a third and fourth ventricle. They are all connected by **foramina** that allow CSF to flow caudally to the spinal cord and through its central canal.

   a. **Lateral:** The two lateral ventricles are the largest of the four. They are symmetrical C shapes and are located at the center of the two cerebral hemispheres. They connect with the third ventricle via two interventricular channels called the **foramina of Monro**.

   b. **Third:** The third ventricle lies on the midline at the level of the thalamus and hypothalamus. It connects with the fourth ventricle via the **cerebral aqueduct** (**of Sylvius**).

   c. **Fourth:** The fourth ventricle is located within the brainstem. The caudal end communicates with the spinal cord's central

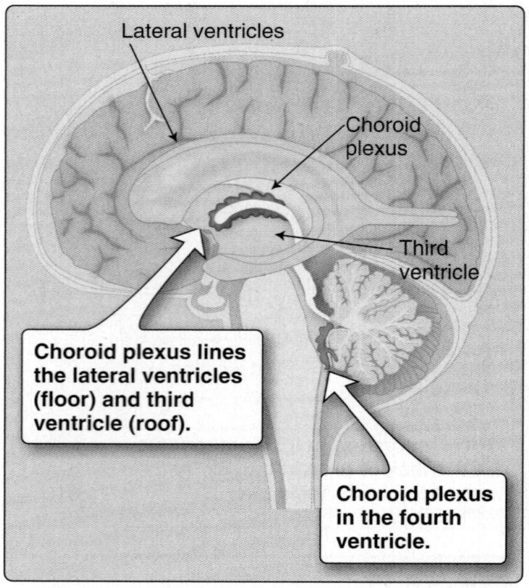

Figure 6.12
Choroid plexuses.

Table 6.1: Plasma and Cerebrospinal Fluid Composition

Solute	Plasma	CSF
Na^+	140	149
K^+	4	3
Ca^{2+}	2.5	1.2
Mg^{2+}	1	1.1
Cl^-	110	125
Glucose	5	3
Proteins (g/dL)	7	0.03
pH	7.4	7.3

Values are approximate and represent free concentrations under normal metabolic conditions. All values (with the exception of protein concentration and pH) are given in mmol/L. CSF = cerebrospinal fluid.

canal. The ventricle also provides a pathway for CSF to flow into the subarachnoid space via three openings. The **foramen of Magendie** is located at the midline. Two **foramina of Luschka** are located laterally.

2. **Location:** The choroid plexuses are localized to specific regions of the ventricles (Figure 6.12). They line the floor of the lateral ventricles and continue through the interventricular channels to line the roof of the third ventricle. In the fourth ventricle, choroid plexus occupies a small portion of the roof.

3. **Structure:** The ventricles and the spinal cord's central canal are lined with **ependymal epithelium**. In the region of choroid plexus, the ependyma gives way to a ciliated **choroid epithelium**, which is responsible for secreting CSF. Choroid epithelial cells contain large numbers of mitochondria, and their apical surfaces are amplified by microvilli, features that are characteristic of an epithelium specialized for high-capacity ion and water transport. The epithelium rests on a basal lamina, which separates it from the vasculature below, and adjacent cells are all coupled by tight junctions. Choroid epithelial activity is supported by a **vascular plexus** comprising a dense network of arteries, capillaries, and veins. The capillaries are large and leaky, and their walls contain fenestrations to facilitate fluid filtration from blood.

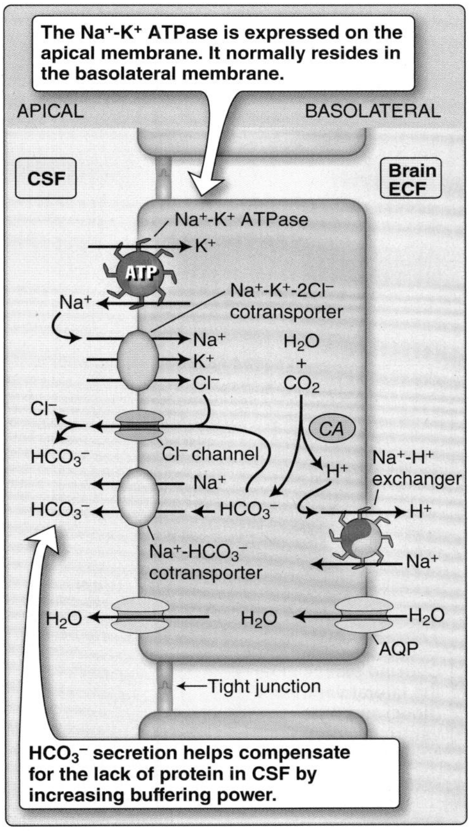

Figure 6.13
Cerebrospinal fluid (CSF) formation. AQP = aquaporin; *CA* = *carbonic anhydrase*; ECF = extracellular fluid.

D. Formation

About 30% of total CSF production can be attributed to secretion by the brain parenchyma. The remaining 70% is produced by the choroid plexus.

1. **Composition:** The basal side of the choroid epithelium is bathed in plasma filtrate, but tight junctions between adjacent epithelial cells create an effective barrier to exchange of ions and other solutes between blood and CSF. The differences between CSF and plasma are notable in several respects (Table 6.1):

   - CSF contains minimal protein or other large molecules. The lack of protein makes the CSF reliant on HCO_3^- for pH buffering.
   - HCO_3^- levels are higher to help buffer acids produced by the CNS.
   - Na^+ and Cl^- levels are higher, which compensates osmotically for the lack of protein.
   - K^+ concentrations are lower.

   The necessary gradients for CSF formation are established at the apical (luminal) surface of the choroid epithelium (Figure 6.13).

2. **Sodium secretion:** The choroid epithelium is highly unusual in that the Na^+-K^+ ATPase is located on the apical membrane rather than the basolateral membrane. The pump drives Na^+ into the CSF.

3. **Potassium absorption:** The Na^+-K^+ ATPase simultaneously removes K^+ from the CSF. More K^+ may be absorbed by an apical Na^+-K^+-$2Cl^-$ cotransporter, using the energy of the Na^+ gradient that favors Na^+ entry into the epithelial cell.

4. **Bicarbonate secretion:** HCO_3^- is generated by *carbonic anhydrase* activity. For every HCO_3^- molecule generated, a H^+ is liberated also. This is released to the vasculature via a Na^+-H^+ exchanger in the basolateral membrane. HCO_3^- is likely secreted into the ventricle via anion (Cl^-) channels and Na^+-HCO_3^- cotransporters.

5. **Chloride secretion:** Cl^- is concentrated within the cells by anion exchangers in the basolateral membrane and then flows across the apical membrane via Cl^- channels.

6. **Water secretion:** Water follows an osmotic gradient generated by secretion of Na^+, HCO_3^-, and Cl^-. Aquaporins provide a pathway for movement.

E. Flow

CSF is produced at prodigious rates (~500 mL/day), flushing the ventricles and CNS surfaces once every 7–8 hrs. High flow rates ensure that byproducts of neuronal activity (inorganic ions, acids, and transmitters spilling over from synapses) are removed in a timely manner before they can build to levels that might interfere with CNS function.

> The choroid plexuses have a mass of only ~2 g. Their ability to generate so much CSF is made possible both by a blood flow that is higher than that of virtually any other tissue (and 10 times that supplying neurons) and by the enhanced surface area for secretion created by the villi and microvilli.

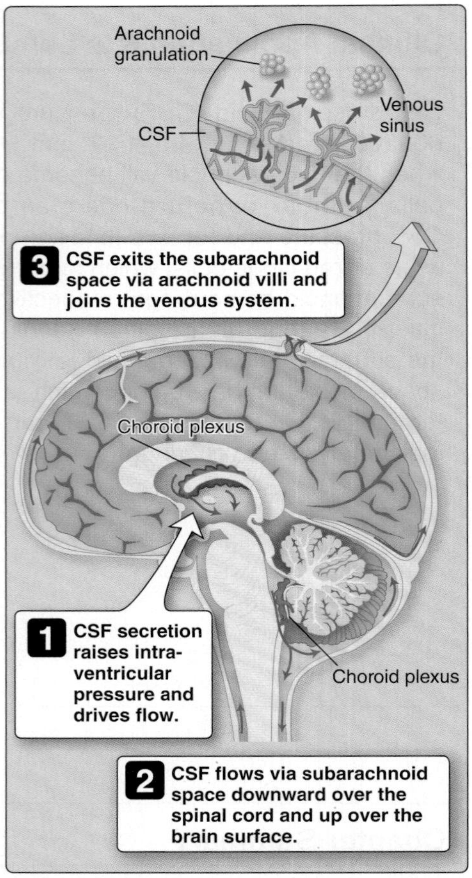
Arachnoid granulation
Venous sinus
CSF
3 CSF exits the subarachnoid space via arachnoid villi and joins the venous system.
Choroid plexus
1 CSF secretion raises intraventricular pressure and drives flow.
Choroid plexus
2 CSF flows via subarachnoid space downward over the spinal cord and up over the brain surface.

Figure 6.14
Pathways for cerebrospinal fluid (CSF) flow over the surfaces of the central nervous system (CNS).

1. **Pathways:** CSF secretion by the choroid plexus increases the pressure within the ventricles by a few millimeters of H_2O, sufficient to drive CSF flow through the ventricles, through the foramina in the fourth ventricle and into the subarachnoid space (Figure 6.14). CSF then percolates through the space and flows over the CNS surfaces, eventually joining venous blood contained within the intracranial sinus. CSF enters the sinuses via **arachnoid villi**, which may be organized into large clumps called **arachnoid granulations**. CSF is transported across the villi via giant vesicles, creating a one-way valve that prevents backflow from sinus to subarachnoid space if CSF pressure drops.

2. **Exchange between extracellular fluids:** CSF and brain ECF are separated in the ventricles by ependymal cells and in other regions by the pia and supporting layer of astrocytic foot processes. Although the pia and astrocytes layers are continuous, the junctions between adjacent cells are leaky and they allow free exchange of materials between CSF and ECF. This allows neuronal and glial waste products to diffuse out of the ECF and be carried away by the CSF.

Clinical Application 6.2: Lumbar Puncture

Cerebrospinal fluid (CSF) pressure is normally within a range of 60–200 mm H_2O (~4.5–14.7 mm Hg), but can rise dramatically when the subarachnoid villi become clogged with bacteria or blood cells. **Lumbar puncture** offers an opportunity both to measure CSF pressure and retrieve fluid samples to test for the presence of white or red blood cells, which might indicate bacterial meningitis or subarachnoid hemorrhage, respectively. Lumbar puncture involves inserting a long, thin (spinal) needle through the dura mater into the subarachnoid space. Fluid is withdrawn from the subarachnoid space in the lumbar region, which is below where the spinal cord terminates. Up to 40 mL of CSF can be withdrawn safely for cytologic analysis and culturing.

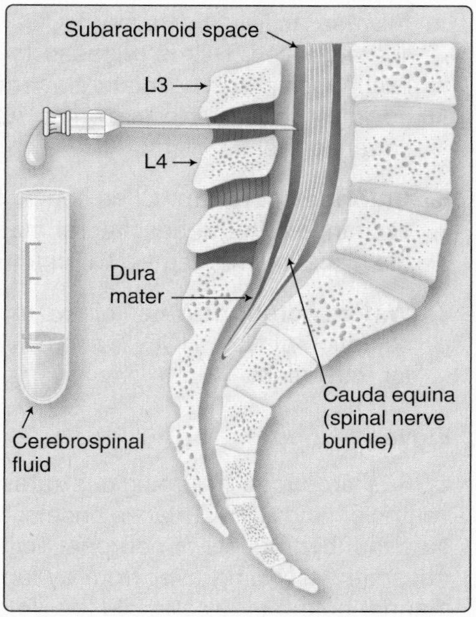

Lumbar puncture.

Chapter Summary

- The **central nervous system** comprises the **spinal cord** and **brain**. The spinal cord contains bundles of nerve fibers organized into **tracts** that relay information between the **peripheral nervous system** (**PNS**) and the brain. **Ascending tracts** relay **sensory** information from the PNS to the brain, whereas **descending tracts** convey **motor** commands to the PNS. The spinal cord also contains intrinsic circuits that facilitate local reflex arcs that do not require input from the brain.
- Peripheral nervous system neurons enter and leave the spinal cord via 31 pairs of **spinal nerves**. **Posterior roots** of these nerves contain **sensory afferent** fibers, whereas **anterior roots** contain **motor efferents**.
- All information flowing between the central and peripheral nervous systems must pass through the brainstem, which contains the **medulla**, **pons**, and **midbrain**. These areas contain **autonomic nuclei** involved in control of respiration, blood pressure, and upper gastrointestinal tract reflexes. The brainstem is associated with 10 **cranial nerves** that innervate the head and neck.
- The **cerebellum** facilitates fine motor control. It integrates sensory information from muscles, joints, and the visual and vestibular systems, and fine-tunes motor commands in anticipation of and during movements.
- The **diencephalon** comprises the **thalamus** and **hypothalamus**. The thalamus processes sensory information, whereas the hypothalamus is an autonomic nervous system control center.
- The **telencephalon** comprises the **basal ganglia**, which are involved in motor control, and the **cerebral cortex**. The cerebral cortex contains sensory, motor, and associative areas and is the seat of higher functioning.
- The central nervous system is protected by five layers, comprising the **pia mater**, a layer of **cerebrospinal fluid**, **arachnoid mater**, **dura mater**, and **bone**.
- Cerebrospinal fluid (CSF) is a colorless, protein-free fluid produced by the **choroid plexus**, a secretory epithelium located within the brain **ventricles**. CSF flows through the ventricles under pressure and then over the surface of both the brain and spinal cord. It drains into venous sinuses located within the dura mater.
- Cerebrospinal fluid (CSF) also acts as a liquid cushion that protects the brain from mechanical trauma and helps distribute its weight evenly within the cranium. CSF is produced at high rates, flushing the ventricles and central nervous system surfaces and carrying away accumulated waste products.

Autonomic Nervous System

7

I. OVERVIEW

Cells erect a barrier around themselves (the plasma membrane) to create and maintain an internal environment that is optimized to suit their metabolic needs. The body similarly is covered with the skin to establish an internal environment whose temperature, pH, and electrolyte levels are optimized for tissue function. Maintaining a stable internal environment (i.e., **homeostasis**) is the responsibility of the **autonomic nervous system** (**ANS**). The ANS is organized similarly to the somatic nervous system and uses many of the same neural pathways. Internal sensory receptors gather information about **blood pressure** (**baroreceptors**), **blood chemistry** (**chemoreceptors**), and **body temperature** (**thermoreceptors**) and relay it to autonomic control centers in the brain. The control centers contain neural circuits that compare incoming sensory data with internal preset values. If comparators detect a deviation from the presets, they adjust the function of one or more organs to maintain homeostasis. The principal organs of homeostasis include the skin, liver, lungs, heart, and kidneys (Figure 7.1). The ANS modulates organ function via two distinct effector pathways: the **sympathetic nervous system** (**SNS**) and the **parasympathetic nervous system** (**PSNS**). The actions of the SNS and PSNS often appear antagonistic, but, in practice, they work in close cooperation with each other.

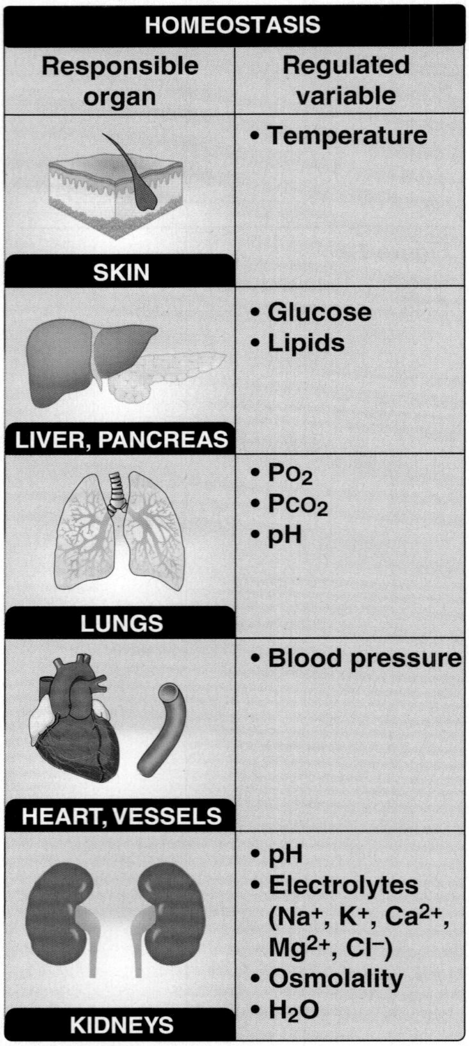

HOMEOSTASIS
Responsible organ
Regulated variable
SKIN
• Temperature
LIVER, PANCREAS
• Glucose
• Lipids
LUNGS
• P_{O_2}
• P_{CO_2}
• pH
HEART, VESSELS
• Blood pressure
KIDNEYS
• pH
• Electrolytes (Na^+, K^+, Ca^{2+}, Mg^{2+}, Cl^-)
• Osmolality
• H_2O

Figure 7.1
Principal homeostatic organs.

II. HOMEOSTASIS

The term "homeostasis" refers to a state of physiologic equilibrium or the processes that sustain such an equilibrium. An individual must maintain homeostatic control over numerous vital parameters to survive and thrive, including arterial P_{O_2}, blood pressure, and extracellular fluid osmolality (see Figure 7.1). Losing control over one or more of these parameters manifests as illness and usually prompts a patient to seek medical attention. It is a physician's task to identify the underlying cause of the imbalance and intervene to help restore homeostasis.

A. Mechanisms

Homeostatic control pathways are seen at both the cellular and organismal level, and they all include at least three basic components that typically form a negative feedback control system (Figure 7.2). There is a sensory component (e.g., a receptor protein) that detects

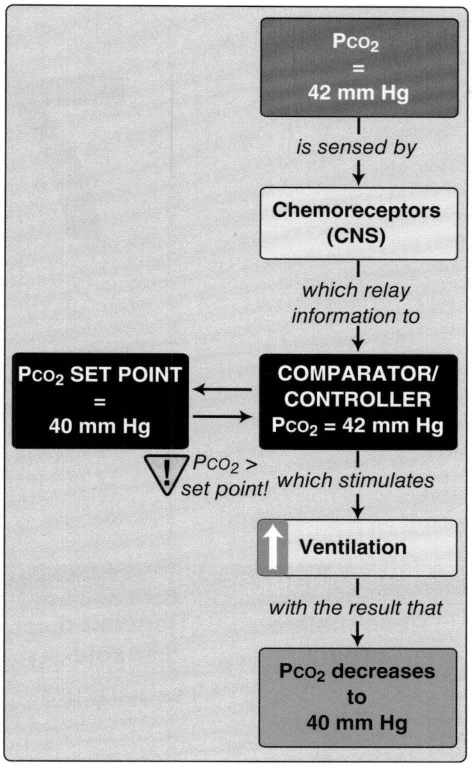

Figure 7.2
Negative feedback control of P_{CO_2}. CNS = central nervous system.

and relays information about the parameter subject to homeostatic control, an integrator (e.g., a neural circuit) that compares incoming sensory data with a system preset value, and an effector component capable of changing the regulated variable (e.g., an ion pump or excretory organ). For example, a rise in arterial P_{CO_2} is sensed by chemoreceptors that feed information to a respiratory control center in the brainstem. The control center responds by increasing respiration rate to expel the excess CO_2. Conversely, a decrease in P_{CO_2} reduces respiration rate. Homeostasis may also involve a behavioral component. Behavior drives intake of salt (NaCl), water, and other nutrients and, for example, impels one to turn on air conditioning or shed clothing if body temperature is too high.

B. Redundancy

Homeostasis at the organismal level typically involves multiple control pathways that are layered and often hierarchical, with the number of layers reflecting the relative importance of the parameter under control. Blood pressure, for example, is controlled by numerous local and central regulatory pathways. Layering creates redundancy, but it also ensures that if one pathway fails, one or more redundant pathways can assert control to ensure continued homeostasis. Layering also allows a very fine degree of homeostatic control.

C. Functional reserve

Organ systems that are responsible for homeostasis typically have considerable **functional reserve**. For example, normal quiet breathing uses only ~10% of total lung capacity, and cardiac output at rest is ~20% of maximal attainable values. Reserves allow the lungs to maintain arterial P_{O_2} and the heart to maintain blood pressure at optimal levels even as body activity level and demand for O_2 and blood flow increases (e.g., during exercise). Functional reserve also allows for progressive decreases in functional capacity, such as occurs with age and disease (see 40·II·A).

III. ORGANIZATION

The ANS, also known as the **visceral nervous system**, is responsible for maintaining numerous vital parameters. Homeostasis must continue when we sleep or our conscious minds are focused on a task at hand, so the ANS operates subconsciously and largely independently of voluntary control. Exceptions include the voluntary interruption of breathing to allow for talking, for example. The ANS is organized along similar principles to those of the somatic motor system. Sensory information is relayed via afferent nerves to the central nervous system (CNS) for processing. Adjustments to organ function are signaled via nerve efferents. The main differences between the two systems relate to efferent arm organization. The ANS employs a two-step pathway in which efferent signals are relayed through ganglia (Figure 7.3).

A. Afferent pathways

ANS sensory afferents relay information from receptors that monitor many aspects of body function, including blood pressure (barore-

ceptors); blood chemistry, that is, glucose levels, pH, Po_2, and Pco_2 (chemoreceptors); skin temperature (thermoreceptors); and mechanical distension of the lungs, bladder, and gastrointestinal (GI) system (mechanoreceptors). Sensory afferent fibers often travel in the same nerves as do autonomic and somatic efferents. Autonomic nerves also contain nociceptive fibers, which provide for visceral pain sensation.

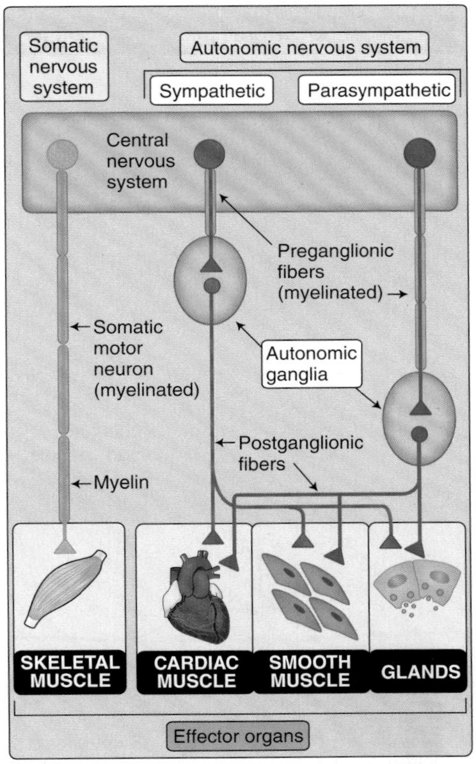

Figure 7.3
Somatic and autonomic nervous system efferent pathways.

B. Efferent pathways

In the somatic motor system, motor neuron cell bodies originate within the CNS (see Figure 7.3). In the ANS, the cell bodies of motor efferents are contained within ganglia, which lie outside the CNS, often in close proximity to their target organs (Figure 7.4; also see Figure 7.3).

1. **Autonomic ganglia:** Ganglia comprise clusters of nerve cell bodies and their dendritic trees. Commands originating in the CNS are carried to ganglia by myelinated **preganglionic neurons**. Unmyelinated **postganglionic neurons** relay the commands to the target tissues.

   a. **Sympathetic:** Sympathetic ganglia are located close to the spinal cord. Therefore, sympathetic preganglionic neurons are relatively short. Postganglionic neurons are relatively long, reflecting the distance between the ganglia and the target cells. There are two types of sympathetic ganglia. **Paravertebral ganglia** are arranged in two parallel **sympathetic chains** located to either side of the vertebral column. The ganglia within the chains are linked by neurons that run longitudinally, which allows signals to be relayed vertically within the chains as well as peripherally. **Prevertebral ganglia** are located in the abdominal cavity.

   b. **Parasympathetic: Parasympathetic ganglia** are located in the periphery near or within the target organ. Thus, parasympathetic preganglionic neurons are much longer than the postganglionic neurons.

2. **Sympathetic efferents:** The cell bodies of sympathetic preganglionic neurons are located in nuclei contained within upper regions of the spinal cord (T1–L3). Neurons located rostrally regulate the upper regions of the body, including the eye, whereas caudal neurons control the function of lower organs, such as the bladder and genitals. Preganglionic neurons leave the spinal cord via a ventral root, enter a nearby paravertebral ganglion, and then terminate in one of several possible locations:

   - within the paravertebral ganglion;
   - within a more distant sympathetic chain ganglion; or
   - within a prevertebral ganglion, a more distal ganglion, or the adrenal medulla.

3. **Parasympathetic efferents:** Preganglionic neurons of the PSNS originate in brainstem nuclei or in the sacral region of the spinal cord (S2–S4). Their axons leave the CNS via cranial or pelvic splanchnic nerves, respectively, and terminate within remote ganglia located close to or within the walls of their target organs.

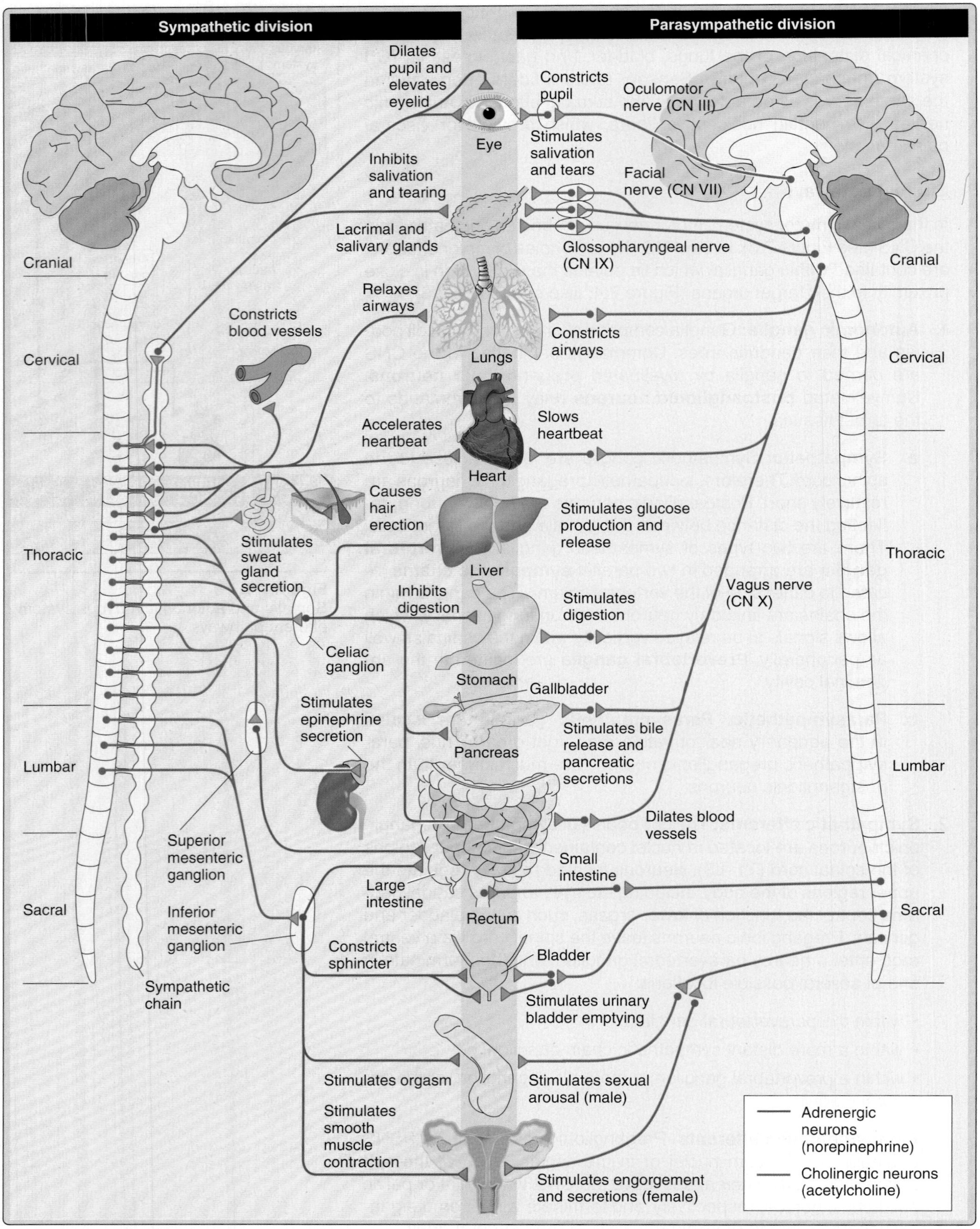

Figure 7.4
Autonomic nervous system organization. CN = cranial nerve.

IV. NEUROTRANSMISSON

The differences between the somatic motor system and the ANS become more apparent when transmitters and synaptic structure are reviewed (Figure 7.5).

A. Preganglionic transmitters

All ANS preganglionic neurons (SNS and PSNS) release acetylcholine (ACh) at their synapses. The postsynaptic membrane bears nicotinic ACh receptors (nAChRs), which mediate Na^+ influx and membrane depolarization when activated, as in skeletal muscle. However, whereas skeletal muscle expresses an N_1-type AChR, ANS preganglionic cell bodies and chromaffin cells in the adrenal medulla express an N_2-type AChR.

> N_1- and N_2-type AChRs have different sensitivities to nAChR antagonists, which makes it possible to inhibit the entire ANS output while leaving the skeletal musculature unaffected, or *vice versa*.[1] Pancuronium is an N_1-type receptor antagonist used in general anesthesia to relax skeletal muscle and aid intubation prior to surgery. It has relatively minor effects on ANS function. Conversely, trimethaphan is an N_2-type antagonist that blocks both arms of the ANS while having little effect on the skeletal musculature.

B. Postganglionic transmitters

Somatic motor neurons act through an ionotropic nAChR and are *always* excitatory. In contrast, ANS effector neurons communicate with their target cells via G protein–coupled receptors and, thus, may have an array of consequences.

1. **Parasympathetic:** All PSNS postganglionic neurons release ACh from their terminals. Target cells express M_1- (salivary glands, stomach), M_2- (cardiac nodal cells), or M_3-type (smooth muscle, many glands) muscarinic AChRs (see Table 5.3).

2. **Sympathetic:** Most SNS postganglionic neurons release norepinephrine from their terminals. Target cells may express α_1- (smooth muscle); β_1- (cardiac muscle); β_2- (smooth muscle); or, less commonly, α_2- (synaptic terminals) adrenergic receptors (see Table 5.2). The exceptions are the SNS efferents that regulate eccrine sweat glands, which release ACh at their terminals and act through an M_3-type AChR (see 16·VI·C·2).

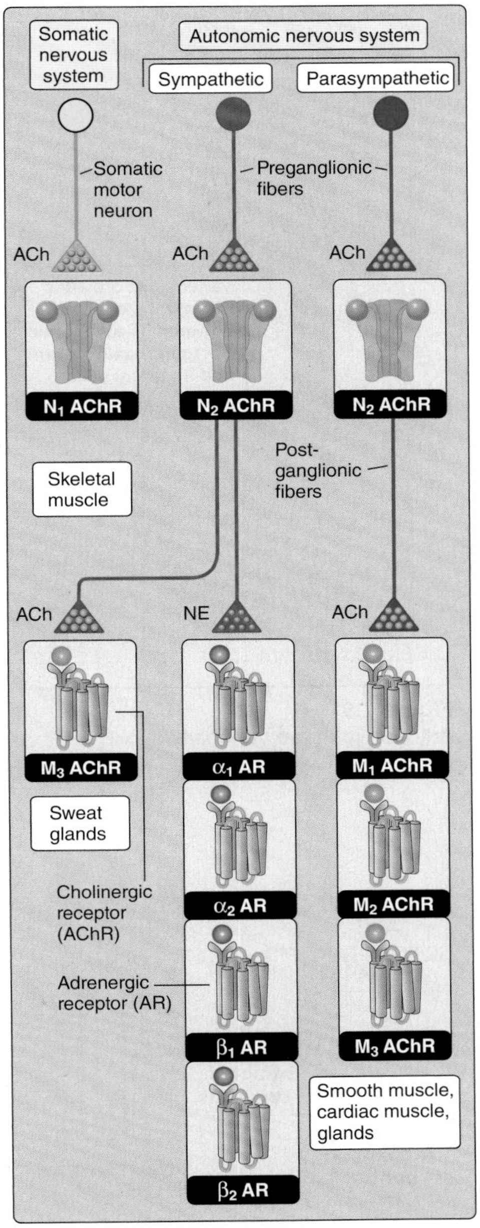

Figure 7.5
Autonomic nervous system neurotransmitters. ACh = acetylcholine; M_1 AChR, M_2 AChR, and M_3 AChR = muscarinic ACh receptors; N_1 and N_2 AChR = nicotinic ACh receptors.

[1]For a more complete discussion of cholinergic antagonists and their actions, see *LIR Pharmacology*, 5e, Chapter 5.

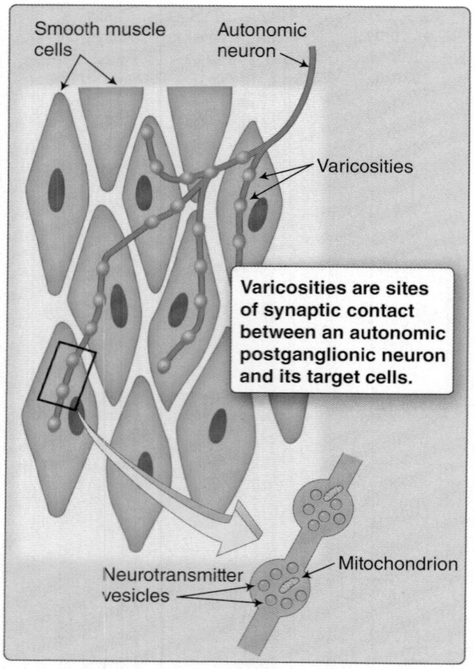

Figure 7.6
Autonomic nerve varicosities.

C. Postganglionic synapses

Somatic motor nerves terminate at highly organized neuromuscular junctions. The site of synaptic contact between an ANS neuron and its target cell is very different. Many postganglionic nerve axons exhibit a string of beadlike varicosities (swellings) in the region of their target cells (Figure 7.6). Each represents a site of transmitter synthesis, storage, and release, functioning as a nerve terminal.

V. EFFECTOR ORGANS

The somatic motor system innervates skeletal musculature. The ANS innervates all other organs. Most visceral organs are innervated by both arms of the ANS. Although the two divisions typically have opposite effects on organ function, they usually work in a complementary rather than antagonistic fashion. Thus, when sympathetic activity increases, output from the parasympathetic division is withdrawn and *vice versa.* The principal targets and effects of ANS control are summarized in Figures 7.1 and 7.4.

VI. BRAINSTEM

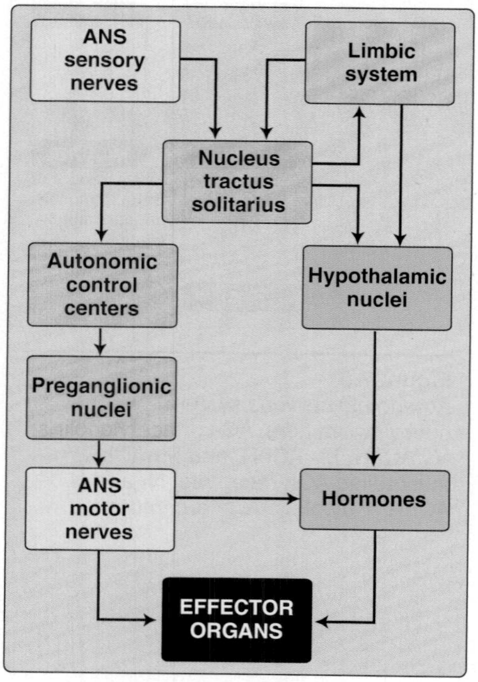

Figure 7.7
Autonomic control centers. ANS = autonomic nervous system.

ANS output can be influenced by many higher brain regions, but the main areas involved in autonomic control include the brainstem, the hypothalamus, and the limbic system. The relationship between these areas is shown in Figure 7.7. The brainstem is the primary ANS control center and can maintain most autonomic functions for several years even after clinical brain death has occurred (see 40·II·C). The brainstem comprises nerve tracts and nuclei. The nerve tracts convey information between the CNS and the periphery. Nuclei are clusters of nerve cell bodies, many of which are involved in autonomic control.

A. Preganglionic nuclei

Preganglionic nuclei are the CNS equivalents of ganglia, comprising clusters of nerve cell bodies at the head of one or more cranial nerves (CNs). Nuclei usually also contain interneurons that create simple negative feedback circuits between afferent and efferent nerve activity. Such circuits mediate many autonomic reflexes, such as reflex slowing of heart rate when blood pressure is too high and receptive relaxation of the stomach when it fills with food (see Clinical Application 7.1). The brainstem contains several important PSNS preganglionic nuclei, including the **Edinger-Westphal nucleus**, **superior** and **inferior salivatory nuclei**, the **dorsal motor nucleus of vagus**, and the **nucleus ambiguus** (Figure 7.8). The nucleus ambiguus contains both glossopharyngeal (CN IX) and vagal (CN X) efferents that innervate the pharynx, larynx, and part of the esophagus. The nucleus helps coordinate swallowing reflexes, and it also contains vagal cardioinhibitory preganglionic fibers.

B. Nucleus tractus solitarius

The nucleus tractus solitarius (NTS) is a nerve tract running the length of the medulla through the center of the solitary nucleus (see Figure 7.8) that coordinates many autonomic functions and reflexes.

Clinical Application 7.1: Autonomic Dysfunction

Disruption of autonomic pathways can result in specific functional deficits or more generalized loss of homeostatic function, depending on the nature of the underlying pathology. **Horner syndrome** is caused by disruption of the sympathetic pathway that raises the eyelid, controls pupil diameter, and regulates facial sweat gland activity. The result is a unilateral **ptosis** (drooping eyelid), **miosis** (inability to increase pupil diameter), and local **anhidrosis** (inability to sweat). More generalized autonomic dysfunctions are common among patients on maintenance dialysis and those with diabetes whose glucose levels are poorly controlled (diabetic autonomic neuropathy, or DAN). DAN can manifest as an inability to control blood pressure following a meal (postprandial hypotension) or upon standing (postural hypotension), gastrointestinal motility disorders (difficulty swallowing and constipation), or bladder dysfunction, among other symptoms.

Tests designed to assess autonomic function include monitoring cardiac responses during changes in posture, hand immersion in ice water (the cold pressor test, designed to induce intense pain), and a Valsalva maneuver.

A Valsalva maneuver involves forced expiration against a resistance, designed to cause intrathoracic pressures to rise to 40 mm Hg for 10–20 s. The pressure increase prevents venous blood from entering the thorax, so cardiac filling is impeded, and arterial pressure falls. In a healthy individual, a fall in arterial pressure is sensed by arterial baroreceptors, initiating a reflex increase in heart rate that is mediated by sympathetic efferents traveling in the vagus nerve. Patients with DAN may have impaired baroreceptor or vagal nerve function and, thus, fail to respond to a Valsalva maneuver with the expected tachycardia.

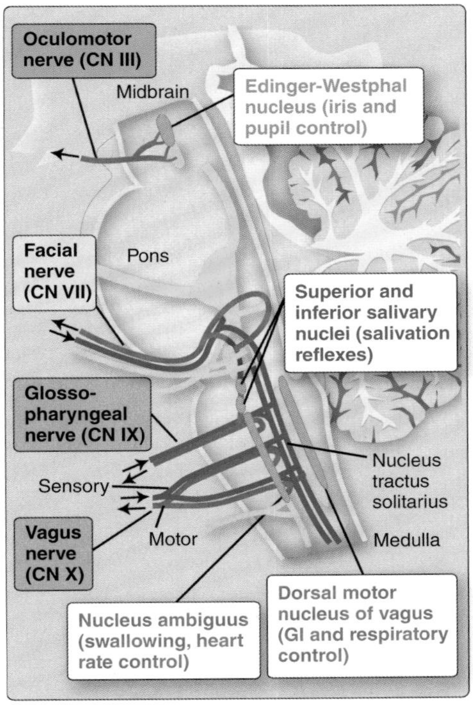

Figure 7.8
Principal autonomic brainstem nuclei. CN = cranial nerve; GI = gastrointestinal.

It receives sensory data from most visceral regions via the vagus and glossopharyngeal nerves (CNs IX and X) and then relays this information to the hypothalamus. It also contains intrinsic circuits that facilitate local (brainstem) reflexes controlling respiration rate and blood pressure, for example.

C. Reticular formation

The reticular formation comprises a collection of brainstem nuclei with diverse functions, including control of blood pressure and respiration (as well as sleep, pain, motor control, etc.). It receives sensory data from the glossopharyngeal and vagal nerves and helps integrate it with effector commands from higher autonomic control centers located in the limbic system and hypothalamus.

D. Control centers

Brainstem areas that have related functions are considered control centers, even if separated spatially. Brainstem control centers include the **respiratory center**, **cardiovascular control center**, and **micturition center** (Table 7.1).

Table 7.1: Brainstem Control Centers

Respiratory Center
Receives sensory information from chemoreceptors that monitor arterial P_{O_2}, P_{CO_2}, and pH.
Controls respiration rate through outputs to the diaphragm and respiratory muscles (see 24·II).
Cardiovascular Center
Receives sensory information from peripheral baroreceptors and chemoreceptors. Controls blood pressure through modulation of cardiac output and vascular tone (see 20·III·B).
Micturition Center
Monitors urinary bladder distension. Facilitates urinary bladder emptying by relaxing the urethral sphincter and contracting the bladder (see 25·VI·D).

VII. HYPOTHALAMUS

The hypothalamus establishes the set point for many internal parameters, including body temperature (37°C), mean arterial pressure (~95 mm Hg), and extracellular fluid osmolality (~290 mOsm/kg). Its influence extends to virtually all of the body's internal systems, belying its tiny size (~4 cm^3, or ~0.3% of brain volume). The ability to establish a set point or narrow operating range minimally requires that the hypothalamus be provided with a way of monitoring the parameters it controls and a way of communicating with the organs that maintain them.

A. Organization

The hypothalamus is located below the thalamus at the base of the brain. It contains several distinct nuclei summarized in Figure 7.9. Note that although some of these nuclei have clearly defined functions, others are organized in functional groups or areas that work cooperatively. In addition to controlling autonomic functions, the hypothalamus can stimulate many behavioral responses, including those associated with sexual drive, hunger, and thirst.

B. Neural pathways

The hypothalamus receives reciprocal innervation from many areas, as might be expected of such a key integrative organ. The major pathways for information flow occur between the hypothalamus and the brainstem as well as the hypothalamus and the limbic system (see Figure 7.7).

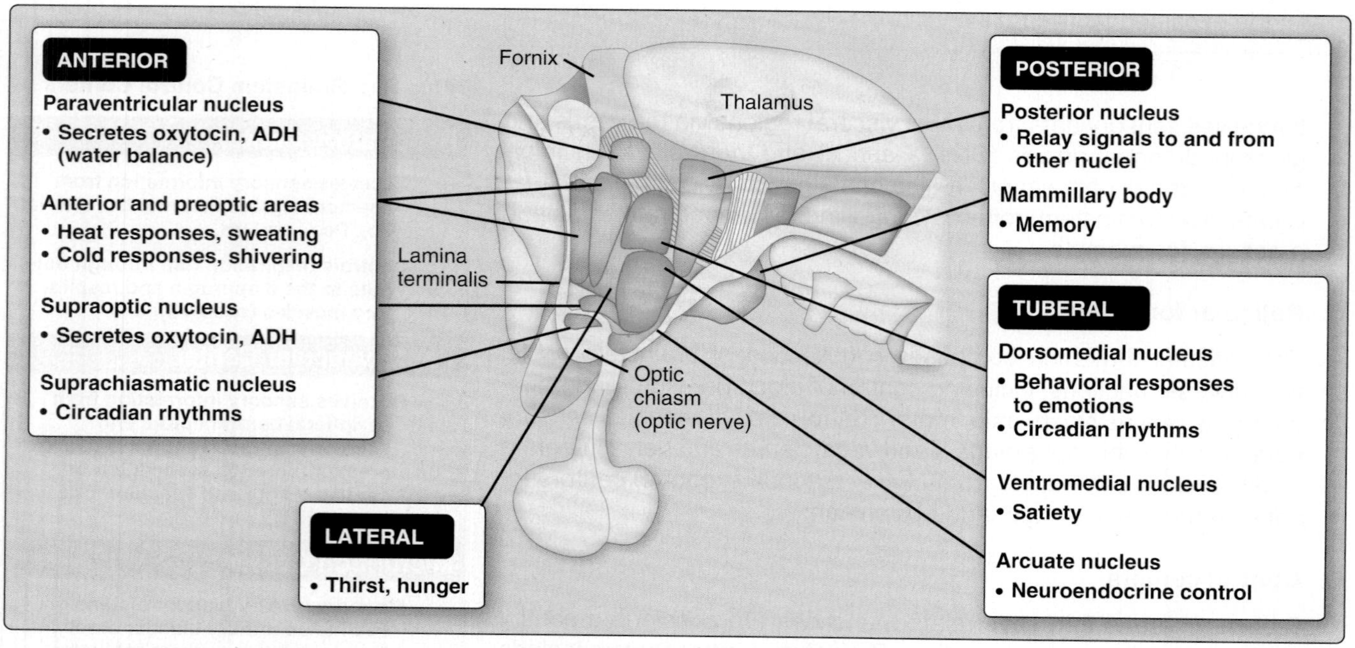

Figure 7.9
Hypothalamic nuclei. ADH = antidiuretic hormone.

C. Circumventricular organs

Any organ that is tasked with maintaining homeostasis needs to be able to monitor the parameters that it controls. In the case of the hypothalamus, this includes ions, metabolites, and hormones. The hypothalamus is a part of the brain, however, meaning that it is isolated from most such factors by the blood–brain barrier (BBB). Although it does receive feedback from peripheral receptors, the information they provide is limited. Therefore, the hypothalamus is provided with windows in the BBB through which it can make observations about blood composition directly. These windows are called **circumventricular organs** (**CVOs**).

1. **Location:** The brain contains six CVOs (Figure 7.10). CVOs comprise specialized brain regions where the BBB is interrupted to allow cerebral neurons to interact with the circulation directly. Some CVOs are sensory, whereas others are secretory.

   a. **Sensory:** Sensory CVOs include the **subfornical organ** and the **organum vasculosum of the lamina terminalis**, both associated with the hypothalamus. The **area postrema** is a brainstem CVO.

   b. **Secretory:** Secretory CVOs include the **median eminence** (part of the hypothalamus), the **neurohypophysis** (**posterior pituitary gland**), and the **pineal gland**.

2. **Structure:** CVOs are designed as interfaces between the brain and the periphery. They are highly vascularized, and blood flows through these regions slowly to maximize time available for exchanging materials between blood and brain. Also, CVO capillaries are fenestrated and leaky, which facilitates movement of ions and smaller proteins between blood and interstitium.

3. **Sensory functions:** Sensory CVOs contain neuronal cell bodies that are sensitive to numerous bloodborne factors (Na^+, Ca^{2+}, angiotensin II, antidiuretic hormone, natriuretic peptides, sex hormones, and feeding and satiety signals). Their axons project to hypothalamic areas that control corresponding variables.

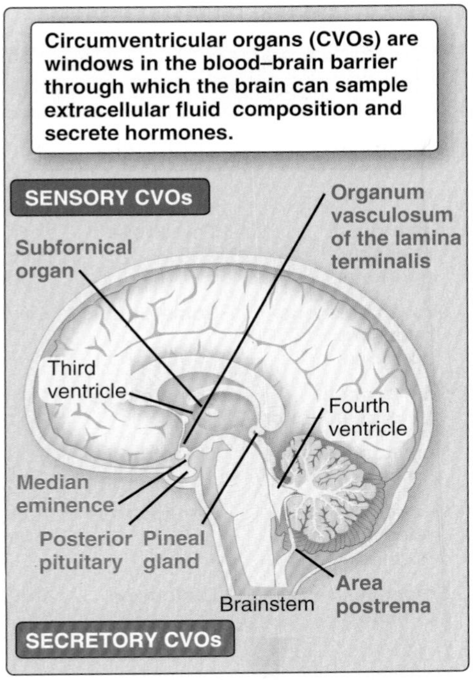

Figure 7.10
Circumventricular organs.

D. Endocrine functions

Most organs in the body are dually regulated by the nervous system and the endocrine system. The hypothalamus's key homeostatic role requires that it be able to influence both systems. It modulates the neural component via nerve tracts and peripheral nerves. It exerts endocrine control using hormones (summarized in Tables 7.2 and 7.3) that are released via the pituitary gland.

1. **Endocrine axes:** The hypothalamus, pituitary, and a dependent endocrine gland together form a unified control system known as an **endocrine axis**. Most endocrine systems are organized into such axes. The advantage of this system is that it allows for both fine and gross control of hormone output. For example, the **hypothalamic–pituitary–adrenal axis** regulates **cortisol** secretion from the **adrenal cortex**. The hypothalamus produces **corticotropin-releasing hormone** (**CRH**), which stimulates

Table 7.2: Anterior Pituitary Hormones

Hypothalamic Hormone	Pituitary Target Cell	Pituitary Hormone	Target Organ (Effects)
Corticotropin-releasing hormone	Corticotrope	Adrenocorticotropic hormone	Adrenal cortex (stress responses)
Thyrotropin-releasing hormone	Thyrotrope	Thyroid-stimulating hormone (TSH)	Thyroid gland (thyroxine release, metabolism)
Growth hormone–releasing hormone	Somatotrope	Growth hormone	Widespread (anabolic)
Somatostatin (release inhibitor)	Somatotrope	Growth hormone	Widespread
Somatostatin (release inhibitor)	Thyrotrope	TSH	Thyroid gland
Gonadotropin-releasing hormone	Gonadotrope	Luteinizing hormone	Gonads (androgen production)
Dopamine (release inhibitor)	Lactotrope	Prolactin	Mammary glands (milk production and letdown)
Gonadotropin-releasing hormone	Gonadotrope	Follicle-stimulating hormone	Gonads (follicle maturation, spermatogenesis)

Table 7.3: Posterior Pituitary Hormones

Pituitary Hormone Released	Pituitary Hormone Target (Effects)
Oxytocin	Uterus (contraction), mammary glands (lactation)
Antidiuretic hormone	Renal tubule (water reabsorption)

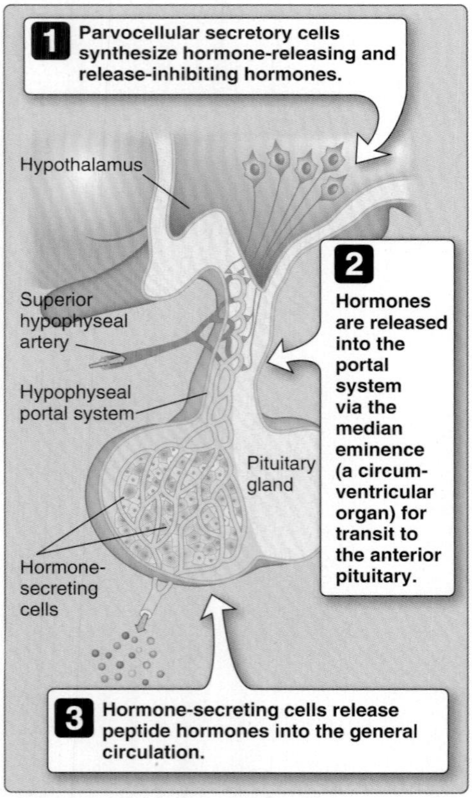
1 Parvocellular secretory cells synthesize hormone-releasing and release-inhibiting hormones.
Hypothalamus
Superior hypophyseal artery
Hypophyseal portal system
Pituitary gland
Hormone-secreting cells
2 Hormones are released into the portal system via the median eminence (a circumventricular organ) for transit to the anterior pituitary.
3 Hormone-secreting cells release peptide hormones into the general circulation.

Figure 7.11
Anterior pituitary.

adrenocorticotropic hormone (**ACTH**) release from the anterior pituitary. ACTH stimulates cortisol production by the adrenal cortex. Cortisol exerts negative feedback control on both ACTH production by the anterior pituitary, and both ACTH and cortisol inhibit CRH synthesis by the hypothalamus.

2. **Pituitary gland:** The pituitary gland (also known as the **hypophysis**) projects from the hypothalamus at the base of the brain and nestles in a bony cavity called the **sella turcica** (Latin for "Turkish saddle"). The hypothalamus and pituitary are connected by the pituitary (or **hypophyseal**) stalk, which contains bundles of neurosecretory axons. The pituitary contains two lobes (Figure 7.11). Although they lie next to each other within a common gland, they have very different embryologic origins and cellular compositions.

   a. **Anterior lobe:** The **anterior lobe** (**adenohypophysis**) has epithelial origins. It comprises a collection of glandular tissues that synthesize and store hormones (see Table 7.2). Hormone release is regulated by the hypothalamus using hormone-releasing or release-inhibiting hormones, which travel from hypothalamus to the anterior pituitary via the **hypophyseal portal system** (see below).

   b. **Hypophyseal portal system:** The hypophyseal portal system directs blood from the hypothalamus to the anterior lobe of the pituitary gland (see Figure 7.11). This unusual serial vascular arrangement is used to carry peptide hormones synthesized by hypothalamic **parvocellular** (small-cell) **neurosecretory**

cells to the anterior pituitary, where they stimulate or inhibit pituitary hormone release. Hypothalamic hormones are synthesized in neurosecretory cell bodies and then transported down their axons to terminals located in the median eminence. The median eminence is a CVO that sits at the head of the pituitary stalk and its portal system. Given an appropriate stimulus, the hormones are released from the nerve terminals into the portal system and carried to the capillaries supplying the anterior lobe's hormone-secreting cells.

c. **Posterior lobe:** The **posterior lobe** (**neurohypophysis**) is neural tissue. Axons from **magnocellular** (large-cell) **neurosecretory cells** in the supraoptic and paraventricular nuclei extend down the pituitary stalk and terminate within a CVO located in the posterior lobe (Figure 7.12). Magnocellular cell bodies synthesize oxytocin (OT) and antidiuretic hormone (ADH), two related peptide hormones (see Table 7.3). The hormones are transported to the nerve terminals via the pituitary stalk and stored in secretory granules (**Herring bodies**) awaiting release. The posterior pituitary is highly vascular and the capillaries fenestrated. When peptides are released, they enter the general circulation directly.

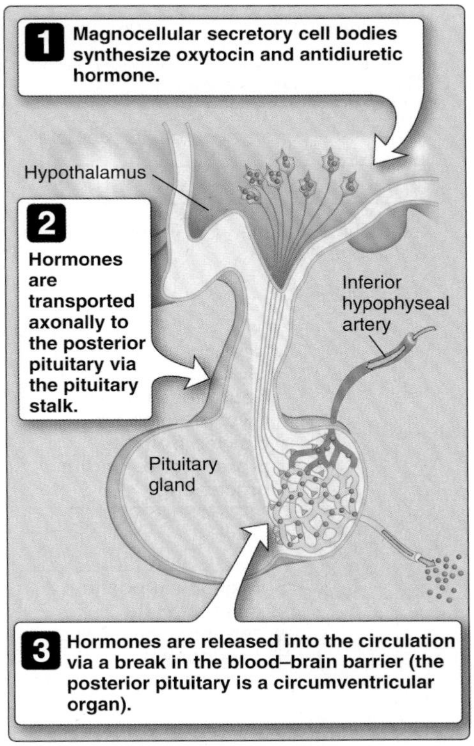

1 Magnocellular secretory cell bodies synthesize oxytocin and antidiuretic hormone.
Hypothalamus
2 Hormones are transported axonally to the posterior pituitary via the pituitary stalk.
Inferior hypophyseal artery
Pituitary gland
3 Hormones are released into the circulation via a break in the blood–brain barrier (the posterior pituitary is a circumventricular organ).

Figure 7.12
Posterior pituitary.

3. **Anterior pituitary hormones:** The anterior pituitary comprises five endocrine cell types (Table 7.4). The hormones they produce can be placed in one of three structurally related groups.

a. **Adrenocorticotropic hormone: ACTH** (**corticotropin**) is synthesized by **corticotropes** as a preprohormone, that is, preproopiomelanocortin (prePOMC). Removing the signal sequence yields POMC, a 241 amino acid–peptide containing ACTH (39 amino acids), melanocyte-stimulating hormone (MSH), and β endorphin (an endogenous opioid). Corticotropes lack the enzymes necessary to generate MSH or β-endorphin, however, so they release ACTH alone.

b. **Glycoprotein hormones: Thyroid-stimulating hormone** (**TSH**), **follicle-stimulating hormone** (**FSH**), and **luteinizing hormone** (**LH**) are related glycoproteins. All three hormones are heterodimers that share a common α subunit called **α-glycoprotein subunit** (**α-GSU**) and a hormone-specific β subunit. TSH is synthesized in **thyrotropes** and comprises a α-GSU–β-TSH dimer. FSH and LH are released by **gonadotropes** and comprise α-GSU–β-FSH and α-GSU–β-LH dimers, respectively. **Human chorionic gonadotropin** (**hCG**) is a related placental hormone comprising an α-GSU–β-hCG heterodimer.

c. **Growth hormone and prolactin: Growth hormone** (**GH**) and **prolactin** are related polypeptides synthesized by **somatotropes** and **lactotropes**, respectively. A related hormone, **human placental lactogen**, is synthesized by the fetal placenta. GH is a single-chain, 191 amino acid–residue polypeptide synthesized and released in several different isoforms. Prolactin, a 199 amino acid–polypeptide, is the only anterior pituitary hormone whose release is under tonic inhibition by the hypothalamus (via **dopamine**).

Table 7.4: Anterior Pituitary Tropic Cell Composition

Cell Type	% of Total
Corticotropes	15–20
Gonadotropes	10
Lactotropes	15–20
Somatotropes	50
Thyrotropes	5

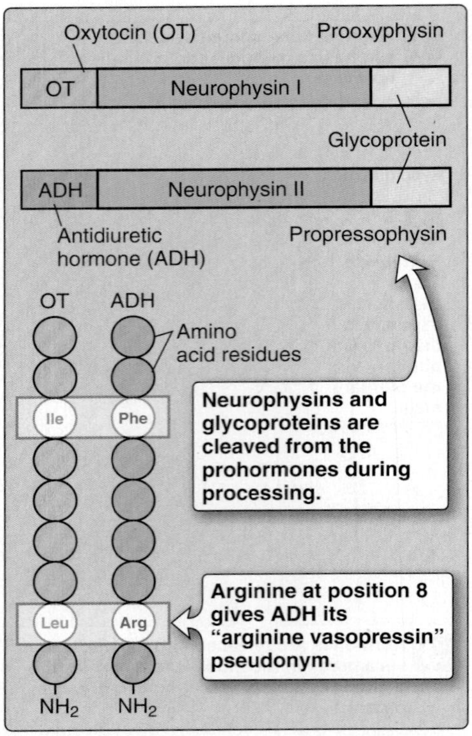

Figure 7.13
Structural similarities between posterior pituitary hormones.

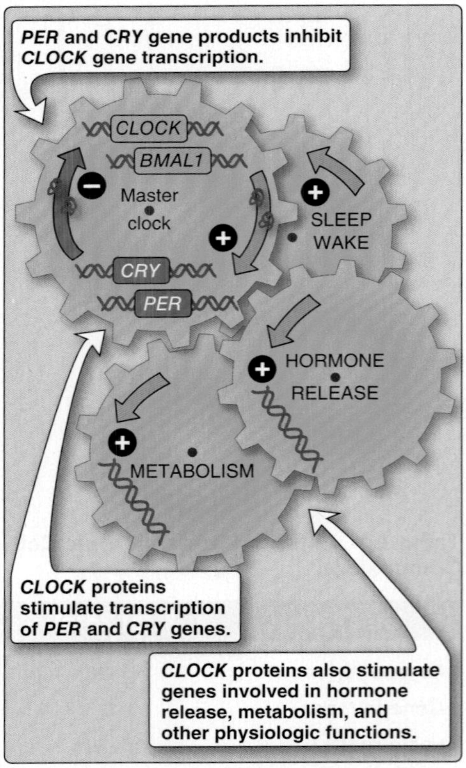

Figure 7.14
The master clock.

4. **Posterior pituitary hormones:** OT and ADH are near-identical nonapeptide hormones (they differ at only two amino acid positions) with a common evolutionary ancestor (Figure 7.13). They are both synthesized as preprohormones that contain a signal peptide, the hormone, a **neurophysin**, and a glycoprotein. The signal peptide and glycoproteins are removed to form prohormones during processing and packaging in the Golgi apparatus. **Prooxyphysin** comprises OT and neurophysin I, whereas **propressophysin** comprises ADH and neurophysin II. The hormones are separated by proteolysis from their respective neurophysins after packaging in neurosecretory vesicles and fast axonal transport to the posterior pituitary. The neurophysins (and glycoproteins) are coreleased along with hormone but have no known physiologic function.

> The structural similarities between OT and ADH causes some functional crossover when circulating hormone levels are sufficiently high. Thus, OT can have mild antidiuretic effects, whereas ADH can cause milk ejection in lactating women.

E. Clock functions

Most bodily functions, including body temperature, blood pressure, and digestion, have daily ("**circadian**," derived from the Latin *circa dies*) rhythms. All cells appear capable of generating such self-sustaining rhythms. The hypothalamus synchronizes these rhythms and entrains them to a circadian cycle established by a master clock. Entrainment allows the body's various physiologic functions to be modified to anticipate coming nightfall or daybreak and optimized to coincide with a sleep–wake cycle. The master clock is located in the **suprachiasmatic nucleus** (**SCN**). It synchronizes bodily functions in part through the endocrine system, with the **pineal gland** acting as a neuroendocrine intermediary.

1. **Molecular clocks:** Although many cortical regions contain circuits that establish seasonal and other rhythms, the master clock responsible for circadian rhythms resides in the SCN (see Figure 7.9). The molecular cogs that make the clock run comprise two sets of genes locked in a negative feedback control system (Figure 7.14). *CLOCK* gene proteins promote transcription of *CRY* and *PER* genes, whose translation products inhibit *CLOCK* gene transcription. The transcription–translation cycle oscillates over a period of ~24 hr.

2. **Setting the time:** Although the master clock oscillates with an inherent periodicity of around 24 hr, the clock is reset daily to entrain it to the light–dark cycle. The clock is set by light falling on a small subset of retinal ganglion cells (~1%–3% of total). These cells express **melanopsin**, a photopigment that allows them to detect and respond to light. Signals from these cells reach the hypothalamus via afferents traveling in the **retinohypothalamic tract** of the optic nerve (Figure 7.15).

3. **Pineal gland:** The SCN synchronizes body functions in part through manipulation of endocrine axes using the pineal gland as an intermediary. The pineal gland is a small (~8 mm) pinecone-shaped (hence the name) gland located at the midline near the posterior wall of the third ventricle (see Figure 7.10). It comprises **pinealocytes** and glial support cells that are similar to **pituicytes** (pituitary glial cells). The SCN communicates with the pineal gland via neural connections to the brainstem and spinal cord, and from there via sympathetic connections to the superior cervical ganglion and pineal gland. The pineal gland is a secretory CVO, which allows melatonin to be released into the circulation directly.

4. **Melatonin:** Melatonin is an indoleamine (N-acetyl-5-methoxytryptamine) synthesized from tryptophan. The synthetic pathway includes *arylalkylamine N-acetyltransferase* (*AA-NAT*), which is regulated by the SCN via adrenergic inputs from the SNS. When light falls on the retina, the sympathetic pathways from the SCN to the pineal gland are activated, and *AA-NAT* activity is inhibited (see Figure 7.15). Melatonin synthesis and secretion fall, as a result, and do not resume until dark (Figure 7.16).

> Individuals with **Smith-Magenis syndrome** (a developmental disorder) have an inverted melatonin secretory response to light. Melatonin levels peak during the daytime and fall at night. These patients have neurobehavioral problems and sleep disturbances, underscoring melatonin's importance in timing CNS function.

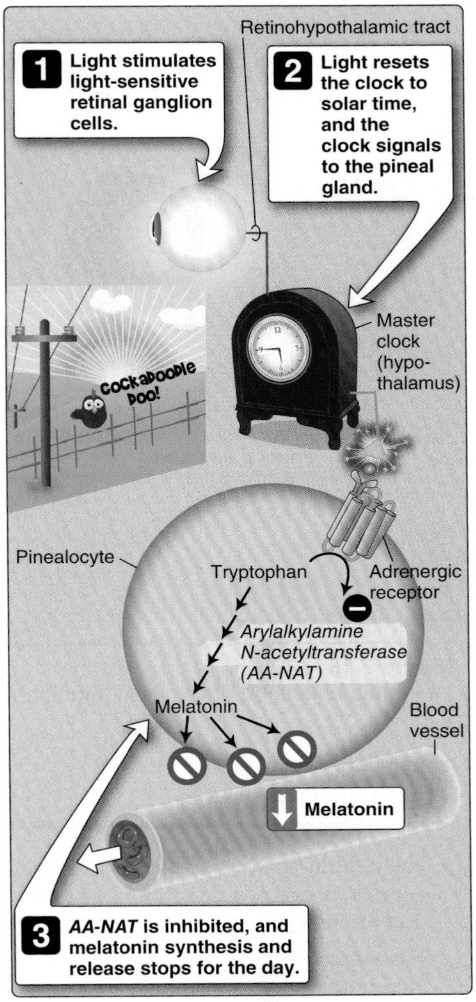

Figure 7.15
Effects of light on melatonin release.

VIII. LIMBIC SYSTEM

The limbic system comprises a collection of functionally related nuclei encircling the brainstem (**hippocampus**, **cingulate cortex**, and **anterior thalamic nuclei**) that strongly influence autonomic activity via connections to the hypothalamus. Many of these nuclei control emotions and motivational drives. These connections explain how emotions such as rage, aggression, fear, and stress can so profoundly exert physiologic effects. Everyone is familiar with the sensations associated with fright: a rapid, pounding heartbeat (increased heart rate and myocardial contractility); rapid breathing (respiratory center); cold, sweaty palms (sympathetic activation of sweat glands); and hairs standing erect on the back of the neck (piloerection).

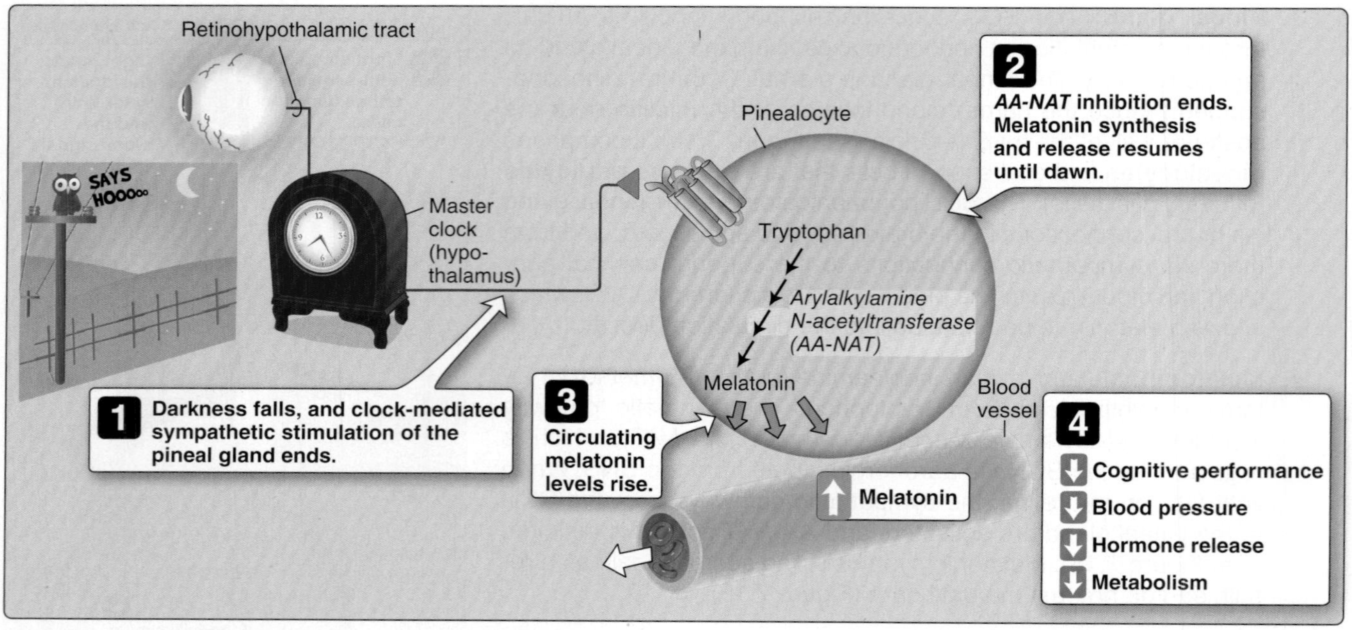

Figure 7.16
Melatonin effects on physiologic function.

Chapter Summary

- The central nervous system comprises the somatic nervous system and **autonomic nervous system** (**ANS**). The somatic nervous system controls skeletal musculature, whereas the ANS controls visceral organ function. The ANS's primary function is to maintain internal **homeostasis**.
- The autonomic nervous system (ANS) operates subconsciously and largely independently of voluntary control. The ANS incorporates two functionally distinct effector pathways (**sympathetic** and **parasympathetic**) that act cooperatively and in a reciprocal fashion to ensure homeostasis.
- The autonomic nervous system receives sensory information from receptors located throughout the body that monitor blood pressure, chemistry, and body temperature. This information is used to modify effector function via local reflexes or higher (central) autonomic control centers.
- Effector commands are relayed from autonomic control centers via **ganglia** that lie outside of the central nervous system. Sympathetic ganglia lie close to the spinal cord, whereas parasympathetic ganglia are located close to or within the walls of their target organs. All preganglionic neurons and parasympathetic effectors release **acetylcholine** at their terminals. Most sympathetic postganglionic motor neurons are adrenergic and release **norepinephrine** at target organs.
- Principal autonomic control centers include the **brainstem** and **hypothalamus**.
- The brainstem contains multiple autonomic control nuclei and control centers. The **nucleus tractus solitarius** and **reticular formation** help integrate autonomic sensory information with effector commands from the hypothalamus and **limbic system**.
- The hypothalamus establishes the **set point** for many vital internal parameters. It exerts homeostatic control through modification of brainstem control pathways and hormonally via the **pituitary gland**.
- The pituitary gland has two lobes: one comprising epithelial glandular tissue (**anterior lobe**), the other neural tissue (**posterior lobe**). Two breaches in the blood–brain barrier (**circumventricular organs**) allow pituitary hormones to be deposited into the general circulation.
- The hypothalamus stimulates release of six peptide (tropic) hormones from the anterior lobe into the circulation using **release-stimulating** or **release-inhibiting hormones**. These hypothalamic hormones reach the pituitary via the **hypophyseal portal system**. Two additional hormones are released from hypothalamic nerve terminals located in the posterior pituitary.
- Sensory circumventricular organs located within the brain allow the hypothalamus to sample the chemistry of extracellular fluid and make adjustments to organ function as necessary to maintain homeostasis.
- The hypothalamus is also the location of the master clock that entrains most organs to a **circadian rhythm**. The master clock resides in the **suprachiasmatic nucleus**, which exerts control both through direct neural connections to organs and through endocrine control. Entrainment of endocrine organs is mediated by the **pineal gland** and **melatonin** release.

Vision

8

I. OVERVIEW

The ability to detect light is common to most organisms, including bacteria, reflecting the importance of the visual sense. Designs for visual organs have arisen multiple times and many remain extant. In humans, photoreception is the purview of the eyes. Each eye comprises a sheet of photoreceptive cells (the **retina**) housed within an optical apparatus (Figure 8.1). The optics project a spatially accurate representation of the visual field onto the photoreceptors, much as a camera lens projects an image onto photographic film or a photosensor array. The simplest cameras use a pinhole as an aperture, which projects an inverted image of the subject onto film. An eye functions similarly, but aperture size (the **pupil**) is variable to control the amount of light falling on the photoreceptors. The inclusion of a variable-focus lens ensures that the projected image stays sharp when the aperture changes. The retina, which is located at the back of the eye, contains two types of photoreceptor cells. One is optimized to function in daylight and provide data that can be used to construct a color image (**cones**). The other is optimized to collect data under minimal lighting conditions, but the data is sufficient only to construct a monochromatic image (**rods**).

II. EYE STRUCTURE

The eye is a roughly spherical organ enclosed within a thick layer of connective tissue (the **sclera**) that is usually white (see Figure 8.1). The sclera is protective and creates attachment points for three pairs of skeletal (**extraocular**) muscles that are used to adjust the direction of gaze, stabilize gaze during head movement, and track moving objects. Because the photoreceptors are located at the back of the eye, photons entering the eye must travel through multiple layers and compartments before they can be detected.

A. Cornea

Light enters the eye via the **cornea**, which is continuous with the sclera. The cornea comprises several thin, transparent layers delimited by specialized epithelia. The middle layers are composed of collagen fibers along with supportive **keratinocytes** and an extensive sensory nerve supply. Blood vessels would interfere with light transmission so the cornea is avascular.

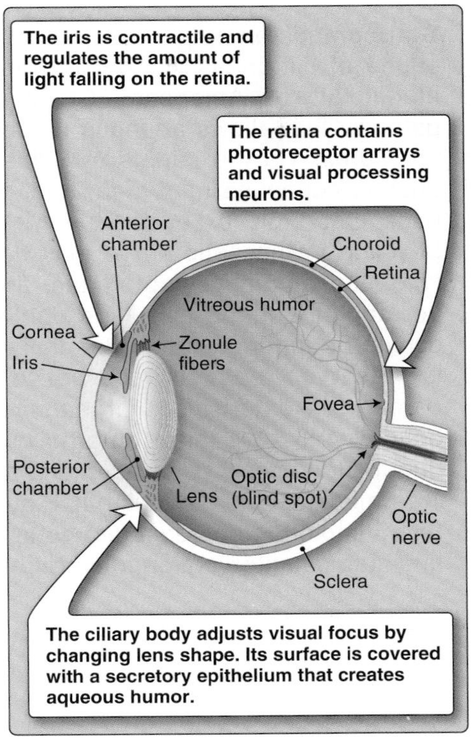

Figure 8.1
Eye structure.

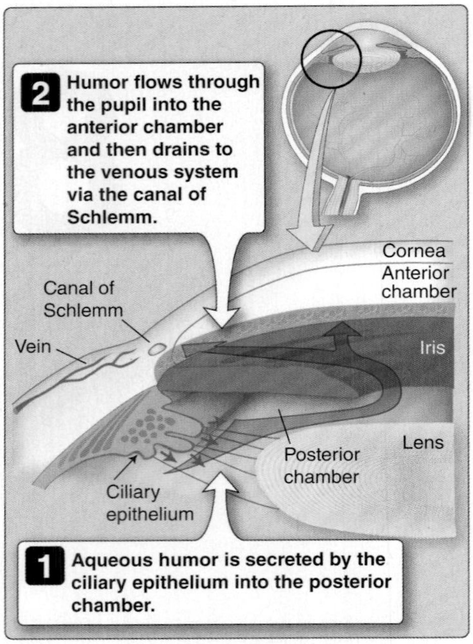

2 Humor flows through the pupil into the anterior chamber and then drains to the venous system via the canal of Schlemm.
Cornea
Anterior chamber
Canal of Schlemm
Vein
Iris
Lens
Posterior chamber
Ciliary epithelium
1 Aqueous humor is secreted by the ciliary epithelium into the posterior chamber.

Figure 8.2
Aqueous humor secretion and flow.

B. Anterior chamber

The anterior chamber is filled with **aqueous humor**, a watery plasma derivative. It is secreted into the **posterior chamber** by a specialized **ciliary epithelium** that covers the **ciliary body**. It then flows through the pupil, into the anterior chamber, and drains via the **canals of Schlemm** to the venous system. Humor is produced continuously to deliver nutrients to the cornea and to create a positive pressure of ~8–22 mm Hg that stabilizes corneal curvature and its optical properties (Figure 8.2).

C. Iris

The iris is a pigmented, fibrous sheet with an aperture (the pupil) at its center that regulates how much light enters the eye. Pupil diameter is determined by two smooth muscle groups that are under autonomic control. Rings of sphincter muscles that are controlled by postganglionic parasympathetic fibers from the ciliary ganglion decrease pupil diameter when they contract (**miosis**) as shown in Figure 8.3. A second group of radial muscles controlled by postganglionic sympathetic fibers originating in the superior cervical ganglion widen the pupil (**mydriasis**). Changes in pupil diameter are reflex responses to the amount of light falling on specialized photosensitive ganglion cells located in the retina (the **pupillary light reflex**). Signals from these

Clinical Application 8.1: Glaucoma

Glaucoma is an optic neuropathy that is the second most common cause of blindness worldwide and a leading cause of blindness among African Americans. Glaucoma commonly occurs when the pathway that allows aqueous humor to pass through the pupil and then drain via the canals of Schlemm is obstructed. Humor production continues unabated, and, thus, intraocular pressure (IOP) rises. Once IOP exceeds 30 mm Hg, there is a danger that axons traveling in the optic nerve may be damaged irreversibly. Patients typically remain asymptomatic, their condition being discovered accidentally during a routine ophthalmic examination. Vision loss occurs peripherally during the initial stages. Because central vision is preserved, patients tend not to notice their deficit until retinal damage is extensive. Ophthalmic examination often shows the optic disc to have taken on a hollowed out or "cupped" appearance due to blood vessel displacement, a finding diagnostic of glaucoma. Treatment includes reducing IOP by using β-adrenergic antagonists (e.g., timolol) to decrease aqueous humor production, for example,[1] and surgical intervention to correct the cause of obstruction.

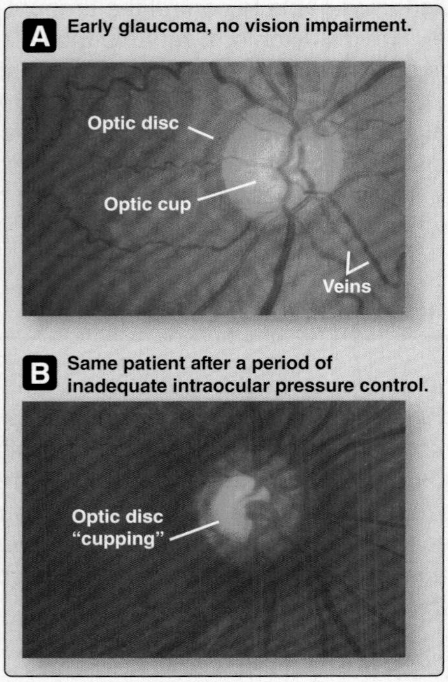

A Early glaucoma, no vision impairment.
Optic disc
Optic cup
Veins
B Same patient after a period of inadequate intraocular pressure control.
Optic disc "cupping"

Glaucoma.

[1]For further discussion of drugs used to treat glaucoma, see *LIR Pharmacology*, 5e, p. 94.

cells travel via the optic nerve to nuclei in the midbrain and then to the Edinger-Westphal nuclei (see Figure 7.8). Here, they trigger a reflex increase in parasympathetic activity via the oculomotor nerve (cranial nerve [CN] III), and the pupil constricts. Pupillary constriction reduces the amount of light entering the eye and helps prevent photoreceptor saturation. Saturation is undesirable in that it functionally blinds an individual. When light levels are low, a reflex pupillary dilation increases the amount of light reaching the retina. The pupillary reflexes elicit identical muscle responses in both eyes, even though light levels may be changing in one eye only.

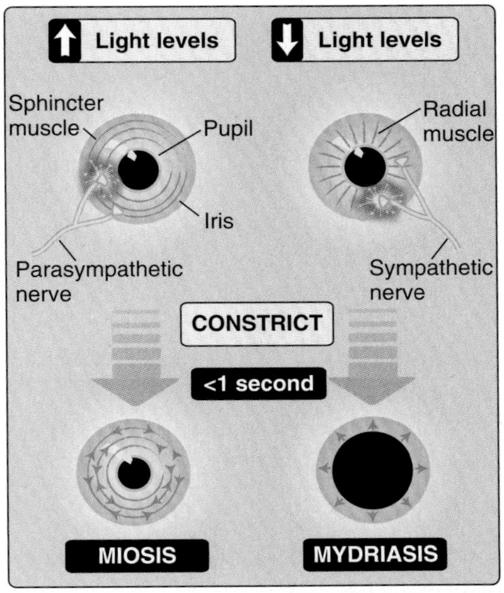

Figure 8.3
Regulation of pupil diameter.

> Pupil diameter always reflects a balance between tonic sympathetic and parasympathetic nerve activity. Thus, when atropine (an acetylcholine-receptor antagonist) is applied topically to the cornea during an ophthalmic examination, the pupil dilates because the balance between sympathetic and parasympathetic influence has been shifted in favor of the sympathetic nervous system.

D. Lens

The **lens** is a transparent, ellipsoid disk suspended in the light path by radial bands of connective tissue fibers (**zonule** fibers), attached to the **ciliary body**. The ciliary body is contractile and functions to modify lens shape and adjust its focus (see below). The lens is composed of long, thin cells that are arranged in tightly packed, concentric layers, much like the layers of an onion. The cells are dense with **crystallins**, proteins that give the lens its transparency and determine its optical properties. The lens is enclosed within a capsule composed of connective tissue and an epithelial layer.

E. Vitreous humor

Vitreous humor is a gelatinous substance composed largely of water and proteins. It is maintained under slight positive pressure to hold the retina against the sclera.

F. Retina

When light reaches the retina, it still has to penetrate multiple layers of neurons and their supporting structures before it can be detected by photoreceptors. The neuronal layers are transparent, so light loss during passage is minimal. The retina contains two specialized regions. The **optic disc** is a small area where the photoreceptor array is interrupted to allow blood vessels and axons from the retinal neurons to exit the eye, creating a **blind spot** (Figure 8.4). Nearby, in the center of the field of vision, is a circular area called the **macula lutea**. At its center is a small (<1-mm diameter) pit called the **fovea**. The neuronal layers separate here to allow light to fall directly on photoreceptors, creating an area of maximal visual acuity (see below).

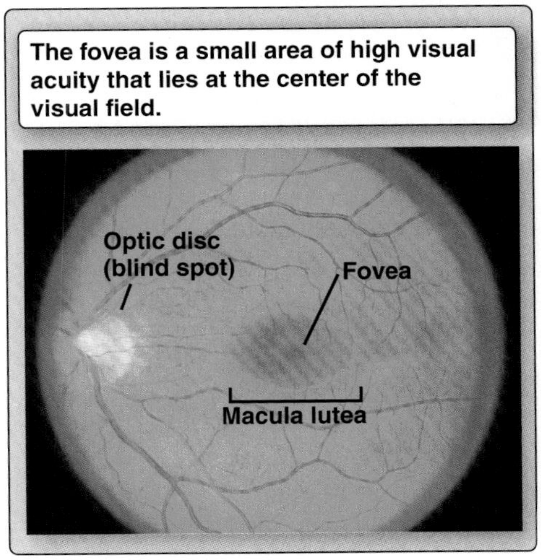

Figure 8.4
Retinal landmarks.

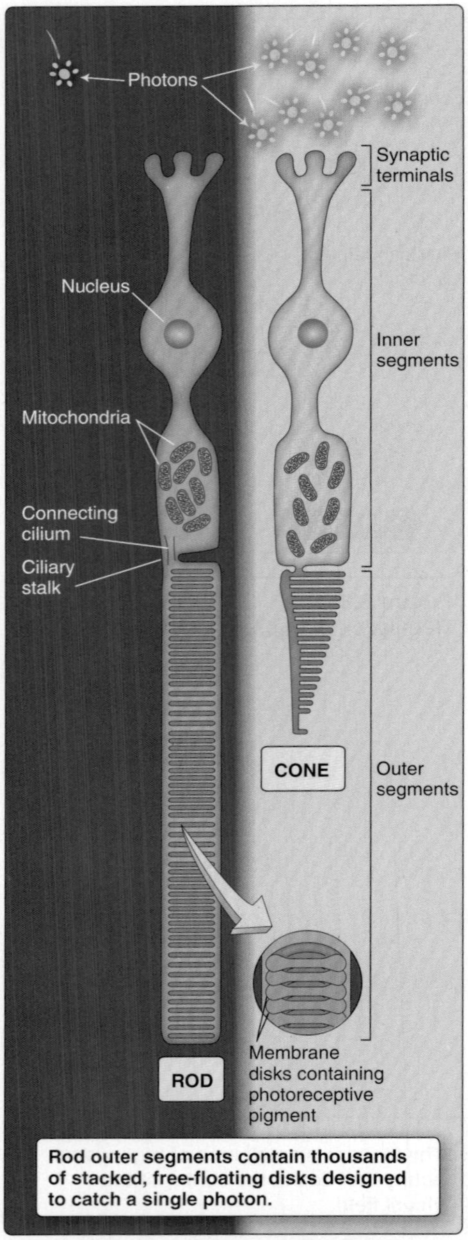

Figure 8.5
Photoreceptor structure.

III. PHOTORECEPTORS

Retinal photoreceptors are arranged in highly regular arrays so that spatial information can be extracted from the photoreceptor excitation patterns. The retina contains two types of photoreceptors that share a similar cellular structure.

A. Types

Rods are specialized to detect single photons of light. They cannot differentiate color but they can generate an image under low-light conditions and thereby facilitate **scotopic vision** (derived from the Greek word for darkness, *skotos*). **Cones** function optimally in daylight and mediate **photopic**, or color, vision.

B. Organization

Photoreceptors are long, thin, excitable cells (Figure 8.5). At the center is a cell body that encloses the nucleus. The cell body extends in one direction to form a short axon that branches into several presynaptic structures. The opposite end of the cell is long and cylindrical and divided into two segments. The **inner segment** contains all the other organelles required for normal cell function, including numerous mitochondria. The inner segment gives rise to a cilium (**connecting cilium**) that is grossly modified to house the phototransduction machinery. This compartment, which is known as the **outer segment**, is connected to the inner segment by a short **ciliary stalk**.

C. Disk membranes

The bloated sensory cilium that comprises the rod outer segment is packed with >1,000 discrete, flattened, membranous disks that are stacked like dinner plates alongside the ciliary axoneme. Cones contain similar but less numerous stacks that are infoldings of the surface membrane. The rod stacks are designed to capture a single photon as it traverses the eye's photosensitive layer. To make this a reality, the disk membrane is so densely packed with photosensory pigment molecules that there is little room left over for lipid!

D. Distribution

The retina lines the inside surface of the eye, covering roughly 75% (~11 cm^2) of its total surface area. Photoreceptors are densely packed within the sheet, with rods outnumbering cones ~20-fold (~130 million rods *versus* 7 million cones). Although both rods and cones are found throughout the retina, their distribution is unequal.

1. **Rods:** Rods dominate the peripheral retina, which optimizes these areas for night vision.

2. **Cones:** Cones are concentrated in the central retina, which imparts this area with a high degree of visual acuity. At its center is the fovea, which contains cones alone (see Figures 8.1 and 8.4). The fovea's lack of rods means that it cannot participate in night vision.

IV. PHOTOSENSOR

The ability to capture the energy of a single photon requires a **chromophore**, a molecule that absorbs certain light wavelengths while reflecting or transmitting others. This property gives the molecule color. The chromophore used in the eye is **retinal**, which can exist in several different conformations. The 11-*cis* conformation is very unstable and, when hit by a photon, immediately flips into a more stable all-*trans* configuration. Transition is rapid (femtoseconds), which makes it an ideal photoreceptive pigment. The task of detecting and reporting the conformational change falls on **opsin**, which is a G protein–coupled receptor. Opsin covalently binds 11-*cis* retinal in the same way that a hormone receptor binds its ligand. The receptor and chromophore combine to create a visual pigment called **rhodopsin**, which has a reddish purple color. When retinal absorbs a photon and transitions, it triggers a change in opsin conformation to generate **metarhodopsin II**. This event initiates a signal cascade that ultimately converts photonic energy into an electrical signal.

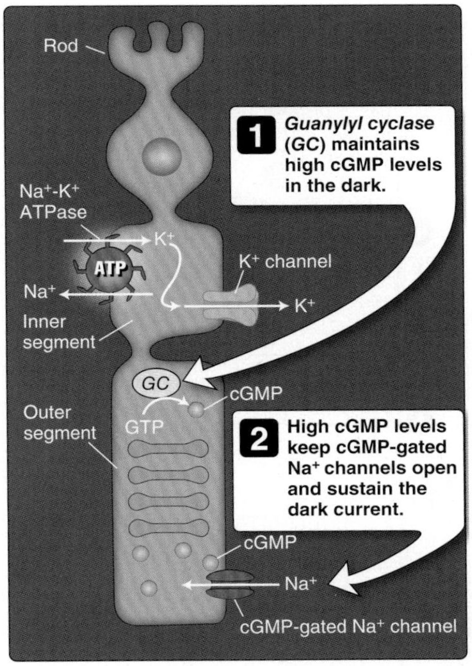

Figure 8.6
Dark current origins. ATP = adenosine triphosphate; cGMP = cyclic guanosine monophosphate; GTP = guanosine triphosphate.

V. PHOTOSENSORY TRANSDUCTION

Phototransduction is highly unusual in that stimulus detection causes receptor *hyper*polarization rather than *de*polarization, as is the case in other sensory systems.

A. Dark current

The rod outer-segment membrane contains a nonspecific cation channel that is gated by cyclic guanosine monophosphate (cGMP) as shown in Figure 8.6. A constitutively active *guanylyl cyclase* (*GC*) maintains high intracellular cGMP levels in the dark, and the channel is always open. Na^+ and small amounts of Ca^{2+} flow into the photoreceptor, creating an inward **dark current**. K^+ leak channels in the inner segment allow K^+ to escape the cell and help offset the current, but membrane potential (V_m) still rests at a relatively shallow −40 mV.

B. Transduction

When a photon hits retinal, rhodopsin contorts and activates **transducin**, which is a G protein ([G_T] Figure 8.7). When activated, the G_T α subunit dissociates and activates a membrane-associated *phosphodiesterase* (*PDE*). *PDE* hydrolyzes cGMP to GMP, and intracellular cGMP levels fall. The cation channel deactivates and closes as a result, and the dark current terminates. The K^+ channel in the inner segment remains open, however, which causes V_m to drift negative. This V_m change constitutes a signal that light has been detected. Although the transduction cascade is relatively slow (tens to hundreds of milliseconds), it does provide for tremendous signal amplification that allows the eye to register single photons.

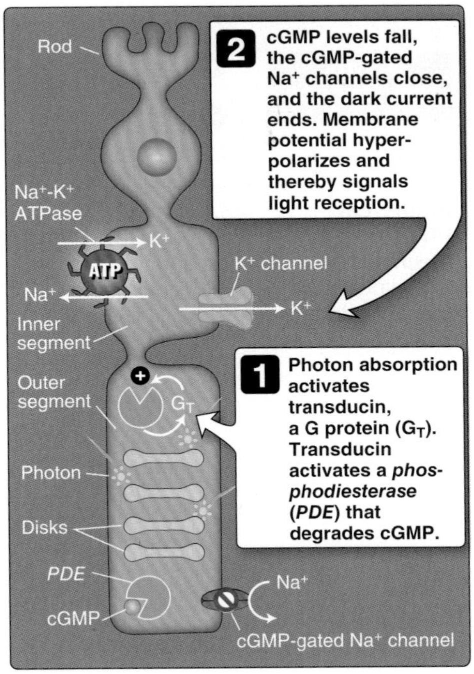

Figure 8.7
Phototransduction under low light conditions. ATP = adenosine triphosphate; cGMP = cyclic guanosine monophosphate.

C. Signal termination

The amplification cascade is so powerful that a rod relies on multiple negative feedback mechanisms to limit and terminate signaling in a timely manner (Figure 8.8).

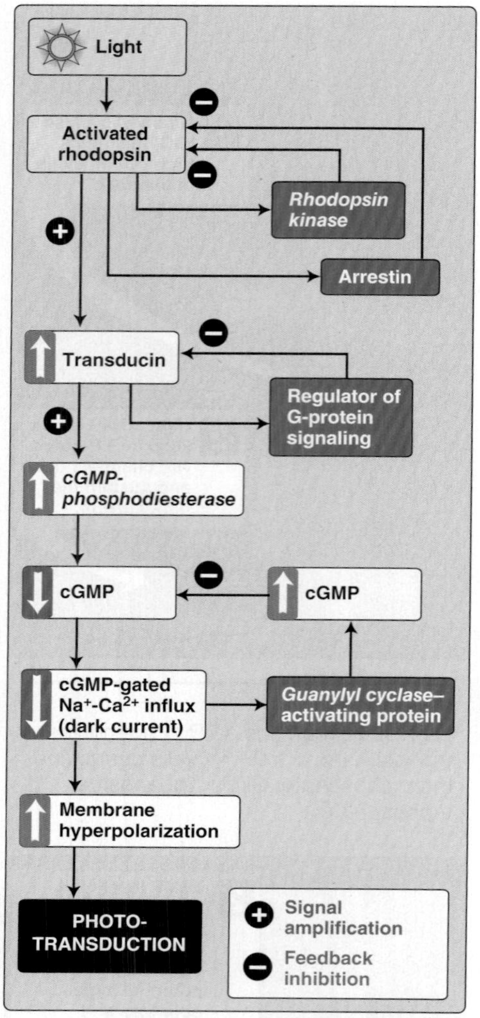

Figure 8.8
Photosensory transduction pathway and mechanisms for limiting and terminating signaling. cGMP = cyclic guanosine monophosphate.

1. **Opsin inactivation:** The active form of opsin is a substrate for *rhodopsin kinase* activity. Opsin has multiple phosphorylation sites, and each successive phosphate transfer further reduces its ability to interact with G_T. Phosphorylation also makes the receptor a favorable target for **arrestin** binding. **Arrestin** is a small protein whose sole function is to block interaction between opsin and transducin and thereby prevent further signaling.

2. **Transducin deactivation:** Rods also contain a **"regulator of G-protein signaling"** protein that enhances the G_T α subunit's GTPase activity and thereby speeds its deactivation.

3. ***Guanylyl cyclase* activation:** The dark current is mediated, in part, by Ca^{2+} influx. Light stops this influx, and intracellular Ca^{2+} concentrations fall. This is sensed by one or more Ca^{2+}-dependent *GC*-activating proteins, which respond by stimulating *GC* activity, which, in turn, counteracts the effects of *PDE*. This pathway is important for helping photoreceptors adapt to light levels that saturate the signaling pathway and also helps restore the dark current once signaling ends.

D. Desensitization

Prolonged exposure to bright light desensitizes the rods. Desensitization is partly an extension of the opsin inactivation process described above. *Rhodopsin kinase* phosphorylates opsin at multiple sites, which increases arrestin-binding affinity and blocks further opsin–transducin interactions. With time, transducin is translocated from the outer segment to the inner segment, effectively breaking the first crucial link in the phototransduction chain and preventing further signaling.

E. Retinal recycling

Retinal is released from opsin shortly after activation, and the pigment turns yellow (**bleaching**). It is then converted to **retinol**, also known as **vitamin A.** Vitamin A is converted to 11-*cis* retinal, which binds to opsin and restores the visual pigment.

> Vitamin A is essential for synthesis of visual pigments. Inadequate dietary intake results in night blindness, characterized by an inability to see in low light due to impaired rod cell function. The condition can be reversed within hours by administering vitamin A.

VI. COLOR VISION

Night vision is monochromatic because rods contain only one visual pigment. They are designed to register small amounts of light, not provide information about its quality. Distinguishing colors requires two or more

pigments that signal maximally at different wavelengths. Color vision employs three cone-cell types, each containing a different visual pigment (Figure 8.9). All use 11-*cis* retinal as a chromophore, and the phototransduction mechanism is the same as that described for rods above. However, the opsins differ in their primary sequence, which shifts pigment sensitivity to different wavelengths on the visible spectrum. **S cones** respond maximally to short wavelengths (violet-blue: ~420 nm), **M cones** to medium wavelengths (green-yellow: ~530 nm), and **L cones** to long wavelengths (yellow-red: ~560 nm). Overlap in pigment absorption spectra means that all three cone types respond to most visible light frequencies, but the intensity of their responses differs according to how close the stimulus is to the cone's optimal range. The brain then extrapolates colors from the data streams emerging from the retina.

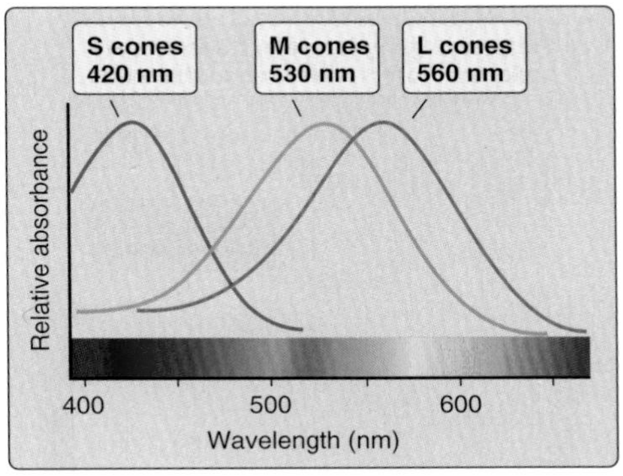

Figure 8.9
Cone spectral sensitivities.

VII. VISUAL PROCESSING

The photoreceptor array is capable of generating immense quantities of visual information.[1] This information is relayed to the CNS visual centers via the optic nerve, whose exit through the retina creates a "blind spot" (see Figures 8.1 and 8.4). If the optic nerve carried raw data, its diameter would have to be increased over a hundredfold to accommodate the required number of axons, and blind-spot size would increase proportionately. Thus, raw visual sensory data is processed extensively before leaving the retina to compress it and minimize the impact of optic nerve diameter on sensory array continuity.

A. Retina structure

The retina is a highly organized structure comprising layers of cells that generate photosensory data (the photoreceptors), process visual signals (**bipolar cells**, **horizontal cells**, **ganglion cells**, and **amacrine cells**), or support neuronal activity (the **pigment epithelium** and neuroglia) as shown in Figure 8.10.

1. **Pigment epithelium:** The innermost retinal layer is a black, pigmented epithelium that absorbs stray photons that might otherwise interfere with imaging and decrease visual acuity. The color is imparted by numerous **melanin** pigment granules. The pigment epithelium also supplies photoreceptors with nutrients, is involved with retinal recycling, and assists with photoreceptor turnover. Photoreceptor membranes are subject to constant damage by photons and, hence, are turned over continually. The entire rod stack is replaced once every ~10 days.

2. **Neuroglia:** Because the retina is an extension of the brain, the photoreceptors and all associated excitable cells receive support from glia. **Müller cells**, which are a retina-specific glial subtype, occupy the spaces between neurons and form a barrier (the **inner delimiting membrane**) that separates the retina from vitreous humor.

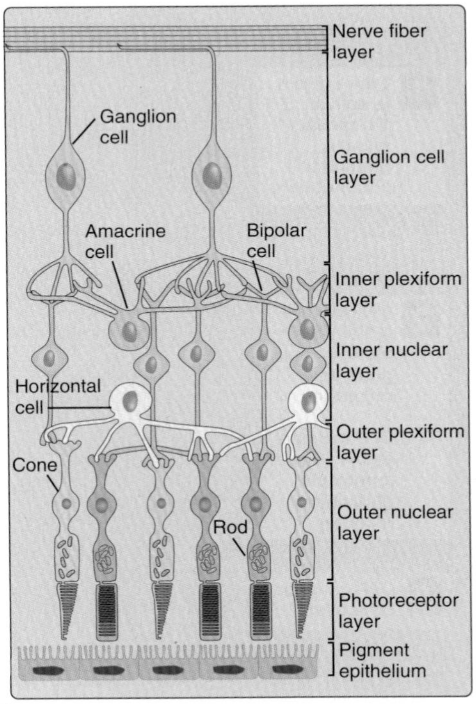

Figure 8.10
Principal retinal cell layers and interconnections.

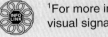

INFO LINK [1]For more information on processing and interpreting visual signals, see *LIR Neuroscience*, Chapter 15.

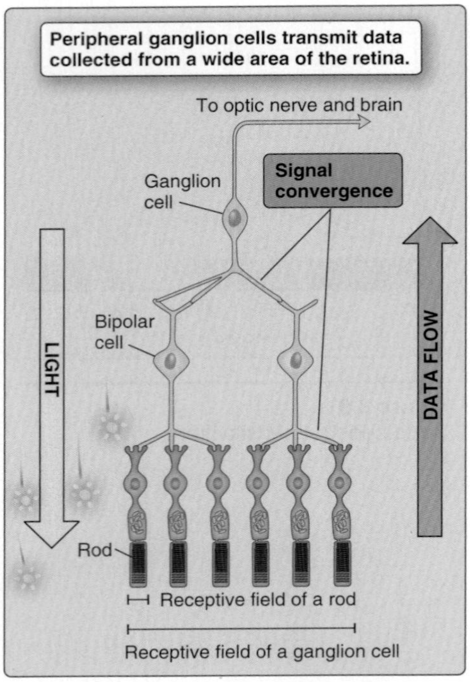

Figure 8.11
Photosensory data flow in the retina.

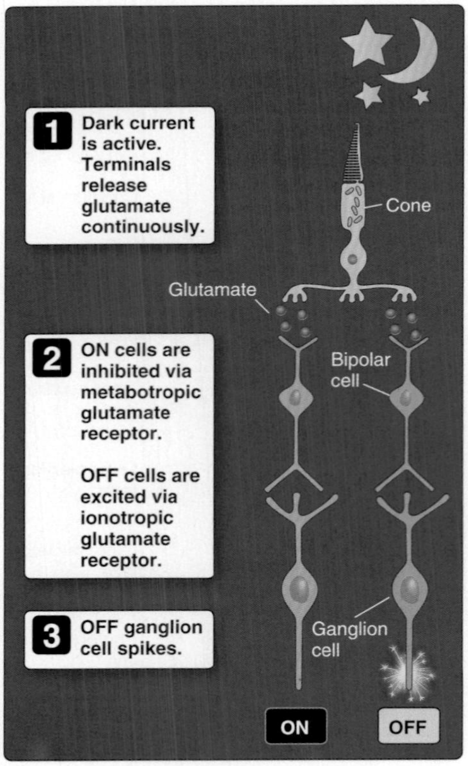

Figure 8.12
ON and OFF bipolar cell activity in the dark.

B. Photoreceptor output

The dark current keeps photoreceptors in a depolarized and excited state, and their presynaptic terminals are specialized to release transmitter (glutamate) continuously. Light reduces this current and causes a hyperpolarization that is graded with stimulus intensity. Photosensory excitation is signaled first to bipolar cells, which is where signal convergence begins (Figure 8.11).

1. **Rods:** Rods function under minimal light conditions in which the goal is simply capturing an image and visual acuity is a lesser concern. Bipolar cells, thus, collect and pool inputs from several spatially distant rods to increase the likelihood of unitary events being transmitted to the cortex. The portion of the visual field that is serviced by a single rod represents its **receptive field**. The receptive field of bipolar cells is, thus, larger because it incorporates the receptive fields of multiple rods (see Figure 8.11).

2. **Cones:** Cone cells operate in daylight. Because all photoreceptors are active under such conditions, pooling inputs covering a large receptive field would degrade the sensory data quality. Thus, cones typically synapse with dedicated bipolar cells that preserve spatial information and, thereby, increase visual acuity.

C. Bipolar cell output

Bipolar cells execute the first data-processing step. Individual cells generally receive inputs from either a group of rods or a single cone, but not both. These dedicated relationships preserve the integrity of the data streams from rods and cones. Bipolar cells are unusual in that they do not generate action potentials but, rather, their output is graded. There are at least 10 different bipolar cell types, but they can be divided into two principal groups: **"ON"** (or, **on-center**) and **"OFF"** (**off-center**) cells. Rods only synapse with ON cells. Most cones synapse with at least one of each.

1. **ON cells:** ON cells employ a metabotropic glutamate receptor (mGluR6) to transduce photoreceptor signals. The glutamate receptor is coupled via an inhibitory G protein to a nonselective cation channel, but, because the relationship is inhibitory, the channel is prevented from opening in the dark (Figure 8.12). When photoreceptors upstream are illuminated, the channel is freed of its inhibitory influence, and the bipolar cell depolarizes in a graded manner (Figure 8.13). These bipolars are known as **sign-inverting** because receptor *hyperpolarization* causes bipolar cell *depolarization*. They are excited when light is turned ON.

2. **OFF cells:** OFF cells express an ionotropic glutamate receptor. Glutamate binding causes a sustained inward current and a membrane depolarization mirroring that of the photoreceptors (see Figure 8.12). When the photoreceptor is stimulated, its membrane hyperpolarizes, and synaptic glutamate release is inhibited. The bipolar cell with which the photoreceptor synapses also hyperpolarizes (see Figure 8.13). When the light is turned OFF, the photoreceptor depolarizes, and synaptic glutamate release resumes, causing bipolar cell excitation and signaling.

D. Ganglion cell output

The vertical data streams running from the photoreceptors to the outer layer of the retina are preserved by the ganglion cells. Unlike most other cells in the retina, ganglion cells generate action potentials that are used to digitally encode visual information for transmission to the brain. ON ganglion cells respond to light with a volley of action potentials (see Figure 8.13), whereas OFF ganglion cells discharge when the light is turned off (see Figure 8.12).

E. Horizontal and amacrine cells

Horizontal and amacrine cells manipulate the sensory data streams as they progress through the retinal layers. Horizontal cells extend their processes laterally within the outer plexiform layer, which allows them to synapse with and collate information from photoreceptors within a wide receptive field. There are several different types of horizontal cell. Their output is usually inhibitory, using either γ-aminobutyric acid or glycine as a neurotransmitter, influencing both photoreceptor and bipolar cell signaling. Their net effect is to increase the contrast between signals received from light and dark areas of the retina. The role of amacrine cells is less well understood.

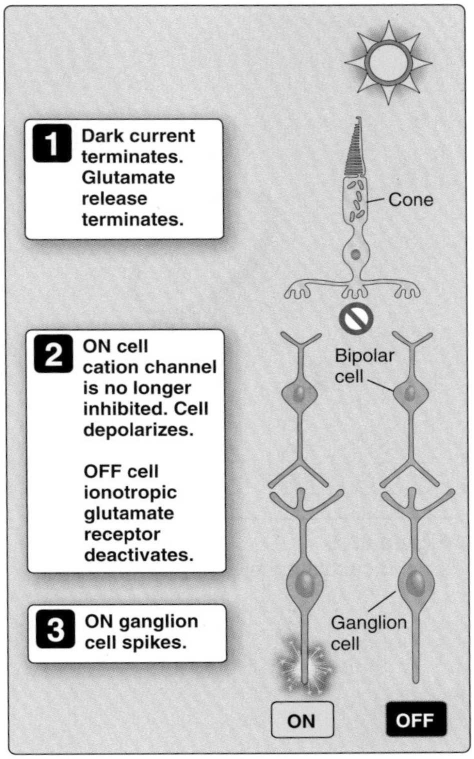

Figure 8.13
ON and OFF bipolar cell responses to light.

VIII. VISUAL PATHWAYS

Ganglion cell axons gather to form the optic nerves (CN II), one per eye, which convey visual signals from the retina to the brain. The optic nerves meet and merge immediately in front of the pituitary gland at a structure called the **optic chiasm** (Figure 8.14). Here, nasal retinal fibers cross the midline and join temporal fibers from the contralateral eye to form **optic tracts** that converge on the **lateral geniculate nucleus** of the thalamus. In practice, this crossing over (decussation) ensures that sensory data from the right visual fields of both eyes is transmitted to the left side of the brain, and *vice versa*. The data streams are then transmitted from the thalamus via **optic radiations** to the primary visual cortex in the occipital lobe for analysis and interpretation.

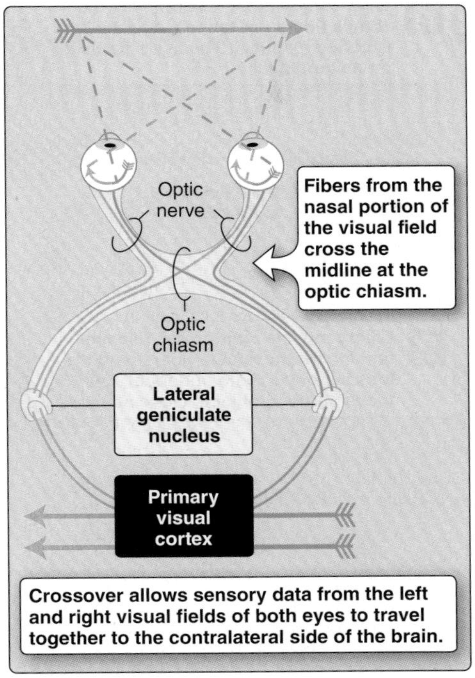

Figure 8.14
Pathways for visual information flow to the brain.

IX. OPTICAL PROPERTIES OF THE EYE

Light entering a room through a window does not form a perfectly focused image of the outside view on the opposite wall. Similarly, light entering the eye does not form a sharp image on the retina unless the rays are manipulated during passage through the eye.

A. Optical principles

Light rays normally travel in parallel lines. The speed at which they travel depends on the medium they are passing through. Their velocity is slowed in air compared with a vacuum and slowed further during passage through water or a transparent solid such as glass. The ratio of light velocity in a vacuum to velocity in a different medium is known as the **refractive index**. Air has a refractive index of about 1.0003, water has an index of 1.33, and the eye's lens and cornea are closer to 1.4. When light transitions at an angle to a medium of different refractive

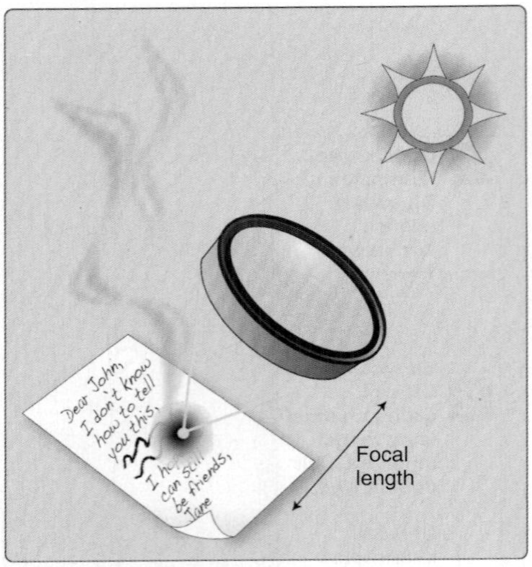
Focal
length

Figure 8.15
Lens focal distance.

index, it bends. The amount that it bends is dependent on the difference in refractive index of the two materials and also the angle of attack.

B. Simple lenses

Simple convex lenses (e.g., in a magnifying glass) have two curved surfaces. Their curvature forces parallel light rays to bend toward a central focal point. Thus, a magnifying glass held above a piece of paper in bright sunlight creates a spot of intense white light on the paper (Figure 8.15). When the glass is held at a distance that equals the **focal length** of the lens, the spot becomes a tightly focused pinhead, and the sun's energy causes the paper to smolder and burn. **Focal power** is the inverse of focal length. Powerful glass lenses are able to bring objects into tight focus within a relatively short distance compared with a weak lens, and this is achieved by increasing the curvature of the surfaces. Focal power is measured in diopters (D) and calculated as the reciprocal of focal length, in meters. A 1D lens brings an object into focus 1 m from the lens. A 10D lens focuses at 0.1 m. The human eye has a maximum power of about 59D.

C. Focal power: Light is forced to change direction several times during its passage from air to the retina. It is refracted at the interfaces between air and cornea, cornea and aqueous humor, aqueous humor and lens, and lens and vitreous humor.

The greatest degree of refraction occurs when light transitions from air to cornea, which, in practice, means that the cornea is the primary determinant of the eye's ability to bring a distant object into focus.

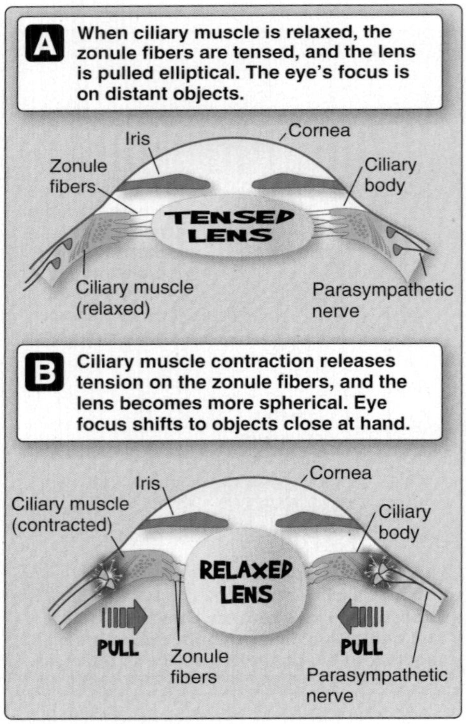
A
When ciliary muscle is relaxed, the zonule fibers are tensed, and the lens is pulled elliptical. The eye's focus is on distant objects.
Iris
Cornea
Zonule fibers
Ciliary body
TENSED LENS
Ciliary muscle (relaxed)
Parasympathetic nerve
B
Ciliary muscle contraction releases tension on the zonule fibers, and the lens becomes more spherical. Eye focus shifts to objects close at hand.
Iris
Cornea
Ciliary muscle (contracted)
Ciliary body
RELAXED LENS
PULL
PULL
Zonule fibers
Parasympathetic nerve

Figure 8.16
Lens accommodation.

D. Accommodation: A simple lens has a fixed focal length. Its surfaces can be shaped to image distant objects or objects close at hand but not both simultaneously. More complex imaging devices, such as telescopes, adjust focus by using two or more simple lenses whose positions relative to each other are varied. The eye is also capable of adjusting focus, but it does so by changing lens shape. This is known as the power of **accommodation**. The lens of a young and healthy individual can adjust focal power by up to 14D in a third of a second.

1. **Resting eye:** A human lens freed from all other influences assumes a near-spherical shape due to the capsule's natural elasticity. The lens is suspended in the light path by numerous radial zonule fibers, which are attached to the ciliary body. In a resting eye, the zonule fibers are tensed by surrounding structures and the lens stretches and flattens into an elliptical shape (Figure 8.16A). Thus, a resting eye focuses on distant objects.

2. **Accommodation mechanism:** Lens shape and focus is determined by ciliary muscle fibers under parasympathetic control (oculomotor nerve, CN III). When the eye is required to focus on close objects, the ciliary muscle ring is excited and contracts around the lens. This sphincter-like movement releases the tension on the zonule fibers and allows the lens to assume a more rounded shape, and focus shifts accordingly (see Figure 8.16B). The lens capsule stiffens with age, decreasing the power of accommodation and increasing the reliance on corrective reading lenses for individuals over age 40 years (**presbyopia**, derived from the Greek for "old eye").

Clinical Application 8.2: Refractive Disorders

An individual is assumed to have normal vision (**emmetropia**) if light from a distant object (>6 m away) traverses the eye optics and forms a focused image on the retina when the ciliary muscle is relaxed. Deviation from normal is very common. **Myopia**, or **nearsightedness**, refers to a condition in which the lens projects an image in front of the retina. It most commonly results from an eye that is longer than normal, but may also be caused by a lens or cornea that is optically more powerful than normal. Placing a lens with concave surfaces in front of the eye can adjust the plane of focus and restore visual acuity. **Hyperopia**, or **farsightedness**, is caused by an eye that is too short or a lens or cornea that projects an image behind the retina. The cause is usually of genetic origin. **Astigmatism** is a common visual defect in which irregularities in the focal power of the cornea or lens cause portions of the projected image to be blurred. Corrective lenses can again restore visual acuity. In recent years, **laser-assisted in situ keratomileusis** (**LASIK**) surgery has gained increasing popularity as a means of correcting myopic vision by providing an alternative to wearing corrective lenses. LASIK surgery involves lifting a corneal flap and then reshaping the cornea using an excimer laser to restore the curvature necessary to focus on distant objects.

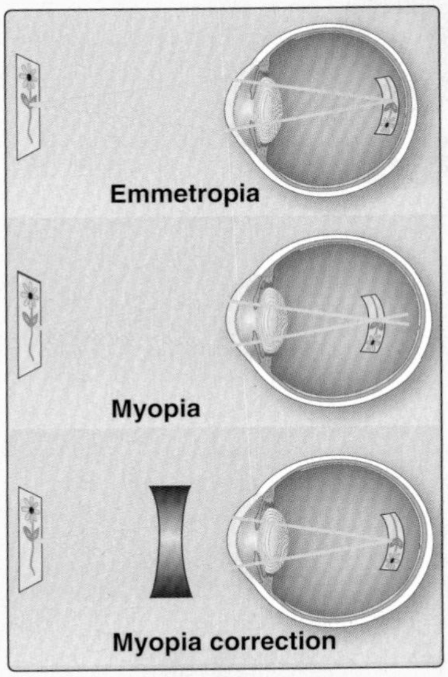

Emmetropia and myopia.

Chapter Summary

- The eye is a visual sensory organ comprising an optical apparatus (a variable aperture and lens) that projects an image of external objects onto a photosensor array. The array makes it possible to capture a digitized representation of the image for transmission to the primary visual cortex.
- Aperture (**pupil**) size is modulated as a way of controlling how much light enters the eye (the **pupillary light reflex**). Pupil size is determined by the iris, which contains two smooth muscle groups that are under autonomic (**oculomotor nerve**) control. A set of **sphincter muscles** decreases pupil size when they contract (**miosis**), whereas radial muscles dilate the pupil (**mydriasis**). The autonomic nervous system also regulates lens shape using **ciliary muscles** to allow the eye to focus on near objects (**accommodation**).
- The retina is a thin, layered, pigmented membrane containing a **photoreceptor array**, **signal processing neurons**, and glia (**Müller cells**) lining the back of an eye. Axons from the visual processing neurons (**ganglion cells**) exit the eye via the **optic nerve**, creating a retinal **blind spot**.
- The retina contains two types of photoreceptor cell. **Rods** are specialized to create monochromatic images in dim light. **Cones** produce color images in daylight. Both photoreceptor types transduce photon detection using modified **sensory cilia** containing stacks of membranes packed with photosensitive pigments.
- Rods are found at highest densities in the peripheral regions of the retina. Cones are concentrated within the center of the visual field. The **fovea** is a small area at the visual field's center that contains only cones and where the neuronal layers part to allow light to fall on photoreceptors directly. These modifications create an area of high visual acuity.
- Photosensory transduction occurs when photons are absorbed by and trigger a conformational change in a **chromophore** (**retinal**). Retinal is associated with a G protein–coupled receptor (**opsin**), together constituting a **visual pigment** (**rhodopsin**).
- Different opsin isoforms attune the visual pigment to particular wavelengths optimized for nighttime or color vision. Color vision relies on cones containing one of three such pigments.
- When light falls on rhodopsin, it activates a G protein (**transducin**), and the α subunit causes resting cyclic guanosine monophosphate (cGMP) levels to fall through activation of a *phosphodiesterase*. This suppresses a cGMP-dependent inward **dark current**, and the membrane hyperpolarizes. This membrane potential change represents a **receptor potential**.
- Photosensory signals are relayed to the cortex via a series of neurons (**bipolar cells** and **ganglion cells**) that begin processing the visual data while maintaining the integrity of the spatial information they contain.

9 Hearing and Balance

I. OVERVIEW

Modern day aquatic vertebrates possess lateral-line sensory systems that detect vibrations and movements in their watery surrounds. Lateral lines comprise lines of pits running down both sides of the body. Movements are transduced by clusters of sensory hair cells embedded in a gelatinous dome that protrudes from each lateral-line pit. When the dome is displaced by local water currents or vibrations, the embedded hairs are displaced also, generating a receptor potential in the hair cell body. Although humans did not retain lateral-line organs during evolution, the hair cell transduction system works so well that it was adapted for use in the inner ear. The inner ear contains two contiguous, hair cell–based sensory systems. The auditory system uses hair cells to transduce vibrations generated by sound waves. The vestibular system uses hair cells to transduce head movements.

II. SOUND

Sounds are atmospheric pressure waves created by moving objects. For example, striking a metal gong causes its metal surface to vibrate back and forth (Figure 9.1). The gong alternately compresses and then decompresses the surrounding air to create a pressure wave that propagates outward at a speed of 343 m/s. We perceive these pressure waves as sounds, the wave's frequency reflecting its pitch. The ability to transduce sound waves (**audition**) allows us to detect objects at a distance and, therefore, has clear survival advantages. If an approaching object represents a threat (e.g., a predator or speeding truck), advance warning of its approach allows time for evasive maneuvers. The ability to hear vocalizations allowed for the development of oral communication and speech. The ability to perceive sound requires that the energy of sound waves be converted into an electrical signal, a process that occurs within the **inner ear** and relies on **sensory hair cells**.

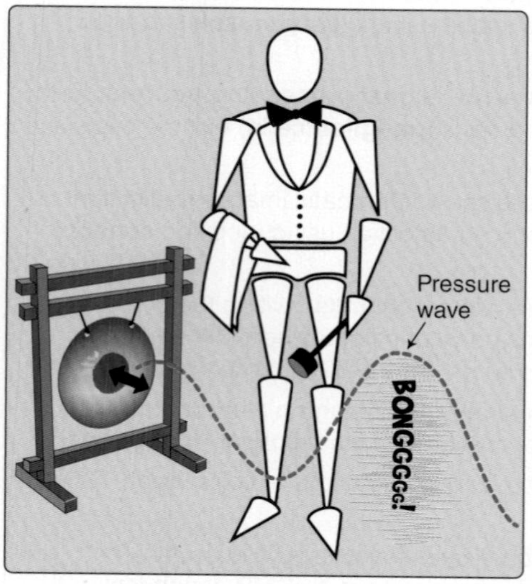

Figure 9.1
Sounds are pressure waves that travel through air.

III. AUDITORY SYSTEM

Designing a system that transduces sound is relatively easy because sound waves vibrate membranes. For example, a dog's bark creates

vibrations in the wall of an empty soda can or milk jug that can easily be sensed by mechanoreceptors in the fingertips. Sounds are usually very complex, however, comprising a series of changing frequencies. The ear decodes such sounds using an array of sensory hair cells embedded within a membrane that is designed to resonate at different frequencies along its length. It combines this with an amplifier to create a remarkably sensitive acoustic analyzer.

A. Structure

The auditory system, or ear, can be divided into three main anatomical components: the **outer**, **middle**, and **inner ear** (Figure 9.2).

1. **Outer ear:** The **pinna** collects and focuses sounds. Sounds are channeled into the ear canal (**external auditory meatus**), which allows them to pass through the skull's temporal bone. The canal ends blindly at the **tympanic membrane** (**eardrum**), which vibrates in response to sound.
2. **Middle ear:** The **middle ear** is an air-filled chamber lying between the eardrum and the inner ear. It connects with the nasopharynx via the **eustachian tube**, which drains fluids and allows the pressure across the eardrum's two surfaces to equalize. Eardrum vibrations are transmitted to the inner ear by an articulating lever system comprising three, small, fragile bones called **ossicles** (Figure 9.3). The bones are known as the **malleus** (hammer), **incus** (anvil), and **stapes** (stirrup), their names roughly reflecting their shapes. The malleus is attached to the eardrum's inner surface and transmits vibrations to the incus. The incus transmits them to the stapes. The stapes' footplate inserts into and is firmly attached to the **oval window** of the inner ear.
3. **Inner ear:** The inner ear contains a convoluted series of fluid-filled chambers and tubes (**membranous labyrinth**). The structures are encased within bone (the **bony labyrinth**) with a thin layer of **perilymph** trapped between bone and membranes. The labyrinth has two sensory functions. The **auditory portion** is called the **cochlea**. The **vestibular portion** contributes to our sense of balance (see section V below). It comprises the **otolith** organs (**utricle** and **saccule**) and three **semicircular canals**.

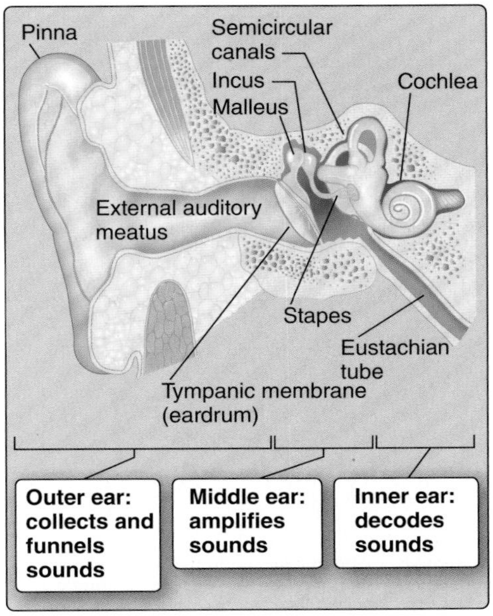

Figure 9.2
Ear anatomy.

B. Impedance matching

The inner ear is filled with fluid that has a high inertia and is difficult to move compared with air. The middle ear's function is, thus, to harness the sound wave's inherent energy and transmit it to the inner ear with sufficient force to overcome the inertia of the fluid contents. This process is called **impedance matching**.

1. **Mechanism:** The ossicles form a lever system that amplifies eardrum movements by ~30% (see Figure 9.3). It also focuses the movements on the stapes' footplate, whose surface area is ~17 times smaller than the eardrum. Amplification and focusing combined increase force per unit area ~22-fold, which allows sounds to be transferred to the inner ear with sufficient force to overcome cochlear fluid inertia.

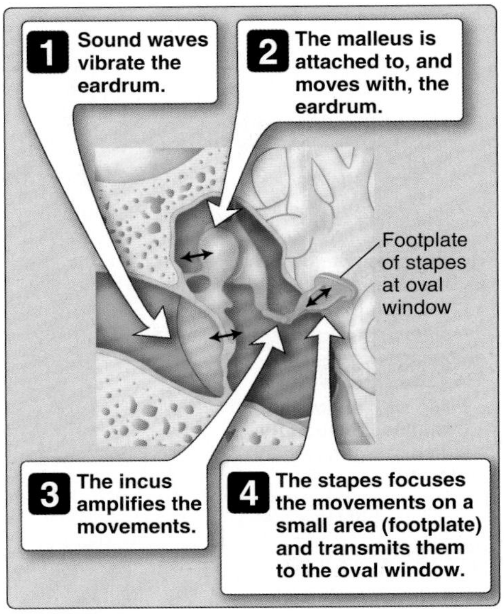

Figure 9.3
Ossicles and their role in impedance matching.

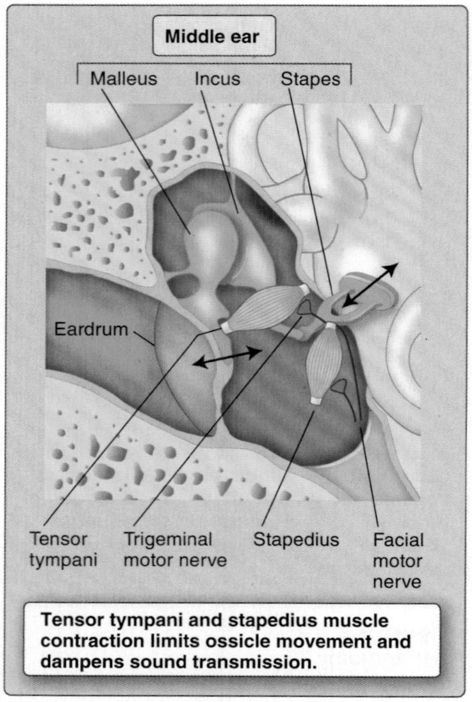

Figure 9.4
Attenuation reflex.

2. **Damping:** Lever system flexibility is modulated to reduce sound amplitude under certain circumstances. The malleus and stapes are attached to two tiny muscles under autonomic control (Figure 9.4). The **tensor tympani** anchors the malleus to the wall of the middle ear and is innervated by the trigeminal nerve (cranial nerve [CN] V). The stapes is anchored by the **stapedius**, which is innervated by the facial nerve (CN VII). When the two muscles contract, the ossicular chain becomes more rigid, and sound transmission is attenuated. The **attenuation reflex** can be triggered by loud sounds but is probably designed to dampen the sound of our own voices when talking.

C. Cochlea

The cochlea is a long (~3 cm), tapered tube containing three fluid-filled chambers that run the length of the tube. The tube is coiled like a snail shell *in vivo*, but the functional architecture is easier to understand when considered uncoiled (Figure 9.5). The three chambers are called the **scala vestibuli**, the **scala media**, and the **scala tympani**.

1. **Scala vestibuli and scala tympani:** The upper and lower chambers are both filled with **perilymph** (a fluid approximating plasma) and are physically connected by a small opening (the **helicotrema**) at the cochlear apex.

2. **Scala media:** The center chamber is separated from the scala vestibuli by the **Reissner membrane** (or **vestibular membrane**) and from the scala tympani by the **basilar membrane**. The scala media terminates short of the cochlea apex and is sealed off from the other two chambers. It is filled with **endolymph**, a K^+-rich fluid produced by the stria vascularis, a specialized epithelium lining one wall of the chamber (see Figure 9.5). The scala media contains the **organ of Corti**, which is the auditory sensory organ.

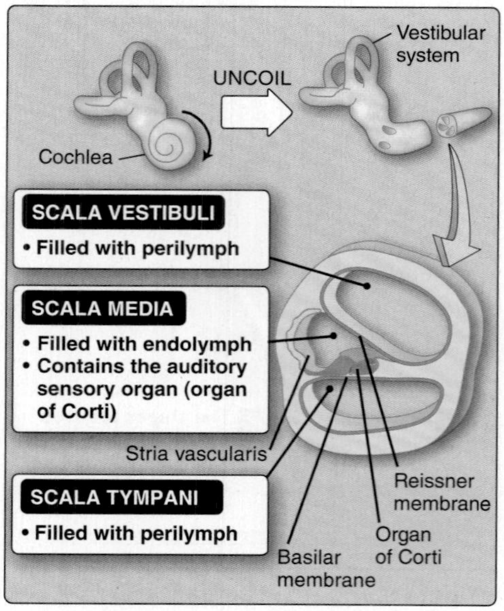

Figure 9.5
Cochlear chambers.

IV. AUDITORY TRANSDUCTION

Sound waves enter the cochlea via the oval window, which forms the basal end of the scala vestibuli (Figure 9.6). Stapes motion sets up a pressure wave in the perilymph that runs down the chamber's length to the apex, passes through the helicotrema, and then pulses back down the scala tympani to the cochlear base. Here, it encounters the **round window**, a thin membrane located between the inner and middle ear. The membrane vibrates back and forth in reverse phase with the wave generated by stapes movement. The stapes would not be able to displace the oval window and set the perilymph in motion if the round window did not exist because the cochlear chamber walls are otherwise rigidly encased in bone. The scala media, which approximates a fluid-filled sac suspended between the two chambers, is buffeted by the pressure wave as it pulses back and forth. Thus, although the sound wave never enters the scala media directly, the entire structure wobbles, much as a waterbed responds when pushed down on hard at one corner. It is this buffeting that is sensed by the organ of Corti.

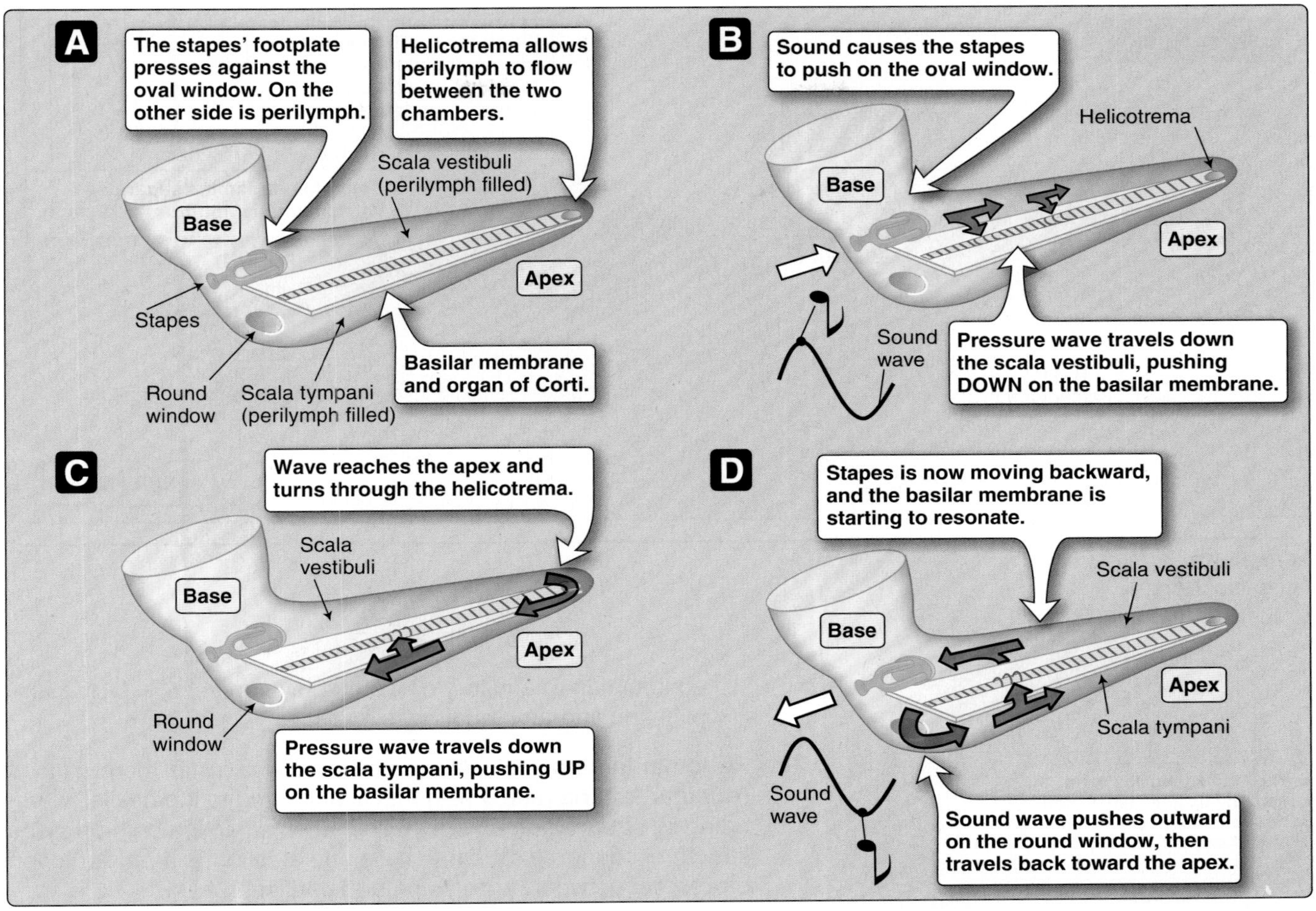

Figure 9.6
Sound wave passage through the cochlear chambers.

A. Organ of Corti

The organ of Corti comprises a sheet of auditory receptor cells (hair cells) and their associated structures, all of which rest on the basilar membrane (Figure 9.7).

1. **Hair cell types:** The hair cells are arranged in rows down the length of the cochlea. Two types of hair cell (**inner** and **outer**) can be distinguished based on their location, innervation, and function.

   a. **Inner:** Inner hair cells ([IHCs] ~3,500 total) form a single row toward the center of the cochlea. They are densely innervated by sensory neurons (up to 20 per cell), whose axons make up the bulk of the cochlear nerve (part of the vestibulocochlear nerve, or CN VIII). IHCs are the ear's primary sound transducers.

   b. **Outer:** There are an additional three rows of outer hair cells (OHCs). Although they number around 20,000 in total, their

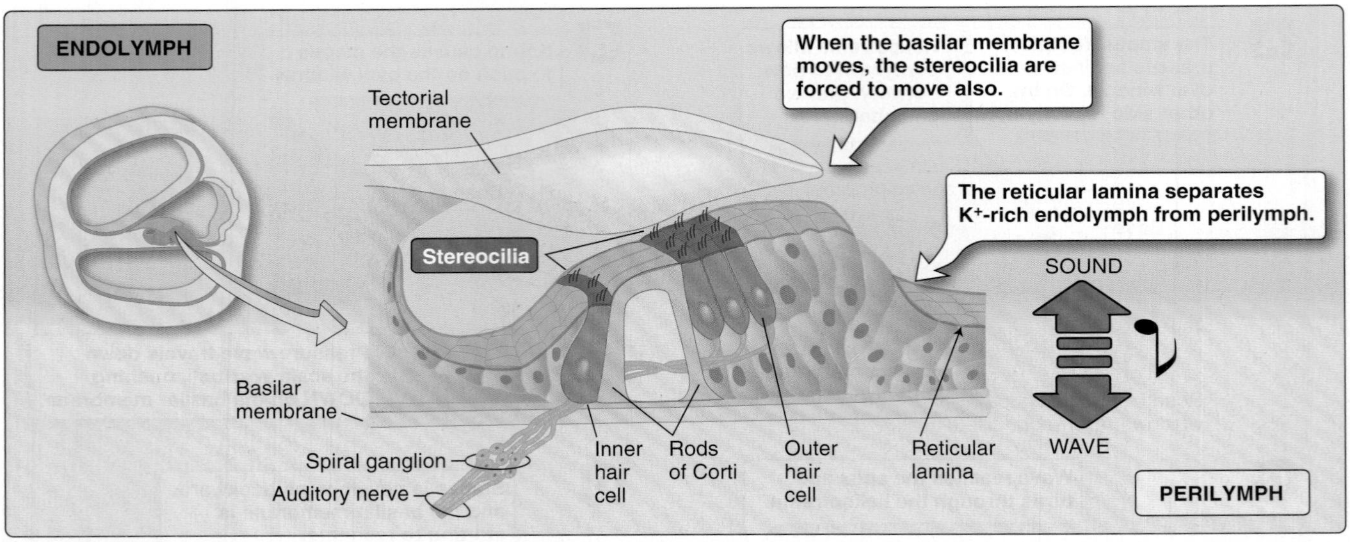

Figure 9.7
Organ of Corti.

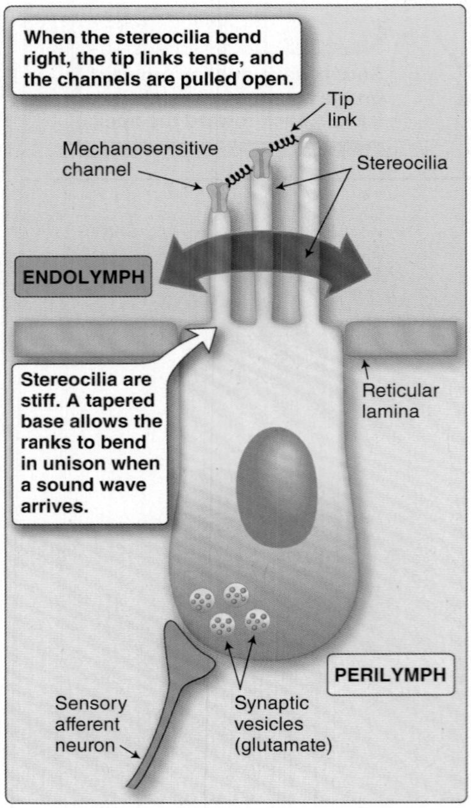

Figure 9.8
Role of stereocilia in mechanosensory transduction.

contribution to auditory nerve output is only ~5%. OHCs amplify and fine-tune auditory signals.

2. **Reticular lamina:** Hair cells are covered with a stiff, membranous **reticular lamina** (see 4·II·A) that is anchored to the basilar membrane by struts (**rods of Corti**). The reticular lamina both provides structural support for the hair cells and also forms a barrier to ion movement between endolymph and perilymph.

3. **Tectorial membrane:** Hair cells bear ~100 sensory stereocilia at their apical surface. They protrude through the reticular lamina into endolymph, a K^+-rich fluid whose unusual ion composition is critical for generating an auditory receptor potential (see below). The tips of OHC stereocilia embed in a gelatinous **tectorial membrane**, which lies over the cells like a blanket. When the basilar membrane is buffeted by sound, the hair cells are dragged back and forth beneath the blanket, and the stereocilia are forced to bend (see Figure 9.7).

B. Hair cell function

In order for sound waves to be perceived as such by the central nervous system (CNS), their energy must be converted to an electrical signal. Movement is sensed at the molecular level by mechanoreceptive ion channels located on sensory hair cells.

1. **Hair cell structure:** Hair cells are polarized, nonneuronal sensory cells. The apical side bears several rows of stereocilia, which are stepped in height to form ranks (Figure 9.8). The basal side synapses with one or more sensory afferent neurons, which the hair cell communicates with using an excitatory neurotransmitter (glutamate) when stimulated appropriately.

2. **Stereocilia:** At birth, hair cells contain a true cilium (**kinocilium**) that may help establish stereociliary orientation. The kinocilium is

not involved in auditory transduction, degenerating shortly after birth. **Stereocilia** are actin filled and rigid. They taper at their base where they meet the hair cell body, creating a hinge that allows for deflection (see Figure 9.8). Cilia are linked at their tips down the ranks by fine elastic protein strands called **tip links**, the lower ends of which are connected to a mechanosensitive or **mechanoelectrical transduction** (**MET**) channel. When the stereocilia bend toward the tallest rank, the tip links tense and pull the MET channel open. This is the mechanosensory transduction step.

C. Mechanotransduction

The hair cells straddle two compartments with strikingly different ion compositions, which favor an unusual *depolarizing* K^+ current when the MET channel opens (Figure 9.9).

1. **Endolymph:** The stereocilia are bathed in endolymph, a unique fluid that is secreted by the **stria vascularis**, a highly vascularized epithelium (see Figure 9.9B). Endolymph is characterized by a K^+ concentration of ~150 mmol/L, far higher than that of perilymph or extracellular fluid (~5 mmol/L). The inside of the hair cell relative to endolymph is −120 mV, which creates a very strong electrochemical gradient for K^+ influx across the stereociliary membrane.
2. **Perilymph:** The basolateral side of the hair cell is bathed in perilymph. This fluid is high in Na^+ and low in K^+, much like extracellular fluid. Perilymph is considered to be at 0 mV, so hair cell membrane potential (V_m) relative to perilymph is −40 mV. The voltage difference between endolymph and perilymph, which approximates 80 mV, is known as the **endocochlear potential**.
3. **Receptor currents:** The mechanosensitive MET channel is relatively nonselective for cations and may be a member of the transient receptor potential channel family (see 2·VI·D). When it opens, K^+ (and Ca^{2+}) flow into the cell and cause a receptor depolarization (see Figure 9.9A). This stands in stark contrast to the usual

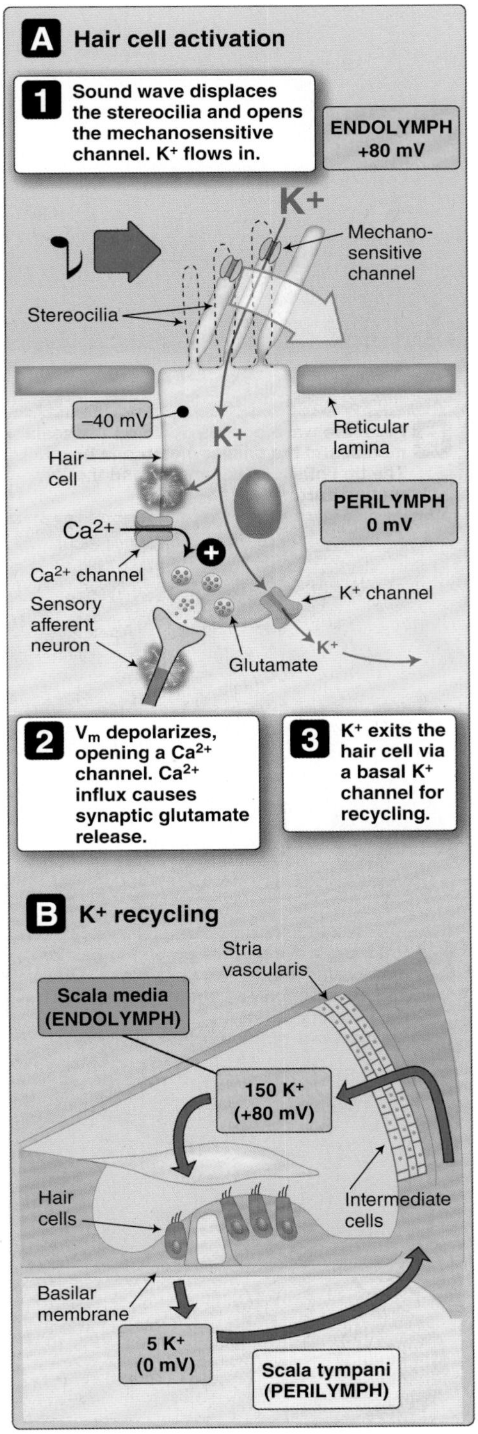

Figure 9.9
Mechanosensory transduction and K^+ recycling. K^+ concentrations are given in mmol/L. V_m = membrane potential.

Clinical Application 9.1: Congenital Hearing Loss

Congenital hearing loss (HL) results from mutations in any of the many genes required for auditory transduction, signaling, and processing. The most common form of HL results from a recessive mutation in the *GJB2* gene (known previously as *DFNB1*), which encodes connexin-26. Connexins are proteins that form gap junction channels between adjacent cells in many tissues, including those of the stria vascularis (see 4·II·F). The *KCNJ10* gene encodes an adenosine triphosphate–gated K^+ channel that is expressed in intermediate cells of the stria vascularis (see Figure 9.9B). These cells are responsible for maintaining high endolymph K^+ concentrations. *KCNJ10* mutation interrupts K^+ recycling, collapses the endocochlear potential that is required for auditory transduction, and causes profound deafness.

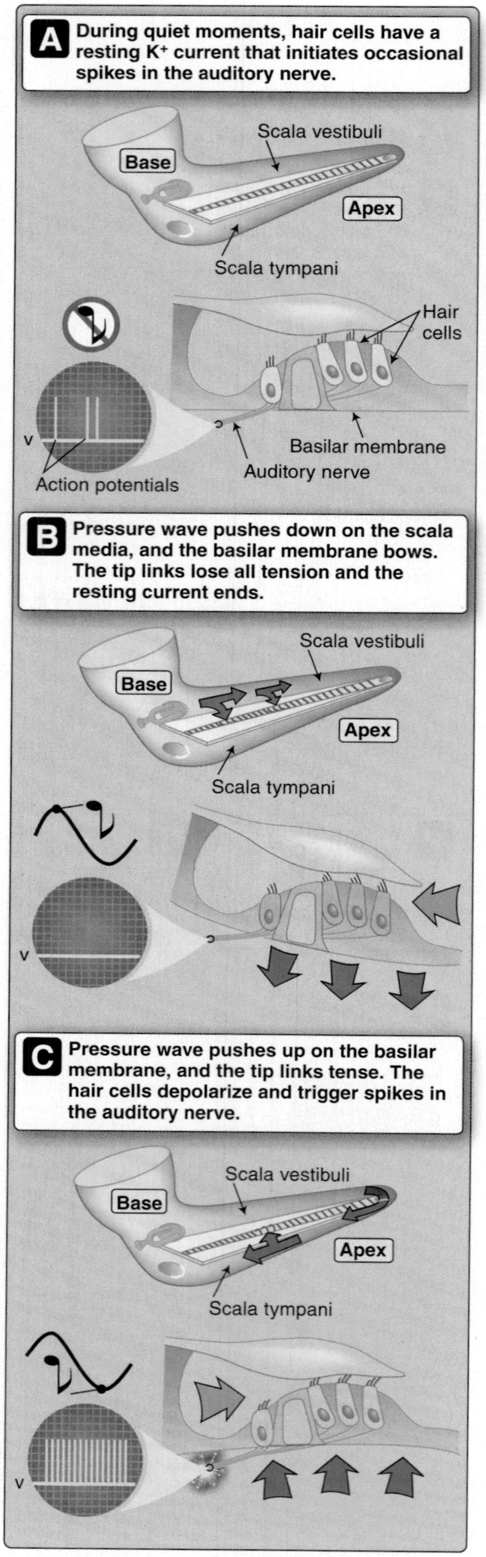

A During quiet moments, hair cells have a resting K+ current that initiates occasional spikes in the auditory nerve.
Scala vestibuli
Base
Apex
Scala tympani
Hair cells
V
Action potentials
Basilar membrane
Auditory nerve
B Pressure wave pushes down on the scala media, and the basilar membrane bows. The tip links lose all tension and the resting current ends.
Scala vestibuli
Base
Apex
Scala tympani
V
C Pressure wave pushes up on the basilar membrane, and the tip links tense. The hair cells depolarize and trigger spikes in the auditory nerve.
Scala vestibuli
Base
Apex
Scala tympani
V

Figure 9.10
Sound-wave effects on hair cells within the organ of Corti. V = voltage.

effects of K^+-channel opening on V_m. Most cells are bathed in low-K^+ extracellular fluid, so K^+-channel opening usually causes K^+ efflux and hyperpolarization.

4. **Potassium recycling:** K^+ exits hair cells via K^+ channels in the basolateral membrane (see Figure 9.9A) and is returned to the endolymph via the stria vascularis (see Figure 9.9B).

D. Auditory encoding

Most sounds are a complex mix of tones of variable intensity. The loudness of a sound correlates with the pressure wave's amplitude. Sound intensities are measured on a logarithmic scale in decibels (dB). Wave frequency (measured in **Hertz** [Hz]) determines whether it is perceived as being a low or high note (i.e., its pitch).

1. **Hearing range:** Normal conversation is held at ~60 dB. Leaves rustle at ~10 dB. Sounds of ~120 dB cause discomfort, and anything louder is perceived as being painful and can cause acute auditory damage. The ability to discriminate sound frequencies above 2,000 Hz decreases with age (**presbycusis**), but most young people can hear sounds within a range of 20–20,000 Hz.

2. **Loudness:** The relationship between sound amplitude and hair cell output is relatively simple.

   a. **Hair cell output:** Loud sounds buffet the scala media to a greater extent than quiet sounds, which displaces the stereocilia to a greater degree and prolongs MET channel open times. Receptor-potential amplitudes and spike frequencies in the sensory afferent neurons are increased proportionally.

   b. **Sensory nerve output:** The stereociliary ranks of IHCs and OHCs all face in the same direction. Directionality is important because the MET channels only open when the cilia are displaced toward the tallest rank, and a similar orientation allows for a unified response to basilar membrane movement. In the absence of sound, the tip links have a resting tension that keeps some MET channels open and generates a basal inward current and spike frequency in the auditory nerve, even when all stereocilia are standing erect (Figure 9.10A). When a sound wave enters the organ of Corti, it buffets the basilar membrane, and the stereocilia rock back and forth beneath the tectorial membrane. When pressure in the scala vestibuli is high, the basilar membrane bows downward, and the stereocilia are moved away from the tallest rank. This motion relieves the resting tip link tension and allows any open MET channels to close. The resting hair cell current and auditory nerve activity ceases (see Figure 9.10B). When pressure in the scala vestibuli is low, the stereocilia are displaced toward the tallest rank. The tip links tense, the MET channels open, and auditory nerve firing rate increases in proportion to pressure wave intensity (see Figure 9.10C). Thus, sound wave's passage causes hair cell V_m to swing from negative to positive, and the sensory afferent responds with volleys of spikes interspersed with quiescent periods.

3. **Frequencies:** The cochlea is organized much like a stringed instrument, such as a piano or harp. At one end, the strings are short and taut, and they resonate at high frequencies. At the other end, the strings are long and produce low-frequency notes. In the cochlea, the basilar membrane serves a similar function to that of strings. The membrane runs the length of the cochlea and is tapered (Figure 9.11). At the base, the membrane is relatively narrow (~0.1 mm) and stiff. Stereocilia here tend to be short and stiff also. In practice, this means that the base resonates, and hair cells are displaced maximally, by high-frequency sounds (see Figure 9.11A,B). The basilar membrane widens to ~0.5 mm and becomes ~100-fold more flexible toward the apex, and the stereocilia become longer and more flexible. The apex is tuned to low frequencies (see Figure 9.11B,C). This provides a way of breaking down a complex sound into its individual frequencies and relaying information about their relative timing and intensity to the CNS. This acoustic analysis is known as **auditory place coding**.

4. **Fine-tuning:** The OHCs respond to sound with receptor potentials, but they contribute minimally to the auditory data stream that flows to the CNS. Instead, OHC receptor potentials are used to modify hair cell shape. Depolarization causes the cell soma to contract, and the stereocilia pull down on the tectorial membrane, whereas hyperpolarization causes cell lengthening. The details are uncertain, but these shape changes are used to amplify sound-induced basilar membrane movements, creating a **cochlear amplifier**. OHC function may, thus, be to fine-tune frequency discrimination by the IHCs.

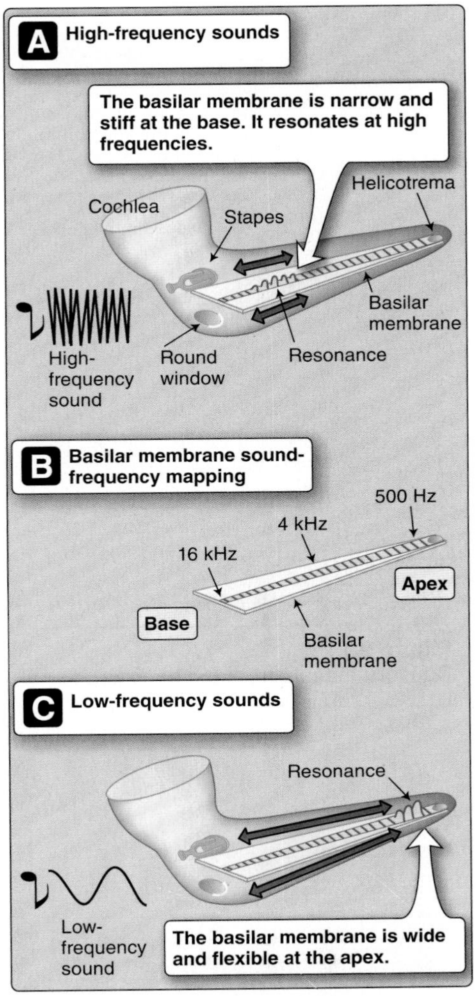

Figure 9.11
Auditory encoding.

> The cochlear amplifier is capable of generating sounds (**otoacoustic emissions**) loud enough to be heard by a bystander! Emissions may occur spontaneously or in response to an applied auditory stimulus. The phenomenon provides a basis for noninvasive testing for hearing defects in newborns and young children. Damage to the inner ear eliminates the emissions.

E. Auditory pathways

The sensory afferents innervating hair cells at the cochlear apex activate at low-frequency sounds, whereas afferents from the base activate at high frequencies (see Figure 9.11), which allows for spatial mapping of sound frequencies (**tonotopy**). Map integrity is preserved during relay back to the auditory cortex for processing and interpretation. Auditory signals are relayed from the cochlea to the CNS via the spiral ganglion, which is located within the bony **modiolus** that forms the center of the cochlear spiral. The cochlear nerve joins the vestibular nerve to become the vestibulocochlear nerve (CN VIII), which projects to cochlear nuclei in the brainstem medulla. Many fibers then cross the midline and travel to the **inferior colliculus** of the midbrain,

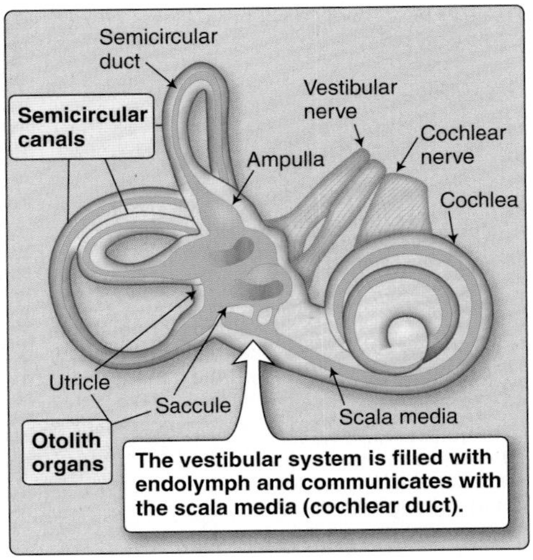

Figure 9.12
The vestibular system and cochlea.

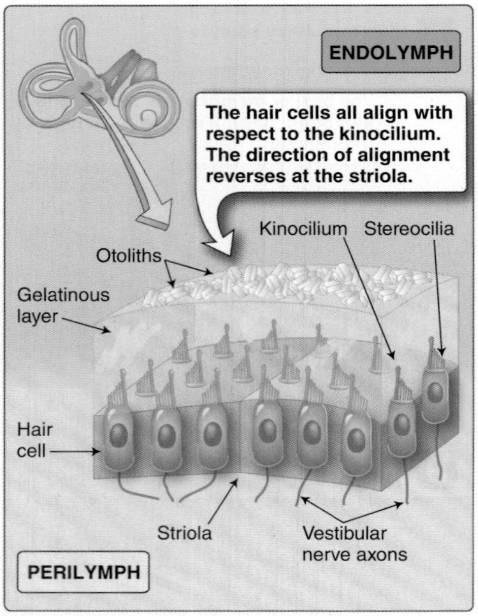

Figure 9.13
Sensory macula from the saccule.

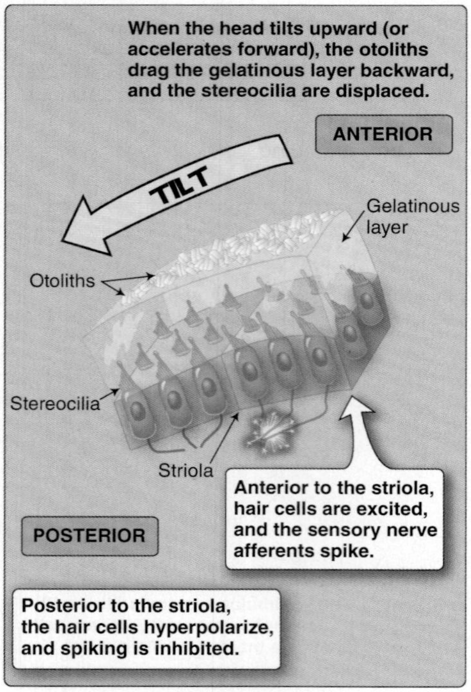

Figure 9.14
Saccular macular response to head tilt.

an area involved in auditory integration. Information is then relayed to the primary auditory cortex and associated areas involved in speech (**Wernicke area**) for interpretation.

V. BALANCE

The ability to stay upright is a feat that we rarely think about until the system is stressed by illness or age or when we place ourselves in situations that force us to pay attention (e.g., walking down a steep trail or stepping into an unstable boat). Maintaining balance requires sensory input from numerous areas that constantly update motor control systems about body position. The most important of these is the vestibular system, a specialized sensory organ that provides rapid and sensitive information about changes in head position. The vestibular system allows us to correct body position before we fall and helps maintain a stable retinal image during head movements.

A. Vestibular system

The vestibular system is a part of the inner ear. The principal components are the otolith organs and the semicircular canals (Figure 9.12). Like the cochlea, the system comprises membranous, labyrinthine structures encased in bone and bathed in perilymph. The interior, which is continuous with the scala media, contains endolymph. Sensory transduction relies on mechanosensory hair cells, which, unlike those of the cochlea, retain their kinocilium.

1. **Otolith organs:** The otolith organs (**utricle** and **saccule**) comprise two chambers near the center of the inner ear. The utricle detects linear acceleration and deceleration. The saccule is oriented to detect movements caused by vertical acceleration (e.g., riding an elevator). Both also respond to changes in head angle.

2. **Semicircular canals:** The semicircular canals detect angular head rotation. As the name suggests, the canals comprise semicircular tubes with a swelling (**ampulla**) at their base. The vestibular system comprises three such canals (**anterior**, **posterior**, and **horizontal**) that are oriented perpendicular to one another to allow detection of movements in any of three dimensions.

B. Otolith organ function

Each otolith organ contains a sensory epithelium (the **macula**) comprising an array of innumerable hair cells and their supporting structures (Figure 9.13). The kinocilium and stereocilia that project from the hair cell apical surface embed in a gelatinous **otolithic membrane** studded with calcium carbonate crystals. These are the **otoliths** ("ear stones") that give the organs their name. Their purpose is to add inertial mass to the membrane.

1. **Mechanotransduction:** When the head tilts forward or backward, the otolithic membrane moves under the influence of gravity and the embedded sensory cilia bend (Figure 9.14). Similar movements occur when the head accelerates or decelerates suddenly. The

mechanotransduction step is identical to that described in Section IV·C above. When the stereocilia move toward the kinocilium, tip links tense and open a mechanosensitive channel in the ciliary tip, the membrane depolarizes, and the sensory nerve afferents fire action potentials. When the stereociliary assembly moves away from the kinocilium, the membrane hyperpolarizes, and vestibular nerve activity decreases.

2. **Hair cell orientation:** Hair cells within both otolith organs are oriented relative to an otolithic membrane ridge (the **striola**). The ridge curves across the width of the macula, meaning that hair cell orientation shifts with the curve also. This allows the hair cells to sense movement in any direction (Figure 9.15). Hair cell orientation also reverses at the striola (see Figures 9.13 and 9.15), which ensures that linear head movements always activate at least some hair cells within both otolith organs, which further enhances sensory discrimination.

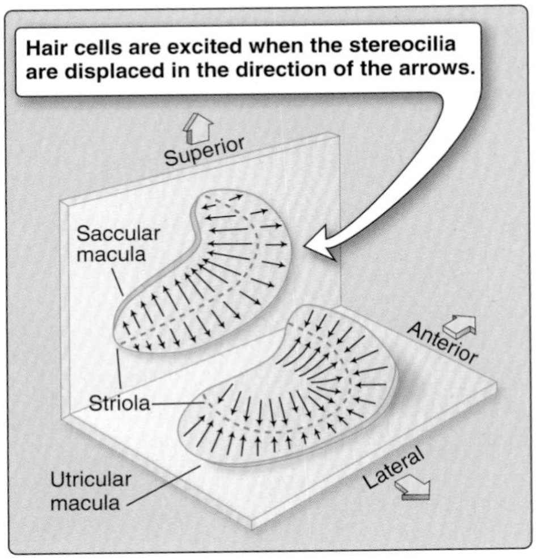

Figure 9.15
Hair cell orientation within the utricular and saccular maculae.

C. Semicircular canal function

The sensory epithelium within semicircular canals is pushed up into a crest (**ampullary crest**, or **crista ampullaris**) covered with sensory hair cells as shown in Figure 9.16. Their cilia are embedded in a gelatinous flap called the **cupula** that occludes the ampulla. The canals are filled with endolymph. When the head moves, the walls of the canal slip past the endolymph, which is held back by inertia. Float an ice cube in a large measuring cup filled with water and take note of its position relative to the spout or handle. If you now rotate in place while holding the cup, you will note that inertia prevents the contents of the cup from moving even though the cup walls are rotating around it. A similar phenomenon happens in the semicircular canals. Within the ampulla, head movement drags the cupula through the endolymph, causing it to deflect. This action tugs on the hair cells and generates a receptor potential, the polarity of which correlates with direction of movement. If the head is rotated continually, the endolymph eventually achieves the same rotational speed, and the receptor potential wanes. Sudden deceleration then initiates a new response. Note that the vestibular system comprises three such canals oriented at right angles to each other so as to detect rotational movements in any direction. The structures of the inner ear are also mirrored on each side of the head. Responses from all six canals and four otolith organs are integrated by brainstem vestibular nuclei.

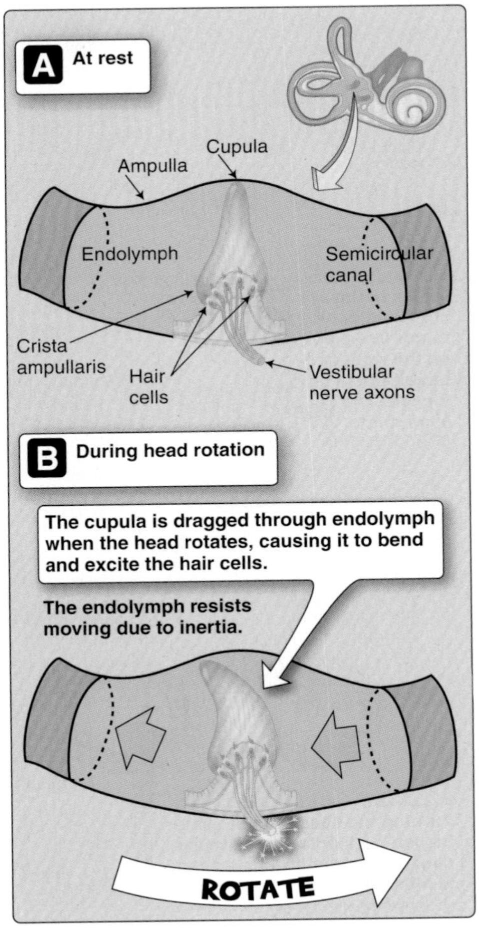

Figure 9.16
Transduction of rotational head movements by the semicircular canals.

D. Vestibular nuclei

The brainstem contains a group of vestibular nuclei that form an integrative control center responsible for balance. Vestibular sensory nerve signals reach the nuclei via the vestibular nerve, vestibular ganglion, and vestibulocochlear nerve (CN VIII). Vestibular sensory afferents also project to the cerebellum. The vestibular nuclei also receive sensory data from the eyes and somatic proprioceptors located in muscles and joints (see 11·II). This information is integrated and then used to execute reflex movements of the eye, head, and the muscles involved in postural control.

Clinical Application 9.2: Vestibular Dysfunction

A normal sense of balance is so important that severe cases of vestibular dysfunction can be disabling. Even mild cases can bring on highly disturbing sensations of **vertigo** and nausea. Vertigo is a sense of spinning in space, or that a room is spinning around a person even when stationary. **Benign paroxysmal positional vertigo (BPPV)** is the most common form of vestibular dysfunction. The symptoms include dizziness, lightheadedness, and vertigo. It is usually brought on by rolling over in bed or when getting out of bed. Tilting the head to look upward may also precipitate an attack. BPPV is caused by otoliths ("ear stones") that have detached from the otolithic membrane and made their way into one of the semicircular canals. They then stimulate the hair cells inappropriately when the head is moved in a particular direction. BPPV usually resolves spontaneously, although episodes may be recurrent. Sequential manipulation of head position by a trained therapist to work the otoliths out of the canals may afford a more permanent solution. Episodes of dizziness and vertigo can also be caused by infection or inflammation of the inner ear (**labyrinthitis**) and is usually accompanied by hearing loss. Appropriate treatment can lead to full recovery of both hearing and balance. **Ménière disease** is an idiopathic inner ear disorder that is believed to result from inadequate drainage of endolymph from the inner ear.

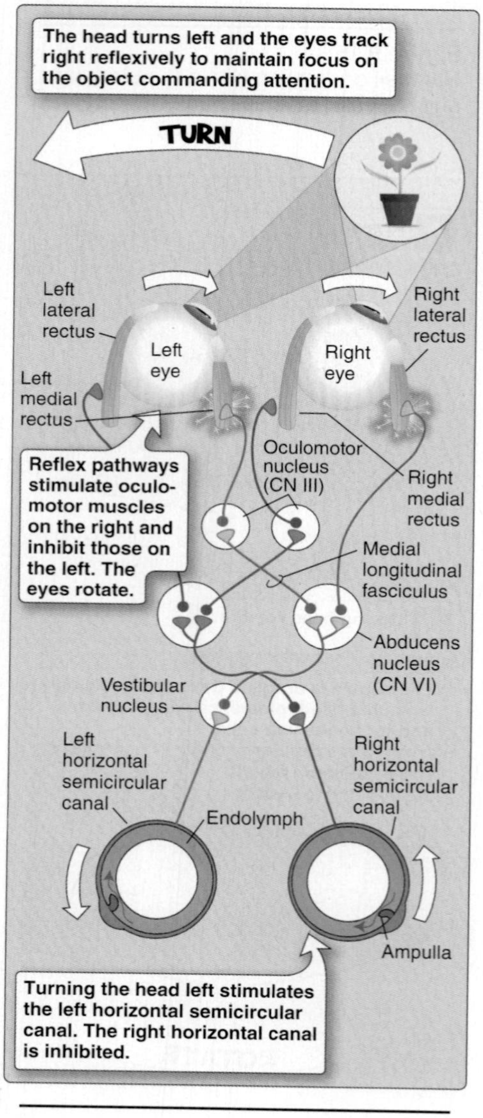

Figure 9.17
The vestibuloocular reflex. CN = cranial nerve.

E. Vestibuloocular reflex

Changes in head position affect eye position also, which is problematic because moving images lack acuity. Camera movement similarly blurs a photographic image. One important vestibular function is, thus, to inform ocular motor-control centers about the direction of head movement so that eye position may be adjusted to maintain a stable retinal image even as the head is moving (a **vestibuloocular reflex** [**VOR**]). Figure 9.17 considers what happens when the head is turned to the left as an example, but similar principles govern responses to head movements in any direction. Turning the head to the left stimulates hair cells in the horizontal semicircular canal on the left side of the head and inhibits output from the horizontal canal on the right. Sensory signals are relayed via the vestibular nuclei to the contralateral **abducens nucleus**, a cranial nerve nucleus (CN VI) located in the brainstem. Excitatory outputs from this nucleus travel directly via the abducens nerve to the left medial rectus eye muscle, one of six muscles that control eye movement. They also travel via the medial longitudinal fasciculus to the oculomotor nucleus (CN III), from where they excite the right lateral rectus muscle. Inhibitory pathways simultaneously suppress left lateral rectus and right medial rectus muscle contraction, and the eyes track right, a movement that is exactly equal and opposite to changes in head position. The retinal image remains centered as a result. The pathways between vestibular system and eyes are very short, and, thus, the reflex is extremely fast.

Clinical Application 9.3: Caloric Reflex Testing

The vestibuloocular reflex (VOR) provides a way for assessing functionality of the vestibular system, brainstem, and oculomotor pathways. During caloric reflex testing, a supine patient's head is inclined 30°, and the external auditory meatus is then irrigated with ~50 mL of ice water from a syringe. The cold water sets up a temperature gradient across the horizontal canal. Endolymph in the section of the canal closest to the external ear becomes denser as it cools, and it sinks under the influence of gravity. This sinking motion displaces endolymph in sections farthest from the cooled ear, causing fluid movement within the canal that mimics the effects of turning the head away from the stimulated ear. The VOR causes both eyes to track slowly toward the cooled ear and then rapidly reset their focus toward the opposite side (the rapid resetting motion is called **nystagmus**). Irrigating the ear with warm water simulates head-turning toward the stimulated ear. The caloric reflex test is a useful means of assessing brainstem function in comatose patients. The VOR is typically abnormal or absent in patients who have suffered brainstem hemorrhage or infarction.

Chapter Summary

- The **auditory system** and **vestibular system** both use **mechanosensory hair cells** to transduce sound waves and head movements, respectively.
- Sounds are collected by the **outer ear** and channeled via the **tympanic membrane** into the middle ear for amplification by an articulating lever system comprising three **ossicles** (**malleus**, **incus**, and **stapes**). The stapes transfers sound via the **oval window** to the **cochlea** of the **inner ear** for auditory transduction.
- The cochlea is a long, tapered tube coiled into a spiral. It contains three fluid-filled tubular chambers. The two contiguous outer chambers (**scala vestibuli** and **scala tympani**) are filled with **perilymph** and provide a pathway for sounds to travel through the system. A central tube (**scala media**) is filled with K^+-rich **endolymph** and contains the auditory sensor.
- Sound is transduced by the **organ of Corti**. It contains four rows of sensory hair cells. **Stereocilia** project from the apical surface of the hair cells into the central chamber and embed in an overlying **tectorial membrane**. The basolateral surface of the cells rests on the **basilar membrane**.
- When sound waves displace the basilar membrane, the stereocilia bend. Adjacent stereocilia are connected by **tip links**, which tense and open mechanoreceptor channels in the stereociliary membrane during passage of sounds. The channels allow K^+ influx from the endolymph, producing a depolarizing receptor potential.
- The basilar membrane is tapered. The base is narrow and stiff and resonates at high frequencies, whereas the apex is wide and flexible and resonates at low frequencies. This allows complex sounds to be broken down to create a **spatial** (**tonotopic**) **map**. Spatial information is preserved during data transmission to the auditory cortex.
- The vestibular system detects head movements. The system comprises three orthogonally arranged **semicircular canals** and two **otolith organs** (**utricle** and **saccule**), all of which are filled with endolymph.
- The semicircular canals detect **head rotation** in any plane. A swelling at the base of each canal (**ampulla**) contains a ridge (**crista ampullaris**) covered with sensory hair cells, each bearing a kinocilium and several stereocilia. The ciliary tips embed in a gelatinous **cupula** that protrudes into the endolymph. Head rotation forces endolymph against the cupula, causing it (and its embedded sensory cilia) to bend. The sensory cilia contain K^+-permeant mechanosensory channels that open during bending to generate a depolarizing receptor potential.
- The otolith organs contain lines of sensory hair cells whose cilia are embedded in a gelatinous membrane encrusted with **otoliths** (calcium carbonate crystals). Shifts in head position or sudden acceleration cause the heavy membrane to move and bend the sensory cilia.

10 Taste and Smell

I. OVERVIEW

The gustatory and olfactory systems are probably the oldest of the senses in evolutionary terms. Both systems allow us to detect chemicals in the external environment and, thus, are usually grouped together. In practice, however, they represent two very different sensory modalities that complement but cannot replace each other. Taste cells are modified epithelial cells, whereas olfactory receptors are neurons. Taste allows us to differentiate between very basic flavors, such as sweet *versus* salty, or savory *versus* sour. Taste is closely linked with appetite and cravings, such as a need to ingest salt (NaCl) or something sweet, and it is also protective. Bitter taste often helps us avoid ingesting toxins, whereas the taste of acid (sour) often indicates food decay. Olfaction allows us to detect and identify thousands of unique chemicals, including pheromones.

Table 10.1: Taste Receptor Type

Taste	Perception	Taste Receptor Type
Salty	Pleasant	I
Sweet	Pleasant	II
Umami	Pleasant	II
Bitter	Aversive	II
Sour	Aversive	III

II. TASTE

There are five basic tastes: **salty**, **sweet**, **umami**, **bitter**, and **sour** (Table 10.1). *Umami* ("good taste" in Japanese) is epitomized by the taste of monosodium glutamate (MSG), which imparts a savory, meaty flavor to food. The taste of fat constitutes a sixth basic taste, but the transduction mechanisms are not fully delineated. The chemical sensations that mimic hot (e.g., the burning sensation associated with chili peppers) and cold (e.g., menthol) are not tastes but rather are mediated by somatosensory pathways located in the oral cavity or nasal passage (see 16·VII·B).

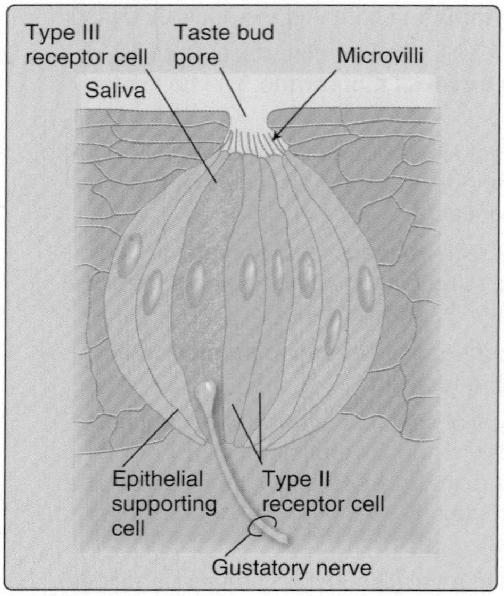

Figure 10.1
Taste bud organization.

A. Taste buds

Taste receptor cells are usually clustered in taste buds, which are distributed throughout the oral cavity. Lingual taste buds are organized like garlic bulbs (Figure 10.1), each containing ~100 elongated neuroepithelial cell "cloves." Adjacent cells are connected apically by tight junctions. Some cells extend microvilli into a taste bud's small central pore, which provides a way for oral fluids (saliva) and their dissolved tastants to enter the taste bud and be sensed. Taste buds contain several different cell types: Type I, II, and III cells are all taste receptors.

B. Type I cells

Type I receptors are nonexcitable cells that transduce salty taste, epitomized by the taste of NaCl (table salt) and, more specifically,

Na^+. Type I cells have **glia-like** properties (see 5·V) and transduce Na^+ sensation using an **epithelial Na^+ channel** (**ENaC**).

1. **Glia-like function:** Type I cells extend membrane processes that surround other cells within the taste bud and may help regulate extracellular K^+ concentrations during excitation. Type I cells also help terminate signaling by hydrolyzing neurotransmitter soon after release (see Section E below).

2. **Transduction:** Type I cells sense salt using ENaC. ENaC is always open, so when salty foods are ingested, Na^+ ions flood into the cells, and the receptor depolarizes (Figure 10.2). Receptor potential amplitude is graded with Na^+ concentration, but the downstream consequences of this receptor potential are still being investigated.

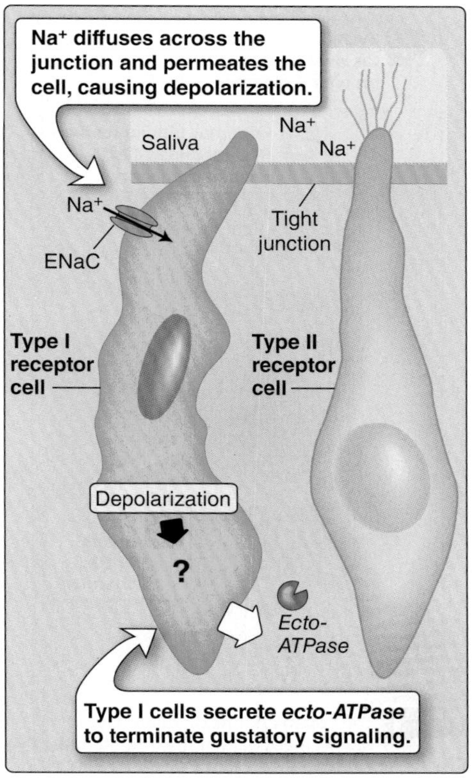

Figure 10.2
Type I receptor cell taste transduction. ENaC = epithelial Na^+ channel.

C. Type II cells

Type II cells are excitable **sensory receptors**. Their membranes contain specific G protein–coupled receptors (GPCRs) that mediate sweet, umami, and bitter tastes, but they do not respond to salty or sour tastants. Individual type II cells are tastant specific.

1. **Tastants:** The three tastant classes detected by type II cells (sweet, umami, and bitter) are perceived as either being pleasant and signaling the presence of food or noxious and indicative of a toxin.

   a. **Sweet:** Sweet tastes are associated with mono- and disaccharides, such as glucose and sucrose. Sugars are a primary energy source, and, therefore, the ability to recognize their presence in food has clear evolutionary advantages. Sugars are sensed by a single GPCR with a T1R2–T1R3 heterodimeric composition.

   > The extracellular domain of the sweet taste receptor is large and shaped like a Venus flytrap. The domain contains binding sites for sugars, but it also recognizes certain proteins as sweet (e.g., monellin). This feature has facilitated the development of a wide range of synthetic peptide sweeteners, including aspartame.

   b. **Umami:** The savory taste is elicited by glutamate, which is released from meat during protein hydrolysis. Some nucleotides (inosine monophosphate and guanosine monophosphate) also produce umami taste. Glutamate elicits at least some of its effects by binding to a T1R1–T1R3 heterodimeric GPCR.

   c. **Bitter:** Many plants, fungi, and some animals produce toxins as a natural defense mechanism. Evolution has helped guide our choice of food by associating many such poisons with a bitter taste. Most bitter tastants are detected by GPCRs. Toxins are such a diverse group that recognizing them as such requires specific receptor proteins. Taste cells that sense bitter

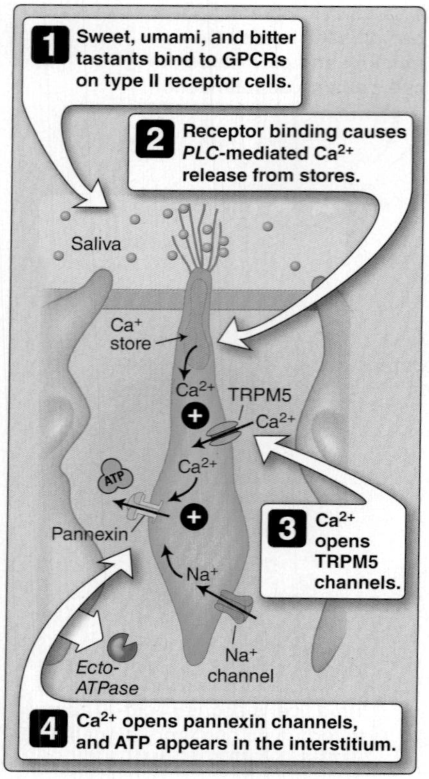

Figure 10.3
Type II receptor cell taste transduction. ATP = adenosine triphosphate; GPCRs = G protein–coupled receptors; *PLC* = *phospholipase C*; TRPM5 = transient receptor potential M5 channel.

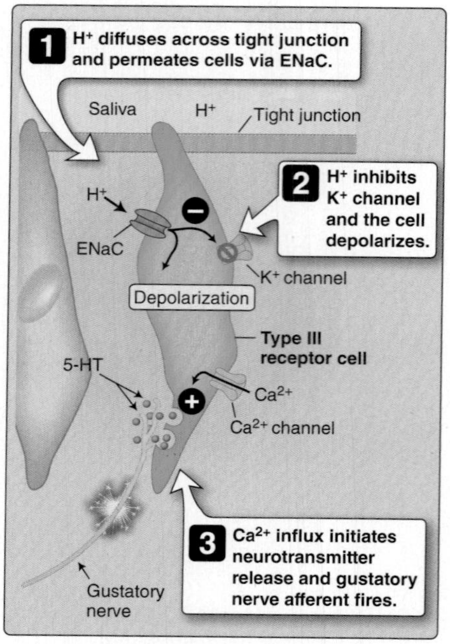

Figure 10.4
Type III receptor cell taste transduction. ENaC = epithelial Na^+ channel; 5-HT = 5-hydroxytryptamine.

stimuli express subsets of >20 GPCR T2R variants. Some are highly tastant specific, whereas others have broad specificity.

> Quinine is a bitter-tasting toxin with antimalarial properties extracted from cinchona tree bark. It blocks most classes of K^+ channel and causes nonspecific membrane depolarization.

2. **Transduction:** Tastant binding to its GPCR activates **gustducin**, a G protein that signals receptor occupancy through adenosine triphosphate (ATP) release into the taste bud interstitium (Figure 10.3).

   a. **Activation:** The gustducin $G_{\beta\gamma}$ unit activates *phospholipase C* (*PLC*) and initiates inositol trisphosphate–induced Ca^{2+} release from intracellular stores. Ca^{2+} then activates influx of Ca^{2+} and other cations via TRPM5 (a transient receptor potential channel; see 2·VI·D), and the receptor cell depolarizes. The gustducin G_α subunit modifies intracellular cyclic adenosine monophosphate (cAMP) levels through modulation of *adenylyl cyclase*, but the consequences are still being investigated.

   b. **Adenosine triphosphate release:** TRPM5-mediated Ca^{2+} influx opens **pannexin** hemichannels in the receptor cell membrane. Pannexins are related to the connexins that form gap junctions between cells (see 4·II·F). When pannexins open, they allow ATP to diffuse out of the cell and enter the interstitium. If a receptor potential generated by tastant binding crosses threshold for action potential formation, the cell then generates a train of spikes mediated by voltage-gated Na^+ channels. Pannexins are also voltage gated, so spiking potentiates ATP release.

D. Type III cells

Type III cells, which are also known as **presynaptic cells,** are the only taste receptor cell class to synapse with a sensory nerve. Type III cells sense sour tastes primarily through H^+-induced membrane depolarization.

1. **Transduction:** Sourness is the taste of acid (H^+). Common dietary examples include acetate (vinegar), citrate (lemons), and lactate (sour milk). H^+ enters cells via ENaC and depolarizes the cell directly, but, once inside the cell, H^+ also reduces K^+ efflux by inhibiting K^+ channels (Figure 10.4). Inhibition further amplifies the depolarization caused by H^+ entry.

2. **Transmission:** If sufficiently large, the H^+-induced receptor potential activates voltage-dependent Na^+ channels in the cell membrane and triggers a spike. Voltage-dependent Ca^{2+} channels then open to allow Ca^{2+} influx and transmitter release (5-hydroxytryptamine [5-HT], also known as serotonin) at a synapse with a sensory afferent neuron.

E. Signal integration

Food contains a mix of different tastants, and, hence, taste bud output usually represents an integrated response to simultaneous stimula-

Clinical Application 10.1: Dysgeusia

Taste is a relatively crude sense that serves primarily as a gatekeeper to the gastrointestinal system, causing us to either accept ingested substances as food or reject potentially injurious substances before they are swallowed. Complete loss of taste (**ageusia**) is uncommon, except in patients with Sjögren syndrome. Sjögren patients suffer from an autoimmune disease that impairs exocrine gland function, including the salivary glands. Saliva is required to carry tastants in dissolved form through the taste bud pore. **Metallic dysgeusia** (a persistent metallic taste) is a common and troublesome side effect of many antibiotics (e.g., tetracycline and metronidazole) and antifungals.[1]

tion of all three taste cell classes. Type III cells are the only receptors that make synaptic contact with an afferent nerve, but type II cells may also stimulate gustatory nerve activity directly through ATP release (the nerve expresses purinoreceptors), and indirectly by stimulating 5-HT release from type III cells. Type I cells influence the output of the other two receptor cell classes by secreting an *ecto-ATPase* that terminates signaling.

F. Taste bud distribution

Taste buds are distributed throughout the oral cavity, although the highest concentrations are located on the dorsal surface of the tongue. Lingual taste buds reside on surface projections called **papillae**. Three types of papillae can be distinguished based on shape and taste bud density: **fungiform**, **foliate**, and **circumvallate** (Figure 10.5).

1. **Fungiform papillae:** The anterior portions of the tongue bear fungiform papillae. Each thumblike projection carries a few taste buds at its tip.
2. **Foliate papillae:** The posterior lateral edge of the tongue bears ridges called foliate papillae. The sides of the papillae are studded with hundreds of taste buds.
3. **Circumvallate papillae:** The largest concentration of taste buds is found on buttonlike circumvallate papillae. They are located in a line across the back of the tongue.

G. Neural pathways

Taste buds are innervated by three different cranial nerves (CNs). The tongue's anterior portions and the palate are innervated by the facial nerve (CN VII). Taste buds on the posterior tongue signal via the glossopharyngeal nerve (CN IX), whereas the vagus (CN X) innervates the pharynx and larynx. All three nerves relay information via the tractus solitarius to a gustatory area within the solitary nucleus (brainstem). Secondary fibers carry gustatory information to the thalamus and primary gustatory cortex.

[1]More information on these drugs can be found in *LIR Pharmacology*, 5e, p. 442.

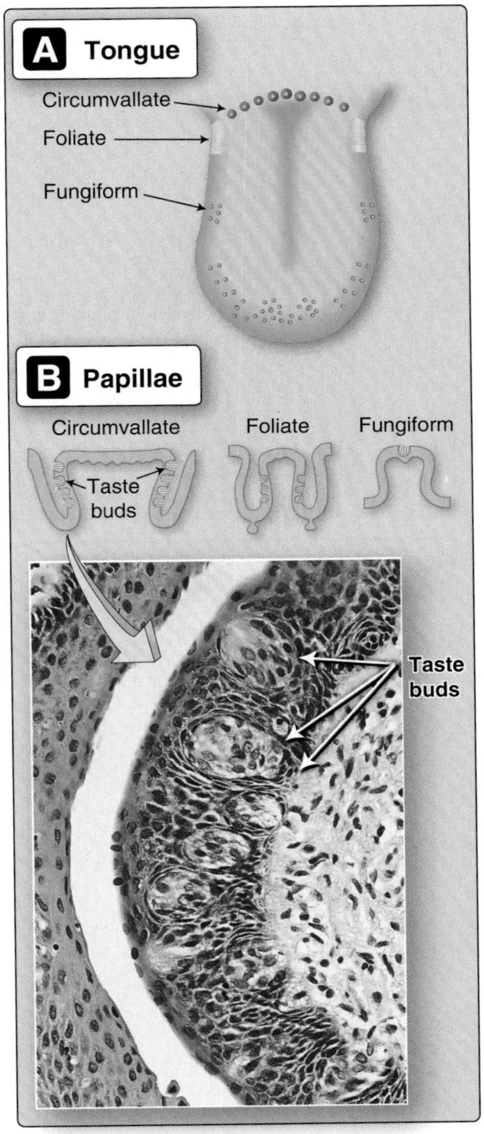

Figure 10.5
Taste bud distribution on the tongue surface.

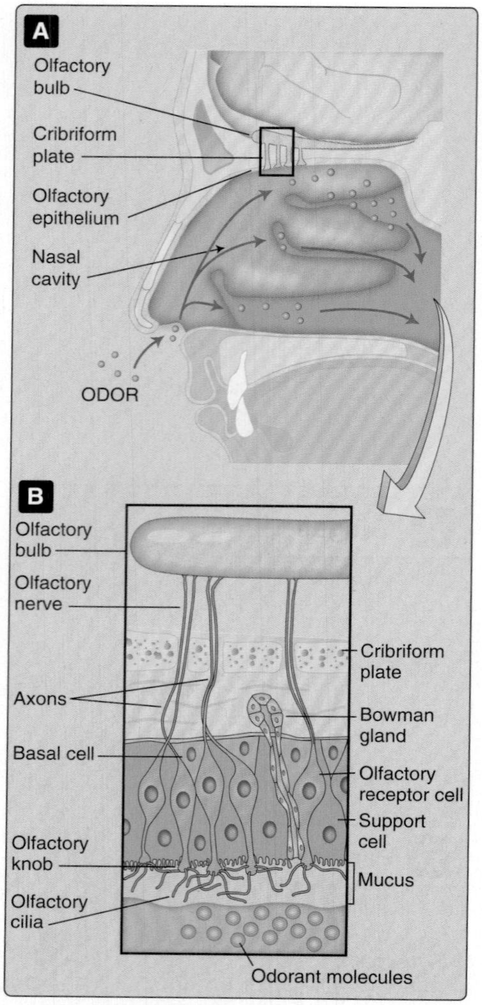

Figure 10.6
Olfactory epithelium.

III. SMELL

The human sense of smell is not as well developed as that in many animals, but, even so, the human olfactory system is capable of distinguishing hundreds of thousands of different odors. Olfactory sensitivity relies on several hundred unique receptors, each encoded by a different gene.

A. Receptors

Odorants are airborne chemicals that are inhaled and carried through the nasal passages during normal breathing or by intentional sniffing. Odors are detected and transduced by chemoreceptors that belong to the GPCR superfamily. The human genome contains ~900 different olfactory receptor genes, of which ~390 are functionally expressed.

B. Receptor cells

Odorant receptors are expressed on cilia that project from sensory neurons contained within a specialized olfactory epithelium that lines the roof of the nasal cavity (Figure 10.6). The sensory neurons are bipolar. The apical portion of the cell body gives rise to a single dendrite that extends toward the epithelial surface and then terminates in a swelling (**olfactory knob**). Each knob supports 10–30 long, nonmotile, sensory cilia which project into a thin layer of watery mucus. The mucus traps passing odorant molecules and allows them to be detected by odorant receptors.

C. Transduction

Odorant binding to its receptor activates an olfactory-specific G protein (G_{olf}) and initiates an *adenylyl cyclase*–mediated rise in intracellular cAMP concentration (Figure 10.7). A cAMP-gated ion channel opens as a result, allowing Na^+ and Ca^{2+} influx. The Ca^{2+} flux, in turn, opens a Ca^{2+}-dependent Cl^- channel, and the combined depolarizing effect of cation influx and anion efflux on membrane potential may be sufficient to trigger a spike in the olfactory neuron.

D. Olfactory epithelium

Olfactory receptor neurons have an average lifespan of ~48 days and then are replaced. Receptor neurons are formed from olfactory

Clinical Application 10.2: Anosmia

Although an inability to detect odors caused by food spoilage can increase the likelihood of food poisoning, loss of the sense of smell (**anosmia**) is not life threatening. Anosmia does significantly impact the quality of life, however. Anosmia markedly impairs food enjoyment and often causes appetite and weight loss, depression, and withdrawal from social events that involve food. Hyposmia commonly occurs during aging and as a result of upper respiratory tract infections. Neurodegenerative diseases (Parkinson and Alzheimer disease) can also impair sense of smell. Anosmia may often follow head trauma as a result of damage to the olfactory cortex or shearing of the olfactory nerves as they pass through the cribriform plate.

epithelial **basal cells**, which are neuroblast stem cells. The epithelium also contains **supporting cells**, which have a glia-like function. The mucus that flows over the epithelium is secreted by **Bowman glands**. Olfactory mucus contains odorant-binding proteins that help ferry hydrophobic odorants to the olfactory receptors. Mucus also contains lactoferrin, lysozyme, and various immunoglobulins that help ensure that pathogens do not gain access to the central nervous system (CNS) via olfactory nerves. The olfactory epithelium is one of the few regions of the body where CNS nerves directly interface with the external environment.

E. Neural pathways

Olfactory receptor cells are primary sensory neurons that project directly to the olfactory bulb, which is an extension of the forebrain. Axons from the sensory neurons form bundles and then pass through foramina in the cribriform plate of the ethmoid bone. Axons travel in the olfactory nerve (CN I) to the glomerulus of the olfactory bulb, where they synapse. Neurons originating in the olfactory bulb project to several brain regions, including the olfactory cortex, the thalamus, and the hypothalamus. Individual olfactory neurons express a single receptor gene. Because the number of odorants that a person can distinguish exceeds receptor gene number by several orders of magnitude, the receptors must recognize specific chemical groups of multiple odorant molecules rather than responding to just one odorant. The brain then extrapolates a unique odor signature based on relative output intensity from each of the different receptor types within the olfactory receptor array.

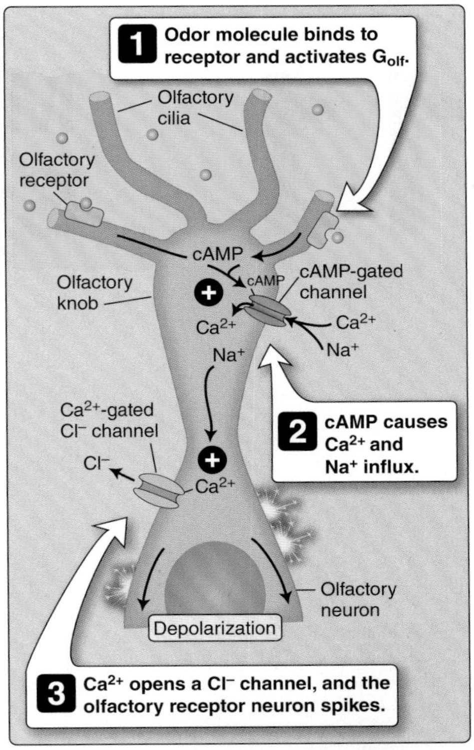

Figure 10.7
Olfactory transduction. cAMP = cyclic adenosine monophosphate; G_{olf} = olfactory-specific G protein.

Chapter Summary

- **Gustation** and **olfaction** (taste and smell) allow us to detect chemicals in food and in inhaled air. Taste is used largely to decide whether or not to swallow ingested food. Olfaction allows for an appreciation of food as well as detection of **pheromones**.
- There are five basic tastes: **salty**, **sweet**, **umami** (savory), **bitter**, and **sour**.
- **Taste receptor cells** reside in **taste buds** found throughout the oral cavity. Type I receptors are **glia-like** cells that transduce salty taste (Na^+). Na^+ excites type I cells by permeating an epithelial Na^+ channel. Type I cells also help terminate gustatory signaling.
- Type II cells express tastant-specific G protein–coupled receptors (GPCRs) that detect sweet, umami, and bitter tastes. The GPCRs act through **gustducin**, a taste cell–specific **G protein**. The activated $G_{\beta\gamma}$ subunit elicits intracellular Ca^{2+} release and depolarization. **Pannexin hemichannels** in the surface membrane then open and allow adenosine triphosphate (ATP) to diffuse out of the cell. ATP stimulates sensory afferent neurons both directly and indirectly by modulating type III cell output.
- Type III cells transduce the sour taste of acid. H^+ permeates an epithelial Na^+ channel and causes receptor depolarization. Type III cells are innervated by gustatory nerves and signal excitation to the afferents via 5-hydroxytryptamine (serotonin) release.
- **Odors** are detected by an **olfactory epithelium** located in the roof of the nasal cavity. **Olfactory receptors** are G protein–coupled receptors expressed on the surface of cilia that project into a mucus layer lining the olfactory epithelium.
- Olfactory receptor cells are bipolar central neurons that relay information to the **olfactory bulb**. Olfactory receptor binding triggers a cyclic adenosine monophosphate (cAMP)-mediated signaling cascade that leads to cAMP-dependent Ca^{2+} influx and Ca^{2+}-dependent Cl^- efflux and causes nerve excitation.

11 Motor Control Systems

I. OVERVIEW

The neural pathways that control muscle activity in humans were developed during early evolutionary history to facilitate directed locomotion. Coordination of muscle groups that move the limbs was accomplished initially using simple neural feedback loops, but, as body complexity and the difficulty of the tasks that it was required to perform increased, so too the muscle control systems evolved. The human body devotes a large percentage of the nervous system to motor control. The simple neural feedback loops were retained during evolution and now function as muscle reflexes, but these pathways have been supplemented with successively higher layers of control (Figure 11.1). Motor regions of the cerebral cortex decide when movements are necessary. Basal ganglia compile motor sequences based on learned experience and then relay these sequences through the thalamus for execution by the primary motor cortex. Control sequences ensure that pairs of muscle groups whose actions typically oppose one another (e.g., **extensors** and **flexors**, **abductors** and **adductors**, and **external** and **internal rotators**) contract and relax in a coordinated fashion to effect smooth limb movements. Motor commands are refined even as they are being executed by the cerebellum, which receives streams of sensory data from muscles, joints, skin, eyes, and the vestibular system. The cerebellum allows the cortex to compensate for unexpected changes in terrain, posture, and limb position.

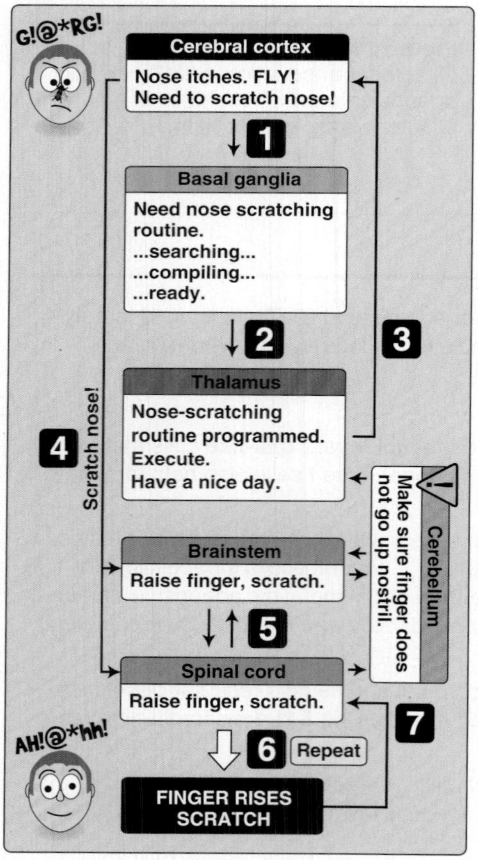

Figure 11.1
Layering of motor control pathways.

II. MUSCLE SENSORY SYSTEMS

Complex motor behaviors (such as walking) require closely coordinated sequences of muscular contractions whose timing and strength is modified constantly during changes in body position and weight distribution. Such coordination is not possible unless the central nervous system (CNS) is informed about limb movements relative to the torso, made possible by the sense of **kinesthesia**. Kinesthesia is a form of **proprioception** and one of the **somatic senses**. Kinesthesia relies primarily on two sensory systems that sense muscle length (**muscle spindles**) and tension (**Golgi tendon organs** [**GTOs**]).

A. Muscle spindles

Skeletal muscle comprises two fiber types. **Extrafusal fibers** (derived from the Latin *fusus* for "spindle") generate the force needed

to move bones. **Intrafusal fibers** are sensory, monitoring muscle length and changes in length. Intrafusal fibers are contained within discrete sensory structures (spindles) that are distributed randomly throughout the muscle body (Figure 11.2).

1. **Structure:** Muscle spindles contain up to 12 intrafusal fibers enclosed within a connective tissue capsule. Each intrafusal fiber comprises a noncontractile portion centered between two weakly contractile regions. The spindles nestle between contractile fibers and are anchored at either end so that contractile and sensory fibers move as one unit. Spindles contain two types of intrafusal fiber: **nuclear bag fibers** and **nuclear chain fibers** (see Figure 11.2). Nuclear bag fibers swell centrally to form a "bag" containing numerous clustered nuclei. Nuclear chain fibers are thinner and more numerous than nuclear bag fibers. Their nuclei form a chain down the fiber's length.

2. **Sensory transduction:** Muscle spindles signal via two types of sensory nerve afferents (**group Ia** and **group II**). Both classes have wide, myelinated axons to maximize signal conduction velocity (Table 11.1; also see 5·III·B). When a muscle is stretched (e.g., by limb extension), the intrafusal fibers are stretched also, causing distortion of the nerves that wrap around them. Stretching activates mechanosensitive cation channels, resulting in depolarization and increased afferent nerve firing frequency.

   a. **Group Ia:** Group Ia fibers coil around the central (equatorial) regions of both nuclear bag and nuclear chain fibers to form **primary muscle spindle receptors**. They yield a **dynamic response** to stretching (Figure 11.3). Type Ia afferents show maximal firing rates when the muscle fibers (and nerve endings) are being stretched actively. Firing rate decreases when the muscle reaches and maintains a new length.

   b. **Group II:** Group II fiber endings are located at the ends of nuclear chain fibers and some nuclear bag fibers (see Figure 11.2). They form **secondary muscle spindle receptors** that yield a **static response** to stretching. Their output is proportional to muscle length, and the nerve fibers continue firing at increased rates if the muscle is held at the new length (see Figure 11.3).

3. **Regulation:** Intrafusal fibers are contractile, but they do not contribute significantly to muscle force development. Instead, the contractile portions serve only to shorten the fiber during muscle excitation and keep the central, sensory portion taut even as the muscle contracts. Maintaining tension allows the intrafusal fibers to continue functioning as stretch sensors throughout the contraction. Intrafusal fibers are innervated by **γ-motor neurons**, which conduct more slowly than the α-motor neurons that stimulate extrafusal muscle contraction (see Table 11.1). The α- and γ-motor neurons fire simultaneously, so that the spindle shortens in parallel with the body of the muscle when the muscle contracts. The combination of muscle spindles and their associated γ-motor neurons constitutes a **fusimotor system**.

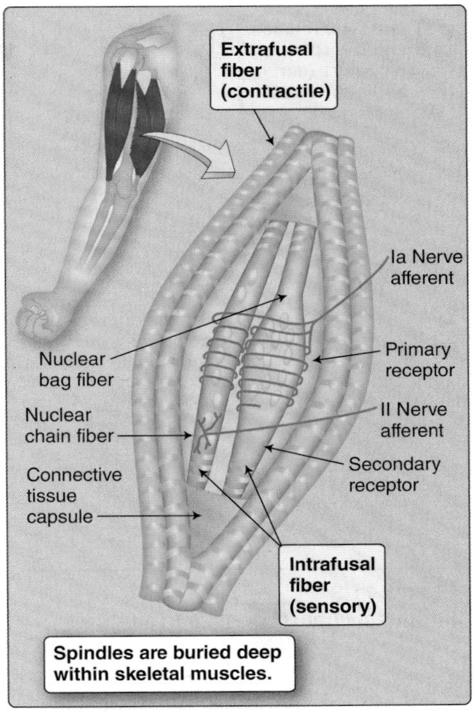

Figure 11.2
Muscle spindle.

Table 11.1: Muscle Nerve Fiber Properties

Class	Innervation	Conduction Velocity (m/s)
Ia Sensory	Muscle spindle (primary endings)	80–120
Ib Sensory	Golgi tendon organs	80–120
II Sensory	Muscle spindle (secondary endings)	35–75
α Motor	Extrafusal muscle fibers	80–120
γ Motor	Intrafusal muscle fibers	15–30

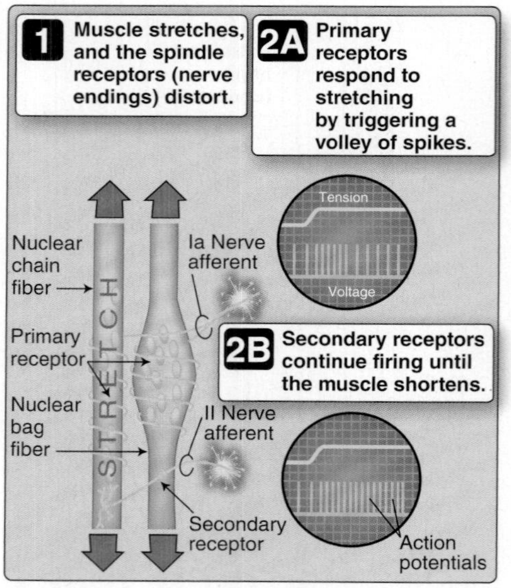

Figure 11.3
Intrafusal fiber responses to stretch.

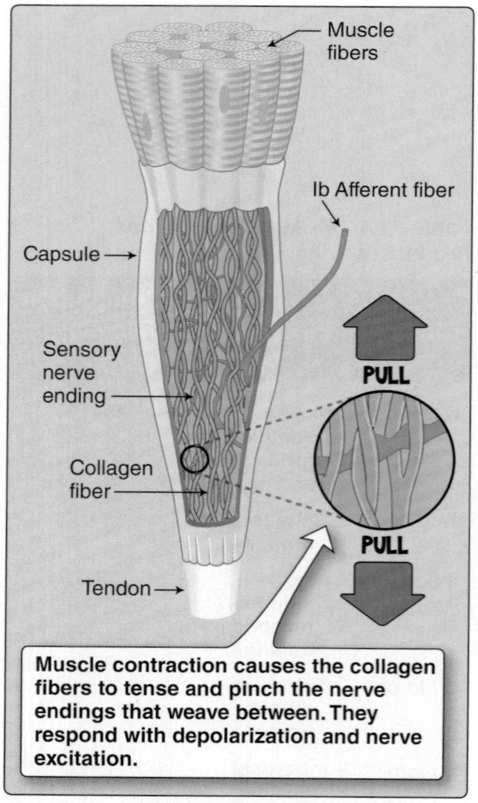

Figure 11.4
Golgi tendon organ.

B. Golgi tendon organs

Each end of a skeletal muscle is attached to a tendon that typically tethers it to a bone. The musculotendinous junction contains GTOs, which are sensory organs that monitor the amount of tension that develops in a muscle when stretched passively or when it contracts (Figure 11.4).

1. **Structure:** GTOs are situated at the junction between a skeletal muscle and a tendon. GTOs comprise a connective tissue capsule filled with collagen fibers that are interwoven with **group Ib sensory nerve endings**. Type Ib nerve afferents are myelinated to increase signal conduction rates (see Table 11.1).
2. **Sensory transduction:** When a muscle is stretched or contracts, the associated GTO is stretched also. The collagen fibers within the GTO tighten and compress the nerve endings that weave between them. Compression opens mechanosensitive channels in the nerve endings, causing depolarization and increasing nerve-firing rates.

C. Joints and skin sensors

Although joints between bones would intuitively seem to be ideal sites for locating receptors that report on limb position, in practice, the role of joint receptors in kinesthesia is minimal. Slow-adapting Ruffini endings in skin do have an important role, however (see 16·VII·A). Skin that covers joints is stretched whenever a limb or digit is retracted, causing Ruffini endings to fire. The importance of sensory data from Ruffini endings is increased in the fingers, where layering of the various muscles and tendons required for execution of fine movements may impede acquisition of sensory information from spindles and GTOs.

III. SPINAL CORD REFLEXES

Walking with an upright gait is, quite literally, a delicate balancing act. A misplaced foot or slight irregularity of terrain can easily upset the balance and precipitate a fall. Avoiding such mishaps requires that a fall be anticipated and an appropriate correction to gait be made with minimal delay. Motor sensory and control neurons are specialized to conduct signals at up to 120 m/s, representing some of the fastest nerve cells in the body (see Table 11.1). This ensures that sensory information is relayed to the CNS and compensatory commands executed in the shortest time possible. Reaction times are enhanced further by using local reflexes mediated by the spinal cord to make many routine adjustments to gait. The neurons involved are relatively short so as to further minimize signal transmission and processing times.

A. Reflex arcs

Reflex arcs are simple neuronal circuits in which a sensory stimulus initiates a motor response directly. Classic examples include withdrawal reflexes triggered by touching a hot stove or stepping on a sharp object. Such arcs are often mediated by the spinal cord, where a sensory neuron synapses with and activates a motor neuron. More complex arcs involve synapses with multiple neurons, at least one of which may be inhibitory. The spinal cord mediates a number of impor-

tant reflex arcs, including the **myotatic reflex**, the **inverse myotatic reflex**, and the **flexion reflex**.

B. Myotatic

A myotatic reflex (also known as a **stretch reflex** or **deep-tendon reflex**) is initiated by stretching a muscle and causes contraction of the same ("homonymous") muscle. Reflex contraction of the thigh (quadriceps) muscles caused by tapping the **patellar ligament** is a familiar example (Figure 11.5).

1. **Response:** Tapping the patellar ligament stretches the quadriceps and activates spindles buried within. Sensory signals are carried by Ia nerve afferents to the spinal cord, where they synapse with and excite α-motor neurons innervating the same muscle. The muscle contracts reflexively, the leg extends, and the foot jerks forward. The myotatic reflex is designed to resist inappropriate changes in muscle length and is important for maintaining posture.

2. **Reciprocal innervation:** Forward foot movement stretches the hamstring muscles at the back of the thigh and stimulates their spindles also. This might be expected to initiate a second reflex that opposes the actions of the first, but the arc is interrupted by a Ia inhibitory spinal interneuron. The Ia interneuron is activated by the same Ia afferent signal that caused the quadriceps to contract. The interneuron synapses with and inhibits the α-motor neurons that innervate the hamstring muscles (e.g., semitendinosus) and, thereby, allows the leg to extend without resistance. This circuitry is referred to as **reciprocal innervation** and is used commonly in situations in which two or more sets of muscles oppose each other around a joint (e.g., flexors and extensors).

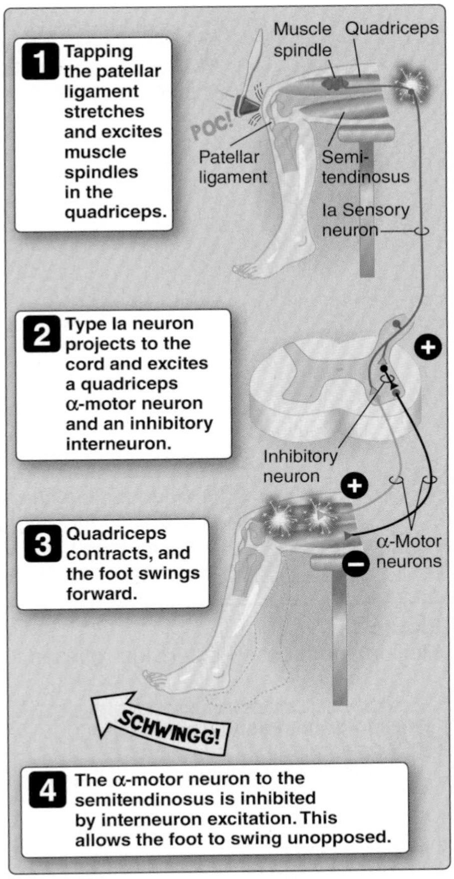

Figure 11.5
Myotatic reflex.

C. Inverse myotatic

The inverse myotatic reflex, also known as a **Golgi tendon reflex**, activates whenever a muscle contracts and GTOs are stretched (Figure 11.6). Group Ib afferents from GTOs synapse with Ib inhibitory interneurons upon entering the spinal cord. When activated, they inhibit α-motor output to the homonymous muscle. Excitatory interneurons simultaneously activate α-motor output to the heteronymous muscle. The Golgi tendon reflex is believed to be important for fine motor control and for maintaining posture, acting synergistically with the myotatic reflex above.

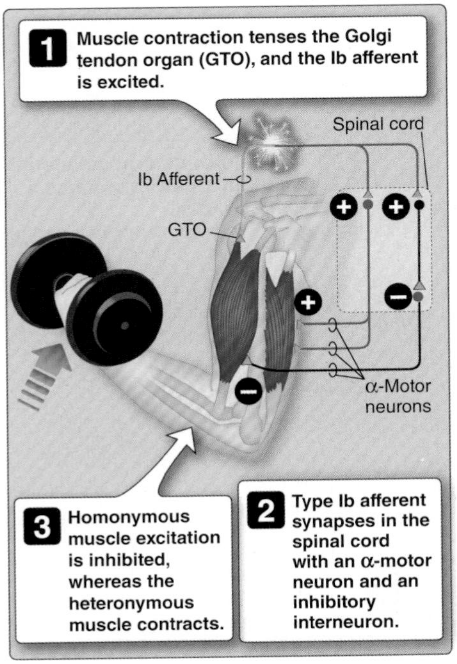

1 Muscle contraction tenses the Golgi tendon organ (GTO), and the Ib afferent is excited.
Spinal cord
Ib Afferent
GTO
α-Motor neurons
3 Homonymous muscle excitation is inhibited, whereas the heteronymous muscle contracts.
2 Type Ib afferent synapses in the spinal cord with an α-motor neuron and an inhibitory interneuron.

Figure 11.6
Inverse myotatic reflex (Golgi tendon reflex).

D. Flexion and crossed-extension

Stepping on a thorn or other injurious object precipitates two urgent actions. The first withdraws the foot from the source of pain (leg flexion). The second braces the opposing limb so that weight can be transferred while still maintaining balance. This complex motion is mediated by **flexion** and **crossed-extension reflexes** (Figure 11.7). Similar reflexes can be induced in the upper limbs. The action sequence can be broken down into three stages: stimulus sensation, wounded (ipsilateral) limb flexion, and then extension of the opposing (contralateral) limb.

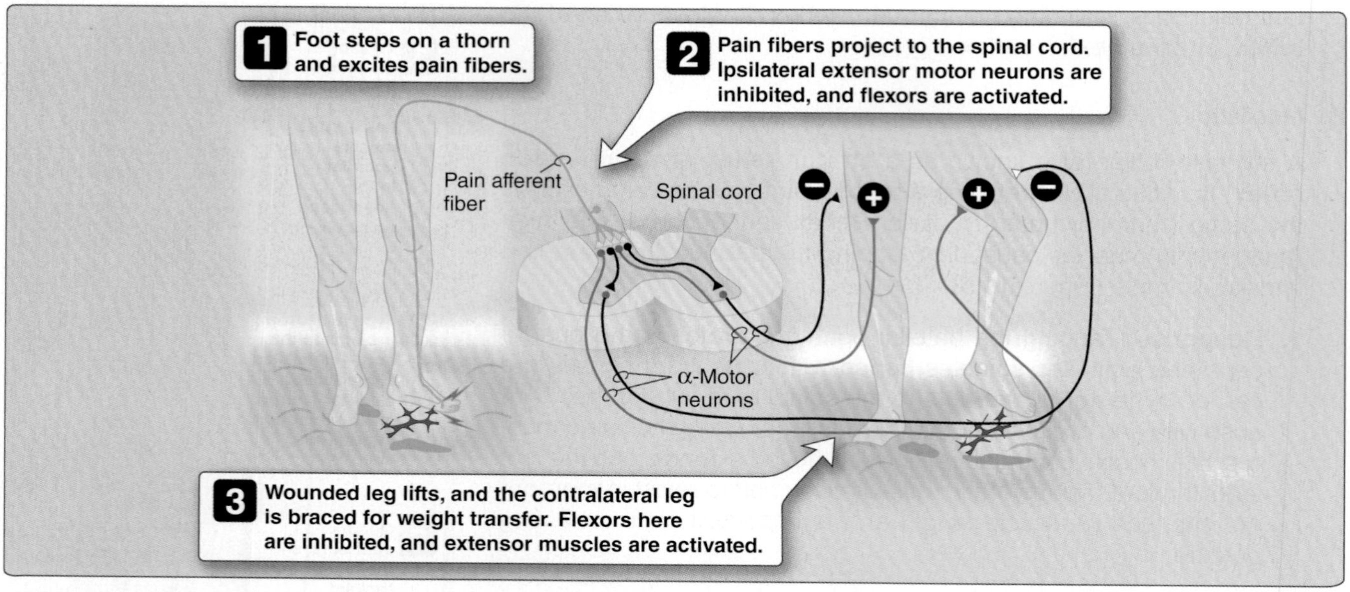

Figure 11.7
Flexion and crossed-extension reflexes.

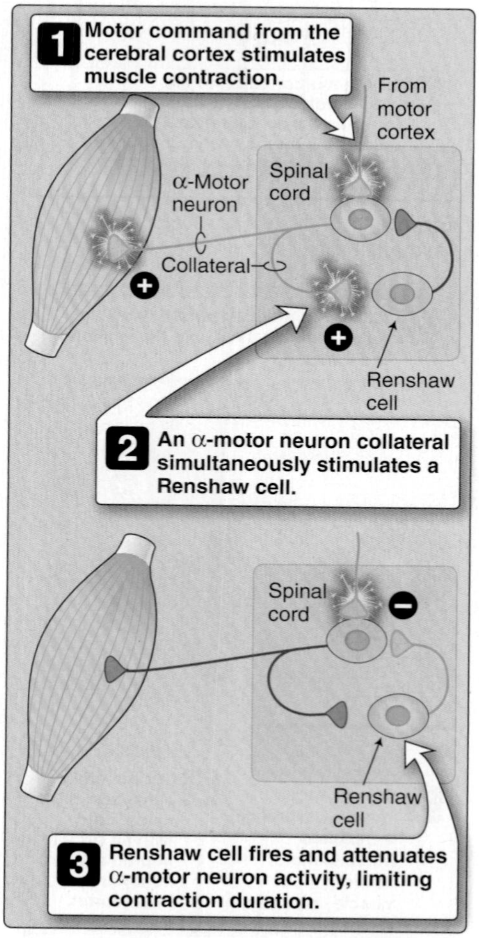

Figure 11.8
Renshaw cells.

1. **Sensation:** Flexion and crossed-extension reflexes are usually initiated as a response to a noxious, painful stimulus. Pain fibers project to and synapse with interneurons in the spinal cord.
2. **Flexion:** The sensory afferents synapse on the ipsilateral side with excitatory motor neurons that innervate flexor muscles. Extensor muscles are inhibited simultaneously, and the limb retracts from the pain source.
3. **Crossed-extension:** Sensory fibers also cross the spinal cord's anterior fissure and synapse with motor neurons controlling contralateral limb movement. Extensors are excited and contract, whereas flexors are inhibited and relax. This is known as a **cross-extension reflex** and braces the contralateral limb for the sudden weight transfer caused by raising the wounded limb.

E. Renshaw cells

Intensely painful stimuli trigger a volley of spikes in the pain afferents that potentially could cause dependent flexor muscles to become tetanic if the reflex circuit were unregulated. Regulation comes in the form of **Renshaw cells**, which are a special class of spinal inhibitory interneuron that are excited by α-motor neuron collaterals (Figure 11.8). Renshaw cells fire whenever a muscle receives a command to contract, but they project back to and inhibit the same α-motor neuron that excited them. Renshaw cells can cause inhibition lasting tens of seconds. Their activity level is tied to that of the motor neuron, so the more intense the command to contract, the greater the degree of α-motor neuron inhibition. Renshaw cells also receive modulating inputs from higher motor control centers that allow for sustained voluntary contractions. They also project to motor neurons innervating opposing and associated muscle groups. These relationships enhance fluidity of limb movement.

F. Central pattern generators

Motor systems execute many repetitive behaviors, such as those associated with locomotion (walking, running, swimming), grooming, bladder control, ejaculation, eating (chewing, swallowing), and breathing (movement of the chest wall and diaphragm). These behaviors do not require conscious thought, although they can be modified by higher control centers. The rhythmic behaviors are established by neuronal circuits known as **central pattern generators** (**CPGs**). CPGs are found in many CNS areas, including the spinal cord. The only requirement is an excitable cell or cell cluster with intrinsic pacemaker activity (e.g., two neurons that sequentially excite each other) and dependent circuits of interconnected neurons that control motor neurons. Walking, for example, involves a repetitive set of motor commands that move one leg forward, shift weight, and then extend the opposite leg. The motor commands and movements are sequential and predictable. Speeding or slowing the pace requires simple adjustments to CPG timing.

IV. HIGHER CONTROL CENTERS

Spinal cord reflexes and CPGs establish stereotyped behaviors, but planning and recalling learned movements requires higher levels of control.

Clinical Application 11.1: Spinal Cord Injury

Tens of thousands of individuals in the United States suffer traumatic spinal cord injury (SCI) every year, most often as a result of a motor vehicle accident, a fall, or violence. Injuries to the spinal cord proper are usually caused by damage to the vertebral column or supporting ligaments, including fractures, dislocation, or disruption or herniation of an intervertebral disk. Acute SCI is often followed by a period of **spinal shock** that lasts 2–6 weeks, characterized by a complete loss of physiologic function caudal to the level of injury. This includes flaccid paralysis of all muscles, absence of tendon reflexes, and loss of bladder and bowel control. Males may often develop priapism. The mechanisms underlying shock are still under investigation.

The extent of SCI is described as being **complete** or **incomplete**. Cord transection causes complete SCI, characterized by a complete loss of sensory and motor function caudal to the trauma site. Incomplete SCI describes injuries in which some degree of sensory or motor function is preserved.

Even in cases of complete SCI, some reflex pathways may recover in time. Because these pathways are now disconnected from higher motor control centers, they can cause inappropriate movements. For example, sudden flexing of the ankle or wrist may trigger prolonged rhythmic contractions caused by unregulated Golgi tendon organ–mediated reflex loops (**clonus**).

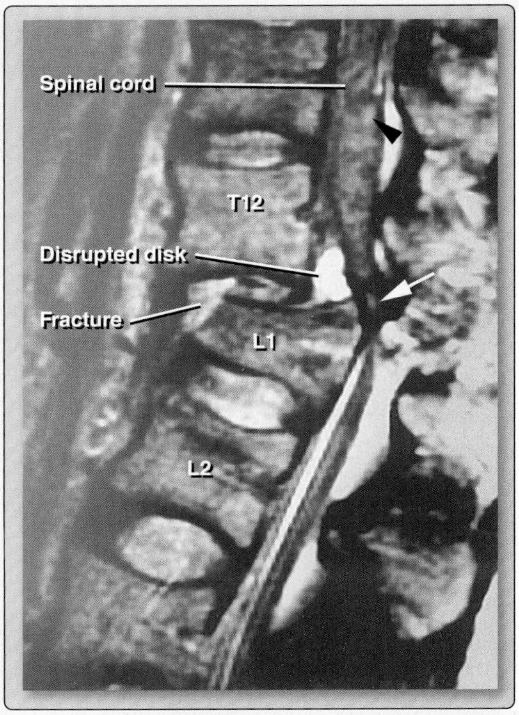

Spinal cord compression (*white arrow*) and hemorrhage (*black arrowhead*) caused by L1 vertebral body fracture and dislocation.

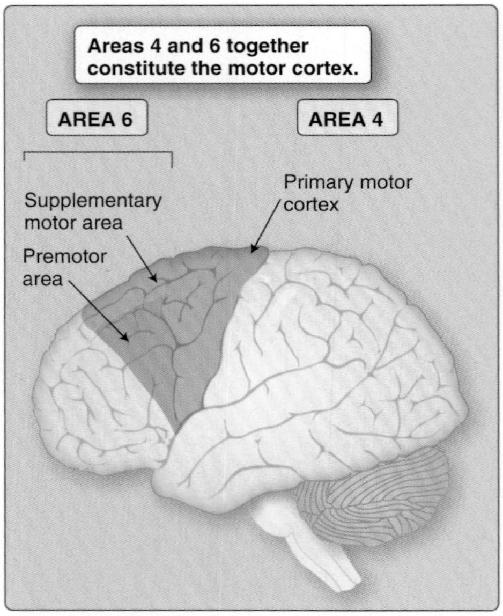

Figure 11.9
Motor cortex organization.

These are added in layers, with each successive layer providing for more sophisticated and finer degrees of motor control (see Figure 11.1).[1] As is the case for most organ systems, the greatest insights into function come from observing what happens when these pathways are damaged.

A. Cerebral cortex

The cerebral cortex is responsible for planning voluntary motor commands. Many different cortical regions are involved in coordinating motor activities, but the most important are in area 4, the **primary motor cortex**, and area 6, which contains the **premotor cortex** and the **supplementary motor area** (Figure 11.9).

1. **Primary motor cortex:** The primary motor cortex sends motor fibers via the **corticospinal tract** to the spinal interneurons that ultimately cause muscle contraction. Commands are executed only after extensive processing by the cerebellum and basal ganglia. Their execution also takes into account information being received simultaneously from various skin and muscle proprioceptors.
2. **Premotor cortex:** The premotor cortex may be responsible for planning movements based on visual and other sensory cues.
3. **Supplementary motor area:** The supplementary motor area retrieves and coordinates memorized motor sequences such as those required for playing the piano.

B. Basal ganglia

The cortex makes decisions about when to move and what tasks need to be accomplished, but execution requires careful planning about the timing of contractile events, the distance that limbs and digits need to move, and the force that needs to be applied. Thus, the sequence of movements required to apply fine paint strokes to a watercolor portrait are very different from those required to apply exterior paint in broad strokes to the side of a house. The task of planning and executing motor commands falls on the basal ganglia. These areas are not absolutely required for motor function, but movements become grossly distorted and erratic if they are damaged.

1. **Structure:** The basal ganglia comprise a group of large nuclei that are located at the base of the cortex in close proximity to the thalamus (Figure 11.10). They work together as a functional unit. The nuclei receive motor commands from the cortex, pass them through a series of feedback loops, and then relay them to the thalamus for return to, and execution by, the primary motor cortex.
   a. **Striatum:** The **striatum** (or **neostriatum**) comprises two nuclei: the **putamen** and the **caudate nucleus**. The striatum is the gateway through which commands from the cortex enter the nuclear complex. The striatum is dominated by GABAergic neurons, and its output is largely inhibitory.

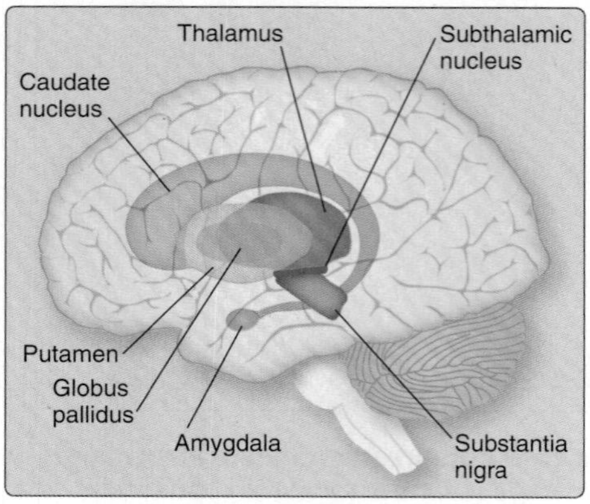

Figure 11.10
Basal ganglia.

INFO LINK

[1]A comprehensive description of the anatomical pathways, structures, and mechanisms involved is beyond the purview of this text but are considered in greater detail in *LIR Neuroscience*.

b. **Globus pallidus:** The **globus pallidus** ([**GP**] also known as the **palladium**) can be divided into two regions (**internal** [**IGP**] and **external** [**EGP**]) based on function. The palladium is composed of inhibitory GABAergic neurons.

c. **Substantia nigra:** The **substantia nigra** (Latin for "black substance") is filled with melanin, a dark pigment that serves as a substrate for **dopamine** formation. Functionally, it can be divided into two areas: the **pars reticulata** and the **pars compacta**. Both regions contain inhibitory neurons. The pars reticulata is primarily GABAergic, whereas the pars compacta contains dopaminergic neurons. Because the pars reticulata and IGP often function together and have a similar anatomical structure, they are often considered as a single functional unit.

d. **Subthalamic nucleus:** The **subthalamic nucleus** is a part of the subthalamus. It is a key link in a basal-nuclei feedback circuit and is the only primarily excitatory (glutamatergic) center within the basal ganglia.

2. **Feedback circuits:** The motor cortex communicates its intent to the striatum. There are two pathways for information flow from the striatum through the basal nuclei: a **direct path** and an **indirect path**. Both end at the thalamus, which is tonically active and stimulates cortical areas that ultimately control the musculature (Figure 11.11).

a. **Direct path:** When the striatum is activated, it inhibits IGP–pars reticulata complex output. These two nuclei are tonically active normally, and their output suppresses the thalamus' tonic output to the motor cortex. Thus, activating the striatum allows the thalamus to stimulate the motor cortex. In practice, the direct path increases motor activity.

b. **Indirect path:** A second pathway involves the EGP and the subthalamic nucleus. Exciting the striatum prevents the EGP from signaling. The EGP normally inhibits the subthalamic nucleus, which would otherwise be tonically active and, therefore, increasing the activity of the IGP. The IGP inhibits the thalamus and prevents it from exciting the motor cortex. In practice, the indirect pathway decreases motor activity.

c. **Biasing output:** When the striatum receives a motor command, the direct and indirect pathways are activated simultaneously, and their effects on the IGP are conflicting and balanced. Any influence that changes this balance might be used to regulate motor output. The substantia nigra pars compacta can potentially have a major influence over motor output because it sends dopaminergic axons back to two areas of the striatum. When active, these neurons increase the activity of the direct pathway via an excitatory dopaminergic (D_1) receptor while simultaneously suppressing the indirect pathway via a dopamine D_2 receptor (see Table 5.2). Both effects favor increased motor activity.

3. **Diseases affecting the basal ganglia:** The balance that exists between the direct and indirect pathways is delicate. Disrupting even a single circuit component can thus have devastating motor consequences. These can include a slowing of movement (**bradykinesia**)

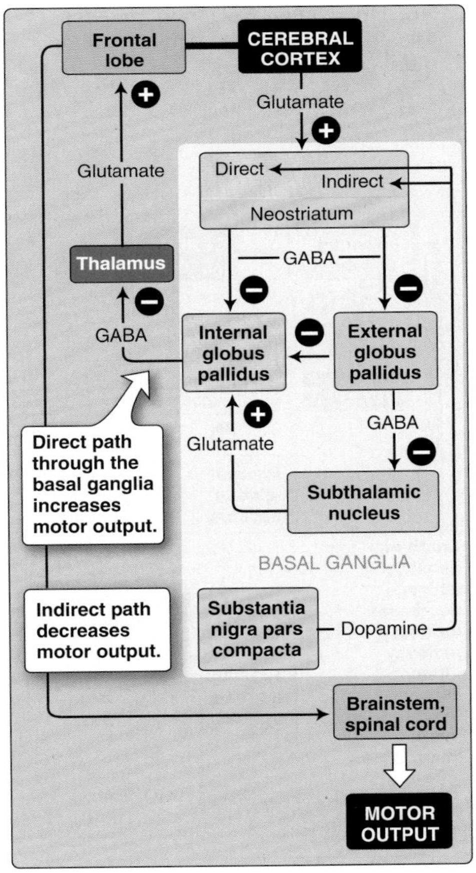

Figure 11.11
Motor function relationships between the basal ganglia. GABA = γ-aminobutyric acid.

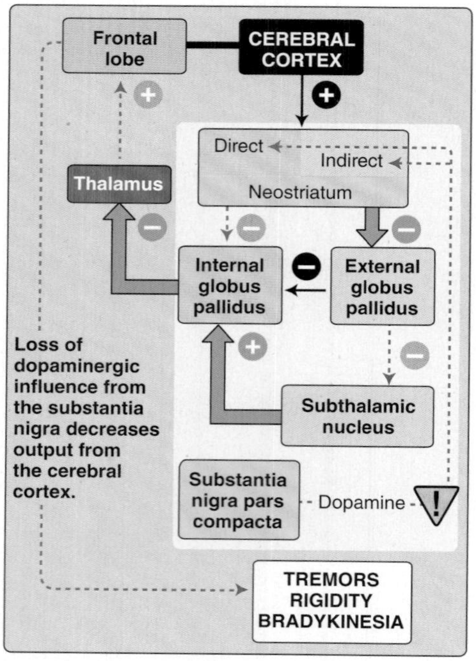

Figure 11.12
Parkinson disease. *Dashed red lines* indicate pathways with diminished influence.

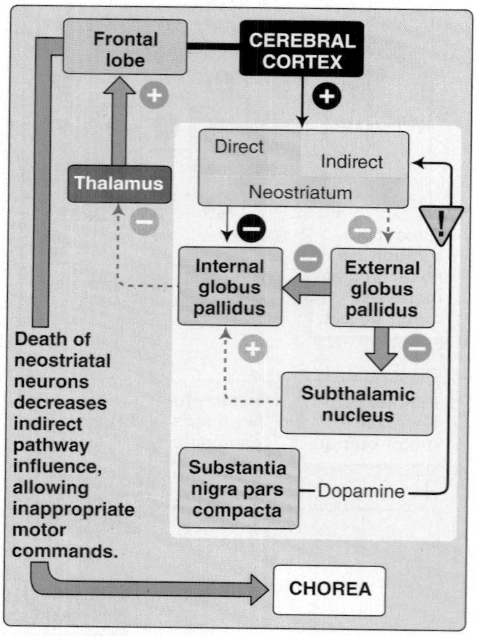

Figure 11.13
Huntington disease. *Dashed red lines* indicate pathways with diminished influence.

or a complete loss of motor control (**akinesia**), rigidity due to increased muscle tone (**hypertonia**), and involuntary writhing movements at rest (**dyskinesia**). The best-studied motor disorders are **Parkinson disease** (**PD**) and **Huntington disease** (**HD**).

a. **Parkinson disease:** The motor disturbances associated with PD result from death of large numbers of dopaminergic neurons within the pars compacta (Figure 11.12). Loss of these neurons causes resting arm and hand tremors; increased muscle tone and limb rigidity; bradykinesia; and, in the later stages, postural instability. Patients also have a slow, shuffling gait. These symptoms all reflect the consequences of losing the dopaminergic feedback loop between the pars compacta and striatum. Directed movement becomes difficult, and the inherent conflicts between direct and indirect pathways become evident. Treatment options currently include drugs that raise dopamine levels, either by providing a substrate for dopamine formation (L-dopa) or by inhibiting its breakdown (*monoamine oxidase* inhibitors[1]; see Figure 5.7 and Table 5.3).

> Tremors are the most common form of movement disorder. A tremor is a rhythmic body movement reflecting an imbalance between the actions of two antagonistic muscle groups. All individuals show physiologic tremors that may be exaggerated by physical stress; hunger; caffeine; and many classes of drugs affecting dopaminergic, adrenergic, and cholinergic neurotransmission.

b. **Huntington disease:** HD is a hereditary disorder affecting **huntingtin**, a ubiquitous protein whose normal function is still not understood fully. Striatal neurons that normally inhibit motor output via the indirect pathway are destroyed due to abnormal protein accumulations, removing the normal constraints on the direct pathway (Figure 11.13). Early disease symptoms include chorea (from the Greek word for "dance"), characterized by involuntary limb muscle contractions that produce abrupt jerking and writhing motions. HD ultimately involves most brain regions, causing severe psychiatric disturbances and dementia. There is no treatment option, and death usually follows diagnosis by ~20 years.

C. Cerebellum

The cerebellum is not essential for locomotion, but it is intimately involved in motor control. It verifies that instructions issued by the cortex are executed as intended and makes corrections as necessary.

INFO LINK [1]For more information on drugs used to treat Parkinson and other neurodegenerative diseases, see *LIR Pharmacology*, 5e, p. 99.

1. **Function:** The full extent of cerebellar involvement in motor control is not known, but principal functions include fine-tuning and smooth execution of movements.

    a. **Fine-tuning:** The cerebellum receives extensive sensory information about body and head position, muscle contractility and length, and tactile information from the skin. It then compares this information with the motor commands that were issued by the higher centers and makes fine motor adjustments as necessary. This prevents a finger from overshooting its target when reaching out to flip a light switch, for example.

    b. **Sequencing:** Activities such as playing the piano involve finger movements that are executed so fast that there is insufficient time for sensory information to be relayed back to the CNS for processing and feedback. Such activities are only possible because the cerebellum anticipates when a particular movement should end and then executes a command that arrests it at a precise moment in time. It simultaneously anticipates and executes a command that ensures a smooth transition to the next motion.

    c. **Motor learning:** The cerebellum is able to anticipate and execute motor commands because it stores and constantly updates information about the correct timing of commands required for complex motor sequences.

2. **Cerebellar dysfunction:** The cerebellum finesses motor commands but is not absolutely required for locomotion. Cerebellar lesions cause varying degrees of coordination loss, depending on the lesion's site and severity.

    a. **Ataxia: Ataxia** refers to a general lack of muscular coordination. Gait may become slow, wide, and swaying. The conscious brain is now forced to think about body position, but the time lag between receipt of proprioceptive information and execution of motor commands means that limbs usually overreach and miss their intended targets. The brain then executes a poorly controlled compensatory movement, resulting in a behavior pattern known as **dysmetria**.

    b. **Intention tremors:** Because the conscious brain has to guide and continually update movements, simple tasks, such as reaching for an object, become slow, and the path taken to the target meanders from side to side (an **intention tremor**). Intention tremors are readily discerned using a simple finger-to-nose test (Figure 11.14).

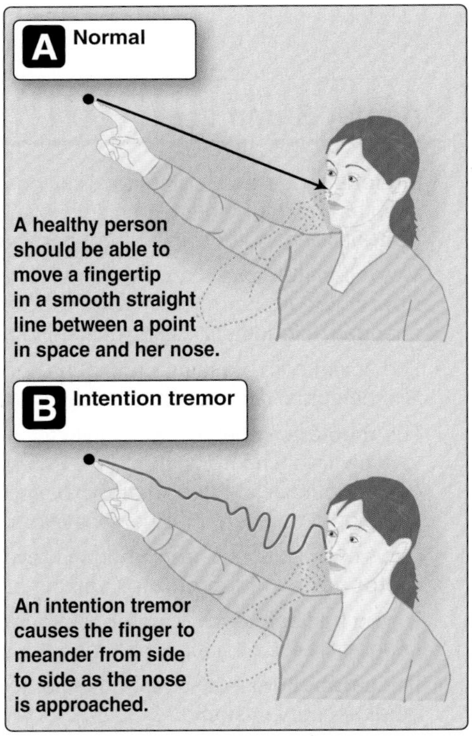

A Normal
A healthy person should be able to move a fingertip in a smooth straight line between a point in space and her nose.
B Intention tremor
An intention tremor causes the finger to meander from side to side as the nose is approached.

Figure 11.14
Intention tremor.

D. Brainstem

Simple neuronal circuits in the spinal cord produce stereotypical behaviors that facilitate walking and other rhythmic movements. The brainstem sets these pathways in motion and coordinates them with reference to sensory information received from the eyes and vestibular system. It also controls eye movement to stabilize visual images during head and body movement. The brainstem contains four important motor control areas: the **superior colliculus** (**tectum**),

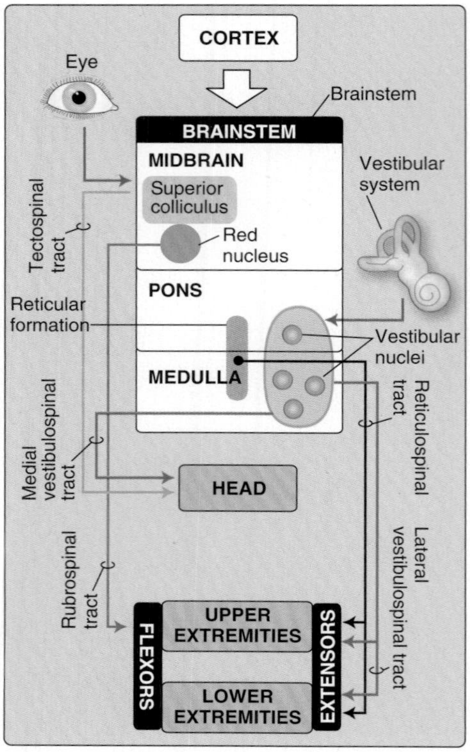

Figure 11.15
Brainstem motor control centers.

the **red nucleus**, the **vestibular nuclei**, and the **reticular formation** (Figure 11.15).

1. **Superior colliculus:** The superior colliculus controls head and neck movements with reference to visual information. Fibers from this area project to the cervical spine via the **tectospinal tract**.
2. **Red nucleus:** The **red nucleus** is located in the midbrain. The red nucleus controls flexor muscles in the upper limbs via the **rubrospinal tract**.
3. **Vestibular nuclei:** There are four **vestibular nuclei**: one in the pons (**superior vestibular nucleus**) and three in the medulla (the **medial**, **lateral**, and **inferior vestibular nuclei**). They receive and integrate information from the inner ear about head and body motion. Output from these regions controls eye movement via the **oculomotor nerve** (cranial nerve III). They also help coordinate head, neck, and body movements via the **vestibulospinal tracts**. The medial vestibulospinal tract arises in the medial vestibular nucleus, which helps stabilize the head during body movements. The lateral vestibulospinal tract projects to all levels of the spinal cord, where it stimulates extensor muscle contraction and inhibits flexors to help control posture during body movements.
4. **Reticular formation:** The **reticular formation** is also involved in many complex motor behaviors. Fibers arising from this area project via the **reticulospinal tract** to all levels of the spinal cord. They influence the activity of both α- and γ-motor neurons to facilitate voluntary body movements originating in the cortex and initiated via the corticospinal tract.

Chapter Summary

- **Skeletal muscles** facilitate locomotion and manipulation of the external environment. Executing complex movements, such as walking, requires multiple levels of coordination, involving the **spinal cord**, **brainstem**, **cerebellum**, **basal ganglia**, and motor areas of the **cerebral cortex**.
- Motor control centers receive sensory data from specialized myofibrils contained within skeletal muscles (**intrafusal fibers**) and from tendons (**Golgi tendon organs** [**GTOs**]). Intrafusal fibers relay information about muscle length and changes in length. GTOs are tension sensors.
- The spinal cord contains **central pattern generators** that sustain rhythmic limb movements during walking, for example. Simple spinal circuits allow for rapid reflex responses to noxious stimuli and unanticipated changes in muscle length.
- The **myotatic reflex** causes muscles to contract when stretched while simultaneously inhibiting opposing muscles to allow free limb movement. The **inverse myotatic reflex** limits muscle contraction and simultaneously activates an opposing muscle. **Flexion** and **crossed-extension reflexes** prepare opposing limbs to brace for transfer of weight when stepping on a sharp or otherwise injurious object, for example.
- Decisions about how and when to move begin in the cortex. Principal motor areas of the cortex include the **primary motor cortex**, the **premotor cortex**, and the **supplementary motor area**.
- Timing and sequencing motor commands is the responsibility of the **basal ganglia**. Motor commands are subjected to a series of feedback loops that hone the sequences and ensure accurate and smooth movements.
- The **cerebellum** fine-tunes movements during execution, referencing information being received from proprioceptors and other sensory systems.
- The **brainstem** executes motor commands and helps coordinate movements with reference to sensory data being relayed from the eyes and vestibular system.

Study Questions

Choose the ONE best answer.

II.1 Autoimmune diseases such as multiple sclerosis cause neurological impairment by affecting axon conduction velocity. Which of the following would slow axonal signal propagation to the greatest extent?

A. Increasing axon diameter
B. Increasing axon length
C. Increasing myelin thickness
D. Decreasing leak-channel density
E. Decreasing depolarization rate

Best answer = E. Axonal conduction velocity is dependent on the rate of membrane depolarization during an action potential, which, in turn, is a function of channel gating kinetics (5·III·B). Conduction velocity would also be reduced by decreasing (not increasing) axon diameter or by demyelination, which would increase current loss via leak channels. Increasing leak-channel density might also be expected to slow axonal conduction velocity. Conduction velocity is independent of axonal length.

II.2 Epilepsy is a common neurologic disorder characterized by spontaneous episodic neuronal firing and seizures. Research indicates that a glial spatial buffering dysfunction may be involved. Spatial buffering's role includes which of the following?

A. Limiting K^+ buildup and nerve hyperexcitability
B. Preventing acidification of brain extracellular fluid
C. Increasing axonal conduction velocity
D. Synaptic neurotransmitter recycling
E. Transferring nutrients from blood to neurons

Best answer = A. Glia take up K^+ from the neuronal interstitium and transfer it via gap junctions to adjacent cells for disposal at a remote site (or to the circulation; 5·V·B). Neural function is highly sensitive to local K^+ concentrations, and buildup could cause hyperexcitability and inappropriate spiking activity. Spatial buffering does not normally play a major role in pH balance. Axonal conduction velocity is enhanced by myelination, which is also a glial function (5·V·A). Glia also participate in neurotransmitter recycling by synaptic uptake and return to neurons (5·V·C), but this is not a spatial buffering function. Nutrient transfer via glial cells is referred to as the "lactate shuttle" which is unrelated to spatial buffering (5·V·D).

II.3 Cerebrospinal fluid (CSF) loss reduces buoyancy, allowing the brain to sag and triggering a "low CSF pressure headache." In addition to buoyancy, what other protective feature does CSF provide?

A. It contains mucin to lubricate the brain.
B. Cerebrospinal fluid volume ≅ 15 mL that forms a cohesive film between brain and cranium.
C. It is enriched in HCO_3^- to buffer pH changes.
D. It is K^+ free to enhance neuronal K^+ efflux.
E. It drains along the olfactory nerve to moisturize the olfactory epithelium.

Best answer = C. Unlike most other body fluids, cerebrospinal fluid (CSF) is protein free (i.e., it contains no mucins). Proteins normally provide a significant defense against pH changes, and, thus, CSF is enriched in HCO_3^- to compensate (6·VII·D). About 120 mL of CSF bathes the central nervous system, floating the brain to prevent compression of cerebral blood vessels and forming a protective cushion between brain and bone (6·VII·B). CSF contains ~3 mmol/L K^+, slightly less than that of plasma. CSF drains into the venous system via an intracranial sinus.

II.4 A 45-year-old woman complains of pain in her fingertips and toes during cold exposure or emotional stress. This "Raynaud phenomenon" is caused by exaggerated sympathetic vasoconstriction in the extremities, producing ischemic pain. Which of the following statements best applies to her condition?

A. Sympathetic ganglia serving the fingers are located in the hand.
B. The sympathetic postganglionic nerve is myelinated.
C. The patient may gain relief from an α-adrenergic inhibitor.
D. Pain may be relieved by an *acetylcholinesterase* inhibitor.
E. The vascular neuromuscular junction contains nicotinic acetylcholine receptors.

Best answer = C. Vasoconstriction is mediated by norepinephrine release from sympathetic nerve terminals. Norepinephrine binds to α-adrenergic receptors on vascular smooth muscle cells, so the patient's vasospasm may be relieved by an α-adrenergic inhibitor (7·IV). Vascular neuromuscular signaling does not involve nicotinic acetylcholine receptors. Sympathetic ganglia are located close to the vertebral column, not peripherally, and postganglionic neurons are unmyelinated. Synaptic transmission within sympathetic ganglia is cholinergic, and, thus, an *acetylcholinesterase* inhibitor would augment sympathetic efferent activity thereby worsening the symptoms.

II.5 A 38-year-old woman is nauseated after receiving cytoxan, an anticancer drug administered to treat breast cancer. Drug-induced nausea is mediated by the area postrema, a sensory circumventricular organ (CVO). Which of the following best describes CVOs' function?

A. Aldosterone and thyroxine are released via circumventricular organs.
B. The hypothalamus monitors plasma composition via circumventricular organs.
C. Circumventricular organs allow blood and cerebrospinal fluid to mix.
D. Circumventricular organ sensory processes extend across the blood–brain barrier.
E. The central chemoreceptor that monitors P_{CO_2} is a circumventricular organ.

Best answer = B. The hypothalamus uses sensory circumventricular organs (CVOs) to monitor plasma composition, which facilitates homeostatic control of Na^+, water, and other body parameters (7·VII·C). A blood–brain barrier (BBB) is interrupted in CVOs, and the capillaries are leaky, allowing fluid to filter from blood for sensing by CVO neurons. CVO sensory neurons do not penetrate the capillary wall and extend across the BBB. Aldosterone and thyroxine are released from the adrenal and thyroid glands, respectively, and they do not contain CVOs. Central chemoreceptors are located behind the BBB. Although CVO capillaries are leaky, blood remains contained in the vasculature by the capillary walls, which prevents blood and brain extracellular fluid or cerebrospinal fluid from mixing.

II.6 A 32-year-old male presents to the emergency department with head trauma after falling from a ladder. An attendant physician shines a flashlight in each eye and observes normal pupillary reflexes. Which of the following best describes such reflexes?

A. They are an example of a vagovagal reflex.
B. Light causes cone receptor depolarization.
C. Reflexive miosis involves ciliary muscle.
D. Miosis results from increased sympathetic stimulation of smooth muscle.
E. Pupillary reflexes are mediated by retinal ganglion cells.

Best answer = E. The pupillary light reflex is triggered by light falling on photosensitive retinal ganglion cells (8·II·C). The pupil constricts reflexively through parasympathetic stimulation of iris sphincter muscles. Light falling on cones hyperpolarizes the receptor cells through a decrease in Na^+ influx through cyclic nucleotide–activated channels. The pupillary light reflex is mediated by the optic nerve and oculomotor nerve, not by the vagus nerve.

II.7 A driver traveling a dark rural road at night is temporarily blinded by the high beams of an oncoming vehicle. Which of the following observations best describes the blinded driver's retinal function?

A. Vision recovery involves rhodopsin dephosphorylation.
B. The channel that mediates night vision also transduces olfaction.
C. The high beams cause blindness through rod depolarization.
D. Light inhibits *guanylyl cyclase*–activating proteins in rods.
E. Temporary blindness is caused by Na^+ channel internalization.

Best answer = A. Rhodopsin activation by light initiates a signal cascade that causes rod signaling, but it also initiates pathways that limit signaling (8·V·C). These include rhodopsin phosphorylation by *rhodopsin kinase*, so recovery necessarily involves rhodopsin dephosphorylation. Rod cell stimulation and desensitization involves membrane hyperpolarization, mediated by a cyclic guanosine monophosphate–dependent Na^+ channel, which is different from the olfactory cyclic nucleotide–gated channel (10·III·C). Na^+ channel internalization is not part of the desensitization process. *Guanylyl cyclase*–activating proteins are stimulated by light.

II.8 A 62-year-old woman with a history of temporal arteritis suffers sudden monocular vision loss caused by retinal artery occlusion and subsequent ganglion cell ischemia. Which of the following statements best describes how these retinal ganglion cells function?

A. They are dedicated to single rods.
B. Light always causes cell depolarization.
C. They signal via the oculomotor nerve.
D. They generate action potentials.
E. They assist photoreceptor recycling.

Best answer = D. Ganglion cells transmit visual information to the brain using action potentials, their axons forming the ocular nerve (the oculomotor nerve controls eye movement). Most other cells in the retina respond to light with graded potentials rather than action potentials (8·VII). Ganglion cells collate data from groups of photoreceptors (not single rods), which gives them a wide receptive field. Ganglion cells activate when light is turned on or off, depending upon where light falls on the retina relative to their receptive field. Pigment cells, not ganglion cells, aid photoreceptor recycling.

II.9 A child with congenital hearing loss is diagnosed with round window atresia (absence of a round window) following computed tomography imaging studies. Atresia impairs hearing by which of the following mechanisms?

A. Pressure across the eardrum cannot equalize.
B. It prevents ossicular chain movement.
C. It impairs impedance matching.
D. It prevents perilymph movement.
E. It stiffens the basilar membrane base.

Best answer = D. The round window allows perilymph to move within the cochlear chambers when the oval window is displaced by the stapes (9·IV). The cochlea is encased in bone, which prevents chamber expansion when the oval window is displaced. Therefore, in the absence of a round window, perilymph movement and basilar membrane flexion cannot occur. Impedance matching is a function of the ossicular chain, and air pressure equalization relies on the eustachian tube, neither of which should be affected by atresia. Basilar membrane stiffening would affect frequency discrimination but should not cause hearing loss.

II.10 A rare inherited disorder that prevents synthesis of tip-link proteins has been observed in animal models. Gene expression might be expected to have which of the following effects on auditory transduction?

A. Endocochlear potential would collapse.
B. Hair cells would lose sensory function.
C. K^+ recycling would be inhibited.
D. Stereocilia would not be displaced by sound.
E. Only vestibular function would be impaired.

Best answer = B. Sounds are transduced by hair cells, which are excited when tip links between adjacent stereocilia tense, and a K^+-permeant mechanoelectrical transduction (MET) channel opens (9·IV·C). The tip links tense when stereocilia are displaced by sound waves passing through the cochlea. If the tip links were missing, the stereocilia would still be displaced by sound, but the hair cells would be unable to generate a receptor potential. Auditory and vestibular hairs cells would be affected similarly. The endocochlear potential and K^+ recycling relies on the stria vascularis to concentrate K^+ within endolymph, which should not be affected by a tip-link disorder.

II.11 Which of the following best describes the properties of the organ of Corti?

A. The apex is attuned to high frequency sounds.
B. The basilar membrane is wider at the apex.
C. The scala media is filled with perilymph.
D. Inner hair cells are sound amplifiers.
E. Stereocilia do not bend toward the kinocilium.

Best answer = B. The basilar membrane resonates at different frequencies along its length (9·IV·D). The membrane is wider at the apex and resonates at low frequencies. The basilar membrane base and the hair cells it supports are attuned to high frequencies. Auditory nerve signals are dominated by output from inner hair cells, which signal when stereocilia bend toward the kinocilium. The outer hair cells are believed to help amplify these signals. The scala media is filled with endolymph, not perilymph.

II.12 The right ear of a comatose patient is irrigated with cold water to assess vestibuloocular reflex (VOR) function. Which of the following statements best describes the VOR or its components?

A. The horizontal semicircular canal detects vertical motion.
B. The vestibuloocular reflex is initiated by otolith displacement.
C. Ear cooling causes receptor-mediated K^+ influx.
D. The vestibuloocular reflex is mediated by thermosensory nerves.
E. Vestibuloocular reflex vestibular nuclei are located in the thalamus.

Best answer = C. The temperature gradient created by irrigating the ear canal with cold water causes endolymph to move within the horizontal semicircular canal (9·V·E). Movement is transduced by mechanosensory channels on sensory hair cells, which open to allow K^+ influx and depolarization. The vestibuloocular reflex does not involve thermoreceptors. The horizontal canal normally detects rotational head movements in a horizontal plane, which relays sensory information to vestibular nuclei in the brainstem, not the thalamus. Otoliths are normally found in the otolith organs, not the semicircular canals.

II.13 A 23-year-old woman with monosodium glutamate (MSG) syndrome complex experiences nausea, palpitations, and diaphoresis after eating food containing MSG. MSG is a food additive that enhances umami taste. Which of the following best describes the MSG sensory transduction mechanism?

A. It is sensed by type I taste receptor cells.
B. The monosodium glutamate receptor is a Na^+ channel.
C. Umami cells release adenosine triphosphate.
D. Monosodium glutamate binds to a domain on "sweet" receptors.
E. Monosodium glutamate activates type III receptor cells.

Best answer = C. Monosodium glutamate (MSG) binds to a G protein–coupled receptor (not a Na^+ channel) on umami-specific type II cells (10·II·C). Receptor binding initiates Ca^{2+} release from intracellular stores, which causes membrane depolarization and opening of pannexin hemichannels. Pannexins allow adenosine triphosphate to diffuse out of the cell and stimulate a gustatory nerve. MSG may have lesser, indirect effects on type I cells (salt) and type III cells (acid) but is detected primarily by type II cells.

II.14 A 32-year-old male presents with anosmia (loss of sense of smell) following accidental inhalation of a volatile chemical at work. Which of the following statements best describes olfactory neuron function?

A. They do not regenerate, so anosmia is permanent.
B. They do not generate action potentials.
C. Olfaction is mediated by *guanylyl cyclase*.
D. Their axons form cranial nerve II.
E. The patient's sense of taste is probably intact.

Best answer = E. Taste and smell are mediated by two different types of sensory cell. Taste receptors are epithelial cells, whereas olfactory receptors are primary sensory neurons (10·III·B). Olfactory neurons are turned over and replaced every ~48 days, so the sensory deficit is probably temporary. Olfactory receptor binding causes a change in *adenylyl cyclase* activity rather than *guanylyl cyclase* activity. If sufficiently intense, a stimulus will cause the neuron to fire an action potential, which is transmitted to the brain via cranial nerve (CN) I, the olfactory nerve (CN II is the optic nerve).

II.15 An 83-year-old man with myasthenia gravis is unable to eat foods such as steak because of bulbar muscle fatigue. Studies of the man's bulbar muscles during contraction might have revealed which of the following compared with normal?

A. Decreased α-motoneuron activity
B. Decreased γ-motoneuron activity
C. Decreased Ia sensory afferent activity
D. Decreased Ib sensory afferent activity
E. Decreased II sensory afferent activity

Best answer = D. The antibodies that are produced in patients with myasthenia gravis destroy the nicotinic acetylcholine receptor, interfering with normal excitation and force development (Clinical Application 12.2). Muscle tension development during contraction is sensed by Golgi tendon organs, which signal via group Ib sensory afferents (11·II·A). Contraction is initiated by α-motoneurons, which might be expected to signal normally, as would the γ-motoneurons that initiate intrafusal fiber contraction. Group Ia and group II afferents relay sensory information from muscle spindles when a muscle is stretched.

II.16 A distracted cook picks up and immediately drops a metal spatula that had become painfully hot to the touch. Which of the following statements best describes such reflexes?

A. They are mediated by local spinal circuits.
B. Pain stimuli are transduced by Ruffini endings.
C. Pain stimuli are transmitted via α-motoneurons.
D. They would be unaffected by demyelination.
E. They are mediated by central pattern generators.

Best answer = A. Reflexive movements triggered by painful stimuli are mediated by spinal reflex circuits (or "arcs"; 11·III·A). Painful stimuli are mediated by pain receptors and transmitted via myelinated sensory afferent fibers to the spinal cord. The short signal path and fibers adapted for high conduction velocity decrease reaction time. The motoneurons are also myelinated, which makes both arms susceptible to demyelinating disease. Ruffini endings sense mechanical stimulation of the skin (16·VII·A; 11·II·C), whereas central pattern generators are involved in establishing rhythmic movements (11·III·F).

UNIT III
Musculoskeletal and Integumental Physiology

12 Skeletal Muscle

I. OVERVIEW

The various organs of the body are housed within compartments (the thorax and abdomen) that are framed and carried by a bony skeleton (bone structure and function is discussed in Chapter 15). Bones also define the limbs, which are used to manipulate objects and for locomotion. Bone movement is facilitated by skeletal muscles, which are usually arranged in antagonistic pairs to create articulating levers. The arm, for example, contains three long bones that form a lever that hinges at the elbow. A pair of antagonistic muscle groups allows for forearm **extension** (triceps) and **flexion** (biceps) as shown in Figure 12.1, but a full range of arm motions uses ~40 more. Skeletal muscles allow the body to sustain itself by hunting and gathering food, but they are also used for nonlocomotory activities, such as maintaining an erect posture and expanding the lungs. Skeletal muscle accounts for approximately 40% of total body mass in an average person, but another 10% of body mass comprises two different types of muscle: cardiac muscle and smooth muscle. The three muscle types use common molecular principles to generate force. Contraction is initiated by a rise in intracellular free Ca^{2+} concentrations, which facilitates interaction between actin and myosin filaments through binding to and activation of a Ca^{2+}-dependent regulatory complex. The two filaments then slide against each other at the expense of adenosine triphosphate (ATP) to generate force. Sliding causes myofilaments and muscle cells (**myocytes**) to contract and shorten. Although the mechanism by which force is generated is similar in all three muscle types, there are significant differences in their organization and in the way they initiate and control contractile force. These differences reflect their unique roles within the human body and will be explored in more detail in Chapters 13 and 14. Skeletal muscle consumes significant amounts of energy (ATP) when maximally active and releases an equally significant amount of heat as a byproduct. It is the integument's (skin's) responsibility to dissipate this heat by transferring it to the external environment (see 38).

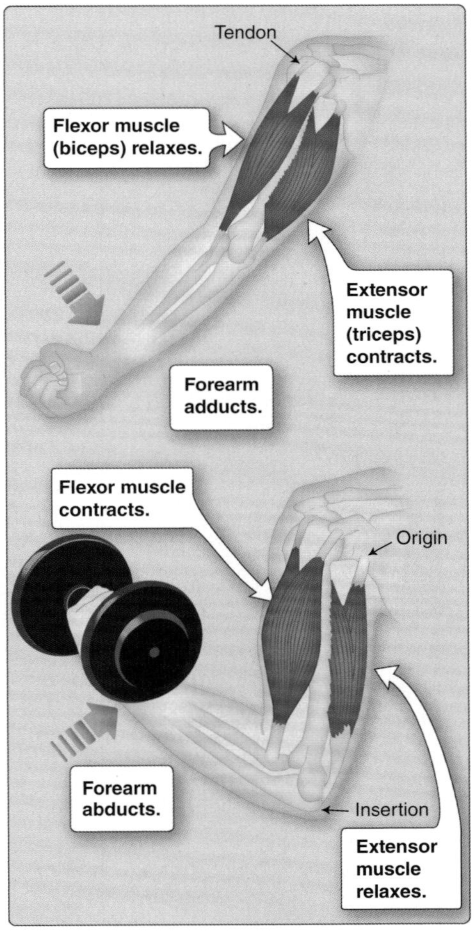

Figure 12.1
Antagonism between the muscles that extend and flex the forearm.

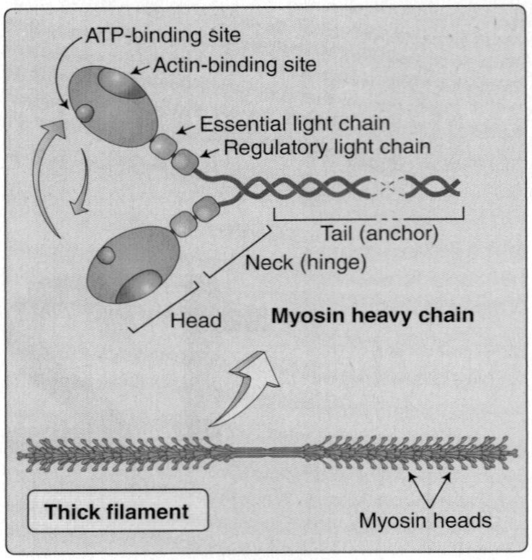

Figure 12.2
Myosin and thick filament structure. ATP = adenosine triphosphate.

II. STRUCTURE

Skeletal, cardiac, and smooth muscle all use the same molecular principle to generate force during contraction. The energy for contraction is provided by ATP hydrolysis. The stimulus for contraction is provided by free calcium ions (Ca^{2+}). The contraction mechanism involves the protein **myosin** attaching to an **actin** thin filament and pulling on it, causing it to slide past the myosin thick filament.

A. Myosin

Myosin is a large hexameric protein (MW = 52 kDa). The main body is composed of two **heavy chains**, each of which is a golf club–shaped polypeptide with a head; neck; and a long, coiled tail (Figure 12.2). The head possesses ATPase activity and contains the actin-interaction site. The neck acts as a hinge that allows the head to pivot and pull during contraction. The tail anchors the protein within a larger filamentous assembly. Each head region associates with two light chains: one **regulatory light chain** and one **essential light chain**. The tails of two heavy chains intertwine to form a coiled coil, and then about 100 such assemblies are bundled like golf clubs crammed in a bag, with the heads protruding in various different directions. Two such bundles are joined tail-to-tail to form a **thick filament**.

B. Actin

Actin forms a molecular "rope" that myosin pulls on during contraction. Actin is synthesized as a globular protein (**G-actin**) and then polymerized to form **F-actin** (Figure 12.3), which contains two beadlike polymers braided in a helical conformation. Actin expresses myosin-binding sites, but these must remain hidden until a signal to contract is received or else the muscle fiber might lock into a rigid state (**rigor**). Access to the binding site is controlled by **tropomyosin** and **troponin** (**Tn**), two regulatory proteins that lie in the grooves of the actin helix near the binding site. Actin, tropomyosin, and troponin together constitute a **thin filament**.

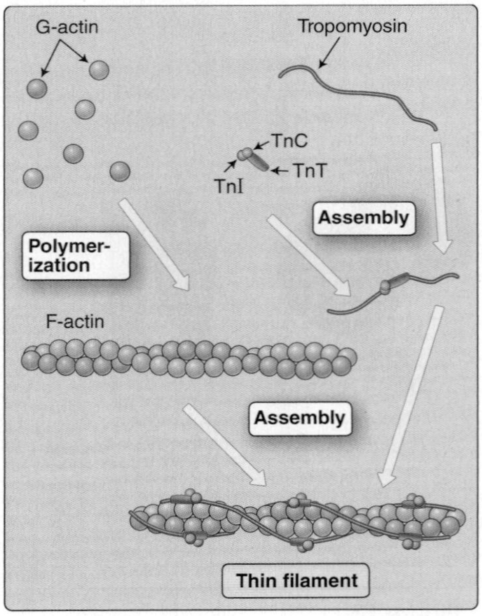

Figure 12.3
Thin filament assembly. TnC = troponin C; TnI = troponin I; TnT = troponin T.

1. **Tropomyosin:** Tropomyosin comprises two identical filamentous subunits that are entwined to form a helix. Tropomyosin molecules lay end to end along the actin filament, concealing the myosin-binding sites beneath.

2. **Troponin:** Troponin is a Ca^{2+}-sensitive protein that uncovers the myosin-binding site when intracellular Ca^{2+} concentrations rise. Troponin is an assembly of three different proteins: **troponin C** (**TnC**), **troponin I** (**TnI**), and **troponin T** (**TnT**).

   a. **Troponin C:** Tn**C** is a **Ca**$^{2+}$-binding protein that senses when a rise in intracellular Ca^{2+} occurs. Two of its four Ca^{2+}-binding sites are occupied normally, which allows Tn to attach to the thin filament. When intracellular Ca^{2+} rises, the two vacant binding sites fill, and Tn changes conformation and pulls tropomyosin deeper into the actin-filament groove. This movement uncovers the myosin-binding site, and myosin binds instantly.

b. **Troponin I:** Tn**I** helps **i**nhibit actin/myosin **i**nteraction until the appropriate time. It also masks myosin-binding sites.

c. **Troponin T:** Tn**T** **t**ethers the **t**roponin complex to **t**ropomyosin.

C. Sarcomere

During contraction, myosin head groups reach along the actin filament from one binding site to the next and then pull. This causes the thin filament to slide past the thick filament, and the muscle shortens. A contracting muscle harnesses the energy from millions of these small movements, but, in order for this to occur, the actin ropes must be tethered end to end and ultimately to the ends of the muscle fiber. The two filaments must also be held tightly against each other within a framework of structural proteins that ensure actin–myosin contact. The result is a contractile unit called a **sarcomere** (Figure 12.4).

> Many descriptive terms that refer to muscle use the "sarco-" prefix. Sarco- is derived from the Greek word for flesh, *sarx*. Flesh refers to the soft tissues that cover bone and are composed primarily of muscle and fat.

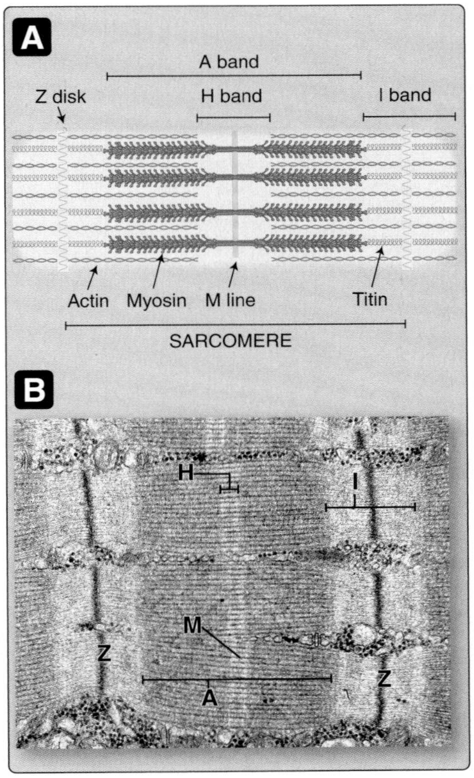

Figure 12.4
Filamentous structure and banding pattern of a muscle sarcomere.

1. **Sarcomeric structure:** A sarcomere is delineated by two **Z disks**, proteinaceous plates that anchor arrays of thin filaments (see Figure 12.4A). The filaments protrude from the disks like bristles from a hairbrush. Thick filaments insert between the thin filaments, so that each thick filament is surrounded by, and may pull simultaneously on, six actin ropes (Figure 12.5). The thick filaments are 60% longer than thin filaments. This allows them to insert into an identical array of thin filaments that extend from the Z disk, which delineates the far end of the sarcomere. This region of overlap is where interactions, or **crossbridges,** between the two types of filament will occur. Overlap between thick and thin filaments produces distinct banding patterns when viewed under polarized light (see Figure 12.4B) that are repeated across the width of a muscle to give it a **striated** appearance.

2. **Structural proteins:** Numerous specialized cytoskeletal proteins rigidly constrain the thick and thin filaments within the framework of a sarcomere and aid in its assembly and maintenance.

   a. **Actinin: α-Actinin** binds the ends of thin filaments to the Z disks (Z disks appear as Z lines in micrographs).

   b. **Titin: Titin** is a massive protein (>3 million MW). One end is attached to a Z disk, the other to the thick filaments. It forms a spring that limits how much the sarcomere can be stretched. It also centers the thick filaments within the sarcomere.

   c. **Dystrophin: Dystrophin** is a large protein associated with Z disks. It helps anchor the contractile array to the cytoskeleton and surface membrane. It also aligns the Z disk with disks in adjacent myofibrils and muscle fibers.

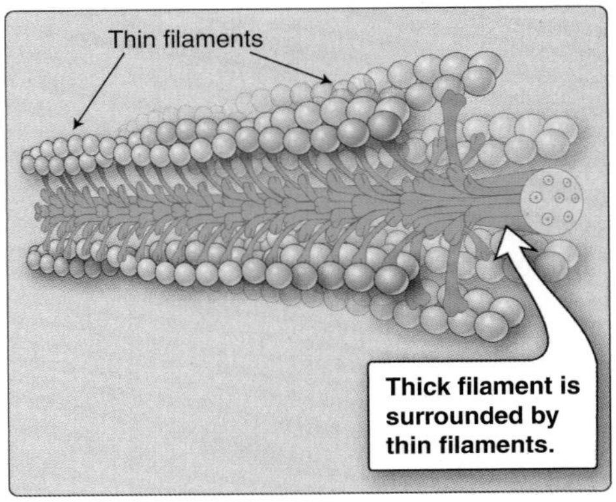

Figure 12.5
Spatial relationship between thick and thin filaments in muscle.

Clinical Application 12.1: Muscular Dystrophy

Muscular dystrophy (MD) refers to a heterogeneous group of inherited disorders that cause muscle wasting. The most common is **Duchenne muscular dystrophy**, which results from a recessive X-linked mutation in the dystrophin gene. Loss of dystrophin function prevents the cytoskeleton and its embedded contractile machinery from attaching to the sarcolemma and the muscle fiber becomes necrotic as a result, causing gradual muscle wasting. MD affects all voluntary muscles. Patients usually do not survive beyond their early 30s, eventually succumbing to respiratory muscle failure.

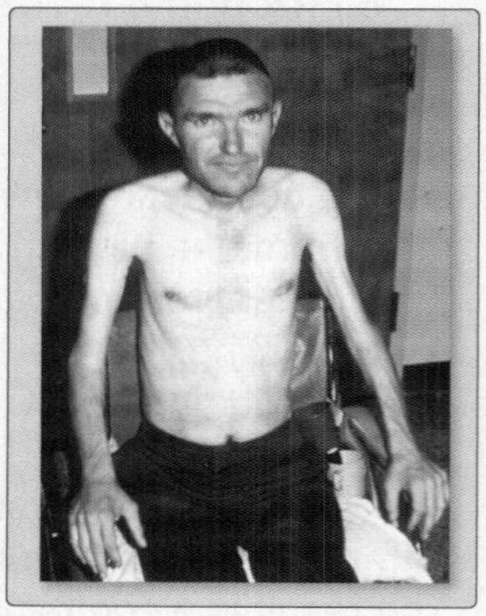

Muscle wasting in a patient with muscular dystrophy.

d. **Nebulin: Nebulin** associates with and extends the length of an actin filament. It is believed to act as a molecular ruler that determines thin-filament length during assembly.

D. Skeletal muscle

The structure of the sarcomere is replicated many times down the length of a muscle to build up a **myofibril** (Figure 12.6). Many hundreds or even thousands of myofibrils are then bundled and wrapped in a **sarcolemma** sheath, which comprises plasma membrane covered by a thin extracellular coat containing numerous collagen fibers to give it strength. The result is a **muscle fiber**. Multiple muscle fibers are bundled and wrapped in connective tissue to form a **fascicle**, and fascicles, in turn, bundled to form a skeletal muscle. Each individual muscle fiber within a fascicle has a **tendon fiber** fused at either end to provide a mechanical link between the muscle and bone. Tendon fibers are made of collagen and are well able to withstand the tension generated by muscle contraction. The fibers gather in interlaced, parallel bundles to form tendons, which are adhered to bone. The two ends of a muscle are referred to anatomically as its **insertion** and **origin** (see Figure 12.1). The insertion attaches to the bone that moves when the muscle contracts and is usually more distal than its stationary origin.

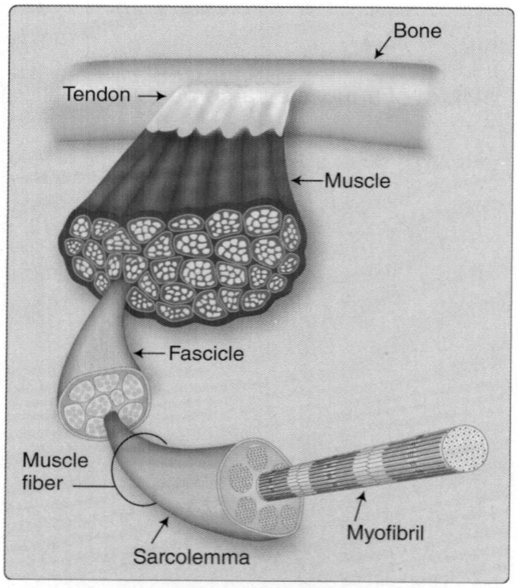

Figure 12.6
Skeletal muscle structure.

E. Sarcoplasm

Sarcoplasm is muscle cell cytoplasm and is rich in Mg^{2+}, phosphates, and glycogen granules. It also contains high levels of **myoglobin**, an oxygen-binding protein that is related to hemoglobin. The sarcoplasm is also dense with mitochondria, which lie tightly alongside the myofibrils to supply the large quantities of ATP used to fuel contraction.

F. Membrane systems

If a skeletal muscle is to work efficiently, its sarcomeres must contract in unison. The stimulus for contraction is a rise in sarcoplasmic Ca^{2+}

levels. Ensuring that every sarcomere receives a signal simultaneously requires two membrane specializations: one external (**transverse [T] tubules**), the other internal (**sarcoplasmic reticulum [SR]**) (Figure 12.7).

1. **Transverse tubules: T tubules** are tubular invaginations of the sarcolemma that extend deep into a muscle fiber's core, branching repeatedly en route so as to contact every myofibril. The tubules align with the ends of the thick filaments, two per sarcomere, and the myofibrils are aligned by their Z disks, so that the tubules have a straight run through the fiber. T tubules carry signals to contract (action potentials) from the cell surface to the structures responsible for releasing Ca^{2+} into the sarcoplasm.

2. **Sarcoplasmic reticulum:** The SR comprises an extensive system of membranous sacs that wrap around the myofibrils. The role of the sacs is to douse the myofibrils with Ca^{2+} and thereby initiate contraction. The SR scavenges Ca^{2+} from the sarcoplasm using pumps (SERCA [sarco/endoplasmic reticulum] Ca^{2+} ATPases) when the muscle is relaxed and then store it temporarily in association with **calsequestrin**, a Ca^{2+}-binding protein that acts like a Ca^{2+} sponge. T tubules communicate the need for Ca^{2+} release to the SR via **terminal cisternae**, two of which from abutting sarcomeres form a junctional structure with a T tubule, which is called a **triad**. Triads are the location of specialized sensors and ion channels that have a key role in **excitation–contraction coupling** (see Section III below).

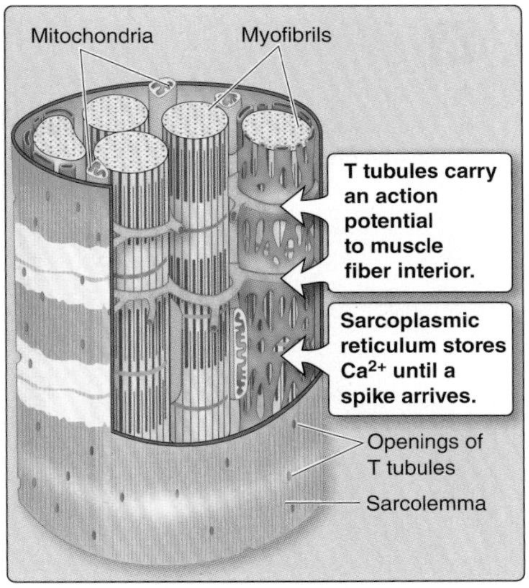

Figure 12.7
Skeletal muscle membrane systems. T tubules = transverse tubules.

G. Neuromuscular junction

All skeletal muscles are under voluntary control. Decisions to initiate contractions originate in the cerebral cortex and are then delivered to the appropriate muscle group by an α-motor neuron that interfaces with the muscle at a **neuromuscular junction (NMJ)** as shown in

Clinical Application 12.2: Myasthenia Gravis

The autoimmune condition **myasthenia gravis (MG)** is the most common disorder affecting neuromuscular transmission. Patients develop circulating antibodies against nicotinic acetylcholine receptors that interfere with normal signaling at the neuromuscular junction. Patients with a more limited form of the disorder, ocular MG, have muscle weakness of the eyelids and extraocular muscles. Generalized MG affects ocular muscles plus bulbar, limb, and respiratory muscles. Although severe respiratory muscle weakness may cause life-threatening respiratory failure, termed a "myasthenic crisis," most MG patients experience milder episodes of weakness that improve with rest. Treatments include anticholinesterase inhibitors (e.g., pyridostigmine), immunosuppressive drugs, immunomodulating therapies (e.g., plasmapheresis), and surgery (thymectomy). The latter may help because many MG patients have either thymic hyperplasia (~60%) or thymomas (15%) and exhibit symptomatic improvement with thymectomy.

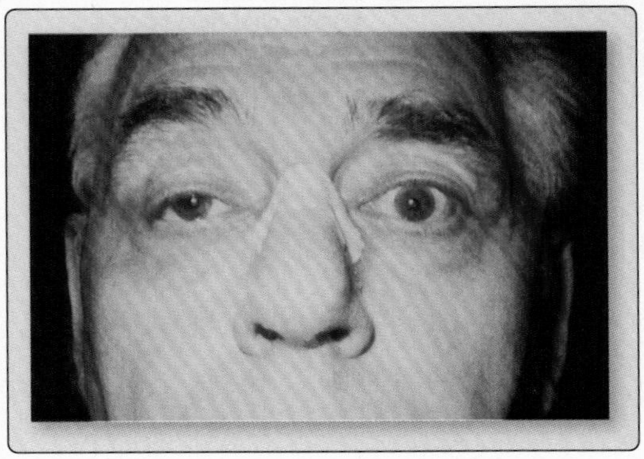

Right eyelid–droop caused by myasthenia gravis.

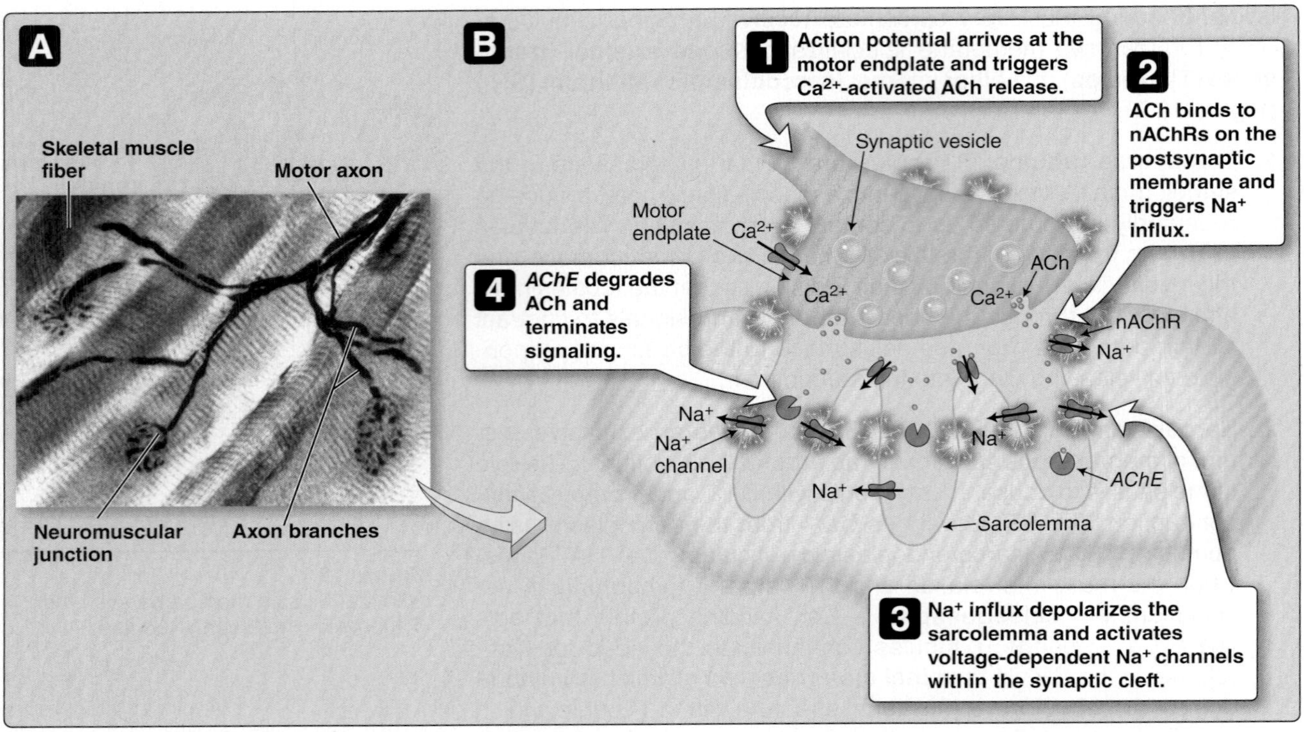

Figure 12.8
Neuromuscular excitation. ACh = acetylcholine; *AChE* = *acetylcholinesterase*; nAChR = nicotinic ACh receptor.

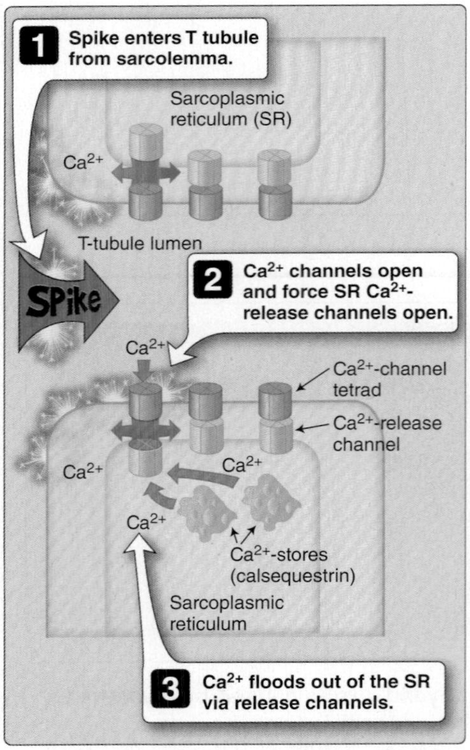

Figure 12.9
Triad structure and excitation. T tubules = transverse tubules.

Figure 12.8A. Motor neurons are flattened in this region to form a **motor endplate** (see Figure 12.8B). The motor nerve terminals are filled with vesicles that release **acetylcholine** (**ACh**) into the synapse in a Ca^{2+}-dependent manner upon excitation. The sarcolemma is deeply folded in the region of the NMJ, and the crests of these folds are studded with **nicotinic ACh receptors** (**nAChRs**). ACh binding to the nAChRs initiates contraction.

1. **Nicotinic acetylcholine receptors:** nAChRs are ion channels that open in response to binding of two ACh molecules to mediate a joint Na^+ and K^+ flux across the sarcolemma. Na^+ influx dominates the exchange, and the membrane depolarizes, a response known as a **motor endplate potential**.

2. **Sodium channels:** The sides of the clefts created by SR folding are dense with voltage-dependent Na^+ channels. The local depolarization induced by the nAChR opens the channels and initiates an action potential that spreads from the clefts over the surface of the muscle fiber.

3. **Signal termination:** The synaptic cleft contains *acetylcholinesterase* (*AChE*), an enzyme that rapidly cleaves ACh and terminates the signal to contract.

III. EXCITATION–CONTRACTION COUPLING

The spike triggered by ACh release at the NMJ propagates over the sarcolemma in a similar manner to that described for nerve cells (2·III·C).

Action potentials are electrical events, whereas contraction is mechanical. Transduction from electrical to mechanical occurs by a process known as excitation–contraction coupling, which begins when the action potential enters the T tubules and encounters a triad.

A. Triad role

The T-tubule membrane in the region of a triad is filled with **L-type Ca^{2+} channels** (**dihydropyridine receptors**) arranged in quadruplets (**tetrads**) as shown in Figure 12.9. They are voltage sensitive, so when the action potential arrives, they open and Ca^{2+} flows into the sarcoplasm from the cell exterior (i.e., the tubule lumen). The amount of Ca^{2+} entering the myocyte via these channels is minute compared with the amount needed for contraction, but the conformational change that accompanies channel opening also forces open **Ca^{2+}-release channels** in the SR (also known as **ryanodine receptors**, or **RyRs**). Ca^{2+} then floods out of the stores and saturates the contractile apparatus.

B. Crossbridge cycling

Dumping of the SR Ca^{2+} stores causes sarcoplasmic Ca^{2+} concentrations to rise from 0.1 μmol/L to ~10 μmol/L within a fraction of a second. Troponin pulls tropomyosin deeper into the groove of the actin filament and exposes the myosin-binding sites, allowing contraction to begin (Figure 12.10).

1. **Myosin binding:** The myosin head immediate latches onto the actin filament, and the two filaments immobilize in a **rigor state**. This step is brief in an active muscle but becomes permanent when ATP is unavailable.
2. **Adenosine triphosphate binding:** ATP binds to the myosin head and decreases its affinity for actin, thereby releasing it.
3. **Adenosine triphosphate hydrolysis:** ATP hydrolysis releases energy that causes the myosin molecule to pivot at the neck, causing the head to reach forward by ~10 nm. ATP hydrolysis reverses the affinity change that occurred during the previous step, and myosin immediately rebinds actin at this new position. The head group is now tensed for action like a cocked trigger, the potential energy from ATP hydrolysis stored primarily in the neck region.

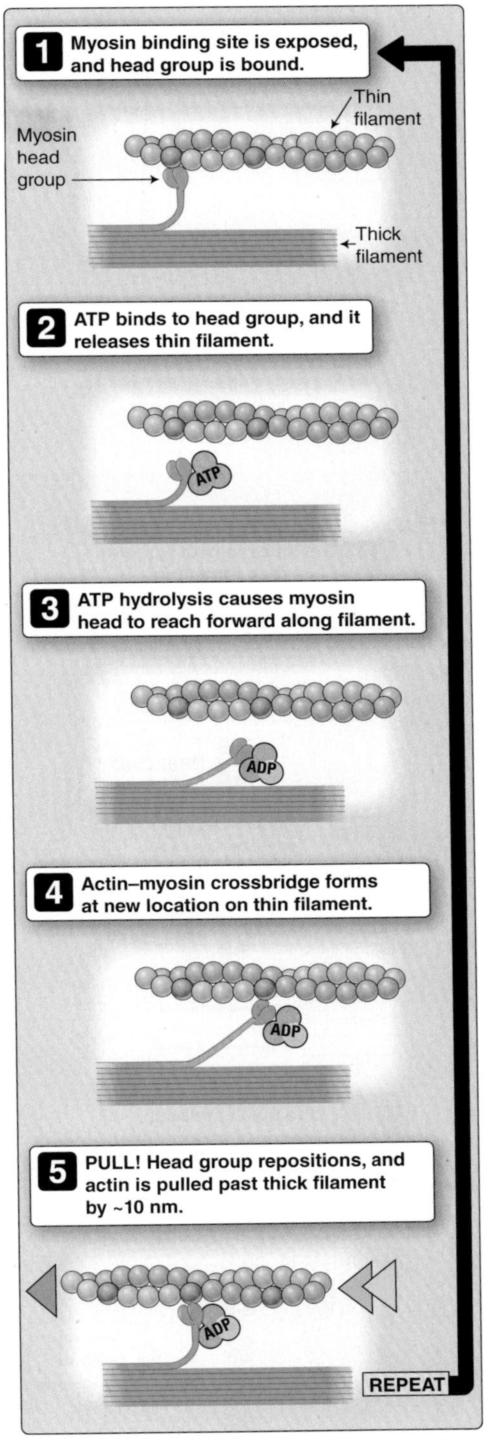

Figure 12.10
The actin–myosin crossbridge cycle. ADP = adenosine diphosphate; ATP = adenosine triphosphate.

Clinical Application 12.3: Rigor Mortis

When the body dies, adenosine triphosphate reserves are depleted rapidly, and the pumps that maintain ion gradients across the membrane stop working. Intracellular Ca^{2+} concentrations rise as a result, causing actin and myosin to bind and lock into a state of **rigor mortis**. Although time of onset can be highly variable depending on ambient temperatures, rigor mortis typically develops within 2 to 6 hours of death. The rigor state persists for 1 or 2 days until the muscle cells deteriorate, and digestive enzymes released from lysosomal lysis finally break the crossbridges, allowing the muscles to relax.

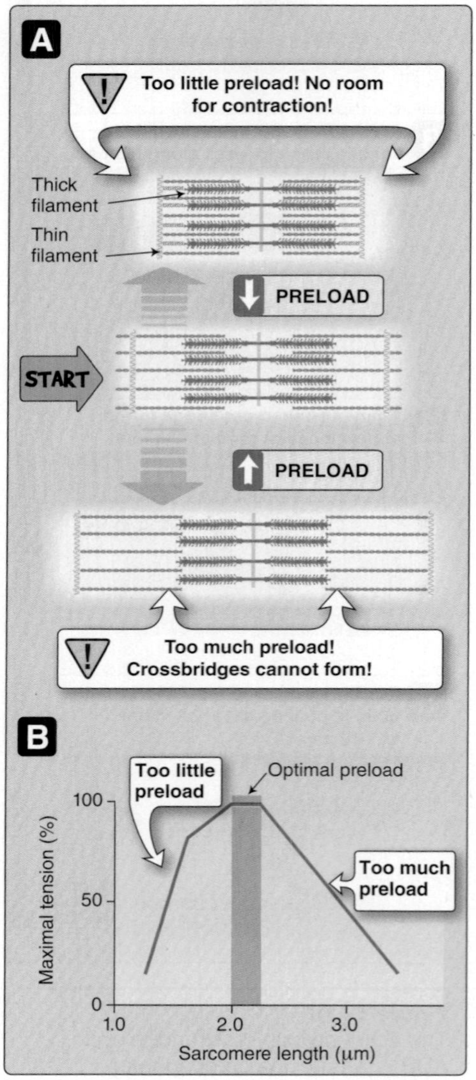

Figure 12.11
Preload effects on force development by skeletal muscle.

4. **Power stroke:** Inorganic phosphate dissociates, and the trigger is released, initiating a conformational change that pulls the head back to its prior position. The head remains attached to actin during this time, so that the entire actin rope is pulled past the thick filament by a distance of ~10 nm.

5. **Adenosine diphosphate release:** Adenosine diphosphate dissociates from myosin, and the crossbridge cycle returns to a rigor state. The cycle then repeats, causing the myosin head to draw in the actin rope ~10 nm at a time.

C. Relaxation

Once the α-motor neuron stops firing, the muscle is allowed to relax, but relaxation requires that the crossbridge cycle be broken. The Ca^{2+} channels and release channels close immediately once the motor neuron stops firing and Ca^{2+} ATPases in the SR and sarcolemma then rapidly clear Ca^{2+} from the sarcoplasm by returning it to the stores or expelling it from the myocyte. Sarcoplasmic free Ca^{2+} concentrations thus fall, causing TnC to release its two labile Ca^{2+} and allowing tropomyosin to slide back into place over the myosin-binding sites. Actin and myosin are prevented from interacting once again, so crossbridge cycling ceases, and the muscle relaxes.

IV. MECHANICS

A skeletal muscle is a complex assembly of contractile and elastic elements. Each crossbridge cycle generates a unit of force that is transferred to a tendon and used to move a bone, lift a load, or tense the diaphragm, for example. Some of the force generated by contraction is wasted because it is expended in tensioning the contractile elements, much as a rubber band has to be tensed before it can be used to lift an attached weight (**passive tension**). An understanding of how a skeletal muscle performs *in vivo* (a discipline called **muscle mechanics**) begins at the sarcomeric level.

A. Preload

Actin filaments are firmly anchored at either end by the Z disks. Myosin tugging on thin filaments draws the Z disks at either end of the sarcomere closer together, thereby shortening it, and the muscle contracts. Shortening continues until the thick filaments run into the Z disks and then stops. If Ca^{2+} release occurs when the thick filaments are already hard up against the Z disks, shortening and contraction cannot occur (Figure 12.11A). Stretching a sarcomere can also prevent contraction if it physically separates the thick and thin filaments and prevents them interacting. Muscle performance peaks when the potential for crossbridge formation reaches a maximum (see Figure 12.11B). Stretching a muscle to optimize actin and myosin interaction is known as **preloading**. Sarcomere length is optimal in resting skeletal muscle, but, in cardiac muscle, increasing preload will further increase force production (see 13·IV).

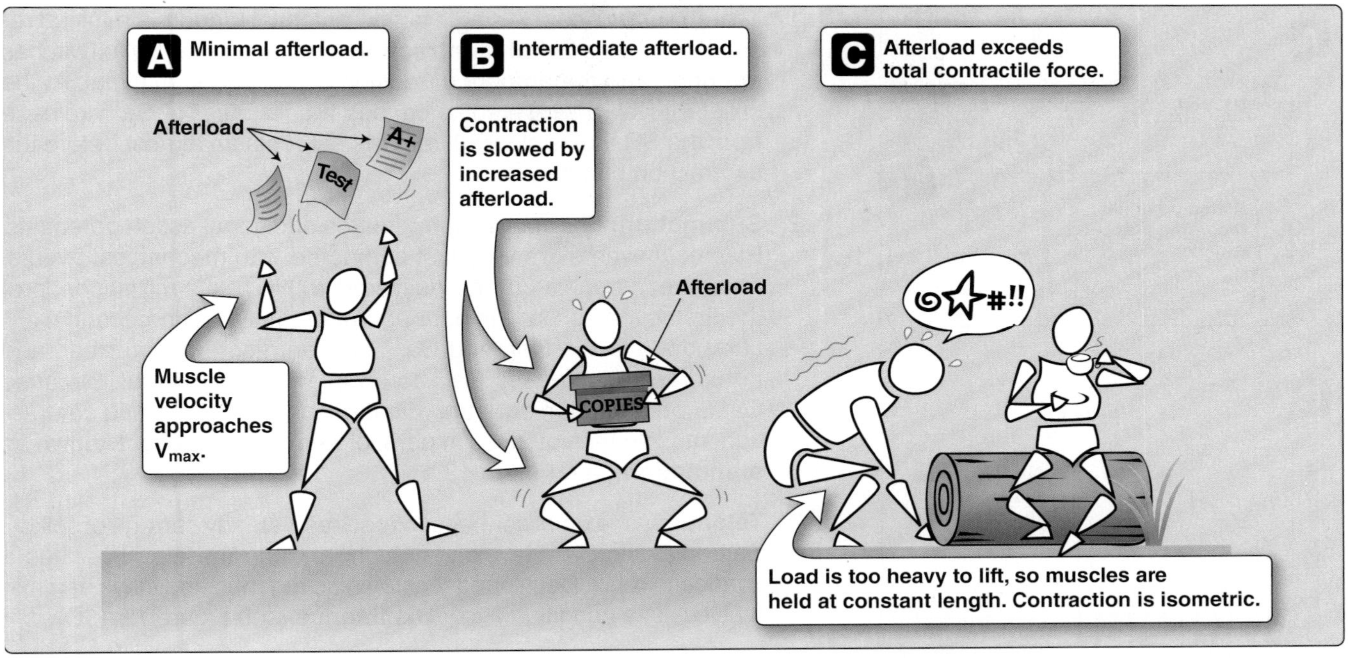

Figure 12.12
Afterload effects on skeletal muscle shortening velocity. V_{max} = maximal velocity.

B. Afterload

Although muscle contraction involves innumerable crossbridge cycles, the amount of force they generate is finite. As everyone knows from personal experience, some loads are too heavy to lift. The load that a muscle is asked to lift determines how much **tension** a muscle must develop in order to move or lift it and is known as the **afterload**. Afterload also determines how fast the muscle contracts during a lift. Minimal loads allow the muscle to contract at a maximal velocity, whereas very heavy loads are lifted slowly (Figure 12.12).

> Functionally, there two types of contraction. An **isometric** contraction occurs when the afterload is too heavy to lift, and the muscle cannot shorten even though crossbridge cycling continues regardless. A muscle undergoing **isotonic** contraction shortens but maintains a constant tension. In practice, most contractions are a mix of both types.

C. Stimulation frequency

Motor neurons usually issue commands to contract using a volley of action potentials rather than just one. The number and frequency of action potentials within a volley determines the amount of force that muscle develops during contraction (Figure 12.13).

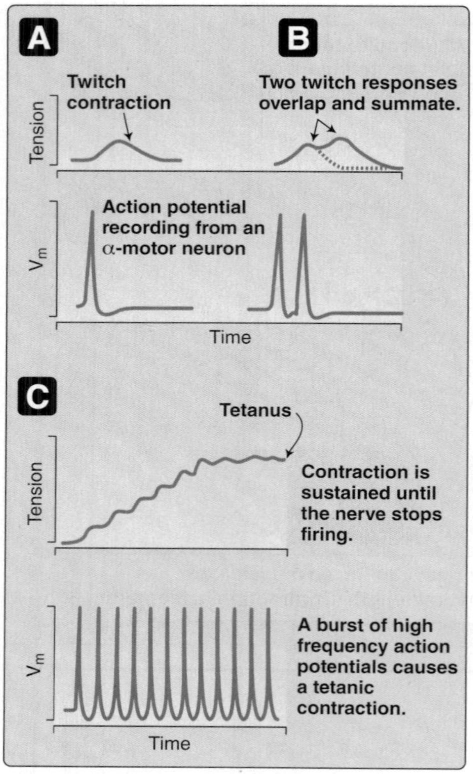

Figure 12.13
Skeletal muscle summation and tetany.

1. **Twitch responses:** A single action potential triggers a minimal response called a **twitch contraction** (see Figure 12.13A). Twitches are brief, and the amount of tension developed is minimal. At the cellular level, a single spike causes a brief burst of Ca^{2+} release from the SR that is cleared from the sarcoplasm almost before the contraction has begun.

2. **Summation:** Full contraction requires additional action potentials. Electrical events are very fast compared with mechanical events, so a second spike can be delivered within a few milliseconds of the first, even as Ca^{2+} is pouring out of the SR. The second Ca^{2+} burst adds to the first, and Ca^{2+} concentration rises further, and a third spike increases Ca^{2+} to an even greater extent. Because Ca^{2+} concentration equates with crossbridge cycling and contraction, muscle tension rises in parallel with Ca^{2+}, an effect known as **summation** (see Figure 12.13B).

3. **Tetanus:** A maximal contraction requires a steady barrage of spikes. Sarcoplasmic Ca^{2+} levels stay high, and the muscle never has a chance to relax. Developed tension also reaches and stays at maximal levels, a condition known as **tetanus** (see Figure 12.13C).

D. Recruitment

Summation is an imprecise method of controlling muscle tension. Some tasks require a finer degree of control than can be achieved using summation alone, and this is made possible by dividing muscles into **motor units**. A motor unit includes a group of fibers that may be spatially separate but that are innervated by the same α-motor neuron (Figure 12.14). Skeletal muscles comprise many such motor units. The multiunit approach allows some units to be tetanic even as others are relaxing. This permits contractile force to be maintained at a constant level and then ramped up more smoothly than can be achieved by summation. It also allows the central nervous system to choose which muscle units to activate depending on their speed (see below) and the task at hand.

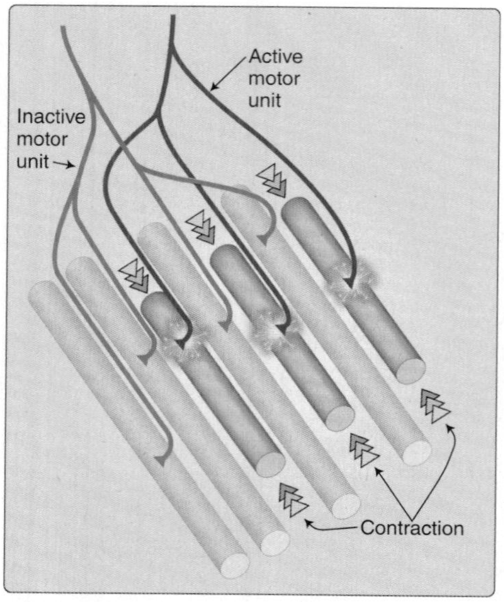

Figure 12.14
Skeletal muscle motor units.

V. TYPES

The body places many demands on skeletal muscle. Some muscles must sustain a contraction for prolonged periods, but response latency is not a concern. The muscles that control posture (muscles in the legs and the back) are classic examples. Conversely, other muscles must respond rapidly but are only ever used for short bursts of activity (e.g., muscles that control eye movement). To deal with these varied needs, the body employs several different classes of myocyte that differ in the amount of force they can develop, their contraction velocity, how quickly they fatigue, and their metabolism. Traditionally, muscle fibers have been placed in one of two groups (Table 12.1). In practice, most muscles are a mix of slow and fast fiber types.

A. Slow twitch (type I)

Slow-twitch fibers contract slowly, but they do not readily fatigue. They are used in maintaining posture, for example. Their energy is derived

primarily from oxidative metabolism, facilitated by a rich vascular system, high levels of oxidative enzymes, and an abundance of myoglobin and mitochondria. The myoglobin gives the slow-twitch fibers a red appearance.

B. Fast twitch (type II)

Fast-twitch fibers contain a myosin isoform that is specialized for rapid movement, but they fatigue more easily. They rely more heavily on glycolysis as a source of ATP than do slow-twitch fibers. Fast-twitch fibers are a diverse group that can be further subdivided into **type IIa** and **type IIx** fibers.

1. **Type IIa:** Type IIa fibers resemble slow-twitch fibers in that they rely primarily on oxidative metabolism, supported by high numbers of mitochondria and the presence of myoglobin (which gives them their red color), yet they also have well-developed glycolytic pathways. These properties allow for wide range of activity types.
2. **Type IIx:** Type IIx fibers are specialized for high-speed contractions of the type used in rapid sprints. They are primarily glycolytic and fatigue easily.

VI. PARALYSIS

Muscle contraction involves several crucial steps, but the NMJ is one of the more critical elements in this pathway because it functions as an on/off switch. If a command to contract is interrupted, the muscle is paralyzed. Nature has devised several potent toxins that target the NMJ to help predators immobilize their prey. The pharmaceutical industry has similarly developed many drugs that act at the NMJ.[1] Two primary targets for both toxins and drugs are the nAChR and *AChE*.

A. Agents that affect nicotinic acetylcholine receptors

nAChR signaling can be blocked both by agonists and antagonists, although the mechanism of action is different.

1. **Agonists: Succinylcholine** comprises two linked ACh molecules that readily binds to the nAChR and initiates contraction. However, whereas ACh is immediately catabolized by *AChE*, succinylcholine is resistant to degradation, allowing it to persist in the synaptic cleft and continue stimulating the receptor. The result is nAChR desensitization and Na^+-channel inactivation. In practice, this means that signaling is blocked after an initial contraction. Succinylcholine is fast acting and, therefore, is often used clinically to paralyze tracheal muscle to permit endotracheal tube insertion. Its effects are reasonably short lived because it readily succumbs to a plasma form of *cholinesterase*.

[1]For a discussion of neuromuscular-blocking drugs see *LIR Pharmacology*, 5e, p. 64.

Clinical Application 12.4: Trismus ("Lockjaw")

The soil microbe *Clostridium tetani* produces tetanus toxin. The bacterium typically enters the body via a cut or deep puncture wound and is then carried by retrograde axonal transport via peripheral neurons to the spinal cord. The spinal cord is home to **Renshaw cells**, inhibitory neurons that form a vital component of a feedback loop that normally limits muscle contractions (see 11·III·E). The toxin interferes with neurotransmitter release by Renshaw cells and thereby allows muscles to contract in an unregulated manner to the point of tetany. Trismus, or "lockjaw," is a classic, early symptom of tetanus toxicity and gives the disease its familiar name. The resulting contractions may be powerful enough to fracture bones. All skeletal muscles are affected, including respiratory muscles, which accounts for the high incidence of fatality associated with this disease. The patient shown here exhibits opisthotonos, caused by tetanic contraction of back muscles.

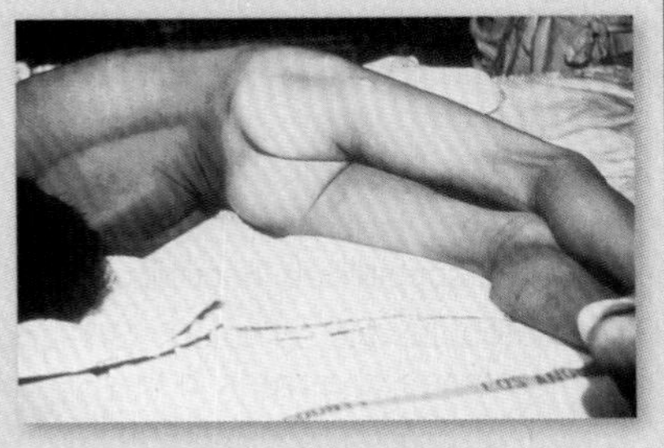

Tetanus-induced opisthotonos.

Table 12.1 Muscle Fiber Types and Properties

	Slow	Fast
Synonyms	Red	White
Myosin ATPase activity	Slow	Fast
Fatigue resistance capacity	High	Low
Oxidative capacity	High	Low
Glycolytic capacity	Low	High
Myoglobin content	High	Low
Mitochondrial volume	High	Low
Capillary density	High	Low

2. **Antagonists:** Two natural toxins bind tightly to the nAChR and prevent ACh from interacting with the receptor. **α-Bungarotoxin** is a potent nAChR antagonist found in cobra venom. **Curare** is synthesized by some plants in the Amazonian region of South America, the active ingredient in which is d-tubocurarine. Native tribal hunters apply curare to arrowheads to help paralyze and fell prey. Anesthesiologists use similar drugs (muscle relaxants) to limit movement during some surgical procedures.

B. Agents that affect acetylcholinesterase activity

Physostigmine is a natural alkaloid that reversibly inhibits *AChE* activity, allowing synaptic ACh concentrations to rise and paralyze the muscle through prolonged stimulation. Synthetic derivates of physostigmine are used to diagnose and treat the symptoms of the autoimmune disease **myasthenia gravis** (see Clinical Application 12.2).

Chapter Summary

- Most **skeletal muscles** attach to and work in conjunction with bone. Together, they create a system of articulating levers that allow the body to move and manipulate objects.
- The tensile force that allows a muscle to contract is generated at the expense of adenosine triphosphate by **myosin head groups**. They cause muscle shortening by binding to and pulling on **actin filaments**.
- Actin–myosin interactions are initiated by Ca^{2+} release via **troponin**. Troponin is a Ca^{2+}-sensitive protein that unmasks myosin-binding sites on the actin filament by repositioning **tropomyosin**.
- Actin and myosin monomers are assembled into **thin** and **thick filaments**, respectively. Overlap between these filaments within a sarcomere gives the banding pattern that is characteristic of **striated muscle**.
- Skeletal muscle is under voluntary control. A command to contract is relayed to the muscle by an **α-motor neuron**. The neuron releases **acetylcholine** (**ACh**) at the neuromuscular junction, which binds to a **nicotinic ACh receptor** on the postsynaptic membrane. Binding initiates an **action potential** that spreads over the surface of the muscle fiber.
- The action potential is distributed to individual sarcomeres by **transverse tubules**. **Excitation–contraction coupling** begins at **triads**, specialized junctions within the tubules where the **sarcolemma** and **sarcoplasmic reticulum** are closely apposed.
- The sarcoplasmic reticulum (SR) contains high concentrations of Ca^{2+} bound to **calsequestrin**. Opening of **Ca^{2+}-release channels** in the SR membrane allows the stores empty out into the sarcoplasm. The resulting rise in sarcoplasmic Ca^{2+} concentration initiates **crossbridge cycling** and **contraction**.
- The force that a muscle generates is dependent on sarcomere length (**preload**). Maximal force generation occurs when there is optimal overlap between thick and thin filaments. Excessive or insufficient preloading reduces the ability to generate force.
- Maximal contraction velocity is observed when a muscle is unloaded. Increasing the **afterload** decreases the velocity of contraction.
- Single action potentials produce **twitch responses** and minimal force generation. Increasing action-potential frequency increases force by **summation**. Constant stimulation results in **tetanus**.
- Most skeletal muscles are a mix of **slow-twitch** and **fast-twitch** fibers. Slow-twitch fibers contract slowly but can produce a sustained contraction. Fast-twitch fibers are capable of rapid response and high contractile force but **fatigue** quickly.
- Interference with one or more steps in excitation–contraction coupling can result in muscle **paralysis**. Pharmaceutical drugs and natural toxins usually paralyze muscle by interrupting signals at the neuromuscular junction.

Cardiac Muscle

13

I. OVERVIEW

Cardiac muscle shares much in common with skeletal muscle. Skeletal and cardiac muscle sarcomeres are organized similarly, so the two muscle types give similar banding patterns when viewed under polarized light (Figure 13.1). The essential principles and molecular components of contraction are the same also. There are some key differences, however, because the tasks required of the two muscles are unique. Skeletal muscle fibers contract independently of each other, each responding to individualized commands from the motor cortex. By contrast, cardiac muscle functions independently of somatic motor control, and the fibers are extensively interconnected so as to form a single functional unit. The impetus for cardiac muscle contraction comes from within the musculature itself and then spreads from myocyte to myocyte, which necessitates well-developed pathways for communication between all muscle fibers within a heart. Cardiac muscle design also incorporates a different mechanism for regulating contractile force compared with skeletal muscle because myocardial excitation *always* involves *all* fibers, and there is no option of recruiting motor units to increase overall contractile force as there is in skeletal muscle. Cardiac muscle contractility is, instead, regulated through changes in membrane Ca^{2+} permeability, which is controlled primarily by the autonomic nervous system (ANS).

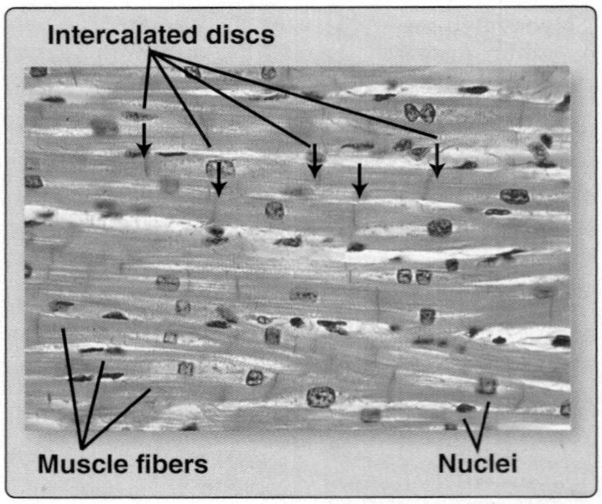

Figure 13.1
Cardiac muscle structure.

II. STRUCTURE

Cardiac myocytes contain contractile arrays that are structurally similar to those of skeletal muscle (see 12·II). The significant differences between skeletal and cardiac muscle begin with the regulatory protein troponin.

A. Troponin

Troponin is a Ca^{2+}-sensitive heterotrimer associated with the actin thin filament. Increases in sarcoplasmic Ca^{2+} concentration cause a conformational change in troponin that rolls tropomyosin away from myosin-binding sites on the thin filament, thereby facilitating actin–myosin interaction. In skeletal muscle, troponin must bind two Ca^{2+} ions before contraction can begin. In cardiac muscle, only one Ca^{2+} ion is required for contraction.

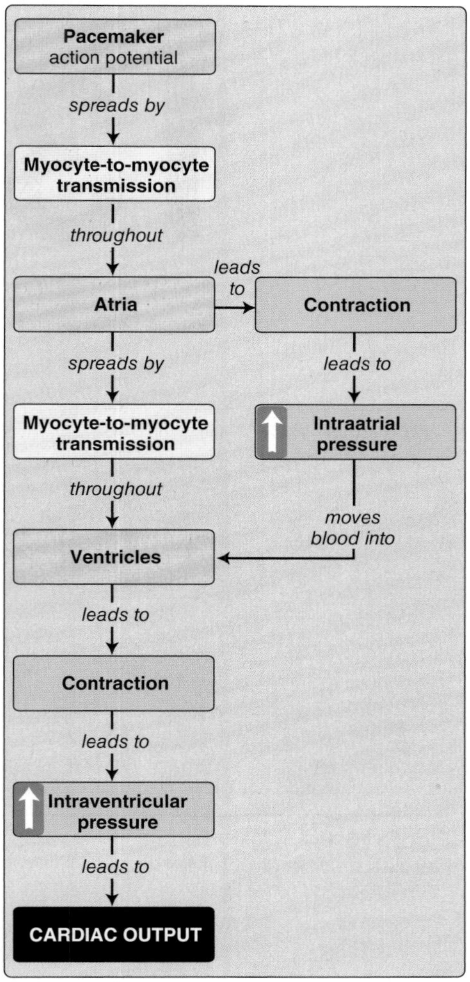

Figure 13.2
Concept map for cardiac function.

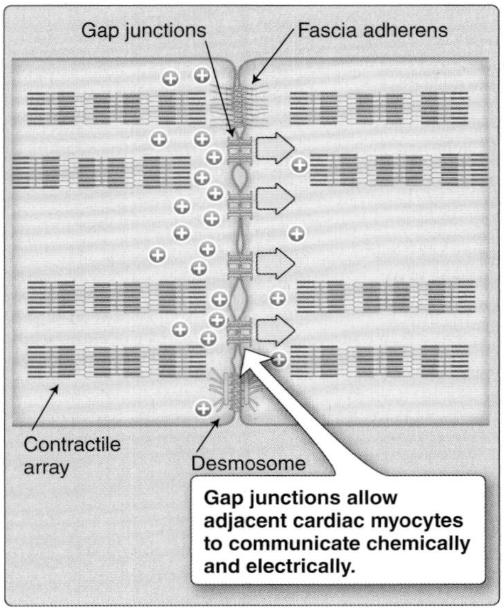

Figure 13.3
Structure of the intercalated disc.

Clinical Application 13.1: Cardiac Markers

Cardiac troponin I (cTnI) and troponin T (cTnT) are troponin isoforms unique to heart muscle. cTnI and cTnT both appear in the circulation within 3 hours of acute myocardial infarction (MI), reaching a peak after 10–24 hours. Their appearance in blood indicates ischemia-induced necrosis and loss of cardiac myocyte integrity and, thus, can be used clinically to detect an acute MI. Blood tests also measure *creatine kinase* (*CK*) levels, but *CK* is not specific to cardiac muscle. Elevated plasma *CK* levels can be caused by cocaine use or exercise-induced skeletal muscle trauma in the absence of any cardiac damage.

B. Communication pathways

Skeletal muscle fibers are under voluntary control and contract only when commanded by an α-motor neuron. Cardiac muscle is *regulated* by the nervous system, but a command to contract normally originates within an area of the myocardium (the **sinoatrial node**) that is specialized to function as a **pacemaker**. Pacemaker cells periodically generate spikes that spread from myocyte to myocyte until every fiber within the organ is involved (Figure 13.2), made possible by gap junctions and extensive cellular branching.

1. **Pacemaker:** The cardiac pacemaker is located high in the wall of the right atrium. Myocytes in this region are noncontractile, and their membranes contain ion channels that conduct a "funny current" ([I_f] see 17·IV·A) that causes them to depolarize spontaneously and generate action potentials ~100 times per minute. Once initiated, an action potential spreads regeneratively across both atria and ultimately involves the ventricles, with contraction following in its wake. The rate at which nodal cells depolarize and initiate spikes is modulated by the ANS as a way of controlling heart rate (see 17·IV·C).

2. **Gap junctions:** Gap junctions link adjacent myocytes and provide pathways for direct electrical and chemical communication (see 4·II·F). They are located at specialized sarcolemmal contact regions known as **intercalated discs** (Figure 13.3), which also contain structural elements (**desmosomes** and **fascia adherens**) that fuse the cells together mechanically and allow them to withstand the tension that is generated during muscle contraction. This contrasts with the complete lack of pathways for communication between adjacent skeletal-muscle fibers.

3. **Cellular branching:** When a command to contract is issued by the pacemaker, it must spread rapidly throughout the entire heart. Whereas skeletal muscle fibers are long, thin, and unbranching, cardiac myocytes branch extensively. The combination of branching cells and gap junctions creates a vast interconnected cellular network that functions as a single unit (a **syncytium**).

Clinical Application 13.2: Hypertrophic Cardiomyopathy

Because the heart is a muscle tissue, inherited disorders of sarcomeric function that cause skeletal muscle wasting may lead to sudden cardiac death (SCD) when cardiac muscle is affected. **Hypertrophic cardiomyopathy** (**HCM**) is an inappropriate myocardial enlargement that is a leading cause of SCD in young athletes. More than half of these deaths result from mutations in genes that encode sarcomeric proteins and, most often, the gene encoding the cardiac isoform of the myosin heavy chain. HCM alleles disrupt the normal organization of myocytes within the myocardium, causing the myocytes to be disarrayed, the fibers hypertrophied, and the cellular structures distorted (compare the micrograph at right with Figure 13.1). Extensive deposition of connective tissue within the interstitium contributes to gross ventricular wall thickening associated with HCM. Patients with HCM are prone to atrial and ventricular arrhythmias, which accounts for the high risk of SCD in affected individuals.

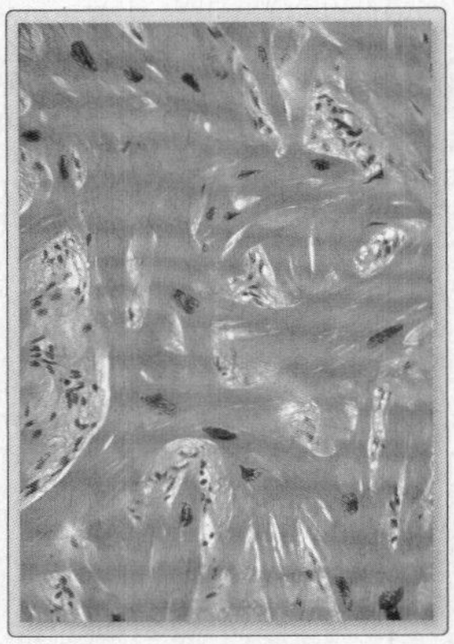

Myocyte disorganization.

C. Transverse tubules and sarcoplasmic reticulum

The action potential generated by the cardiac pacemaker is a signal that is conveyed into the core of each muscle fiber by **transverse (T) tubules** and then transmitted to the sarcoplasmic reticulum (SR) for **electromechanical transduction**. Although the SR provides the bulk of the Ca^{2+} needed for contraction, the membrane system is less extensive in cardiac muscle than in skeletal muscle. There are two other notable differences in cardiac T-tubule and SR design compared with skeletal muscle.

1. **Location:** In skeletal muscle, the T tubules align with the ends of the thick filaments, two per sarcomere. In cardiac muscle, there are fewer tubules, and they run along the Z line. T tubules tend to be wider in cardiac muscle and branch less extensively.
2. **Diads:** T tubules carry action potentials to the SR. In skeletal muscle, the tubules interface with two SR cisternae at junctional complexes known as triads. In cardiac muscle, tubules associate with a single extension of the SR at an analogous structure called a **diad** (see below).

III. CONTRACTILITY REGULATION

The molecular mechanisms underlying contraction are essentially the same in skeletal and cardiac muscle (see 12·III). The key difference between skeletal and cardiac muscle relates to the degree to which sarcoplasmic Ca^{2+} rises during excitation because Ca^{2+} availability is used to regulate contractility in cardiac myocytes.

A. Calcium-induced calcium release

When an action potential arrives at a diad, it encounters and activates L-type Ca^{2+} channels (also known as **dihydropyridine receptors**) in the T-tubule membrane (Figure 13.4). In skeletal muscle, this opening event forces Ca^{2+}-release channels (also known as **ryanodine receptors**) in the SR to open also. Ca^{2+} then floods into the sarcoplasm,

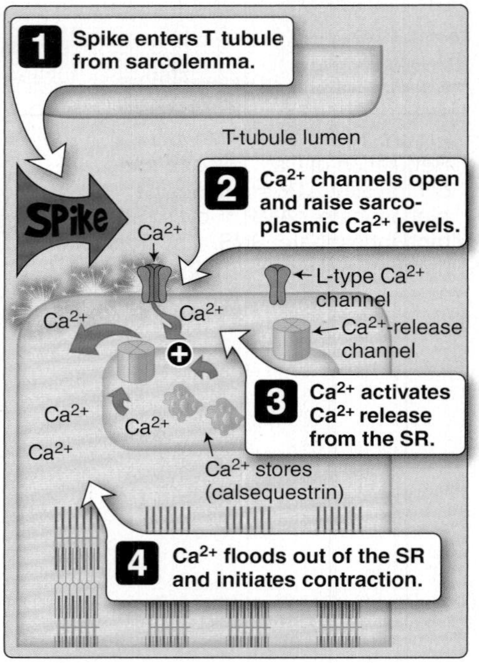
1 Spike enters T tubule from sarcolemma.
T-tubule lumen
2 Ca2+ channels open and raise sarcoplasmic Ca2+ levels.
Spike
Ca2+
L-type Ca2+ channel
Ca2+-release channel
3 Ca2+ activates Ca2+ release from the SR.
Ca2+ stores (calsequestrin)
4 Ca2+ floods out of the SR and initiates contraction.

Figure 13.4
Ca^{2+}-induced Ca^{2+} release. SR = sarcoplasmic reticulum; T tubule = transverse tubule.

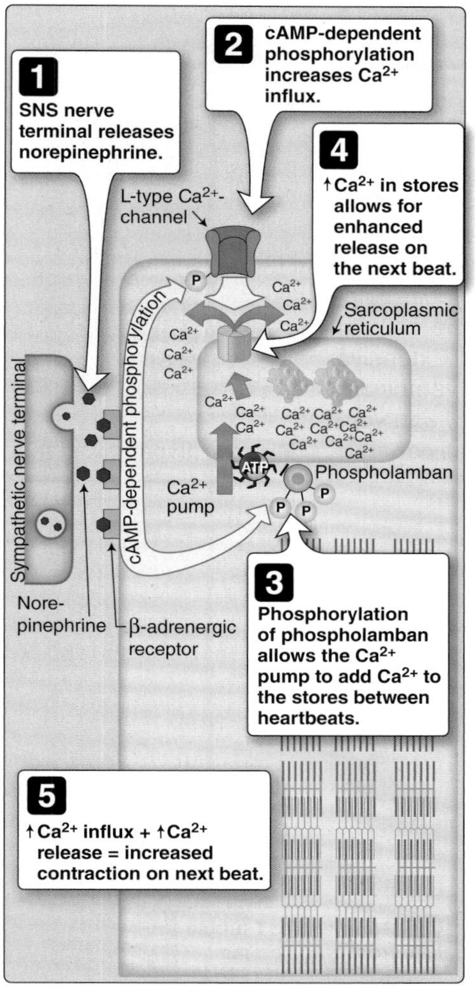

Figure 13.5
Regulation of Ca^{2+} release and contractility. ATP = adenosine triphosphate; cAMP = cyclic adenosine monophosphate; SNS = sympathetic nervous system.

and contraction ensues. In cardiac muscle, L-type Ca^{2+}–channel opening creates a Ca^{2+} **trigger flux** that is alone responsible for activating **Ca^{2+}-induced Ca^{2+} release** (**CICR**) from the SR. There are far fewer CICR channels in cardiac muscle than in skeletal muscle, reflecting the differences between the principal way that force is regulated in the two muscle types. In skeletal muscle, excitation always triggers maximal Ca^{2+} release from the SR, and, therefore, the amount of force developed is always maximal also. In cardiac muscle, the size of the trigger flux and the amount of force generated is regulated by the ANS. An approximate 1:1 relationship between the number of L-type Ca^{2+} channels and dependent release channels permits fine control over sarcoplasmic Ca^{2+} concentration and contractility.

B. Contractility regulation

The ANS controls both the rate and force with which cardiac muscle contracts. Rate control involves the coordinated actions of both the sympathetic and parasympathetic branches of the ANS (see 17·IV·C), but contractility is regulated primarily by the sympathetic nervous system (SNS). The two main targets of this regulation are the L-type Ca^{2+} channels and the Ca^{2+} ATPase that sequesters Ca^{2+} within the SR (SERCA; Figure 13.5).

1. **Sympathetic activation:** The SNS typically activates when changes in tissue activity increase the need for arterial blood supply. In practice, such needs are met by SNS-stimulated increases in cardiac output (CO). CO is increased by increasing heart rate (HR) and by making the heart a more effective pump (i.e., increased contractility). SNS terminals release norepinephrine at the cardiac neuromuscular junction and stimulate epinephrine release from the adrenal medulla. Epinephrine travels via the circulation to the heart. Neurotransmitter and hormone both bind to β_1-adrenergic receptors on cardiac sarcolemma and increase protein phosphorylation by *protein kinase A* through activation of *adenylyl cyclase* and cyclic adenosine monophosphate formation.

2. **Calcium channels:** L-type Ca^{2+} channel phosphorylation increases channel open probability and thereby increases the size of the Ca^{2+} trigger flux. Increasing the trigger flux increases the magnitude of CICR from the SR and increases the number of Ca^{2+} ions available to bind troponin and initiate crossbridge cycling.

3. **Calcium pump:** Cardiac muscle relaxation relies on two transporters to clear Ca^{2+} from the sarcoplasm. SERCA returns Ca^{2+} to the SR, whereas a Na^+-Ca^{2+} exchanger in the sarcolemma transports Ca^{2+} out of the cell. SERCA is associated with a small integral SR membrane protein called **phospholamban**, which functions as a pump limiter. When phospholamban is phosphorylated, its ability to inhibit the pump is curtailed, which allows the pump to cycle faster. There are two important consequences: faster relaxation times and increased amounts of Ca^{2+} stored for release on the next heartbeat.

   a. **Relaxation:** When the pump runs faster, intracellular Ca^{2+} concentration drops to resting levels (0.1 μmol/L) more quickly. A rapid decrease in sarcoplasmic Ca^{2+} concentration allows

the contractile machinery to relax faster, which prolongs the time available for ventricular filling during HR increases (see 39·V·B).

b. **Calcium release:** When the SR pump runs faster, more Ca^{2+} is stored in the SR, and less is returned to the extracellular space compared with previously. Stocking the stores with additional Ca^{2+} makes more available for release on the next contraction, and contractile force increases as a result.

IV. PRELOAD DEPENDENCE

Skeletal and cardiac muscle show a similar inverted U-shaped relationship between sarcomere length and force generation. Skeletal muscle is designed such that thick- and thin-filament overlap is optimized when the muscle is at rest, and further stretching decreases contractility. Cardiac muscle is designed to take advantage of the length–force relationship to match its performance to the volume of blood entering its chambers.

A. Cardiac output

Cardiac muscle makes up the walls of the heart's four chambers (Figure 13.6). When the heart fills with blood, the chambers and cardiac myocytes contained within their walls stretch (**preload**), as shown in Figure 13.7. Cardiac muscle is designed such that in the absence of preload (i.e., an empty heart), sarcomeric length is minimal, and the possibility for further shortening and force development is small. Preloading pushes the sarcomere progressively higher up the left arm of the inverted U. In practice, this means that if the amount of blood returning from the vasculature between beats is high, the chamber walls and sarcomeres are stretched to a greater extent. Stretching increases the amount of force that the muscle is able to generate on the next beat, but this force is actually *needed* to expel the additional blood volume (preload) on the next beat. The length–tension relationship thereby provides the heart with a near-perfect way of matching contractile force to the volume of blood contained within its chambers. This phenomenon is known as the **Frank-Starling law of the heart**.

B. Length-dependent activation

The relationship between preload and force has traditionally been explained in terms of optimizing overlap between thick and thin filaments. However, preloading also causes **length-dependent activation** of the contractile apparatus. Activation sensitizes the crossbridge cycle to Ca^{2+}, allowing for increased force generation even though the Ca^{2+} trigger flux and amount of Ca^{2+} released from the SR remains constant. The velocity of contraction increases during activation also. The molecular mechanisms of length-dependent activation are uncertain at present.

[1]For a discussion of beta blockers and Ca^{2+}-channel blockers see *LIR Pharmacology,* 5e, p. 222.

Clinical Application 13.3: Beta Blockers and Calcium-Channel Blockers

Beta blockers and **Ca^{2+}-channel blockers** are two important classes of drug used to control cardiac function and blood pressure.[1] Beta blockers (e.g., propranolol) are β-adrenergic receptor antagonists that prevent norepinephrine- and epinephrine-mediated increases in myocardial Ca^{2+} concentration, thereby reducing heart rate and contractility and decreasing blood pressure. Ca^{2+}-channel blockers prevent Ca^{2+} influx through L-type Ca^{2+} channels. Verapamil and diltiazem are both relatively myocardium-specific. Nifedipine is a dihydropyridine Ca^{2+}-channel blocker that reduces blood pressure by relaxing vascular smooth muscle.

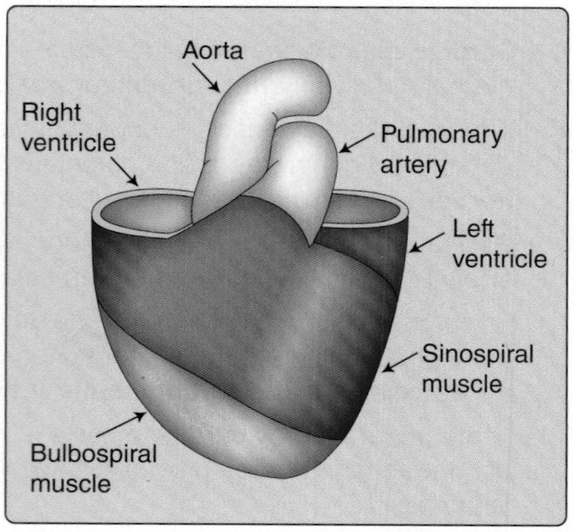

Figure 13.6
Bands of cardiac muscle.

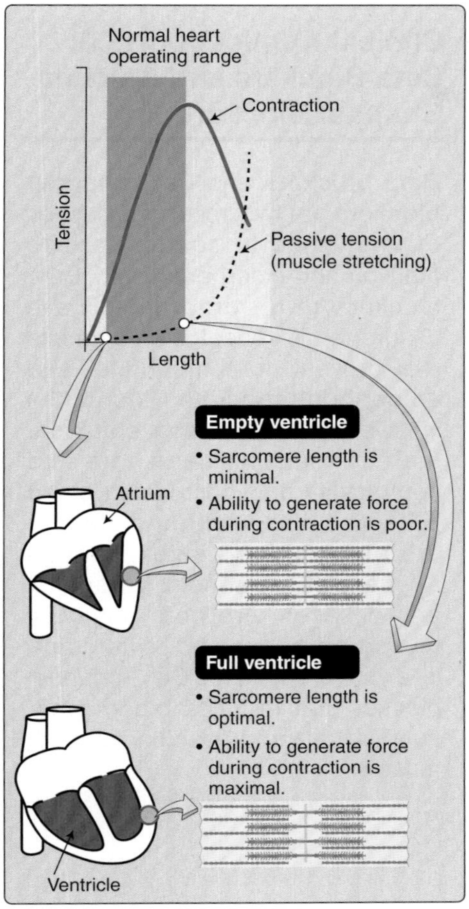

Figure 13.7
Preloading effects on the ventricular myocardium.

C. Preload limits

Excessive preloads could potentially stretch the sarcomere to the point where force generation is impaired and CO is compromised. Thus, cardiac muscle is designed with structural components that strongly curtail stretching beyond an optimal range (i.e., the peak of the inverted "U" in Figure 13.7). Structural enhancements include intracellular elastic elements, increased amounts of connective tissue in the extracellular matrix, and a fibrous sac around the heart (**pericardium**) that limits its filling volume.

V. ENERGY SOURCE

Cardiac muscle contracts about once a second for 80 or more years, but the contractions are always brief (<1 s). This contrasts with skeletal muscle in which isometric contractions can last for minutes and can be sustained to the point of fatigue. A heart capable of even one prolonged (tetanic) contraction would quickly kill its owner! Thus, cardiac muscle has lost the ability to sustain adenosine triphosphate (ATP) production and contraction for more than a few seconds. It maintains modest ATP stores that support short contractions and then regenerates these stores using aerobic pathways when relaxed. The limited anaerobic capability creates a high dependence on O_2. If O_2 supply is limited due to reduced arterial blood supply, a creatine phosphate pool can sustain ATP levels for several tens of seconds, and then lactate acid begins to be produced. Prolonged O_2 deprivation (minutes) causes irreversible hypoxic muscle damage and **myocardial infarction**.

Chapter Summary

- **Cardiac muscle** shares many features in common with skeletal muscle. Both have highly organized and aligned sarcomeres that give them a **striated appearance** under polarized light.
- Whereas skeletal muscle is under voluntary control, cardiac muscle functions **autonomously**. A specialized **pacemaker** periodically generates action potentials that spread from myocyte to myocyte to involve the entire heart. The **autonomic nervous system** regulates the timing and force of contraction but does not initiate it.
- **Gap junctions** that connect adjacent myocytes facilitate action-potential propagation. The myocytes also branch extensively to maximize interactions with adjacent cells.
- Force generation is regulated by the autonomic nervous system via **β-adrenergic receptors** and cyclic adenosine monophosphate–dependent phosphorylation by ***protein kinase A***. *Kinase* targets include the **L-type Ca^{2+} channel** in the sarcolemma and **sarcoplasmic reticular Ca^{2+} pump**. Contractile force rises as a consequence.
- Force generation by cardiac muscle is highly **preload dependent**. Increasing sarcomeric length increases the force of contraction, thereby providing a way for the heart to increase contractility when required to expel increased filling volumes.
- Cardiac muscle relies on aerobic pathways to supply the adenosine triphosphate needed for contraction. The limited anaerobic capability makes heart muscle vulnerable to decreasing O_2 availability during interruptions in blood supply.

Smooth Muscle

14

I. OVERVIEW

Skeletal and cardiac muscle are both designed to contract rapidly. Rapid kinetics facilitate locomotion (skeletal muscle) and sustain a cardiac output (cardiac muscle) that supports flow to dependent organs, even at rest. However, the human body performs many other functions that require muscular involvement on a less urgent time scale. The tasks are varied but are ably fulfilled by the highly adaptable smooth muscle. Smooth muscle is found in all regions of the body. It is layered within the walls of blood vessels and airways (Figure 14.1), functioning to adjust lumen diameter to regulate blood and air flow, respectively. Smooth muscle mixes and propels food and secretions through the gastrointestinal (GI) tract. Smooth muscle also regulates pupil diameter as a way of regulating the amount of light that enters the eye, and it regulates lens shape to adjust visual focus. Some of smooth muscle's diverse functions are summarized in Table 14.1. The varied demands made of smooth muscle have necessitated many organ-specific adaptations to its structure and function. For example, whereas striated muscle is activated by a handful of neurotransmitters and hormones, smooth muscle is modulated by hundreds of chemical signals. Contractile filaments within resting striated muscle typically have a standard length that does not vary over time. Contractile filaments in smooth muscle do not appear to have a standard length and may be disassembled and reassembled even as a contraction ensues. Plasticity of smooth muscle sarcomeres and cytoskeletal framework is necessary to maintain contractility during changes in hollow organ luminal volume (e.g., urinary bladder, GI tract, gallbladder, and uterus). Many details of smooth muscle structure and function are currently unresolved, but, like striated muscle, its core function is to generate force, accomplished using the same basic principles as in skeletal and cardiac muscle. Contraction begins when intracellular Ca^{2+} concentrations rise, and force is generated when head groups extending from a **myosin** thick filament bind to and pull on **actin** thin filaments.

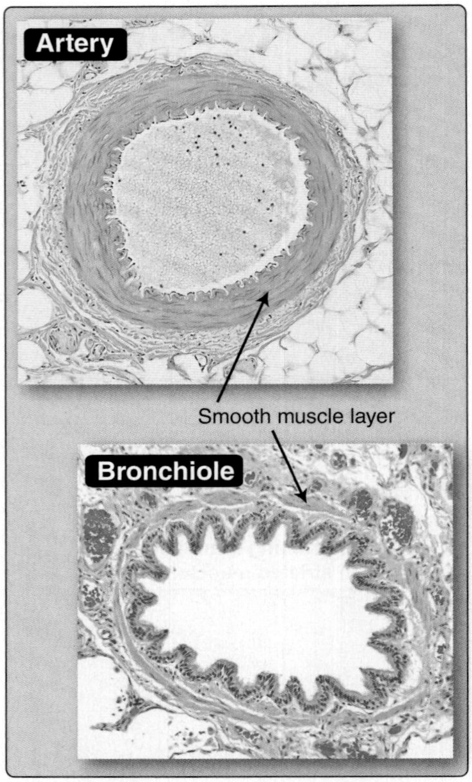

Figure 14.1
Arterial and airway smooth muscle.

II. STRUCTURE

Smooth muscle develops force through actin–myosin interactions, but the contractile filaments are organized less formally than they are in striated muscle.

Table 14.1: Smooth Muscle Functions

Location	Function
Blood vessels	Control diameter, regulate blood flow
Lung airways	Control diameter, regulate air flow
Gastrointestinal tract	Tone, motility, sphincters
Urinary system	Propel urine through ureter, bladder tone, internal sphincter
Male reproductive tract	Secretion, propel semen
Female reproductive tract	Propulsion (fallopian tube), childbirth (uterine myometrium)
Eye	Control of pupil diameter (iris muscle) and lens shape (ciliary muscle)
Kidney	Regulate blood flow (mesangial cells)
Skin	Hair erection (pili muscles)

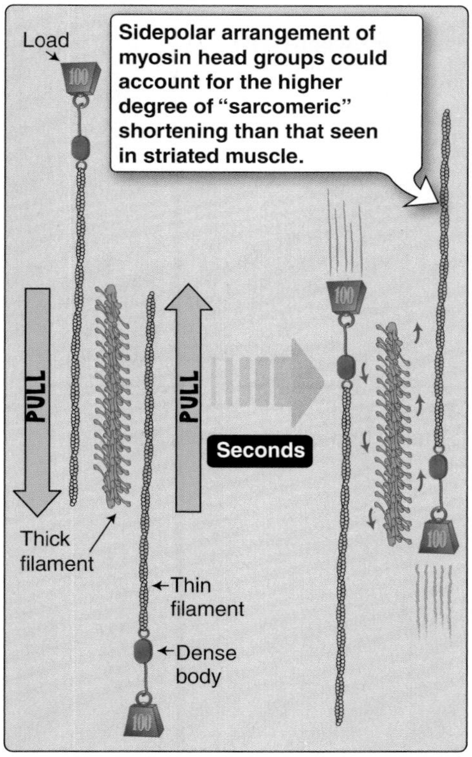

Figure 14.2
A possible model for smooth muscle contraction.

A. Contractile units

Smooth muscle cells are densely packed with actin and myosin filaments, which are organized into contractile units that have been referred to as "minisarcomeres." However, because the contractile units are not aligned by Z disks as they are in skeletal and cardiac muscle, there are no visible striations when viewed under polarized light. The lack of striations gives "smooth muscle" its name.

1. **Thick filaments:** Smooth muscle myosin has a similar tertiary structure to that of striated muscle, but its amino acid sequence is quite distinct. Smooth muscle myosin is hexameric like that of other muscle types, containing two **heavy chains** that include head and neck regions and two pairs of **light chains** (a 17–kDa **essential light chain** and a 20–kDa **regulatory light chain**, also known as **MLC_{20}**). The myosin heads contain ATPase activity and an actin-binding site. Although the ultrastructural basis for smooth muscle contraction has not been elucidated, one suggestion is that the myosin head groups within a thick filament have a sided or "**sidepolar**" arrangement that allows two actin filaments to be pulled simultaneously in different directions (Figure 14.2). Such an arrangement could account for observations that smooth muscle myocytes can shorten to a far greater degree than can striated muscle fibers.

2. **Thin filaments:** Thin filaments in striated muscle are associated with troponin, which imparts Ca^{2+} dependence to contraction. Smooth muscle does not contain troponin, but it does associate with two muscle-specific regulatory proteins called **caldesmon** and **calponin**. Their properties have been studied extensively *in vitro*, but their roles *in vivo* are still debated.

    a. **Caldesmon:** Caldesmon is an actin-associated myosin ATPase inhibitor. Inhibition is relieved by high **Ca^{2+}-calmodulin** (**CaM**) concentrations or by phosphorylation by *Ca^{2+}-CaM–dependent protein kinase* (or several other endogenous *kinases*).

    b. **Calponin:** Calponin is an abundant actin-associated myosin ATPase inhibitor that is regulated by *Ca^{2+}-CaM–dependent protein kinase*–dependent phosphorylation.

3. **Assembly:** For a muscle cell to exert a useful force, its contractile units have to align roughly with the long axis of the cell. In smooth muscle, this seems to involve parallel assemblies of 3–5 thick filaments, each surrounded by numerous (10–12) thin filaments. Both thick and thin filaments appear to have variable lengths. This contrasts with striated muscle, where filament length is standardized and strictly controlled. The filaments are anchored by α-actinin–rich **dense bodies** (**desmosomes**) found scattered throughout the sarcoplasm (Figure 14.3) and believed to be the functional equivalent of Z disks in striated muscle (see 12·II·C·1). Dense bodies do not align as they do in striated muscle, but the general arrangement of thin filaments, thick filaments, and dense bodies does resemble a sarcomere.

B. Organization

Smooth muscle cells do not contain myofibrils as such, but long assemblies of contractile filaments and associated dense bodies appear to stretch the length of the cell. Contractile arrays are suspended

within the cytoskeletal framework by a network of intermediate filaments comprising **vimentin** and **desmin**, the relative amounts depending on smooth muscle type. The arrays are tethered to the sarcolemma by **dense plaques**, which are related to dense bodies. The plaques are distributed over the entire cell surface and link adjacent cells mechanically (**adherens junctions**; see 4·II·G and Figure 14.3) so that smooth muscle may exert directed force upon contraction.

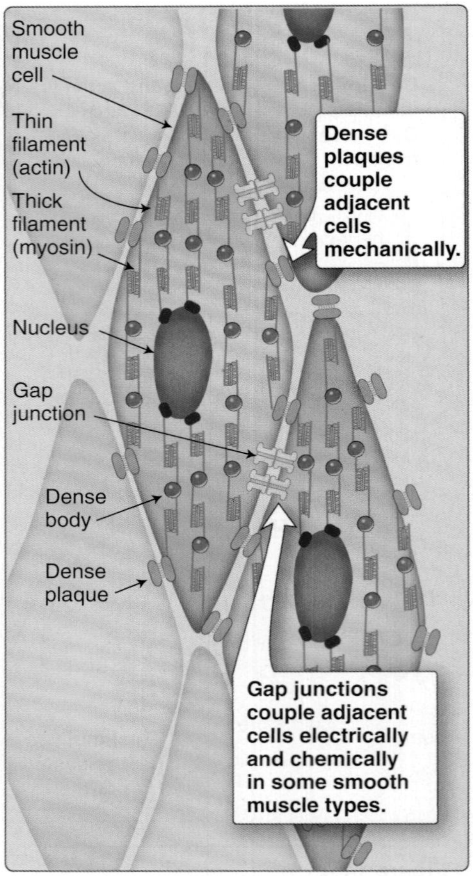

Smooth muscle cell
Thin filament (actin)
Thick filament (myosin)
Nucleus
Gap junction
Dense body
Dense plaque
Dense plaques couple adjacent cells mechanically.
Gap junctions couple adjacent cells electrically and chemically in some smooth muscle types.

Figure 14.3
Suggested contractile filament alignment within a myocyte.

C. Membrane systems

Crossbridge cycling begins when intracellular Ca^{2+} concentrations rise above rest (0.1 μmol/L). In striated muscle, this involves **L-type Ca^{2+} channels** (**dihydropyridine receptors**) in transverse (T)-tubule membranes and **Ca^{2+} release** from the **sarcoplasmic reticulum** (**SR**) via **ryanodine receptors** (see 12·III). Although the structural relationships between contractile arrays and SR is less well organized in smooth muscle compared with striated muscle, smooth muscle employs similar mechanisms to initiate contraction.

1. **Caveolae:** Smooth muscle cells do not contain T tubules that carry action potentials deep into the body of a muscle fiber, but they do show linear arrays of **caveolae** that may have a related function. Caveolae are 50–100 nm flask-shaped sarcolemmal pockets that form narrow junctions (15–30 nm) with the underlying SR. Caveolae are enriched in L-type Ca^{2+} channels, suggesting that the junctions serve a role similar to that of diads and triads in striated muscle.

2. **Sarcoplasmic reticulum:** Smooth muscle cells contain extensive tubular networks of SR that store Ca^{2+} until contraction begins. Unlike striated muscle, smooth muscle SR contains two types of Ca^{2+}-release channels, one activated by Ca^{2+}, the other by **inositol trisphosphate** (**IP_3**) (Figure 14.4).

   a. **Calcium-induced calcium-release channels:** Ca^{2+}-induced Ca^{2+} release (CICR) channels are opened by Ca^{2+} entering the myocyte via voltage-dependent Ca^{2+} channels in the surface membrane. Although CICR is the primary means by which contraction is initiated in striated muscle, its role in smooth muscle is less clear.

   b. **Inositol trisphosphate–gated calcium channels:** Smooth muscle SR also contains an IP_3-gated Ca^{2+} channel. IP_3 is a second messenger that communicates binding of one or more chemical signals, including many hormones and neurotransmitters (e.g., norepinephrine and acetylcholine), at the cell surface.

D. Neuromuscular junction

Most smooth muscle types are controlled by the **autonomic nervous system** (**ANS**). Depending on function and location within the body, smooth muscle may receive inputs from the **sympathetic nervous system**, **parasympathetic nervous system**, and **enteric nervous system**. The ANS neuromuscular junction is less developed than in skeletal muscle, but pre- and postsynaptic structures are arranged similarly. ANS efferents may contact multiple smooth muscle cells via a series of **varicosities** (swellings) spaced along the length of an axon, each of which is the site of a neuromuscular junction (see Figure 7.6).

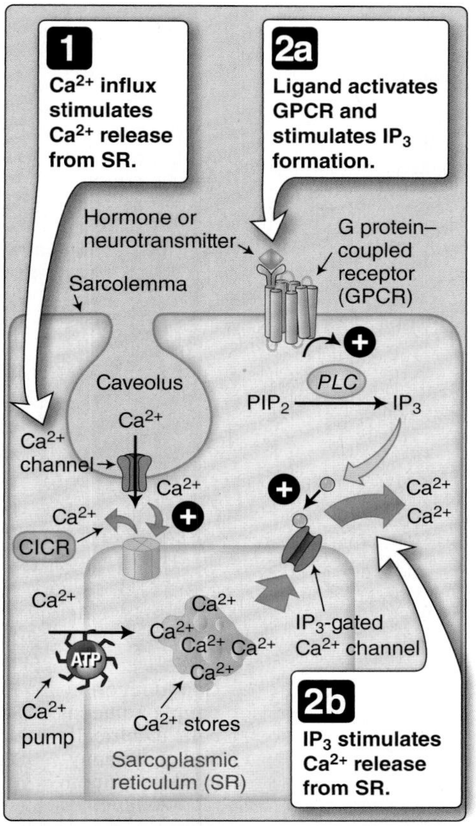

Figure 14.4
Excitation–contraction coupling in smooth muscle. ATP = adenosine triphosphate; CICR = Ca^{2+}-induced Ca^{2+} release; IP_3 = inositol trisphosphate; PIP_2 = phosphatidylinositol 4,5-bisphosphate; *PLC = phospholipase C.*

Clinical Application 14.1: Irritable Bowel Syndrome

Irritable bowel syndrome is a gastrointestinal (GI) disorder associated with intestinal cramping, increased flatulence, and altered bowel habits. It has no known cause or cure. Treatment options are limited, but include oral antispasmodics such as L-hyoscamine, an atropine analog.[1] Atropine is an alkaloid cholinergic receptor antagonist that blocks parasympathetic nervous system–induced increases in smooth muscle contractility in the GI tract and urinary bladder. The muscle relaxes as a result, which decreases cramping-associated pain. Natural remedies include peppermint oil, which has a similarly relaxing effect on smooth muscle. It is believed to act by blocking Ca^{2+} channels in the smooth muscle plasmalemma, thereby reducing contractility and relaxing the muscle. Because peppermint oil relaxes sphincters also, it is contraindicated in patients with gastroesophageal reflux disease (GERD). GERD is associated with lower esophageal sphincter (LES) dysfunction. Because the LES is composed entirely of smooth muscle, peppermint oil further impairs sphincter contractility and exacerbates GERD symptoms.

III. EXCITATION–CONTRACTION COUPLING

The number of competing signals that vie for control of smooth muscle function is immense (hundreds). Some of these signals may promote contraction and others relaxation, but, ultimately, most converge on Ca^{2+}, just as in other muscle types. The main difference between smooth muscle and striated muscle is that the Ca^{2+}-sensitive step has been transferred from the thin filament (via troponin and tropomyosin) to the thick filament (via myosin phosphorylation).

A. Calcium source

Contraction begins when sarcoplasmic Ca^{2+} concentrations rise. There are three potential mechanisms by which this can occur in smooth muscle: Ca^{2+} influx across the sarcolemma, CICR from the SR, and IP_3-mediated Ca^{2+} release from the SR. The relative contributions of these pathways to the net rise in intracellular Ca^{2+} concentration varies according to smooth muscle type.

1. **Calcium influx:** Smooth muscle cells express at least two types of Ca^{2+} channels. L-type Ca^{2+} channels that are found concentrated in caveolae are voltage-gated. The channels open in response to membrane depolarization to mediate Ca^{2+} influx and the upstroke of an action potential commonly seen in detrusor muscle and visceral muscle, for example. Most smooth muscle cells

[1]For a discussion of antispasmodics, see *LIR Pharmacology,* 5e, p. 61.

also express one or more **receptor-operated Ca^{2+} channels** (**ROCCs**). For example, visceral smooth muscle expresses muscarinic ROCCs, whereas vascular smooth muscle expresses adrenergic ROCCs. Although ROCC-mediated Ca^{2+} fluxes are relatively minor and generally insufficient to support contraction on their own, they do depolarize cells to potentiate (or initiate) Ca^{2+} influx and contraction via L-type Ca^{2+} channels.

2. **Calcium-induced calcium release:** CICR can be visualized as **"Ca^{2+} sparks"** in fluorescent imaging studies of smooth muscle, but their role is still being investigated. In some smooth muscle types (e.g., detrusor muscle), CICR may potentiate contraction, whereas in others (e.g., vascular smooth muscle), Ca^{2+} release relaxes the muscle by opening Ca^{2+}-dependent K^+ channels that hyperpolarize the membrane and decrease L-type Ca^{2+} channel open probability.

3. **Inositol trisphosphate:** Most smooth muscle types express a rich variety of **G protein–coupled receptors** (**GPCRs**) that modulate contraction through *phospholipase C* and IP_3 formation. When intracellular IP_3 concentrations rise, the IP_3-gated Ca^{2+} channel in the SR opens, and the Ca^{2+} stores are released (see Figure 14.4). IP_3-mediated Ca^{2+} release and smooth muscle contraction can (and frequently does) occur independently of an action potential or other membrane potential change and is known as **pharmacomechanical coupling**. The IP_3 pathway is a primary means of initiating contraction in smooth muscle.

B. Contraction

In striated muscle, rising intracellular Ca^{2+} concentrations prompt troponin to move tropomyosin away from myosin-binding sites on the actin filament (see 12·III·B). Striated muscle myosin has a high intrinsic ATPase activity, and, with the binding sites exposed, contraction proceeds rapidly. In smooth muscle, the ATPase remains inactive until the MLC_{20} regulatory light chain is phosphorylated. A rise in sarcoplasmic Ca^{2+} is sensed by CaM, which then activates *myosin light-chain kinase* (*MLCK*) as shown in Figure 14.5. *MLCK* phosphorylates MLC_{20} and, in the presence of ATP, the myosin head group is now able to reach forward and attach to a binding site on the actin thin filament. Crossbridge cycling then proceeds by essentially the same mechanism as described for striated muscle (see 12·III·B). Because myosin phosphorylation is required for smooth muscle contraction, it is said to be **thick-filament regulated** as opposed to **thin-filament regulated** in striated muscle. The myosin isoform's kinetics and the multistep nature of the activation process means that smooth muscle contracts 10–20 times slower than does skeletal muscle.

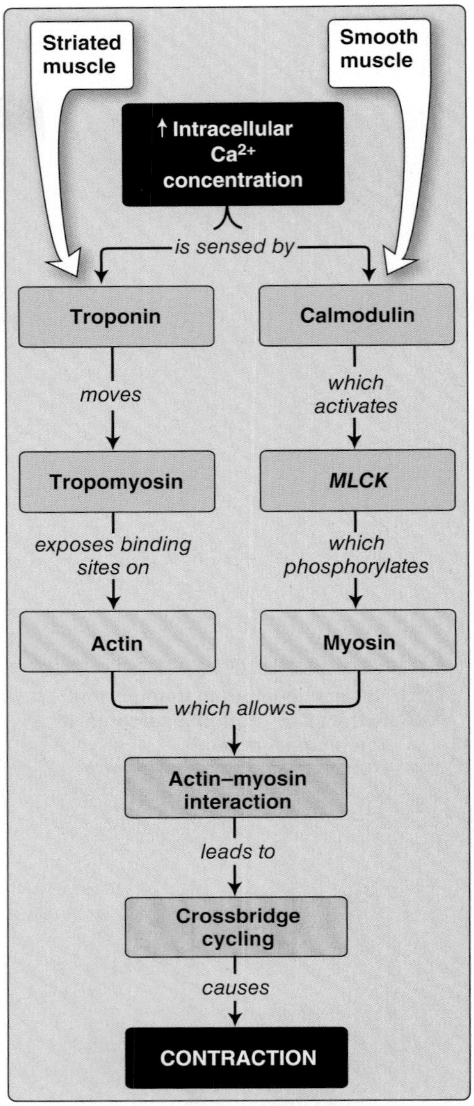

Figure 14.5
Differences between events facilitating crossbridge cycling in striated and smooth muscle. *MLCK = myosin light-chain kinase.*

C. Relaxation

Muscle relaxation occurs when sarcoplasmic Ca^{2+} concentrations renormalize. Because smooth muscle contraction involves a *kinase* and a phosphorylation step, relaxation requires a *phosphatase*.

1. **Calcium renormalization:** When excitatory signaling ends, Ca^{2+} is expelled from the cell by Ca^{2+} ATPases and Na^+-Ca^{2+} exchangers,

Clinical Application 14.2: Erectile Dysfunction

Penile erection during male sexual arousal results from increased blood flow and engorgement of three sinusoidal tissue cavities within the penis (the **corpora cavernosae** and **corpus spongiosum**). Blood flow to these cavities is regulated in part by parasympathetic nerves that signal via nitric oxide (NO) release. NO relaxes intracavernosal trabeculae, which comprise smooth muscle cells lined by endothelium. Relaxation allows the erectile tissue to fill with blood in preparation for sexual activity. NO is a short-lived signaling molecule that exerts its effects through activation of *guanylyl cyclase*, cyclic guanosine monophosphate (cGMP) formation, and phosphorylation of proteins involved in regulation of intracellular Ca^{2+} concentration by *cGMP-dependent protein kinase*. Men who have difficulty achieving or sustaining an erection have benefited from the availability of drugs such as sildenafil and tadalafil, which are *phosphodiesterase* (*PDE*) inhibitors.[1] *PDEs* degrade NO and terminate signaling. Inhibiting *PDEs* increases local NO levels, allowing an erection to be sustained.

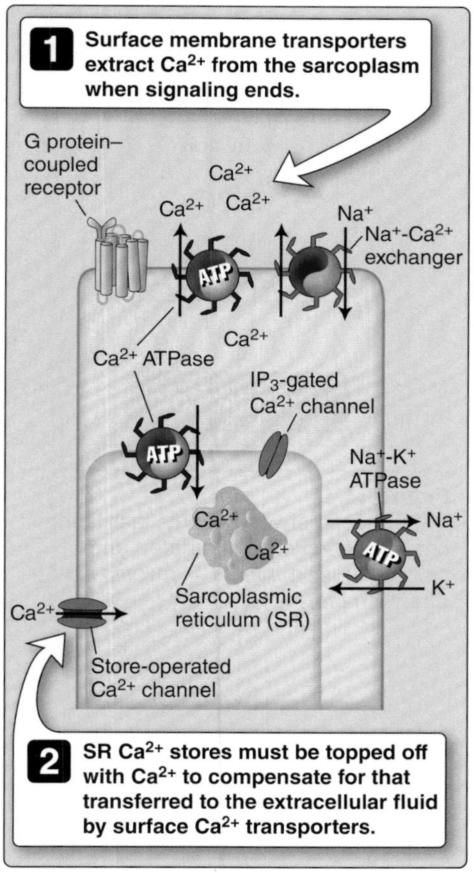

Figure 14.6
Ca^{2+} handling during smooth muscle relaxation. ATP = adenosine triphosphate; IP_3 = inositol trisphosphate.

and returned to the SR by a sarco(endo)plasmic reticulum Ca^{2+} ATPase ([SERCA] Figure 14.6). *MLCK* then deactivates.

> Compounds that modulate smooth muscle contractility through *protein kinase A* and *protein kinase G*–dependent pathways (e.g., nitric oxide) do so by influencing SERCA and L-type Ca^{2+}-channel activity.

2. **Store refill:** The plasmalemmal Ca^{2+} transporters compete with SERCA for available Ca^{2+} during relaxation, so the stores have to be topped off with Ca^{2+} from the outside of the cell via a **store-operated Ca^{2+} channel** (**SOC**). This is a critical step that must be completed for continued smooth muscle contractility. SOCs are common to many cell types, including lymphocytes, where a need for topping off is sensed by **Stim1**, a Ca^{2+} sensor, and requires **Orai**, a Ca^{2+} channel or Ca^{2+}-channel subunit.

3. **Dephosphorylation:** Once MLC_{20} is phosphorylated, cross-bridge cycling continues for as long as ATP is available to power contraction. Relaxation of smooth muscle is, therefore, dependent on *myosin phosphatase*, which is constitutively active and always undoing the work of *MLCK*. *Myosin phosphatase* is a protein trimer comprising a *protein phosphatase* catalytic domain (PP1c), a myosin binding subunit (MYPT1), and a small subunit of unknown function. When sarcoplasmic Ca^{2+} concentrations fall and *MLCK*

[1]For a discussion of *phosphodiesterase* inhibitors, see *LIR Pharmacology*, 5e, p. 363.

deactivates, *myosin phosphatase* quickly strips MLC_{20} of phosphate groups and the muscle relaxes.

D. Latchbridge formation

If myosin is dephosphorylated when still attached to actin, it locks or "latches" in place. Latchbridges have an intrinsically slow cycling rate that allows blood vessels, sphincters, and hollow organs such as the bladder to sustain contractions for prolonged periods with minimal ATP use (~1% of the amount required by skeletal muscle to achieve the same effect). Because latchbridges are contractile events, they can only occur if intracellular Ca^{2+} levels remain minimally elevated above background. Although many details of latchbridge formation, termination, and regulation have yet to be elucidated, the ability to form latchbridges is a unique and fundamental property of smooth muscle.

E. Regulation

In striated muscle, force regulation usually occurs through control of intracellular Ca^{2+} concentration. In smooth muscle, force is regulated via changes in MLC_{20} phosphorylation state. Force development is dependent on crossbridge formation, which can only occur when MLC_{20} is phosphorylated. Because MLC_{20} phosphorylation state is dependent on both *MLCK* and *myosin phosphatase*, there are multiple potential control points. Two principal regulatory pathways involve *Rho-kinase* (*ROCK*) and *protein kinase C* (*PKC*) as shown in Figure 14.7.

1. ***Rho-kinase:*** *ROCK* is a *serine–threonine protein kinase* regulated by RhoA, a GTP-binding protein. RhoA is activated indirectly following GPCR binding by, for example, norepinephrine, angiotensin II, endothelin, or any of a number of other ligands. *ROCK* has a several targets, including the MYPT1 *myosin phosphatase* myosin binding subunit. Phosphorylation inhibits *myosin phosphatase* activity and thereby promotes contraction. *ROCK* also has *MLCK*-like activity and stimulates contraction through direct effects on MLC_{20}. Note that this pathway acts independently of any changes in intracellular Ca^{2+} concentration.

2. ***Protein kinase C:*** Agonists that activate *phospholipase C* and promote contraction via IP_3 simultaneously activate *PKC* via diacylglycerol release. *PKC* also phosphorylates many proteins that regulate contraction, including CPI-17. CPI-17 is a endogenous 17-kDa protein that becomes a potent *myosin phosphatase* inhibitor when phosphorylated, thereby promoting contraction. *ROCK* also phosphorylates CPI-17 *in vitro*, but the possible significance for smooth muscle contraction function *in vivo* is unresolved.

> The *ROCK*- and *PKC*-mediated contractility increases account for a phenomenon known as **"Ca^{2+} sensitization."** Ca^{2+} sensitization refers to an observed shift in the Ca^{2+}-dependence of contractility toward lower intracellular free Ca^{2+} values following hormone or transmitter binding.

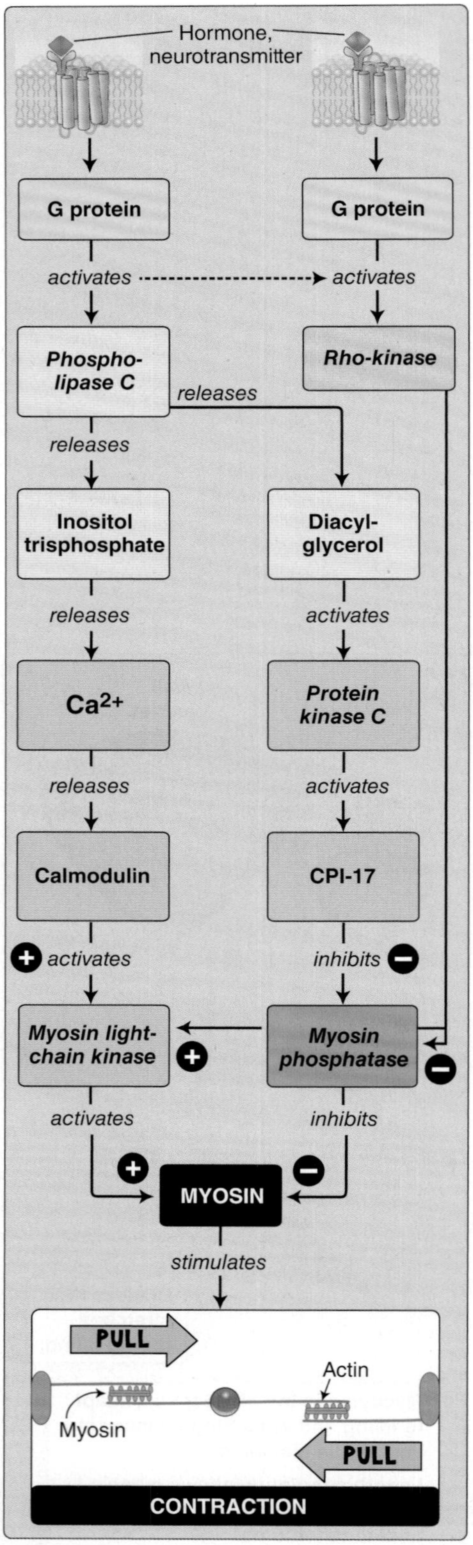

Figure 14.7
Pathways regulating smooth muscle contraction.

Clinical Application 14.3: *Rho-kinase* and Hypertension

Essential hypertension is a vascular smooth muscle disease. The underlying cause and cellular pathways involved are still largely unknown, despite hypertension's prevalence and cost, both financially and in terms of its effects on the health of the individual. High blood pressure reflects an inappropriate systemic vasoconstriction, forcing the left ventricle to generate higher arterial pressures to force blood through the narrowed vessels. Because *rho-kinase* (*ROCK*) is a primary regulator of smooth muscle contractility, the *ROCK* signaling pathway has come under scrutiny as a possible cause of and means of treating hypertension. Fasudil is a *ROCK*-specific *kinase* inhibitor currently used in Japan to treat the cerebral vasospasm that commonly follows subarachnoid hemorrhage. The inhibitor relaxes most smooth muscle types and is currently being investigated as a possible treatment option for essential and pulmonary hypertension.

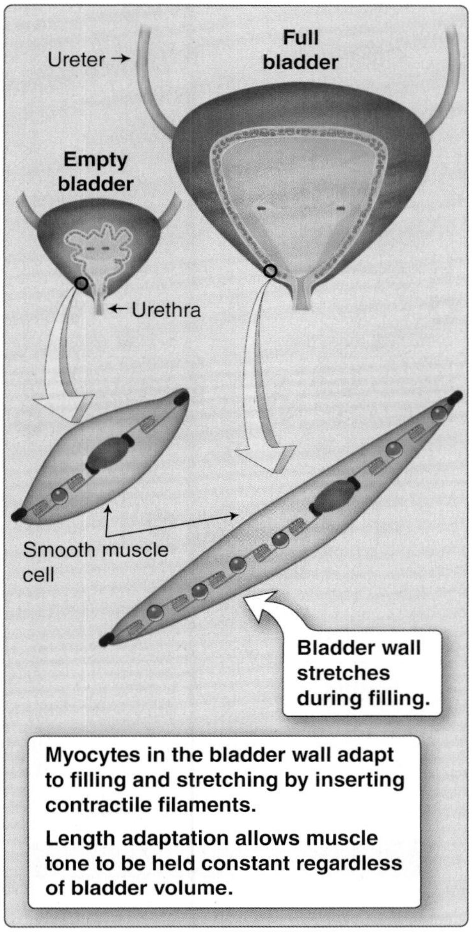
Ureter
Full bladder
Empty bladder
Urethra
Smooth muscle cell
Bladder wall stretches during filling.
Myocytes in the bladder wall adapt to filling and stretching by inserting contractile filaments.
Length adaptation allows muscle tone to be held constant regardless of bladder volume.

Figure 14.8
Length adaptation during urinary bladder filling.

IV. MECHANICS

Striated muscle and smooth muscle use the same basic principles for force generation so their mechanics are similar. Optimizing the degree of thick and thin filament overlap (preloading) maximizes force development, and contraction velocity slows as the afterload increases. There are notable differences, however, as discussed below.

A. Contraction rate

Smooth muscle is capable of contracting rapidly when the need arises; when eye pupil size must be decreased to reduce the amount of light falling on the retina, for example. The myosin isoform found in smooth muscle has inherently slow ATPase activity, however, which limits maximum contraction velocity to only a fraction of that seen in skeletal muscle.

B. Length adaptation

Smooth muscle regulates wall tone in many hollow organs whose lumen size is variable. For example, airways and blood vessels rhythmically dilate and constrict in time with the respiratory and cardiac cycles, respectively, yet the smooth muscle contained within their walls maintains a constant tone throughout the cycle. Similarly, urinary bladder luminal volume increases from ~6 to ~500 mL when full, yet the detrusor muscle within its walls can contract and expel urine any point during filling cycle. During bladder filling, the muscle "minisarcomeres" are stretched to their physiologic limits. Stretching reduces the degree of actin–myosin overlap and limits contractility, as in other muscle types, but smooth muscle is unique in that it adapts to this stress over a period of minutes and regains full contractility even at an increased length. This phenomenon is known as **length adaptation** and is a fundamental property of smooth muscle. The pathways involved have not been well delineated, but the phenomenon

represents one of the key differences between striated muscle and smooth muscle. Whereas the length and number of contractile filaments in striated muscle is largely fixed and the contractile arrays ordered so as to optimize force production along a constant vector, smooth muscle is a fundamentally plastic cell type. When stretched, myocytes are believed to replicate contractile units and insert them in series with existing assemblies (Figure 14.8). Preexisting filaments are probably lengthened also. When the muscle returns to normal length, it again adapts, probably by removing the additional contractile units or shortening the contractile arrays. Note that the entire cytoskeletal framework must necessarily be remodeled to accommodate such changes. The pathways and proteins involved may be the same as those involved in initiating and regulating contraction.

V. TYPES

Smooth muscle is a diverse tissue type that can be classified in many ways, but one of the more useful is based on function. **Phasic smooth muscle** contracts transiently when stimulated. Examples include muscles that make up the walls of the GI tract (stomach, small intestine, large intestine), and urogenital tract (ureters, urinary bladder, vas deferens, fallopian tube, uterus). **Tonic smooth muscle** is capable of sustained contractions, a feature that is often used to maintain a constant muscular tone. Examples include vascular and airway muscle, sphincters (e.g., lower esophageal sphincter, pyloric sphincter), and eye ciliary and iris muscles.

> Most smooth muscle types are a blend of phasic and tonic, which allows them to respond to a range of stimuli.

A. Phasic

Phasic smooth muscle often functions like cardiac muscle. Specialized pacemaker cells generate action potentials that spread via gap junctions from myocyte to myocyte until the whole tissue is involved.

1. **Pacemakers:** Some smooth muscle cells are capable of generating spontaneous membrane potential (V_m) changes. In the GI tract, these may manifest as "slow waves" with a periodicity of ~3–5 per minute. If the waves achieve an amplitude sufficient to cross threshold, action potentials may be initiated that spread via gap junctions throughout the entire muscle body. A wave of contraction follows in its wake (Figure 14.9A). Because all myocytes within phasic muscle are excited as a single unit, phasic muscle may also be referred to as **"unitary"** smooth muscle.

2. **Action potentials:** Smooth muscle action potentials are characteristically slow and their form and time course highly variable. The upstroke is mediated by L-type Ca^{2+} channels, as is the upstroke of cardiac nodal cell "slow" action potential (see 17·IV·B). Ca^{2+} influx simultaneously initiates contraction and opens Ca^{2+}-activated K^+ channels ("BK" channels). Ca^{2+}-channel inactivation and BK-mediated K^+ efflux together help repolarize the membrane and yield the action potential downstroke.

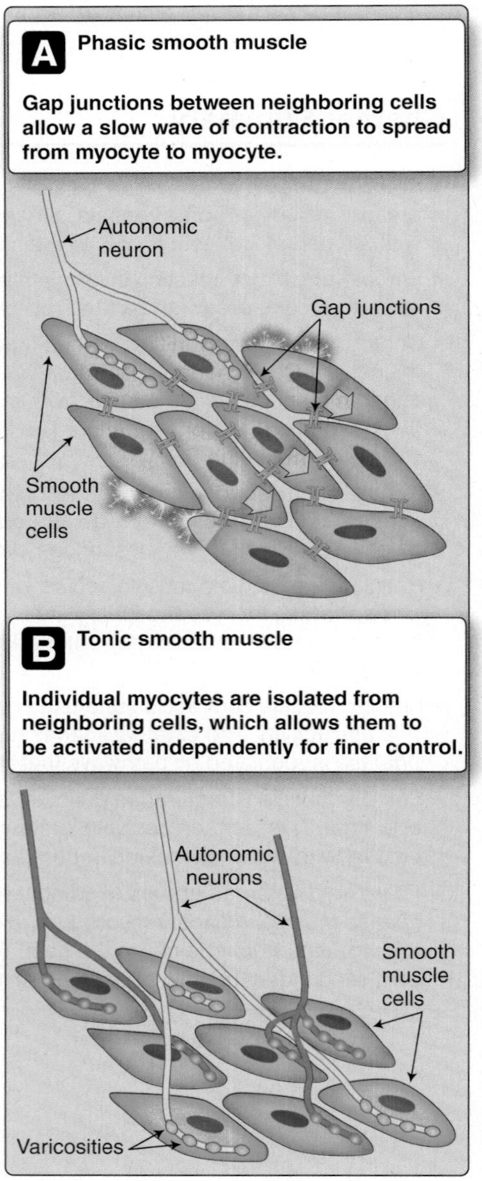
A
Phasic smooth muscle
Gap junctions between neighboring cells allow a slow wave of contraction to spread from myocyte to myocyte.
Autonomic neuron
Gap junctions
Smooth muscle cells
B
Tonic smooth muscle
Individual myocytes are isolated from neighboring cells, which allows them to be activated independently for finer control.
Autonomic neurons
Smooth muscle cells
Varicosities

Figure 14.9
Contractility control in phasic and tonic smooth muscle.

B. Tonic

Tonic smooth muscle (also known as **multiunit** smooth muscle) resembles skeletal muscle in that individual myocytes or groups of myocytes function independently of their neighbors (see Figure 14.9B). This feature allows for fine control over movements, which is advantageous for precise control of pupillary diameter and eye lens shape, for example. Multiunit control also allows force to be ramped up through recruitment, much as skeletal muscle force is controlled using combinations of discrete motor units (see 12·IV·D). Tonic smooth muscle typically does not generate action potentials and, indeed, contraction may occur in the absence of any V_m change through pathways described in Section III above.

Chapter Summary

- **Smooth muscle** serves many diverse functions in essentially all areas of the body. It is found in the walls of many hollow organs, including **blood vessels, airways**, **intestines**, and **urogenital tract**. Smooth muscle contracts slowly compared with striated muscle but is able to maintain a steady tone with minimal energy expenditure.
- Smooth muscle contraction involves interaction between actin and myosin, but the thick and thin filaments are loosely organized compared with striated muscle.
- Contraction is usually initiated when intracellular Ca^{2+} concentrations rise. Ca^{2+} may originate extracellularly and enter the cell via Ca^{2+} channels, or from Ca^{2+} stores in the **sarcoplasmic reticulum** (**SR**). The SR releases its Ca^{2+} stores either in response to Ca^{2+} influx from the outside of the cell (Ca^{2+}-induced Ca^{2+} release) or a rise in intracellular inositol trisphosphate (IP_3) concentration. The latter occurs as a result of surface receptor binding and acts via an **IP_3-gated Ca^{2+} channel** in the SR.
- Control of crossbridge cycling in smooth muscle is through myosin phosphorylation. Rising intracellular Ca^{2+} concentrations cause Ca^{2+}–calmodulin-dependent activation of ***myosin light-chain kinase*** (***MLCK***). *MLCK* phosphorylates the myosin head group, and crossbridge cycling begins.
- Relaxation of smooth muscle occurs when Ca^{2+} levels fall, and *myosin light-chain kinase* deactivates. ***Myosin phosphatase*** then dephosphorylates myosin and allows relaxation to occur. When Ca^{2+} levels are barely above baseline, smooth muscle may enter a **latch state** in which muscle tone is maintained for prolonged periods with minimal energy use.
- Smooth muscle contractile state usually represents a balance between *myosin light-chain kinase* and *myosin phosphatase* activity, and external ligands are able to modulate contractility by manipulating this balance. *Myosin phosphatase* is regulated by pathways that include ***protein kinase C*** and ***Rho-kinase***, both of which potentiate contraction.
- Smooth muscle is required to maintain a steady tone in hollow organs whose internal volume changes appreciably over time. This is made possible by **length adaptation**, a process that allows the length–tension relationship to shift in parallel with organ expansion or contraction. Length adaptation may involve sarcomeric and cytoskeletal remodeling.
- There are two broad groups of smooth muscle. **Phasic** (**unitary**) smooth muscle functions as a single unit, much like cardiac muscle. Adjacent myocytes are connected via gap junctions, which allows for waves of excitation and contraction to propagate from one cell to the next. **Tonic** (**multiunit**) smooth muscle is composed of myocytes that function and are controlled independently of each other, an organization reminiscent of skeletal muscle.

Bone

15

I. OVERVIEW

The brain and many other soft tissues of the body are enclosed within a protective framework made of bone. Bone also fashions limbs that, together with skeletal muscles, facilitate locomotion and allow objects to be manipulated. The 206 bones that make up the human **skeleton** work in conjunction with cartilage, ligaments, tendons, and skeletal muscle. These tissues help maintain skeletal structure and control bone movement. Bone is a connective tissue (see 4·IV·A) that has been mineralized to give it high resilience to stress and trauma. The mineral component of bone persists long after the body has died and the soft tissues have decomposed, but bones are not lifeless structures. Bone is honeycombed with tunnels and cavities that are teeming with cells and that provide channels through which blood vessels and nerve fibers permeate the matrix. Many bones also contain a central chamber filled with marrow that manufactures blood cells and stores fat (Figure 15.1). Finally, bone is a highly dynamic tissue that is constantly being remodeled and turned over. Remodeling is partly an adaptive response to mechanical stress, but it also reflects bone's immense repositories of calcium and phosphate (>99% and 80% of body total, respectively) that can be mobilized when plasma concentrations fall below optimal.

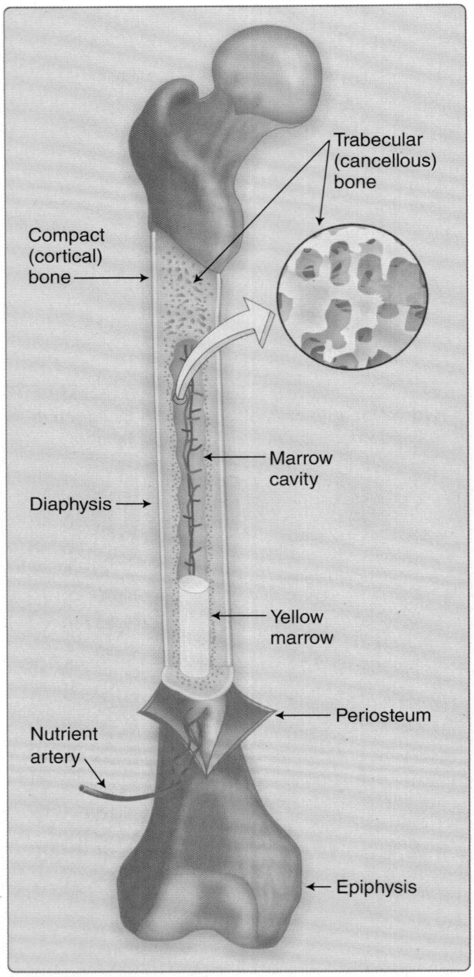

Figure 15.1
Bone structure.

II. FORMATION

Bone structure reflects two competing needs. Bones must be strong to protect the soft tissues from mechanical trauma and to support body weight during locomotion. Bone's resilience is afforded in part by minerals, which makes bones heavy. On the other hand, bones must be sufficiently light to allow for rapid responses to external threats (e.g., from predators). Light bones are prone to fracture, however. Bone breakage is a potentially lethal event (through predation, hemorrhage, or circulatory shock) and, therefore, must be avoided at all cost. Thus, bone design represents a compromise between strength and weight. In practice, bones contain just enough minerals to withstand normal mechanical stress limits plus an added safety margin.

Table 15.1: Bone Composition

Component (% by wt)	Composition
Organic (30%)	Cells (~2%) Type I collagen (~93%) Ground substance (~5%)
Inorganic (70%)	Ca^{2+} and PO_4^{3-} crystals

Limb bones typically fracture when subjected to stresses that cause them to deform by three to four times that experienced during normal physiologic activity. If stress levels increase chronically (e.g., during weight training), bone remodels to compensate for the added stress (the Wolff law) and to reestablish normal safety margins.

Most of a bone's mineral content and strength is concentrated in a thin, highly compacted outer layer. This construction is similar to the hollow steel tubing used to fashion the legs of chairs and tables. Bone's interior is not air filled but is composed of a light, highly porous matrix. Resistance to compression and mechanical shear is accomplished using a resilient mix of minerals and proteins (Table 15.1).

A. Mineral component

Resistance to compression is achieved using thin, tabular **hydroxyapatite** crystals measuring ~50 nm long. Hydroxyapatite is a mineral comprising calcium and phosphate (Ca_{10} $[PO_4]_6$ $[OH]_2$) that occurs naturally in stalagmites and mineral crusts.

B. Protein scaffolding

Hydroxyapatite crystals do not readily compress, but they do shear. Taking advantage of their natural properties requires that they be cemented and tethered within collagen fibers (Figure 15.2). The cement is made from mucopolysaccharides and is rich in Mg^{2+} and Na^{+} (~25% of total body Na^{+} is contained within bone). Collagen is the principal component of tendons, tissues notable for their flexibility and high resistance to tensile (stretching) and shear stress. When the collagen fibers with their embedded hydroxyapatite crystals are cemented together by **ground substance** (see 4·IV·B·3), they create a material that is able to support heavy loads and resist mechanical impact yet is flexible enough to torque and bend without fracturing. Similar construction techniques are used to make adobe (a mix of straw and mud) and reinforced concrete (a mix of steel rods or "rebar" and cement).

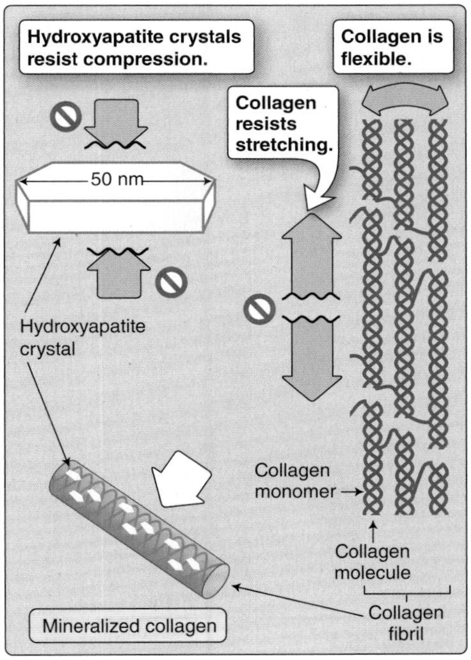

Figure 15.2
Mineralized collagen formation.

C. Assembly

Bone is formed by **osteoblasts**, which are related to fibroblasts (see 4·IV·B). Bone formation can be divided into four steps: collagen deposition; secretion of ground substance; crystal seeding; and, finally, maturation.

1. **Collagen deposition:** Bone formation begins with the creation of **collagen** scaffolding. Collagen molecules comprise chains of tropocollagen subunits, each composed of three polypeptide chains braided into a helix. Monomers join end to end and then spontaneously assemble into **collagen fibrils**, which are the cellular equivalent of steel rebar in concrete. Monomers within the fibrils are crosslinked extensively and staggered like bricks in a wall for additional stability and strength.

Clinical Application 15.1: Osteogenesis Imperfecta

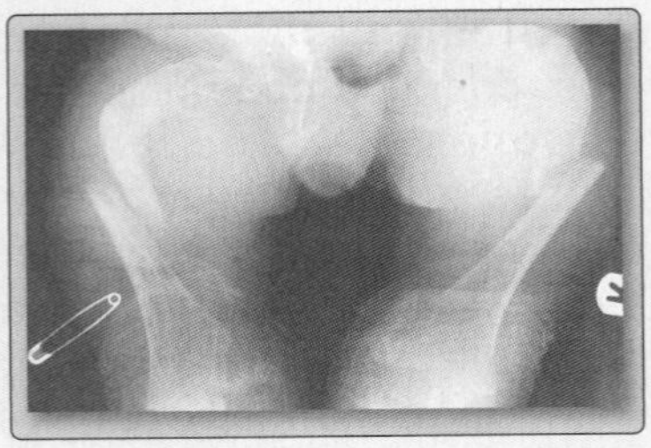

Fractures in an infant with osteogenesis imperfecta.

Osteogenesis imperfecta, or "brittle bone disease," results from a group of inherited defects in two type I collagen genes. The most common is a point mutation that replaces glycine with a bulkier amino acid at a site where the three strands of the helix normally come close together. The result is a molecular "bleb" that interferes with normal fibril formation and packing of hydroxyapatite crystals. Bones formed with such collagen are weak and prone to fracture.[1]

2. **Ground substance:** Ground substance is an amorphous, gel-like matrix of glycosaminoglycans, proteoglycans, glycoproteins, salts, and water that fills the space between cells in all tissues. Bone ground substance differs from that found in other tissue in that it is saturated with Ca^{2+} and PO_4^{3-}. The combination of collagen fibrils and ground substance is called **osteoid** (Figure 15.3).

3. **Crystal seeding:** Osteoblasts continue secreting Ca^{2+} and PO_4^{3-} into ground substance until it becomes supersaturated, at which point the minerals begin precipitating as calcium phosphate crystals. Osteoblasts also secrete seed crystals that are cemented to collagen fibrils to provide nucleation sites for continued growth. The fibrils slowly become encrusted with amorphous mineral deposits.

4. **Maturation:** In succeeding months, the calcium phosphate crystals are remodeled by osteoblasts to form mature hydroxyapatite. The tabular hydroxyapatite crystals are tethered to collagen fibrils by proteoglycans, and the fibrils themselves become extensively interlinked to give bone the tensile strength that approaches that of structural steel.

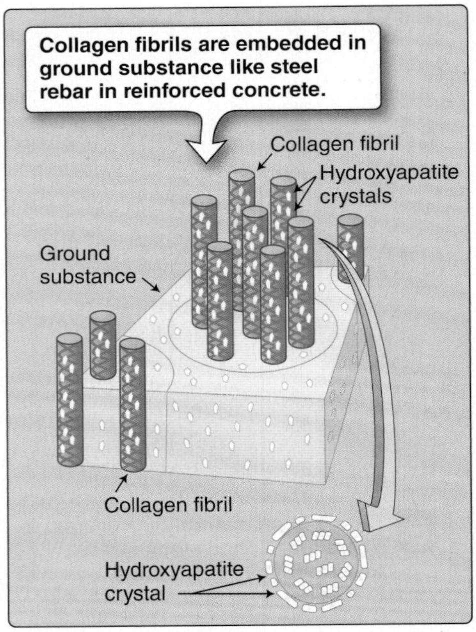
Collagen fibrils are embedded in ground substance like steel rebar in reinforced concrete.
Collagen fibril
Hydroxyapatite crystals
Ground substance
Collagen fibril
Hydroxyapatite crystal

Figure 15.3
Osteoid.

D. Immature

Bone deposition is, by nature, a very slow process, which presents a problem when bone is damaged and must be repaired. Even though bone fracture is an uncommon event in healthy people, most individuals break a bone at some point in their lives. Breaking a proximal phalanx in a finger is painful and inconvenient, but breaking a major limb bone (e.g., a femur) is more serious because it impairs motion. In the wild, breaking a leg can leave an animal extremely vulnerable to predation, so deposition of new bone usually occurs with a view to speed rather than strength. The result is **woven bone**, which,

INFO LINK [1]For additional discussion of osteogenesis imperfecta, see *LIR Biochemistry*, 5e, p. 49.

Clinical Application 15.2: Paget Disease

Paget disease is the second most common bone disorder after osteoporosis. The underlying causes are unknown. Paget disease manifests as inappropriately high levels of osteoclast activity, often affecting solitary bones. Osteoclasts normally digest bone during remodeling. Osteoblasts compensate for the resulting bone loss by laying down new, woven bone, but the rate of turnover in affected bones is so high that it never has time to mature and strengthen. Most Paget disease patients remain asymptomatic. Others present clinically with bone deformities, arthritis, bone pain, symptoms caused by peripheral nerve compression, and increased incidence of fractures.

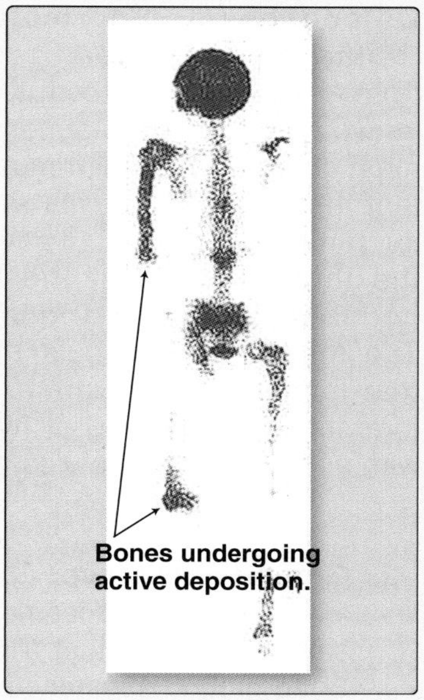

Bones undergoing active deposition.

Bone scan of a patient with Paget disease.

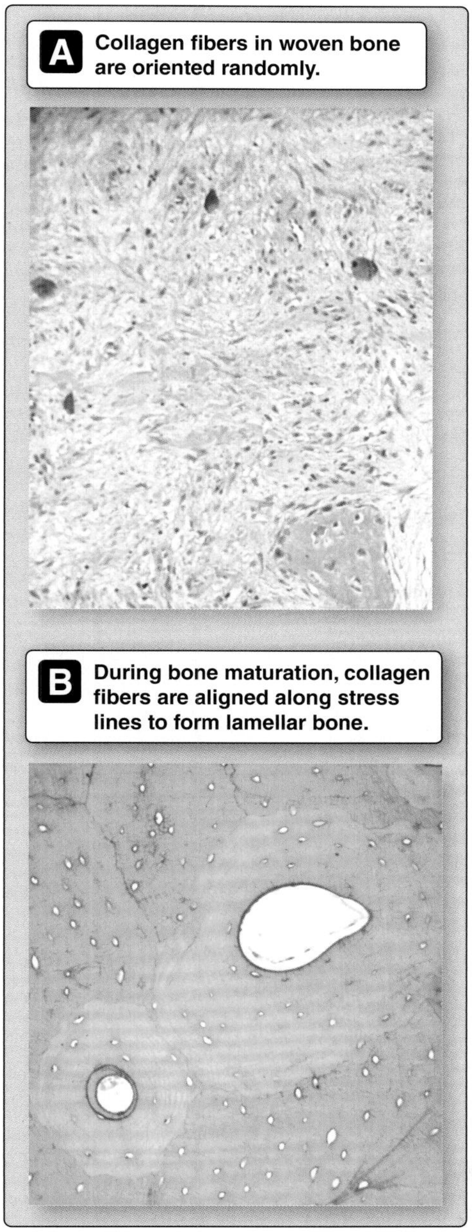

A Collagen fibers in woven bone are oriented randomly.
B During bone maturation, collagen fibers are aligned along stress lines to form lamellar bone.

Figure 15.4
Woven and lamellar bone.

although weaker than the mature form, allows breaks to heal in weeks rather than months (Figure 15.4A). In time, woven bone is replaced through remodeling with the mature **lamellar** form (see Figure 15.4B). Woven bone has the appearance of fabric when viewed in section, reflecting the fact that osteoblasts deposit collagen fibers at random within osteoid. The random orientation provides resistance to stress in all directions and, therefore, is an excellent all-purpose patch for a broken bone. Woven bone is also found at bone growth plates.

E. Mature

Bone takes up to 3 years to mature fully. During this time, collagen fibers are aligned with predominant stress lines to provide maximal strength. Lamellar bone is laid down in 10–30 concentric rings that form cylinders ~200 μm wide and a few millimeters long (~1–3 mm) known as **osteons**, or **Haversian systems** (Figure 15.5). At the center of each cylinder is a **Haversian canal** that provides a thoroughfare for blood vessels and nerve fibers. Two types of lamellar bone can be distinguished based on density and porosity, **compact bone** and **trabecular bone**.

1. **Compact:** Compact bone is extremely dense and is configured for strength. Also known as **cortical** bone, it is found at the periphery of all bones.

2. **Trabecular:** Trabecular bone (**cancellous**, or **spongy bone**) lines the marrow cavity at the center of a bone. It has a lacy, porous structure that gives it a very high surface area. When plasma Ca^{2+} levels fall, trabecular bone is the first to be sacrificed in order

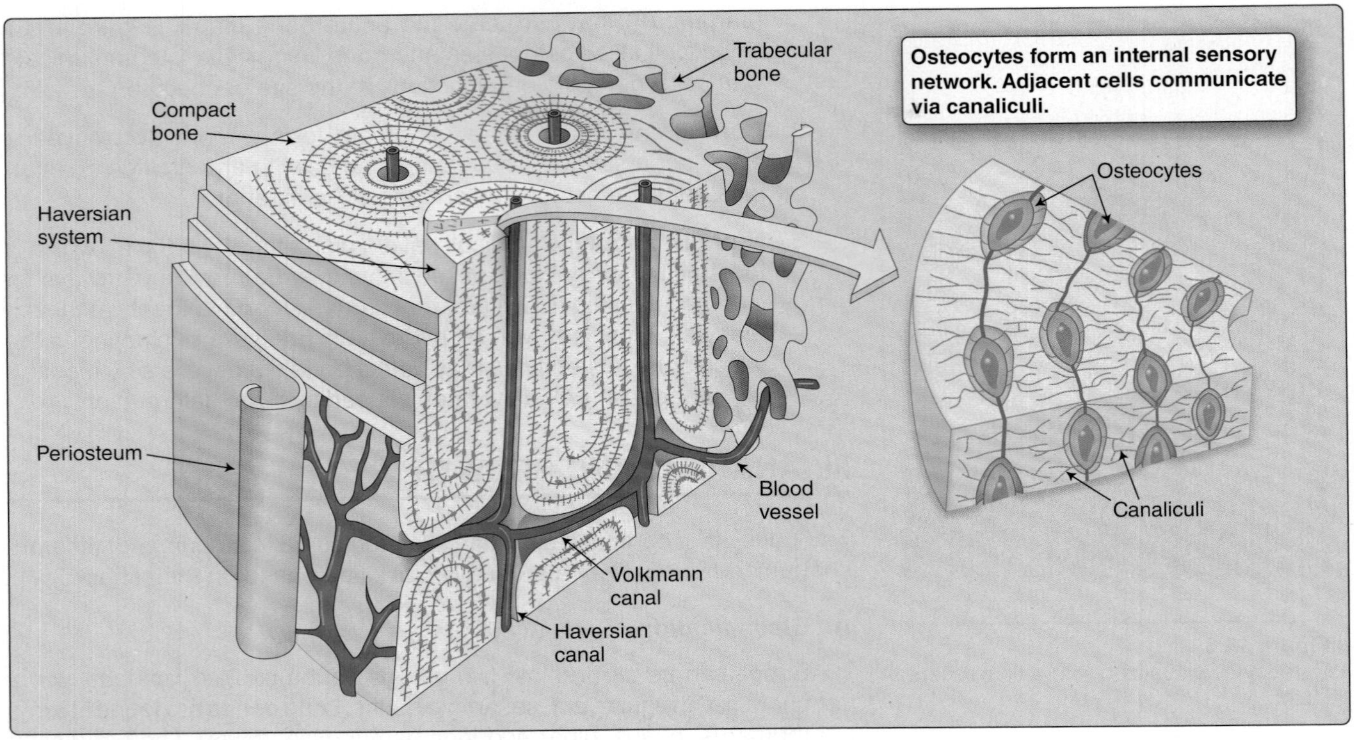

Figure 15.5
Haversian systems in compact bone.

to release its mineral content to the circulation. When Ca^{2+} and PO_4^{3-} levels renormalize, the bone is rebuilt. Trabecular bone also provides critical mechanical support, particularly in the vertebrae.

F. Vasculature

Bone is a living tissue that must be supplied with blood. Blood vessels reach bone by way of the **periosteum**, a fibrous membrane that covers the nonarticulating surfaces of bones and serves as an attachment point for blood vessels and nerves. Supply arteries penetrate the bone cortex and terminate in the medulla. Smaller arterial vessels course through the marrow cavity and then reenter bone to supply the cells within. Vessels travel longitudinally in **Haversian canals** and outward toward the cortex via **Volkmann canals** (see Figure 15.5).

G. Stress-monitoring system

When bone formation is complete, osteoblasts either undergo apoptosis, or they persist as either **osteocytes** or **bone lining cells**. Together, these two cell types form a vast sensory network that monitors bone stress and integrity.

1. **Osteocytes:** Some osteoblasts become entombed in the osteoid of their own making during bone formation and persist, potentially for years, as osteocytes. Osteocytes reside in small (~10–20 μm) cavities called **lacunae**, where they remain to monitor bone stress levels and signal the need for remodeling if microfractures appear (see Figure 15.5). The cells extend long, thin processes (**dendrites**) in all directions through microscopic channels

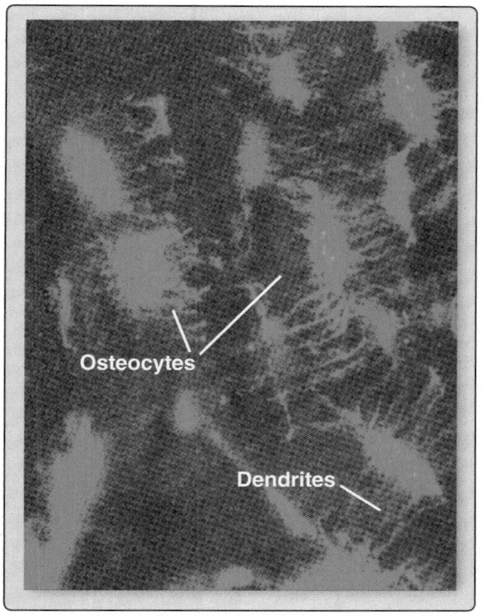

Figure 15.6
Osteocytes visualized with a fluorescent stain.

(**canaliculi**) that permeate the entire bone matrix (Figure 15.6). Canaliculi allow osteocytes and bone lining cells to communicate with each other via gap junctions at the site of contact.

2. **Bone lining cells:** Bone lining cells form a monolayer covering every bone surface. They communicate with and provide an interface between osteocytes and the bone exterior.

3. **Nutrient supply:** Canaliculi have a diameter of <0.5 μm, which is too small to carry blood vessels, but they do provide pathways for extracellular fluid (ECF) flow that carries O_2 and nutrients to the osteocytes. Flow is driven by hydrostatic pressure originating in the vasculature. Osteocytes may detect bone stress levels by monitoring changes in ECF flow rates caused by bone deformation.

III. ANATOMY AND GROWTH

The bones that comprise a human skeleton come in a number of different shapes and sizes. They are usually classified based on their shape.

A. Classification

Bones can be divided into five groups on the basis of anatomy: long, short, flat, irregular, and sesamoid bones. Long bones are found in arms (humerus, radius, ulna) and legs (femur, tibia, fibula). The bone shaft (**diaphysis**) is a long, thin tube of cortical bone with trabecular bone at the center (see Figure 15.1). They typically widen (the **metaphysis**) toward the end (**epiphysis**) to form a site of articulation with another bone. The ends are broadened to spread the load, and they are filled with trabecular bone that acts as a shock absorber during locomotion. In children, the region between the metaphysis and epiphysis contains a growth plate that is the active site of new bone formation, also called the physis, or epiphyseal plate.

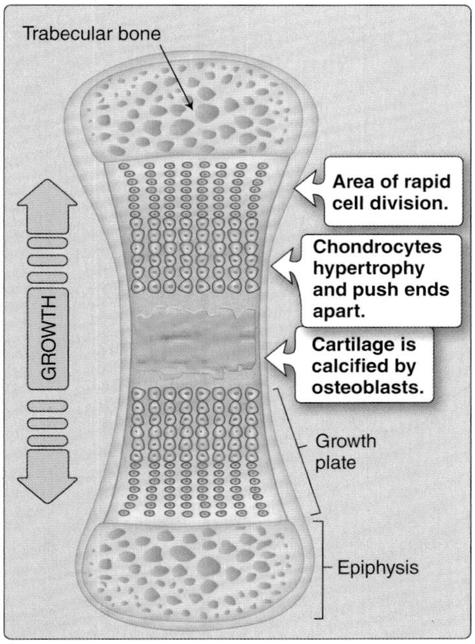

Figure 15.7
Bone growth.

B. Growth

Bone growth (widening and lengthening) during fetal development and throughout childhood is effected by **chondrocytes**. Chondrocytes are derived from the same mesenchymal stem cell line that gives rise to osteoblasts and are arranged in 10–20 columns within a growth plate. Chondrocytes nearest the ends of the bone divide rapidly (Figure 15.7). Further down the column, the chondrocytes enlarge and push the ends apart. Chondrocytes also secrete cartilage, which becomes mineralized by osteoblasts and forms a template for further ossification. Older chondrocytes eventually undergo apoptosis, leaving spaces within the matrix that are invaded by nerves, blood vessels, and additional osteoblasts, which complete the task of bone maturation. When skeletal growth is complete, the growth plate dwindles, and the two epiphyses are united with the shaft (**epiphyseal closure**).

C. Bone marrow

Bone marrow is a soft tissue located in the center of some bones. There are two types: red and yellow.

1. **Red:** Red marrow is the source of virtually all blood cells, including red cells, most white cells, and platelets. Blood cells are formed from hematopoietic multipotent stem cells. Marrow is also the

location of mesenchymal stem cells, which give rise to osteoblasts and chondrocytes, among others.

2. **Yellow:** Yellow marrow stores fat. Yellow marrow appears and accumulates in long bones during adulthood. It is derived from and can be converted back to red marrow if there is a need to increase blood cell production.

IV. REMODELING

Bone is a dynamic tissue that is constantly being turned over at a rate of about 20% per annum in young adults, slowing to 1%–4% per annum in older adults. Remodeling is partly a response to mechanical stress but also reflects bone's vital role as a Ca^{2+} and PO_4^{3-} repository. Remodeling involves four bone cell types that, together, comprise a **basic multicellular unit** (**BMU**). Osteocytes signal the need for remodeling, bone lining cells facilitate and coordinate remodeling, **osteoclasts** digest old bone, and osteoblasts lay down new bone (Figure 15.8). BMUs are roving demolition and construction crews replacing bone constantly for the life of an individual.

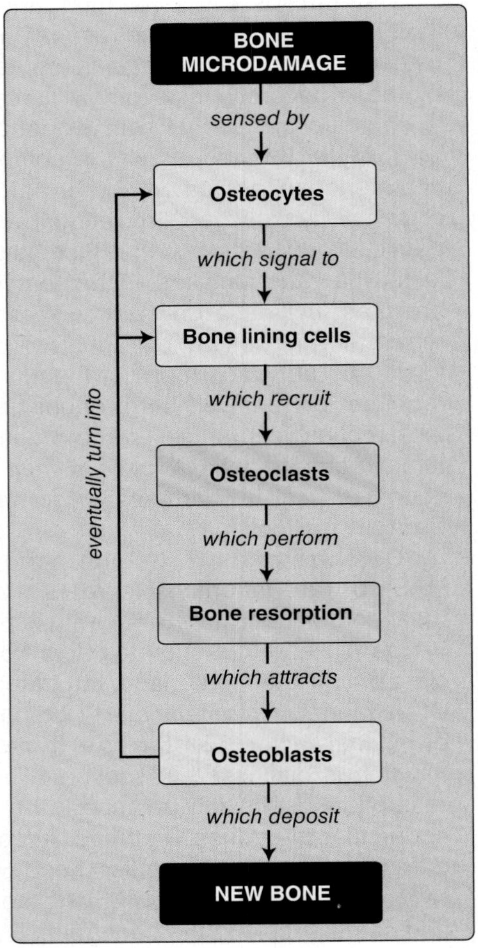

Figure 15.8
Concept map for the basic multicellular unit.

A. Causes

There are three main forces driving bone remodeling: mechanical stress, microdamage, and Ca^{2+} and PO_4^{3-} homeostatic needs.

1. **Mechanical stress:** Many bones are subject to repeated mechanical stress associated with lifting and carrying weight. For example, the arm bones form a lever system powered by skeletal muscles. When the muscles contract to lift a weight, the levers are stressed. Bones are designed to withstand such stresses within a normal physiologic range, but a muscle that is exercised repeatedly grows stronger and increases the stress on the levers. Thus, bones are also designed to sense mechanical stress and lay down additional bone mass to compensate if necessary. Conversely, when the stress on bones is reduced, they lose mass.

> The stroke (playing) arm of professional tennis players is subjected to years of repeated mechanical stress. The bones of the forearm respond by increasing bone density, diameter, and length. Individuals who have been freed from gravity and its associated mechanical stresses lose bone mass. Astronauts who remain in space for prolonged periods exercise daily to help offset the effects of bone unloading, but they still lose pelvic bone mass at a rate of 1%–2% per month.

2. **Microdamage:** Bones constantly develop microscopic damage as a result of mechanical stress, either acutely or as the result of normal actions that are performed repeatedly over time. The organic component of bone also deteriorates with time, which increases the likelihood of microfissures and microscopic cracks forming.

Clinical Application 15.3: Remodeling Disorders

At any one time, around 1 to 2 million basic multicellular units are at work remodeling bone in the adult skeleton. In the absence of external influences, such as mechanical stress or disease, total bone mass remains constant. This exact matching of bone deposition to bone resorption requires tight functional **coupling** between osteoblasts and osteoclasts. A disturbance in the balance between the activities of the two cell types causes loss of bone mass or abnormal bone deposition.

Osteoporosis is a common term for a group of disorders in which the balance between bone resorption and formation is tipped in favor of osteoclasts. The osteoblasts fail to keep up with osteoclast activity, and the bone becomes increasingly porous and fragile as a result. Fracture is common in patients with osteoporosis. Bone loss associated with aging (type II osteoporosis) is common in both women and men, but postmenopausal women are at particular risk. Estrogen limits osteoclast activity, so when circulating levels of this hormone fall after menopause, osteoclasts become increasingly active, and bone is resorbed faster than it is replaced. Osteoporosis affects all bones, but the effects are most dramatic on trabecular bone, which is the primary site of remodeling in healthy individuals. Excessive resorption thins all trabeculae and truncates many of them, which destroys the template that is required for renewed bone deposition. It also seriously compromises their mechanical functions and greatly increases the likelihood of fracture. Treatment options for both men and women include oral bisphosphonates (e.g., alendronate, trade name Fosamax), which inhibit bone mineral breakdown.[1]

Osteopetrosis, or "stone bone," results from a heterogeneous group of rare inherited disorders that impair osteoclast function. The most common form (~60%) results from a mutation in a subunit of the V-type ATPase that secretes acid from the ruffled border of an osteoclast onto the bone surface, but other mutations affect genes encoding Cl^- channels, intracellular H^+ pumps, and RANK (receptor activator of nuclear factor κB); the role of these proteins in normal osteoclast function is discussed in section C below. The mutations tip the bone resorption–deposition balance in favor of the osteoblasts, resulting in bones that are dense yet brittle. Affected individuals may show skeletal deformities, an increased likelihood of fractures, and secondary effects related to incursion of bone into the marrow space and the vascular and nerve supply to the bone matrix.

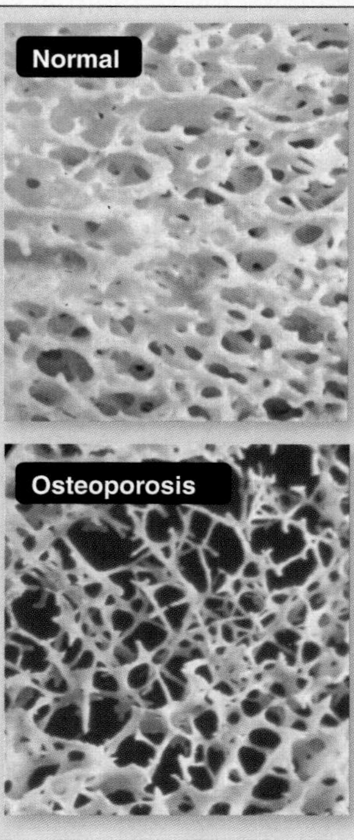

Comparison of normal and osteoporetic trabecular bone.

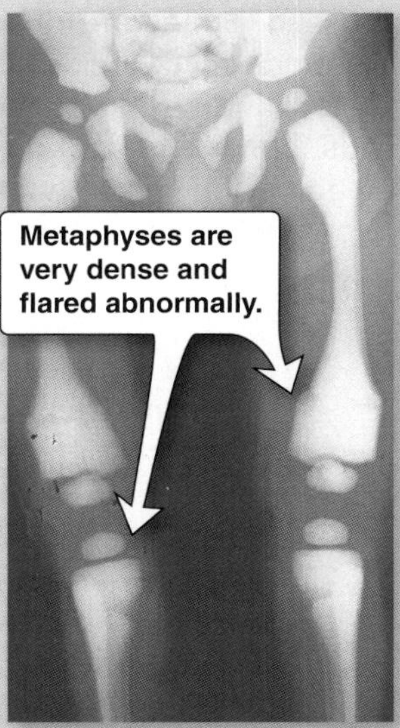

Abnormal bone density in an infant with osteopetrosis.

[1]For a discussion of agents used to treat osteoporosis, see *LIR Pharmacology*, 5e, p. 365.

Because fissures and cracks can ultimately result in fracture, damaged areas are replaced with new bone through remodeling.

3. **Hormones:** Bone contains immense reservoirs of Ca^{2+} and PO_4^{3-} that can be mobilized and circulated if plasma levels fall. The balance between bone resorption and deposition is controlled by two hormones (parathyroid hormone [PTH] from the parathyroid gland and calcitonin from the thyroid) and by vitamin D. The pathways involved are summarized briefly in Section V below and are considered in more detail in Chapter 35.

B. Signaling

Bone remodeling involves extensive signaling between the various cellular participants. Few of the pathways or signals involved have been characterized fully, although there are many candidates. When microdamage occurs, osteocytes at the fissure site undergo apoptosis, and bone lining cells then initiate the remodeling sequence (Figure 15.9, step 1).

C. Remodeling sequence

The bone remodeling sequence lasts ~200 days in total. The sequence can be divided into four phases: activation, resorption, reversal, and formation.

1. **Activation:** During the activation phase, bone lining cells recruit osteoclast precursors to the remodeling site, expose underlying mineral, and then form a canopy over the BMU worksite.

   a. **Osteoclast precursors:** Bone is absorbed by osteoclasts, a blood cell line related to macrophages ("-clast" is derived from the Greek word *klastos*, meaning "broken"). Osteoclasts are large, multinucleate, phagocytic cells formed by fusion of many hematopoietic precursors. The precursors are called to action from the vasculature by chemoattractant chemokines, including macrophage colony–stimulating factor (see Figure 15.9, step 2).

   b. **Mineral exposure:** Osteoclasts digest osteoid very slowly, so bone lining cells lend assistance by releasing *collagenase* and other enzymes onto the bone surface to expose the mineral.

   c. **Canopy:** Bone lining cells then lift off the bone surface as a single sheet and form a canopy over the worksite. This canopy creates a **bone-remodeling compartment** (**BRC**) whose environment can be optimized for remodeling. The canopy becomes highly vascularized, which allows for recruitment of osteoclast and osteoblast precursors from the vasculature and from marrow (see Figure 15.9, step 3).

2. **Resorption:** The resorption phase takes about 2 weeks to complete. Osteoclast precursors fuse to form mature osteocytes and then digestion and resorption begins.

   a. **Fusion:** Osteoclast precursors express a receptor on their surface that is related to the tumor necrosis factor receptor (**RANK** [receptor activator of nuclear factor κB]). When precursors

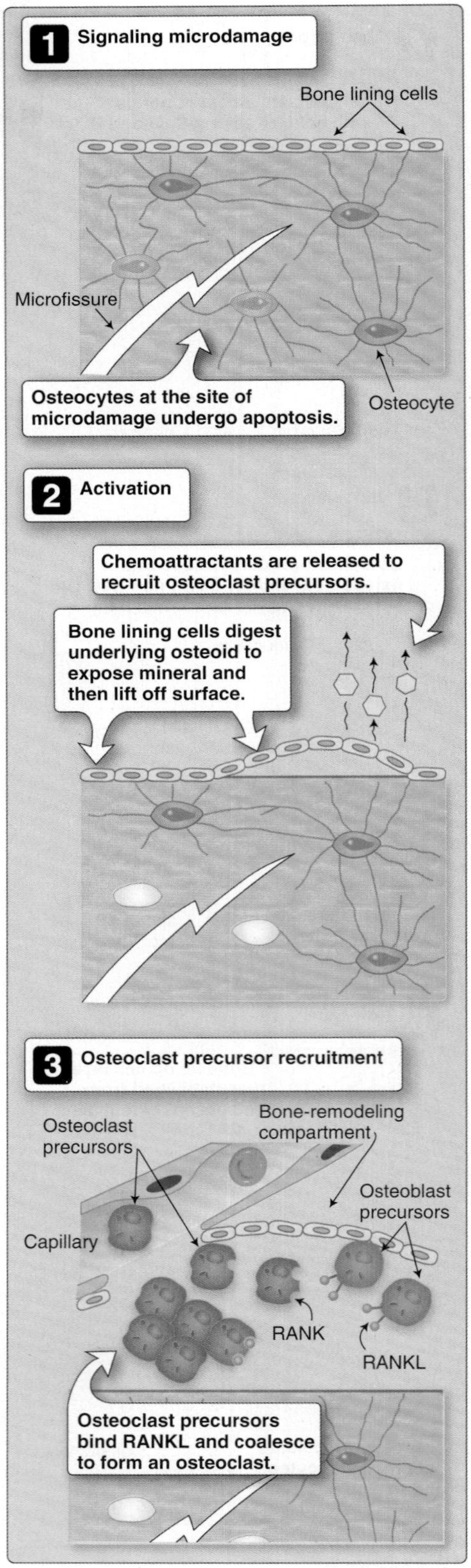

Figure 15.9
Bone remodeling cycle. RANK = receptor activator of nuclear factor κB; RANKL = RANK ligand.

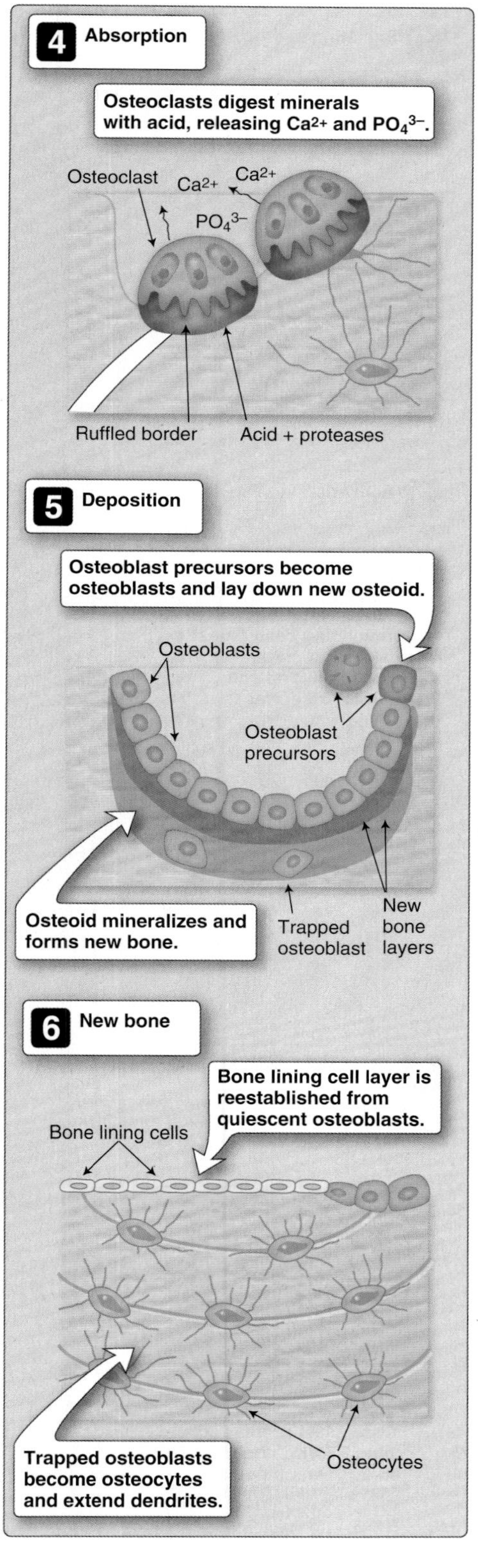
4 Absorption
Osteoclasts digest minerals with acid, releasing Ca2+ and PO43–.
Osteoclast
Ca2+
Ca2+
PO43–
Ruffled border
Acid + proteases
5 Deposition
Osteoblast precursors become osteoblasts and lay down new osteoid.
Osteoblasts
Osteoblast precursors
Osteoid mineralizes and forms new bone.
Trapped osteoblast
New bone layers
6 New bone
Bone lining cell layer is reestablished from quiescent osteoblasts.
Bone lining cells
Trapped osteoblasts become osteocytes and extend dendrites.
Osteocytes

Figure 15.9
(continued)

arrive at the BRC, they encounter osteoblast precursors (bone marrow stromal cells), which express **RANKL** (RANK ligand) on their surface. Contact between the two sets of precursors allows RANK–RANKL binding, and an intracellular cascade is then initiated within the osteoclast precursors that culminates in the synthesis of a variety of fusion proteins. Four or five osteoclast precursors then fuse to form large, multinucleate osteoclasts (see Figure 15.9, step 4).

b. **Digestion:** The osteoclasts seal to the exposed bone matrix at their periphery and polarize. The surface area of the apical membrane increases to form a **ruffled border**, which pumps acid onto the bone surface using a vacuolar-type H^+ ATPase. Lysosomes fuse with the apical membrane and empty their contents onto the bone surface also. Constituents include acid and *cathepsin K*, a *protease* that specifically digests collagen and other connective tissue components in osteoid. Acid degrades hydroxyapatite and releases Ca^{2+} and HPO_4^{3-} for transfer to the circulation. Osteoclasts usually create a simple pit on the surface of trabecular bone, but in cortical bone, they work deep below the surface, creating tunnels that run for several millimeters through the matrix. These tunnels are eventually replaced with a new Haversian system.

3. **Reversal:** Once the resorption phase is complete, bone digestion stops, and the worksite must be refilled. The reversal phase involves several concurrent steps, including osteoclast apoptosis, surface cleaning, and recruitment of osteoblast precursors.

   a. **Apoptosis:** Osteoblasts determine when sufficient bone has been removed, at which point they release **osteoprotegerin** (**OPG**). OPG is a decoy receptor that binds to and masks RANKL on the osteoblast surface, thereby preventing activation of additional osteoclast precursors. The osteoclasts detach and undergo apoptosis.

   b. **Cleaning:** Mononuclear cells arrive at the worksite and clean it with *proteases* in preparation for new bone deposition.

   c. **Osteoblast precursor gathering:** Bone resorption liberates numerous growth factors from the matrix, including insulin-like growth factor (IGF). The growth factors recruit osteoblast precursors from blood and bone marrow. The growth factors are incorporated into the bone matrix by osteoblasts during bone formation. Precursors arriving at the worksite migrate into the BRC and differentiate into osteoblasts (see Figure 15.9, step 5).

4. **Formation:** Bone regrowth is slow compared with resorption, and it takes several months for the cavity to be refilled. The steps involved in osteoid deposition and mineralization are outlined in Section II·C above. Once the cavity has been filled with osteoid, the osteoblasts cease work and either die or differentiate into bone lining cells or remain in place as osteocytes (see Figure 15.9, step 6). The osteocytes inhibit further bone deposition by releasing **sclerostin** into the canalicular system. Sclerostin diffuses to the surface and prevents osteoblast formation by blocking receptors that mediate osteoblast differentiation.

V. REGULATION

The balance between bone resorption and formation is regulated locally and by hormones. The pathways involved are complex and remain undefined. Local regulatory factors include nitric oxide, prostaglandins, and IGF. The resorption–formation cycle is also influenced by hormones involved in Ca^{2+} and PO_4^{3-} homeostasis and by estrogens and androgens.

A. Calcium and phosphate homeostasis

Plasma Ca^{2+} and PO_4^{3-} levels are regulated through the concerted actions of vitamin D, calcitonin, and PTH (see Chapter 35). Vitamin D and calcitonin have minimal direct effects on bone remodeling. Chronic elevation in plasma PTH levels causes bone resorption, thereby increasing Ca^{2+} and PO_4^{3-} availability. PTH binds to osteoblasts and stimulates release of factors that recruit and activate osteoclast precursors.

B. Sex hormones

Estrogens and androgens are both required for an individual to maintain a stable skeletal mass. Circulating levels of these hormones decline with age in both men and women, and, hence, both sexes may develop osteoporosis in later years (see Clinical Application 15.3). However, because males generally achieve a greater bone mass during development than do females, age-related effects have less of an impact on skeletal integrity in men. These hormones stimulate OPG synthesis and reduce RANKL expression by osteoblasts, which limits osteoclast precursor activation and bone resorption. Estrogen also directly increases osteoclast apoptosis, which further favors bone mass retention.

Chapter Summary

- Bone is a connective tissue comprising **mineralized collagen fibers** cemented within an **amorphous ground substance**. Calcium phosphate crystals (predominantly hydroxyapatite) give bone strength, whereas collagen imparts flexibility and resistance to tensile stress that allows bones to torque and bend without fracturing.
- Bone is formed by **osteoblasts**. Bone is created by embedding collagen fibers within a matrix that is supersaturated with Ca^{2+} and PO_4^{3-}. Osteoblasts seed the fibrils with **hydroxyapatite crystals**, which then become nucleation points for further crystal growth.
- Newly formed bone (**"woven"** bone) is disorganized and takes several years to mature. Woven bone is gradually replaced with a **lamellar** form in which the collagen fibers are realigned along predominant stress lines to maximize strength.
- Once bone formation is complete, osteoblasts either undergo apoptosis or persist as **osteocytes** and **bone lining cells**. Osteocytes reside in small cavities (**lacunae**) located throughout the bone matrix, whereas bone lining cells cover the surface. Osteocytes and bone lining cells communicate via **dendrites**, together forming a **sensory network** that monitors bone stress levels and integrity.
- Remodeling is initiated by bone lining cells, which recruit **osteoclast precursors** to a worksite and then raise a canopy over the site to create a compartment whose microenvironment can be optimized for remodeling.
- Osteoclast precursors fuse to become multinucleate **osteoclasts**. Osteoclast formation is initiated by osteoblast precursors via the **RANK–RANKL** signaling pathway.
- Osteoclasts digest bone using acids and *proteases*. The eroded cavity is then cleaned by mononuclear cells, and new bone is laid down by osteoblasts.
- The remodeling cycle is regulated primarily by **parathyroid hormone** (**PTH**), a key Ca^{2+} and PO_4^{3-} homeostatic hormone. PTH stimulates bone resorption when circulating Ca^{2+} levels are low.

16 Skin

I. OVERVIEW

Individual cells erect a membrane around their periphery to create a barrier between the extracellular and intracellular environments, which allows them to regulate their cytoplasmic composition. The body similarly encloses its tissues within skin, a multilayered covering comprising **epidermis**, **dermis**, and a functionally linked **hypodermis** (Figure 16.1). Skin forms a physical barrier that excludes microorganisms and other foreign substances, while simultaneously helping the body retain vital fluids. Skin has several important functions. The outermost layers protect underlying tissues from abrasion and other mechanical insults, chemicals, pathogens, and ultraviolet (UV) light. It contains active microbial defense mechanisms that bolster its barrier function when breached. Skin also serves a vital thermoregulatory role. It secretes aqueous solutions that enhance heat loss by evaporation. Also, the amount of blood traveling through the skin is modulated as a way of conserving or offloading body heat to the environment. Finally, skin is a sensory organ that contains a variety of nerves and specialized receptors that collect sensory information about the external environment and interactions with foreign bodies. Skin and its associated appendages form the **integument**, or body covering.

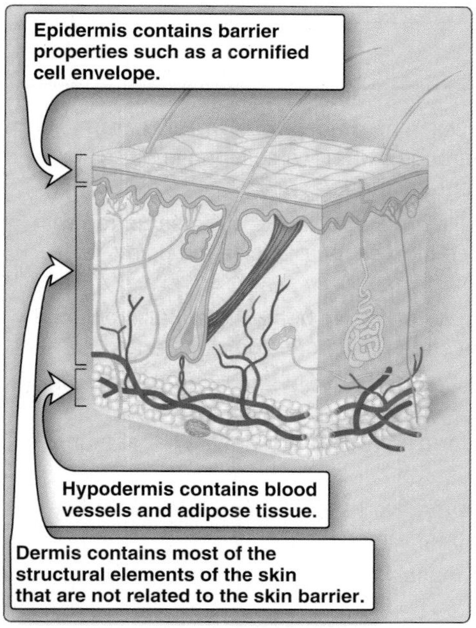

Epidermis contains barrier properties such as a cornified cell envelope.
Hypodermis contains blood vessels and adipose tissue.
Dermis contains most of the structural elements of the skin that are not related to the skin barrier.

Figure 16.1
Skin structure and associated appendages.

II. ANATOMY

Skin comprises two anatomical layers: a superficial epidermis and the dermis. Combined epidermal and dermal thickness varies from 0.5–5 mm, depending on location. This measurement does not include **subcutaneous tissue** (the hypodermis), which is often considered a part of the skin (see Section V below). Skin is the body's largest organ, accounting for 15%–20% of total body mass. Functionally, it is helpful to distinguish between hairy skin (**nonglabrous**) and areas without hair follicles (**glabrous** skin [derived from *glaber*, meaning "bald" or "hairless" in Latin]).

A. Hairy skin

Most areas of the body are covered with hairy skin. Although these areas do contain sensory receptors related to touch, pressure, temperature, and pain, their density is lower than that of glabrous skin. Nonglabrous skin is important for thermoregulation.

B. Glabrous skin

Glabrous skin is smooth and hairless. Examples include the lips, soles of the feet, palms of the hand, and fingertips. The palms and soles of the feet are contact points for grasping and locomotion. Hair here would interfere with motor functions and would decrease the ability to discern the texture and temperature of surfaces. Glabrous skin contains a high density of sensory fibers to aid taction and thick structural adaptations to help protect against abrasions. Glabrous skin does not play a large role in thermoregulation, but skin blood flow does both increase and decrease as well as become erythemic due to elevated estrogen levels in pregnancy and chronic alcoholic liver disease (Figure 16.2).

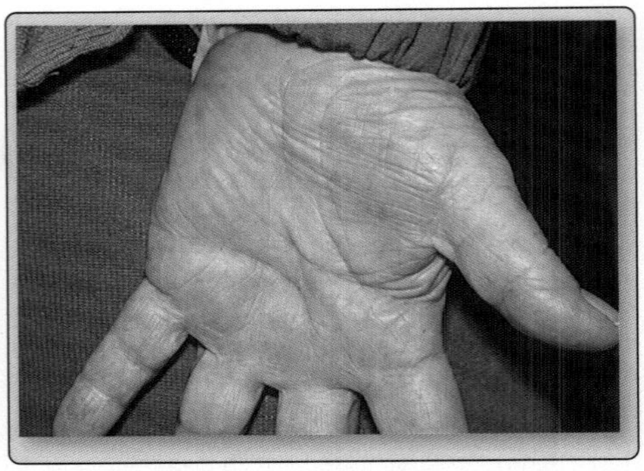

Figure 16.2
Palmar erythema.

III. EPIDERMIS

The external environment is inherently hostile. It contains a number of elements capable of causing tissue damage, so the body must erect a barrier to protect itself. This barrier is called the **epidermis**, a reinforced skin layer located at the interface with the external world. Its function is primarily protective, but it also helps minimize water loss from underlying tissues.

A. Structure

The epidermis is a stratified squamous epithelium that is shed and renewed constantly. It lacks blood vessels, obtaining nutrients by diffusion from the deeper layers. It is composed mainly of **keratinocytes**, but it also contains **melanocytes**, **Langerhans cells**, and **Merkel cells** (Figure 16.3).

1. **Keratinocytes:** Keratinocytes are the most abundant cells in the epidermis. They are created by the division of basal cells in the **stratum basale**. They differentiate as they progress toward the surface and ultimately become inert (the **stratum corneum**). The primary products of keratinocytes are keratins and lipids.

   a. **Keratin:** Keratin is a resilient, fibrous protein that eventually fills (**keratinizes**) the cells during a process in which the nucleus and organelles are removed. A cornified cell layer develops (the **stratum corneum**, composed of **corneocytes**) that has a turnover rate of ~14 days.

   b. **Lipids:** Keratinocytes synthesize and secrete a lipid mixture that contains cholesterol, fatty acids, and ceramides. Lipid precursors are deposited and stored in lamellar bodies prior to secretion.

2. **Melanocytes:** Melanocytes produce **melanin**, a photoprotective pigment that is synthesized in membrane-bound organelles called **melanosomes**. Melanin is then passed to keratinocytes and other adjacent cells through **pigment donation**. Melanin biosynthesis is regulated through melanocortin receptors via α-melanocyte-stimulating hormone (α-MSH) and adrenocorticotropic hormone (ACTH). The amount and type of melanin determines skin hue.

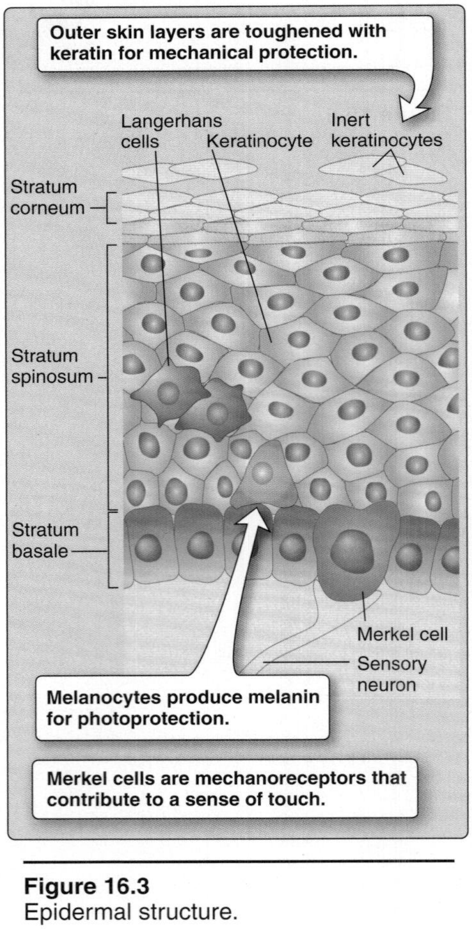

Figure 16.3
Epidermal structure.

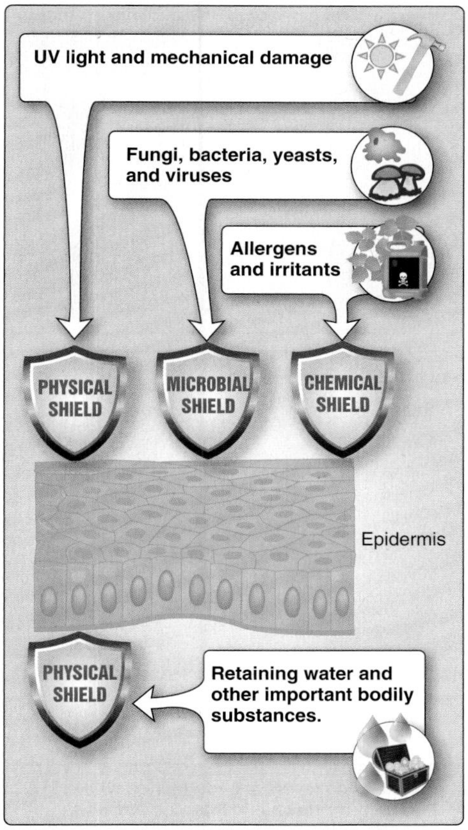

Figure 16.4
Skin barrier functions. UV = ultraviolet.

3. **Langerhans cells:** Langerhans cells are antigen-presenting cells that are integral to adaptive immune responses. They ingest foreign matter, digest it, and then present fragments of the material on the cell surface. Presentation allows other immune cells to recognize the fragments. Langerhans cells are also important in delayed (type IV) hypersensitivity reactions.

4. **Merkel cells:** Merkel cells are slow-adapting mechanoreceptors located in skin areas that have a high tactile sensitivity. Merkel cells are also located at the base of hair follicles to further aid in touch sensation. Merkel cells associate with a nerve terminal to form a Merkel disk receptor (see Section VII below).

B. Barrier functions

The epidermis forms a barrier that both protects tissues from damage and minimizes evaporative water loss. The epidermis shields the body in four principal areas: it is resistant to mechanical and other physical assaults, photoprotective, antimicrobial, and also water repellent (Figure 16.4).

1. **Physical shield:** The superficial layers of the epithelium are keratinized, which creates a tough, multilayered, physical barrier around the body. The barrier resists mechanical abrasion and mild penetrating insults that inevitably occur during physical contact with solid objects. The barrier also resists chemical attack and prevents underlying tissue exposure to toxins and allergens.

2. **Photoprotective shield:** UV radiation naturally originates from the sun. UV light can be highly deleterious to biologic tissues because it breaks chemical bonds and disrupts the structure of DNA and protein. Melanocytes synthesize and donate melanin to adjacent cells to create a photoprotective barrier that absorbs UV radiation and causes it to dissipate safely as heat. Repeated exposure to the sun or other sources of UV radiation can stimulate melanocyte proliferation and melanin production, causing the skin to darken, which increases the level of photoprotection.

3. **Antimicrobial shield:** The superficial keratinized layers provide a physical barrier to microbes. If the physical barrier is breached, Langerhans cells and other immune components provide a rapid response to microbial invasion, warding off infection until the barrier can be repaired.

4. **Water-resistive shield:** The cornified cell envelope of keratinocytes creates a water-repellent shield that serves a dual function.

   a. **Structure:** The shield is composed of lipids that are synthesized by keratinocytes and then secreted onto the surface. The lipid mixture forms a coating approximately 5-nm thick that is connected to the cell membrane via ester bonds. The cellular component is thicker (>15 nm) and is composed of cross-linked, insoluble proteins, including loricrin and keratin.

   b. **Function:** The shield functions much like car wax. It causes water to bead on the epithelial surface and thereby prevents solutes from being washed out of the layers below. The waxy shield also minimizes evaporative water loss from underlying tissues.

Clinical Application 16.1: Abrasions and Burns

Abrasions and burns caused by ultraviolet radiation, heat, and fire can impair one or more skin barrier functions. For example, burn patients have increased transepithelial water loss from the skin that can challenge fluid homeostasis if a sufficiently large area of the body surface is involved. Burn patients are also at an increased risk of developing infections, which is why burned areas are often wrapped with bandages to supplement the physical barrier, and why topical antibiotics are applied as an antimicrobial shield.

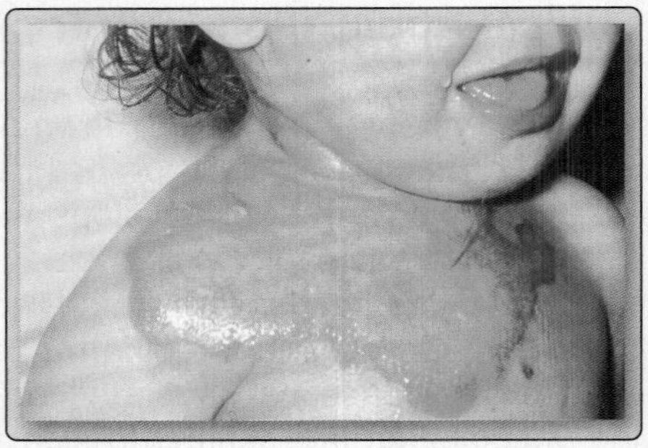

Scalding Burn.

IV. DERMIS

The epidermal barrier is supported and maintained by the dermis. The dermis does not contribute to barrier function directly, but it gives skin strength and elasticity. It also contains immune cells that react to pathogens that may have breached the barrier.

A. Structure

The dermis comprises a meshwork of connective tissue. Primarily, type 1 collagen fibers provide structural support to the skin, whereas elastin fibers provide elasticity (see 4·VI·B·2). Within this matrix are nerve roots and sensory receptors, the cutaneous vasculature, and most skin specializations.

B. Cellular components

The principal cellular components of the dermis include mast cells, macrophages and dermal dendritic cells, and fibroblasts.

1. **Mast cells:** Mast cells are involved in both immune and inflammatory responses. Activated mast cells release histamine, prostaglandins, leukotrienes, cytokines, and chemokines (Figure 16.5). These agents increase cutaneous blood flow and capillary permeability.

2. **Macrophages and dermal dendritic cells:** Macrophages are phagocytic and aid in a number of immune-related responses. Dermal dendritic cells are antigen-presenting cells similar to the Langerhans cells of the epidermis. Dermal dendritic cells are integral to cutaneous adaptive immunity responses.

3. **Fibroblasts:** Fibroblasts are responsible for both the synthesis and degradation of fibrous and nonfibrous connective tissue proteins. These cells are also important in wound healing and scarring.

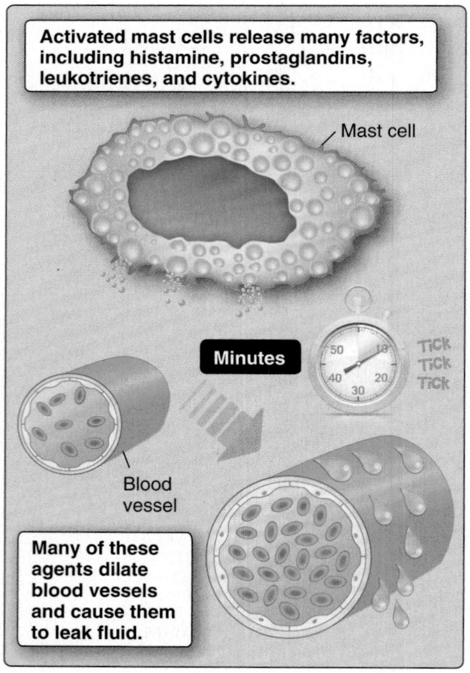

Figure 16.5
Mast cell activation effects.

V. HYPODERMIS

The hypodermis lies beneath the dermis (see Figure 16.1). It consists primarily of subcutaneous fat, blood and lymph vessels, and nerves. Around

Clinical Application 16.2: Triple Response

Wheal, erythema, and flare (known as a "**triple response**") denote a classic reaction to abrasion or histamine-releasing stimuli. First, an erythemic (red) spot develops that spreads outward for a few millimeters, reaching maximal size in about 1 minute. Second, a brighter flush spreads slowly in an irregular flare around the original spot. Third, an edemic wheal forms over the original spot. Mast-cell histamine release can account for the triple response, mediating vasodilation and fluid extrusion into the interstitial space and stimulating nerve endings to give the sensation of itch. The triple response to histamine is often used as a positive control for a skin prick allergy test. Histamine is pricked into the skin followed by a row of other potential allergens such as pet dander, dust mites, and pollens.

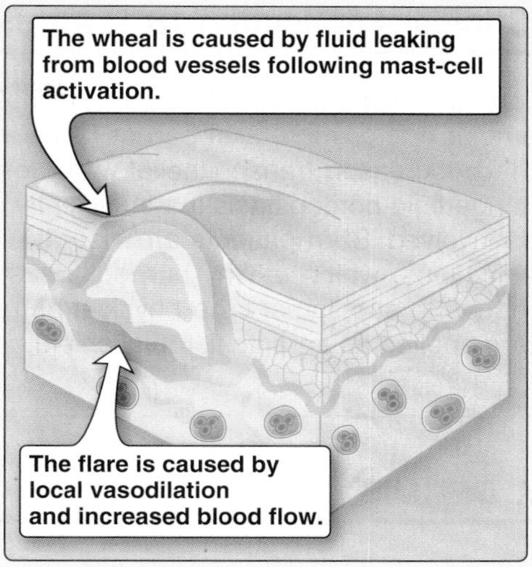

The wheal is caused by fluid leaking from blood vessels following mast-cell activation.
The flare is caused by local vasodilation and increased blood flow.

Triple response.

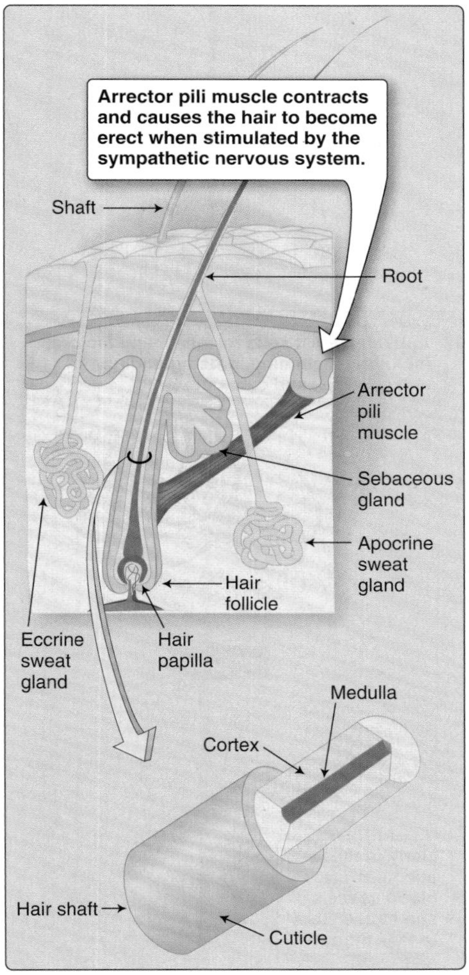

Figure 16.6
Hair follicle and associated structures.

50% of total body fat is located within the hypodermis of an average person. Thus, it is possible to take caliper measures of skin-fold thickness to estimate peripheral fat stores. The hypodermis cushions the skin, allows it to slide over underlying structures, and anchors skin to the tissues below.

VI. SPECIALIZED SKIN STRUCTURES

Skin contains several specialized glands and appendages that are protective or that participate in thermoregulation. An "appendage" is traditionally defined as a structure that protrudes from the body. Dermatologists use the term to indicate any specialized skin structure, including glands that originate in the dermis or hypodermis and then protrude through the epidermis to the skin surface. Appendages include hairs, sebaceous glands, sweat glands, and nails.

A. Hair

The base of each hair is attached to a piloerector muscle that is controlled by the sympathetic nervous system ([SNS] Figure 16.6). When stimulated to contract, piloerector muscles cause hair erection and produce "goose bumps."

1. **Structure:** Hairs are constructed from three layers of fused keratinized cells (see Figure 16.6). The outermost protective layer (**cuticle**) is colorless. A middle layer (**cortex**) imparts strength and contains two types of melanin, the relative proportions of which give hair its natural color. Larger hairs also contain an inner **medulla**. The portion that protrudes beyond the epidermis is known as the **hair shaft**. The shaft emerges from a **hair follicle**, a specialized skin structure containing the hair bulb, keratinocytes, and associated glands (apocrine and sebaceous).

> Some forms of cancer can be treated using chemotherapy, which targets rapidly dividing cells. The treatment is nondiscriminatory, however, also affecting rapidly proliferating keratinocytes, which causes hair thinning and loss.

2. **Hair cycle:** The hair follicle cycle consists of growth, rest, regression, and shedding. Growth-phase duration determines hair length and can vary from one area of the body to the next, which explains why head hairs are typically much longer than in other regions.

3. **Associated structures:** Hair follicles may also contain sebaceous and sweat glands (described below). Follicles also contain a sensory nerve fiber network (**root plexus**) that provides information about touch, pressure, and pain (described in Section VII below).

B. Sebaceous glands

Sebaceous glands produce sebum, a lipid-based secretion. The gland comprises keratinocytes and sebocytes, the latter of which are responsible for sebum synthesis. The gland duct empties sebum directly into the hair follicle, coating the shaft and then flowing onto the epidermal surface. Sebum functions are likely related to its antioxidant, antimicrobial, and hydration properties. Sebum secretion is continuous, but gland output is modulated by sex hormones. Androgens and growth hormone increase secretion rate, whereas estrogens inhibit it.

C. Sweat glands

Sweat glands secrete fluids of variable composition at the epidermal surface.

1. **Types:** There are three classes of sweat glands: apocrine, eccrine, and apoeccrine (Figure 16.7).

   a. **Apocrine:** Apocrine sweat glands are restricted to the axillae and perineum. They activate in response to emotional stimuli and secrete a viscous, milky fluid into the follicle. Bacterial action on these secretions produces odors that may be involved in pheromonal signaling and are the reason that underarm antiperspirants and deodorants were developed. The pathologically foul-smelling sweat in these areas is known as **bromhidrosis**.

   b. **Eccrine:** Eccrine sweat glands are widespread, the greatest concentrations being found on the palms of the hands and soles of the feet. The glands secrete a hypotonic fluid directly onto the skin surface. Fluid evaporation cools the skin and is important for thermoregulation. The skin is capable of producing copious amounts of sweat (1.0–1.5 L/hr on the low end and more than 3 L in heat-acclimated individuals). Such high levels of fluid loss from the body can compromise fluid balance and cardiovascular function.

Clinical Application 16.3: Acne

Acne is a common adolescent disorder that may also occur in adults. Closed comedo acne pimples (whiteheads) occur when skin cells block a hair follicle's external opening. Sebum becomes trapped but continues to be produced. Bacteria may colonize the accumulating sebum, causing local inflammation. Drugs that reduce sebum secretion (e.g., retinoids) help control acne occurrence and spreading.

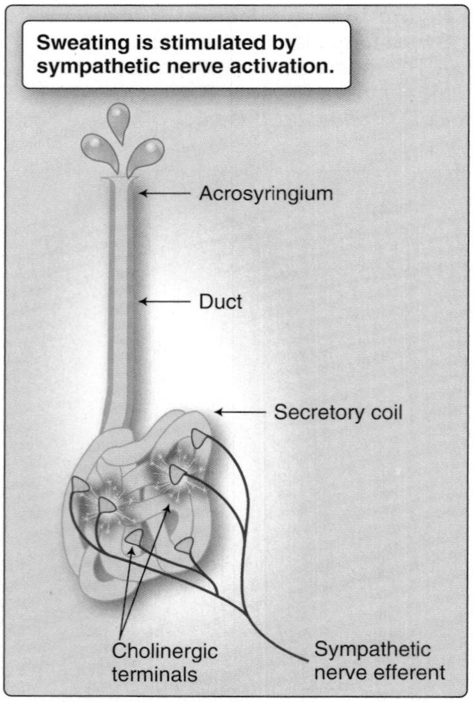

Figure 16.7
Eccrine sweat gland activation.

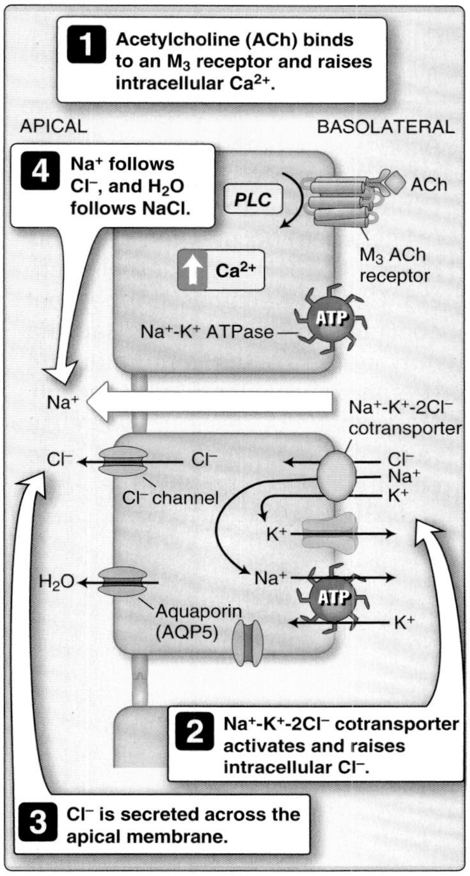

Figure 16.8
Sweat formation by clear cells.
ATP = adenosine triphosphate;
M_3 = muscarinic type-3 receptor;
PLC = phospholipase C.

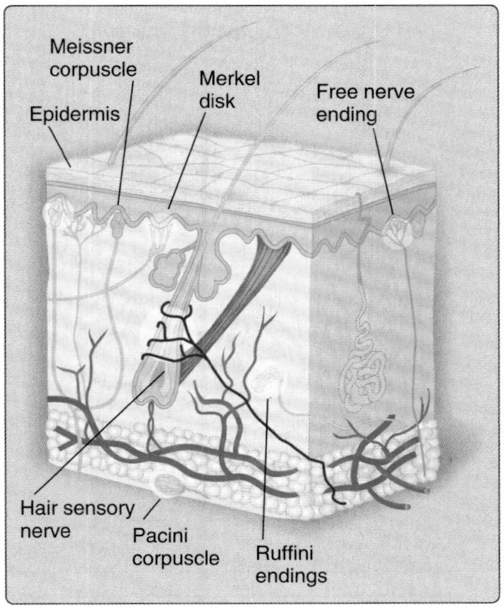

Figure 16.9
Cutaneous tactile receptors.

c. **Apoeccrine:** Apoeccrine sweat glands are located primarily in the axillae, where they make up 50% of the total sweat gland number, and perineum. These glands are of a mixed archetype and function: Their pores typically open into the hair follicle, but their sweat composition is comparable to that of an eccrine gland. Apoeccrine glands tend to produce copious amount of sweat when stimulated appropriately in a continuous rather than pulsatile fashion.

2. **Structure:** Sweat glands can be functionally subdivided into a pore opening (**acrosyringium**), a **secretory coil**, a **duct**, and a layer of **myoepithelial cells** that allows sweat to be discharged onto the skin surface.

a. **Secretory coil:** When active, sweat glands secrete a precursor fluid into the secretory coil lumen that comprises a protein-free plasma filtrate. Osmotic forces drive the fluid through the duct toward the skin surface. Sweating is stimulated by SNS postganglionic cholinergic nerves (see Figure 16.7). Acetylcholine (ACh) binds to muscarinic type-3 receptors (G protein–coupled receptor superfamily) on the secretory cells (**clear cells**), which increases cytosolic Ca^{2+} from both extracellular and endoplasmic reticular sources to stimulate Na^+-K^+-$2Cl^-$ cotransporter activity. The resultant cation influx is balanced by K^+ leakage and Na^+ extrusion by a basolateral Na^+-K^+ ATPase. Apical membrane Cl^- permeability increases also by a less understood mechanism, thereby increasing luminal Cl^- concentrations and causing paracellular Na^+ transport (Figure 16.8). The combined increase in Na^+ and Cl^- in the coil lumen osmotically draws water into the lumen via aquaporin-5 channels.

> Clear cells also express adrenergic receptors, which causes them to be activated by catecholamines during SNS activation.

b. **Reabsorbing duct:** Ions, principally Na^+ and Cl^-, are reabsorbed as the precursor fluid passes through the duct by channels such as the epithelial Na^+ channel (ENaC) and the cystic fibrosis transmembrane conductance regulator (CFTR). The resultant fluid is sweat, which is hypotonic. Disorders such as cystic fibrosis reduce Cl^- reabsorption through the CFTR, leading to increased ion loss in sweat.

c. **Myoepithelial cells:** The secretory coil is surrounded by myoepithelial cells, which contract upon cholinergic stimulation. Contraction does not force sweat out of the coil but rather provides the structural support that allows high osmotic forces (up to 500 mm Hg) to develop within the coil. These forces eventually propel fluid to the skin's surface in pressure pulses.

D. Nails

Nails are hard, scaly epidermal extensions that shield the posterior fingertips. The nail (known as the **nail plate**) is a hardened keratinized structure that mechanically protects the underlying skin (**nail bed**). Nail keratin contains a high number of disulphide bonds, which

give it its strength and rigidity. Skin keratin is softer, reflecting fewer disulphide bonds.

VII. CUTANEOUS NERVES

The cutaneous senses are a part of the **somatosensory system**. Every square millimeter of skin represents an opportunity to interact with and analyze the external environment, and, thus, it is dense with sensory nerve fibers. Not only do these nerves provide us with a sense of touch (**taction**), they also sense pain (**nociception**), itch (**pruritoception**), and temperature (**thermoreception**; see 38·II·A·2).

A. Touch

Physical contact can take many forms. Sometimes it might be a light touch, such as dragging a feather across the skin. Other times it might be the intense pressure of holding a plastic grocery bag full of cans. The ability to sense such disparate stimuli requires mechanoreceptors that are attuned to varying aspects of stimulus intensity, frequency, and duration. Their depth below the skin surface partly determines the size of their **receptive field**. A receptive field defines the area that a sensory receptor monitors. Receptors that collect stimuli over a wide receptive field have an increased chance of recording events but are unable to locate the source of the stimulus precisely. Receptors with small receptive fields are able to pinpoint the source of the stimulus with a high degree of accuracy and are usually clustered in large numbers to ensure adequate coverage over a wide surface area.

1. **Tactile receptors:** Skin contains several different types of mechanoreceptors that transduce tactile stimuli (Figure 16.9). Transduction occurs when a sensory nerve ending is deformed. The endings may be bare or encased in accessory structures that modify their sensitivity and responsiveness to different types of stimuli. Glabrous skin contains rapidly adapting Pacini and Meissner corpuscles. Hairy skin contains slowly adapting **Merkel disks** and **Ruffini endings** as well as rapidly adapting hair plexus neurons and sensory fibers around hairs.

    a. **Pacini corpuscles:** Pacini corpuscles are rapidly adapting mechanoreceptors ~1 mm long that sense high-frequency vibrations in glabrous skin (Figure 16.10A). They reside deep in the skin, and their receptive field is wide. The corpuscles consist of a sensory nerve ending wrapped in numerous layers of fibrous tissue with gelatinous fluid between, so that they resemble an onion in cross section. The entire structure is then wrapped in a connective tissue capsule. The gelatinous layers cushion the nerve so that only transient stimuli are able to deform and excite the nerve membrane. The afferent nerves are myelinated for most of their length, which allows for rapid relay of sensory signals.

    b. **Meissner corpuscles:** Meissner corpuscles also adapt rapidly (see Figure 16.10B). They are exquisitely sensitive to touch and low-frequency vibrations, which produce a fluttering sensation. They are smaller than Pacini corpuscles, but their construction and skin distribution is similar, in that a sensory nerve ending

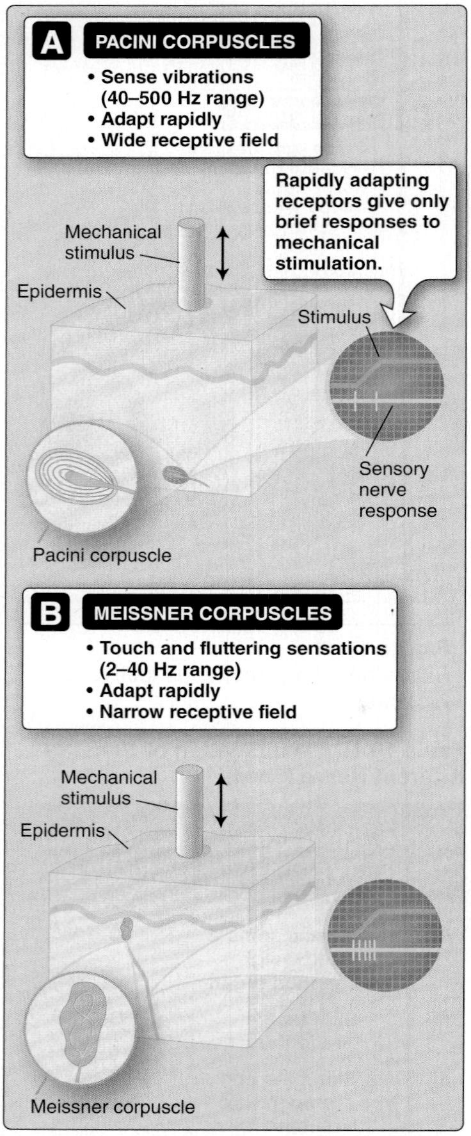

Figure 16.10
Rapidly adapting cutaneous tactile receptors.

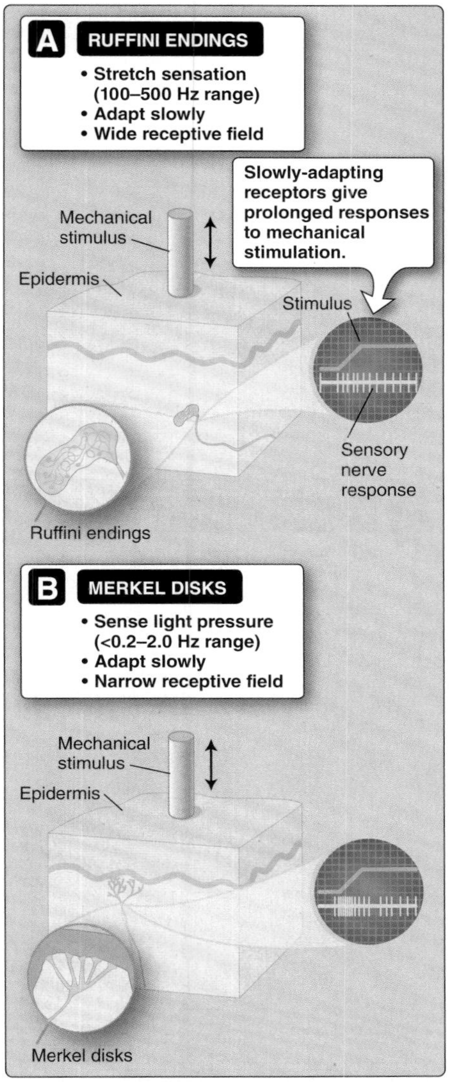

Figure 16.11
Slowly adapting cutaneous tactile receptors.

Table 16.1: Classification of Sensory Afferent Nerve Fibers

Nerve Class	Receptor Type	Conduction Velocity (m/s)
Aα	Muscle spindle and Golgi tendon organ	80–120*
Aβ	Skin tactile receptors	33–75*
Aδ	Skin pain and temperature receptors	5–30*
C	Skin pain, temperature, and itch receptors	0.5–2.0

*Myelinated.

meanders between stacked layers of flattened support cells, all enclosed within a capsule.

c. **Ruffini endings:** These afferents are slowly adapting receptors (Figure 16.11A) located in the deeper layers of the skin. The nerve endings branch and weave between bundles of collagen fibers to form a long, thin, spindle-shaped structure. The fibers are enclosed within a connective tissue capsule that is firmly tethered to surrounding tissues. When skin is stretched, the capsule and structures within are distorted also.

d. **Merkel disks:** Merkel disks are also slowly adapting receptors (see Figure 16.11B) that respond best to low-frequency stimulation and light touch. They lie just below the skin surface, which gives them a very narrow receptive field. Fingertips are endowed with very large numbers of Merkel disks, which allows for fine discrimination of form and texture.

e. **Hairs:** Every hair on the body surface functions as a mechanosensor, made possible by the presence of a sensory nerve that wraps around its follicle (see Figure 16.9). When the hair bends, the nerve ending distorts and signals.

f. **Free nerve endings:** Free sensory nerve endings can be found throughout the skin. They can also contribute to taction, in addition to other sensations including pain, itch, and temperature.

2. **Mechanosensory transduction:** Deformation of a tactile nerve ending opens a Na^+ channel, causing a depolarizing receptor potential (also known as a **generator potential**). The Na^+ channel may be an ENaC family member.

3. **Sensory nerve fibers:** Sensory nerve afferents are classified according to how fast they relay signals to the central nervous system ([CNS] Table 16.1). All of the tactile sensory afferents are myelinated (type Aβ) and conduct at relatively high velocity. These signals then travel via the spinal cord to the CNS for processing (see 6·II). The area perceived by a particular mechanoreceptor is dependent on receptor type and body area. The hands are more discriminatory than the upper arms, for example (Figure 16.12).

B. Pain

Mechanical and thermal stimuli that are innocuous or even pleasurable at low intensities can cause significant cellular damage at higher levels. The function of pain is to alert the CNS of local damage and to initiate a motor reflex that causes the body to either avoid or pull back from the source of stimulation (see 11·III·D).

1. **Nociception:** A number of nociceptor types transduce painful stimuli into a membrane potential change.

   a. **Mechanical:** Mechanical nociceptors respond to intense pressure or mechanical deformation of the skin. They also respond to sharp objects that stab or cut the skin. These receptors are likely very high threshold mechanoreceptors that only respond to mechanical stimuli when they reach noxious levels.

b. Thermal: Temperature extremes (freezing cold and burning heat) cause tissue damage. Cold stimuli become noxious at ~20°C, with intensity of perceived pain increasing linearly to ~0°C. Cold responses are also sensitive to the rate of cooling, with rapid cooling producing more intense responses. The threshold for noxious heat sensation is ~43°C.

c. Chemical: Because nociceptive fibers are free nerve endings, they are accessible to chemicals that cross the epidermal barrier or that are released by damaged tissues (Table 16.2). Capsaicin, the active ingredient in hot chili peppers, produces a burning sensation via activation of nociceptors when applied topically.

d. Polymodal: A subpopulation of nociceptors is sensitive to two or more stimuli and is known as **polymodal**.

2. Nociceptive stimulus transduction: The precise mechanisms by which nociceptive stimuli are sensed and signaled are not understood fully, but transduction of many noxious stimuli involves transient receptor potential (TRP) channel family members (see 2·VI·D).

a. Receptors: Heat activates TRPV1, a member of the vanilloid class of TRP receptors. It is also activated by capsaicin. Skin cooling activates TRPM8. Activating either TRP channel class results in Na^+ and Ca^{2+} influx and excitation. Hydrogen ions excite nociceptive neurons by permeating an acid-sensing channel of the ENaC family. Other channels may be involved in pain sensation also.

b. Nociceptive fibers: Nociceptor activation is relayed to the CNS by fast (myelinated) Aδ fibers and slower C fibers. The Aδ fibers mediate sensations of sharp, intense, pricking pain (**first pain**), followed by a more prolonged dull, throbbing, burning pain associated with C-fiber activation (**second pain**).

C-fibers are particularly sensitive to lidocaine, a local anesthetic that is applied topically to relieve skin itching and pain.[1] It blocks the Na^+ channel that mediates the nerve action potential. Lidocaine is also commonly injected to anesthetize teeth prior to dental surgery or is combined with prilocaine (a related Na^+-channel blocker) in an ointment. Lidocaine is sometimes combined with a vasoconstrictor to reduce local blood flow and thereby reduce drug washout effects. Local anesthesia is prolonged as a result.

[1]For additional discussion of amide local anesthetics such as lidocaine, see *LIR Pharmacology*, 5e, pp. 147–149.

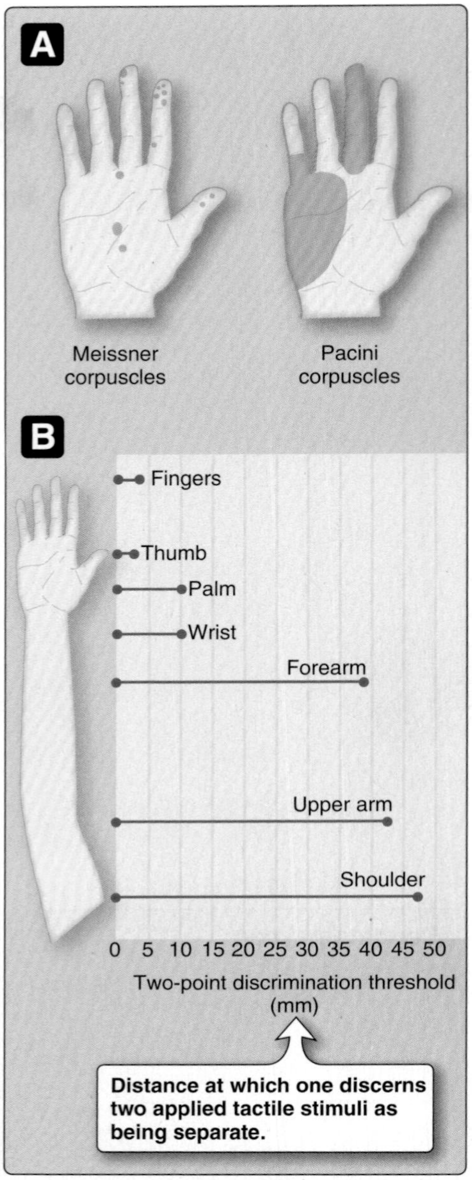

Figure 16.12
Receptive fields of two receptor types in the hand and sensory discrimination along the arm.

Table 16.2: Nociceptor-Activating Chemicals

Source	Chemical
Mast cells	Histamine
Mast cells, traumatized skin cells	Prostaglandins
Stressed skin cells	K^+ Bradykinin H^+
Sensory afferents	Substance P
Cholinergic efferents	Acetylcholine

3. **Sensitization:** Tissue damage initiates a chain of events that sensitizes surrounding afferent nerve endings to innocuous stimuli, thereby causing them to be perceived as painful (**hyperalgesia**). Sensitization initially remains localized to the site of damage (**primary hyperalgesia**) but spreads within minutes to involve surrounding areas (**secondary hyperalgesia**). Sensitization follows the progress of swelling and inflammation and involves many of the common inflammatory mediators. Its effects may persist for months after recovery from the initial injury.

C. Itch

Pruritis (derived from *prurire*, the Latin word for "itch") is the most recently recognized member of the cutaneous senses. Itches appear designed to trigger reflex scratching or rubbing to remove an insect or other irritant. The sensation is mediated by two populations of C-type nerve fibers. One fiber type responds optimally to histamine, whereas the other (nonhistamine) type is activated by a wide range of pruritogens, such as prostaglandins, interleukins, *proteases*, and ACh. Itch sensations can be suppressed by painful stimuli (such as scratching) and by antihistamines and potentiated by analgesics. The mechanisms by which painful and pruritic sensations interact are not understood fully.

D. Dermatomes

Sensory information from skin receptors is relayed by the CNS via afferent nerves. The nerves have a limited area of coverage, which can be mapped onto the body surface as a series of discrete bands called **dermatomes** (Figure 16.13). Each band corresponds to a single

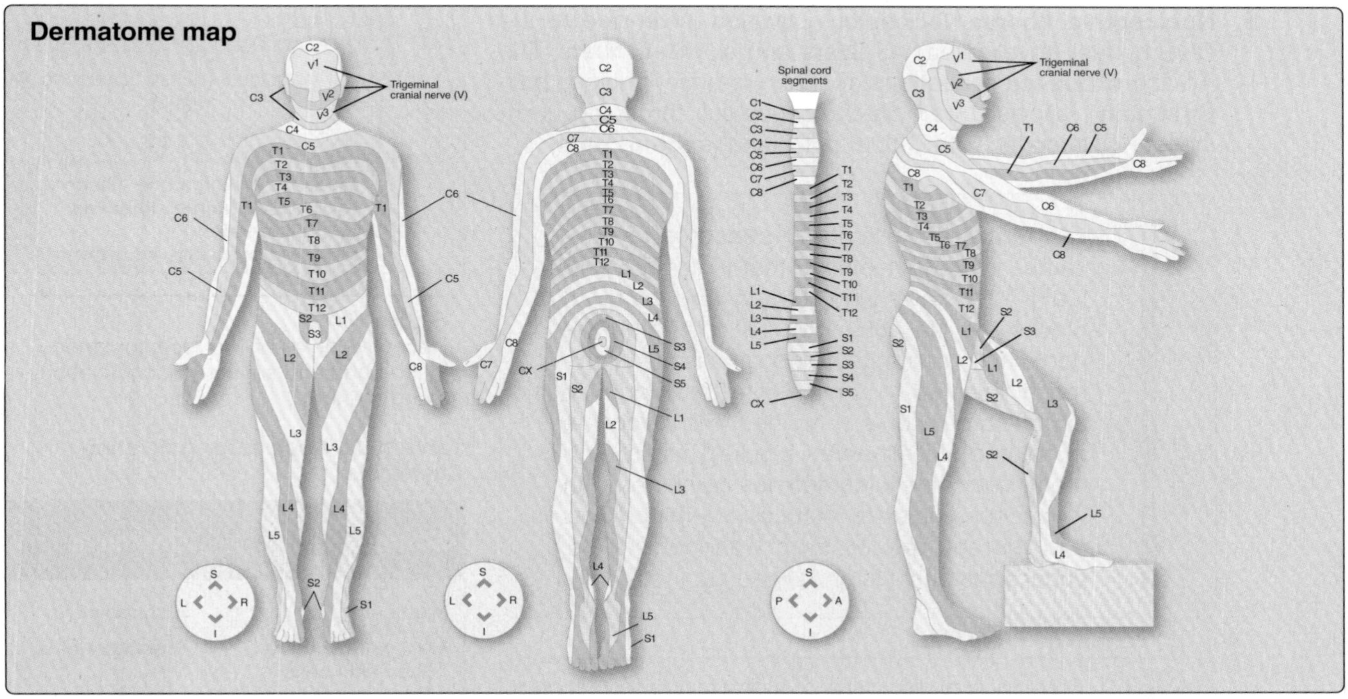

Figure 16.13
Dermatomes.

spinal segment. There is overlap in coverage between the bands, so that cutting a single pair of posterior nerve roots does not result in complete sensory loss in the corresponding dermatome. Pain that localizes to a particular dermatome may be helpful in identifying the site of spinal cord injury, for example.

Chapter Summary

- The **epidermis** provides the majority of skin barrier functions. The water barrier is provided by a combination of a thin lipid layer and a thicker protein layer. The ultraviolet (UV) barrier is provided by **melanin**, which absorbs a portion of UV radiation.
- The **dermis** is the location of the majority of the functional aspects of the skin and is home to most **skin appendages**. The dermal layer contains **mast cells**, which are involved in local inflammatory responses.
- **Hair follicles** provide a pore for glandular secretions of both sebaceous and apocrine sweat glands. Hair growth is a complex process involving active growth, regression, and then shedding of the hair.
- **Eccrine sweat glands** participate in thermoregulation and involve the formation of isotonic precursor sweat in the secretory coil and then ion reabsorption in the duct. This produces a hypotonic solution that can evaporate depending on environmental conditions.
- The **cutaneous senses** are a part of the **somatosensory system**, which monitors events occurring within or at the surface of the body.
- **Tactile** receptors provide information resulting from physical contact with objects. They activate in response to touch, the sensations being transduced by mechanical deformation of a sensory neuron. Some receptors (**Pacini** and **Ruffini**) respond best to high-frequency vibration or stretch. Others (**Meissner** and **Merkel** receptors) are more sensitive to pressure and low-frequency events.
- Sensory afferents from tactile receptors are myelinated and conduct impulses at high speed.
- Pain receptors activate in response to intense and noxious mechanical, thermal, or chemical stimuli.

Study Questions

Choose the ONE best answer.

III.1 Which of the following cytoskeletal proteins functions like a spring, limiting the extent to which the sarcomere can be stretched?

A. α-Actinin
B. Dystrophin
C. Nebulin
D. Titin
E. Z disk

Best answer = D. Titin is a massive, thick filament–associated structural protein that limits sarcomere length when a muscle is stretched (12·II·C). Thin filament–associated proteins do not act as springs but, rather, provide structural integrity. For example, α-actinin binds the ends of thin filaments to Z disks (structural plates that serve as attachment points for thin filaments); dystrophin anchors the contractile array within the cytoskeletal framework; and nebulin, which extends the length of the actin filament, is believed to establish thin-filament length.

III.2 When two acetylcholine molecules bind to a nicotinic receptor on skeletal muscle, the channel opens and allows transmembrane passage of ions. The resulting ion flux is dominated by which of the following choices under normal physiologic conditions?

A. Ca^{2+}
B. Mg^{2+}
C. H^+
D. Cl^-
E. Na^+

Best answer = E. The nicotinic acetylcholine receptor (nAChR) is a nonspecific cation channel that allows passage of Na^+, K^+, and Ca^{2+} (2·VI·B). The Ca^{2+} flux is small and not physiologically significant. The electrochemical gradients for Na^+ and K^+ are normally configured such that the nAChR supports simultaneous Na^+ influx and K^+ efflux. Na^+ influx dominates this exchange, however, and the myocyte depolarizes (12·II·G). Membrane depolarization then opens voltage-dependent Na^+ channels, which allows an action potential to propagate across the sarcolemma and down the T-tubule system to initiate muscle contraction. The nAChR does not support significant H^+, Cl^-, or Mg^{2+} fluxes under normal, physiologic conditions.

III.3 A 22-year-old woman receives botulinum toxin type A (a cholinergic presynaptic release inhibitor) injections to treat palmar hyperhidrosis (excess sweating). Her grasp is weakened by the treatments, through a decrease in the synaptic levels of what substance?

A. *Acetylcholinesterase*
B. Acetylcholine
C. Calsequestrin
D. Myoglobin
E. Nicotinic receptors

Best answer = B. Botulinum toxin is a protease that prevents exocytosis and release of neurotransmitters from nerve terminals (5·IV·C). Botulinum toxin type A is commonly used clinically to inhibit acetylcholine (ACh) release at the neuromuscular junction, which reduces synaptic ACh levels. It reduces eccrine sweat gland activity through a similar mechanism. *Acetylcholinesterase*, which normally degrades ACh and terminates neuromuscular signaling (12·II·G), would not be affected by the toxin. Myoglobin is a sarcoplasmic O_2 storage and transport molecule, whereas calsequestrin is a Ca^{2+}-binding protein found in the sarcoplasmic reticulum. Neither is directly involved neuromuscular transmission.

III.4 Phospholamban is a regulatory protein associated with the cardiac sarcoplasmic reticulum Ca^{2+} ATPase. Phospholamban phosphorylation would most likely increase the rate of which of the following events?

A. Relaxation
B. Ca^{2+} influx
C. Crossbridge cycling
D. Electrical conduction
E. Nodal cell depolarization

Best answer = A. Phospholamban normally acts as a rate limiter on sarcoplasmic reticulum (SR) Ca^{2+} ATPase (SERCA) function (13·III·B). Phospholamban phosphorylation reduces its inhibitory effects, allowing the pump to speed up. SERCA normally helps remove Ca^{2+} from the sarcoplasm following excitation. Increasing pump speed causes sarcoplasmic free Ca^{2+} levels to fall faster than normal, promoting decreased relaxation times. Ca^{2+} influx occurs during excitation and probably would not be appreciably affected by changes in SERCA. Crossbridge cycling rate is dependent on actin–myosin interactions. Electrical conduction between myocytes is dependent on gap junction function. Although faster relaxation times do facilitate heart rate increases, the rate of nodal cell depolarization is controlled through ion channel modulation.

III.5 Cardiac muscle contraction is dependent on a rise in sarcoplasmic Ca^{2+} concentration. The bulk of the Ca^{2+} required for full force generation flows through which of the following Ca^{2+} channel types?

A. Dihydropyridine receptors
B. Ryanodine receptors
C. Inositol trisphosphate–gated channels
D. Transient receptor-potential channels
E. Stretch-activated channels

Best answer = B. Full force development by a cardiac myocyte relies on Ca^{2+} release from stores in the sarcoplasmic reticulum ([SR] 13·III·A). Release is mediated by Ca^{2+}-induced Ca^{2+} release (CICR) channels, also known as ryanodine receptors. Dihydropyridine receptors are L-type Ca^{2+} channels that mediate voltage-gated Ca^{2+} fluxes across the T-tubule membrane. Ca^{2+} influx via this pathway acts as a trigger for CICR. Inositol trisphosphate mediates Ca^{2+} release from the SR in smooth muscle. Transient receptor-potential channels are found in many tissues, often mediating cellular sensory stimulus transduction (2·VI·D). Stretch-activated channels are also widespread but ryanodine receptors are the principal pathway for Ca^{2+} fluxes during contraction.

III.6 What type of smooth muscle Ca^{2+} channels localize to plasma membrane caveolae and are gated primarily by membrane potential change?

A. Ca^{2+}-induced Ca^{2+} release channels
B. Receptor-operated Ca^{2+} channels
C. Store-operated Ca^{2+} channels
D. Inositol trisphosphate–gated Ca^{2+} channels.
E. L-type Ca^{2+} channels

Best answer = E. L-type Ca^{2+} channels are voltage gated, opening in response to membrane depolarization. They are found in many cell types, including smooth muscle, where they are concentrated within plasma membrane pockets called caveolae (14·II·C). Receptor-operated Ca^{2+} channels open when a ligand binds to the associated receptor, rather than a voltage change. Ca^{2+}-induced Ca^{2+} release channels and inositol trisphosphate–gated Ca^{2+} channels are located on the sarcoplasmic reticulum membrane and mediate Ca^{2+} store release. Store-operated Ca^{2+} channels are used to top off intracellular Ca^{2+} stores with extracellular Ca^{2+} during muscle relaxation (14·III·C). Channel opening is controlled by a store Ca^{2+} sensor (Stim1).

III.7 A pharmaceutical company is intent on developing a drug that decreases smooth muscle–induced vasospasm. Which of the following enzymes normally antagonizes smooth muscle contraction and might, thus, make a suitable target for modulation (stimulation or upregulation) by a pharmaceutical product?

A. *Rho-kinase*
B. *Myosin phosphatase*
C. *Myosin light-chain kinase*
D. *Protein kinase C*
E. *Phospholipase C*

Best answer = B. *Myosin phosphatase* normally dephosphorylates a myosin regulatory light chain (MLC_{20}) to inhibit myosin ATP-ase activity, thereby blocking smooth muscle contraction (14·III·C). Upregulating this enzyme would decrease contractility and potentially reduce vasospasm. *Myosin phosphatase* activity is regulated by at least two different pathways. *Rho-kinase* and *protein kinase C* are components of separate pathways that normally inhibit *myosin phosphatase* activity and promote contraction. *Myosin light-chain kinase* (*MLCK*) phosphorylates and activates smooth muscle myosin, thereby facilitating contraction. *Phospholipase C* also promotes contraction through a pathway that simultaneously stimulates *MLCK* and inhibits *myosin phosphatase.*

III.8 What are the mineral crystals that resist compression and give bones their characteristic strength and resilience?

A. Urate
B. Hydroxyapatite
C. Glycosaminoglycan
D. Creatinine
E. Calcium oxalate

Best answer = B. Hydroxyapatite is a crystalline mineral containing calcium and phosphate (15·II·A). Hydroxyapatite crystals are cemented within collagen fibers and then bundles of mineralized fibers embedded in ground substance to create a material that has a high resistance to compression and tensile stress. Urate, creatinine, and calcium oxalate are found at high concentrations in urine. When sufficiently concentrated, they form crystals that may be observed in urine sediments. Glycosaminoglycan is a mucopolysaccharide found in ground substance, which fills the spaces between all cells, including bone.

III.9 What type of precursor cells found in bone express RANKL (receptor activator of nuclear factor κB ligand) on their surface to facilitate bone resorption?

A. Osteoblast
B. Osteoclast
C. Osteocyte
D. Bone lining cell
E. Hematopoietic

Best answer = A. Bone resorption and remodeling involves several cell types that work together within a bone-remodeling compartment (15·IV·C). Osteoblast precursors express receptor activator of nuclear factor κB (RANK) ligand (RANKL) on their surface. RANKL binds to RANK, a receptor expressed on the surface of osteoclast precursors (these are hematopoietic precursors), causing several such cells to fuse and form large multinucleated osteoclasts. Osteoclasts digest bone and release its mineral content for return to blood. Osteocytes are cells embedded in the bone matrix that monitor stress and integrity. Bone lining cells are found on the bone surface and signal a need for remodeling as needed.

III.10 A 36-year-old woman has a parathyroid hormone (PTH)-secreting tumor. Which of the following might be expected to increase as a result of chronic PTH elevation?

A. Bone resorption
B. Bone deposition
C. Ca^{2+} excretion from the intestines
D. PO_4^{3-} absorption from the intestines
E. PO_4^{3-} reabsorption from the kidneys

Best answer = A. Parathyroid hormone (PTH) release from parathyroid glands is normally regulated by plasma Ca^{2+} levels (15·V·A; 35·V·B). When plasma Ca^{2+} (or Mg^{2+}) levels decrease, PTH is secreted to stimulate Ca^{2+} resorption from bone. PTH's effects are mediated by PTH receptors on osteoblasts, which then recruit osteoclasts to a bone-remodeling site. PTH does not stimulate bone deposition. PTH causes Ca^{2+} reabsorption (not excretion) by the renal distal tubule (27·III·C). PTH also inhibits PO_4^{3-} reabsorption by the renal proximal tubule, thereby increasing excretion rates (26·VI·A).

III.11 A 4-year-old boy with a family history of cystic fibrosis has been presenting with mild respiratory and gastrointestinal symptoms. If cystic fibrosis is suspected, his sweat composition might best be described as which of the following, compared with that of a healthy boy his age?

A. Hypotonic
B. Isotonic
C. Hypertonic
D. Copious
E. Scant

Best answer = C. The cystic fibrosis (CF) transmembrane conductance regulator (CFTR) is an adenosine triphosphate–binding cassette transporter that functions as a Cl^- channel in many epithelia. CFTR defects prevent Cl^- secretion by respiratory and gastrointestinal epithelia. Secretion creates an osmotic gradient that is used to draw water onto the apical surface, so CF patients typically form mucus that is thick, which makes it difficult to expel from the lungs, for example. In sweat glands, CFTR is used to reabsorb Cl^- from the ductal lumen during sweat's passage to the skin surface, causing sweat to become hypotonic (16·VI·C·2·b). In CF patients, a CFTR defect prevents Cl^- (and Na^+) reabsorption, so their sweat is hypertonic. CFTR defects do not cause major changes in sweat volume.

III.12 A 42-year-old jackhammer operator presents with decreased high-frequency vibration sensitivity in the glabrous skin of the hands. Which receptor is most likely being affected?

A. Ruffini endings
B. Merkel disks
C. Free nerve endings
D. Pacini corpuscles
E. Hair sensory fibers

Best answer = D. Pacini corpuscles are rapidly adapting tactile receptors responsible for sensing vibrations in the 40–500 Hz range (16·VII·A). Glabrous skin does not have hair and, therefore, no hair sensory fibers. Ruffini endings are slowly adapting tactile receptors that sense skin stretch rather than vibrations. Merkel disks sense light skin pressure. Glabrous skin also contains free nerve endings, but these are less sensitive to vibration than are Pacini corpuscles.

UNIT IV

Cardiovascular System

17 Cardiac Excitation

I. OVERVIEW

The cardiovascular system is responsible for making oxygen and nutrients available to every cell in the body. It also carries away metabolic waste products, including heat. The principal components of the cardiovascular system are blood, blood vessels, and the heart (Figure 17.1). Blood carries materials to and from the tissues, blood vessels are conduits that bring blood close to cells, and the heart is used to create the pressure that is needed to propel blood around the system. The heart is a multichambered organ comprising cardiac muscle. When the muscle contracts, chamber diameter decreases and blood is forced under pressure from chamber to chamber and then out into the vasculature. The unique design of the heart means that contraction of its various parts must be sequenced for output efficiency. Sequencing is effected using a wave of excitation whose rate of progress is modulated to allow time for blood to move between its chambers. Movement of this wave through time and space creates electrical gradients within the surrounding tissues that can be detected and recorded at the body surface as an electrocardiogram (ECG). Because the timing and pattern of excitation varies little from one individual to the next, an ECG can be used to detect abnormalities in the excitation pathway or in the overall structure of the heart.

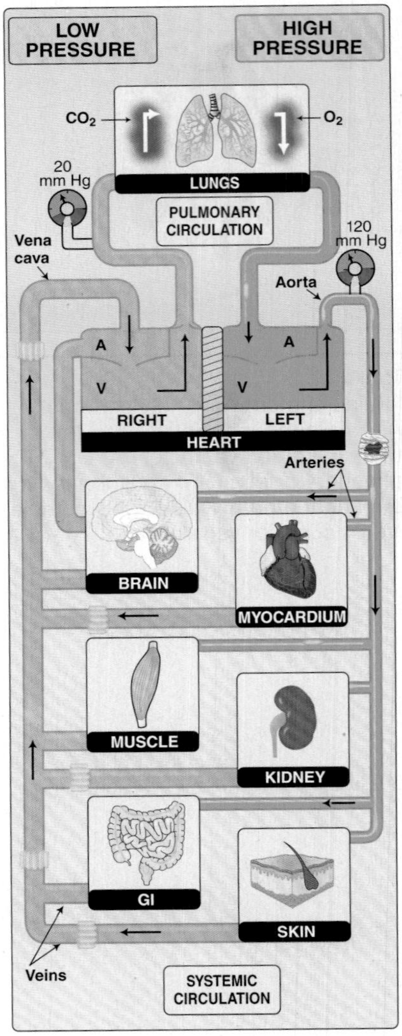

Figure 17.1
Cardiovascular circuitry. A = atrium; GI = gastrointestinal; V = ventricle.

II. CARDIOVASCULAR CIRCUITRY

The cardiovascular system is a closed loop through which blood cycles continuously throughout the life of the individual. The loop incorporates two structurally and functionally distinct circulations. The **systemic circulation** supplies all body organs with nutrients and carries away waste products. The **pulmonary circulation** conveys the contents of the vasculature through the lungs for exchange of CO_2 and O_2.

A. Heart structure

Blood is propelled around the vasculature by two muscular pumps, one on each side of the heart (Figure 17.2A). Each pump contains two chambers: one **atrium** and one **ventricle**. The left heart pumps

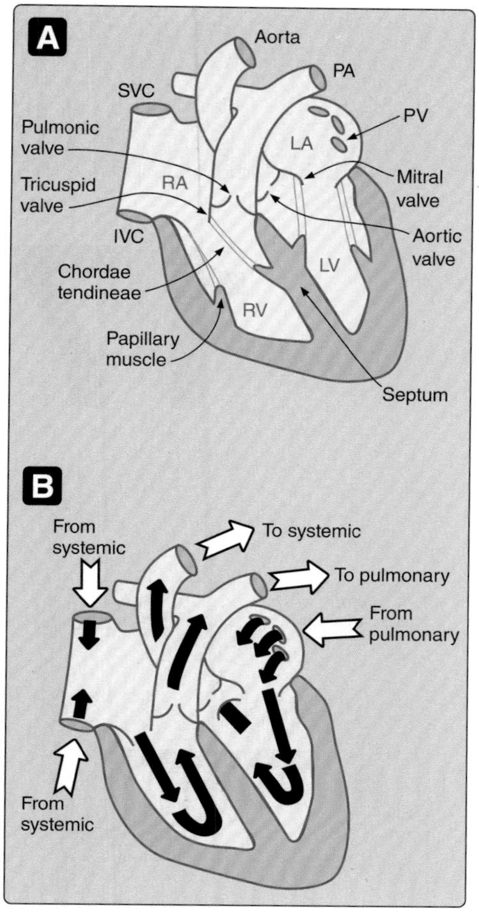

Figure 17.2
Cardiac anatomy and flow patterns. IVC = inferior vena cava; LA = left atrium; LV = left ventricle; PA = pulmonary artery; PV = pulmonary vein; RA = right atrium; RV = right ventricle; SVC = superior vena cava.

blood through the **aorta** to the organs of the systemic circulation. It returns to the heart via the **vena cavae**. The right side of the heart perfuses the pulmonary circulation. Blood exits the right ventricle via the **pulmonary artery**, passes through the lungs, and then enters the left heart via the **pulmonary vein**.

B. Cardiac chambers

Atria and ventricles serve different functions, which is reflected in the amount of cardiac muscle contained within their walls.

1. **Atria:** Atria act as holding tanks for blood collected from the venous system during ventricular contraction. Accumulated blood is transferred to the ventricles by atrial contraction at the beginning of each cardiac cycle. Minimal amounts of pressure are required to push blood into the ventricles and, therefore, atrial walls contain relatively small amounts of muscle and are thin.

2. **Ventricles:** Ventricles drive blood at high pressure through vast networks of vessels, made possible by chamber walls that are thick with cardiac muscle. The left ventricle (LV) typically generates peak pressures of 120 mm Hg. The right ventricle (RV) pumps blood through a system of relatively low-resistance vessels and, therefore, its walls are less muscular than those in the LV. The RV generates peak pressures of about 20 mm Hg.

B. Valves

One-way valves situated between atria and ventricles (**atrioventricular [AV] valves**) and between ventricles and their outlets (**semilunar valves**) help ensure that flow around the cardiovascular system is unidirectional (see Figure 17.2B).

1. **Atrioventricular:** The **tricuspid** (right side) and **mitral** (left side) valves allow blood to pass from atrium to ventricle and close when ventricular contraction begins. **Chordae tendineae** are filaments attached to the edges of the valve leaflets. The chordae work in conjunction with **papillary muscles** to brace the valves and prevent them from being everted by high pressures generated within the ventricles during contraction (see Figure 17.2A).

2. **Semilunar:** The **pulmonary** (right side) and **aortic** (left side) valves prevent backflow from the arterial system into the ventricles. The semilunar valves are subject to high shear stress associated with high-velocity ventricular outflow, so are thicker and more resilient than AV valve leaflets.

III. SEQUENCING CONTRACTIONS

The heart's unique architecture requires that its various regions contract in an orderly sequence for maximal pumping efficiently. Sequencing is accomplished using a wave of depolarization that spreads from myocyte to myocyte until it envelops the entire heart. This is made possible by **gap junctions** (see 4·II·F), which create electrical connections between every cell in the heart and transform it into a **syncytium**. Once a wave of depolarization is initiated, it involves and engulfs adjacent myocytes with the inevitability of a line of toppling dominoes (Figure 17.3).

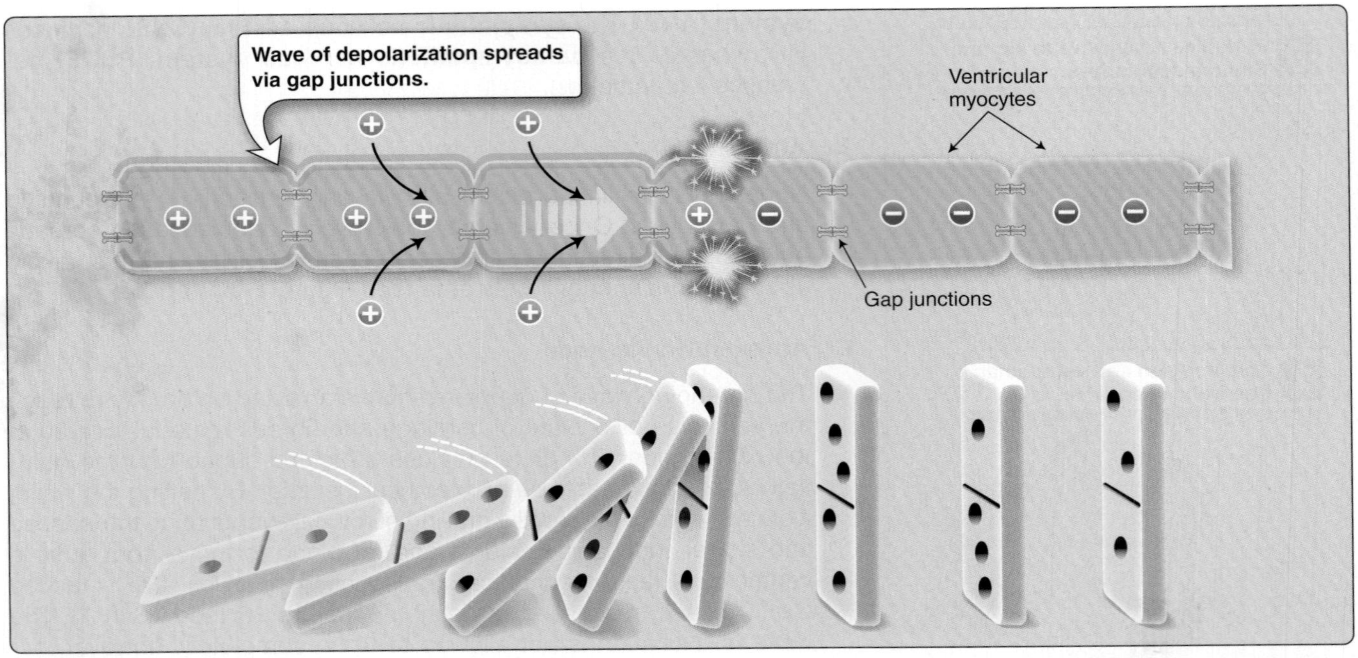

Figure 17.3
Signal propagation in the myocardium.

> The current record for number of dominoes toppled in one cascade is just shy of 4.5 million, a record set in the Netherlands on Domino Day, 2009. It took a team of 89 builders 2 months to set up! Because a misplaced foot or finger could easily trigger the entire array, it had to be designed with breaks that were completed immediately before the competition began. Because the myocardium does not have such safety features, every cell in the heart has the potential to become the primary pacemaker should its membrane become unstable. "Ectopic" pacemakers are a constant and potentially lethal threat to cardiac function.

The wave of depolarization that drives myocardial contraction and cycling of the pump originates in the **sinoatrial (SA) node** (Figure 17.4). Nodal cells have an unstable membrane potential (V_m) that drifts slowly positive over time. Once V_m crosses the **threshold for action potential (AP) formation** ([V_{th}] see 2·III·B·1), the cell spikes and a wave of excitation begins.

A. Pacemaker

The SA node comprises a group of specialized cardiac myocytes located near the superior vena cava in the wall of the right atrium (see Figure 17.4 step 1). These myocytes have lost most of their contractile elements and their function, instead, is to generate spontaneous APs. The rate at which APs are initiated and, thus, heart rate (HR), is under simultaneous control of both arms of the **autonomic nervous**

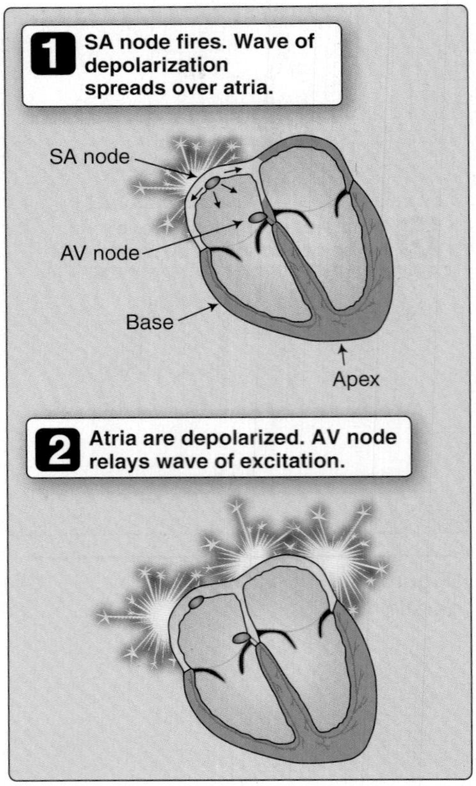

Figure 17.4
Cardiac excitation cycle.
AV = atrioventricular; SA = sinoatrial.

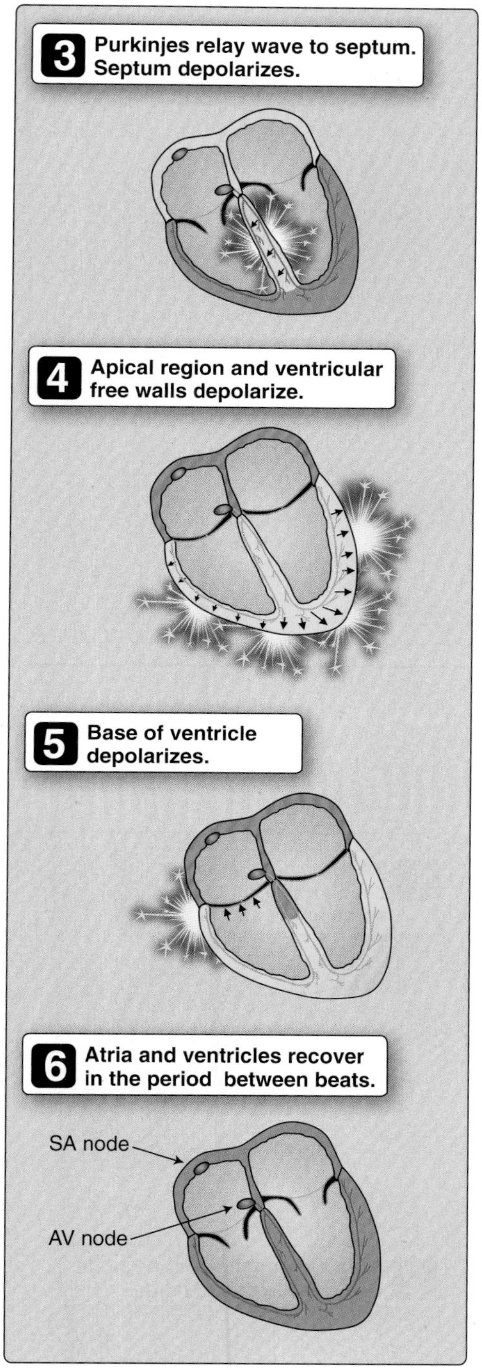

3 Purkinjes relay wave to septum. Septum depolarizes.
4 Apical region and ventricular free walls depolarize.
5 Base of ventricle depolarizes.
6 Atria and ventricles recover in the period between beats.
SA node
AV node

Figure 17.4
(continued)

system (**ANS**). The **sympathetic nervous system** (**SNS**) increases HR, whereas the **parasympathetic nervous system** (**PSNS**) decreases it (see below).

B. Atria

Nodal cells are linked electrically via gap junctions to surrounding atrial myocytes. Once initiated, the wave of depolarization spreads outward in all directions with a conduction velocity of ~1 m/s, taking about 100 ms to engulf both atria (see Figure 17.4 step 2).

C. Atrioventricular node

The spreading wave of depolarization is arrested before it can reach the ventricles by a plate of cartilage and fibrous material located at the AV junction. The plate provides structural support for the heart valves, but it also acts as an electrical insulator. By halting the wave, it allows time for the rapidly moving electrical events to be transduced into slower mechanical events and for blood to move from atria to ventricles. Electromechanical transduction involves Ca^{2+}-induced Ca^{2+} release, discussed in detail in Chapter 13 (see 13·III·A). The wave of excitation is not allowed to fizzle completely, however, because there is an electrical bridge between the atria and ventricles. At the entrance to this bridge is the AV node, a patch of noncontractile cardiomyocytes that are specialized to conduct signals slowly (0.01–0.05 m/s). It takes about 80 ms for the electrical "spark" to traverse the AV node, just long enough for blood to be pushed by atrial contraction through the AV valves.

D. His–Purkinje system

Once the wave of excitation migrates through the AV node, the ventricular walls must be stimulated to contract in a sequence that squeezes blood upwards toward the outlets: **septum** → **apex** → **free walls** → **base**. This is made possible using tracts of tissue comprising myocytes that are specialized to deliver the wave of depolarization at high speeds to the different regions of the ventricles.

> The His–Purkinje system is often compared to a system of interstate highways or motorways that allow vehicles to travel at high speed through a city's heart to a remote location. The slower myocyte-to-myocyte conduction pathway is equivalent to taking back streets to the same location, a route that is generally much slower.

The pathway to the ventricles begins with the **common bundle of His**, a tract of specialized myocytes that emerges from the AV node and then sweeps downward into the interventricular septum (see Figure 17.4 step 3). Here it separates into left and right bundle branches, which then branch again to deliver the excitation signal to all regions of the left and right ventricles, respectively. High-speed **Purkinje fibers** (conduction velocity ~2–4 m/s) carry the wave of depolarization to the contractile myocytes (see Figure 17.4, step 4).

E. Ventricles

Ventricular myocytes are similar to atrial myocytes, conducting the wave of depolarization from cell-to-cell via gap junctions at 1 m/s. Excitation of both ventricles is essentially complete within 100 ms (see Figure 17.4 step 5), although the slower mechanical events will take another 300 ms to complete.

IV. ELECTROPHYSIOLOGY

The mechanism by which the wave of excitation is accelerated or delayed during its travel through the various regions of the heart is elegant in its simplicity. The heart is, in essence, a large sculpted muscle. The rate at which different myocytes within this muscle conduct an electrical signal is dependent on how fast they depolarize, which, in turn, is governed by the relative mix of ion channels contained within the sarcolemma.

A. Ion channels

Cardiac myocytes are all excitable cells regardless of location. They all express a Na^+-K^+ ATPase that generates and maintains Na^+ and K^+ gradients across the membrane and establishes V_m. Ion-selective channels then open and close to manipulate V_m. Cardiac function is dependent on five principal conductances, a **voltage-dependent Na^+ channel** common to neurons, a **voltage-dependent Ca^{2+} channel**, two types of **K^+ channels**, and a **hyperpolarization-activated, cyclic nucleotide–dependent nonspecific channel** (**HCN**) common to all cells with intrinsic pacemaker activity (Figure 17.5).

1. **Sodium channel:** The Na^+ channel opens in response to membrane depolarization and is extremely fast activating (0.1–0.2 ms). It mediates an inward Na^+ flux (I_{Na}) that drives V_m to zero and several tens of millivolts positive (**overshoot**). It then rapidly inactivates (~2 ms) and stays inactivated until V_m returns to –90 mV. I_{Na} generates the upstroke of the AP in atrial and ventricular myocytes.

2. **Calcium channel:** L-type Ca^{2+}channels also open in response to membrane depolarization, but they activate more slowly than Na^+ channels (~1 ms). Once open, they mediate a Ca^{2+} influx that both depolarizes the cell (I_{Ca}) and initiates contraction. Ca^{2+} channels also inactivate but very slowly (~20 ms).

3. **Potassium channels:** K^+ channels mediate K^+ efflux and are used to repolarize a membrane after excitation. There are two principal types of K^+ channel found in myocytes. A fast-activating K^+ current mediates a minor transient outward current (I_{to}) in atrial and ventricular muscle. Membrane repolarization is the responsibility of a voltage-gated K^+ current (I_K) that activates slowly (~100 ms) and after a considerable delay following membrane depolarization.

4. **HCN-gated channel:** HCN mediates a **"funny current"** (**I_f**). The current is so named because it has peculiar properties. It is a nonspecific cation channel activated by hyperpolarization that supports simultaneous K^+ efflux and Na^+ influx. Na^+ dominates

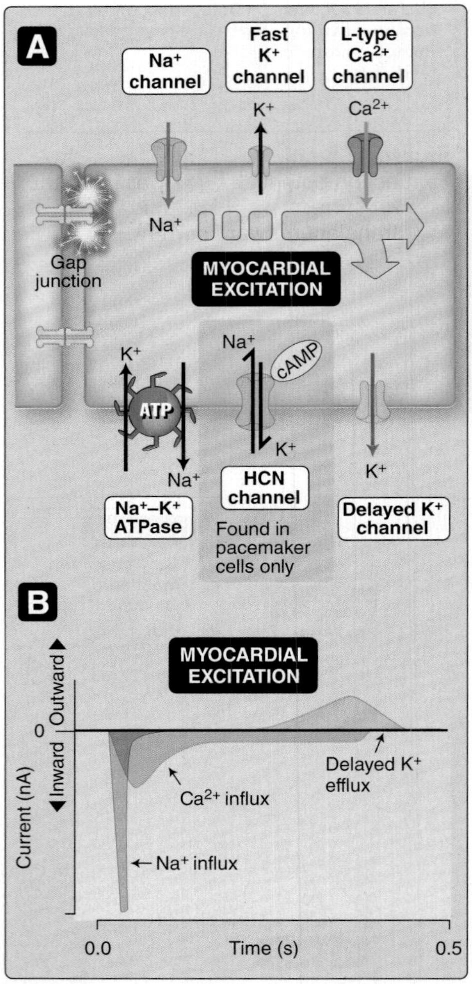

Figure 17.5
Principal ion channels and currents in cardiac myocytes. ATP = adenosine triphosphate; cAMP = cyclic adenosine monophosphate; HCN = hyperpolarization activated, cyclic nucleotide–dependent Na^+ channel.

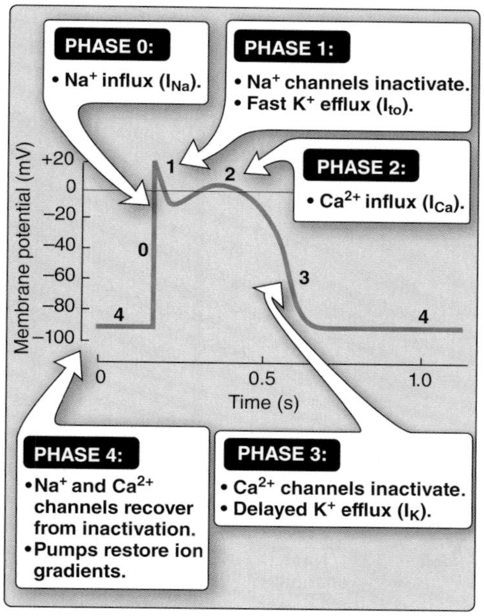

Figure 17.6
Fast action potential. I_{Ca} = Ca^{2+} current; I_K = K^+ current; I_{Na} = Na^+ current; I_{to} = transient outward current.

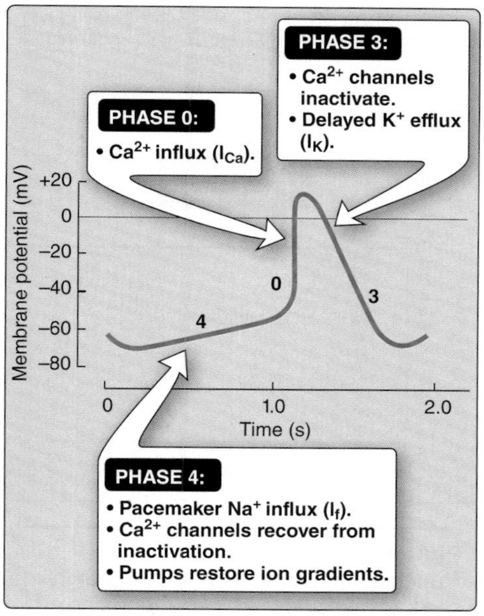

Figure 17.7
Slow action potential. I_{Ca} = Ca^{2+} current; I_f = funny current; I_K = K^+ current.

the exchange, and the membrane depolarizes. I_f is a pacemaker current, discussed in more detail below.

B. Action potentials

The rate at which myocytes conduct electrical signals is dependent on the mix of ion channels involved. Excitation of contractile myocytes is dominated by voltage-gated Na^+ channels. Nodal-cell excitation is dominated by L-type Ca^{2+} channels. The consequences are most apparent in the shape of APs recorded from the two myocyte classes. Atrial and ventricular myocytes express a fast AP that rises rapidly during I_{Na} activation. Nodal cells express slow APs that rely on I_{Ca} to provide the upstroke.

1. **Fast:** A fast AP has five distinct phases (Figure 17.6). The currents that underlie the different phases overlap in time, but pharmaceuticals used to treat dysrhythmias and other cardiac disorders are often grouped according to which AP phase they affect primarily and, thus, are identified here.

   a. **Phase 0: Phase 0**, the AP upstroke, is caused by Na^+ channel opening. The sarcolemma of atrial and ventricular myocytes is rich in Na^+ channels, and they open rapidly once the wave of excitation arrives. The result is a massive Na^+ influx that drives V_m toward the equilibrium potential for Na^+ (+60 mV).

   b. **Phase 1: Phase 1** reflects Na^+-channel inactivation, which brings V_m closer to 0 mV. Phase 1 repolarization is aided by I_{to}.

   c. **Phase 2:** The AP plateau is maintained by Ca^{2+} influx through L-type Ca^{2+} channels. I_{Ca} inactivates slowly during the AP, but a few Ca^{2+} channels remain open to prolong the plateau and ensure that Ca^{2+} release and contraction completes before excitation terminates.

   d. **Phase 3:** Membrane repolarization is mediated by delayed activation of I_K.

   e. **Phase 4:** The interval between APs is used to return Ca^{2+} to the intracellular stores and to pump Na^+ out of the cell in exchange for K^+. Return to a resting V_m (–90 mV) also allows Na^+ and Ca^{2+} channels to recover from their inactivated state, a process that takes several tens of milliseconds.

2. **Very fast:** Purkinje cells are designed to conduct the wave of excitation at high speed. Their membranes contain more Na^+ channels and fewer Ca^{2+} channels than ventricular myocytes, meaning that phase 0 more closely follows I_{Na}. It is the rate of phase 0 depolarization that determines conduction velocity. Purkinje cells are also three to four times thicker than ventricular myocytes, which allows for faster conduction velocities (see 5·III·B·2).

3. **Slow:** Nodal cells express slow APs dominated by I_{Ca} (Figure 17.7). The primary reason is that nodal cells have a resting V_m that is significantly more positive than contractile myocytes (–65 mV *versus* –90 mV). Na^+ channels are inactivated and cannot be opened at –65 mV, forcing nodal cells to rely on the slower Ca^{2+} channels to provide the AP upstroke.

> Nodal cell membranes contain functional Na^+ channels that will support a Na^+ current if they are allowed to recover from inactivation. Recovery involves holding V_m at –90 mV under controlled conditions.

a. **Phase 0:** The upstroke of a slow AP is driven by Ca^{2+} influx through L-type Ca^{2+} channels, which activate an order of magnitude more slowly than Na^+ channels. Slow APs propagate very slowly as a result.

b. **Phase 3:** Phase 3 repolarization is mediated by I_K.

c. **Phase 4:** Phase 4 corresponds to a period of recovery, but nodal cells are notable in that phase 4 is unstable, and it drifts slowly positive with time. This drift is caused by I_f and is the key to automaticity and pacemaker function.

C. Pacemakers

Pacemaker ability is conferred on cells by HCN. When open, HCN channels cause V_m to slide gradually toward the threshold for AP formation. The cyclic adenosine monophosphate (cAMP) dependence of this channel also provides the ANS with a way of regulating the rate of phase 4 depolarization, which, in turn, regulates HR. When intracellular cAMP levels rise, HCN open probability increases, and V_m depolarizes at an accelerated rate. Falling cAMP levels decrease HCN opening, and the rate of phase 4 depolarization slows. Because maintaining a regular heartbeat is critical for survival, three different cell types are granted the ability to function as pacemakers: SA nodal cells, AV nodal cells, and Purkinje fibers.

1. **Funny current:** HCN is activated by the hyperpolarization that occurs at the end of phase 3 (Figure 17.8A). The ensuing depolarization usually takes several hundred milliseconds to reach V_{th}, at which point a spike and a new wave of excitation is initiated. The spike inactivates HCN until the end of phase 3, at which point the cycle repeats.

2. **Other pacemakers:** The SA node is the heart's primary pacemaker. It has an intrinsic rate of ~100 beats/min, but HR is usually lower because the PSNS reduces HR when the prevailing need for cardiac output (CO) is low (see Figure 17.8B). Should the SA node be damaged and fall silent, then the AV node takes over as pacemaker. The AV node is normally subservient to the SA node because its intrinsic rate is 40 beats/min. It takes ~1.5 s for AV nodal phase 4 to reach V_{th} (see Figure 17.8C), but the new wave of excitation originating in the SA node typically arrives well before this time (**overdrive suppression**). Purkinje cells are tertiary pacemakers. Their intrinsic rate is very low (~20 beats/min), partly because V_m is about 25 mV more negative in Purkinje cells than nodal cells, and, thus, it takes much longer for V_m to reach and cross threshold from this more negative level (see Figure 17.8D).

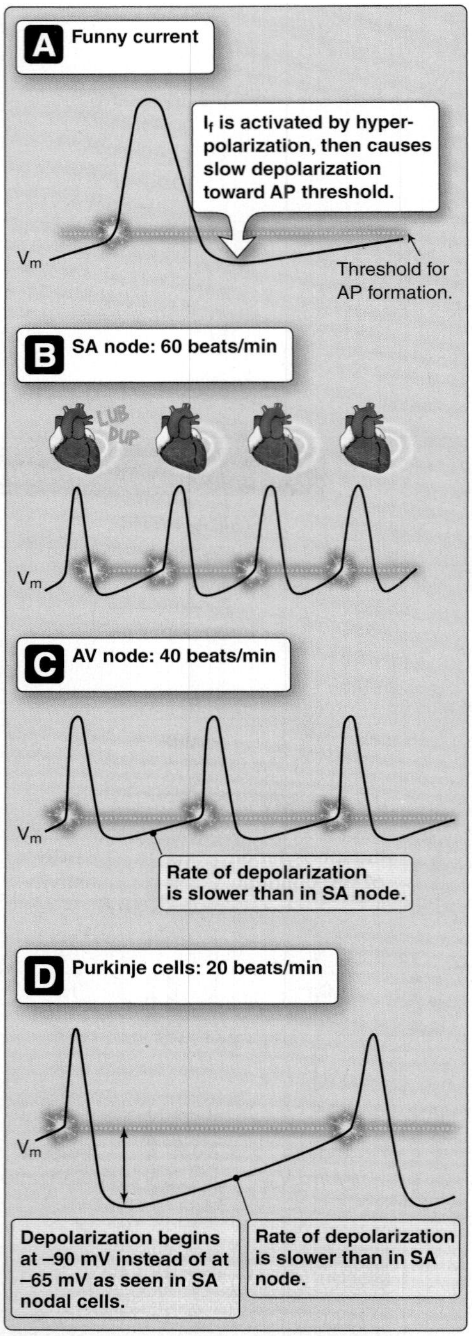

Figure 17.8
Pacemakers. AP = action potential; AV = atrioventricular; I_f = funny current; SA = sinoatrial; V_m = membrane potential.

3. **Regulation:** Because HR is a primary determinant of CO, the SA node is heavily regulated by the ANS. The SNS increases HR by releasing norepinephrine onto β_1-adrenergic receptors on nodal cells. These are G protein–coupled receptors (GPCRs) that increase *adenylyl cyclase* (*AC*) activity and intracellular cAMP concentration. cAMP binds to and increases HCN open probability and accelerates the rate of phase 4 depolarization (Figure 17.9). HR thus increases (**positive chronotropy**). PSNS terminals release acetylcholine (ACh) onto nodal cells. ACh binds to muscarinic type-2 receptors, which are also GPCRs that depress *AC* activity and decrease cAMP formation. The rate of phase 4 depolarization slows, and HR decreases (**negative chronotropy**).

4. **Other currents:** The rate of phase 4 depolarization is also influenced by a Ca^{2+} current and a ligand-gated K^+ current, both of which are regulated by the ANS (see Figure 17.9).

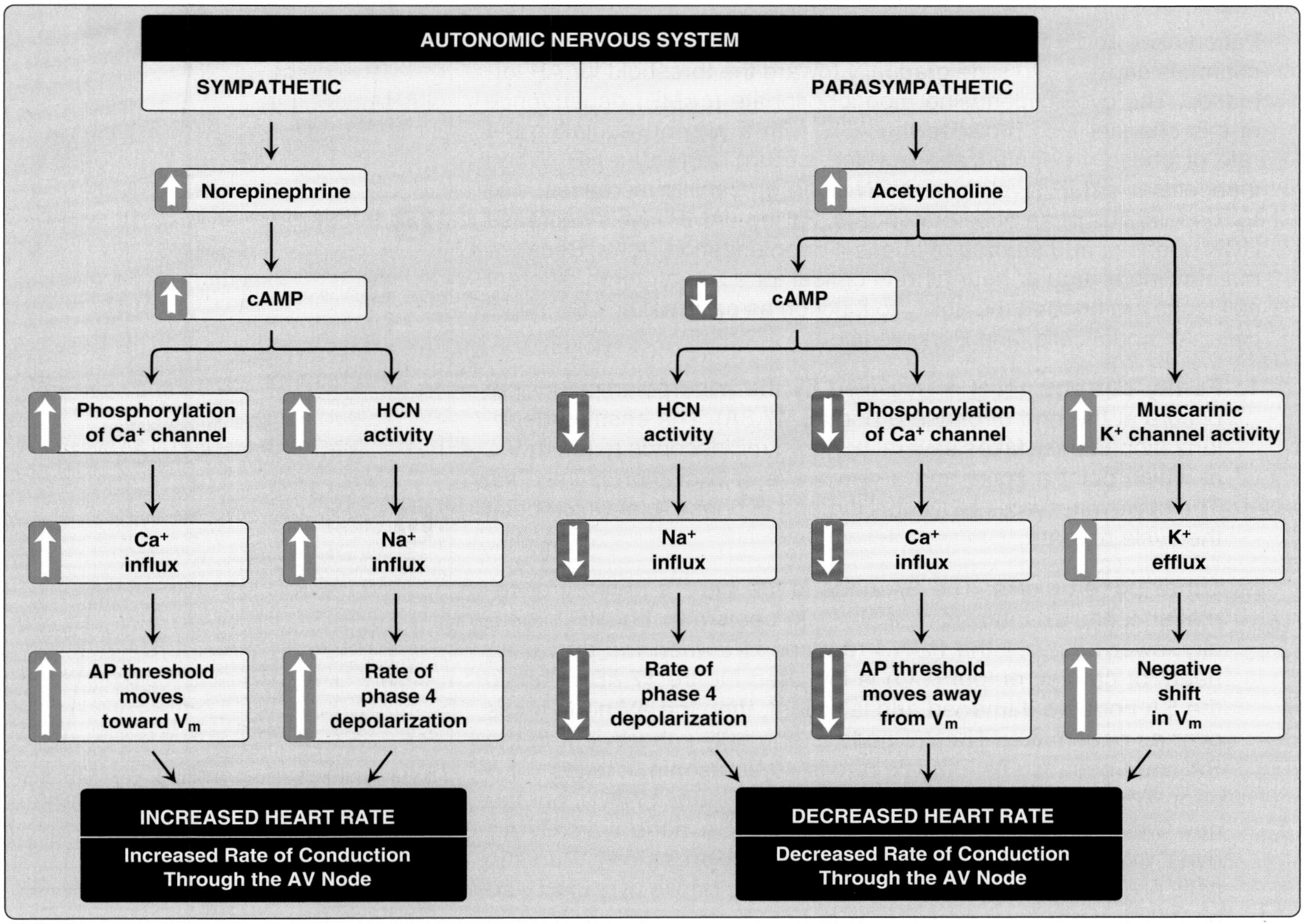

Figure 17.9
Autonomic regulation of nodal cells. AP = action potential; AV = atrioventricular; cAMP = cyclic adenosine monophosphate; HCN = hyperpolarization activated, cyclic nucleotide–dependent Na^+ channel; V_m = membrane potential.

a. **Calcium current:** Catecholamine-induced increases in intracellular cAMP concentration increase Ca^{2+} channel activity via *protein kinase A* (*PKA*)-dependent phosphorylation. This contributes to the increased rate of phase 4 depolarization in SA nodal cells, and it also moves V_{th} closer to V_m. PSNS activation decreases *AC* and *PKA* activity and, thus, moves I_{Ca} away from V_{th}. The ANS has similar effects on Ca^{2+} channels in the AV node, but here they manifest as a change in conduction velocity (**dromotropy**).

b. **Potassium current:** The PSNS has an additional level of control through activation of an ACh-activated K^+ current that causes nodal cell membranes to become more negative. Because phase 4 begins at a more hyperpolarized level, it takes longer to reach V_{th} and HR slows.

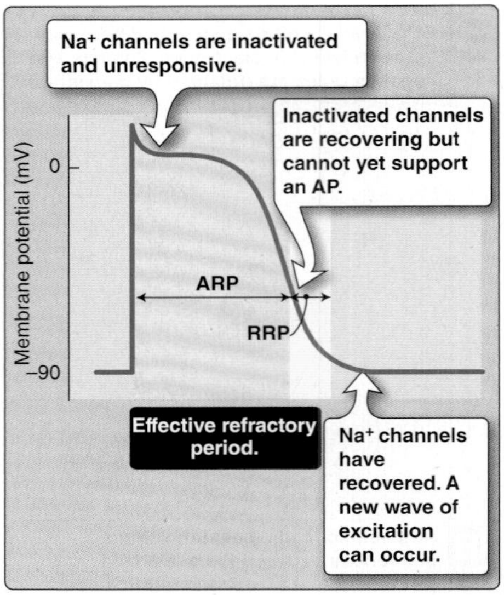

Figure 17.10
Refractory periods. AP = action potential; ARP = absolute refractory period; RRP = relative refractory period.

D. Refractory periods and arrhythmias

Because every cell in the myocardium is electrically connected via gap junctions, the heart is vulnerable to pacemakers located within the contractile portions of the myocardium (**ectopic pacemakers**). These may have intrinsic rates that are so fast that the heart is unable to function as a pump. Fortunately, I_{Na} inactivates during depolarization, lessening this possibility by creating an **absolute refractory period** (**ARP**), a time during which a myocyte is insensitive to new waves of excitation (Figure 17.10). Once the current begins recovering from inactivation, the myocyte progresses to a **relative refractory period** (**RRP**), a time during which it may be possible to elicit a small response from the cell, but not one that propagates. The ARP and RRP together constitute an **effective refractory period**. Refractory periods also insure against the possibility of tetanic contraction.

V. ELECTROCARDIOGRAPHY

The heart is a three-dimensional organ. It takes around a third of a second for the various regions to activate fully, during which time there are waves of electrical activity racing through its internal structures. The **ECG** captures a series of one-dimensional snapshots of these electrical events to create a remarkably detailed picture about their timing, direction, and the mass of tissues involved.

A. Theory

An ECG records extracellular potentials using electrodes adhered to the body surface. The potentials are generated by current flowing through surrounding tissues from depolarized areas of the heart to polarized regions (**electrical dipole**) as shown in Figure 17.11. Current intensity is directly proportional to size of the dipole. Three ECG electrodes are arranged in a triangle (**Einthoven triangle**) around the heart and connected to an ECG recorder. The recorder systematically compares voltage differences between pairs of electrodes and generates a moving paper record. These comparisons, which are facilitated by rapid switching within the ECG recorder, are known as

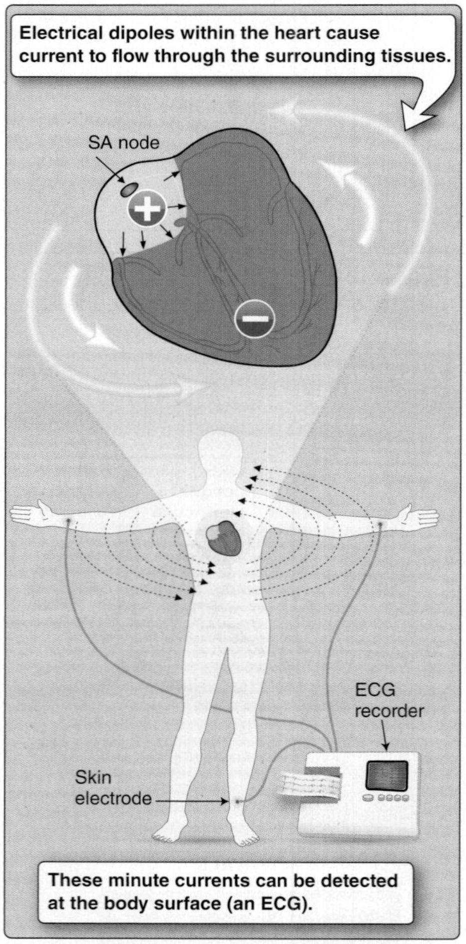

Figure 17.11
Electrocardiography. ECG = electrocardiogram; SA = sinoatrial.

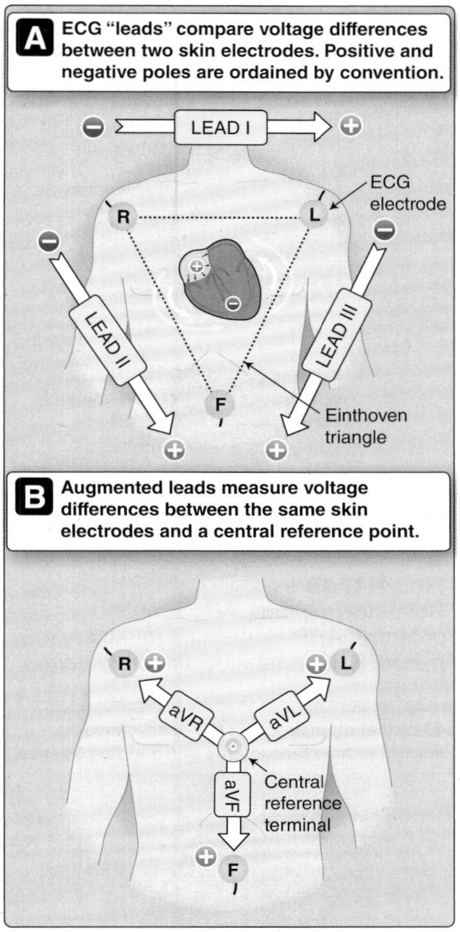

Figure 17.12
Electrocardiogram (ECG) limb leads.

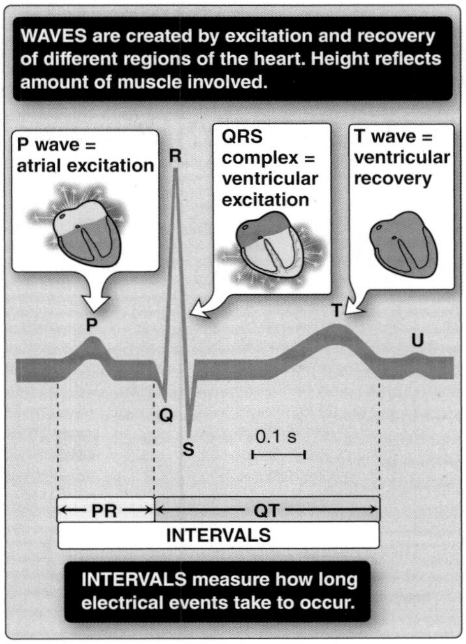

Figure 17.13
The electrocardiogram.

ECG leads (Figure 17.12). There are two general types of leads: limb leads and precordial leads.

1. **Bipolar limb leads:** There are three **bipolar limb leads**, which are created by comparing voltage differences between each of the three ECG electrodes (see Figure 17.12A). Lead I records voltage differences between the right and left shoulders, lead II compares the right shoulder and the left foot, and lead III compares the left shoulder and left foot. By convention, the left shoulder is designated the positive pole of lead I, whereas the foot is designated the positive pole of leads II and III.

2. **Augmented limb leads:** Three **unipolar leads** compare voltage differences between skin electrodes and a common reference point (**central terminal**) that is held close to zero potential (see Figure 17.12B). Leads **aVL**, **aVR**, and **aVF** measure voltage differences between this point and the left shoulder, right shoulder, and foot, respectively. The skin electrodes are considered to be the positive pole in each case.

3. **Precordial leads: Precordial** or **chest** leads compare voltage differences between the common reference point and six additional skin electrodes placed in a line directly above the heart ($\mathbf{V_1}$ through $\mathbf{V_6}$).

B. Electrocardiograph

All ECG recordings are standardized so that their interpretation becomes a simple matter of pattern recognition to a trained eye. By convention, when a wave of depolarization is moving through the heart toward the positive pole of a lead, it causes an upward (positive) deflection on the ECG recording. Movement toward the negative pole causes a downward (negative) deflection. Depolarization of a large muscle mass generates a larger dipole than a smaller mass, so it generates a larger deflection on the record.

C. Normal electrocardiogram

A typical ECG recording comprises five waves, P through T, that correspond to the sequential excitation and recovery of the different regions of the heart (Figure 17.13).

1. **P wave:** The myocardium rests between beats, and the ECG pen rests at the **isoelectric line**. Excitation begins with the SA node, but the current that it generates is too small to record at the body surface. The wave of depolarization then spreads across the atria, registering as the **P wave**. When both atria are depolarized fully, the pen returns to baseline. A normal P wave has a duration of 80–100 ms.

2. **QRS complex:** The P wave is followed by a brief period of quiet during which the wave of excitation moves slowly through the AV node and crosses from atria to ventricles via the bundle of His. This progression does not register on the recording. Ventricular depolarization produces the **QRS complex**. The three components reflect excitation of the intraventricular septum (**Q wave**), the apex and the free walls (**R wave**), and finally the regions near

Table 17.1: Timing of Electrocardiogram Waveforms

Name	How Measured	Significance	Time
PR interval	From start of the P wave to start of QRS complex	Time for wave of excitation to traverse atria and AV node	120–200 ms
PR Segment	From end of P wave to start of QRS complex	Time for wave of excitation to traverse AV node	50–120 ms
QT interval	From start of QRS complex to end of T wave	Duration of myocardial excitation and recovery	300–430 ms

AV = atrioventricular.

the base (**S wave**). The recording returns to baseline when the entire ventricular myocardium is depolarized, roughly coinciding with phase 2 of the ventricular AP. The entire complex lasts 60–100 ms.

3. **T wave:** Ventricular repolarization registers on the ECG recording as the **T wave**. On rare occasions, the T wave may be followed by a small **U wave**, thought to represent papillary muscle repolarization.

The time intervals between the waves are also named and can provide important insights into cardiac function (Table 17.1).

D. Rhythms

The heart of a healthy individual at rest beats with a **normal sinus rhythm** of 60–100 beats/min. Some individuals have normal rates that fall below this range (**sinus bradycardia**), whereas strenuous physical activity typically causes the normal rate to exceed 100 beats/min (**sinus tachycardia**). The "sinus" prefix to both rhythms indicates that the rate is established by the SA node. Abnormal rhythms (**dysrhythmias** and **arrhythmias**) can originate from virtually any part of the myocardium (Figure 17.14).

Clinical Application 17.1: Long QT Syndrome

Long QT syndrome (LQTS) refers to a set of related inherited and acquired disorders that delay phase 3 membrane repolarization and manifest as a prolongation of the QT interval on the electrocardiograph. LQTS patients are at risk of developing **torsades de pointes** ("twisting around the points"), a characteristic ventricular tachycardia in which the QRS complex rotates about the isoelectric line. Torsades is of concern because it often precipitates sudden cardiac death (SCD). Phase 3 is delayed by reducing I_K or by prolonging I_{Na} or I_{Ca}. The most common form of LQTS is caused by *LQT1* K^+-channel gene mutation, which reduces I_K. *LQT3* mutations prevent the Na^+ channel from inactivating fully during depolarization and is a particularly lethal form of LQTS. Arrhythmic events may be precipitated by any of a number of factors, including exercise and abrupt sounds. Patients typically suffer palpitations, syncope, seizures, or SCD as a result.

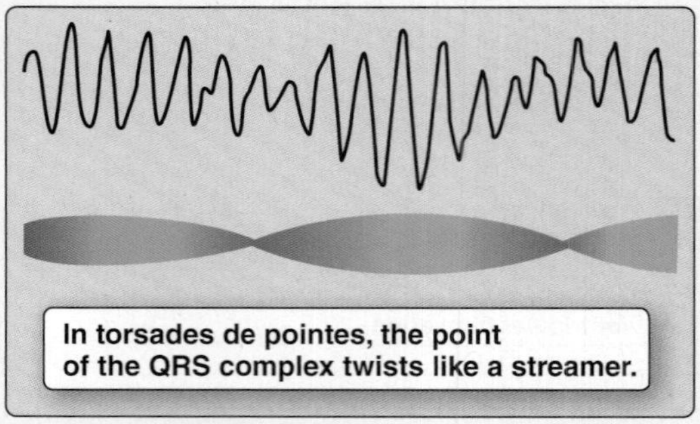

In torsades de pointes, the point of the QRS complex twists like a streamer.

Torsades de pointes.

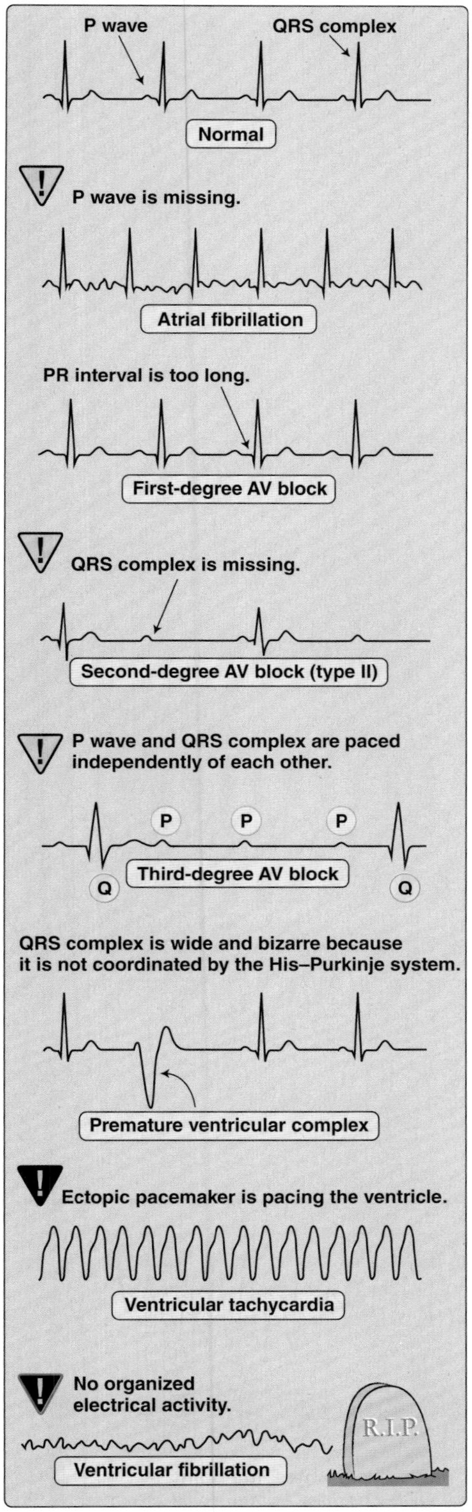

Figure 17.14
Normal and abnormal cardiac rhythms. AV = atrioventricular.

1. **Atrial arrhythmias: Atrial fibrillation** (**AF**) is an arrhythmia caused by one or more extranodal atrial pacemakers that typically cycle at several hundred times per minute. AF is relatively common, especially among older adults and in patients with heart failure. Loss of atrial pump function reduces CO, and patients typically present with fatigue, dyspnea, and lightheadedness as a result. The AV node acts as a filter that usually protects the ventricles from arrhythmias of atrial origin. The node is re-excited by the chaotic electrical activity running through the atria whenever it emerges from its refractory period so HR may be irregular and tachycardic, but the QRS is normal because the wave of excitation is still coordinated by the His–Purkinje system.

2. **Atrioventricular block:** Functional and anatomic defects in the AV node can delay or interrupt transmission of signals to the ventricles, a condition known as **AV nodal block**. Block occurs during the PR interval, because this is the time when the wave of excitation propagates from atria to ventricles. AV block is generally described as being first, second, or third degree, according to severity.

    a. **First degree: First-degree block** is characterized by a lengthening of the PR interval ($>$0.2 s). It is usually benign and asymptomatic.

    b. **Second degree:** Two types of **second-degree block** are recognized. **Möbitz type I** (also known as **Wenckebach block**) describes a rhythm in which the PR interval lengthens gradually until a complete block occurs, at which point the ventricles fail to excite, and the ECG recording drops a QRS complex. **Möbitz type II** block is characterized by ECG recordings in which the QRS complex is dropped with no prior warning. Type I is usually benign. Type II may progress rapidly to third-degree block.

    c. **Third degree: Third-degree block** is caused by a defect in the AV node or conduction system that completely prevents electrical signals from reaching the ventricles. In absence of guidance from the SA node, pacemakers located in the bundle of His or Purkinje network often take over the responsibility for driving ventricular contraction. The ECG typically shows a normal, regular P wave, and a QRS complex that may also be regular but temporally disconnected from the sinus rhythm.

3. **Ventricular dysrhythmias:** Dysrhythmias can also have ventricular origins. **Ectopic ventricular rhythms** originating in the contractile portions of the myocardium propagate via gap junctions until the entire heart is involved. Because the wave of excitation spreads via the myocardial equivalent of slow back streets rather than the Purkinje highway system, the resulting QRS complexes are broad. The excitation sequence is abnormal also, so the QRS complex is highly atypical. Occasional ($<$6/min) **premature ventricular contractions** (**PVCs**) of ectopic origin are common and usually benign (see Figure 17.14). Ectopic pacemakers have the potential to pace the myocardium at high rates

Clinical Application 17.2: Myocardial Infarction

If a region of the myocardium is deprived of adequate blood flow, it becomes **ischemic**. Prolonged or extreme ischemia causes muscle death, an event known as **myocardial infarction** (**MI**). MI is usually precipitated by stenosis or complete occlusion of a coronary supply artery by atherosclerotic plaque. Depending on the severity of flow impairment, MI may prove immediately fatal or may be limited to focal ventricular wall necrosis. Patients suffering an acute MI typically present with intense ischemic pain. Diagnosis of an MI can be confirmed by measuring circulating levels of cardiac biomarkers, such as troponins, and may often manifest on an electrocardiogram (ECG) as **ST-segment elevation**.

ST-segment elevation occurs because injured and dying cells leak K^+ into the extracellular space. All cells maintain high intracellular K^+ concentrations using the ubiquitous Na^+-K^+ ATPase. When cells die, their membranes disrupt, and K^+ is released. All cells also rely on a steep transmembrane K^+ gradient to maintain V_m at normal resting levels, so the appearance of K^+ extracellularly causes healthy myocytes peripheral to the ischemic event to depolarize. The ischemic area thereby creates an electrical dipole within the resting myocardium that generates an **injury current**. The current flows in the period between beats and causes a baseline offset on an ECG recording. An observer only becomes aware of the offset during the ST segment, a time during which the entire myocardium is depolarized and the dipole and its dependent current disappear. In practice, the injury current *tricks* the eye into believing that the ST segment is elevated. Damaged areas eventually necrose and are replaced with scar tissue, at which point the dipole and the injury current disappear.

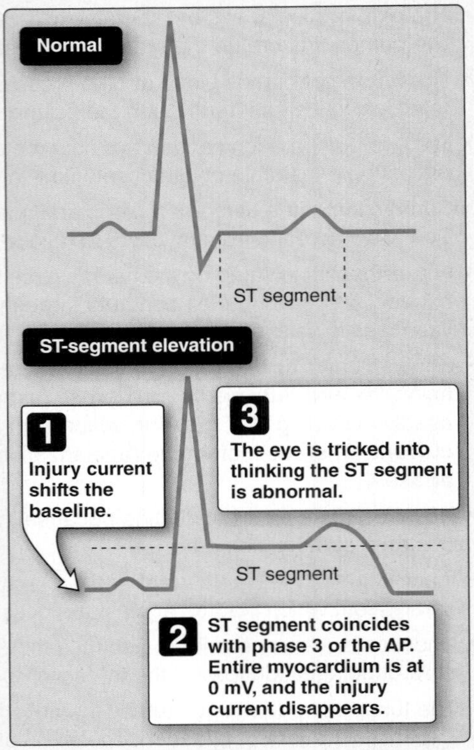

ST-segment elevation. AP = action potential.

(~300 beats/min), a rhythm known as **ventricular tachycardia** (**V-tach**). The onset of V-tach is a grave event because pacing the heart at such high rates disrupts pump function to the point where CO drops to zero. A myocardium deprived of O_2 supply quickly degenerates into **ventricular fibrillation** (**V-fib**) and sudden cardiac death.

E. Mean electrical axis

The **mean electrical axis** (**MEA**) averages the many electrical vectors generated by the wave of excitation as it moves through the heart. It provides a single value that indicates which region of the heart dominates the electrical events (Figure 17.15). In a healthy individual, the LV dominates because it contains the largest mass of tissue. By convention, a circle is drawn around the heart in the plane of the limb leads, and the left side taken to be at 0°, the right side to be at 180°, the feet to be at +90°, and the head to be at −90°. In a normal, healthy individual, the MEA lies between −30° and +105°. If the MEA is less than −30°, the left side of the heart must be contributing to the MEA to a greater extent than normal. This is known as **left axis deviation** and is usually an indication of left ventricular hypertrophy. An MEA of greater than +105° (**right axis deviation**) indicates right ventricular hypertrophy.

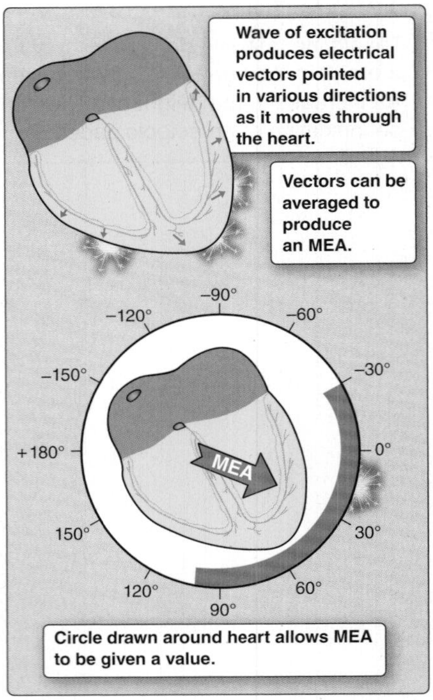

Figure 17.15
Mean electrical axis (MEA).

Chapter Summary

- The cardiovascular system contains two vascular circuits that are connected in series to form a loop. The heart contains two pumps that create the pressure necessary to drive blood through the two circuits.
- The **left heart** pumps blood at high pressure through the **systemic circulation**. The **right heart** drives blood at a relatively low pressure through the **pulmonary circulation**.
- The two pumps each contain two chambers: one **atrium** and one **ventricle**. One-way valves at the ventricular inlets and outlets help maintain unidirectional flow through the circuits.
- Contraction of the different regions and chambers within the heart is carefully coordinated by a wave of depolarization. The wave spreads from myocyte to myocyte throughout the myocardium via **gap junctions**.
- A heartbeat is initiated by the heart's pacemaker, the **sinoatrial node**, located in the wall of the right atrium. Atria contract first, forcing their contents through the **atrioventricular** (**AV**) **valves** toward the ventricles. Excitation of the ventricles is delayed by the **AV node** and then coordinated by high-speed **Purkinje fibers**.
- Conduction velocity through the different regions of the heart is related to the shape of the action potential. Contractile myocytes and Purkinje fibers express **fast action potentials** comprising five phases. **Phase 0** (the upstroke) is due to a rapidly activating **Na^+ current**, which then **inactivates** (**phase 1**). The plateau (**phase 2**) is due to a slow-activating **Ca^{2+} current**, whereas membrane repolarization (**phase 3**) is effected by a **K^+ current**. The interval between beats is termed **phase 4**.
- Nodal cells are slow conducting because they lack the phase 0 Na^+ current, leaving the Ca^{2+} current to drive the upstroke of the action potential.
- Inactivation of the Na^+ current and Ca^{2+} current during depolarization causes myocytes to become **refractory** to further stimulation, which insures against the possibility of tetanus.
- Nodal cells are **pacemakers** because they express a **"funny current"** that is activated during phase 4 and that causes slow depolarization toward the threshold for spike formation.
- The funny current is regulated by the **autonomic nervous system**. The **sympathetic** branch enhances the current, increases the rate of phase 4 depolarization, and speeds heart rate. **Parasympathetic** nerve stimulation has the opposite effect.
- The waves of excitation moving through the heart generate currents that can be recorded at the body surface to produce an **electrocardiogram** (**ECG**). A typical ECG recording comprises several distinct waveforms corresponding to excitation of the atria (**P wave**), ventricular septum, apex, and free walls (**Q**, **R**, and **S**) and then repolarization of the ventricles (**T wave**).
- The timing, magnitude, and form of these waves can be used to diagnose defects in heart function. When there is a break in the normal conduction pathways, the time between P wave and QRS complex is prolonged (**"block"**). An increase in the height of a P wave or QRS complex is indicative of **hypertrophy**. Widening of the QRS complex may be indicative of an ectopic pacemaker. Displacement of a segment may be indicative of **ischemia** and **infarction**.

Cardiac Mechanics

18

I. OVERVIEW

The heart's function is to generate pressure within the arterial compartment. Pressure is required to drive blood flow through the vasculature, delivering O_2 and nutrients to all cells in the body. The amount of blood that the heart pumps out on each cycle is precisely matched to metabolic needs. At rest, the needs of the various tissues are modest and cardiac output (CO) approximates 5–6 L/min in an average person. Any increase in tissue activity (digesting a meal, walking, climbing stairs) requires that CO increase to support the active tissue's increased needs. Raising CO is achieved in part by increasing cycle frequency (heart rate [HR]), but the heart, unlike a conventional pump, has the unique ability to increase the amount of blood it ejects on each stroke, and to do so with increased force and efficiency. These features allow a fit athlete's heart to ramp up output by as much as fivefold to sixfold during strenuous exercise. Matching CO to tissue needs is the responsibility of the autonomic nervous system (ANS), which regulates beat frequency, the extent of ventricular filling prior to contraction, and contractile force.

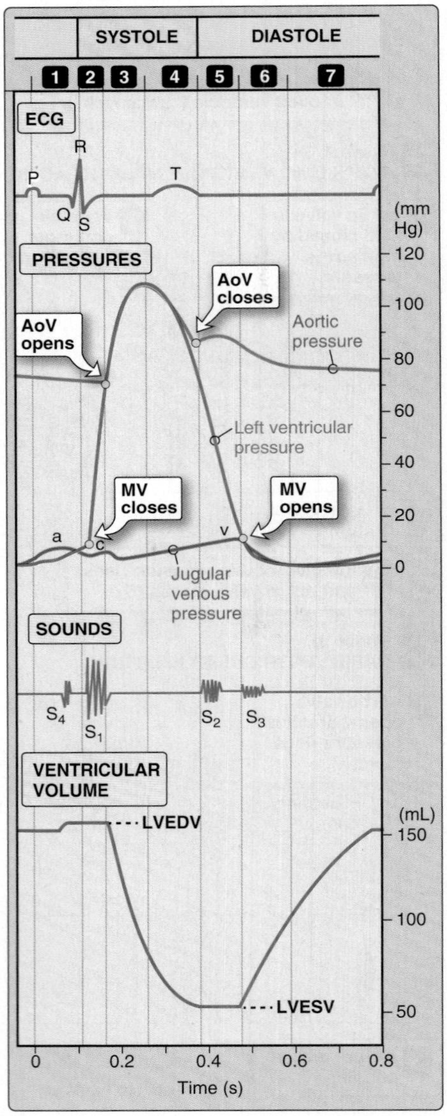

Figure 18.1
The cardiac cycle. AoV = aortic valve; ECG = electrocardiogram; LVEDV = Left ventricular end-diastolic volume; LVESV = left ventricular systolic volume; MV = mitral valve.

II. CARDIAC CYCLE

The cardiac cycle consists of alternating periods of contraction (**systole**) and relaxation (**diastole**). When describing the cycle, it is useful to correlate four measures of activity: the electrical events that initiate and coordinate contraction (recorded as an **electrocardiogram** [**ECG**]), pressure within various parts of the system, volume changes, and the sounds associated with blood moving between the various compartments. These four indices are collated in Figure 18.1. This diagram focuses on the left ventricle (LV), but the right ventricle (RV) functions similarly, albeit at lower ejection and filling pressures.

A. Phases

The cardiac cycle can be subdivided into seven discrete phases (numbers corresponding with phases indicated at the top of Figure 18.1).

1. **Atrial systole:** The cardiac cycle begins with atrial systole, which is initiated by atrial excitation and follows the crest of the P wave on the ECG.

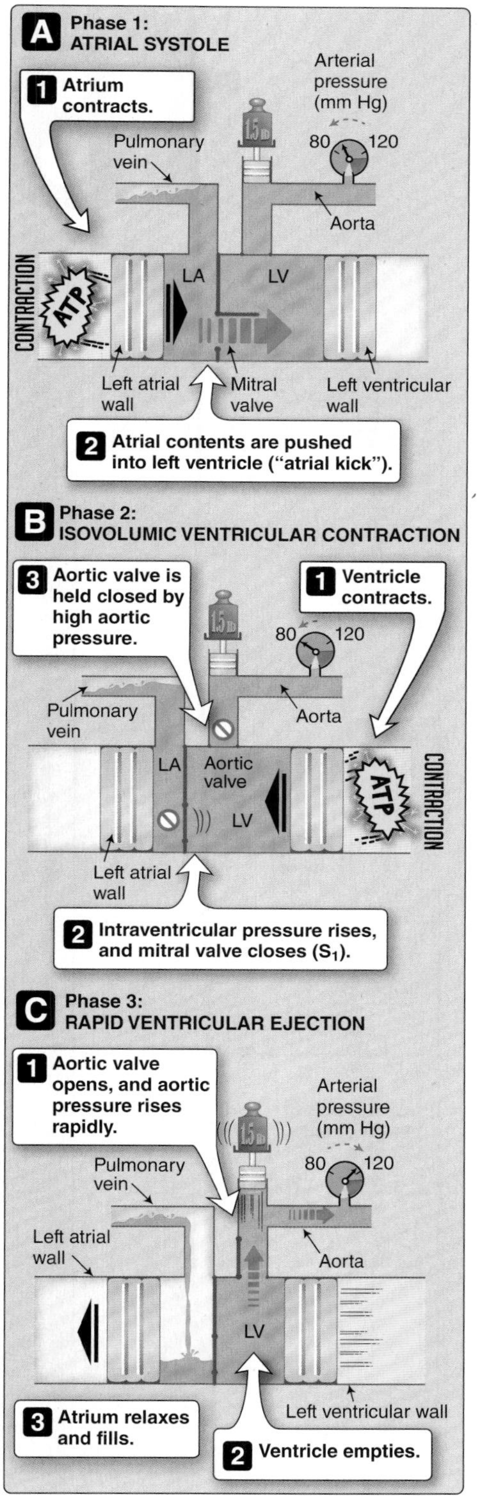

Figure 18.2
The cardiac cycle in a model heart.
ATP = adenosine triphosphate;
LA = left atrium; LV = left ventricle;
LVP = left ventricular pressure.

2. **Isovolumic ventricular contraction:** Ventricular systole begins with mitral valve closure, which occurs during the QRS complex. It takes around 50 ms for the ventricle to develop sufficient pressure to force the aortic valve open, during which time the myocytes are contracting around a fixed volume of blood. This phase is, thus, known as **isovolumic** (also **isovolumetric** or **isometric**) **contraction**.

3. **Rapid ventricular ejection:** The aortic valve finally opens, and blood exits the ventricle and enters the arterial system at high velocity (**rapid ejection**).

4. **Reduced ventricular ejection:** Ejection velocity decreases as ventricular systole nears completion (**reduced ejection**). Aortic valve closure marks the end of this phase.

5. **Isovolumic ventricular relaxation:** With the ventricle once again a sealed vessel, there follows a period of **isovolumic relaxation**.

6. **Rapid ventricular filling:** When the mitral valve opens, blood that had been damming in the atrium during systole surges forward into the ventricle. The **rapid passive filling** phase signals the beginning of diastole.

7. **Reduced ventricular filling:** The cardiac cycle ends with reduced filling. This phase, which is also known as **diastasis**, typically disappears when HR increases because cycle length is shortened largely at the expense of diastole.

B. Ventricular pressure and volume

Figure 18.2 provides a schematic representation of the left heart to illustrate the following descriptions of blood flow and pressure development during a single cardiac cycle. The numbered phases in Figure 18.2A–F correlate with phases indicated in Figure 18.1.

1. **Diastole:** The LV refills with pulmonary venous blood by way of the left atrium during diastole. By the end of diastole, the ventricle has neared capacity, but atrial contraction forces a small additional blood bolus into the chamber lumen (**"atrial kick"**; see Figures 18.1 and 18.2A), and left ventricular pressure (LVP) rises to around 10–12 mm Hg. LV **end-diastolic volume** (**EDV**) then stands at ~120 mL in an individual at rest, although the range of values considered normal is wide (70–240 mL).

> In a healthy person at rest, atrial systole boosts LV volume by around 10%, but this amount can climb to 30%–40% when HR is high, and the time available for filling is reduced. Patients in heart failure can become so dependent on atrial contribution to resting output that loss of atrial systole due to fibrillation can compromise CO and arterial pressure.

2. **Systole:** LVP climbs at maximal rates during isovolumic contraction (see Figure 18.2B). Once LVP meets and exceeds aortic pressure, the aortic valve opens, and blood is ejected into the arterial system (see Figure 18.2C). Pressures continue to climb even though blood is being ejected because the LV myocytes are still actively contracting (see Figure 18.1). The rapid ejection phase accounts for ~70% of total ventricular output and drives aortic pressure toward a peak of ~120 mm Hg. The ventricular myocytes now begin to repolarize, contraction wanes, and LVP falls rapidly. The kinetic energy imparted to blood by LV contraction continues to drive ventricular outflow for a brief period, but the rapid fall in LVP soon causes the pressure gradient across the aortic valve to reverse and the aortic valve slams shut (see Figure 18.2D).

3. **Diastole:** Once intraventricular pressure dips below atrial pressure, the mitral valve opens, and filling begins (see Figure 18.2E). The LV's helical muscle bands cause it to shorten, twist, and wring blood out through the valves during systole. With relaxation, the myocardium rebounds through natural elasticity (discussed in more detail below), and pressure continues to decrease rapidly even as blood surges in from the atrium (rapid, passive filling) as shown in Figure 18.2F. The cycle then repeats itself.

4. **Ventricular efficiency:** The LV does not empty completely during systole, and **end-systolic volume** (**ESV**) is usually around 50 mL. Subtracting ESV from EDV gives **stroke volume** (**SV**), which defines the amount of blood transferred from the LV to the arterial system during systole. SV should be >60 mL in a healthy person. Dividing SV by EDV yields **ejection fraction** (**EF**), normally ~55%–75%. EF is an important measure of cardiac efficiency and health and, therefore, is used clinically to assess cardiac status in patients with heart failure, for example.

> EF is typically estimated noninvasively using two- or three-dimensional echocardiography. An echocardiogram can measure ESV and EDV, wall thickness, muscle shortening velocity, and blood flow patterns during contraction and relaxation.

C. Aortic pressure

The arterial system is comprised of small-bore vessels that do not stretch easily (see 19·II·C). In practice, this means that arterial pressure rises sharply when the LV forces blood into the system, but this pressure readily dissipates when blood flows out of the system and enters capillary beds.

1. **Systole:** Aortic pressure is insensitive to intraventricular events so long as the aortic valve is closed. Once LVP climbs above aortic pressure and the aortic valve opens, aortic pressure then rises and falls in near synchrony with LVP (see Figures 18.1 and 18.2C).

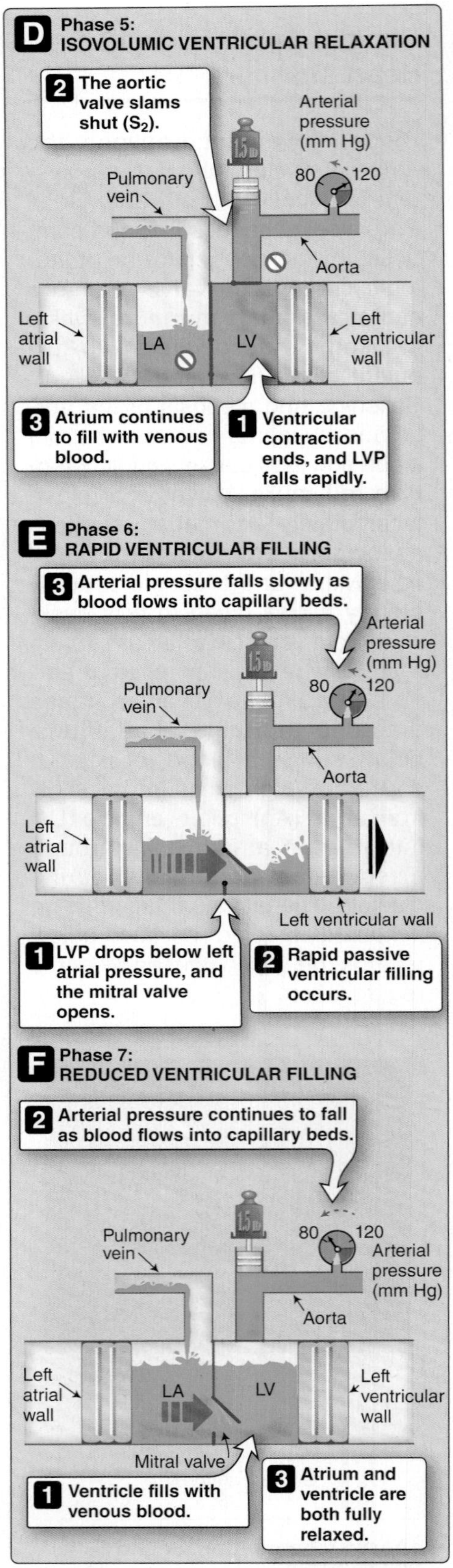

Figure 18.2
(continued)

Clinical Application 18.1: Heart Murmurs

Blood movement between the atria, ventricles, and arterial system is usually smooth and silent. Physiologic changes in blood composition and congenital or pathologic changes in valve structure can create **heart murmurs**, which are sounds associated with turbulent blood flow (see 19·V·A·4). Stiffening and stenosis of the aortic and pulmonary valves obstruct blood flow and cause systolic **ejection murmurs**. Valves that close incompletely also cause murmurs associated with backward blood flow. Mitral and tricuspid **insufficiency** allows blood to flow backward from ventricle to atrium during systole, whereas incompetent aortic and pulmonary valves cause diastolic murmurs. The figure below shows a blood jet (*arrow*) caused by high-pressure backflow from aorta (Ao) to left ventricle (LV) into the left atrium (LA) through a regurgitant aortic valve during diastole. The jet was imaged using echocardiography. The trace below records the sound caused by aortic regurgitation.

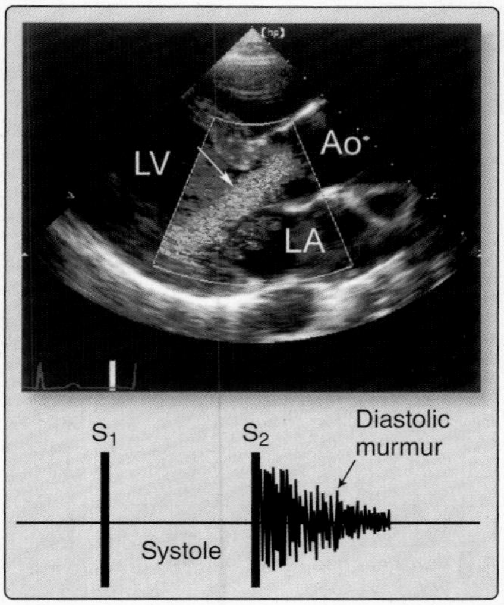

Aortic regurgitation. S_1 and S_2 = first and second heart sounds.

2. **Diastole:** Aortic pressure dips briefly immediately following aortic valve closure, creating a characteristic **dicrotic notch** or **incisura** in the aortic pressure curve. The notch is caused by the aortic valve bulging backward into the LV under the weight of aortic pressure when it closes. Aortic pressure declines slowly throughout diastole, reflecting blood draining out of the arterial system and into the capillary beds (see Figures 18.1, 18.2E, and 18.2F).

D. Venous pressure

The lack of valves between atria and the venous system allows intra-atrial pressure changes to be transmitted backward into the veins lying close to the heart. Internal jugular venous pressure (a prominent neck vein) displays three distinct pressure waves during the cardiac cycle (see Figure 18.1).

1. **a wave:** Right atrial contraction generates a pressure wave that forces blood forward into the right ventricle and also creates an **a wave** in a jugular venous recording.
2. **c wave:** Ventricular contraction causes intraventricular pressure to rise sharply, and the atrioventricular (AV) valves bulge backward into the atria as a consequence. Backward deflection of the tricuspid valve generates a jugular venous pressure pulse known as the **c wave**.
3. **v wave:** During ventricular systole, blood continues to flow from the venous system into the right atrium and dam against the closed tricuspid valve. Pressure builds as the atrium fills, registered as the upslope of the **v wave** in a jugular venous recording. The downslope correlates with rapid atrial emptying when the tricuspid valve opens, blood surging forward into the right ventricle (a similar surge occurs on the left side of the heart; see Figure 18.2E).

E. Heart sounds

There are four **heart sounds** associated with the cardiac cycle. The first two are valve-closing events, and the second two are sounds caused by blood entering the LV.

1. **First:** The first heart sound (**S_1**) occurs at the beginning of ventricular systole. LVP develops rapidly during this time, causing blood to start moving backward toward the atria. The movement immediately catches the leaflets of the AV valves and causes them to snap shut (see Figure 18.2B). Valve closure and reverberations within the LV wall registers as a low rumbling **"lub"** sound, which lasts ~150 ms.
2. **Second:** Aortic and pulmonary valve closure is associated with the second heart sound (**S_2**), which can be heard as a brief **"dup."** S_2 often splits into two distinct aortic (A_2) and pulmonary (P_2) components that reflect slight asynchronous closure of the two valves. Splitting is most acute during inspiration, when a drop in intrathoracic pressure enhances the pressure gradient driving blood flow from the systemic veins into the RV. The RV receives more blood than the LV as a result. A large blood volume takes longer to eject

than a small volume and, thus, RV systole is prolonged compared with LV systole. Pulmonary valve closure is delayed to the extent that it can be heard as a separate sound (P_2).

3. **Third:** Auscultation of children and thin adults may reveal a low-intensity rumbling heart sound (**S_3**) during early diastole. S_3 is caused by blood rushing into the LV (i.e., rapid passive filling) and causing turbulence that makes the LV walls reverberate and rumble.

4. **Fourth:** The fourth heart sound (**S_4**) is associated with atrial contraction. The force of atrial systole is too weak to be detectable by ear in healthy individuals. A ventricle that requires filling assistance can stimulate atrial hypertrophy, however, and then S_4 may become apparent as a brief, low-frequency sound. It reflects blood being forced into the ventricle at high pressure, causing reverberations within the ventricular wall.

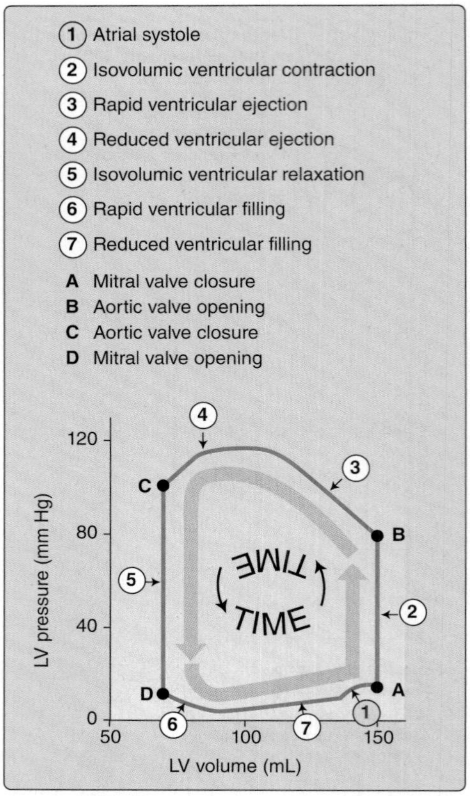

Figure 18.3
Pressure–volume loop. LV = left ventricle.

III. CARDIAC OUTPUT

The body's need for O_2 and nutrients changes constantly with changing activity level. Ensuring that these needs are met requires that CO be adjusted in parallel. To appreciate how and why CO is regulated *in vivo*, it is important to understand the various factors that influence LV output.

A. Determinants

CO (L/min) is calculated from the product of HR (beats/min) and SV (mL):

$$CO = HR \times SV$$

1. **Heart rate:** HR is established by the sinoatrial (SA) node, the cardiac pacemaker. HR is dependent on the autonomic nervous system, which controls the rate at which the pacemaker generates a wave of excitation. Sympathetic nervous system (SNS) activation raises HR, whereas parasympathetic stimulation decreases it.

2. **Stroke volume:** SV is dependent on LV **preload**, **afterload**, and **contractility**.

   a. **Preload:** Preload refers to a load that is applied to a myocyte and establishes muscle length before contraction begins. In the LV, preload equates with the volume of blood entering the chamber during diastole (EDV), which is dependent on end-diastolic pressure (EDP).

   b. **Afterload:** Afterload is the load against which a myocyte must shorten. In a healthy individual, the principal component of LV afterload is aortic pressure.

   c. **Contractility:** Contractility is a measure of a muscle's ability to shorten against an afterload. In practice, contractility equates with sarcoplasmic free Ca^{2+} concentration.

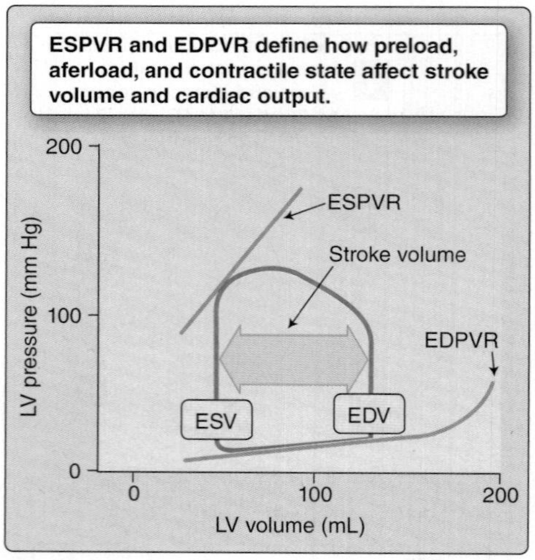

Figure 18.4
The end-systolic pressure–volume relationship (ESPVR) and the end-diastolic pressure–volume relationship (EDPVR). EDV = end-diastolic volume; ESV = end-systolic volume; LV = left ventricle.

B. Pressure–volume loop

The **pressure–volume** (**PV**) loop (Figures 18.3 and 18.4) examines the relationship between blood volume and pressure within the LV

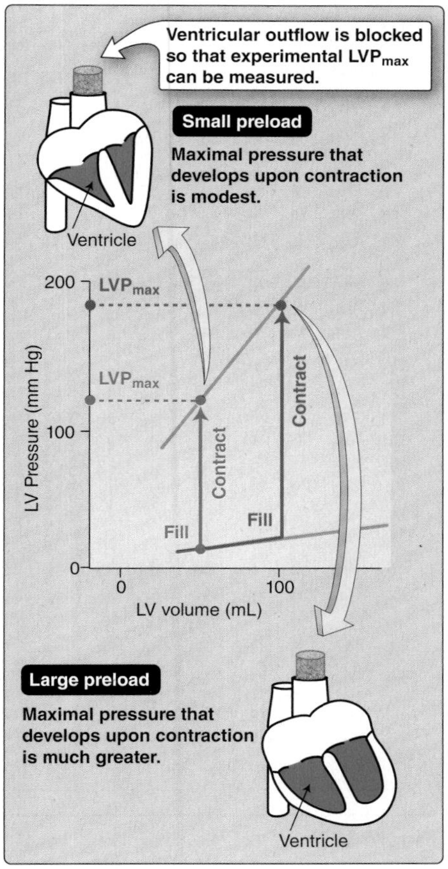

Figure 18.5
Preload effects on maximal left ventricular pressure (LVP_{max}) development. LV = left ventricle.

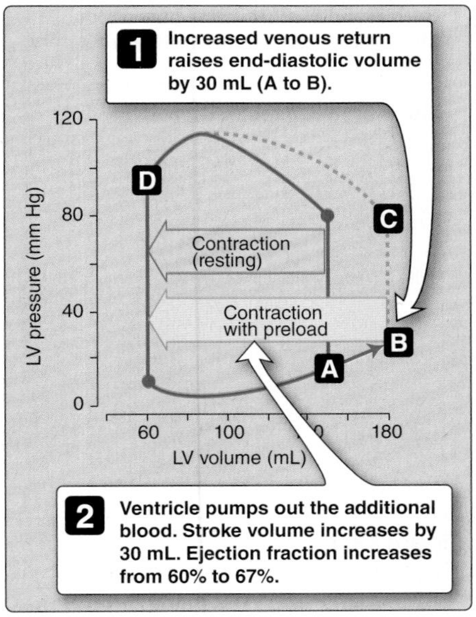

Figure 18.6
Effects of increasing preload on the pressure–volume loop. LV = left ventricle.

during a single cardiac cycle. It is particularly useful in demonstrating how preload, afterload, and contractility affect cardiac performance. The loop replots the LVP trace from Figure 18.1 and folds it back on itself in time. All seven phases are reproduced in Figure 18.3, and the four points labeled A through D represent valve events.

C. Pressure–volume relationships

The PV loop is constrained by two curves, the **end-diastolic** and **end-systolic pressure–volume relationships** (**EDPVR** and **ESPVR**, respectively), that define how a ventricle performs when presented with any given preload, afterload, or contractile state (see Figure 18.4).

1. **End-diastolic pressure–volume relationship:** The EDPVR describes passive pressure development during ventricular filling. During early diastole, the ventricle swells with blood relatively easily. The sharp upturn in the EDPVR between 150 and 200 mL reflects the ventricle reaching capacity. Any further volume increases require that significant pressure be applied to stretch the myocardium.

2. **End-systolic pressure–volume relationship:** The ESPVR defines the maximum pressure that the LV develops at any given filling volume (LVP_{max}). LVP_{max} is an experimental value, determined by initiating a contraction after the aorta has been clamped to prevent outflow during pressure development (Figure 18.5). The ESPVR demonstrates one of the key properties of the myocardium, namely that increasing cardiac filling volume increases contractile force and pressure development.

D. Preload

LV preload is determined by EDP, but commonly used surrogates include **right atrial pressure** and **central venous pressure** because systemic venous pressures drive flow through the right side of the heart and into the LV.

1. **Effect on the sarcomere:** Stretching (preloading) a cardiac muscle fiber within physiologic limits increases sarcomere length from 1.8 μm to 2.2 μm and, in so doing, sharply increases the amount of tension that develops upon contraction, called "length-dependent activation" (see 13·IV·B). Strong elastic elements within the myocardium (elastin and collagen) resist lengthening beyond 2.2 μm and account for the abrupt upturn in the whole-heart EDPVR discussed above (also see Figure 13.7).

2. **Effect on a pressure–volume loop:** Increased ventricular preloading increases EDV (A moves to B in Figure 18.6). Preloading has no direct effect on arterial pressure, so the point at which the aortic valve opens is unchanged (see Figure 18.6, point C). Myocytes contract down to the same absolute length regardless of preload, so ESV remains unchanged (see Figure 18.6, point D). Both SV and EF have increased, however, because preloading causes length-dependent sarcomeric activation (see 13·IV·B).

3. **Preload in practice:** A ventricle's innate ability to respond to increased preloading with an increased SV is known as the **Starling**

law of the heart or the **Frank-Starling relationship** (see Figure 13.7). Changes in preload are used for both short- and long-term adjustments in CO. Short-term changes involve the SNS, which constricts veins to force their contents forward toward the ventricle. Long-term management of CO is effected through sustained increases in circulating blood volume and preload and involves fluid retention by the kidneys (see 20·IV).

E. Afterload

Afterload is the force against which the ventricle must work in order to eject blood into the arterial system. Under normal circumstances, afterload equates with mean arterial pressure (MAP).

1. **Effects on the sarcomere:** The effects of increasing afterload on sarcomeric function are easiest to demonstrate using a single muscle fiber and a series of weights, or loads (Figure 18.7). The lightest weight will be lifted with little difficulty and at maximum velocity. As the load increases, the rate of contraction slows. At the heavier end of the scale, shortening is both very slow, and the height to which the weight is lifted is reduced.

2. **Effects on a pressure–volume loop:** Increasing arterial pressure has no direct effect on preload, so EDV remains unchanged (Figure 18.8). Increasing MAP does increase the amount of pressure that the LV must develop in order to force the aortic valve open, however, as indicated by the upward movement of point A to point B in Figure 18.8. It takes time to develop additional pressure, so isovolumic contraction is prolonged (not apparent from the loop because there is no time index). Ejection is truncated (see Figure 18.8, points B to C) because the extent to which the myocytes can shorten at such high pressures is limited, as defined by the ESPVR. In simpler terms, myocytes have a limited amount of adenosine triphosphate (ATP) available for developing force during each contraction. If they use more ATP developing pressures necessary to force the aortic valve open against a higher afterload, less is subsequently available to sustain ejection. The net result is that the aortic valve closes prematurely (see Figure 18.8, point C). Thus, even though EDV is unchanged, SV and EF both decrease.

3. **Afterload in practice:** Arterial pressure is changing constantly, but the cardiovascular system can compensate by parallel changes in myocardial contractility. In the short term, this involves the SNS and modulation of Ca^{2+} release (discussed below). Chronic changes in afterload invoke pathways that are involved in myocardial remodeling.

F. Inotropy

Inotropy refers to the ability of a muscle cell to develop force and is synonymous with contractility. The ventricular myocardium, unlike skeletal muscle, is unusual in that its contractile state can be varied as a way of altering cardiac performance. Agents with a positive inotropic effect (e.g., epinephrine, norepinephrine, digoxin) cause the myocardium to contract faster, develop higher peak systolic pressures, and then relax faster. Negative inotropes (e.g., beta blockers and Ca^{2+}-channel blockers) have the opposite effect.

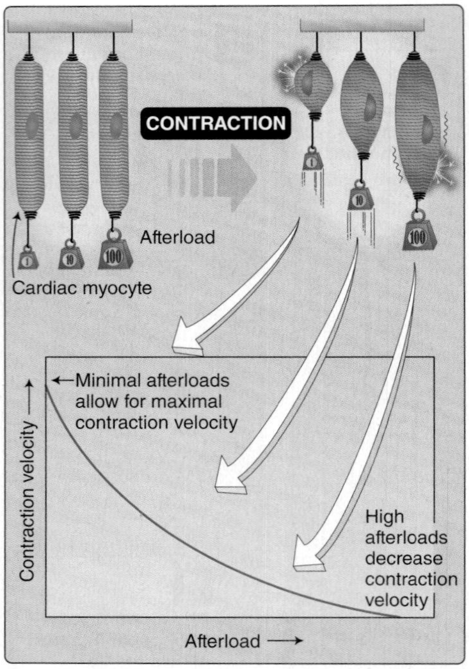

Figure 18.7
Afterload effects on contraction velocity in cardiac myocytes.

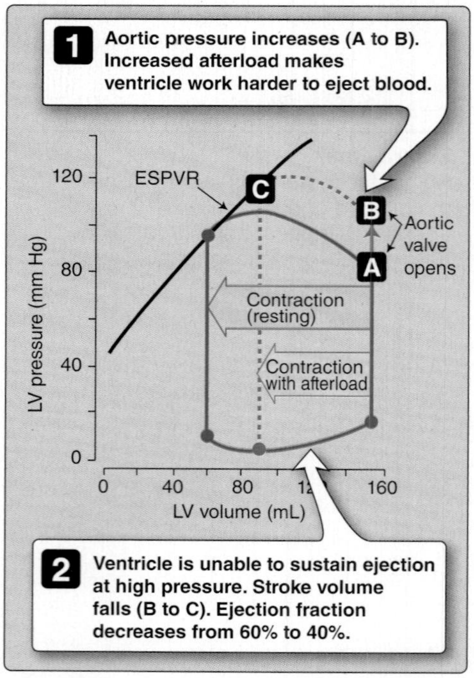

Figure 18.8
Effects of increasing afterload on the pressure–volume loop. ESPVR = end-systolic pressure–volume relationship; LV = left ventricle.

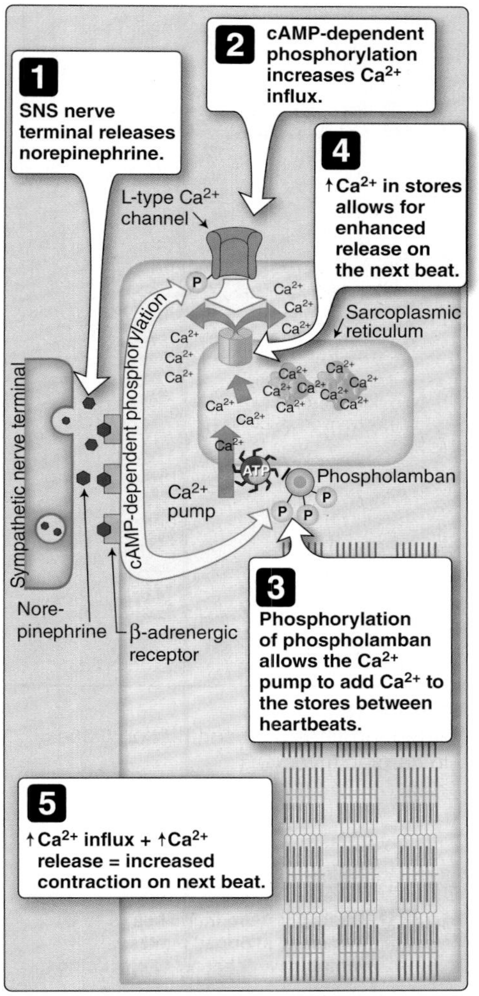

Figure 18.9
Sympathetic nervous system (SNS) modulation of contractility. ATP = adenosine triphosphate; cAMP = cyclic adenosine monophosphate.

1. **Effects on the sarcomere:** Contractility equates with intracellular free calcium concentration. Ca^{2+} binds to and activates troponin to expose myosin-binding sites on actin (see 13·II·A). Myosin then binds to and pulls on the actin filament at the expense of one ATP molecule. Thus, a single free calcium ion equates with a single unit of contractile force. Contractility is regulated by the SNS. SNS nerve terminals release norepinephrine onto myocardial β_1-adrenergic receptors, which activates *protein kinase A* (*PKA*) via the cyclic adenosine monophosphate (cAMP) transduction pathway. *PKA* has three principal targets: **L-type Ca^{2+} channels**, **Ca^{2+}-release channels** in the sarcoplasmic reticulum (SR), and a **Ca^{2+} pump** that refills the stores (Figure 18.9).

   a. **Calcium channels:** Myocardial excitation causes "trigger Ca^{2+}" influx through L-type Ca^{2+} channels in the sarcolemma. *PKA*-dependent phosphorylation of these channels increases their Ca^{2+} permeability and increases the size of the trigger flux.

   b. **Calcium release:** The Ca^{2+} trigger flux opens Ca^{2+}-release channels in the SR, allowing Ca^{2+} to flood out of the stores and into the sarcoplasm. Increasing the size of the trigger flux increases Ca^{2+} release from the stores. *PKA* further magnifies this effect by sensitizing the release channels to a rise in sarcoplasmic Ca^{2+} concentration.

   c. **Calcium pump:** During diastole, Ca^{2+} is returned to the SR by a SERCA (SR Ca^{2+} ATPase) pump. Phosphorylation stimulates SERCA and increases Ca^{2+} sequestration during diastole, allowing for an even greater Ca^{2+} release on the next beat. The phosphorylated pump also clears Ca^{2+} from the sarcoplasm more efficiently, allowing for faster myocardial relaxation following excitation, facilitating an increase in HR.

> A failing heart can be aided using cardiac glycosides such as digitalis.[1] These drugs inhibit the Na^+-K^+ ATPase and raise intracellular Na^+ concentration. This, in turn, reduces the driving force for Ca^{2+} efflux by the Na^+-Ca^{2+} exchanger, making more Ca^{2+} available for the next contraction. Contractility is increased as a result.

2. **Effect of inotropy on a pressure–volume loop:** Increasing Ca^{2+} availability allows the myocardium to contract with greater speed and force at any given filling pressure. The ESPVR shifts upward

[1]For a discussion of the actions and side effects of digitalis glycosides, see *LIR Pharmacology*, 5e, p. 201.

and to the left (Figure 18.10). Changes in inotropy have no immediate effect on preload or MAP, so the filling and isovolumic contraction phases of the PV loop are largely unchanged. However, because the myocardium is now working more efficiently, a greater volume of blood is squeezed out of the ventricle compared with previously, and ESV falls (see Figure 18.10, points A to B). SV and EF are both increased.

3. **Inotropy in practice:** Changes in inotropy are an important mechanism by which CO and blood pressure are regulated. The underlying biochemical changes occur rapidly, meaning that control can be exerted on a beat-by-beat basis. In practice, changes in inotropy do not occur in isolation, because the SNS increases HR and preload simultaneously (Figure 18.11). All three variables act in concert to ensure that CO matches demand (see 20·III·D).

> Inotropic state is a vital indicator of cardiac well-being, so it is important to be able to measure contractility in a clinical setting. The best indicator is the rate at which LVP rises during early isovolumic contraction, but this has to be measured invasively by a catheter-tip manometer threaded through a peripheral vein into the ventricle. Noninvasive alternatives include Doppler ultrasound techniques that estimate myocardial-shortening velocity or blood ejection velocity through the aortic valve.

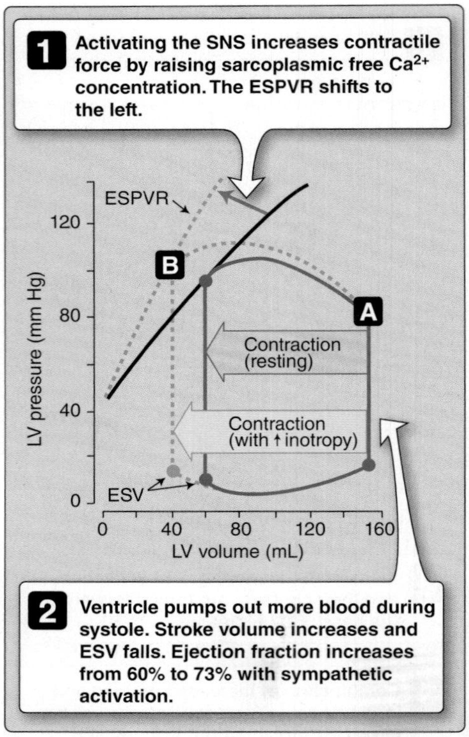
1 Activating the SNS increases contractile force by raising sarcoplasmic free Ca^{2+} concentration. The ESPVR shifts to the left.
ESPVR
LV pressure (mm Hg)
120
80
40
0
B
A
Contraction (resting)
Contraction (with ↑ inotropy)
ESV
0
40
80
120
160
LV volume (mL)
2 Ventricle pumps out more blood during systole. Stroke volume increases and ESV falls. Ejection fraction increases from 60% to 73% with sympathetic activation.

Figure 18.10
Effects of increasing contractility on the pressure–volume loop. ESPVR = end-systolic pressure–volume relationship; ESV = end-systolic volume; LV = left ventricle; SNS = sympathetic nervous system.

IV. CARDIAC WORK

The heart performs work when it moves blood from veins to arteries, and, thus, any changes in cardiovascular performance that affect output (i.e., changes in preload, afterload, inotropy, or HR) necessarily affect cardiac workload also. The various determinants of CO are unequal in the way they place demands on cardiac workload.

> A simple way of estimating workload on a heart is using the rate–pressure product, in which HR is multiplied by systolic blood pressure. Although imprecise and contraindicated when there is evidence of aortic stenosis, it suffices in a clinical situation.

A. Components

The heart performs two kinds of work: **internal work** and **external work**.

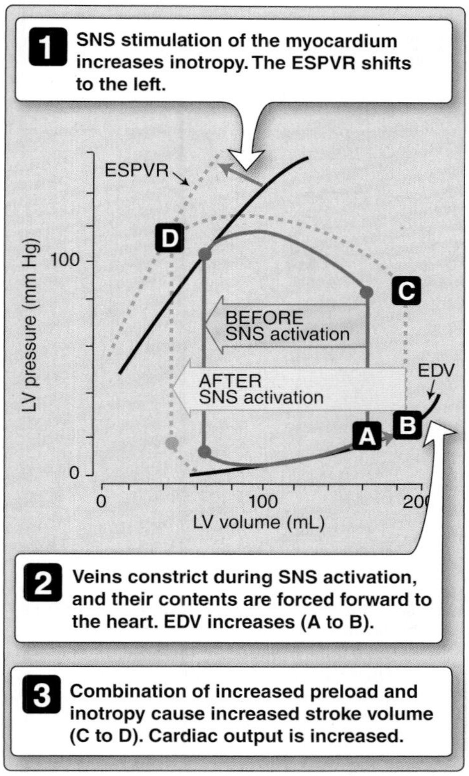

Figure 18.11
Combined effects of sympathetic nervous system activation on the pressure–volume loop. EDV = end-diastolic volume; ESPVR = end-systolic pressure–volume relationship; LV = left ventricle; SNS = sympathetic nervous system.

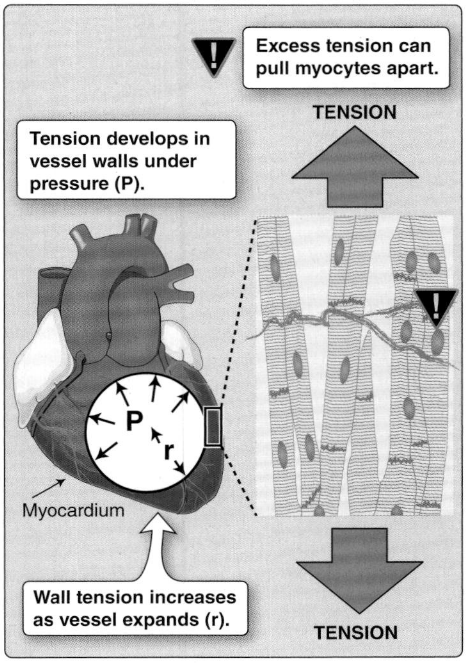

Figure 18.12
Left-ventricular wall tension caused by increasing intraventricular pressure.

1. **Internal:** Internal work accounts for >90% of total cardiac workload. Internal work is expended in isovolumic contraction, which generates the force necessary to open the aortic and pulmonary valves. The amount of energy consumed in internal work can be quantified by multiplying the amount of time spent in isovolumic contraction by ventricular wall tension (see below).

2. **External:** External work, or **pressure–volume work**, is work expended in transferring blood to the arterial system against a resistance. External work accounts for <10% of total cardiac workload, even at maximal levels of output. External work (or **minute work**) can be determined from:

$$\text{Minute work} = \text{MAP} \times \text{CO}$$

External work is represented graphically by the area contained within a PV loop. At rest, about 1% of this work is expended in imparting kinetic energy to blood, but the kinetic component can increase to as much as 50% of total at the high levels of output (e.g., during strenuous exercise).

B. Ventricular wall tension

Wall tension is a significant determinant of cardiac workload. Tension is a force that develops within the walls of pressurized chambers (Figure 18.12) and is counterproductive because it tugs on the ends and sides of the myocytes and contributes to afterload. Wall tension can be quantified using the **law of Laplace**:

$$\sigma = \text{LVP} \times \frac{r}{2h}$$

where σ is wall stress, r is ventricular radius, and h is myocardial wall thickness. The Laplace law helps illustrate how differently changes in preload, afterload, and HR impact myocardial performance.

1. **Preload:** The volume of a sphere is proportional to radius cubed ($V = 4/3 \times \pi \times r^3$). Thus, if EDV (preload) were to double, intraventricular radius would rise by ~26% and wall tension would rise by an equivalent amount. In practice, preloading a heart is a relatively efficient way of increasing CO and minimizing the effects on cardiac workload.

2. **Afterload:** If aortic pressure doubled, LVP would have to rise by a similar amount in order to eject blood. Wall tension would rise by 100%. Changes in afterload stress the myocardium to a much greater extent than do changes in preload.

3. **Heart rate:** If HR doubles, the amount of time spent in systole and isovolumic contraction doubles also. In effect, doubling HR increases wall tension and cardiac workload by ~100%.

Clinical Application 18.2: Hypertensive Myocardial Hypertrophy

Hypertension (HTN) is a leading risk factor for numerous disorders such as myocardial infarction, heart failure, intracerebral hemorrhage, and chronic kidney disease. Studies suggest that >90% of the population will develop HTN in later years (> age 55 years). HTN is defined as a systolic BP of ≥140 mm Hg and a diastolic BP of ≥90 mm Hg. HTN represents increased afterload to the left ventricle (LV), forcing it to generate higher pressures to eject blood into the arterial system. Chronic increases in afterload initiate compensatory pathways that remodel the LV myocardium to increase its contractile strength. New myofibrils are laid down alongside existing ones, causing the LV wall to widen. Increasing ventricular wall thickness decreases lumen capacity and makes the myocardium less compliant and difficult to fill, which increases the likelihood of diastolic failure (see 40·V·A).

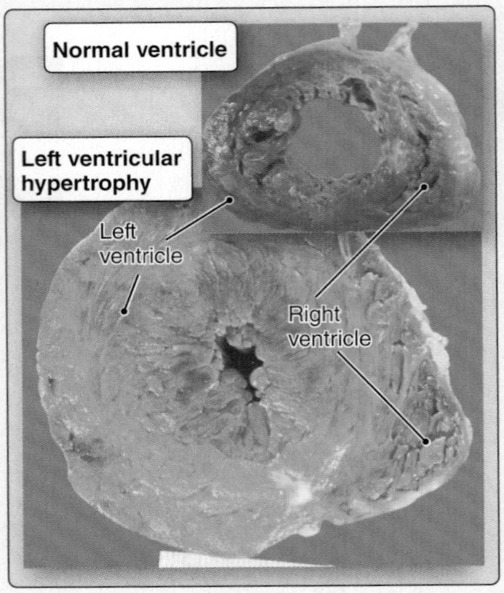

Hypertensive ventricular hypertrophy.

Chapter Summary

- The heart repeatedly contracts (**systole**) and then relaxes and fills (**diastole**).
- The cardiac cycle can be divided into several discrete phases. It begins with **atrial systole**, which pushes blood forward into the ventricle and completes preloading.
- **Ventricular systole** follows **atrial systole**. Intraventricular pressure rises rapidly and forces the mitral valve closed (focusing on the left side of the heart, but the right functions similarly). Contraction is initially **isovolumic**, but once luminal pressure exceeds aortic pressure, the aortic valve opens, and **rapid ejection** begins. Aortic pressure, which had been declining during diastole, climbs when blood is forced into the arterial system.
- Rapid ejection gives way to **reduced ejection**. The ventricle then begins to relax, and luminal pressure falls rapidly. **Isovolumic relaxation** begins when the aortic valve is forced closed by high aortic pressures.
- When ventricular pressure drops below atrial pressure, the mitral valve reopens, and there is **rapid passive ventricular filling**, aided by the influx of blood that has dammed up against the mitral valve during diastole.
- Venous pressure records exhibit an **a wave** during atrial contraction, a **c wave** during ventricular contraction, and a **v wave** caused by venous blood damming within the atria during diastole.
- There are four **heart sounds**: S_1 correlates with atrioventricular valve closure. S_2 is associated with semilunar valve closure. S_3 is a sound of ventricular filling commonly heard in children and in adults with a failing ventricle. S_4 is a pathologic sound associated with contraction of a hypertrophied atrium.
- **Cardiac output** is the product of **stroke volume** and **heart rate**. Stroke volume is determined by ventricular **preload**, **afterload**, and **contractility** (**inotropy**). All of these parameters are influenced by the autonomic nervous system.
- **Preload** is determined by end-diastolic pressure and volume. Preload increases stroke volume through length-dependent activation of the sarcomere (the **Frank-Starling mechanism**). **Afterload** is the force that must be overcome for a ventricle to eject blood, and this usually equates with **arterial pressure**. Increases in afterload decrease stroke volume. **Inotropy** (contractility) is directly correlated with **sarcoplasmic Ca^{2+} concentration**. Positive inotropes, such as epinephrine and norepinephrine, increase cardiac contractility by increasing both the rate of pressure development and peak systolic pressure.
- A heart must perform both internal and external work. Moving blood from ventricle to arterial system is termed **external** or **pressure–volume work**.
- Most of the energy used by a heart during the cardiac cycle is consumed during isovolumic contraction (**internal work**). A significant portion of this work is used in overcoming wall tension.

19 Blood and the Vasculature

I. OVERVIEW

The functions and design of the cardiovascular system are, in many ways, similar to that of a water utility in a modern city. A water utility is tasked with distributing clean water to its many consumers. The distribution network is vast, and pumping stations are required to ensure that water arrives at sufficiently high pressure for adequate flow from faucets and showerheads (Figure 19.1). Waste water is collected and returned to treatment plants under low pressure by an elaborate system of drains. The cardiovascular system similarly distributes blood at high pressure to ensure adequate flow to many consumers (cells). Waste (venous) blood travels back to the heart at low pressure for "treatment" by the lungs. Water utilities distribute water, a Newtonian fluid whose flow characteristics behave predictably under pressure. The cardiovascular system circulates blood, a viscous non-Newtonian fluid comprising water, solutes, proteins, and cells. Considerable pressure must be applied to blood in order to make it flow through the vasculature at rates sufficient to meet the needs of the tissues. The capillaries used to deliver blood to individual cells are extremely leaky, unlike the copper pipe used in household water-distribution systems. Leakiness means that the pressure used to drive flow through the system also drives fluid out of the vasculature and into the intercellular spaces. Lastly, the pipe-work used to distribute and collect blood from cells is composed of biologic tissue that stretches and causes the vessels to distend when pressure is applied. Distensibility poses a threat to system function because there is always the potential that the entire vascular contents might become trapped in the pipes, thereby allowing the vascular faucets to run dry.

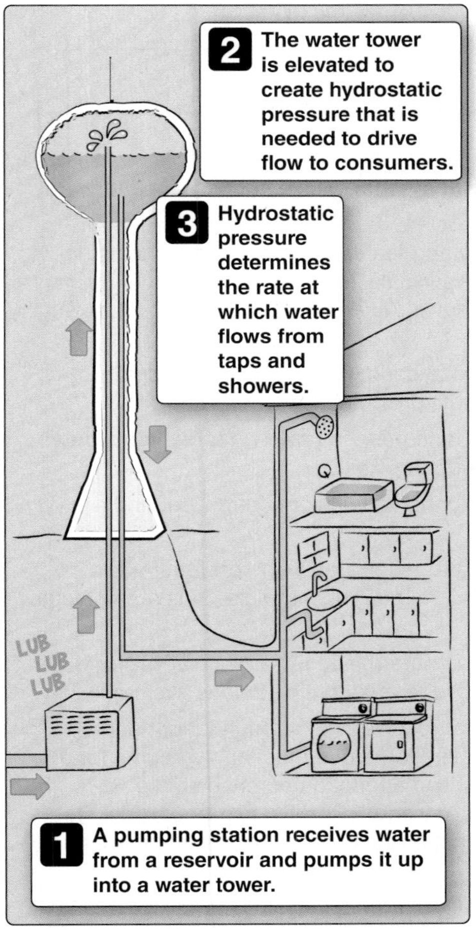

2 The water tower is elevated to create hydrostatic pressure that is needed to drive flow to consumers.
3 Hydrostatic pressure determines the rate at which water flows from taps and showers.
LUB
LUB
LUB
1 A pumping station receives water from a reservoir and pumps it up into a water tower.

Figure 19.1
Hydrostatic pressure drives flow through plumbing systems.

II. VASCULATURE

The systemic vasculature comprises a vast network of blood vessels that channel O_2-rich blood to within a few microns of every cell in the body. Here, O_2 and nutrients are exchanged for CO_2 and other metabolic waste products, and then blood is returned to the heart for reoxygenation by the lungs and redistribution to the tissues. Figure 19.2 provides an overview of the systemic vasculature and its various components.

A. Organization

The human body contains ~100,000,000,000,000 cells, every one of which must be supplied with blood. Creating a vascular

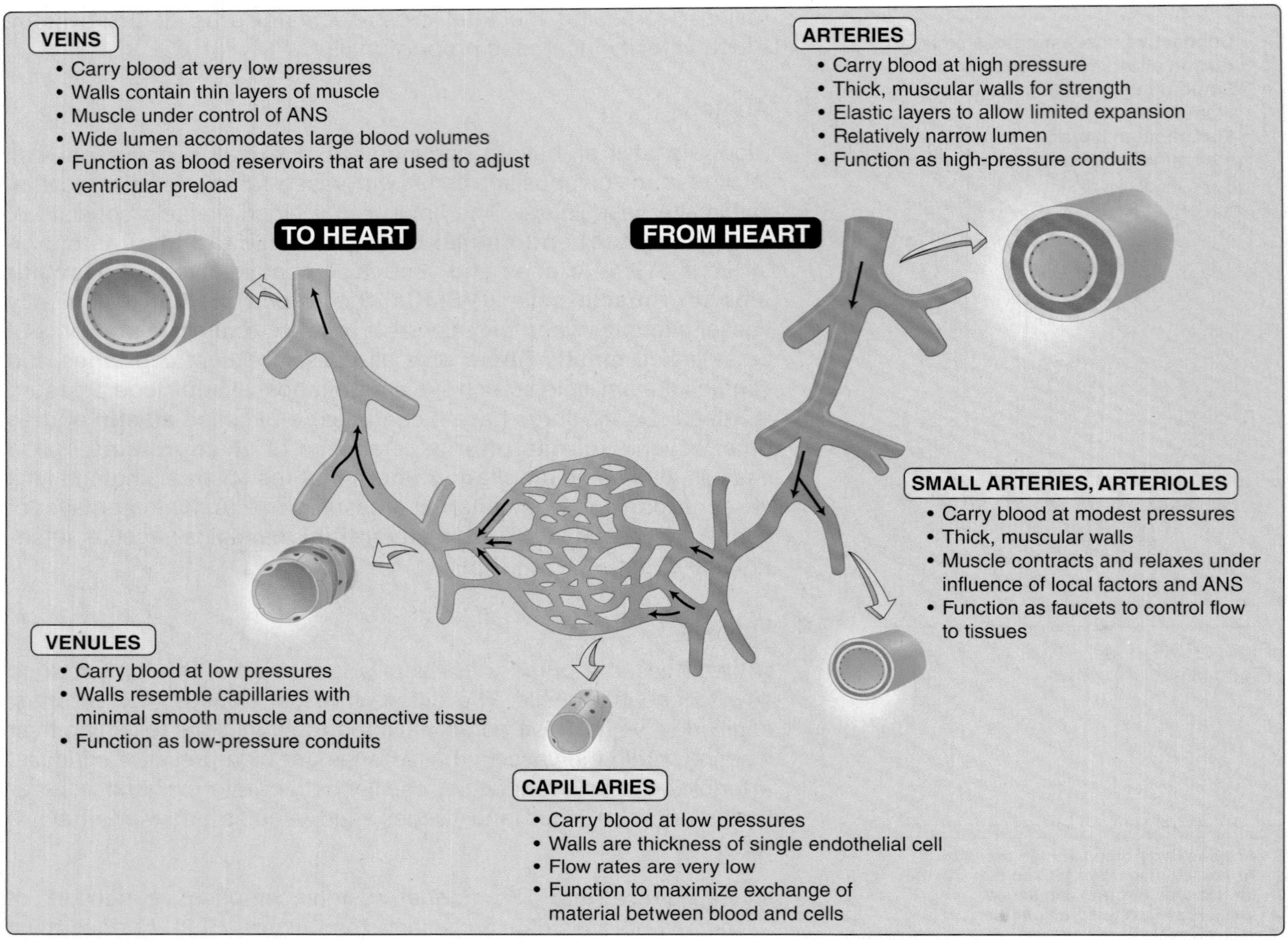

Figure 19.2
Properties and functions of the vessels comprising the systemic vasculature. ANS = autonomic nervous system.

distribution network that is capable of such a task requires extensive branching of the vascular tree. Thus, blood leaves the left ventricle (LV) via a single large-diameter vessel (the aorta), which then branches repeatedly to yield ~10,000,000,000 tiny capillaries. The branching pattern greatly increases vascular cross-sectional area, from ~4 cm^2 (aorta) to ~4,000 cm^2 total at the level of the capillaries (Figure 19.3). Blood flow velocity drops proportionally. Blood exits the LV at up to ~50 cm/s, but velocity has dropped to <1 mm/s by the time blood reaches the capillaries. A low flow rate greatly increases the time available for exchanging materials between blood and tissues during blood's passage through capillary beds. Note that whereas arteries and veins are arranged in series with each other, capillaries are organized in parallel circuits (see Figure 19.2). This vascular arrangement has important physiologic consequences, as discussed in Section IV below. Blood travels back to the heart via venules, which join and fuse to form veins. Smaller veins merge to form larger veins, each

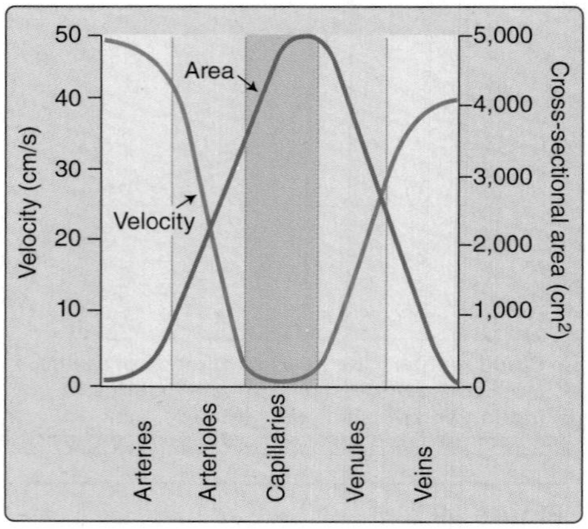

Figure 19.3
Blood velocity and vessel cross-sectional area across the systemic vasculature.

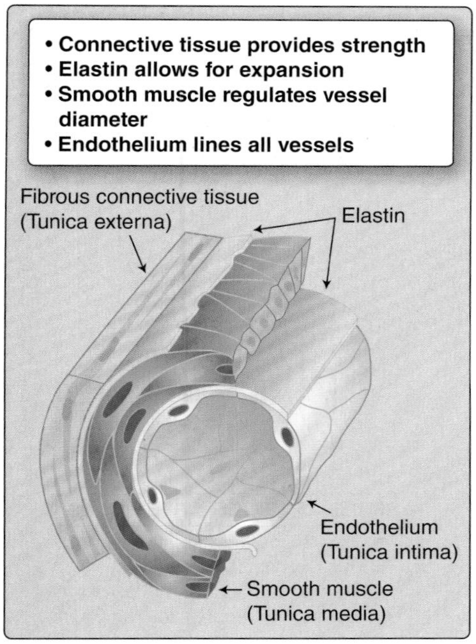
• Connective tissue provides strength
• Elastin allows for expansion
• Smooth muscle regulates vessel diameter
• Endothelium lines all vessels
Fibrous connective tissue (Tunica externa)
Elastin
Endothelium (Tunica intima)
Smooth muscle (Tunica media)

Figure 19.4
Blood vessel structure.

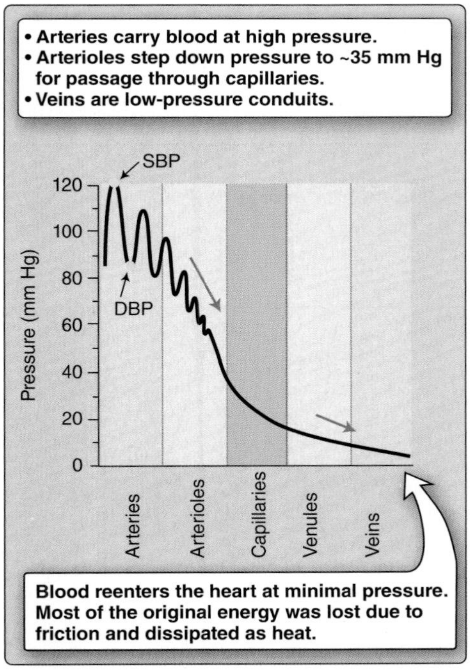
• Arteries carry blood at high pressure.
• Arterioles step down pressure to ~35 mm Hg for passage through capillaries.
• Veins are low-pressure conduits.
SBP
DBP
Pressure (mm Hg)
120
100
80
60
40
20
0
Arteries
Arterioles
Capillaries
Venules
Veins
Blood reenters the heart at minimal pressure. Most of the original energy was lost due to friction and dissipated as heat.

Figure 19.5
Perfusion pressures across the systemic vasculature. DBP = diastolic blood pressure; SBP = systolic blood pressure.

fusion decreasing the total cross-sectional area of the system. Blood velocity increases proportionally.

B. Anatomy

Blood vessels all have a common structure, although vessel wall thickness and composition varies with vessel function and its location within the vasculature. The lining of all blood vessels consists of a single layer of **endothelial cells** (the **tunica intima**) as shown in Figure 19.4. Arteries and veins also contain layers of **vascular smooth muscle cells** ([**VSMCs**] the **tunica media**) that modify vessel diameter when they contract or relax. Extensive networks of cross-linked **elastic fibers** give all vessels except capillaries and venules the ability to stretch like a rubber hose when blood pressure is raised. Elastic fibers have a central core of coiled **elastin** and an outer covering of **microfibrils** composed of **glycoproteins**. Blood vessels also contain **collagen fibers** that resist stretching and limit vessel expansion when internal pressures rise. A thin outer layer of connective tissue (the **tunica adventitia**) maintains vascular integrity and shape.

C. Vessels

A blood vessel's primary function is to provide a conduit for blood flow to and from cells. The different vessel classes (i.e., arteries, capillaries, veins) have additional important functions reflecting their location within the vasculature. Arteries are high-pressure conduits, arterioles are flow regulators, capillaries facilitate exchange of materials between blood and tissues, and venules and veins have a reservoir function.

1. **Large arteries:** The arterial system comprises a network of narrow-bore distribution vessels (see Figure 19.2). Arteries must carry blood at high pressure (Figure 19.5), so their walls are thick and their lumens narrow, which limits arterial system capacity. The walls of the larger arteries (also called **elastic arteries**) contain smooth muscle layers and are rich in elastin fibers. The muscle layers have a resting tone, which limits arterial distensibility and helps maintain the pressure of the blood within.

2. **Small arteries and arterioles:** The walls of the smallest arteries and arterioles are dominated by their smooth muscle layers. Collectively known as **resistance vessels**, they act as faucets or stopcocks to control blood flow to capillaries (Figure 19.6). When tissue demand for O_2 and nutrients is high, the VSMCs relax, and flow to the tissues increases. A decreased demand for blood or intervention by the central nervous system constricts the muscular "faucets," and flow to the tissues is reduced (see 20·II).

3. **Capillaries:** Capillaries bring blood to within 30 μm of virtually every cell in the body. They are designed to keep the blood contained within the vasculature while simultaneously maximizing the opportunity for exchange of materials between blood, interstitium, and tissues. Their walls are the thickness of a single endothelial cell plus the **basal lamina**. In some tissues, capillaries permit direct communication between blood and cells via

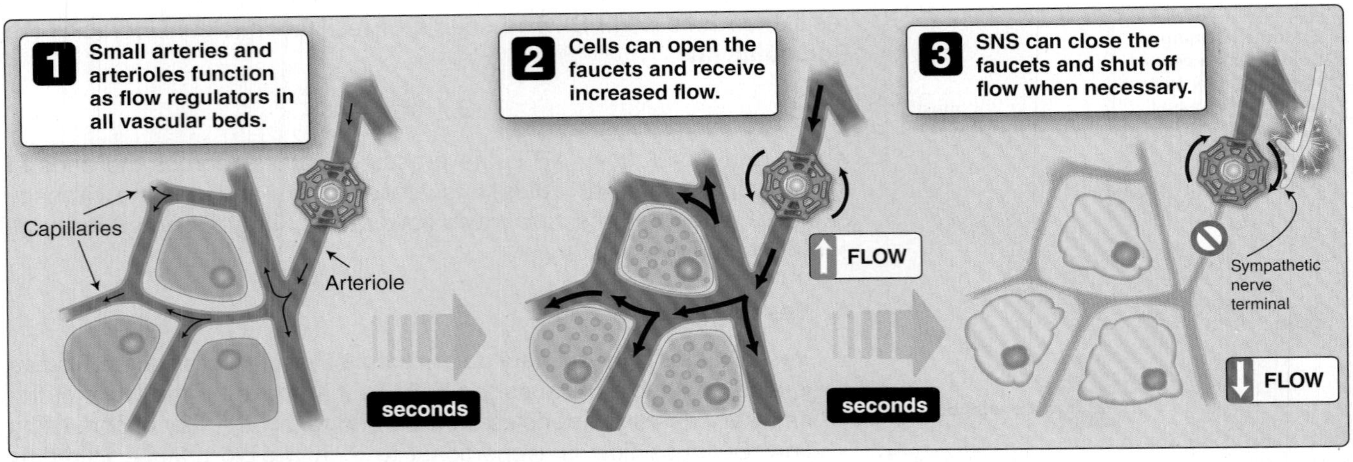

Figure 19.6
Vascular faucets regulate blood flow. SNS = sympathetic nervous system.

transmural **pores** (**fenestrations**) and junctional **clefts** between adjacent cells (Figure 19.7).

> Although capillaries average only 1 mm in length and 8–10 μm in diameter, they provide a total surface area for exchange of 500–700 m^2 in an average adult.

4. **Venules and veins:** Venules and veins are low-pressure conduits for blood to make its way back to the heart. Smaller venules are almost indistinguishable from capillaries, which allows them to participate in fluid and metabolite exchange. Venules widen and fuse with each other as they progress toward the heart. Larger venules contain VSMCs within their walls, but far fewer than those seen in vessels of equivalent size in the arterial system. The paucity of muscle means that vein walls are thin (see Figure 19.2), making them highly distensible and able to accommodate large volumes of blood. Under resting conditions, ~65% of total blood volume resides in the venous compartment, creating a reservoir that is used to sustain ventricular preload when the need arises (see 20·V). The larger veins contain valves that help maintain unidirectional flow through the system and counter the tendency of blood to be pulled downward under the influence of gravity.

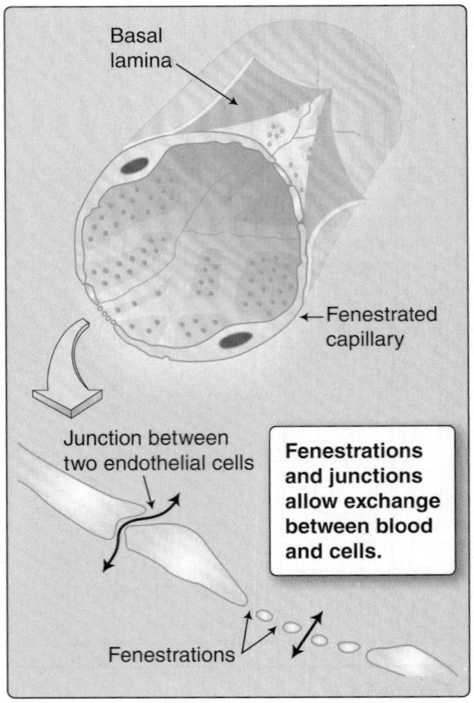

Figure 19.7
Capillary wall structure.

III. BLOOD FLOW DETERMINANTS

Blood will not flow through the vasculature unless it is forced to do so by application of pressure, which is required to overcome a **resistance** to flow. Understanding the origins of this resistance is important clinically because it is a primary indicator of cardiovascular status and health. For example, whereas resistance usually goes up during circulatory shock, septic shock is associated with a profound decrease in vascular

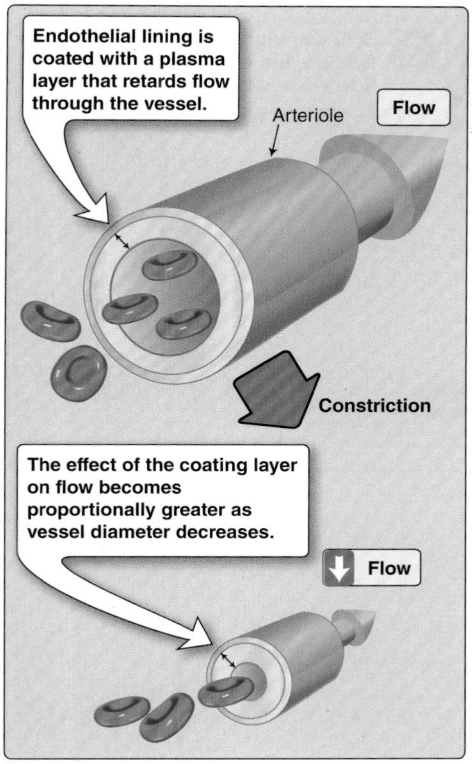

Figure 19.8
Effects of constriction on flow through a resistance vessel.

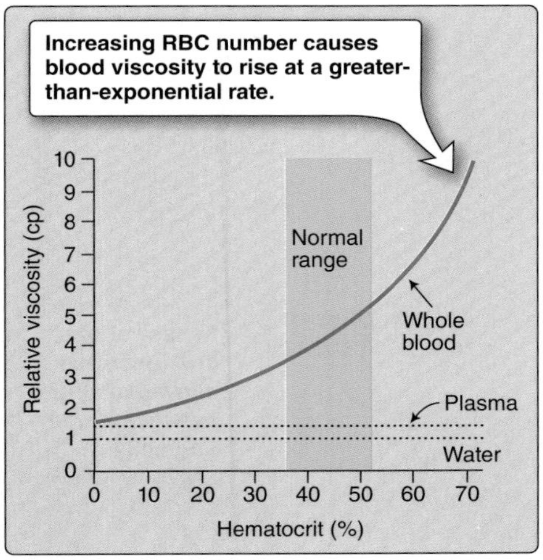

Figure 19.9
Relationship between hematocrit and blood viscosity. cp = centipoise; RBC = red blood cell.

resistance (see 40·III·C). The source of resistance is considered in the **Poiseuille law**:

$$Q = \frac{\Delta P \times \pi r^4}{8L\eta}$$

where Q denotes flow, ΔP is the pressure gradient across the ends of the vessel, r is internal radius, L is vessel length, and η is the viscosity of blood. Whereas pressure drives flow, radius, length, and viscosity all contribute to flow resistance.

A. Vessel radius

Vessel radius is the primary determinant of vascular resistance. Radius is also a variable because the VSMCs that make up the walls of the small arteries and arterioles contract and relax as a way of controlling flow. Because flow is proportional to r^4, a twofold change in radius causes a 16-fold change in flow. The potency of radius's effect on flow relates to a layer of plasma that clings to the inner surface of all vessels. The layer forms through interactions between blood and the vascular endothelium and, in so doing, impedes flow. Although the depth of the coating layer is essentially the same in all vessels irrespective of bore size, its contribution to total cross-sectional area is much greater in a small-diameter vessel than in a large one, and, therefore, resistance to flow through a small vessel is correspondingly larger (Figure 19.8).

B. Vessel length

Blood flow through a vessel is inversely related to vessel length, again reflecting blood's tendency to interact with the vascular endothelium. Vessel length does not change physiologically and is not considered further.

C. Blood viscosity

Blood is a complex fluid, whose viscosity varies with flow. A fluid's viscosity is measured relative to water. Adding electrolytes and organic molecules (including proteins) to water raises viscosity from 1.0 cp to ~1.4 cp. Cells, principally red blood cells (RBCs), have the greatest impact, with viscosity rising at a greater-than-exponential rate with hematocrit (Figure 19.9).

1. **Hematocrit:** Hematocrit measures the percentage of whole blood volume that is occupied by RBCs. Hematocrit is determined clinically by centrifuging a tube containing a small blood sample to separate cells from plasma. Hematocrit can then be estimated from the height of the packed RBC layer within the tube, which is dependent on both RBC number and volume. Normal values for hematocrit range between 41%–53% for males and 36%–46% for females.

> A hematocrit that falls below a normal range indicates **anemia**. Anemia is more usually defined in terms of hemoglobin (Hb) levels, however. Normal Hb values for males range from 13.5–17.5 g/dL, 12.0–16.0 g/dL in females.

2. **Flow resistance:** RBCs increase flow resistance by rubbing up against the vessel wall. RBCs travel through capillaries measuring only 2.5 μm in diameter, which is surprising given that these blood cells are typically pictured as 8-μm diameter disks (Figure 19.10). RBCs readily distort, however, which allows them to slip through narrow vessels. They also travel in the center of the vessel, which minimizes interactions with the endothelium. Even so, the heart must still expend significant energy to overcome the resistance associated with friction between RBCs and vessel walls.

> Cells can be likened to plastic bags filled with water. Like bags, most cells rupture when deformed mechanically. RBCs more closely resemble plastic bags that are only partially filled with water. They deform easily, which allows them to squeeze through narrow vessels and pores. RBCs swell and deform less readily as they age, which allows the body to target them for destruction. Disposal occurs in the spleen, where RBCs are forced through a filter made of connective tissue fibrils (**splenic cords**). Young, flexible RBCs pass through the filter with relative ease, but older RBCs become trapped and are then phagocytosed.

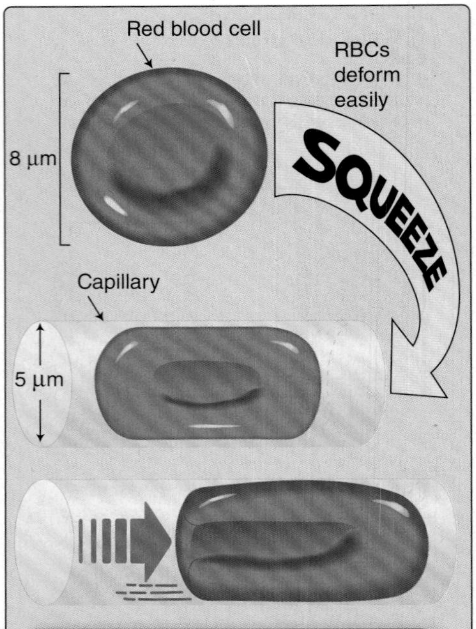

RBCs look like mushroom caps when traveling through capillaries. Friction between the RBC and vessel wall creates flow resistance.

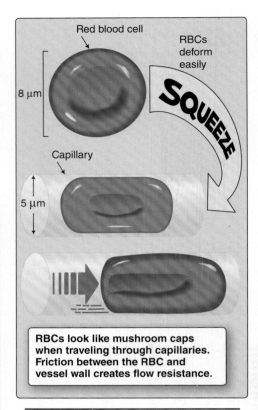

Figure 19.10
Red blood cell flexibility.

3. **Anemia:** Anemia is associated with a decrease in RBC number. Blood viscosity and flow resistance decreases also. **Physiologic anemias** occur when blood volume is expanded faster than RBC production, such as during pregnancy (see 37·IV·C) or exercise training.

4. **Polycythemia:** Increasing RBC numbers increases blood viscosity and resistance to flow. People living at high altitude demonstrate a physiologic **polycythemia** stimulated by reduced atmospheric O_2 levels. Although increased RBC production helps compensate for reduced O_2 availability, the tradeoff is increased workload on the heart, limiting the altitude at which humans can comfortably exist to around 5,000 m.

IV. HEMODYNAMIC OHM LAW

Vascular resistance represents afterload to the LV and determines how hard it must work to generate output. If resistance increases, the heart is forced to work harder to compensate. A healthy heart is well equipped to meet the demands placed on it by changes in vascular resistance under normal conditions, but increased resistance can seriously stress a diseased heart. For this and other reasons, it is important to be able to quantify vascular resistance in a clinical setting. Identifying and summing all of the individual components that make up vascular resistance

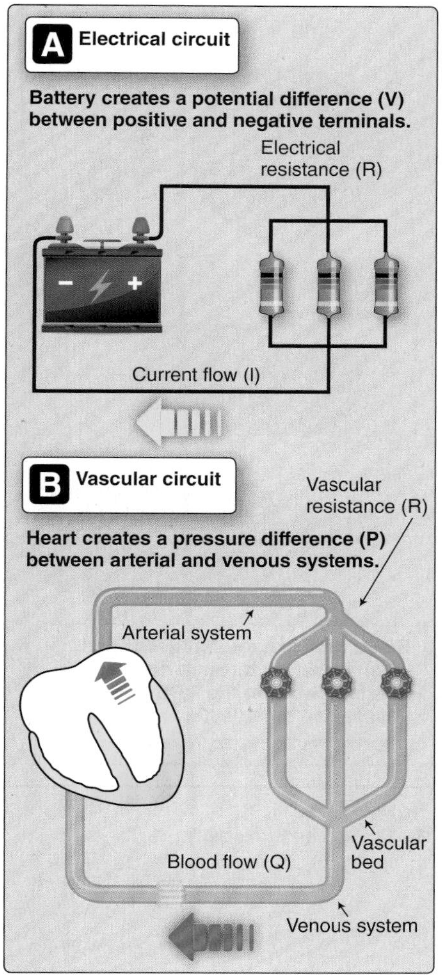

Figure 19.11
The Ohm law.

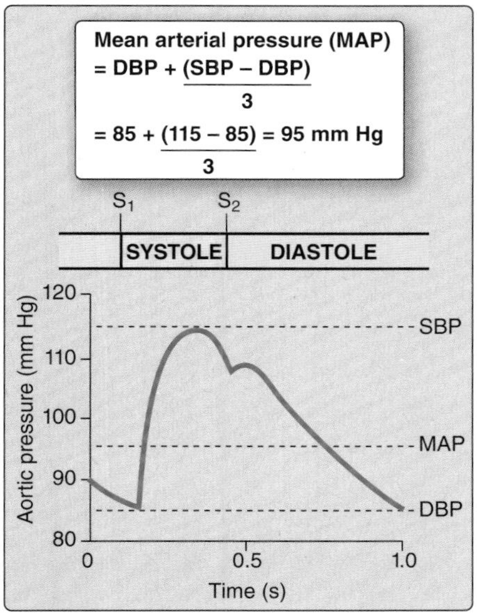

Figure 19.12
Derivation of mean arterial pressure.
DBP = diastolic blood pressure;
SBP = systolic blood pressure.

Clinical Application 19.1: Polycythemia Vera

Polycythemia vera is a neoplastic disorder affecting myeloid red blood cell (RBC) precursors. The disease causes uncontrolled RBC production, and hematocrit (HCT) climbs accordingly. Polycythemia is defined by a HCT of >48% in females and >52% in males.

Once HCT reaches around 60%, the RBCs are so closely packed that they collide with each other and begin forming aggregates and clots. Cohesion is dependent on fibrinogen and other large plasma proteins that coat the RBC surface. Viscosity and vascular resistance increases to such a degree that the left ventricle is unable to generate sufficient pressure to maintain even basal flow rates. Patients typically present with headaches, weakness, and dizziness associated with decreased cerebral perfusion. A common complaint is relentless pruritis (skin itching) after taking a warm bath.

Left untreated, the median survival time is 6–18 months. This improves to >10 years if treated with serial phlebotomy, optimally reducing HCT to <42% in females and <45% in males. The main risk factors are thrombotic events (i.e., stroke, deep vein thromboses, myocardial infarction, and occlusion of peripheral arteries).

in a typical human is not feasible. As an alternative, vascular resistance can be estimated with relative ease from knowledge of pressure and flow using a modified version of the **Ohm law** (Figure 19.11A). The Ohm law describes the effects of electrical resistance (R) on current flow (I) in a DC circuit:

$$I = \frac{V}{R}$$

where V is the voltage drop across the resistance. The hemodynamic form of the Ohm law is thus:

$$Q = \frac{P}{R}$$

Where Q is blood flow, P is the pressure gradient across a vascular circuit, and R is vascular resistance (see Figure 19.11B). As discussed above, R is defined by vessel radius, length, and blood viscosity ($R = 8L\eta \div \pi r^4$). The hemodynamic form of the Ohm law makes it possible to calculate R for any vessel or vascular circuit, regardless of its size, from measurements of pressure and flow.

A. Systemic vascular resistance

The largest circulation in the body and the one with the greatest resistance is the systemic circulation. The value of **systemic vascular resistance** ([**SVR**] also known as **total peripheral resistance**) is calculated as

$$SVR = \frac{MAP - CVP}{CO}$$

where MAP − CVP represents the pressure difference between the aorta (**mean arterial pressure** [**MAP**]) and vena cavae (**central venous pressure** [**CVP**]). MAP is a time-averaged value that recognizes that arterial pressure rises and falls in step with the cardiac cycle (Figure 19.12). MAP is calculated as

$$MAP = DBP + \frac{(SBP - DBP)}{3}$$

where **SBP = systolic blood pressure** and **DBP = diastolic blood pressure** (SBP − DBP is also known as **pulse pressure**). Using typical normal values for MAP (95 mm Hg), CVP (5 mm Hg), and CO (6L/min), SVR is calculated to be 15 mm $Hg{\cdot}min{\cdot}L^{-1}$.

> SVR typically varies between 11 and 15 mm $Hg{\cdot}min{\cdot}L^{-1}$ in an average person. SVR may also be expressed clinically in units of $dyn{\cdot}s{\cdot}cm^{-5}$, calculated by multiplying the values above by 80. Thus, a normal SVR range is between ~900 and 1,200 $dyn{\cdot}s{\cdot}cm^{-5}$.

Pulmonary vascular resistance (**PVR**) can be calculated in a similar manner (by using mean pulmonary artery and left atrial pressures), amounting to ~2–3 mm $Hg{\cdot}min{\cdot}L^{-1}$ (150–250 $dyn{\cdot}s{\cdot}cm^{-5}$) in an average person.

B. Serial and parallel circuits

Hemodynamic circuits are treated in the same way as electrical circuits when calculating the combined resistance of multiple individual components (Figure 19.13). The total resistance (R_T) of a circuit containing three resistors ($R_1 - R_3$) arranged in series is equal to the sum of the individual components. If each of the resistors below has a resistance of 10 units, $R_T = 30$ units.

$$R_T = R_1 + R_2 + R_3 = 10 + 10 + 10 = 30$$

Calculating the total resistance of the same three resistors arranged in parallel requires that the reciprocal of each component be summed (see Figure 19.13):

$$\frac{1}{R_T} = \frac{1}{R_1} + \frac{1}{R_2} + \frac{1}{R_3} = \frac{1}{10} + \frac{1}{10} + \frac{1}{10}; R_T = 0.3$$

Note that R_T of the parallel circuit is 3.3 units, significantly less than that of any individual component. Thus, even though the systemic circulation contains approximately 10^{10} capillaries that individually have a very high resistance to flow, their parallel arrangement means that their *combined* resistance is relatively low.

> Adding capillaries to a vascular circuit causes SVR to decrease, not increase, because they provide additional pathways for blood flow (see Figure 19.13).

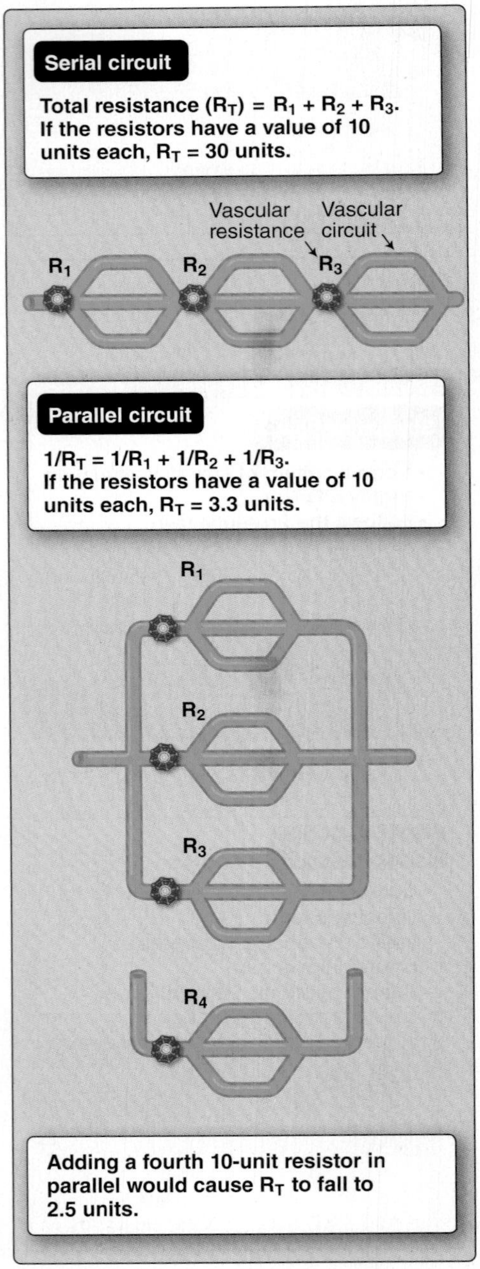
Serial circuit
Total resistance (RT) = R1 + R2 + R3.
If the resistors have a value of 10 units each, RT = 30 units.
Vascular resistance
Vascular circuit
R1
R2
R3
Parallel circuit
1/RT = 1/R1 + 1/R2 + 1/R3.
If the resistors have a value of 10 units each, RT = 3.3 units.
R1
R2
R3
R4
Adding a fourth 10-unit resistor in parallel would cause RT to fall to 2.5 units.

Figure 19.13
Calculating the resistance of vascular circuits.

V. LIMITS TO THE POISEUILLE LAW

The Poiseuille law helps identify the sources of flow resistance in the cardiovascular system, but system complexity limits its application to smaller arterial vessels and to capillaries. Confounding cardiovascular design features include the predilection for turbulent flow, the fact that blood viscosity is velocity dependent, and the compliance of blood vessels.

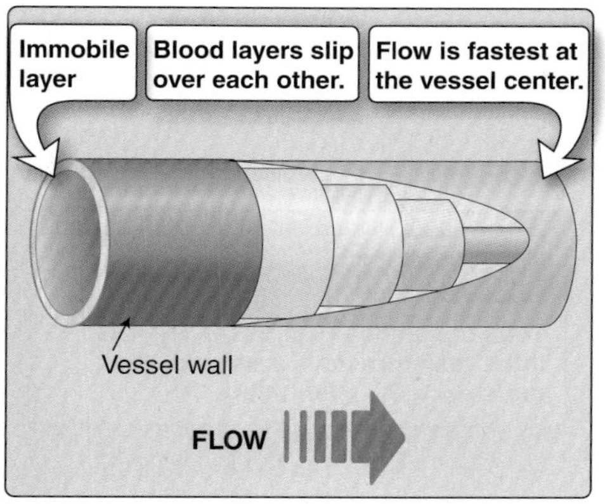

Figure 19.14
Laminar blood flow.

A. Turbulence

When blood flows through the vasculature, it experiences drag caused by its various components interacting with the vessel wall. As discussed above, the vascular endothelium is coated with a layer of immobilized plasma. This coating exerts drag on blood that is flowing closer to the center of the vessel, creating another slowed layer that exerts its own drag, and so on toward the center of the vessel. Thus, flow through vessels occurs in concentric layers that slip over each other, with fastest flow at the center and slowest up against the walls of the vessel. This flow pattern is referred to as **laminar** or **streamline flow** (Figure 19.14). Laminar flow is observed in most regions of the cardiovascular system, and the Poiseuille law is valid for as long as it is maintained. When streamline flow is disrupted, kinetic energy is squandered on chaotic motion, a pattern known as **turbulence** (Figure 19.15).

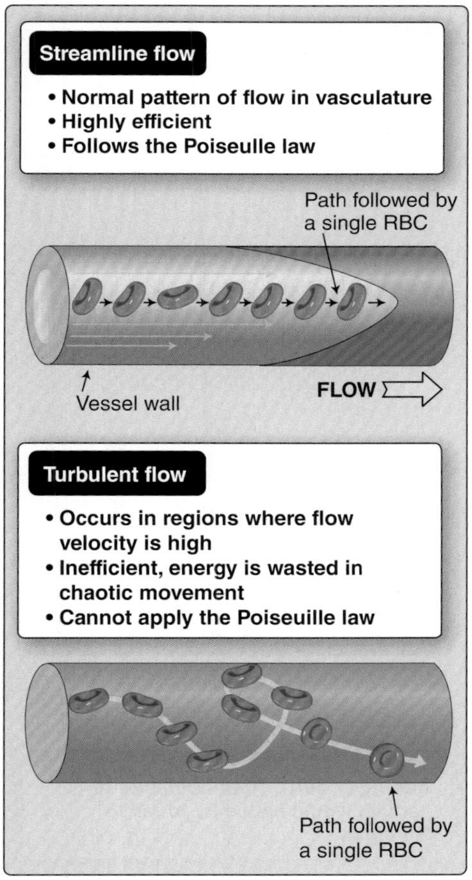

Figure 19.15
Streamline and turbulent flow.
RBC = red blood cell.

1. **Reynolds equation:** The likelihood of turbulence can be predicted using the **Reynolds equation:**

$$N_R = \frac{v \times d \times \rho}{\eta}$$

where N_R is Reynolds number, v is mean blood velocity, d is vessel diameter, ρ (rho) is blood density, and η is blood viscosity. Blood density does not change within the parameters of normal human physiology. Many blood vessels constrict and relax and hence their internal diameter changes constantly, but not to the extent that they cause turbulence *in vivo*. Velocity and viscosity are both physiologically relevant variables, however.

2. **Blood velocity effects:** Turbulence is most likely to be observed within the heart chambers or within the vessels that enter and leave the heart. These are regions where large volumes of blood are moving at high flow velocities. Turbulence occurs once a certain critical velocity is achieved, causing orderly streamline flow to become chaotic and inefficient.

3. **Equation of continuity:** Congenital and pathologic heart valve defects are common causes of turbulence. The aortic valve is located in a high-pressure, high-velocity region of the cardiovascular system where it is subject to constant wear and tear. It is not uncommon for the valve leaflets to calcify and stiffen with age, or perhaps fuse along their commissures as a result of repeated inflammation. Such changes reduce the cross-sectional area of the valve orifice and obstruct outflow. Because CO has to be maintained at a basal 5–6 L/min regardless of circumstance, left

ventricular pressure increases and drives flow at higher velocity through the narrowed outlet (Figure 19.16). The extent to which velocity is increased by stenosis is defined by the **equation of continuity:**

$$Q = v_{nl} \times A_{nl} = v_s \times A_s$$

where Q is flow, v_{nl} and v_s are flow velocities through normal and stenotic valves, respectively, and A_{nl} and A_s are valve cross-sectional areas. If Q is constant and A_s is reduced, velocity must increase to compensate.

4. **Sounds:** Turbulent flow creates crosscurrents and eddies and causes kinetic energy to be expended when blood impacts the vessel wall. Impacts cause vibrations that travel to the body surface where they can be heard as sounds. Common examples encountered clinically include **murmurs** and **Korotkoff sounds**.

   a. **Murmurs:** Blood being forced at high velocity through a stenotic aortic or pulmonary valve yields a **systolic murmur**. Valves that fail to close completely also produce murmurs. Such murmurs are caused by blood being forced backward through an incompetent valve and impacting blood contained within the atria or ventricles (see Clinical Application 18.1).

   b. **Korotkoff sounds:** Turbulence can be induced artificially for diagnostic purposes. Partial occlusion of the brachial artery with a pressure cuff causes murmurs that reflect blood being ejected at high velocity through the compressed area and impacting the column of blood beyond. These murmurs (Korotkoff sounds) can be heard with a stethoscope placed downstream of the cuff. Sounds are first heard when cuff pressure drops just below SBP, allowing small amounts of blood to jet through the occluded artery. The murmurs typically disappear when cuff pressure drops below DBP and the artery is fully patent. These sounds thereby provide a convenient way of approximating SBP and DBP.

5. **Hematocrit:** Because blood velocity is inversely related to viscosity and hematocrit, anemia can also increase the likelihood of turbulence. For example, the physiologic anemia that accompanies pregnancy causes **functional murmurs**, sounds associated with ejection of blood at high velocity through a normal valve (see 37·IV·C).

6. **Occurrence of turbulence *in vivo*:** In an ideal system, turbulence can be expected when N_R exceeds 2,000. When N_R is below 1,200, laminar flow prevails. The cardiovascular system is less than ideal. Many factors, especially the extensive branching that is inherent to the vascular tree, lower the threshold for turbulence to around 1,600. Vessel branches disrupt laminar flow and create foci for local eddy currents to form.

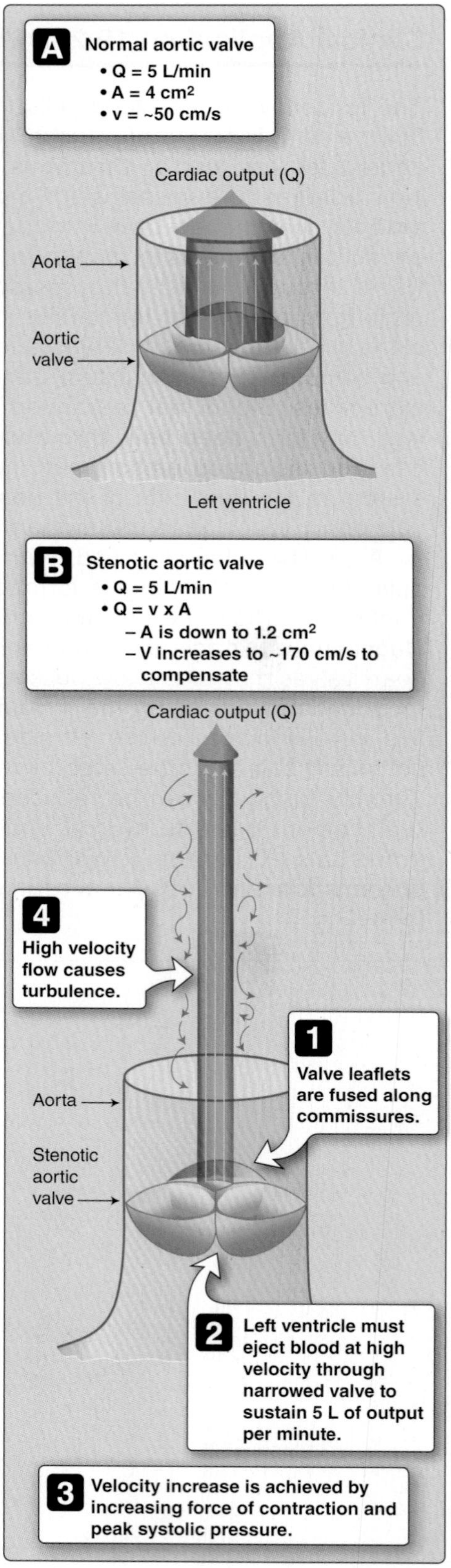
A
Normal aortic valve
• Q = 5 L/min
• A = 4 cm2
• v = ~50 cm/s
Cardiac output (Q)
Aorta
Aortic valve
Left ventricle
B
Stenotic aortic valve
• Q = 5 L/min
• Q = v x A
– A is down to 1.2 cm2
– V increases to ~170 cm/s to compensate
Cardiac output (Q)
4 High velocity flow causes turbulence.
1 Valve leaflets are fused along commissures.
Aorta
Stenotic aortic valve
2 Left ventricle must eject blood at high velocity through narrowed valve to sustain 5 L of output per minute.
3 Velocity increase is achieved by increasing force of contraction and peak systolic pressure.

Figure 19.16
Effect of aortic stenosis on ventricular ejection velocity. A = area; Q = flow; v = ejection velocity.

Clinical Application 19.2: Thrombi and Anticoagulation Therapy

The tendency for red blood cells to aggregate in regions where flow velocity is low is of serious clinical concern because such aggregates can lead to **thrombus** formation. Thrombi are blood clots adhered to a vessel wall. Thrombi may break loose to form **emboli**, which then travel through the vasculature until they encounter and become lodged in a vessel that is too small to travel through. In a healthy person, thrombi can form during prolonged periods of immobility such as during long-haul airplane flights. Cramped cabins and hard seats restrict mobility and compress the vasculature that returns blood from the lower extremities. Blood that is trapped within the deep veins of the legs may form **deep vein thromboses**. Disembarkation restores flow and dislodged emboli may then travel through the venous system to the right side of the heart and become lodged in the pulmonary vasculature (**pulmonary embolism**).

Atrial fibrillation (AF) and heart valve replacement also puts patients at risk of thrombus formation. AF prevents orderly contraction and flow through the atria, creating regions of blood stagnation within the affected chamber. Prosthetic ("mechanical") heart valves may also allow pockets of stagnation to form behind their leaflets, increasing the incidence of thrombus formation. If this valve is located on the left side of the heart, a freed embolus potentially can enter the cerebral vasculature and cause a stroke. Thrombi formation can be reduced in AF and mechanical valve replacement patients by oral anticoagulants such as warfarin (trade name Coumadin).[1] Warfarin is a vitamin-K antagonist that prevents formation of several coagulation factors required for clot formation.

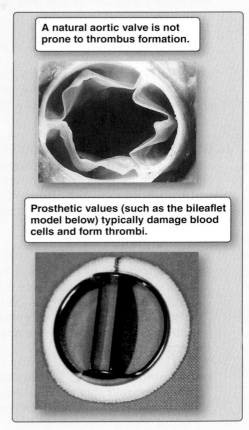

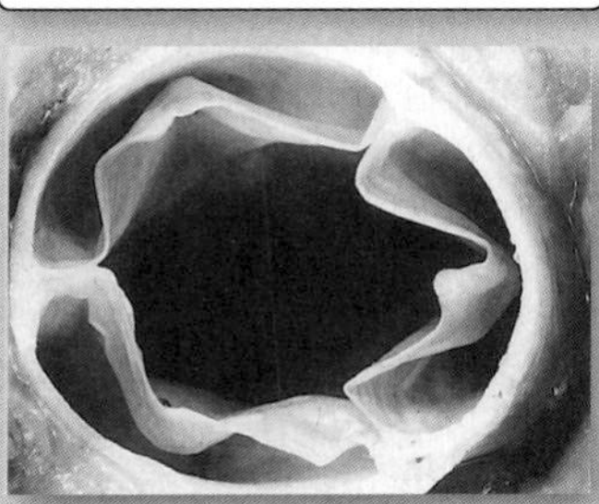

A natural aortic valve is not prone to thrombus formation.

Prosthetic values (such as the bileaflet model below) typically damage blood cells and form thrombi.

Natural and prosthetic aortic valves.

B. Velocity effects

When blood is stationary or barely moving, RBCs have time to adhere to each other and form aggregates that resemble stacks of coins (**rouleaux**). Aggregates require more effort to move through the circulation than do individual cells, thereby increasing resistance to flow. Rouleaux begin to break apart as flow velocity increases, and viscosity falls in parallel (Figure 19.17).

C. Vessel compliance

The Poiseuille law assumes that blood vessels are rigid tubes. Although smaller vessels (capillaries, arterioles, and small arteries) are relatively nondistensible, most veins and larger arteries expand when internal pressure rises (Figure 19.18). Distensibility

[1]For more information on anticoagulants, see *LIR Pharmacology*, 5e, p. 251.

invalidates application of the Poiseuille law, but it does provide cardiovascular benefits.

1. **Venous reservoir:** Distensibility defines the ease with which a vessel swells when filling pressure rises. Recall that veins have thinner walls than arteries. Therefore, although a pressure increase of 1 mm Hg might cause a 1-mL increase in arterial volume, it would cause a similarly sized vein to swell by 6–10 mL. The distensibility of veins means that the venous system as a whole has a much higher **compliance** or **capacitance** than does the arterial system, which allows it to function as a blood reservoir. Compliance is a measure of a vessel's ability to accommodate volume (V) when filling pressure (P) increases:

$$\text{Compliance} = \frac{\Delta V}{\Delta P}$$

2. **Arterial pump:** During systole, the ventricle ejects blood into the arterial tree faster than it can be passed on to capillaries. The distensibility of large arteries allows them to expand to accommodate a full ventricular stroke volume (Figure 19.19) and then transmit it to the capillary beds during diastole when the ventricle is relaxing and the aortic valve is closed (**diastolic runoff**). The energy driving flow during diastole was stored in the elastic elements of the arterial wall by the ventricle during systole. This storage and runoff effect (known as a **windkessel**) is advantageous in that it evens out pressure and flow through the vasculature over time even though ventricular ejection is overtly phasic.

> *Windkessel* is German for "air chamber" and refers to a feature of early fire engine design. Prior to invention of the internal combustion engine, fire engines were horse drawn and hand operated. In the event of fire, water was pumped by hand into the air chamber, causing pressure to develop within. Pressurized water from the windkessel was then directed from a hose at the fire. The inclusion of the windkessel in the design ensured a continuous, steady stream of water from the hose, even when fire personnel were unable to pump.

3. **Age effects:** Aging is associated with a stiffening of the vessel walls due to calcification and collagen deposition (**arteriosclerosis**). Loss of distensibility reduces the amount of blood that can be stored in the arterial system during systole for subsequent diastolic runoff. The LV is forced to make up for this deficit by generating higher pressures to drive increased flow during systole. This pressure manifests as an essential hypertension that is common in older adults.

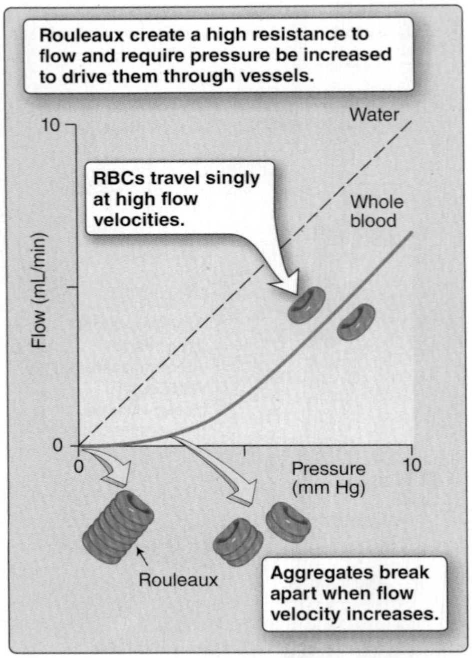

Figure 19.17
Effects of rouleaux on the pressure required to induce blood flow. RBCs = red blood cells.

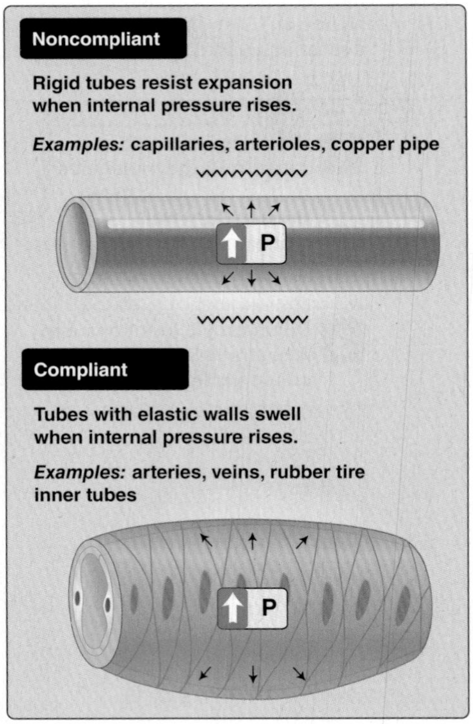

Figure 19.18
Relative compliance of blood vessels.

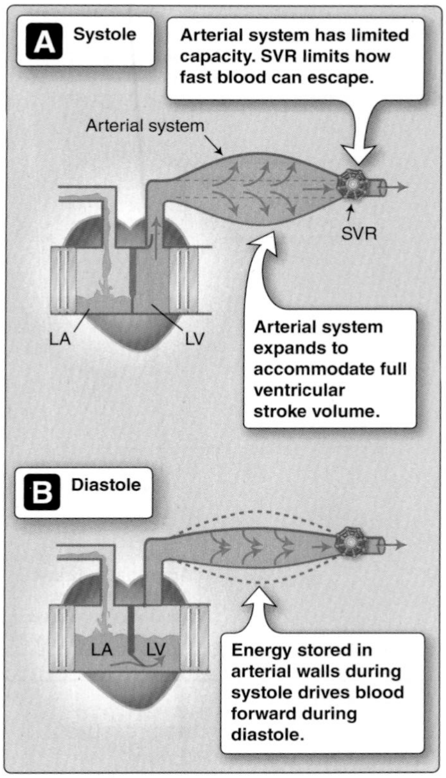

Figure 19.19
Arterial walls expand during systole and then drive flow during diastole. LA = left atrium; LV = left ventricle; SVR = systemic vascular resistance.

VI. EXCHANGE BETWEEN BLOOD AND TISSUES

The primary function of the cardiovascular system is to deliver O_2 and nutrients to all cells in the body. Blood is distributed by capillaries, vessels with exceptionally thin walls (~0.5 μm, or the width of an endothelial cell) designed to facilitate diffusional exchange between blood and cells. Blood moves through capillaries very slowly (~1 mm/s), which maximizes the opportunity for exchange during passage. In discussions of how exchange occurs at the cellular level, it is helpful to recognize four general mechanisms: pinocytosis, bulk flow, diffusion via pores, and diffusion via cells (Figure 19.20).

A. Pinocytosis

Pinocytotic vesicles form when the plasma membrane invaginates and pinches off to capture and internalize an extracellular fluid sample. The vesicles then migrate across the vessel wall and release their contents on the opposite side. Pinocytosis is not a major pathway for exchange, but it does provide for transit of large, charged molecules such as antibodies.

B. Bulk flow

Capillaries are typically very leaky, with **fenestrations** in their walls and **clefts** between adjacent cells that provide ready pathways for exchange of ions and solutes. Blood entering capillaries is pressurized, so these same pathways allow water and anything dissolved in it to be driven out of the vasculature and into the interstitium. This **bulk flow** of fluid is not completely unregulated, however. Intercellular junctions typically contain a proteinaceous barrier that both cements

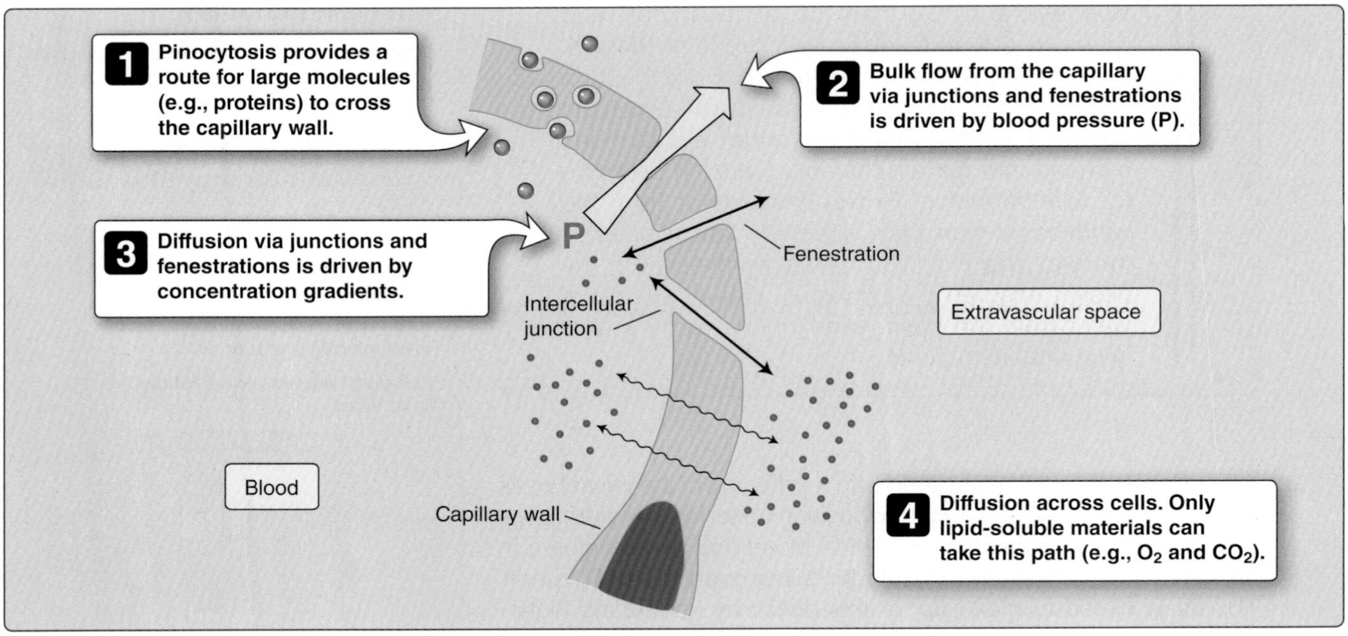

Figure 19.20
Four general mechanisms by which materials are exchanged across the capillary wall.

cells together and filters fluid as it leaves the blood. Proteins are too large to escape via junctions or pores and remain trapped within the vasculature.

C. Diffusion via fenestrations and pores

The same pathways that allow for bulk flow also provide pathways for simple **diffusion** of water and other small molecules. Movement is driven by chemical concentration gradients between blood, interstitium, and cells.

D. Diffusion across endothelial cells

Lipid-soluble materials cross between blood and interstitium by simple diffusion across endothelial cells and their plasma membranes. This is the primary means by which O_2 and CO_2 are exchanged.

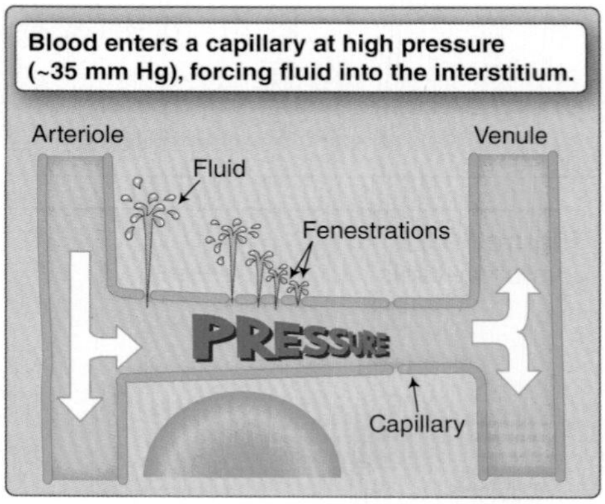

Blood enters a capillary at high pressure (~35 mm Hg), forcing fluid into the interstitium.
Arteriole
Venule
Fluid
Fenestrations
PRESSURE
Capillary

Figure 19.21
Pressure-induced fluid loss from the vasculature.

VII. FLUID MOVEMENT

The leakiness of capillaries is problematic. Blood has to enter capillaries under pressure to ensure that it has sufficient energy to traverse the capillaries and veins and return to the heart, yet this same pressure also drives fluid out of the vasculature (Figure 19.21). The severity of the problem is such that, in the absence of any counter measure, we would lose our entire blood volume to the interstitium within an hour or two!

A. Vascular water retention

The main force holding water in the blood stream is an osmotic potential that is generated by proteins that are trapped in the blood stream by virtue of their size. **Albumin** (60,000 MW) is the principal plasma protein (~80% of total), although **globulins** (140,000 MW) are important also. Proteins help blood retain water through direct osmotic effects, but their many negatively charged groups secondarily attract and concentrate osmotically active cations such as Na^+ and K^+ (a **Donnan effect**).

B. Interstitial water retention

The space between blood vessels and cells (the **interstitium**) contains collagen fibers that provide structural support to tissues, but the bulk of the space is occupied by a dense network of fine proteoglycan filaments (see 4·IV·B·3). Fluid that filters from the bloodstream becomes trapped by these filaments, much as water is trapped by filaments in gelatin (Figure 19.22). The interstitial gel normally contains ~25% of total body water, creating an invaluable fluid reservoir that can be recruited to reinforce vascular volume should the need arise.

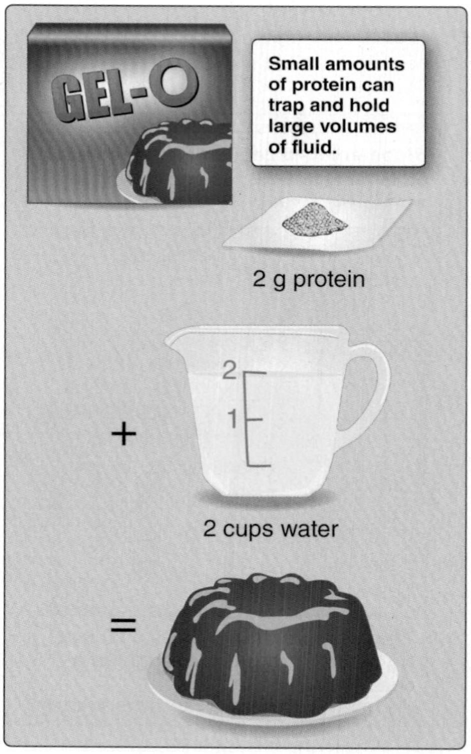

Figure 19.22
Contents of a gelatin dessert.

C. Lymphatic system

Blood loses several liters of fluid to the interstitium on a typical day, far more than the proteoglycan gel can absorb. It is the lymphatic system's responsibility to retrieve excess fluid and return it to the circulation, along with any proteins that may have escaped the vasculature.

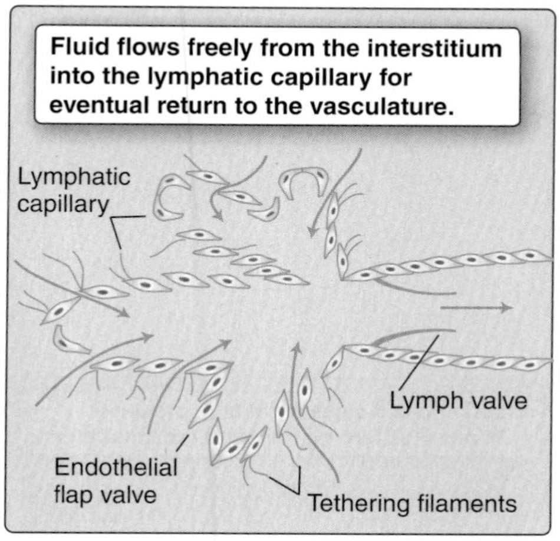

Figure 19.23
Lymph vessels.

1. **Structure:** Lymphatic capillaries are simple, blind-ended endothelial cell tubes that arise in the interstitium (Figure 19.23). Adjacent endothelial cells overlap to create flap valves that allow fluid influx but discourage retrograde flow, and protein filaments tethered to the cell margins maintain vessel patency

2. **Flow:** The larger lymphatic vessels are structurally similar to veins. They contain valves that help maintain unidirectional flow, and their walls contain smooth muscle layers that contract spontaneously in response to rising fluid pressure within. Contraction propels lymph onward and simultaneously creates a slight negative pressure within the lymphatic capillaries that allows them to suction fluid and protein from the interstitium. The lymphatics ultimately drain into the left and right subclavian veins.

D. Starling forces

Maintaining a balance between the forces governing fluid filtration and reabsorption from the vasculature is vital for continued health. Excess filtration causes edema, whereas an inability to recover filtered fluid can compromise LV preload and MAP. There are four principal **Starling forces** governing fluid movement, which are related in the **Starling law of the capillary**:

$$Q = K_f [(P_c - P_{if}) - (\pi_c - \pi_{if})]$$

where Q is net fluid flow across the capillary wall, K_f is a filtration coefficient that recognizes that total surface area and permeability of capillary beds varies from tissue to tissue, P_c is capillary hydrostatic pressure, P_{if} is interstitial fluid pressure, π_c is plasma colloid osmotic pressure, and π_{if} is interstitial colloid osmotic pressure.

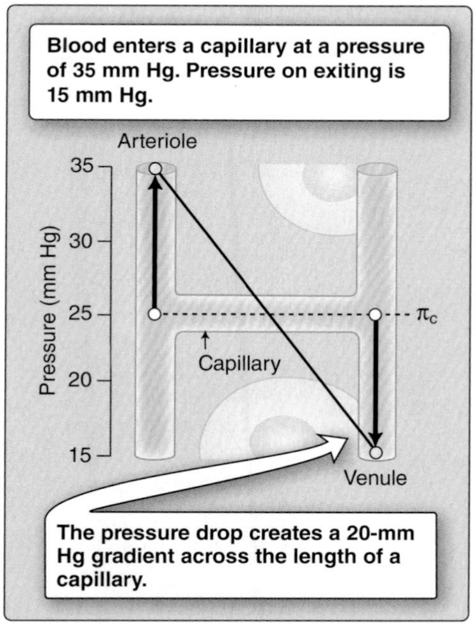

Figure 19.24
Hydrostatic pressure gradient across the length of a capillary. π_c = plasma colloid osmotic pressure.

1. **Capillary hydrostatic pressure:** Blood enters capillaries at a pressure of ~35 mm Hg. Blood exits capillaries and enters veins at a pressure of ~15 mm Hg (Figure 19.24). Mean capillary hydrostatic pressure (P_c) is typically closer to venous pressure than it is to arteriolar pressure but still is usually a positive pressure that drives fluid out of the capillary and into the interstitium.

2. **Plasma colloid osmotic pressure:** The main force opposing P_c is the osmotic pressure created by plasma proteins. Values for π_c typically average ~25 mm Hg.

3. **Interstitial fluid pressure:** P_{if} is typically between 0 and −3 mm Hg normally, due largely to lymphatic suctioning. The lymphatic system does have a finite capacity for fluid removal, however, and if fluid filters from the vasculature faster than it can be removed, the tissue swells. Tissues that are enclosed within skin, bone, or another physical boundary have limited opportunity for expansion, so P_{if} climbs and can become a significant force driving fluid back into the capillary.

4. **Interstitial colloid osmotic pressure:** The interstitium always contains a small amount of protein that creates an osmotic pressure of <5 mm Hg favoring fluid movement out of the capillary. The lymphatic system removes proteins along with fluid, but

Clinical Application 19.3: Lymphatic Filariasis

Lymphatic filariasis results from infection by one of three parasitic nematodes, most commonly *Wuchereria bancrofti* (>90% of total). Also known familiarly as **elephantiasis**, infection can cause gross disfiguration of the legs, arms, and genitalia. The infection is suggested to affect as many as 120 million individuals worldwide and is endemic in the developing regions of Asia, Africa, and South America. Infection occurs by way of a mosquito bite, which injects its host with larval nematodes. The larvae migrate to and establish themselves in lymphatic vessels, where they mature, mate, and breed to produce microfilariae (larvae). The presence of larvae within the lymphatic vessels interferes with drainage and causes pitting edema. Patients typically are infected in childhood but do not become symptomatic until adulthood after they have accumulated large numbers of parasites during repeated infections. Treatment involves prolonged (>1 year) dosing with antihelminthic drugs such as ivermectin.

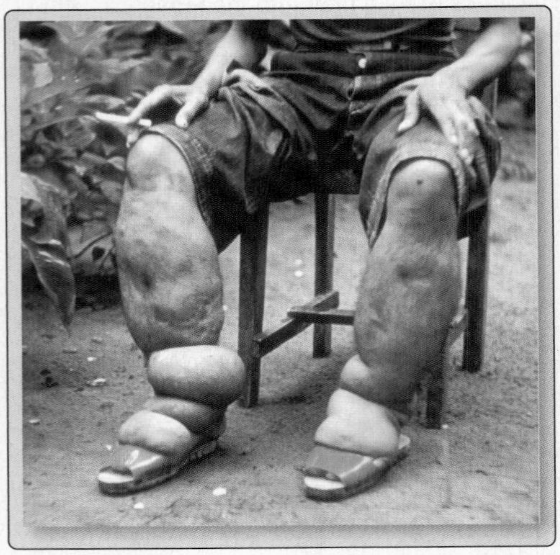

Elephantiasis.

capillaries continually leak proteins via larger fenestrations and intercellular clefts.

E. Starling equilibrium

The balance of forces governing fluid movement across the capillary wall is so perfect that net flow is close to zero in most tissues. Any excess filtrate is returned to the circulation by the lymphatics, which collect <4 L daily. This figure disguises the fact that an additional 16–18 L leaves and is then reabsorbed by capillaries daily. This fluid turnover occurs because of local imbalances between P_c and π_c. At the arteriolar end of the capillary, P_c exceeds π_c by ~10 mm Hg, causing fluid to filter from the capillary and enter the interstitium. By the time that blood has traversed the capillary, P_c has dropped below π_c. Absorption is now favored, and most of the filtered fluid is recovered. The near-precise balancing between filtration and reabsorption across the capillary wall has been termed the Starling equilibrium (Figure 19.25).

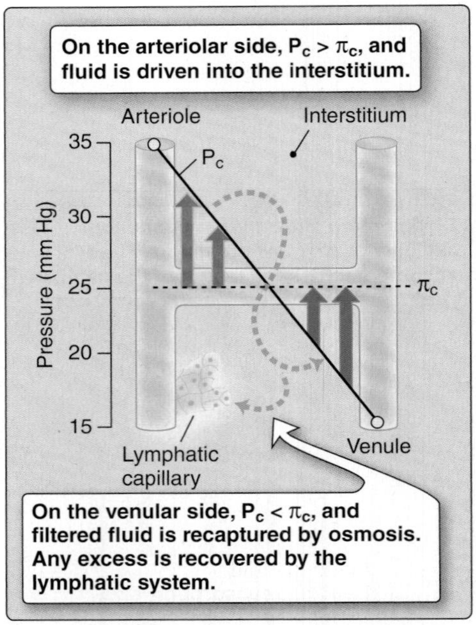
On the arteriolar side, Pc > πc, and fluid is driven into the interstitium.
Arteriole
Interstitium
Pc
πc
Pressure (mm Hg)
35
30
25
20
15
Lymphatic capillary
Venule
On the venular side, Pc < πc, and filtered fluid is recaptured by osmosis. Any excess is recovered by the lymphatic system.

Figure 19.25
The Starling equilibrium. P_c = capillary hydrostatic pressure; π_c = plasma colloid osmotic pressure.

VIII. STARLING EQUILIBRIUM DISTURBANCES

Because water traverses the capillary wall so easily, disturbances in the Starling equilibrium can rapidly cause large amounts of fluid to leave the bloodstream and enter the interstitium, or *vice versa*. This feature is put to good use in several aspects of cardiovascular design.

A. Renal circulation

Kidneys cleanse blood of surplus water, electrolytes, and various waste products. As shown in Figure 19.26, blood arrives at the glomerulus, a specialized renal capillary network, at pressures that greatly exceeds π_c (P_c = ~60 mm Hg; see 25·III·A). The pressure excess causes 180 L/day of protein- and cell-free fluid to filter into the Bowman space. Most of the

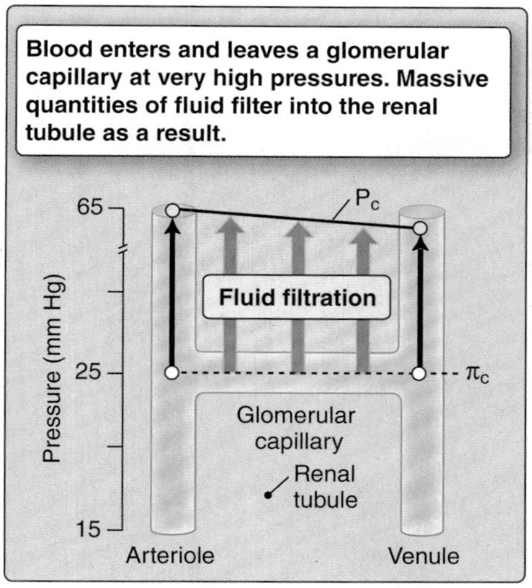
Blood enters and leaves a glomerular capillary at very high pressures. Massive quantities of fluid filter into the renal tubule as a result.
65
25
15
Pressure (mm Hg)
Pc
Fluid filtration
πc
Glomerular capillary
Renal tubule
Arteriole
Venule

Figure 19.26
Fluid filtration from renal glomerular capillaries. P_c = capillary hydrostatic pressure; π_c = plasma colloid osmotic pressure.

water and essential ions and other solutes are later recovered, leaving waste products concentrated in urine.

B. Pulmonary circulation

Mean pulmonary vascular pressures are far lower than in the systemic circulation. P_c for pulmonary capillaries averages 7 mm Hg (compare with ~25 mm Hg in the systemic circulation). π_{if} tends to be higher (~14 mm Hg), but the net driving force for fluid movement is still directed inward (Figure 19.27). This is advantageous because it ensures that the lungs remain relatively fluid free. Fluid accumulating in the pulmonary interstitium and alveolar sacs would interfere with O_2 and CO_2 exchange.

C. Decreased blood volume

The interstitium contains 10 L of fluid on average. This represents a readily accessible fluid reservoir that can be recruited by the vasculature to support CO when circulating blood volume decreases. Causes include hypohydration (hypohydration occurs when fluid intake is insufficient to replenish the amount lost through sweating, for example) and hemorrhage. The heart maintains MAP by drawing down the venous blood reservoir. Hydrostatic pressure on the venular side of the capillary falls as a result, causing the pressure gradient across the capillary to steepen. With π_c now dominating along much of the capillary's length, the Starling forces favor fluid recovery from the interstitium. Circulating blood volume rises as a result (Figure 19.28).

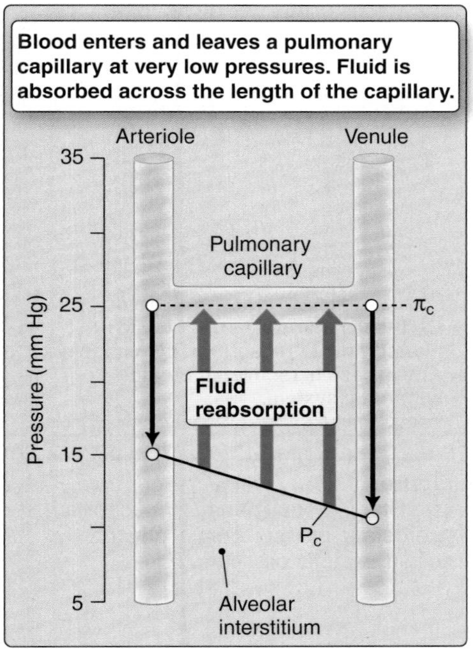
Blood enters and leaves a pulmonary capillary at very low pressures. Fluid is absorbed across the length of the capillary.
Arteriole
Venule
35
25
15
5
Pressure (mm Hg)
Pulmonary capillary
πc
Fluid reabsorption
Pc
Alveolar interstitium

Figure 19.27
Fluid absorption by pulmonary capillaries. P_c = capillary hydrostatic pressure; π_c = plasma colloid osmotic pressure.

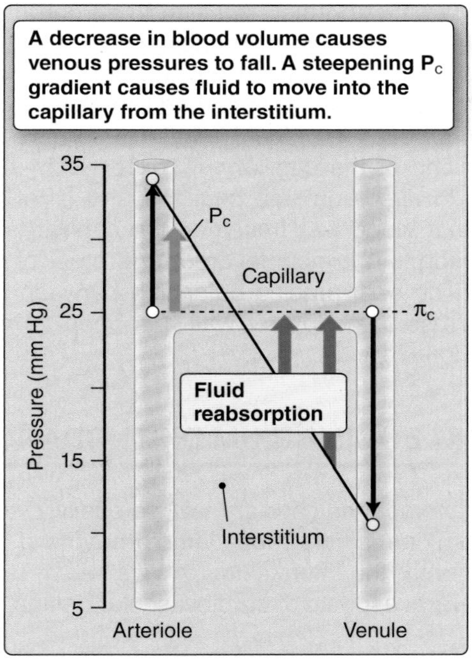
A decrease in blood volume causes venous pressures to fall. A steepening Pc gradient causes fluid to move into the capillary from the interstitium.
35
25
15
5
Pressure (mm Hg)
Pc
Capillary
πc
Fluid reabsorption
Interstitium
Arteriole
Venule

Figure 19.28
Use of Starling forces to recruit fluid from the interstitium. P_c = capillary hydrostatic pressure; π_c = plasma colloid osmotic pressure.

Clinical Application 19.4: Congestive Heart Failure

Edema is encountered frequently in a clinical setting, and there are many causes. One of the most common is **congestive heart failure**. Left ventricular failure presents as an inability to maintain arterial pressure at levels that ensure adequate tissue perfusion. The body compensates by retaining fluid (see 20·IV) to increase circulating blood volume and raise central venous pressure (CVP). The enhanced preload helps compensate for the failure-induced decrease in output through the Frank-Starling mechanism, but raising CVP flattens the hydrostatic pressure gradient across the capillary. The lymphatic system helps compensate for large quantities of fluid that now filter from the vasculature into the interstitium, but the tendency for edema (tissue **congestion**) is increased greatly. Initially, this may manifest as swelling of the feet and ankles, but in the later stages of failure, pulmonary edema may occur also.

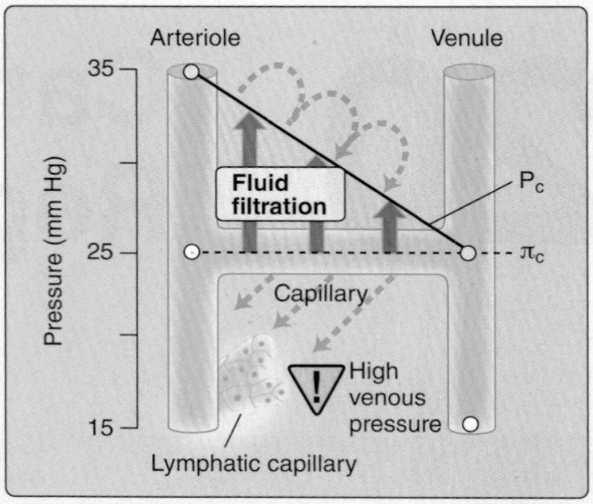

Excess fluid filtration causes edema during congestive heart failure.
P_c = capillary hydrostatic pressure;
π_c = plasma colloid osmotic pressure.

Chapter Summary

- Blood travels through several different classes of vessel once it leaves the heart.
- **Arteries** and **arterioles** are thick-walled, narrow-bore vessels designed to carry blood under high pressure. The smaller arteries and arterioles contract and relax in order to modulate flow to the capillaries.
- **Capillaries** are simple tubes of endothelial cells designed to facilitate exchange of materials between blood and tissues.
- **Veins** are low-pressure drainage vessels with thin walls and a high capacity that allows them to function as a blood reservoir.
- The heart pumps blood at high pressure to overcome several factors that resist flow. These are summarized in the **Poiseuille law**, which states that flow is proportional to the pressure gradient driving flow, vessel radius to the fourth power, and the inverse of vessel length and blood viscosity.
- Vessel radius is the main determinant of vascular resistance, which is why the smaller arteries and arterioles (**resistance vessels**) control flow so effectively through contraction and relaxation. Blood viscosity is largely a reflection of **hematocrit**.
- Estimates of flow resistance recognize that several factors can invalidate the Poiseuille law, including the fact that blood **viscosity changes** with flow velocity, the occurrence of **turbulence**, and vessel **capacitance**.
- Viscosity increases when red-cell density is high or flow rates are low, and cells are given an opportunity to aggregate. Aggregates increase resistance to flow. **Turbulent flow** is less efficient than **streamline flow** because kinetic energy is dissipated through chaotic motion. It usually only occurs in regions of the cardiovascular system where flow velocities are high such as when blood is forced through a heart valve. **Vessel capacitance** relates to a tendency to distend when filling pressure is increased. **Distensibility** allows arteries to store blood under pressure during systole and then release it to the capillary beds during diastole (**diastolic runoff**). The high capacitance of veins allows them to function as a blood reservoir for use when cardiac output increases or to help sustain arterial pressures when circulating blood volume decreases.
- Capillaries leak fluid continually through pores, fenestrations, and junctions between adjacent endothelial cells, driven by hydrostatic pressure. Some of this fluid is trapped by a **proteoglycan gel** that fills the interstitium. The gel gives up fluid to support circulating blood volume when needed. Most filtered fluid is returned to the circulation by osmotic forces associated with plasma proteins (**albumin** and **globulins**) that are trapped in the vasculature by virtue of their large size. Excess filtrate is returned to the vasculature by the **lymphatic system**.
- Disturbances in the forces that control fluid movement across the capillary wall (the **Starling equilibrium**) can have serious repercussions. Increases in venous pressure can raise net capillary hydrostatic pressure to the point where filtered fluid overwhelms the lymphatics. The result is **edema**.
- Decreases in capillary hydrostatic pressure allow fluid to be pulled from the interstitium. This provides a means of supporting cardiac output during a circulatory emergency.

20 Cardiovascular Regulation

Figure 20.1
Pressure is required to drive flow through vessels.

I. OVERVIEW

The volume of blood contained within the cardiovascular system represents only around 20% of its total capacity. Carrying a limited blood volume has clear energetic advantages, but such frugality requires that flow to the different organs be carefully metered with an eye to the needs of the system as a whole if cardiovascular catastrophe is to be avoided. The threat stems from the cardiovascular system's absolute dependency on pressure to drive flow. Just as a broken main in a city's infrastructure can deprive consumers of fresh water, uncontrolled flow through a low-resistance circuit (e.g., an exercising muscle) can cause perfusion pressure and flow through the vasculature to drop precipitously (Figure 20.1). Because some tissues (e.g., the brain and heart) are highly dependent on sustained arterial blood flow for normal function, loss of arterial pressure is a potentially fatal event. Thus, although the cardiovascular system includes flow regulators (**resistance vessels**) that can be operated by the tissues themselves if their nutrient needs increase, it also incorporates mechanisms by which the central nervous system (CNS) can monitor and maintain arterial pressure by reapportioning flow for the benefit of the system as a whole.

II. VASCULAR CONTROL

The cardiovascular system functions very much like a public water utility. Pumping stations ensure that there is always sufficient volume and pressure in the mains to supply consumer needs. Consumers typically do not leave taps running, but rather turn them on and then off as their need to bathe or fill kettles has been satisfied. Consumers recognize that fresh water is a valuable commodity and that the supplies are limited. Similarly, resistance vessels allow tissues (the consumers) to draw arterial blood from the cardiovascular system on the basis of metabolic need. Resistance vessels are located at key positions within the vasculature and, hence, they are subject to a multitude of controls. Four general control mechanisms can be recognized: **local**, **central** (neural), **hormonal**, and **endothelial**.

A. Local

All tissues are able to regulate their blood supply through local control of resistance vessels. Flow is tied to tissue need. Increased activity

causes the resistance vessels to dilate, and blood flow increases proportionately. If supply exceeds prevailing needs, the resistance vessels constrict reflexively. There are two broad classes of local control mechanisms: **metabolic** and **myogenic**.

1. **Metabolic:** Cells continually release various metabolic byproducts, including adenosine, lactate, K^+, H^+, and CO_2. When tissue activity increases, metabolites are produced in greater quantities, and interstitial concentrations rise (Figure 20.2). Resistance vessels lie close to the cells they serve and are sensitive to the appearance of these metabolites in the extracellular fluid. Some metabolites act directly on **vascular smooth muscle cells (VSMCs)**, whereas others act through endothelial cells, but all cause the VSMCs to relax and the vessel to dilate. Blood flow increases as a result, simultaneously providing the tissues with the nutrients they need and also carrying away metabolites (Figure 20.3). When activity ceases, metabolite concentrations fall, and a reflex vasoconstriction again matches flow with need.

2. **Myogenic:** Resistance vessels in many circulations constrict reflexively when intraluminal pressures rise. Contraction is mediated by stretch-activated Ca^{2+} channels in the VSMC membranes and may protect capillaries from surges in arterial pressure. Postural changes can cause sudden gravity-induced pressure spikes of >200 mm Hg in the pedal vasculature, for example.

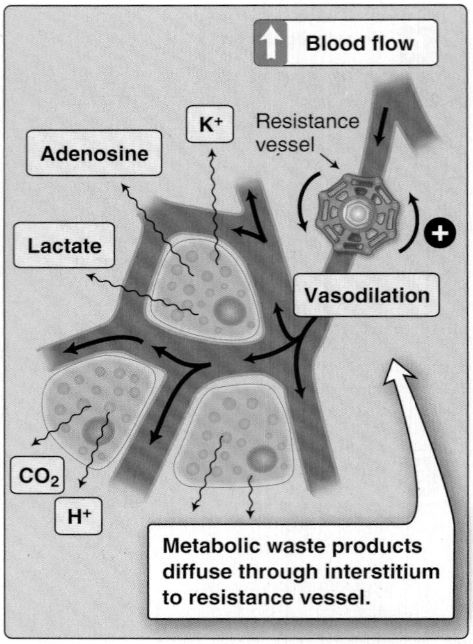

Figure 20.2
Metabolic control of resistance vessels.

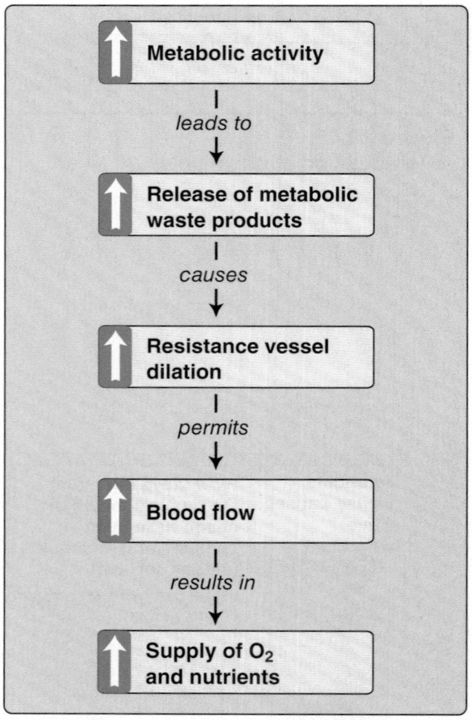

Figure 20.3
Metabolic control concept map.

B. Physiologic consequences

Local and myogenic control mechanisms operate independently of external influence, which frees the CNS from having to micromanage circulatory control. This autonomy manifests in several ways, including flow **autoregulation** and **hyperemia**.

1. **Autoregulation:** Autoregulation is the intrinsic ability of an organ to maintain stable blood flow in the face of changing perfusion pressures (Figure 20.4). If arterial pressure increases suddenly (e.g., during a pressure spike), flow increases also. Metabolites are washed away faster than they are produced, and the resistance vessels constrict reflexively. The myogenic response potentiates this effect, so that over the course of several seconds, flow rates are restored to levels approximating those observed prior to the pressure change. Conversely, a sudden drop in arterial pressure leads to reflex vasodilation, and flow is restored within a few seconds. When presented graphically (see Figure 20.4B), pressure extremes are seen to overwhelm the resistance vessels' autoregulatory powers, but flow remains relatively stable over a wide range of pressures.

2. **Hyperemia: Active hyperemia** is a normal vasodilatory response to increased tissue activity (Figure 20.5). Muscles also demonstrate **postexercise hyperemia**, a period of increased blood flow that persists even after activity has ceased. This reflects a time during which metabolite levels are still high, and the muscles are repaying oxygen debts that were accumulated during exercise (see 39·VI·C).

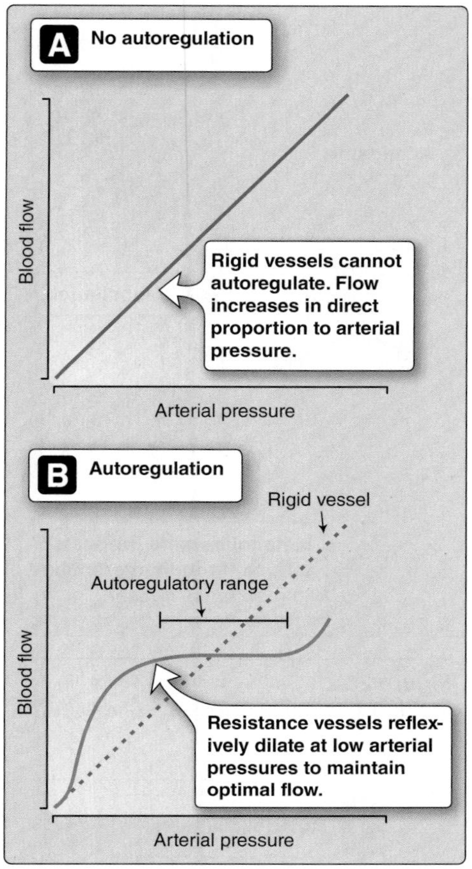

Figure 20.4
Autoregulation of blood flow.

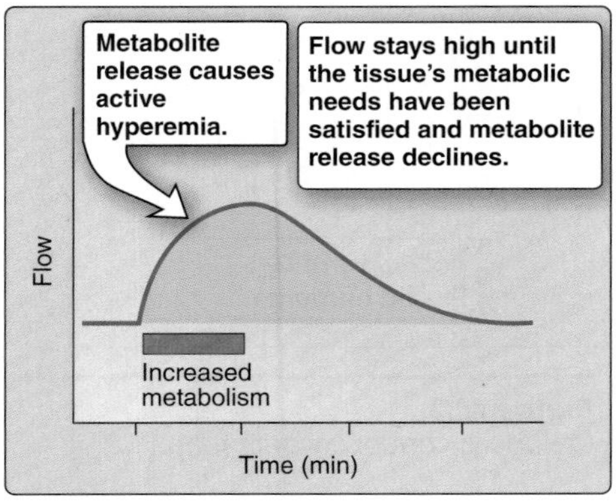

Figure 20.5
Active hyperemia.

C. Central control

All resistance vessels are innervated by the sympathetic nervous system (SNS). When arterial pressure falls, SNS nerve terminals release norepinephrine onto the VSMCs, causing them to contract. Contraction is mediated by α_1-adrenergic receptors via the inositol trisphosphate (IP_3) transduction pathway, causing Ca^{2+} release from the sarcoplasmic reticulum (see 1·VII·B·3).

D. Hormonal control

Many circulating hormones modulate resistance vessels, including **antidiuretic hormone** (**ADH**), **angiotensin II** (**Ang-II**), and **epinephrine**.

1. **Antidiuretic hormone:** ADH, also known as **arginine vasopressin**, is released from the posterior pituitary when tissue osmolarity rises or blood volume decreases (see 28·II·C). Its principal role is in extracellular fluid volume regulation through control of renal water retention, but if circulating levels are sufficiently high (e.g., during hemorrhage), it can vasoconstrict also. ADH affects VSMCs directly via ADH V_1 receptors.

2. **Angiotensin II:** Ang-II is a potent vasoconstrictor. It appears in the bloodstream when renal artery pressure falls, although sympathetic activity can trigger Ang-II release also (see Section IV·C below). Ang-II affects VSMCs directly via AT_{1A} receptors.

3. **Epinephrine:** Epinephrine is produced and released by the adrenal medulla during SNS activation. Its primary effect is to increase myocardial contractility and heart rate (HR), but it also binds to α_1-adrenergic receptors on VSMCs to potentiate the direct SNS-mediated vasoconstriction. Resistance vessels in some circulations (e.g., skeletal) express a β_2-adrenergic receptor that mediates epinephrine-mediated vasodilation. In the skeletal vasculature, this pathway may facilitate increased blood flow to muscles during "flight-or-fight" responses.

E. Endothelial control

The endothelial lining of resistance vessels acts as an intermediary for a number of vasoactive compounds, including **nitric oxide** (**NO**), **prostaglandins** (**PGs**), **endothelium-derived hyperpolarizing factor** (**EDHF**), and **endothelins** ([**ETs**] Figure 20.6).

1. **Nitric oxide:** NO is a potent vasodilator that acts on both arteries and veins. Also known as **endothelium-derived relaxing factor** (**EDRF**), it is synthesized by a constitutive endothelial *NO synthase* (*eNOS*, or *Type III NOS*), following a rise in intracellular Ca^{2+} concentrations. NO is a gas with a half-life of less than 10 s *in vivo*, meaning that its actions remain highly localized. It diffuses through the endothelial cell membrane to adjacent VSMCs and then binds to and activates a soluble *guanylyl cyclase*. Rising cyclic guanosine monophosphate (cGMP) levels cause *cGMP-dependent protein kinase* to phosphorylate and inhibit ***myosin light-chain kinase***. It also phosphorylates and increases the activity of a **SERCA** (**SR Ca^{2+} ATPase**) **pump** and causes intracellular Ca^{2+} concentration to fall. The net result is

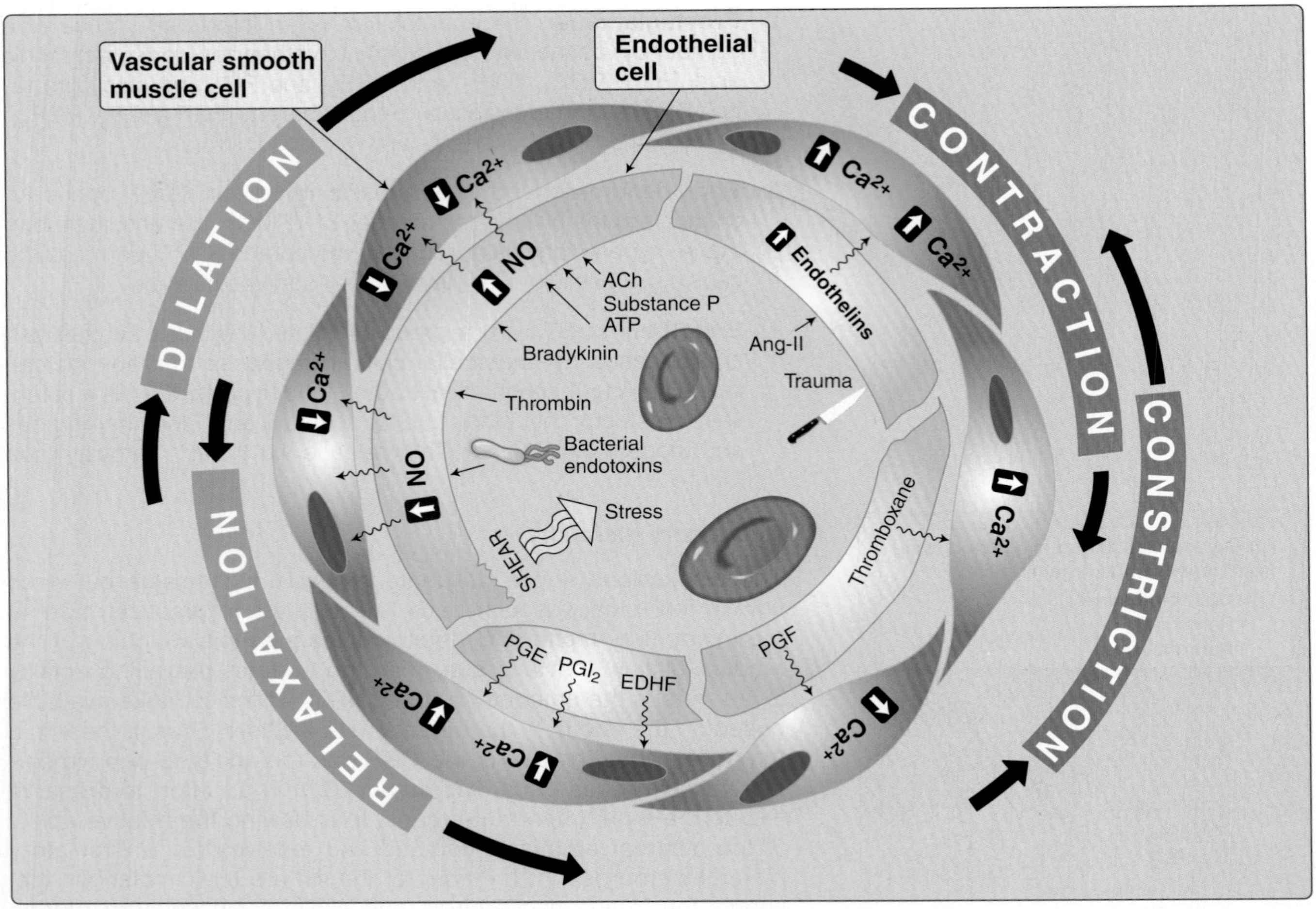

Figure 20.6
Role of endothelium in controlling resistance vessels. ACh = acetylcholine; Ang-II = angiotensin II; ATP = adenosine triphosphate; EDHF = endothelium-derived hyperpolarizing factor; NO = nitric oxide; PGE = prostaglandin E; PGF = prostaglandin F; PGI_2 = prostaglandin I_2.

vasodilation and increased blood flow. NO mediates the actions of many vasodilators, including **local modulators**; **neurotransmitters**, such as acetylcholine, substance P, and adenosine triphosphate; **bradykinin**; **thrombin**; flow-induced **shear stress**; and the bacterial **endotoxins** that cause **septic shock**.

Nitroglycerin and related nitrates are commonly used to relieve the pain of **angina pectoris**. Angina is caused by inadequate myocardial O_2 supply, typically because coronary arteries have become narrowed by plaque (**atherosclerosis**). Nitrates break down to release NO *in vivo*, causing arterial and venous vasodilation to lower ventricular afterload and preload, respectively. Reducing cardiac workload restores the balance between oxygen demand and supply and relieves the angina.

2. **Prostaglandins:** The endothelium is an important source of a number of vasoactive PGs, which it synthesizes from arachidonic acid. PGE (PGE_1, PGE_2, and PGE_3) and PGI_2 (prostacyclin) relax VSMCs in many vascular beds, whereas PGF (PGF_1, $PGF_{2\alpha}$, $PGF_{3\alpha}$) and thromboxane A_2 are vasoconstrictors.

3. **Endothelium-derived hyperpolarizing factor:** EDHF opens K^+ channels in the plasma membranes of VSMCs. The ensuing membrane hyperpolarization reduces membrane Ca^{2+} permeability, causing intracellular Ca^{2+} levels to fall and vasodilation.

4. **Endothelins:** ETs are a group of related peptides synthesized and released by endothelial cells in response to many factors, including Ang-II, mechanical trauma, and hypoxia. ET-1 is a potent vasoconstrictor that binds to ET_A receptors on VSMC membranes and triggers intracellular Ca^{2+} release via the IP_3 pathway (see Figure 20.6).

During exercise, blood is directed away from nonessential organs:
• Gastrointestinal
• Renal
• Reproductive
Redirected flow is used to support essential organs:
• Brain
• Heart
• Skeletal muscle

Figure 20.7
Blood flow redistribution during exercise.

F. Circulatory hierarchy

The discussions above delineate several mechanisms by which flow to individual vascular beds is regulated. In practice, most of the moment-by-moment control involves a simple weighing of the amount of flow that a tissue needs to support prevailing activity levels versus the amount that the CNS is willing to make available based on the needs of the organism as a whole. Thus, if there is a threat to arterial pressure, the CNS has the ability to deprive certain vascular beds of cardiac output (CO) in an effort to preserve flow to the more important organs. In reviewing the relative ability of the different organs to demand and receive flow, a circulatory hierarchy emerges. At the head of the list are the circulations that supply the brain, myocardium, and skeletal musculature (during exercise). Here, local control mechanisms dominate, and central controls have little or no effect. At the bottom of the hierarchy are organs such as the gut, kidneys, and skin that receive blood flow under optimal conditions, but flow is sacrificed if there is a need to preserve arterial pressure.

The rationale for such a hierarchy can best be understood in evolutionary terms. One of the greatest challenges that the cardiovascular system faces involves intense physical activity, of the kind that might be required for chasing after prey or running away from predators (Figure 20.7). Maintaining optimal flow to the three organs at the top of the hierarchy is critical for sustaining such activities. Meeting the challenge requires that the CNS temporarily divert flow away from organs at the bottom of the hierarchy in order to support the needs of the skeletal musculature (see 39·V). Fortunately, these same organs also have relatively low metabolisms so that their sacrifice does not threaten survival.

III. ARTERIAL PRESSURE CONTROL

Survival of the individual requires that pressure be maintained in the arterial system at all times. Because all organs of the body have the ability to demand increased flow, they could easily cause arterial pressure to collapse if their arteriolar supply vessels were not strictly regulated.

The cardiovascular system includes two distinct pathways for monitoring and maintaining arterial pressure. The first is fast activating and helps compensate for short-term pressure changes. Known as a **baroreceptor reflex** (**baroreflex**), it employs simple feedback loops that include **sensors** to monitor pressure and flow, an **integrator** to compare current with preset pressure values, and **effector mechanisms** that make any necessary adjustments. A second, slow-activating system manipulates **mean arterial pressure** ([**MAP**] see 19·IV·A) through changes in circulating blood volume by modifying renal function (discussed in Section IV below).

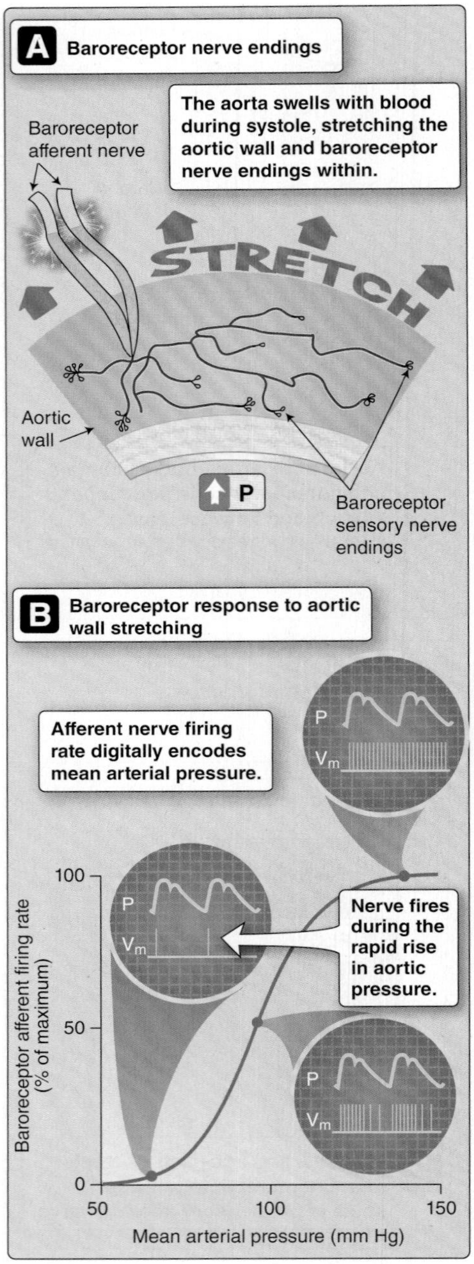

Figure 20.8
Arterial baroreceptors. P = pressure; V_m = membrane potential.

A. Sensors

Three main groups of sensors provide the integrator (located in the brainstem medulla) with information about pressure and flow in the cardiovascular system: **high-pressure arterial baroreceptors** located in the **aortic arch** and **carotid sinus**, low-pressure **cardiopulmonary receptors**, and **chemoreceptors**.

1. **Arterial baroreceptors:** The aortic and carotid baroreceptors are the primary means of detecting changes in MAP. They monitor pressure indirectly by responding to arterial wall stretch.

    a. **Anatomy:** Baroreceptors are clusters of bare sensory nerve endings buried within the elastic layers of the aorta and the carotid sinus (Figure 20.8A). Information from the former is relayed to the brain via sensory afferents traveling in the **aortic nerve** and the **vagus nerve** (**cranial nerve** [**CN**] **X**). Afferents from the carotid sinus travel in the **sinus nerve**, which joins with the **glossopharyngeal nerve** (**CN IX**) en route to the brainstem.

    b. **Function:** In the absence of stretch, the baroreceptors are inactive. When MAP increases, the walls of the aorta and carotid sinus expand, and the embedded nerve endings are stretched. The nerves respond with graded receptor potentials. If the degree of deformation is sufficiently high, the receptor potentials trigger spikes in the sensory nerve (see Figure 20.8B). Baroreceptors are especially sensitive to *changes* in pressure, responding to the sharp rise in pressure that occurs during rapid ejection with strong depolarization and a train of high-frequency spikes. During reduced ejection and diastole, the depolarization abates and spike frequency drops to a new steady-state level that reflects diastolic pressure.

    c. **Sensitivity:** Stretch-sensitivity varies from one nerve ending to the next, thereby allowing for responsiveness over a wide pressure range (see Figure 20.8B). The carotid baroreceptors have a response threshold of around 50 mm Hg and saturate at 180 mm Hg. The aortic baroreceptors operate over a range of 110–200 mm Hg.

2. **Cardiopulmonary receptors:** A second set of baroreceptors is found in low-pressure regions of the cardiovascular system. They provide the CNS with information about the "fullness" of the vascular system, and their principal role is in modulating renal

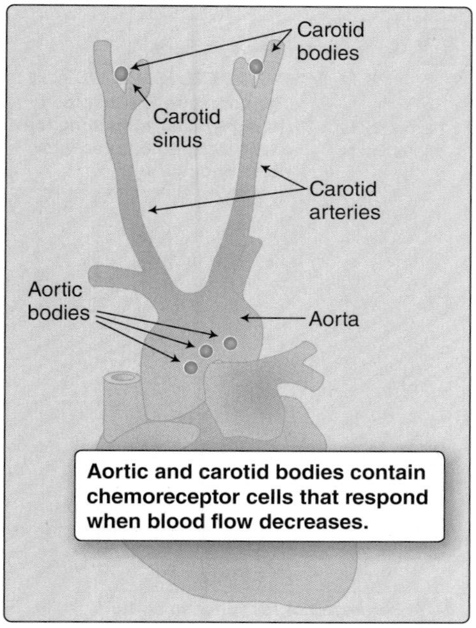

Figure 20.9
Peripheral chemoreceptors.

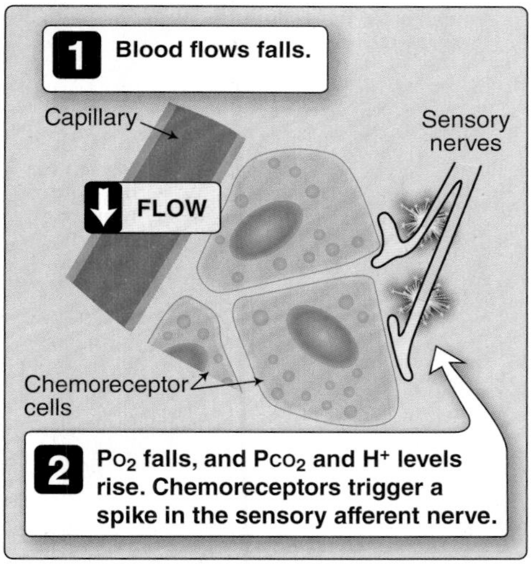

Figure 20.10
Chemoreceptor function.

function. However, because fullness correlates with ventricular preload, they also have a role in maintaining MAP.

a. **Anatomy:** The receptors are similar to those found in the arterial system: bare sensory nerve endings embedded in walls of the vena cavae, the pulmonary artery and vein, and the atria. They relay information back to the CNS via the vagal nerve trunk.

b. **Function:** Atria contain two functionally distinct populations of baroreceptors. **A receptors** respond to tension that develops in the atrial wall during contraction. **B receptors** are sensitive to atrial wall stretching during filling. B receptors are also involved in raising HR when central venous pressure (CVP) is high, a response known as the **Bainbridge reflex**.

3. **Chemoreceptors:** Chemoreceptors monitor local metabolite levels, which reflect adequacy of perfusion pressure and flow.

a. **Anatomy:** There are two groups of chemoreceptors, one located in the brainstem medulla, the other peripheral. Peripheral chemoreceptors are discrete, highly vascularized glomus cell clusters lying close to the aortic arch and carotid sinus (the **aortic** and **carotid bodies**, respectively) as shown in Figure 20.9. Sensory fibers from the aortic bodies travel in the vagus nerve, whereas nerves from the **carotid bodies** travel with the sinus nerve and join the glossopharyngeal trunk en route to the medulla.

b. **Function:** Peripheral chemoreceptors activate when arterial O_2 levels fall (<60 mm Hg) or when Pco_2 or H^+ levels rise (Pco_2 > 40 mm Hg or pH <7.4) as shown in Figure 20.10. Medullary chemoreceptors are sensitive to the pH of brain interstitial fluid, which is dependent on arterial Pco_2. The chemoreceptors seem designed to monitor lung function and are principally involved in respiratory control (see 24·III), but hypercapnia and acidosis can also reflect low perfusion pressures.

B. Central integrator

The sensory afferents converge on the **medulla oblongata**. Here, arterial pressures are compared with preset values and decisions then made about the nature and intensity of a compensatory response.

1. **Control centers:** The medulla contains a collection of nuclei that together comprise a **cardiovascular center**. Some cells in this area cause vasoconstriction when active and are known as the **vasomotor center**. Another group comprises a **cardioacceleratory center**, which increases HR and myocardial inotropy when activated. A third group slows HR when active (the **cardioinhibitory center**). The three control centers are interlinked extensively so as to generate a unified response to changes in arterial pressure (Figure 20.11).

2. **Feedback loops:** Arterial pressure is a product of CO and **systemic vascular resistance** (**SVR**) (MAP = CO × SVR), and the control centers adjust both parameters simultaneously. Control is exerted using simple feedback loops (Figure 20.12). Sensory

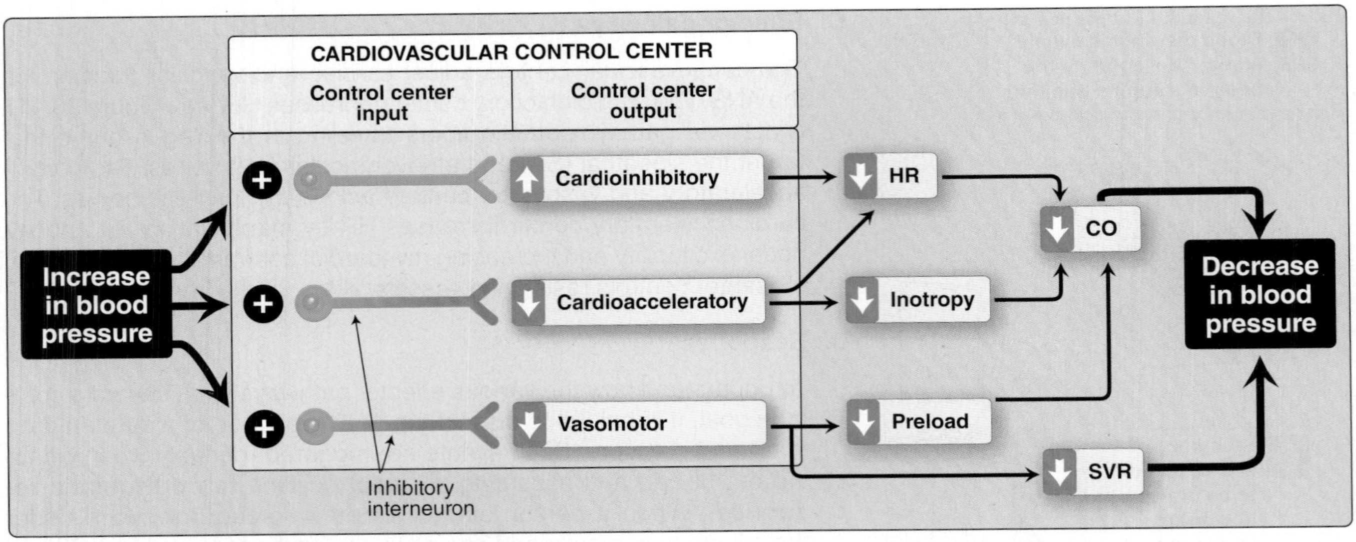

Figure 20.11
Organization and output of the medullary cardiovascular control center. CO = cardiac output; HR = heart rate; SVR = systemic vascular resistance.

afferents project to the **nucleus tractus solitarus** within the medulla and synapse with interneurons that, in turn, project to the three control centers (see Figure 20.11). The sensory afferents are all excitatory, but the interneurons may be either excitatory (glutamatergic) or inhibitory (GABAergic). The cardioinhibitory center receives inputs from excitatory interneurons, so this area is excited when MAP is high (a **positive feedback loop**). The cardioacceleratory and vasomotor centers are innervated by inhibitory interneurons. When MAP is high, they suppress the activity of the nerves that they innervate. Inhibition is required because the cardioacceleratory and vasomotor centers control sympathetic nerves that are tonically active in the absence of external input. This arrangement creates a **negative feedback loop** between MAP and SNS output.

3. **Integration with other central and peripheral pathways:** There are numerous inputs to the cardiovascular center from other regions of the brain and periphery.

   a. **Brainstem:** The **brainstem** also contains a respiratory center that controls breathing. The cardiovascular and respiratory centers work in close cooperation with each other to maintain optimal arterial P_{O_2} and P_{CO_2}.

   b. **Hypothalamus: Hypothalamic control centers** help coordinate vascular responses to changes in external and internal body temperatures.

   c. **Cortex: Cortical control centers** account for changes in cardiovascular performance induced by emotions (fainting or anticipatory changes associated with exercise, for example).

   d. **Pain centers: Pain centers** can precipitate profound changes in blood pressure by manipulating cardiovascular center output.

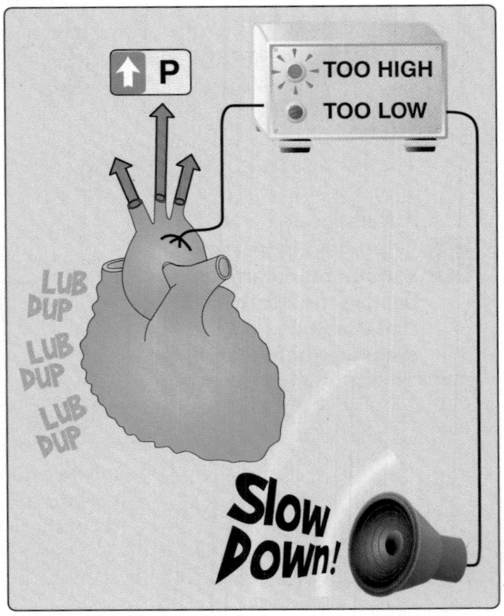

Figure 20.12
Cardiovascular feedback loop.

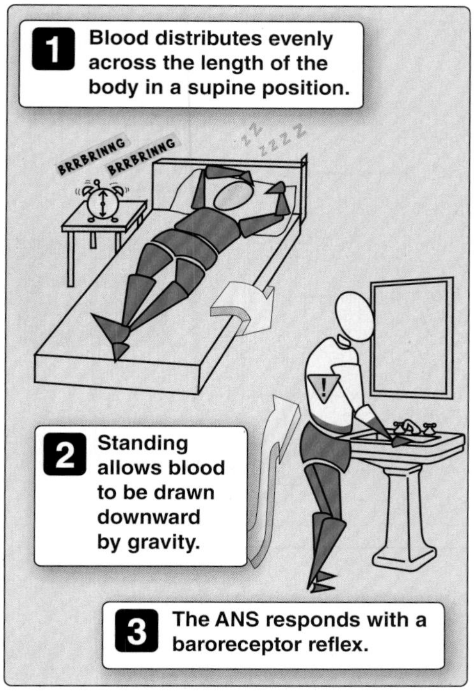

Figure 20.13
Orthostatic reflex. ANS = autonomic nervous system.

C. Effector pathways

The cardiovascular centers adjust cardiac and vascular function via the ANS. The cardioinhibitory center depresses HR (see Figure 20.11). It acts via parasympathetic fibers traveling in the vagus nerve that target the sinoatrial (SA) and atrioventricular (AV) nodes. The cardioacceleratory and vasomotor centers act via sympathetic nerves. The cardioacceleratory center increases HR by manipulating SA and AV nodal excitability and increasing myocardial contractility. The vasomotor center controls resistance vessels, veins, and adrenal glands.

D. Response

To understand how the various effector pathways work toward a common goal, it is helpful to dissect the ANS response to a sudden drop in arterial pressure. Such events are triggered routinely when climbing out of bed and assuming an erect position (an **orthostatic response**). When a person stands, blood is forced downward under the influence of gravity and begins to pool in the legs and feet (Figure 20.13). Venous return (VR), CVP, and ventricular preload all fall as a consequence. Stroke volume (SV), CO, and MAP all follow (Figure 20.14). The baroreceptor nerve endings are stretched to a lesser degree and spike frequency in the afferent arm of the reflex pathway falls. Within the cardioinhibitory center, the loss of excitatory input causes withdrawal of parasympathetic output to the SA node and myocardium. Within the two pressor regions of the cardiovascular

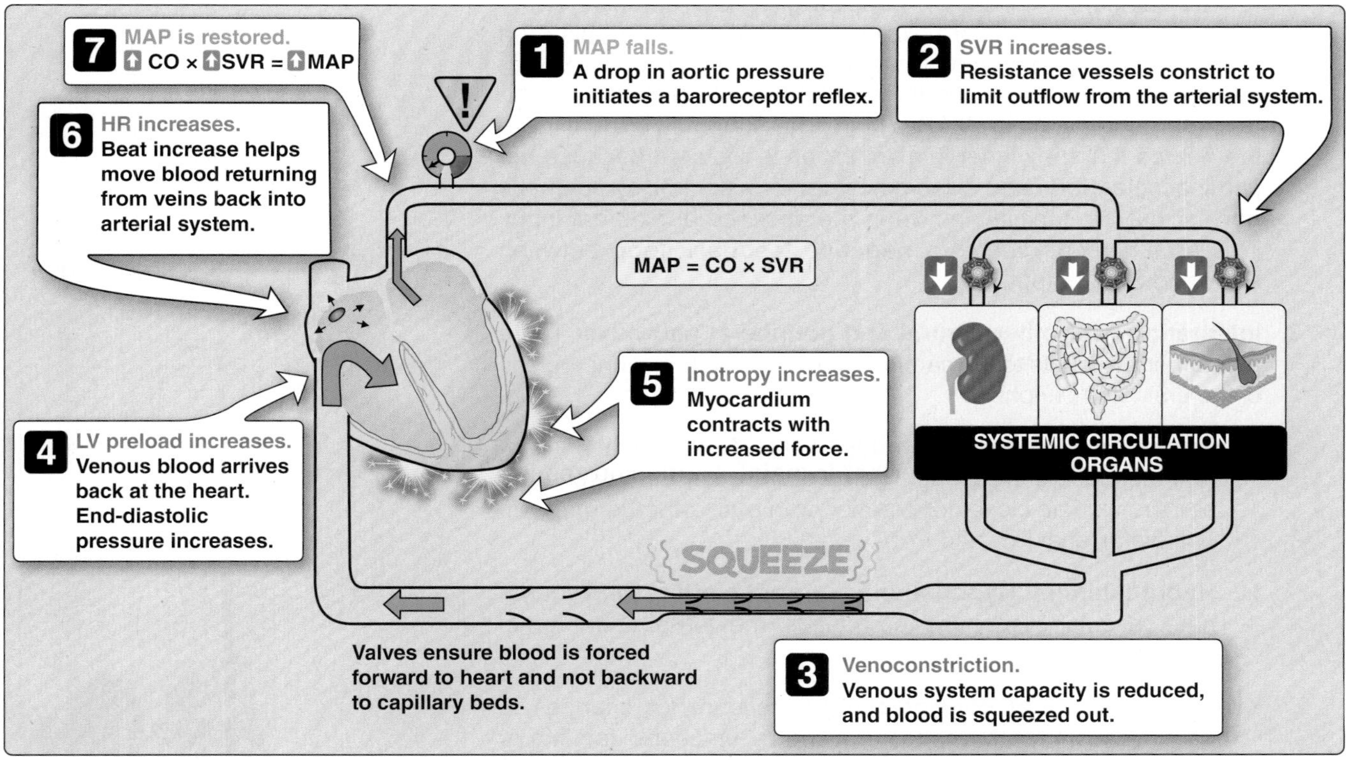

Figure 20.14
Baroreceptor reflex. CO = cardiac output; HR = heart rate; LV = left ventricle; MAP = mean arterial pressure; SVR = systemic vascular resistance.

center, decreased baroreceptor output weakens inhibitory interneuron influence on SNS effector pathways. With the brakes removed, sympathetic nerves now drive up arterial pressure by constricting resistance vessels and veins, increasing myocardial contractility, and increasing HR.

1. **Resistance vessels:** All resistance vessels are innervated by SNS nerve terminals that cause vasoconstriction when active. SVR increases, and outflow from the arterial tree is reduced as a result.
2. **Veins:** Veins and larger venules contract when the SNS is active, reducing the capacity of the venous reservoir and causing intravenous pressures to rise. Valves ensure that this pressure increase forces blood forward toward the heart. Here, it increases left ventricular preload (end-diastolic pressure) and increases SV on the next beat.
3. **Myocardium:** SNS activation increases myocardial contractility by increasing intracellular Ca^{2+} release. The myocardium now works with increased efficiency and contributes to the increased SV caused by preloading. The SNS also speeds myocardial relaxation rate by increasing the rate at which Ca^{2+} is released from the contractile machinery and then removed from the sarcoplasm. Faster relaxation times make more time available for preloading during diastole and thereby facilitate a concurrent increase in HR.
4. **Nodes:** SA and AV nodes are innervated by both parasympathetic and sympathetic nerve terminals, both of which are active at rest. A drop in MAP simultaneously causes parasympathetic activity to

Clinical Application 20.1: Orthostatic Hypotension

Orthostatic or postural hypotension is a common complaint of older adults. The condition describes a 10–20 mm Hg drop in arterial pressure that accompanies standing from a sitting position. The resulting decrease in cerebral perfusion pressure leads to momentary lightheadedness, dizziness, weakness, or darkening of the vision. In extreme cases, patients may not be able to rise from a supine position without fainting (**syncope**). Orthostatic hypotension can be caused by low circulating blood volume, but aging is also associated with a decrease in baroreceptor sensitivity caused by hardening of the arteries (**arteriosclerosis**). Hardening results from deposition of collagen and other fibrous materials in the arterial wall, which lowers its compliance. Atherosclerosis is a form of arteriosclerosis associated with lipid deposition and plaque formation, which thickens the arterial walls and encroaches on the arterial lumen. The baroreceptor reflex relies on aortic and carotid artery distensibility in order to transduce changes in arterial pressure. Atherosclerosis prevents the sensory nerve endings from detecting the drop in pressure that accompanies translocation of blood to the lower extremities upon standing, and compensatory responses are thus delayed for several seconds.

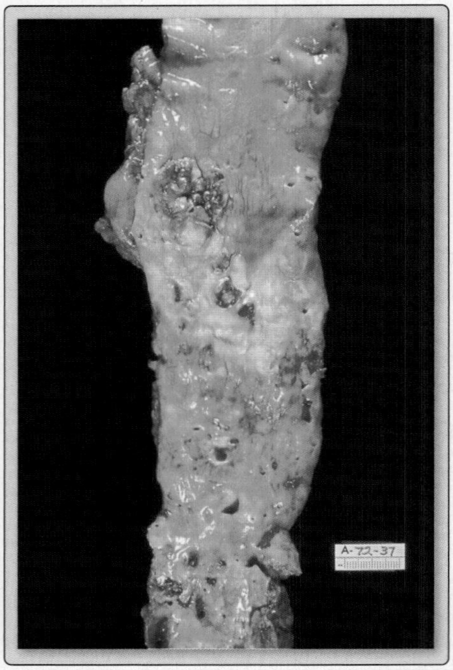

Intima of a sclerotic aorta scarred by lesions.

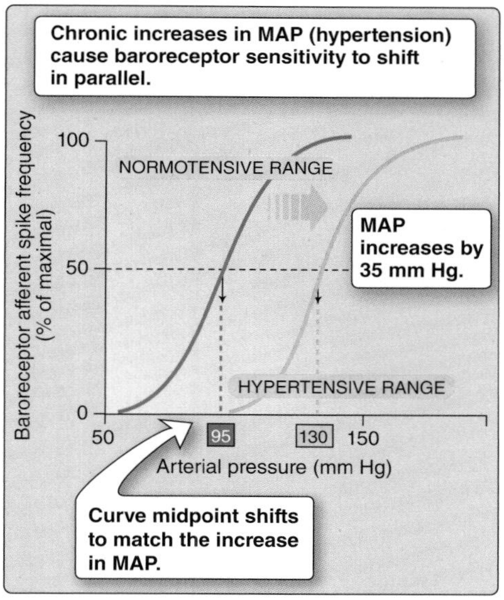

Figure 20.15
Shifts in baroreceptor sensitivity. MAP = mean arterial pressure.

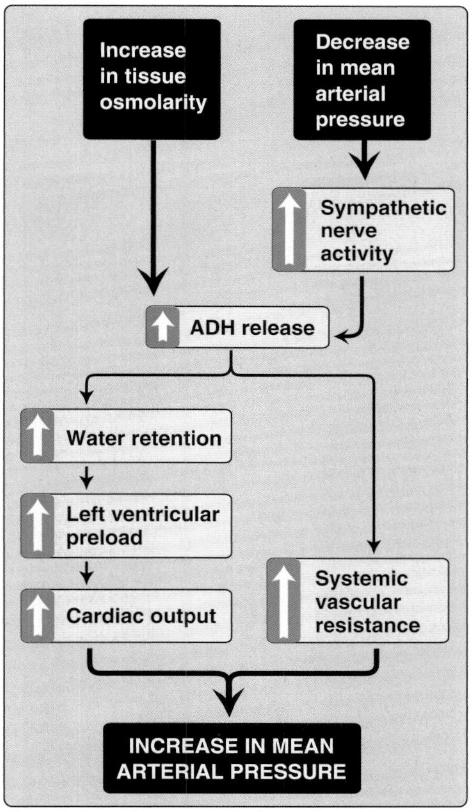

Figure 20.16
Antidiuretic hormone (ADH) effects on blood pressure.

be withdrawn and sympathetic activity to increase, accelerating the rate at which SA nodal cell membrane potential slides toward the threshold for action potential formation (phase 4 depolarization; see 17·IV·C). HR and CO are both increased as a result.

5. **Adrenal glands:** SNS activation causes the adrenal glands to secrete epinephrine into the circulation. Epinephrine binds to the same receptors on blood vessels and the myocardium as does neurally derived norepinephrine.

E. Baroreceptor reflex limitations

The baroreceptor reflex is an extremely effective short-term control mechanism, but pressure changes that are sustained for more than a few minutes cause a parallel shift in system sensitivity (Figure 20.15). The advantage to the shift is that it allows the system to retain responsiveness over a wide range of pressures, even if MAP is increased to a level that would previously have saturated the system. The disadvantage is that although the reflex is ideal for moment-by-moment adjustments in arterial pressure, it cannot be used for long-term pressure control (>1–2 days).

IV. LONG-TERM CONTROL PATHWAYS

A drop in arterial pressure activates the baroreceptor reflex outlined above, but it also initiates pathways that require 24–48 hr to become fully effective. These pathways converge on the kidney, which is responsible for long-term control of blood pressure through regulation of vascular fullness (**circulating blood volume**). Because blood is principally water, this necessarily involves regulation of water output and intake, but it also requires regulation of Na^+ levels because this is the ion that governs how water partitions between the intracellular and extracellular compartments. These concepts are discussed in more detail in 3·III·B and 28·II and III.

A. Water output

Water output is controlled by ADH, a peptide that is synthesized by the hypothalamus and then transported to the posterior pituitary for release. It stimulates water reabsorption by the renal collecting tubule and collecting ducts. At high concentrations, ADH also increases SVR by constricting resistance vessels (Figure 20.16). Several sensors and pathways regulate ADH release including **osmoreceptors**, baroreceptors, and Ang-II.

1. **Osmoreceptors:** The brain contains a number of regions that have the potential to monitor plasma osmolality, including areas surrounding the third ventricle in close proximity to the hypothalamus (see 7·VII·C). Tissue osmolarity is a reflection of total body water and salt concentration. When osmolarity exceeds 280 mOsm/kg, the receptors cause ADH to be released into the circulation.

2. **Baroreceptors:** A decrease in circulating blood volume causes CVP to fall, which is sensed by the cardiopulmonary receptors. Loss of preload also causes arterial pressure to fall and triggers a baroreceptor reflex. The CNS cardiovascular control centers

respond by increasing sympathetic activity and promoting ADH release.

3. **Angiotensin II:** Activating the ***renin*-angiotensin-aldosterone system** (**RAAS**) causes circulating Ang-II levels to rise. The list of target organs for Ang-II includes the hypothalamus, where it stimulates ADH release.

B. Water intake

Water enters the body along with food, but the bulk of liquid intake occurs through drinking, driven by thirst. The sensation is triggered by decreasing blood volume and arterial pressure, suggesting a prominent role for the cardiovascular control center.

C. Sodium output

Osmoreceptors control water retention and excretion, but they sense the "saltiness" of body fluids rather than water *per se*. Thus, if tissue osmolality remains high, they will urge retention of water regardless of total accumulated volume. The primary determinant of circulating blood volume is Na^+ concentration, which is regulated through RAAS, as described below (Figure 20.17).

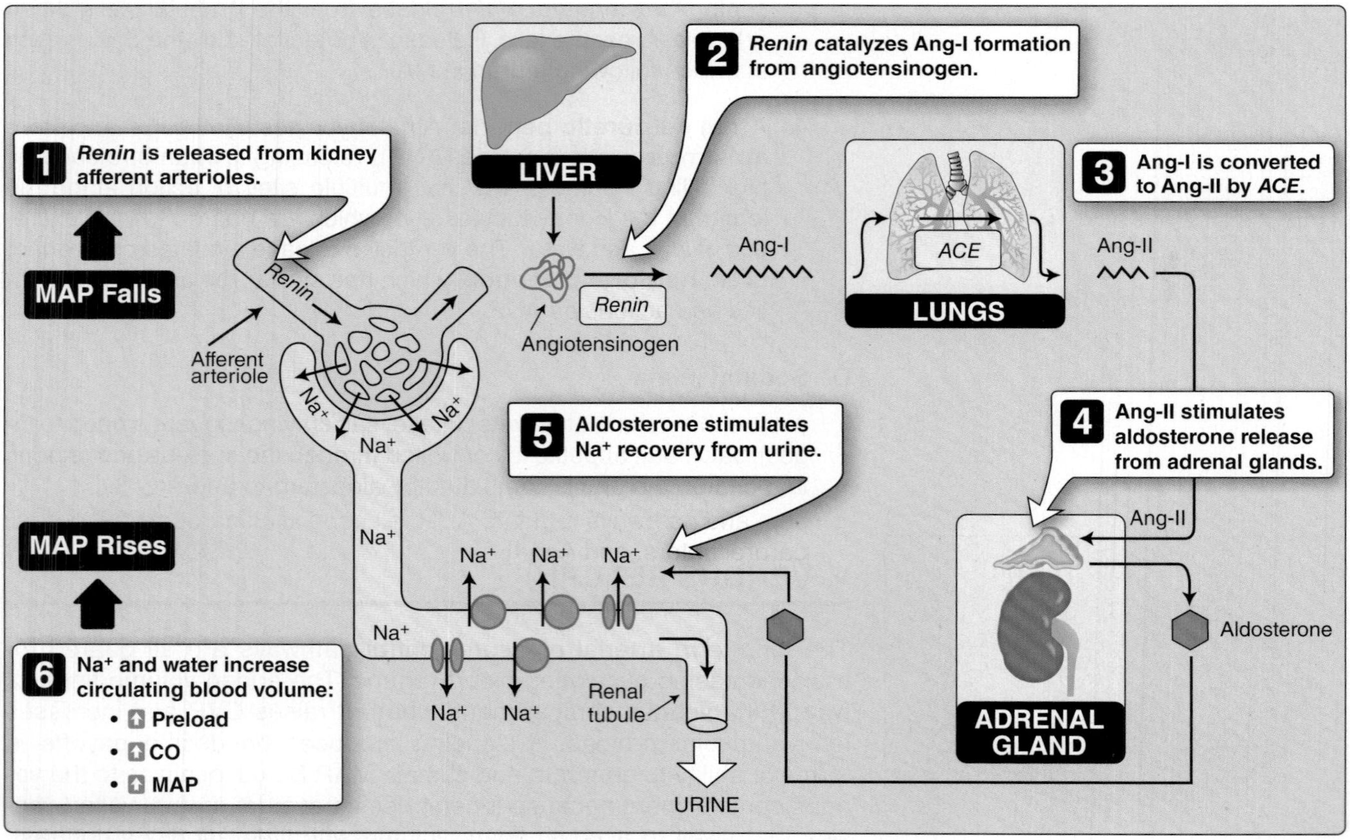

Figure 20.17
Renin-angiotensin-aldosterone system (RAAS). *ACE* = *angiotensin-converting enzyme*; Ang-I = angiotensin I; Ang-II = angiotensin II; CO = cardiac output; MAP = mean arterial pressure.

1. ***Renin*-angiotensin-aldosterone system:** *Renin* is a proteolytic enzyme synthesized by granular cells in the wall of glomerular afferent arterioles (see 25·IV·C). The cells form a part of the juxtaglomerular apparatus (JGA), which senses and regulates Na^+ recovery by the renal tubule. When the JGA is stimulated appropriately, it releases *renin* into the bloodstream. Here, *renin* breaks down **angiotensinogen** (a circulating plasma protein formed in the liver), to release **angiotensin I**. The latter serves as a substrate for ***angiotensin-converting enzyme (ACE)***. *ACE* is expressed by many tissues, including the kidney, but conversion largely occurs during transit through the lungs. The product is Ang-II, which constricts resistance vessels, stimulates ADH release from the posterior pituitary, stimulates thirst, and promotes aldosterone release from the adrenal cortex.

2. **Aldosterone:** Aldosterone targets **principal cells** in the renal collecting tubule epithelium (see 27·IV·B). It has multiple actions, all of which promote recovery of Na^+ and osmotically obligated water from the tubule. Aldosterone acts by modifying expression of genes that encode Na^+ channels and pumps, which is why it takes up to 48 hours for this pressure control pathway to become maximally effective.

3. ***Renin*:** The afferent arteriole of the renal glomerulus is a baroreceptor that triggers *renin* release from the granular cells when arteriolar pressure falls. Release is potentiated by the SNS, which activates following a drop in MAP.

4. **Atrial natriuretic peptide:** Atrial myocytes synthesize and store **atrial natriuretic peptide** (**ANP**), releasing it when stretched by high filling volumes. ANP has multiple sites of action along the length of the kidney tubule, all of which are geared toward excretion of Na^+ and water. The ventricles release a related compound, **brain natriuretic peptide**, which has similar release characteristics and actions as ANP.

D. Sodium intake

Just as thirst stimulates water intake, salt craving triggers a need to ingest NaCl. Salt appetite is controlled through the nucleus accumbens in the forebrain and is stimulated by aldosterone and Ang-II.

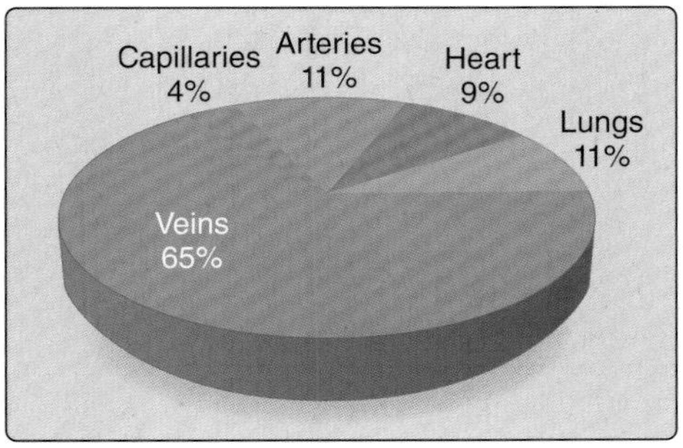

Figure 20.18
Blood distribution within the cardiovascular system.

V. VENOUS RETURN

The long-term arterial pressure-control pathways are all geared toward increasing circulating blood volume. The added volume finds its way to the venous compartment, where it raises CVP and increases left ventricular preload. Preloading produces handsome payoffs in terms of ability to generate and sustain MAP. Blood localizes to the venous compartment because it comprises a system of thin-walled vessels that swell to accommodate volume with little effort. By contrast, the arterial system comprises a series of high-pressure, narrow-bore tubes that have a very limited capacity (~11% of total blood volume) as shown in Figure 20.18. Capillaries are numerous but hold even less

Clinical Application 20.2: Coronary Artery Bypass Grafts

Coronary heart disease (CHD) is a leading cause of death in the western hemisphere. Patient usually present with angina or myocardial infarction caused by coronary artery occlusion, usually as a result of atherosclerosis. Treatment options may include revascularization with coronary artery bypass grafting (CABG or, familiarly, "cabbage") surgery. Although the walls of veins are much thinner and less muscular than those of arteries, they do have considerable strength. This makes it possible to use them as surrogate coronary supply vessels during CABG. Surgery involves removing a donor vein (usually the greater saphenous leg vein) and grafting a segment between the aorta and a site distal to the occlusion. A grafted vein is reversed in orientation to allow blood to flow freely through the valves.

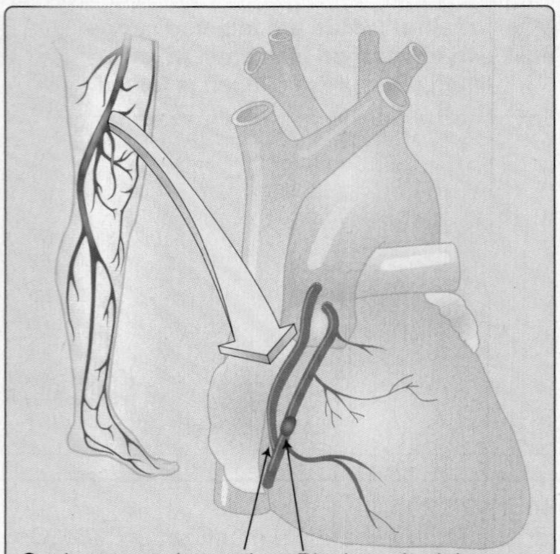
Saphenous vein used to bypass blockage
Blockage in right coronary artery

Use of veins for coronary artery bypass grafting.

blood than do arteries (~4% of total). The cardiopulmonary system also has a very limited capacity. The venous system typically contains >65% of total blood volume (the **unstressed blood volume**), creating an invaluable reservoir that can be mobilized by venoconstriction for use elsewhere should the need arise. However, the features that make veins a good reservoir also allow them to trap blood and limit VR under certain circumstances. When VR is reduced, CO is reduced also. Thus, any consideration of how the cardiovascular system functions as a unit must include an understanding of the role and limitations of the venous system.

A. Venous reservoir

Veins have thin walls, which allow these vessels to collapse easily when intraluminal pressures fall (Figure 20.19). Increasing venous pressures by a few mm Hg then causes veins to swell with minimal resistance. Once the system reaches capacity, the vessels have to be stretched to accommodate additional volume. The pressures required to do this are not achieved physiologically.

B. Venoconstriction

Vein walls contain layers of VSMCs that are innervated by and contract during sympathetic activation. However, whereas arteries are capable of contracting down to the point of occlusion, venoconstriction is limited by a vein's unique microanatomy.

1. **Vein anatomy:** The VSMCs contained within the walls of veins are attached in series with collagen filaments. The filaments are folded and coiled in a relaxed vein, slowly unfolding and tensing as the vessel fills (Figure 20.20). The filaments effectively

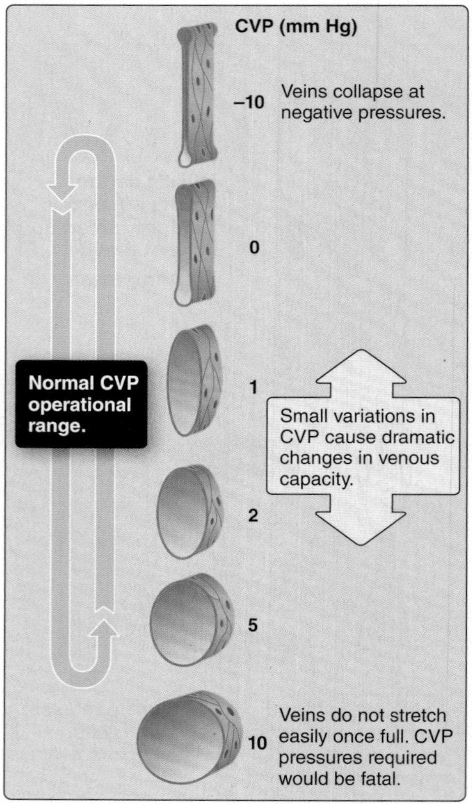
CVP (mm Hg)
–10
Veins collapse at negative pressures.
0
1
Normal CVP operational range.
Small variations in CVP cause dramatic changes in venous capacity.
2
5
10
Veins do not stretch easily once full. CVP pressures required would be fatal.

Figure 20.19
Effects of filling pressure on venous capacity. CVP = central venous pressure.

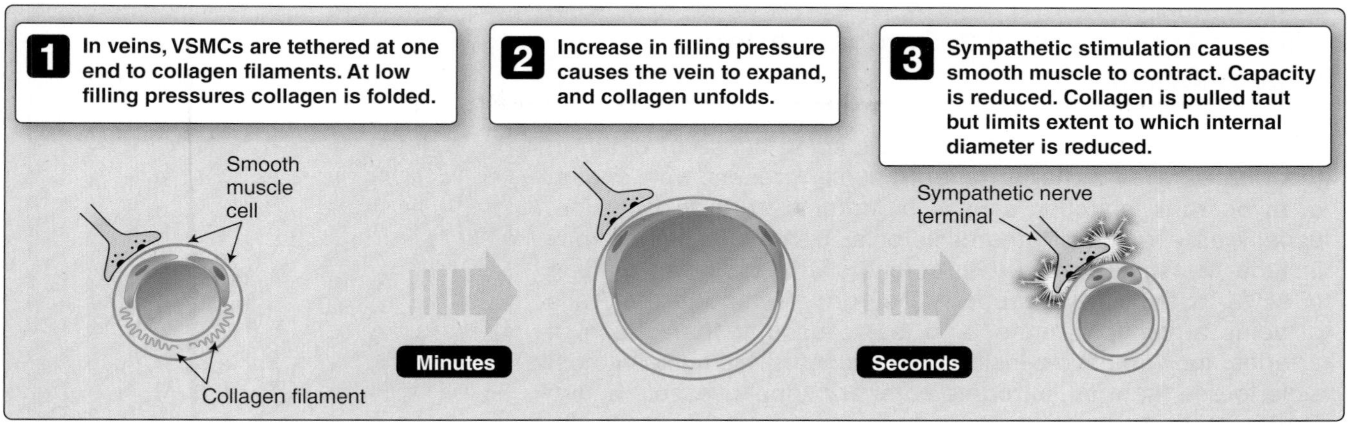

Figure 20.20
Venous reservoir and its mobilization. VSMC = vascular smooth muscle cell.

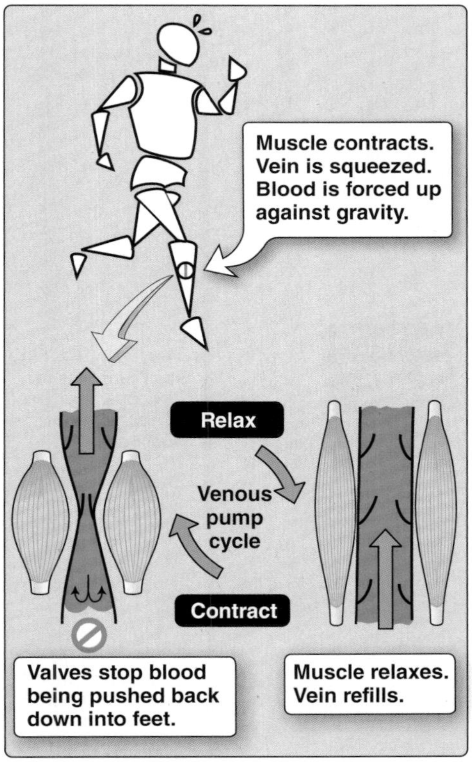

Figure 20.21
Venous pump.

limit the extent to which internal diameter can be reduced by venoconstriction.

2. **Effects of venoconstriction:** Venoconstriction has three principal effects: it mobilizes the blood reservoir, reduces overall capacity, and decreases transit time. It has minimal effect on flow resistance.

   a. **Mobilization:** Venoconstriction raises venous pressure by a few mm Hg and drives blood out of the reservoir. Valves ensure that blood is forced forward toward the heart, where it preloads the left ventricle and increases CO through the Frank-Starling mechanism.

   b. **Capacity:** Venoconstriction decreases the internal diameter of veins and thereby decreases system capacity. Blood that had previously resided in veins is ultimately transferred to the capillary beds that supply active tissues.

   c. **Transit time:** Reducing system capacity reduces the amount of time that blood takes to traverse the system and thereby increases the rate at which it can be reoxygenated and forwarded to active tissues.

   d. **Resistance:** The venous system remains a low-resistance pathway even after sympathetic activation, and there is no significant effect on flow resistance.

C. Venous pump

The high capacity of veins and their tendency to swell in response to even mild filling pressures means that large volumes of blood can easily become trapped in the lower extremities under the influence of gravity. Venoconstriction can reduce system capacity, but, in the absence of any additional motive force, the resultant loss of VR can eventually threaten CO and the ability to maintain arterial pressure. Blood pooling in veins is normally prevented by a **venous**

pump. Whenever skeletal muscles contract, they compress the blood vessels running between the fibers (**extravascular compression**). Intravenous pressures are low, so compression readily collapses the veins and expels their contents. Valves ensure that resultant flow is in the direction of the heart (Figure 20.21). Rhythmic contraction and relaxation of leg muscles effectively pumps blood upward against gravity, simultaneously "milking" blood from the pedal vasculature and ensuring continued VR.

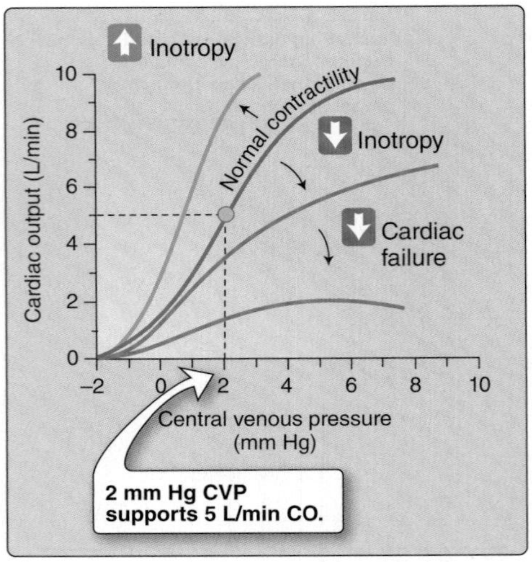

Figure 20.22
Cardiac function curves. CO = cardiac output; CVP = central venous pressure.

D. Cardiac output and venous return

The example above demonstrates that CO is limited by the rate at which blood traverses the vasculature. Movement through the vasculature is, in turn, dependent on CO. Thus, a true understanding of how the cardiovascular system functions *in vivo* requires that the interdependence between CO and VR be appreciated.

1. **Return supports output:** The preload-dependence of CO is defined by the **cardiac function curve** (Figure 20.22). Increases in left-ventricular filling pressure increase CO through length-dependent activation, and filling pressure is dependent on CVP. Changes in ventricular inotropy modify this relationship: Positive inotropes shift the curve upward and to the left, whereas negative inotropes shift the curve downward and to the right.

2. **Output creates return:** Quantifying how CO affects CVP requires that the heart and lungs be replaced with an artificial pump whose output can be controlled (Figure 20.23A). Prior to turning the pump on, normal circulating blood volume (5 L) must be restored. The vasculature stretches when accommodating this much blood, creating a pressure of approximately 7 mm Hg (see Figure 20.23B) known as **mean circulatory filling pressure** (**MCFP**). MCFP is

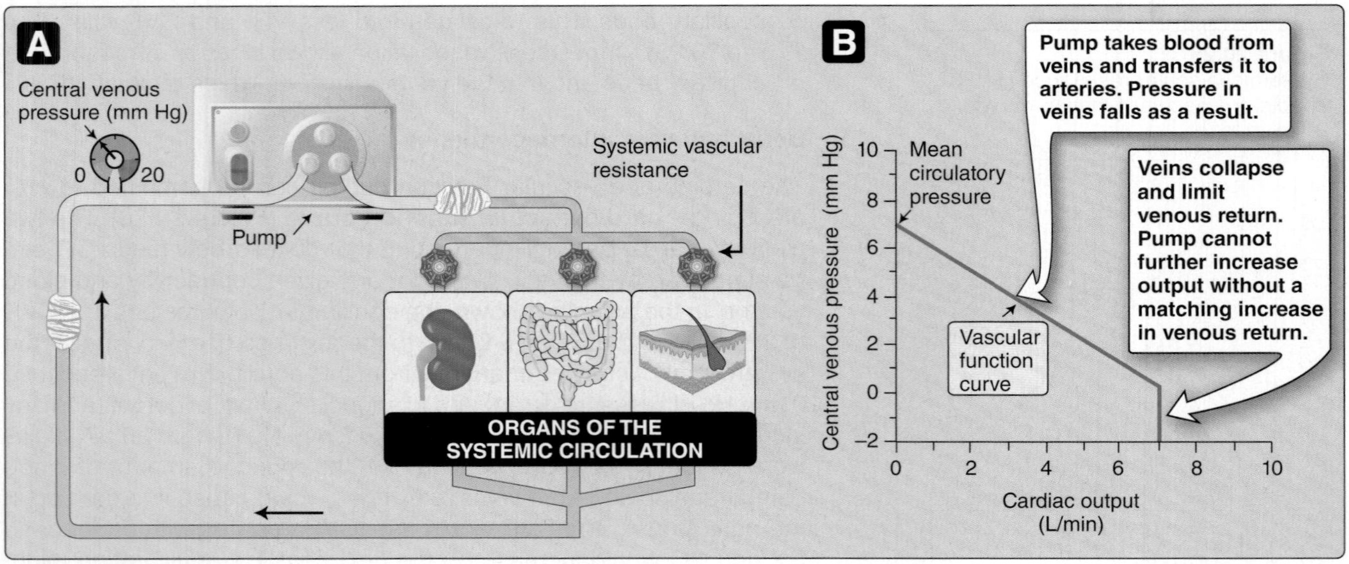

Figure 20.23
Dependence of central venous pressure on cardiac output.

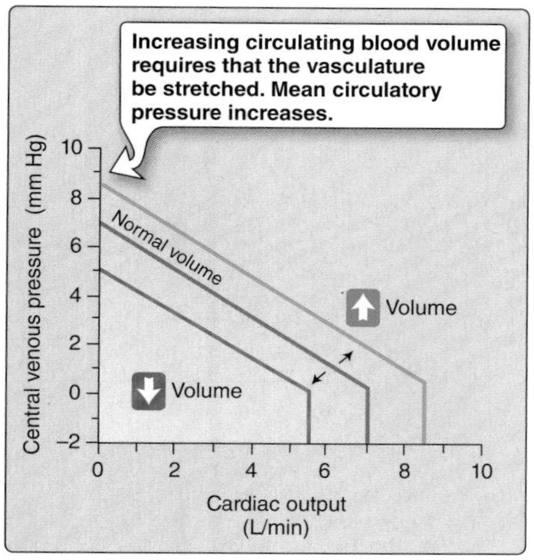

Figure 20.24
Vascular function curves.

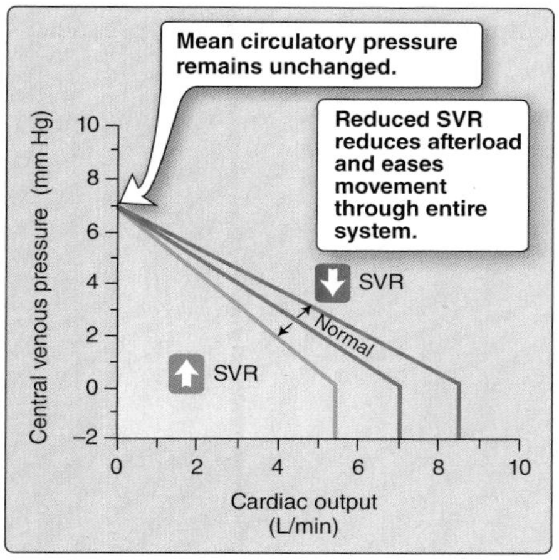

Figure 20.25
Systemic vascular resistance (SVR) effects on vascular function curves.

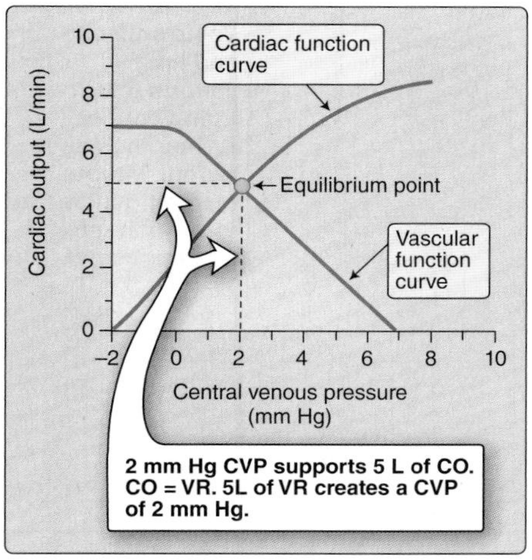

Figure 20.26
Cardiovascular function curve. CO = cardiac output; CVP = central venous pressure; VR = venous return.

defined as the pressure that exists in the vasculature when the heart is arrested, and all parts of the system have come into equilibrium. When the pump is turned on, it translocates blood from the veins to the arteries. Because the arterial compartment has a relatively small volume and outflow is limited by resistance vessels, translocation generates significant pressure within the arterial system. It simultaneously causes CVP to fall because blood is being withdrawn. Driving the pump faster causes CVP to fall further until it finally becomes negative (see Figure 20.23B). At this point, the great veins collapse and limit any additional increases in CO. The plot shown in Figure 20.23B is known as a **vascular function curve**.

3. **Circulating blood volume:** The vascular function curve is dependent on circulating blood volume (Figure 20.24). If blood volume increases, then MCFP necessarily increases also because the vasculature stretches to a greater degree to accommodate the extra volume. When the pump is turned on, CVP falls as before, but, because overall pressure in the system is higher, collapse of the large veins is delayed. Conversely, if circulating blood volume is decreased, then MCFP decreases, and collapse of the great veins occurs at lower output levels.

4. **Venous capacity:** Venoconstriction and venodilation caused by changes in SNS activity produce similar effects to changes in circulating blood volume. Initiating a baroreflex reduces venous system capacity and MCFP rises. Venodilation reduces system capacity, and MCFP falls.

5. **Systemic vascular resistance:** Constriction and relaxation of resistance vessels has little or no affect on MCFP because the contribution of the small arteries and arterioles to overall vascular capacity is small. Changes in SVR do impact CVP, however. When resistance vessels constrict, they reduce flow through the capillary beds. This translates into less VR, and CVP falls (Figure 20.25). Conversely, vasodilation allows blood to surge through capillary beds and into the venous system, which raises CVP.

E. Heart and vein interdependence

The cardiac and vascular function curves can be combined to create a single **cardiovascular function curve** (Figure 20.26). The two plots overlap at an equilibrium point that defines how much CO can be supported by the vasculature for any given contractility and blood volume. In the example shown, the equilibrium point resides at a CVP of 2 mm Hg and a CO of 5 L/min. In the absence of any changes, the system cannot stray permanently from this equilibrium point because 2 mm Hg of pressure is required to support 5 L/min of output, and any increase in CO would drop CVP below 2 mm Hg (Figure 20.27). If HR were suddenly slowed to reduce CO, the reduced amount of blood being translocated from veins to arteries would cause it to dam up in the right atrium, and CVP would rise. CVP equates with preload, so SV and CO would increase on the next beat. Re-equilibration might require several beats to accomplish, but eventually CO and CVP would settle back to 5 L/min and 2 mm Hg.

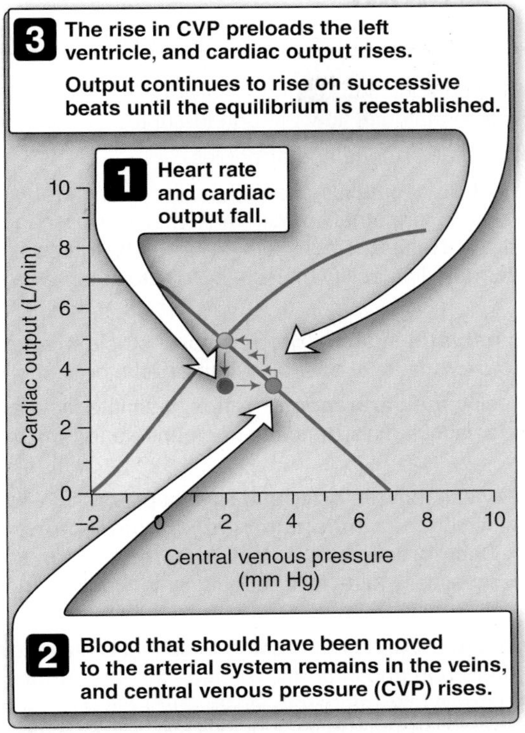

Figure 20.27
Changes in output cannot be sustained away from equilibrium.

F. Moving the equilibrium point

Sustained increases and decreases in CO require that the cardiac function curve and/or vascular function curve be modified to establish a new equilibrium point. The former is accomplished through changes in inotropy, the latter through changes in circulating blood volume.

1. **Inotropy:** Increasing myocardial inotropy allows the ventricle to pump out more blood on every stroke, even though CVP falls as a consequence. A new equilibrium point is created, as shown in Figure 20.28. Increases in inotropy are typically seen during exercise, for example. Conversely, an infarcted myocardium translocates less blood from the right atrium to the arterial system on every stroke. The new equilibrium point settles at a higher CVP.

2. **Circulating blood volume:** Transfusing a subject with blood, for example, raises CVP and increases preload, permitting a higher CO in the absence of any change in inotropy. Conversely, hemorrhage reduces preload, and the equilibrium point shifts to a lower value for CO (see Figure 20.28). Similar effects can be achieved acutely with venoconstriction and venodilation, respectively.

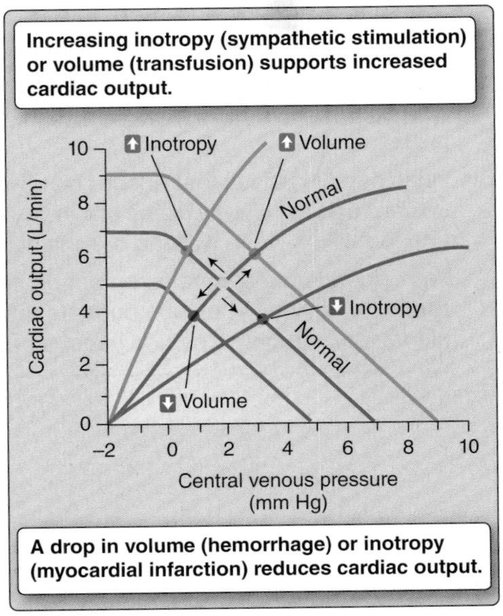

Figure 20.28
Cardiovascular function curves.

Chapter Summary

- The capacity of the cardiovascular system greatly exceeds its contents. Blood flow to individual organs must be carefully metered to maintain arterial pressure at sufficiently high levels to maintain flow through the entire system.
- Blood flow to the tissues is regulated by **resistance vessels** (small arteries and arterioles).
- All tissues can demand additional flow to support increased activity. Control is effected locally through release of **metabolites** and **paracrine factors** (e.g., endothelin, prostacyclin, endothelium-derived hyperpolarizing factor, and nitric oxide) which cause resistance vessels to constrict or dilate. Resistance vessels are also **stretch sensitive**, allowing for reflex vasoconstriction during arterial pressure surges. The ability of tissues to match their own blood supply to prevailing needs is called **autoregulation**.
- Resistance vessels are also innervated by the **sympathetic nervous system**, which vasoconstricts when active. Sympathetic innervation allows the autonomic nervous system to override local controls when arterial pressure is threatened.
- The autonomic nervous system maintains arterial pressure using the **baroreceptor reflex**, a simple feedback pathway that monitors arterial pressure and adjusts cardiac output and vascular resistance to compensate for any changes that might occur.
- There are three main groups of pressure sensors. The primary sensors are **arterial baroreceptors**, located in the wall of the aorta and carotid sinus. Secondary pressure sensors include the **cardiopulmonary baroreceptors**, which are located in the walls of atria and in the pulmonary circulation. **Chemoreceptors** residing in aortic and carotid bodies in the periphery and within the central nervous system monitor pressure indirectly through flow-induced changes in H^+, CO_2, and O_2 levels.
- The various sensors relay information to a group of **cardiovascular control centers** located in the brainstem. The **cardioinhibitory** and **cardioacceleratory centers** control heart rate and inotropy, whereas the **vasomotor center** controls blood vessels and adrenal glands. When blood pressure is low, the cardioacceleratory center increases heart rate and myocardial contractility. The vasomotor center increases vascular resistance and also forces blood out of the veins by **venoconstriction**. Blood pressure rises as a consequence.
- The baroreceptor reflex is used for immediate and short-term adjustments in arterial pressure. Long-term pressure control involves modulating total body water and Na^+. The kidneys play a central role in both cases.
- The kidneys' ability to retain water is controlled by **antidiuretic hormone** (**ADH**). ADH is released from the **posterior pituitary** in response to increases in tissue osmolarity and sympathetic activation.
- Na^+ retention by the kidney is controlled locally. Baroreceptors located in the walls of renal glomerular afferent arterioles stimulate ***renin*** release when blood pressure is low. *Renin* proteolyses **angiotensinogen** to form **angiotensin I**, which is then converted to **angiotensin II** (**Ang-II**) by ***angiotensin-converting enzyme***. Ang-II promotes Na^+ recovery from the kidney tubule, stimulates antidiuretic hormone release, and constricts resistance vessels.
- Na^+ and water retention increases **circulating blood volume**. The additional volume collects in the venous system and raises **central venous pressure** (**CVP**). Increased CVP increases left ventricular preload, cardiac output, and arterial pressure.
- Veins contain valves that ensure blood is forced toward the heart during venoconstriction. Valves also help move blood upward against gravity. Veins are easily compressed by skeletal muscles as they contract. Rhythmic compression of leg and feet veins when walking or running, effectively milks blood out of the lower extremities (**venous pump**).
- The dependence of cardiac output on **venous return** can be presented graphically in the form of **cardiovascular function curves**. They demonstrate that cardiac output can be raised or lowered physiologically through modulation of inotropy, but only if central venous pressure is sufficient to support the induced level of output.

Special Circulations

21

I. OVERVIEW

The peripheral vasculature serves a variety of organs whose functional diversity has required specialization of circulatory design and control (Figure 21.1). The heart and brain have minimal anaerobic capabilities, which makes them highly dependent on their blood supply for normal activity. Coronary and cerebral vascular control is, thus, dominated by local regulatory mechanisms, which facilitate accurate matching of O_2 supply with tissue demand. By contrast, splanchnic circulatory control is dominated by central regulatory mechanisms. The organs that comprise the gastrointestinal (GI) system require significant amounts of blood to support their digestive and absorptive functions when active (see Figure 21.1). Digestion involves secreting copious amounts of fluid into the GI tract lumen. This fluid is derived from blood that must be supplied by the splanchnic vasculature. The digestive organs are able to communicate their need for increased blood flow through local control mechanisms, but the central nervous system (CNS) retains the ability to shut off splanchnic blood flow completely if there is an urgent need for blood elsewhere in the cardiovascular system. Regulation of blood supply to skeletal muscles is unusual in that the relative dominance of central and local controls changes according to the needs of the muscle relative to those of body as a whole.

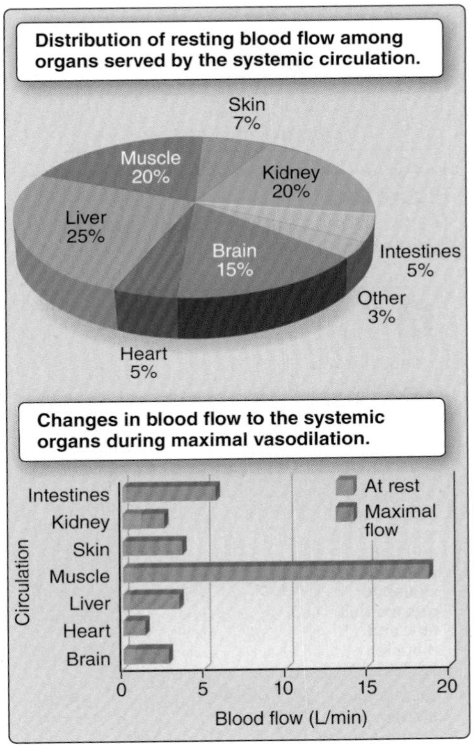

Figure 21.1
Blood flow to organs of the systemic circulation.

II. CEREBRAL CIRCULATION

The cerebral circulation supplies the brain, an organ that represents only 2% of body weight yet commands 15% of resting cardiac output (CO). This demand for blood flow reflects the brain's high rate of metabolism. Brain tissue has few metabolic stores and is highly dependent on oxidative pathways for energy production. In practice, this dependence means that the brain is highly reliant on the cerebral circulation for normal function.

A. Anatomy

Four major arterial branches supply the brain: the left and right vertebral arteries and the left and right carotid arteries (Figure 21.2). The vertebral arteries join to form the basilar artery, which travels via the brainstem to the base of the brain, then splits again and links via communicating arteries to the internal carotids, creating the **circle of Willis**. Extensive interconnections between adjacent arteries allow for continued flow around potential sites of blockage.

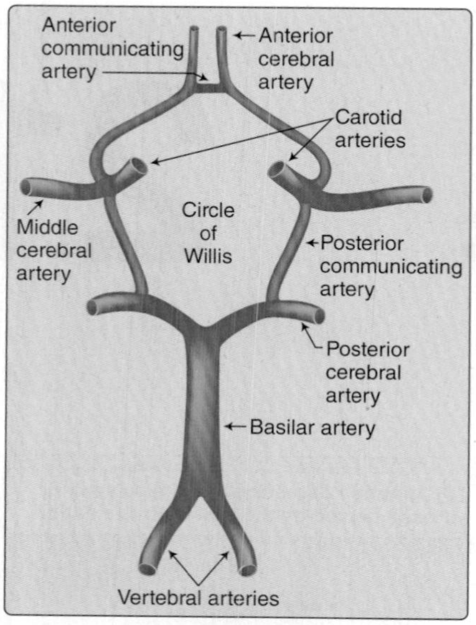

Figure 21.2
Major arteries supplying the cerebral circulation.

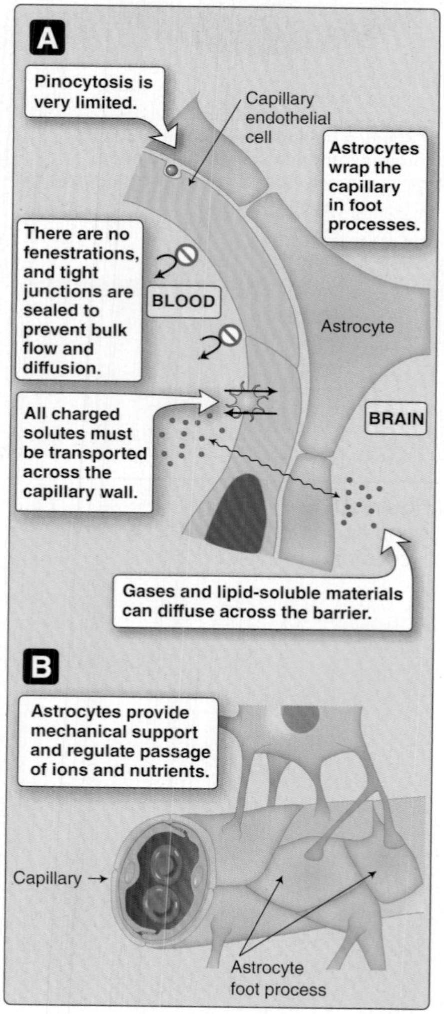

Figure 21.3
The blood–brain barrier.

B. Blood–brain barrier

Blood contains a variety of chemicals that could adversely affect brain function. These include many hormones and neurotransmitters, such as epinephrine, glycine, glutamate, and adenosine triphosphate. The brain is protected from these and other chemicals by the **blood–brain barrier** (**BBB**) that provides three layers of defense: physical, chemical, and cellular.

1. **Physical barrier:** Capillaries in most circulations are relatively leaky, allowing materials to exchange across their walls by four general mechanisms (see 19·VI). Cerebral capillaries are uniquely modified to deny passage by most of these routes. Cerebral capillary endothelial cells rarely form pinocytotic vesicles, which are a major route for transport in skeletal muscle vasculature, for example. Cerebral capillary walls contain no fenestrations, and adjacent endothelial cells are fused together by impermeant tight junctions (Figure 21.3A). This effectively blocks bulk flow and diffusion of ions and water. Lipid-soluble molecules such as O_2 and CO_2 can readily diffuse across the capillary wall, but all other materials must be carried across. Cerebral endothelial membranes are dense with transporters that deliver glucose, amino acids, choline, monocarboxylic acids, nucleotides, and fatty acids to the brain. Movement of ions and protons across the BBB is regulated by channels, exchangers, and pumps. Aquaporins allow water to migrate between blood and brain in response to changes in osmolarity.

2. **Chemical barrier:** Cerebral capillary endothelial cells contain *monoamine oxidase*, *peptidase*, *acid hydrolase*, and a variety of other enzymes that are capable of degrading hormones, transmitters, and other biologically active molecules. They provide a chemical barrier to bloodborne factors.

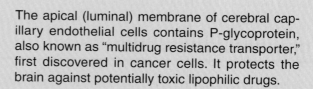

> The apical (luminal) membrane of cerebral capillary endothelial cells contains P-glycoprotein, also known as "multidrug resistance transporter," first discovered in cancer cells. It protects the brain against potentially toxic lipophilic drugs.

3. **Supporting structures:** Cerebral capillaries have a thick basal lamina and are supported mechanically by astrocytes (see Figure 21.3B). Astrocytes also maintain tight-junction integrity and regulate exchange of material between blood and brain (see 5·V).

> The impenetrability of the BBB presents a logistical problem when trying to treat intracranial infections or tumors. In such cases, it may be necessary to shock the endothelium osmotically in order to create a temporary breach that allows passage of antibiotics and chemotherapy agents.

C. Regulation

The cerebral circulation maintains steady flow rates when mean arterial pressure is varied from ~60–130 mm Hg. This is a significantly wider pressure range than observed in other vascular beds. Autoregulation is effected primarily though local control mechanisms.

1. **Local controls:** Cerebral resistance vessels dilate in response to the same metabolic factors that allow for local control in other circulations, but they are especially sensitive to changes in P_{CO_2} (Figure 21.4). Small increases in P_{CO_2} cause profound vasodilation, whereas decreases in P_{CO_2} cause vasoconstriction.

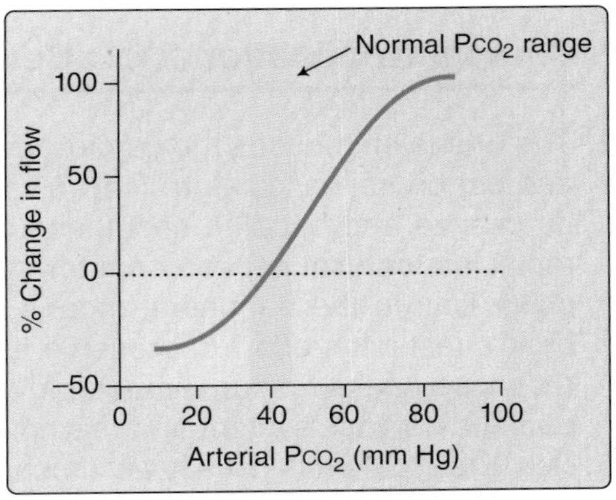

Figure 21.4
Cerebral blood flow dependence on P_{CO_2}.

> P_{CO_2} sensitivity explains why hyperventilating can cause loss of consciousness. When a person breathes at rates beyond his or her physiologic needs, blood P_{CO_2} falls. Loss of vasodilator influence causes the cerebral resistance vessels to constrict, and the person becomes lightheaded. Normal cerebral flow can be restored by rebreathing expired air to increase the CO_2 content of alveolar air. Arterial P_{CO_2} increases as a result, causing the cerebral vessels to dilate, and flow renormalizes.

2. **Central controls:** Cerebral resistance vessels are innervated by both branches of the autonomic nervous system (ANS). The nerves are active, but their effects on the vascular smooth muscle cells are typically minor compared with responses to metabolites.

D. Regional flow patterns

The cranium affords mechanical protection to the brain, but it also physically constrains cerebral blood flow and volume. All tissues of the body swell when metabolic activity rises, reflecting an increased volume of blood flowing through the vasculature. In the brain, demand for blood flow similarly increases and decreases as mental focus shifts, and the neurons become more or less active. The brain resides within the skull, however, which prevents any changes in intracranial volume. Brain tissues also resist volume changes because they are largely composed of fluid, which is incompressible. Thus, regional increases in cerebral flow that accompany increased activity are typically matched by opposing changes in a different brain area (Figure 21.5).

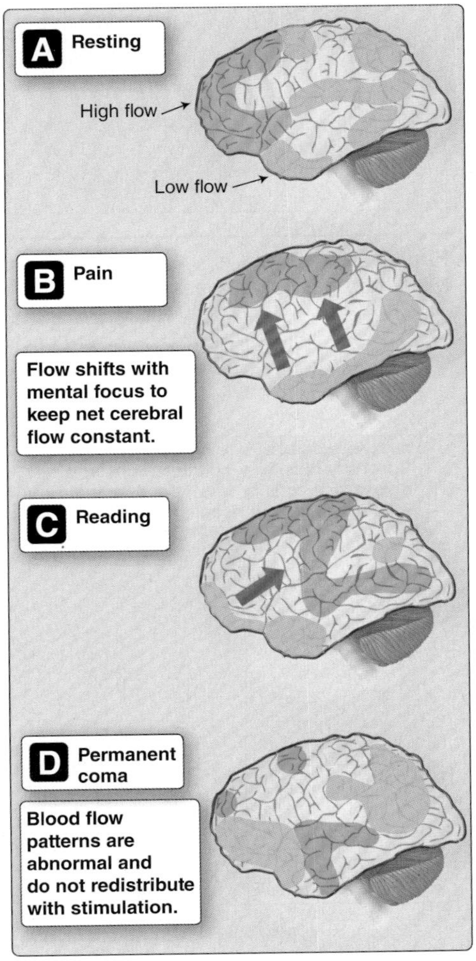

Figure 21.5
Shifting regional cerebral flow patterns.

E. Flow interruption

The brain has a very low tolerance for ischemia. A 20%–30% decrease in cerebral flow causes lightheadedness. A 40%–50% decrease causes fainting (**syncope**). Complete interruption of flow for >4–5 minutes can cause organ failure and death. Cerebral vessels that have narrowed with age or disease may cause a **transient ischemic attack** (**TIA**), a localized reduction in flow and loss of cerebral function lasting minutes or hours. Interruptions in cerebral flow (**cerebrovascular accidents**, or **strokes**) occur when a cerebral vessel is occluded. Such events cause infarction and more permanent neurologic defects.

Clinical Application 21.1: Central Nervous System Ischemic Response

The brainstem houses the cardiovascular control center, a region that has autonomic connections to the heart and peripheral vasculature (see 7·VI). Loss of blood flow to the cardiovascular center causes the neurons to depolarize and become spontaneously active as their ion-pumps fail and their ion gradients dissipate. The result is a **central nervous system** (**CNS**) **ischemic response**, a massive sympathetic outpouring that shuts down flow to the peripheral organs and causes blood pressure to rise to maximal levels. It represents the brain's last-ditch effort to preserve its own supply. The **Cushing reaction** is a special kind of CNS ischemic response caused by an increase in intracranial pressure. The hypothalamus responds by activating the sympathetic nervous system and raising arterial pressure, but this initiates a baroreflex-mediated slowing of heart rate. Bradycardia is usually accompanied by systolic hypertension and respiratory depression (**Cushing triad**) and is an indication of extreme intracranial pressure and impending death.

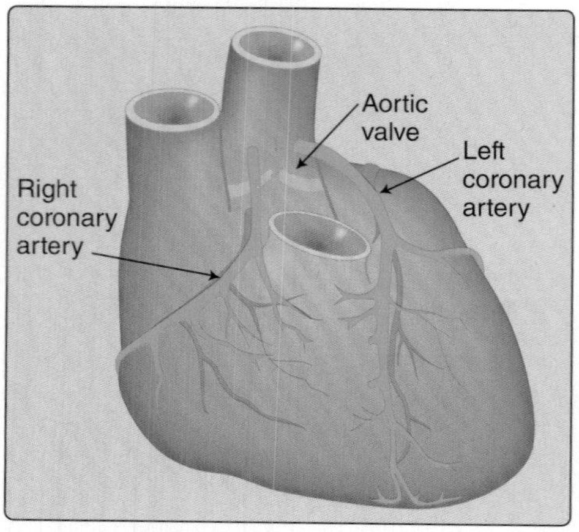

Figure 21.6
Major coronary blood vessels.

III. CORONARY CIRCULATION

The coronary circulation supplies the myocardium, a tissue that rivals the brain in terms of its nutritional demands and the critical importance of continued flow for normal function.

A. Anatomy

The myocardium is supplied by left and right coronary arteries that originate from the root of the ascending aorta immediately above the aortic valve (Figure 21.6). The right coronary artery generally supplies the right heart, whereas the left coronary artery supplies the left. The arteries course over the heart's surface and then dive down through the muscle layers. The vasculature is notable for numerous collaterals connecting adjacent arteries and also for the presence of precapillary sphincters (see below).

B. Regulation

At rest, the coronary circulation receives ~5% of CO. Cardiac muscle extracts >70% of available O_2 from blood, and it has a very low capacity for anaerobic metabolism, much like the brain. This O_2 dependence means that any increase in work must be matched by an increase in coronary flow, achieved entirely through local control mechanisms.

1. **Local controls:** Coronary resistance vessels are exceptionally sensitive to adenosine. Local control mechanisms allow for a fourfold to fivefold increase in coronary flow when CO increases, a phenomenon called **coronary reserve** (Figure 21.7).
2. **Central controls:** Coronary resistance vessels are innervated by both branches of the ANS, but their influence is overridden by local controls.

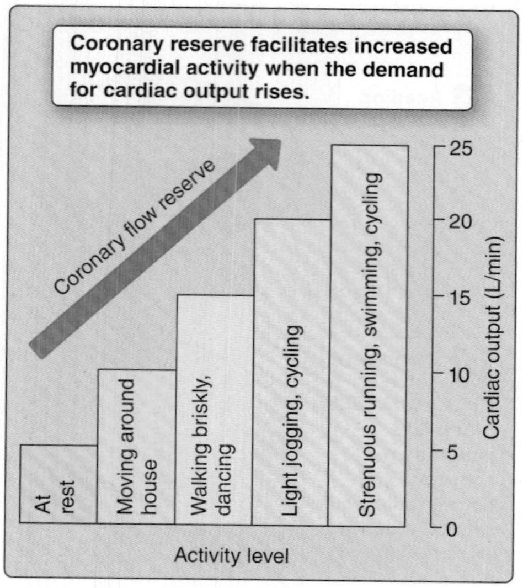

Figure 21.7
Coronary reserve.

Some individuals have resistance vessels that are abnormally susceptible to sympathetic vasoconstrictor influence, causing their coronary arteries to spasm during high sympathetic drive. The ischemia that results from these temporary flow interruptions is experienced as pain, known as **Prinzmetal**, or **variant**, **angina**.

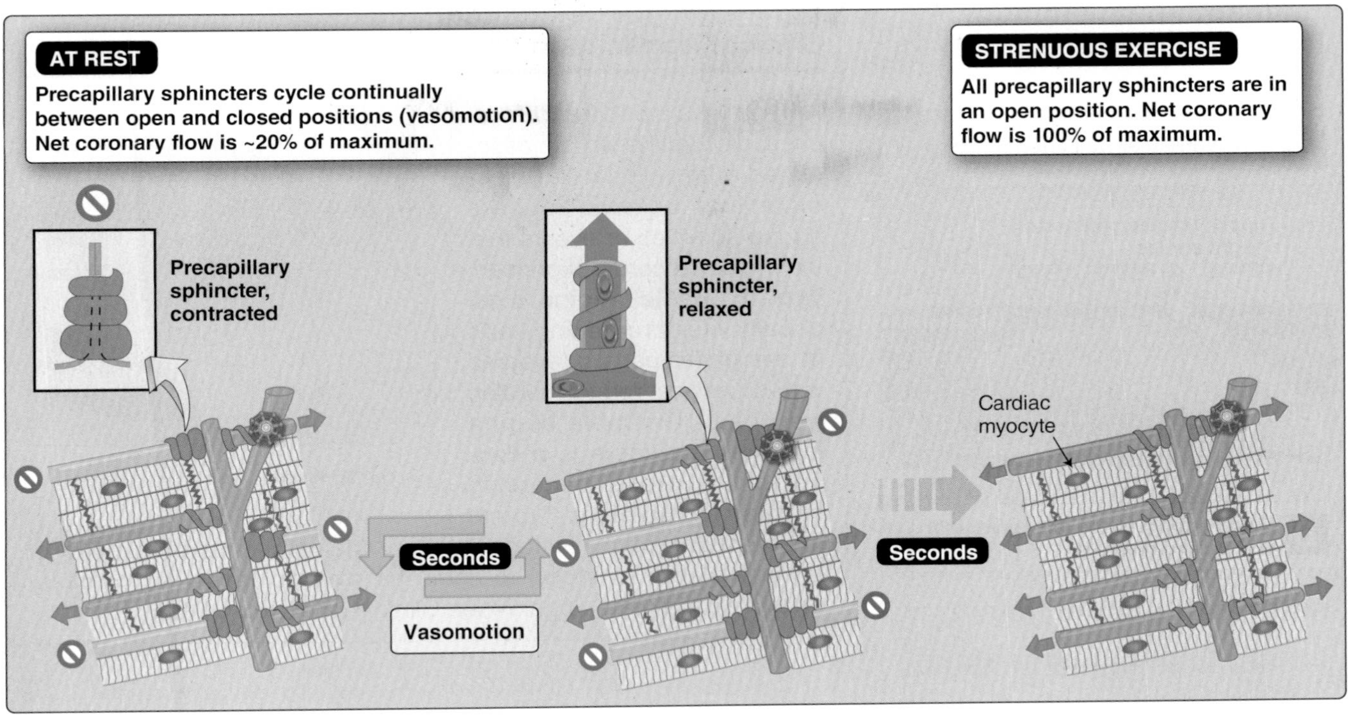

Figure 21.8
Vasomotion and basis of coronary reserve.

C. Precapillary sphincters

Precapillary sphincters comprise single smooth muscle cells wrapped around the inlets to individual capillaries (Figure 21.8). They contract and relax with changes in local metabolite concentrations and function as on/off switches to capillary flow. When CO is minimal, most sphincters are contracted ("off"), and flow is inhibited. They relax intermittently as local metabolite levels rise but again contract when the increased flow washes the metabolites away. At rest, only a small proportion (~20%) of sphincters is relaxed, and capillaries are actively perfused, but the pattern of capillary flow shifts continually (**vasomotion**). When cardiac workload increases, levels of metabolic waste products rise, and the sphincters spend a much greater percentage of time in the "on" position. At maximal levels of CO, all sphincters are open all the time, and coronary flow rises to maximal levels also (see Figure 21.8).

D. Extravascular compression

Blood flow through most systemic vascular beds follows the aortic pressure curve, rising during systole and falling during diastole. Flow through the left coronary artery drops sharply during systole and then rises sharply with the onset of diastole (Figure 21.9). This unique flow pattern occurs because ventricular myocytes collapse the arterial supply vessels as they contract (**extravascular compression**) as shown in Figure 21.10. The effect is felt strongest during early systole because aortic pressure, the main force maintaining vascular patency, is at a low point. During diastole, the compressive forces are

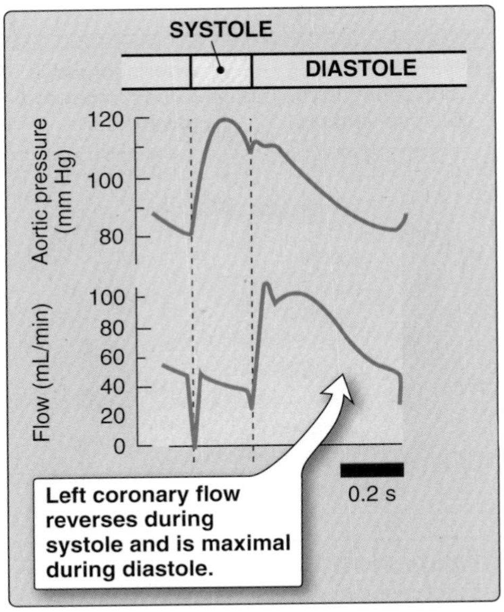

Figure 21.9
Left coronary blood flow.

Clinical Application 21.2: Angina

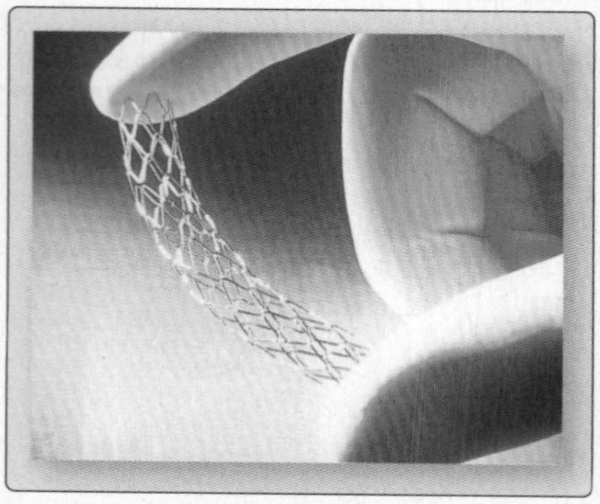

A metal stent.

Angina is a specific form of chest discomfort or pain associated with myocardial ischemia that usually appears during activities or events that increase myocardial workload. **Typical** (**stable**) **angina** is not usually life threatening, and its symptoms can be reversed with drugs that reduce cardiac workload.[1] **Unstable angina** indicates that there is a danger of the vessel becoming blocked completely. Typical interventions might include balloon angioplasty or implanting of a metal sleeve (a **stent**) to open the stenotic vessel, or coronary artery bypass graft surgery (see Clinical Application 20.2).

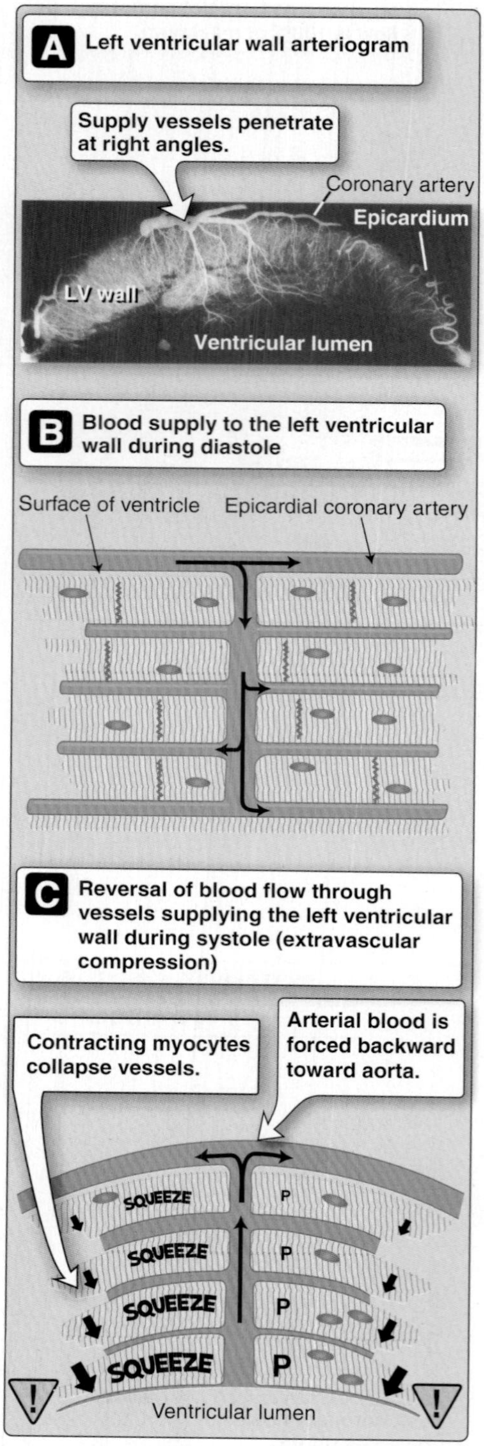

Figure 21.10
Extravascular compression in the left ventricular wall. LV = left ventricle; P = pressure.

removed, and blood surges through the musculature at peak rates (see Figure 21.9).

E. Flow interruption

Because the ventricular myocytes extract such high levels of O_2 from the blood, a delicate balance exists between myocardial workload and coronary supply. If the balance is disturbed, then myocytes become ischemic and infarcted. Most commonly, this occurs due to **atherosclerosis** and **coronary artery disease**.

1. **Atherosclerosis:** Atherosclerotic lesions appear at an early age in the populations of most Western countries. They evolve to become complex plaques of lipids, hypertrophied myocytes, and fibrous material. Plaques enlarge at the expense of the vascular lumen and impair blood flow. This causes an imbalance between coronary supply and myocardial demand, resulting in ischemia. Ischemic myocytes release large quantities of vasoactive compounds, such as adenosine, but vasodilators have no effect on plaque. As the O_2 deficit continues, the myocytes release lactic acid, which stimulates pain fibers within the myocardium and causes **angina pectoris**.

2. **Collaterals:** Collaterals are vessels ~100 μm in diameter that connect adjacent arterioles. They are usually constricted in a healthy heart, but, if a supply vessel becomes occluded, they dilate in response to rising metabolite levels. Flow through collaterals

INFO LINK [1]For a discussion of antianginal drugs, see *LIR Pharmacology*, 5e, Chapter 18.

may prevent **infarction** if the occluded vessel is small. In time, these channels enlarge to provide near-normal flow to the ischemic area.

IV. SPLANCHNIC CIRCULATION

The splanchnic circulation serves the liver, gallbladder, spleen, pancreas, and the length of the gut. It is the largest systemic circulation and commands 20%–30% of CO, even at rest (see Figure 21.1).

A. Anatomy

Blood reaches the splanchnic circulation via the celiac and the mesenteric arteries. It then passes through the spleen, pancreas, stomach, and the small and large intestines and drains via the portal vein into the liver (Figure 21.11). This is one of the few body regions where two organs are arranged in series with each another, but the circulatory design is functional because it allows the liver to filter and begin processing absorbed nutrients before the blood is returned to the general circulation.

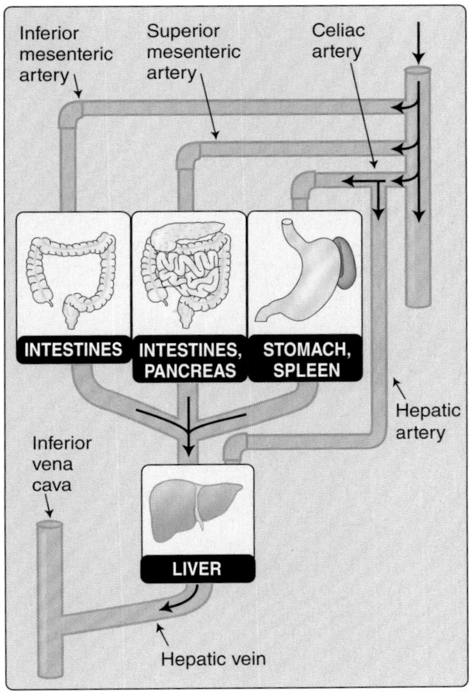

Figure 21.11
Splanchnic circulation.

B. Regulation

Blood flow in the splanchnic circulation is controlled locally and by the ANS, but the enteric nervous system is involved also.

1. **Local controls:** The mechanisms by which the gut autoregulates are poorly delineated. Increases in flow during a meal may be triggered by metabolites, gastrointestinal hormones, vasodilator kinins released from the intestinal epithelia, bile acids, and by the products of digestion.

2. **Central controls:** The splanchnic vasculature is regulated by both ANS divisions. The parasympathetic nervous system increases blood flow both in anticipation of and while digesting a meal (a classic "rest-and-digest" response). The sympathetic nervous system (SNS) constricts all splanchnic vascular beds during "fight-or-flight" responses, thereby shunting blood away from the GI tract for use elsewhere in the circulation.

C. Splanchnic reservoir

At rest, the splanchnic circulation holds ~15% of total circulating volume, representing a significant reservoir that can be tapped by the SNS should there be a more urgent need for blood elsewhere. The consequences of sympathetic activation on splanchnic flow depend on stimulation intensity.

1. **Mild sympathetic activation:** Mild SNS activation curtails splanchnic flow, but circulation renormalizes within minutes by reflexive resistance vessel dilation caused by rising metabolite levels, a phenomenon known as **autoregulatory escape** (Figure 21.12). Moderate SNS activation (e.g., as seen during moderate exercise) produces a more persistent curtailment of splanchnic flow.

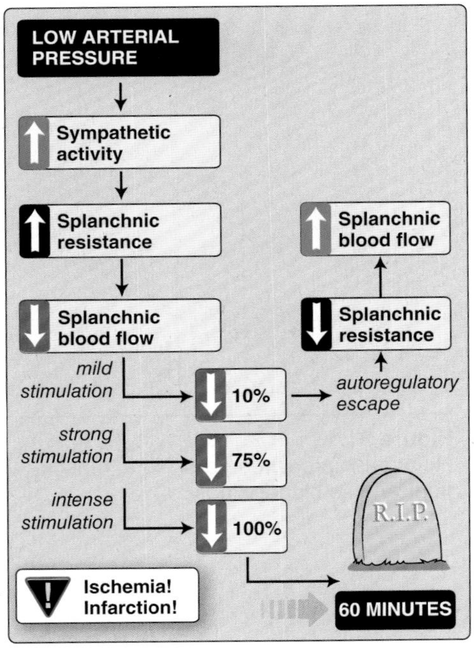

Figure 21.12
Sympathetic dominance over splanchnic blood flow.

2. **Maximal sympathetic activation:** Vigorous exercise places intense demands on the cardiovascular system. Strong SNS stimulation of splanchnic resistance vessels reduces total flow to <25% of basal values, whereas venoconstriction forces 200–300 mL of blood from the splanchnic vasculature. Intestinal tissues compensate for reduced flow by increasing O_2 extraction from the residual supply.

D. Circulatory shock

Severe hemorrhage and other forms of circulatory shock precipitate extreme levels of sympathetic activity that reduce splanchnic flow to minimal levels for prolonged periods (see Figure 21.12). If flow is not restored within an hour or so, the epithelial lining the small intestine infarcts and begins to disintegrate. Breakdown allows toxic materials from the gut (bacterial enterotoxins and endotoxins) to enter the blood stream, resulting in toxemia and septic shock (see 40·IV·C)

V. SKELETAL MUSCLE CIRCULATION

Resting flow to skeletal muscle is modest considering the mass of tissue that it serves (~20% of CO). Such modesty belies the profound intrinsic vasodilatory capabilities of muscle during exercise, however.

A. Anatomy

When skeletal muscles are active, they are highly dependent on the vasculature to deliver O_2 and nutrients and to carry away heat, CO_2, and other metabolic waste products. These functions are facilitated by an unusually dense capillary network. The skeletal vasculature is supplied by superficial feed arteries that branch multiple times within the muscle groups until they become terminal arterioles. Each arteriole gives rise to numerous capillaries that travel parallel to individual muscle fibers within a fascicle. Each fiber is typically associated with three or four capillaries, which reduces the range over which O_2 must diffuse to reach the innermost myofibrils to ~25 μm (Figure 21.13).

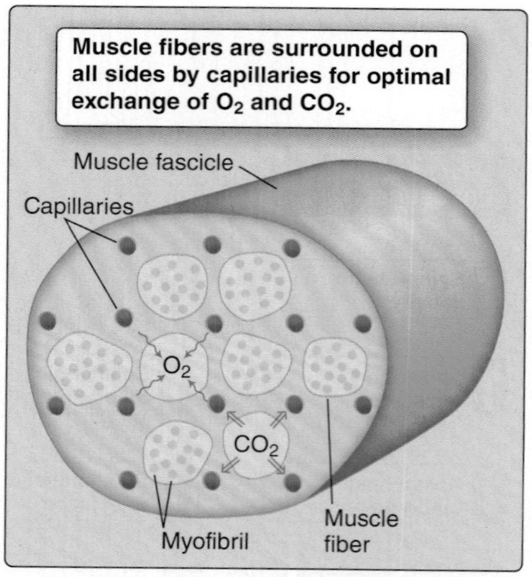
Muscle fibers are surrounded on all sides by capillaries for optimal exchange of O2 and CO2.
Muscle fascicle
Capillaries
O2
CO2
Muscle fiber
Myofibril

Figure 21.13
Relationship between a skeletal muscle fiber and its blood supply.

B. Regulation

Local and central control mechanisms affect the skeletal vasculature with equal potency. These mechanisms can yield dramatic flow extremes depending on circumstance.

1. **Local controls:** In resting muscle, only a small percentage of capillaries are actively perfused because the terminal arterioles (resistance vessels) that feed them are constricted. When the muscle becomes active, metabolite concentrations rise, the arterioles dilate and previously inactive capillaries now course with blood (**capillary recruitment**).

2. **Central controls:** Skeletal resistance vessels are richly innervated by SNS fibers whose resting tone maintains flow at minimal levels governed largely by a muscle's metabolic needs. When arterial pressure falls, SNS activity increases as a normal

part of a baroreceptor reflex. The consequences depend on whether or not the muscle is being exercised at the time.

a. **At rest:** SNS activation decreases flow to a resting muscle. The increased resistance contributes to the rise in systemic vascular resistance that accompanies a baroreceptor reflex.

b. **During exercise:** During exercise, local metabolite concentrations dominate vascular control. The potency of these mechanisms is such that flow through the skeletal vasculature may rise to 25L/min during intense exercise, which amounts to 500% of resting CO!

C. Extravascular compression

Contracting muscles compress the blood vessels that run between the fibers and cause temporary interruptions in flow. Isometric contractions (e.g., induced by lifting a weight) can inhibit flow for several tens of seconds and are followed by a vigorous **reactive hyperemia** upon relaxation. Isotonic exercises, such as jogging and swimming, involve rhythmic cycles of contraction and relaxation and produce a phasic flow pattern (Figure 21.14). Note that flow oscillates between two extremes with each contraction, but flow is increased overall (**active hyperemia**).

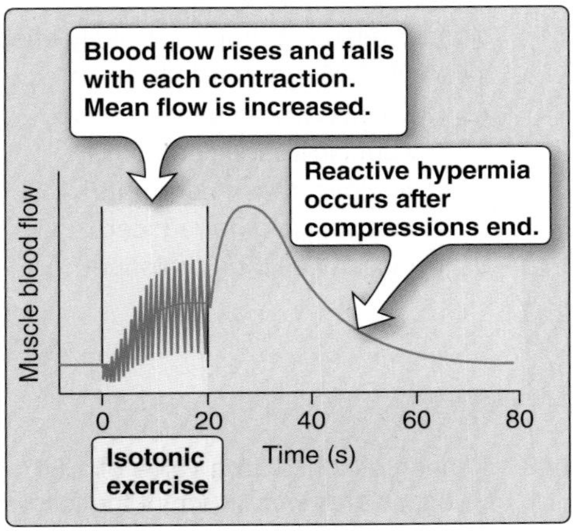

Figure 21.14
Phasic blood flow pattern in skeletal vasculature during aerobic exercise.

Chapter Summary

- The circulations that supply the various organs of the body contain unique features tailored to support organ function.
- The **cerebral circulation** supplies the brain, a tissue notable for a high O_2 demand and its dependence on continued blood flow for function. Loss of flow initially causes fainting, with irreversible cellular changes occurring following within minutes.
- The brain is protected from bloodborne agents by the **blood–brain barrier**. **Tight junctions** between adjacent endothelial cells create a physical barrier to ions and other water-soluble chemicals. Endothelial enzymes that degrade potentially threatening molecules provide a chemical barrier, and **astrocytes** provide mechanical support. All substrates needed for cerebral metabolism must be transported across the barrier.
- Flow through the cerebral vasculature is regulated by local metabolite concentrations. Cerebral resistance vessels are exceptionally sensitive to **P_{CO_2}**. Flow volume is limited by the cranium, so flow increases in one area of the brain are balanced by decreases in another area to keep overall volume constant.
- The **coronary circulation** similarly supplies tissues that are highly dependent on O_2 for continued function and any increase in myocardial workload must be matched by increased flow. Coronary resistance vessels are regulated by local factors, primarily **adenosine**. Sympathetic nerve terminals have no physiologic effect.
- Flow through the left coronary arteries decreases during early systole, reflecting the effects of **extravascular compression** during ventricular contraction. The left ventricle receives the bulk of its supply during diastole. If flow through the coronary vasculature becomes limiting, lactic acid concentrations rise and cause pain (**angina**). Interruptions in coronary flow cause ischemia and may result in **myocardial infarction**.
- The splanchnic circulation supplies all of the organs involved in digestion, including the liver, which is **arranged in series** with the other splanchnic organs.
- The **splanchnic vasculature** holds 15% of total circulating blood volume at rest. Gastrointestinal organs have low O_2 requirements so flow can be diverted elsewhere during a hypotensive crisis without short-term risk of ischemia.
- The sympathetic nervous system dominates splanchnic vascular control. Prolonged splanchnic vasoconstriction ($>$1 hr) can cause death of the intestinal epithelium and breaches in the barrier separating blood and intestinal contents.
- The **skeletal vasculature** is controlled by both central and local factors. When muscle activity increases, rising metabolite levels can increase flow 25 times. When muscle is inactive, sympathetic constrictor influence can divert flow for use elsewhere.

Study Questions

Choose the ONE best answer.

IV.1 A 65-year-old man with a history of hypertension is prescribed a Ca^{2+}-channel blocker to help reduce his blood pressure. What is the likely effect of this drug on the ventricular myocardium?

A. It would have no effect.
B. It would increase contractility.
C. It would increase heart rate.
D. Phase 2 would be reduced.
E. Phase 1 would be prolonged.

Best answer = D. Ca^{2+}-channel blockers reduce Ca^{2+} influx via L-type Ca^{2+} channels during phase 2 of the ventricular action potential, thereby reducing phase 2 and decreasing contractility (17·IV·A). Phase 1 is mediated by Na^{+} and K^{+} channels (17·IV·B) and would not be affected by a Ca^{2+}-channel blocker. Heart rate is determined by the rate of phase 4 depolarization in sinoatrial nodal cells, which is governed in part by L-type Ca^{2+} channels (17·IV·C·4). Ca^{2+}-channel blockers would be expected to decrease heart rate.

IV.2 A mean electrical axis value of $-60°$ would most likely be associated with which of the following conditions?

A. Pulmonary hypertension
B. Premature ventricular contractions
C. Aortic stenosis
D. Pulmonary edema
E. Left-ventricular infarction

Best answer = C. Aortic stenosis forces the left ventricle (LV) to work harder and generate higher peak systolic pressures to sustain cardiac output (40·V·A). Over time, this causes LV hypertrophy, which manifests on an electrocardiogram as left-axis deviation (normal range = $+105°$ to $-30°$; 17·V·E). Pulmonary hypertension and edema promote right ventricular hypertrophy. A left-ventricular myocardial infarction would likely shift the axis to the right, not left, because viable LV muscle mass is reduced. Premature ventricular contractions do not directly alter the mean electrical axis.

IV.3 A 50-year-old woman reports "thumping" sensations in her chest. An electrocardiogram records occasional wide, premature QRS complexes. Which of the following most likely explains the origin of these complexes?

A. Atrial fibrillation
B. Ventricular fibrillation
C. An irritable ectopic focus
D. First-degree heart block
E. Myocardial ischemia

Best answer = C. Premature ventricular contractions (PVCs) are characterized by wide and abnormally shaped QRS complexes. They reflect waves of excitation that travel through the myocardium via the slow myocyte-to-myocyte route, rather than the fast conduction His–Purkinje system (17·V·D·3). PVCs are typically triggered by irritable foci located in the ventricular myocardium (i.e., ectopic) rather than the sinoatrial node. Atrial fibrillation manifests as loss of a P wave, whereas first-degree heart block prolongs the PR interval. A fibrillating ventricle shows no organized electrocardiogram waveforms. Ischemia may affect the ST segment, but QRS complexes still occur in normal position.

IV.4 Which of the following electrocardiogram events coincides with the "reduced ventricular ejection" phase of the cardiac cycle?

A. P wave
B. PR interval
C. QRS complex
D. ST segment
E. T wave

Best answer = E. The T wave corresponds to ventricular repolarization (17·V·C), which occurs during reduced ejection (18·II). The P wave coincides with atrial systole, which continues during the PR interval. The QRS complex is caused by ventricular excitation, which is followed by isovolumic contraction and rapid ejection. The ST segment encompasses isovolumic contraction and persists through rapid ejection.

IV.5 A 7-year-old boy of normal height and build for his age is undergoing a routine physical. The family physician notes a third heart sound (S_3) during auscultation. Which of the following statements best describes the cause of the S_3 in this boy?

A. It coincides with rapid ventricular ejection.
B. It indicates atrial hypertrophy.
C. An electrocardiogram would show right-axis deviation.
D. It is caused by aortic valve regurgitation.
E. It is the sound of ventricular filling.

Best answer = E. The third heart sound (S_3) occurs during ventricular filling and is caused by sudden tensing and reverberation of the ventricular walls (18·II·E). Although usually a sign of underlying pathology in adults, it is a normal finding in children. Ventricular filling and S_3 occur during diastole, not rapid ejection. Atrial hypertrophy would yield an S_4, whereas a right-axis deviation does not necessarily correlate with a heart sound. Regurgitant valves produce murmurs, not heart sounds (Clinical Application 18.1).

IV.6 A 44-year-old woman is diagnosed with dilated cardiomyopathy, a condition caused by impaired ventricular contractility and compensatory fluid retention. What is the advantage to fluid retention and preloading?

A. It increases ventricular wall tension.
B. It increases ventricular stroke volume.
C. It decreases ventricular afterload.
D. It reduces cardiac workload.
E. It reduces the need for resting cardiac output.

Best answer = B. Increased preloading stretches the myocardium, thereby increasing the amount of force developed upon contraction through length-dependent activation of the sarcomere (18·III·D). Preloading increases stroke volume and ejection fraction, which helps compensate for reduced ventricular contractility. The disadvantage to enhanced preloading is that it increases ventricular radius and wall tension, thereby increasing afterload and overall workload (law of Laplace; 18·IV·B). Resting cardiac output is determined by the metabolic needs of the tissues, not by preload.

IV.7 An 11-year-old boy's dentist gives him nitrous oxide gas (N_2O) via a face mask to anesthetize him. Which among the following choices is the main route by which the N_2O reached the brain?

A. Endocytosis across the capillary wall
B. Specialized endothelial transporters
C. Bulk flow through fenestrations
D. Diffusion through intercellular junctions
E. Diffusion through the endothelial cells

Best answer = E. Nitrous oxide gas is a small, highly soluble molecule that, like O_2 and CO_2, easily diffuses through endothelial cell membranes (19·VI). Endocytosis is used primarily as a means of moving large proteins between the bloodstream and tissues, whereas transporters are typically used to transport charged molecules against a concentration gradient. Fenestrations and intercellular junctions provide routes for passage of water and any dissolved ions.

IV.8 Kwashiorkor is a severe form of childhood malnutrition seen mostly in developing countries. Symptoms include hepatomegaly and pitting edema of the lower extremities. The pitting edema is most likely due to which of the following?

A. A plasma protein deficit
B. Inadequate cardiac output
C. Excessive fluid retention
D. Reduced interstitial pressure
E. Decreased hematocrit

Best answer = A. Plasma proteins create an osmotic potential (plasma colloid osmotic pressure) that holds fluid in the vasculature (19·VII·D). Kwashiorkor results from inadequate protein intake, which impairs the liver's ability to synthesize proteins. Fluid filters into the interstitium and causes edema as a consequence. Reduced cardiac output and pressure would reduce fluid filtration. Excessive fluid retention would exacerbate edema caused by the plasma protein deficit. Interstitial pressure is increased by and counters fluid filtration, whereas changes in hematocrit have no effect.

IV.9 Skeletal muscle metabolism increases dramatically during physical activity, sustained by equally dramatic perfusion increases. Which of the following mechanisms facilitates activity-induced increases in muscle blood flow?

A. Flow-induced nitric oxide release
B. Norepinephrine-induced vasodilation
C. Rising metabolite levels
D. Antidiuretic hormone release
E. Histamine release

Best answer = C. Increased blood flow to active tissues ("active hyperemia") is mediated by local accumulation of metabolic byproducts, including CO_2, H^+, and adenosine, which cause reflexive dilation of resistance vessels (20·II·A). Norepinephrine (from sympathetic nerve terminals) and antidiuretic hormone (from the posterior pituitary) both constrict resistance vessels. Flow-induced nitric oxide release may contribute to increased flow at high levels of cardiac output (20·II·E), but this effect is secondary to the influence of metabolic byproducts. Histamine can cause vasodilation but usually only as a part of an allergic reaction.

IV.10 An 11-year-old girl wrestling with her younger brother causes him to faint when inadvertently applying pressure on his left carotid sinus. Syncope most likely occurred as a result of which of the following?

A. Occlusion of his carotid artery
B. Occlusion of his jugular vein
C. Cerebral vasculature vasoconstriction
D. Carotid baroreceptor stimulation
E. Carotid chemoreceptor stimulation

Best answer = D. Applying pressure in the area of the carotid sinus stimulates the baroreceptors within the vessel walls, thereby mimicking the effects of increased blood pressure (20·III·A). This promotes a reflex decrease in cardiac output and systemic vascular resistance (SVR). Arterial pressure falls as a result, causing cerebral hypotension and syncope. Occlusion of only one of the cerebral arteries or veins is unlikely to decrease cerebral perfusion or change cerebral CO_2 levels sufficiently to cause syncope. Carotid body chemoreceptors do not directly sense blood pressure, but, when stimulated, they increase SVR and blood pressure.

IV.11 A 45-year-old woman faints when standing up following a 90-minute Medical Physiology lecture. Which of the following variables increases in a healthy compensating person after rising to an upright position?

A. Systemic vascular resistance
B. Left ventricular preload
C. Right ventricular preload
D. Cerebral venous pressure
E. Aortic baroreceptor firing rate

Best answer = A. When a person stands upright, blood pools in the lower extremities (20·III·D). Decreased venous return causes right and left ventricular preload to fall, which reduces ventricular stroke volume and cardiac output. Arterial pressure begins to fall as a result, which is sensed via a decrease in arterial (aortic and carotid sinus) baroreceptor firing rate. If the pressure drop is severe, cerebral blood flow can be compromised. Normally compensating individuals tolerate the upright position by initiating a baroreceptor reflex, which includes an increase in systemic vascular resistance.

IV.12 A 55-year-old man with severe angina is scheduled for quadruple bypass surgery. The coronary arterioles downstream of the stenotic regions during the anginal episodes are most likely dilated fully. What is the primary cause of this vasodilation?

A. Parasympathetic activity
B. Norepinephrine
C. Adenosine
D. Lactic acid
E. High velocity flow

Best answer = C. Coronary resistance vessels are controlled by the needs of the myocardium via changes in local metabolite concentrations, especially adenosine (21·III·B). Lactic acid also causes vasodilation but to a lesser degree than adenosine. The shear stress caused by high velocity flow can cause vasodilation via nitric oxide release but is unlikely in the setting of decreased perfusion and angina. The parasympathetic nervous system does not have any significant role in coronary vessel regulation, whereas norepinephrine constricts blood vessels.

UNIT V
Respiratory System

Lung Mechanics

22

I. OVERVIEW

All cells generate adenosine triphosphate (ATP) to fuel their many activities. The preferred pathway for ATP formation is aerobic glycolysis, which requires a constant supply of molecular oxygen (O_2) and carbohydrates. O_2 and glucose metabolism yields water and CO_2 (**internal respiration**). CO_2 dissolves in water to form carbonic acid, which must be continually expelled from the body. The task of supplying cells with O_2 and glucose falls on the cardiovascular system, as does the task of removing waste products such as CO_2. O_2 is available from the atmosphere and will readily enter the circulation if blood is brought into close proximity. CO_2 is also a gas that can be discharged to the atmosphere at the same time O_2 is being taken up. The primary function of the lung is to facilitate exchange of these gases between the blood and atmosphere (**external respiration**). Lungs contain a respiratory epithelium that creates a large **blood–gas interface**. Total surface area of the interface is ~80 m^2, or roughly half the size of a singles' tennis court. The interface is extremely thin to facilitate rapid gas exchange between blood and inspired air. These features together ensure that O_2 and CO_2 rapidly equilibrate across the interface as blood circulates through the pulmonary vasculature. Air is pumped in and out of the lungs through rhythmic contraction and relaxation of respiratory muscles. The air pump flushes CO_2 out of the lungs and replenishes O_2, ensuring that the gradients driving diffusion of both gases between blood and atmosphere remain optimal.

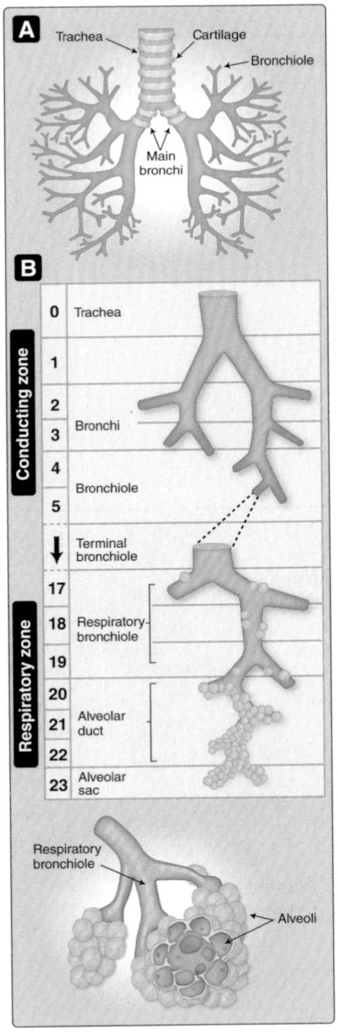

Figure 22.1
Branching structure of the lung airways.

II. AIRWAY ANATOMY

Creating a blood–gas interface within the thorax with a surface area sufficient to meet the demands of internal respiration requires an elaborate system of branching tubes (**airways**) and air sacs (**alveolar sacs**) as shown in Figure 22.1A. The airways channel air from the external atmosphere to the blood–gas interface. The airways begin with the trachea (generation 0) and then branch repeatedly to yield a bronchial tree. The tree contains ~23 branch **generations** (see Figure 22.1B) and comprises two functionally distinct zones, a conducting zone and a respiratory zone.

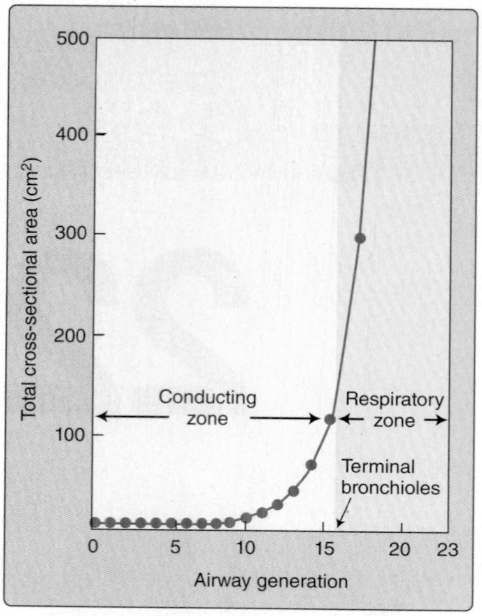

Figure 22.2
Amplification of lung surface area.

A. Conducting zone

Airways in the conducting zone do not participate in gas exchange, they simply channel airflow. The larger airways (generations 0 through ~10) are supported structurally with cartilage to help maintain patency. Generations 10 through 16 are called **bronchioles**, with the terminal bronchioles (~generation 16) demarcating the end of the conducting zone. The conducting zone is lined with a mucus-secreting, ciliated epithelium. The cilia beat constantly, sweeping mucus and trapped particulates up and out of the lungs (the **mucociliary escalator**).

> Tobacco smoke impairs respiratory cilia function. Ciliary arrest allows bacteria and other inhaled particulates to accumulate in the lungs, causing local irritation and epithelial inflammation. Smokers are prone to coughing episodes and bronchitis as a result. Ciliary function is typically restored with tobacco cessation.

B. Respiratory zone

The **respiratory zone** (generations 17–23), comprising the respiratory bronchioles, alveolar ducts, and alveolar sacs, is characterized by a tremendous amplification of cross-sectional area even as the passages narrow (Figure 22.2). The respiratory zone is the location of the blood–gas interface.

C. Alveolar sacs

Alveoli are thin-walled, polyhedral sacs with internal diameters of 75–300 μm (Figure 22.3). The lungs contain ~300 million alveoli, interconnected via **pores of Kohn**. The alveolar lining separates atmospheric air from the vasculature. It comprises two types of respiratory epithelial cell, or **pneumocyte**.

1. **Type I pneumocytes: Type I pneumocytes** are thin and flat. They make up the bulk of alveolar surface area (~90%).
2. **Type II pneumocytes: Type II**, or **granular**, **pneumocytes** are present in equal numbers but are more compact and, therefore, occupy less area. They are filled with numerous **lamellar inclusion bodies** that contain **pulmonary surfactant**. Type II cells are capable of rapid division, which allows them to repair alveolar wall damage. They subsequently transform into type I cells, which divide rarely.
3. **Blood–gas interface:** Pulmonary capillaries meander between adjacent alveolar sacs. Their density is so great that they create a near-continuous sheet of blood covering alveolar surfaces. The distance separating red cells from atmospheric air approximates the width of a capillary endothelial cell plus a pneumocyte (~300 nm total).

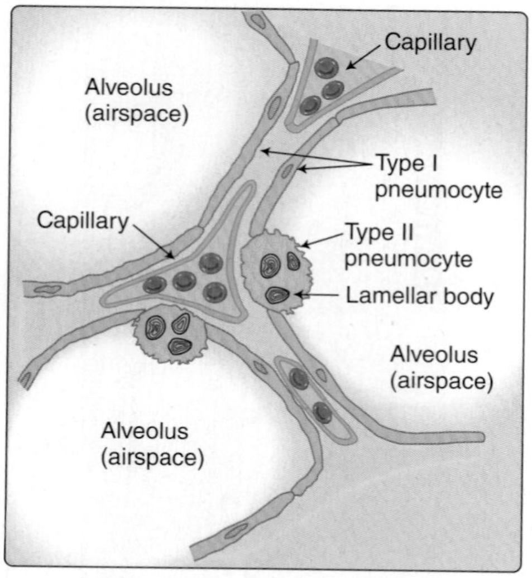

Figure 22.3
Alveolar wall structure.

III. BLOOD SUPPLY

The lung receives blood from two different sources: the **pulmonary** and **bronchial** circulations.

A. Pulmonary circulation

The **pulmonary circulation** brings O_2-poor venous blood from the right ventricle via pulmonary arteries to the blood–gas interface for gas exchange. Pulmonary veins then carry O_2-rich blood to the left side of the heart for delivery to the systemic circulation. The pulmonary circulation has a low vascular resistance, and, thus, mean pulmonary arterial pressures are low (~16 mm Hg). The pulmonary circulation receives the entire output of the heart (~5 L/min at rest, ~25 L/min during strenuous exercise).

B. Bronchial circulation

The **bronchial circulation** is a systemic vascular bed that supplies the conducting airways with O_2 and nutrients. Bronchial arteries arise from the aorta and feed capillaries that drain either via bronchial veins or via anastomoses with pulmonary capillaries into veins of the pulmonary circulation. These connections allow small amounts of deoxygenated blood to bypass the blood–gas interface and reenter the systemic circulation without being oxygenated. This **venous admixture** represents a **physiologic shunt** that decreases pulmonary vein O_2 saturation by 1%–2%.

IV. SURFACE TENSION AND SURFACTANT

Subdividing the lung into 300 million alveoli creates a surface area that is sufficiently large to supply the needs of internal respiration, but there is a significant tradeoff. Each alveolus is moistened with a thin film of **alveolar lining fluid**. The fluid generates **surface tension**, which has significant consequences for lung performance.

A. Surface tension

Water molecules are much more strongly attracted to each other than to the air. This attraction creates surface tension as the individual molecules within alveolar lining fluid draw close to each other and away from the air–water interface. Surface tension always minimizes the area of an exposed surface, which is why soap bubbles or falling raindrops become roughly spherical (Figure 22.4). The moisture film within an alveolus behaves much like a bubble, even though it maintains a connection with the pulmonary lumen during normal breathing. Surface tension is such a powerful force that alveoli (indeed, the entire lung) would collapse unless provided with a means of diminishing its effects (Figure 22.5A).

B. Surfactant

Type II pneumocytes synthesize and release pulmonary surfactant specifically to counter the effects of surface tension. Surfactant is a complex mix of lipids and proteins.

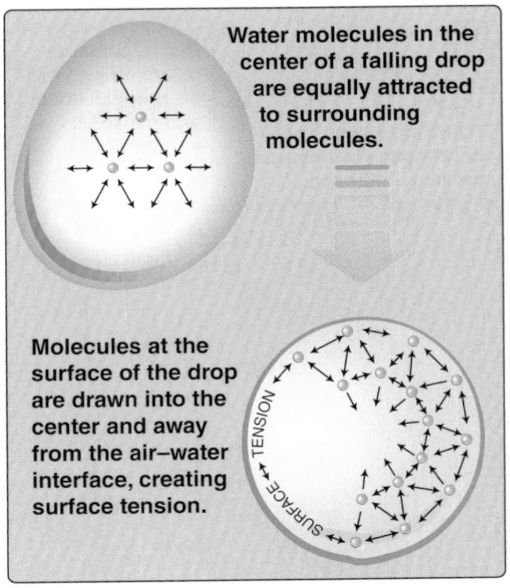

Figure 22.4
Origins of surface tension.

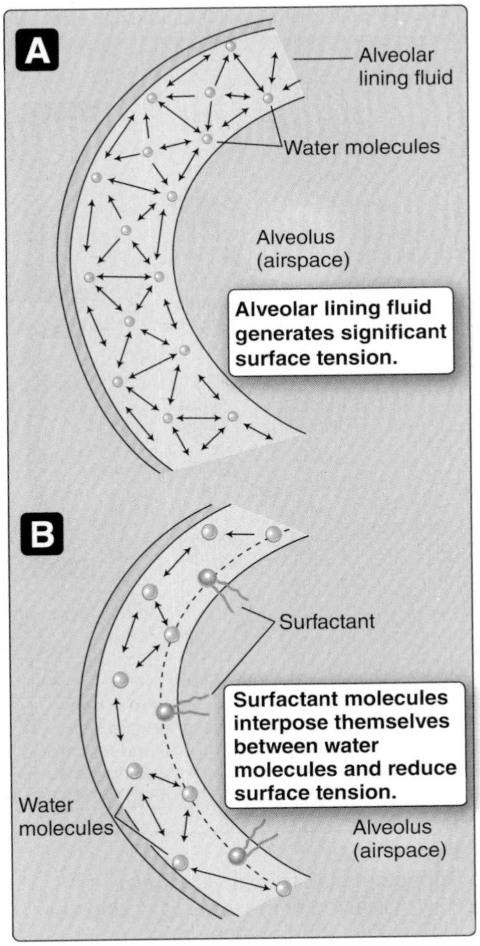

Figure 22.5
Surfactant effects on surface tension created by alveolar lining fluid.

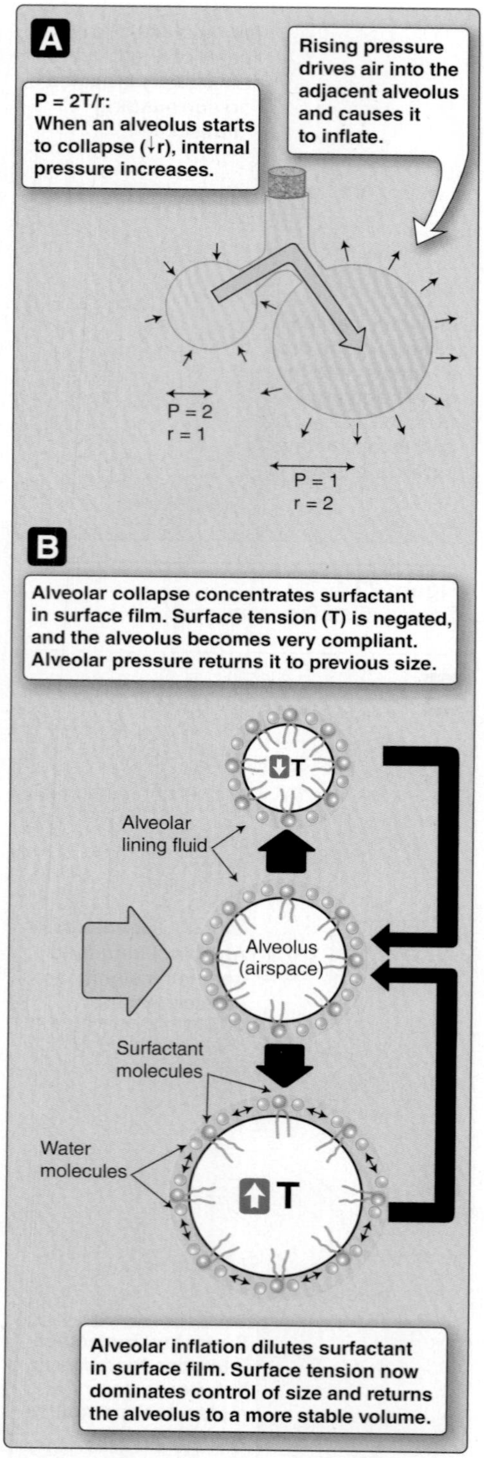

Figure 22.6
Surfactant stabilizes alveolar size.
P = intraalveolar pressure, r = alveolar radius; T = surface tension.

1. **Composition:** Surfactant's principal component is dipalmitoyl phosphatidylcholine (DPPC), a phospholipid. Other components alter secretion rate, aid its distribution within the surface film, or defend the lung against pathogens. Surfactant is stored in lamellar bodies and exocytosed onto the alveolar surface as needed.

2. **Effects on surface tension:** DPPC and other surfactant phospholipids have hydrophilic head groups and hydrophobic tails. When secreted onto the alveolar surface, the molecules localize to the air–water interface, where they spread to form a monolayer (see Figure 22.5B). Their tail groups orient toward the air-filled alveolar lumen, whereas the head groups remain immersed in the superficial aqueous layer. The polar nature of the head groups allows them to interact with and interpose themselves between adjacent water molecules, thereby weakening surface tension. The intensity of surfactant's effects increases in direct proportion to the density of molecules in the surface film.

3. **Functions:** Surfactant's importance in pulmonary function cannot be overstated. There are three main functions; stabilizing alveolar size, increasing compliance, and keeping lungs dry.

   a. **Stabilizing alveolar size:** When two bubbles of unequal size are connected, the smaller bubble collapses and the larger one inflates (Figure 22.6A). This phenomenon is explained by the **Laplace law:**

   $$P = \frac{2T}{r}$$

   Where P is pressure, T is surface tension, and r is bubble radius.

   The Laplace law predicts that pressure within a sealed bubble rises when its radius is reduced. If the collapsing bubble communicates with a larger bubble, rising pressures within the small bubble drive air into the larger bubble. Alveoli approximate bubbles (although their exact shape is more polyhedral), and all alveoli are interconnected via the pulmonary lumen. The Laplace law predicts sequential collapse of all but one alveolus! Although alveolar collapse (**atelectasis**) occurs with regularity *in vivo*, surfactant greatly reduces its extent. A decrease in alveolar volume decreases surface area, which concentrates surfactant molecules within the surface film (see Figure 22.6B). Concentrating the molecules further weakens the forces that create surface tension, thereby preventing collapse. Conversely, alveolar expansion decreases surfactant molecule density and allows surface tension to dominate control of alveolar volume. Surfactant keeps alveolar diameter relatively stable throughout the lung.

   b. **Increasing compliance:** Lung compliance is a measure of the amount of pressure needed to inflate lungs to a given volume ($\Delta V/\Delta P$; see also 19·V·C·1). Surface tension decreases lung compliance and thereby increases the effort required for inflation. Surfactant reduces surface tension's adverse effect on compliance and thereby makes the lungs easier to inflate.

Clinical Application 22.1: Infant Respiratory Distress Syndrome

Infants born prematurely have underdeveloped lungs that are incapable of producing levels of surfactant necessary for stabilizing alveolar volume. Atelectasis is common, as are hyperexpanded regions of the lung. Collapsed alveoli cannot participate in gas exchange and the infant becomes cyanotic as a result. The infant's lungs also have low compliance, which increases the work of breathing. **Infant respiratory distress syndrome** (**IRDS**) is characterized by hypoxia, tachypnea, tachycardia, and exaggerated breathing movements. Ventilatory failure is a likely outcome in the absence of medical intervention. IRDS infants are supported with mechanical ventilation and by delivering surfactant to the lungs until the lungs are sufficiently developed to produce adequate surfactant.

Figure 22.7
Expanding and contracting an accordion's bellows creates airflow and musical notes.

c. **Keeping lungs dry:** The collapsing fluid bubble within an alveolus exerts a negative pressure on the alveolar lining. This pressure creates a driving force for fluid movement from the interstitium onto the alveolar surface. The presence of fluid within an alveolar sac interferes with gas exchange and negatively impacts lung performance. Surfactant reduces the pressure gradient and thereby helps keep lungs fluid free.

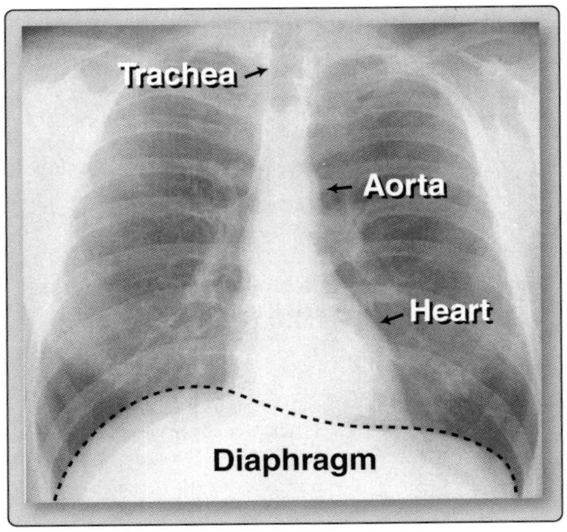

Figure 22.8
Normal posteroanterior chest x-ray.

V. MECHANICS OF BREATHING

The blood–gas interface is separated from the external atmosphere by a distance of ~30 cm (i.e., the length of the trachea and other intervening airways). O_2 cannot diffuse over such a distance fast enough to meet the demands of internal respiration, so air must be drawn into the lungs by an air pump.

A. Pump structure

The pump functions much like an accordion, a musical instrument comprising a bellows operated using two handholds ("manuals") as shown in Figure 22.7. When the manuals are drawn apart, the bellows expand, and pressure inside them drops. This creates a pressure gradient that drives airflow over a set of reeds that give the instrument its familiar sound. Lung tissue (the pulmonary equivalent of bellows) is too fragile to be attached to the muscles and tendons that might function as manuals. Instead, they are hermetically sealed to the lining of the thoracic cavity. This allows the chest wall and diaphragm to expand the bellows while keeping the force per unit area applied to the lungs minimal (Figure 22.8). The seal relies on **pleurae** and pleural fluid.

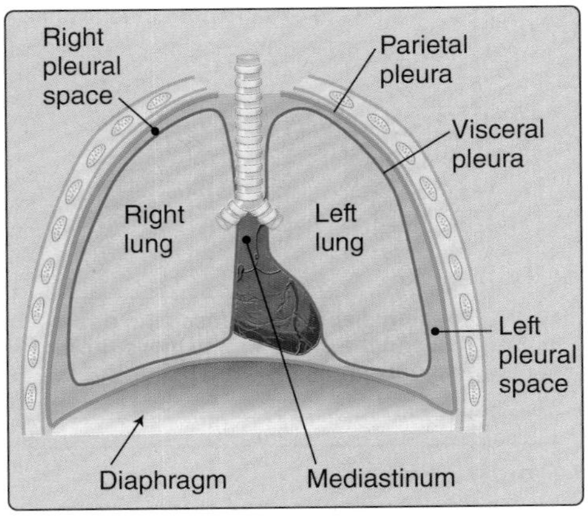

Figure 22.9
Pleurae.

B. Pleurae

Pleurae are thin, serous membranes that cover the lungs. Similar membranes cover the heart (the **pericardium**) and viscera (the **peritoneum**). Pulmonary pleurae have two essential air-pump functions: creating the hermetic seal and secreting pleural fluid.

1. **Hermetic seal:** The lungs are enveloped in **visceral pleura** (Figure 22.9). Each lung is enclosed within its own pleural cavity,

and there is no connection between the two. The chest wall, diaphragm, and mediastinum (heart, larger blood vessels, airways, and associated structures) are covered with **parietal pleura**. The visceral and parietal pleurae are physically attached to their respective underlying structures but not to each other, and the two membranes are separated by the **intrapleural space**. The pleurae effectively exclude air from the intrapleural space to hermetically seal the lungs to the diaphragm and rib cage.

2. **Pleural fluid:** The parietal pleura is innervated and vascular. It is believed to be the source of a viscous **pleural fluid** that is secreted into the intrapleural space. Pleural fluid has two important functions: lubrication and cohesion.

    a. **Lubrication:** Pleural fluid lubricates the pleural surfaces and allows the lungs to slide freely over the chest wall and diaphragm during normal breathing movements.

    b. **Aiding inspiration:** Pleural fluid is secreted and reabsorbed constantly. The volume contained within the intrapleural space at any one time is ~10 mL total, but it spreads to create a thin film that covers all surfaces, making the two pleura almost inseparable under physiologic circumstances. The same cohesive force makes two glass microscope slides difficult to separate when a drop of water is caught between them. Cohesion allows forces generated by movement of the chest wall and diaphragm to be transferred directly to the lung surface.

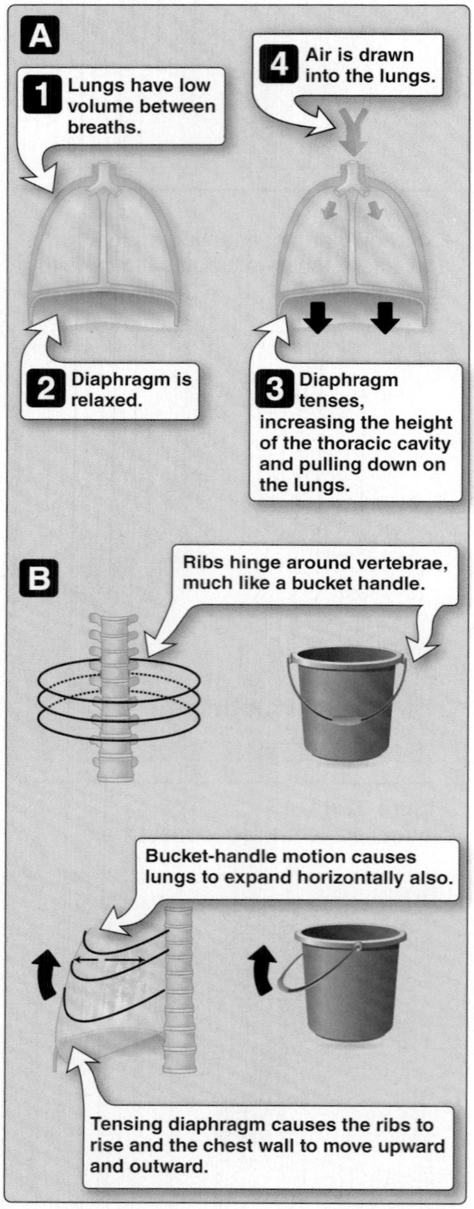

Figure 22.10
Thoracic volume changes during inspiration.

C. Pump cycling

Respiration involves repeated cycles of inspiration and expiration. Inspiration draws air into the lungs and increases O_2 availability at the blood–gas interface.

1. **Inspiration:** The air pump is operated by skeletal muscles (Table 22.1). Most important of these is the diaphragm (see Figure 22.9), a dome-shaped muscle that separates the thoracic and abdominal cavities that is innervated by the phrenic nerve. When the muscle contracts, intrathoracic volume increases.

    a. **Vertical dimensions:** Diaphragm contraction pushes downward on the abdominal contents and increases the vertical dimensions of the thoracic cavity by between 1 and 10 cm, depending on activity level (Figure 22.10A).

    b. **Cross-sectional area:** Contraction also increases cross-sectional area by pulling upward on the lower ribs (see Figure 22.10B). The ribs rise in a bucket-handle fashion, a motion aided by contraction of the external intercostal muscles.

2. **Expiration:** Expiration is generally passive and is driven both by the effects of surface tension on alveolar volume and energy stored in a lung's elastic elements during inspiration (**elastic recoil**). Elasticity reflects an abundance of elastin and collagen fibers in both airways and alveoli. Expiratory muscles are typically

Table 22.1: Muscles Used in Breathing

Inspiration	
Diaphragm	**Costal fibers** are attached to the ribs; **crural** fibers course around the esophagus and are tethered via ligaments to the vertebrae. Contraction pushes downward on the abdominal contents and raises the chest wall.
External intercostals	Connect adjacent ribs and are angled forward so that contraction raises the chest wall.
Accessory muscles	Accessory muscles are used during forced inspiration and exercise. **Scalenes** elevate the first two ribs, **sternomastoids** lift the sternum, and muscles in the upper part of the respiratory tract dilate the upper airways.
Expiration	
Abdominal muscles	Muscles of the abdominal wall (**rectus abdominis**, **transversus abdominis**, and **internal** and **external oblique muscles**) contract during forced expiration to compress the abdominal cavity and push the diaphragm upward. These muscles are also activated during vomiting, coughing, and defecation.
Internal intercostals	Connect adjacent ribs. They pull the ribs downward and inward when they contract.

used during exercise or when airway resistance is increased by disease, for example (see Table 22.1).

VI. STATIC LUNG MECHANICS

Surface tension's influence on lung volume is tempered by surfactant, but it still remains a significant force that impacts lung behavior during normal breathing.

A. Forces acting on a static lung

A healthy lung at rest is subject to two equal and opposing forces, one directed inward and the other outward.

1. **Inward:** As discussed previously, a lung's elasticity and surface tension effects generate an inwardly directed force that favors smaller lung volumes (Figure 22.11A).

2. **Outward:** The muscles and various connective tissues associated with the rib cage also have elasticity. At rest, the elastic elements favor outward movement of the chest wall.

3. **Net effect:** The two opposing forces create negative pressure within the intrapleural space (**intrapleural pressure**, or $\mathbf{P_{pl}}$). P_{pl} is measured relative to the atmosphere and averages several centimeters of water, depending on vertical position within the lung (discussed below). If either pleura is breached, air rushes into the pleural space, driven by the pressure difference between atmosphere and pleural space (**pneumothorax**) as shown in Figure 22.11B. P_{pl} falls to zero, elastic elements in the chest wall cause it to spring outward, and the lung collapses.

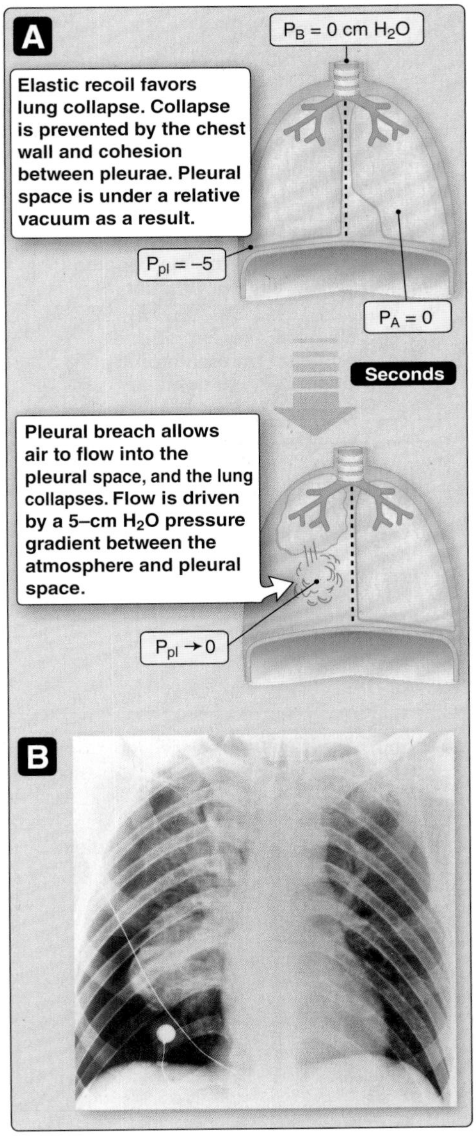
A
P_B = 0 cm H_2O
Elastic recoil favors lung collapse. Collapse is prevented by the chest wall and cohesion between pleurae. Pleural space is under a relative vacuum as a result.
P_{pl} = −5
P_A = 0
Seconds
Pleural breach allows air to flow into the pleural space, and the lung collapses. Flow is driven by a 5–cm H_2O pressure gradient between the atmosphere and pleural space.
P_{pl} → 0
B

Figure 22.11
Pneumothorax. P_A = alveolar pressure; P_B = barometric pressure; P_{pl} = intrapleural pressure.

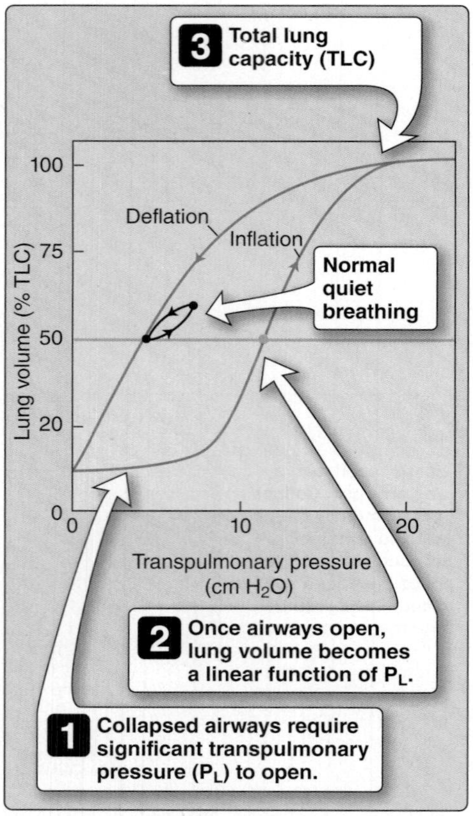

Figure 22.12
Pulmonary pressure–volume loop.

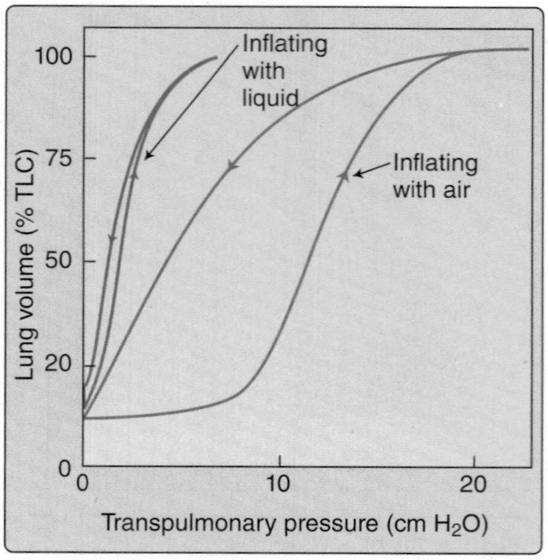

Figure 22.13
Pressure–volume loop for a fluid-filled lung. TLC = total lung capacity.

The human thorax contains two pleural cavities. In practice, this means that injury resulting in pneumothorax usually affects only one lung at a time. Thus, although pneumothorax is still a serious condition, it is not immediately fatal. By contrast, the American buffalo (bison) contains a single pleural cavity that is relatively easy to breach with a gunshot or arrowhead. This immense animal once roamed the plains of North America in large numbers, but its vulnerability to pneumothorax allowed hunters and farmers to systematically decimate herds during the late 1800s.

B. Pressure–volume curves

A collapsed lung can help us understand how much effort is required to inflate it during normal breathing.

1. **Inflation:** A collapsed lung can be inflated in one of two ways, both of which modify **transpulmonary pressure (P_L)**. P_L is the difference between **intraalveolar pressure (P_A)** and P_{pl}:

$$P_L = P_A - P_{pl}$$

A mechanical positive-pressure ventilator can be used to raise pressure inside the lung ($P_A > P_{pl}$), inflating it much as one inflates a balloon. Alternatively, air can be withdrawn from the pleural space to create a negative pressure outside the lung ($P_{pl} < P_A$). Both maneuvers raise P_L. Lung collapse allows the airways to close off and seal with fluid. Restoring patency requires that the surface tension seal be broken, which requires considerable effort. P_L must be increased by several cm H_2O before any significant increase in volume occurs (Figure 22.12, phase 1). Once P_L exceeds ~7–10 cm H_2O, airways pop open, and volume increases linearly with inflation pressure (phase 2). At ~20 cm H_2O, the lung reaches its maximal volume, known as **total lung capacity (TLC)**, as shown in Figure 22.12, phase 3.

2. **Deflation:** A lung that is allowed to deflate from TLC yields a different pressure–volume curve from that seen during inflation (a phenomenon known as **hysteresis**). This is because surfactant is recruited from pneumocytes to the alveolar surface film during lung inflation. The surfactant decreases elastic recoil and thereby resists lung deflation. Note that the **hysteresis loop** begins and ends at a positive volume (normally ~500 mL; see Figure 22.12). This is because the larger airways collapse at zero pressure and trap air within the more distal regions.

3. **Normal breathing:** Normal breathing involves changes in lung volume that are only a fraction of total, but the hysteresis is still evident (see Figure 22.12). Note also that inspiration normally begins at ~50% of TLC. When the lung is resting between breaths, the chest wall prevents it from collapsing, and, at 50% TLC, all

alveoli are patent. The chest wall also positions the resting lung on the steepest portion of the pressure–volume curve, meaning that the increase in P_L during inspiration is maximally effective in increasing alveolar volume.

4. **Surface tension effects:** Surface tension's influence on the pressure–volume loop can be estimated by filling lungs with fluid to eliminate air–water interfaces (Figure 22.13). Fluid-filled lungs are more compliant, and the hysteresis associated with surface tension disappears.

C. Gravitational effects

Lungs and blood have mass and, thus, are subject to the influence of gravity. Gravity causes significant regional differences in P_L and alveolar volume.

1. **Apex:** When the thorax is positioned vertically, a lung within can be imagined to hang suspended by its apical pleura. Suspension creates a strongly negative P_{pl} (and strongly positive P_L) locally and causes apical alveoli to inflate to ~60% of their maximal volume (Figure 22.14). Gravity similarly stretches the coils at the top of a Slinky (the famous toy) farther apart than those at the base (Figure 22.15). In practice, gravitational influences force the lung apex to function near the top of the pressure–volume curve, where the opportunity for further expansion during inspiration is very limited.

2. **Base:** The lung base supports the mass of pulmonary tissue above it. Alveoli in this region are compressed, much like coils at the base of the Slinky. The weight of tissue above also pushes outward against the chest. P_{pl} and P_L both approach zero (see Figure 22.14). In practice, this means that alveoli at base of the lung respond to increases in P_L with large changes in volume because they occur over the lower, steepest part of the pressure–volume curve.

D. Lung compliance

The amount that lung volume increases in response to changes in P_L is a measure of its **compliance**. Lungs are highly compliant organs, increasing volume by ~200 mL for every cm H_2O of transpulmonary pressure. Compliance is governed both by surface tension and the elastic properties of the lungs and chest wall. Lungs become less compliant with age due to connective tissue deposition.

VII. LUNG DISEASES

Obstructive and **restrictive pulmonary diseases** are two broad disease groups that cause significant changes in the static lung properties. The two groups are typified, respectively, by **emphysema** and **pulmonary fibrosis**. We will revisit these diseases frequently to help illustrate the mechanical principles involved in normal breathing.

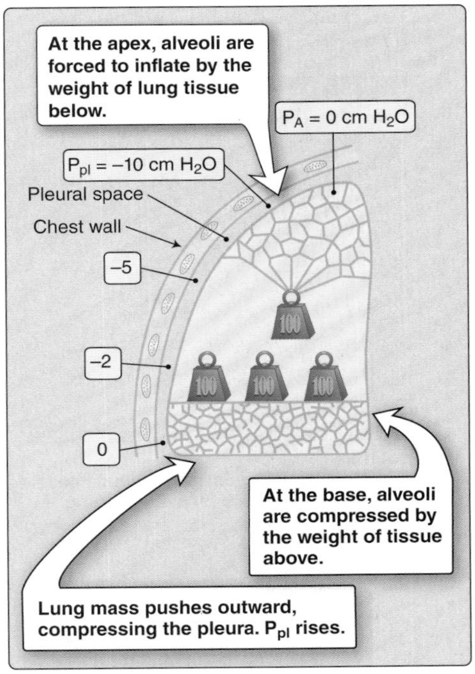

P_A = 0 cm H_2O
P_{pl} = −10 cm H_2O

Lung mass pushes outward, compressing the pleura. P_{pl} rises.

Figure 22.14
Gravitational effects on alveolar volume. P_A = intraalveolar pressure; P_{pl} = intrapleural pressure.

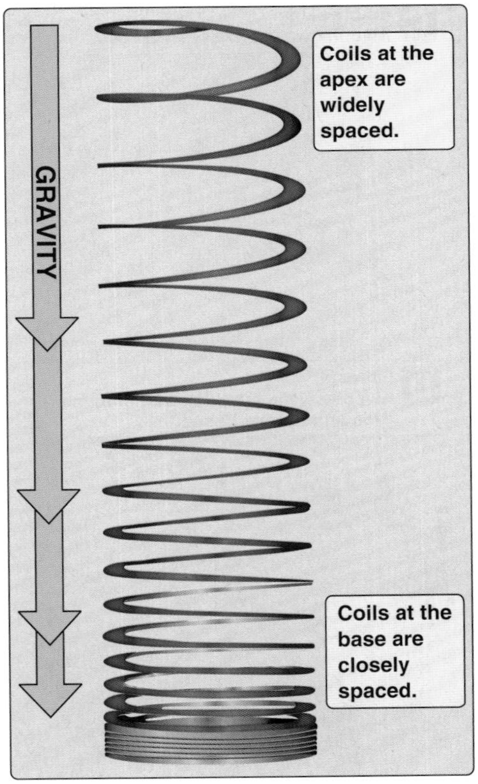

GRAVITY
Coils at the apex are widely spaced.
Coils at the base are closely spaced.

Figure 22.15
Gravitational effects on a Slinky.

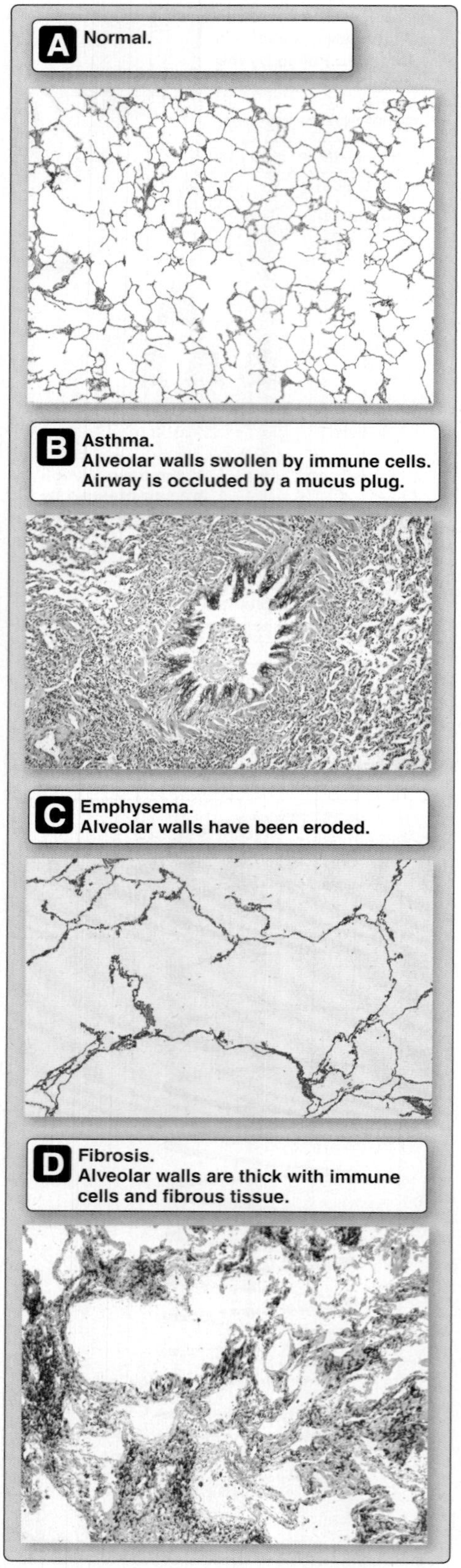

Figure 22.16
Normal and diseased lungs.

A. Obstructive pulmonary disease

Emphysema, chronic bronchitis, and asthma all represent **obstructive pulmonary diseases**, which increase airway resistance to airflow. Because the former two examples often coexist and may be difficult to distinguish clinically, they are commonly grouped and discussed as **chronic obstructive pulmonary disease** (**COPD**). COPD is extremely common and has become the fourth leading cause of death in the United States. There are three general obstructive mechanisms: airway occlusion, wall thickening, and loss of mechanical tethering.

1. **Airway occlusion:** Airways may be occluded by foreign bodies or, more commonly, by secretions that are excessive or difficult to expel (Figure 22.16B). Occlusive diseases include **chronic bronchitis**, **asthma**, and **bronchiectasis**.

2. **Wall thickening:** When the airway wall hypertrophies or becomes edematous, it encroaches on the lumen and reduces its cross-sectional area (see Figure 22.16B).

3. **Loss of mechanical tethering:** All structures in the lung are linked mechanically. Together, they form a dependent network, much like the fabric of a nylon stocking (**interdependence**). Interdependence maintains airway patency when external forces may favor collapse. Emphysema develops when alveolar walls (the fabric of the lung) erode, allowing surrounding airways to collapse and obstruct airflow during normal breathing (see Figure 22.16C). Emphysema is commonly caused by heavy smoking.

> **Emphysema** denotes anatomic tissue loss, although the term is sometimes used to describe smoking-related lung disease. It is a finding evident on computed tomography imaging of the lung. It is also a pathologic finding seen at autopsy or lung tissue biopsy. Postmortem examination of an emphysematous lung shows enlarged cystic air spaces replacing normal lung. Alveolar loss reduces elastic recoil, increases pulmonary compliance, and reduces the surface area available for O_2 uptake.

B. Restrictive pulmonary disease

Pulmonary fibrosis is a **restrictive pulmonary disease**. Others include pleural diseases and problems affecting breathing muscles, all of which limit lung expansion. Pulmonary fibrosis (scarring) results from any one of a number of **interstitial lung diseases**. Scarring typically begins with an injury to the alveolar epithelium. Causes include any of several compounds inhaled in the workplace (e.g., asbestos, beryllium, coal dust, sawdust), circulating drugs (e.g., antibiotics and chemotherapeutic agents), systemic diseases (e.g., rheumatoid arthritis, lupus, scleroderma, and sarcoidosis), or may be idiopathic. The initial insult causes the alveolar wall to thicken and the alveolar space to fill with an exudate containing lymphocytes, platelets, and other immune effector cells (see Figure 22.16D). The space is then

infiltrated by fibroblasts, which lay down bundles of collagen and other fibers between the alveolar sacs. Scar tissue is relatively noncompliant, so the lung becomes stiff and expands with difficulty during inspiration. Diseased lungs decrease O_2 uptake, and hypoxemia may develop as scarring progresses.

VIII. DYNAMIC LUNG MECHANICS

During inspiration, air moves from the external environment through a set of branching tubes of ever-decreasing diameter. Flow is driven by the pressure difference between the external atmosphere and alveolus ($\Delta P = P_B - P_A$). Flow is inversely proportional to airway resistance (R):

$$\dot{V} = \frac{\Delta P}{R}$$

where $\dot{V}$ is airflow (volume ÷ unit time).

A. Pressures driving airflow

Airflow occurs in response to pressure gradients set up between the alveoli and the external atmosphere. The body has no way of controlling P_A directly. Instead, the diaphragm and other respiratory muscles manipulate intrapleural pressure. When P_{pl} falls, P_L rises, and the alveoli expand. P_A becomes negative because the product of pressure and volume of a fixed number of air molecules remains constant (as per the **Boyle law**).

$$P_{A1}V_{A1} = \downarrow P_{A2}\uparrow V_{A2}$$

Where P_{A1} and P_{A2} denote alveolar pressure before and after alveolar expansion (V_{A1} and V_{A2}). Alveolar expansion thus creates a $P_B > P_A$ pressure gradient that drives airflow into the lungs (Figure 22.17). Because flow occurs against a resistance, it takes time for air to move in or out of the lungs and for the pressure gradient to dissipate, particularly at the points farthest removed from the site of highest resistance.

B. Resistance to airflow

In a healthy individual, breathing is usually an effortless and unconscious act, so it seems surprising that airways can offer resistance to flow. The resistance has several origins. Resistance is proportional to airway length (l) and the viscosity (η) of the gas moving through it and inversely proportional to the fourth power of airway radius (r), as stated in the **Poiseuille law**:

$$R = \frac{8L\eta}{\pi r^4}$$

Airway radius has the greatest influence on resistance, although viscosity and turbulence should be considered also.

1. **Airway radius:** Airway radius decreases with each successive generation within the bronchiolar tree. Decreasing radius increases resistance, but the negative impact on net airflow through the lung is more than offset by the gain in airway numbers with each successive generation. In other words, although individual bronchioles have a very high resistance, their combined resistance is almost negligible (calculated from the sum of reciprocals;

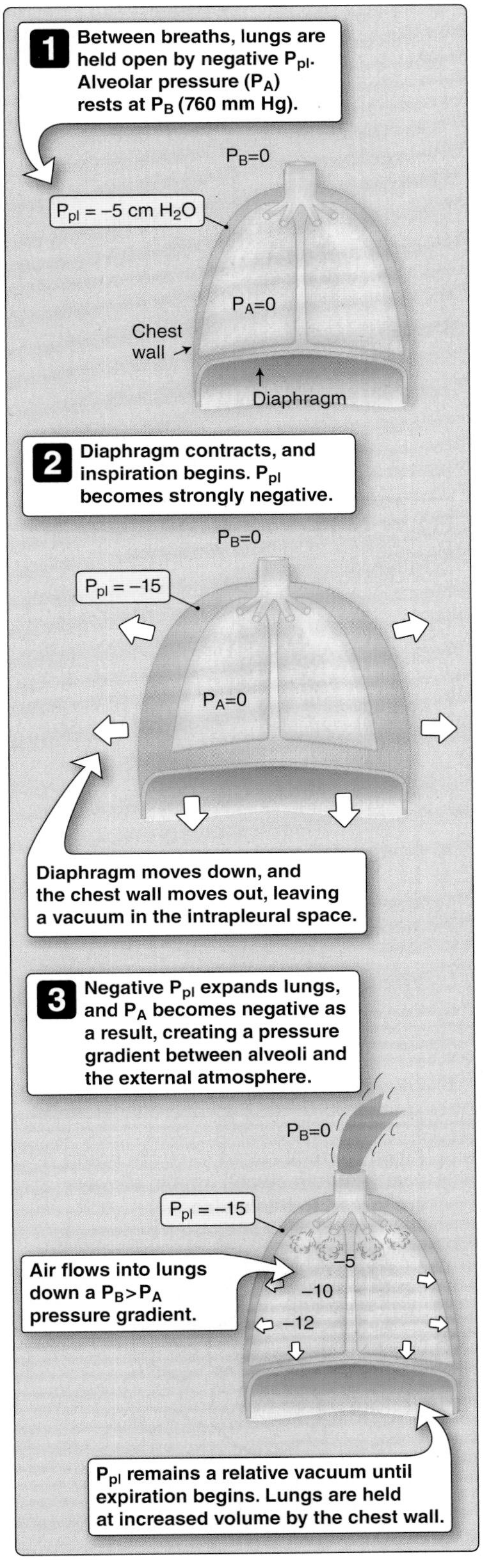

Figure 22.17
Pressure gradients driving airflow during inspiration. P_B = barometric pressure; P_{pl} = intrapleural pressure. All values are given in cm H_2O.

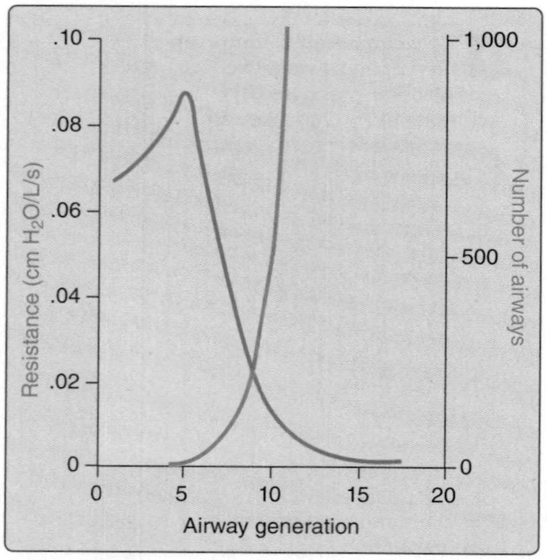

Figure 22.18
Resistance to airflow within the bronchial tree.

see also 19·IV·B). The site of greatest resistance in the lung is in the pharynx and larger airways (generations 0 through ~7) as shown in Figure 22.18.

2. **Air viscosity:** Air viscosity is dependent on air density. Air density increases when compressed, as during a deep-sea dive, for example. Increasing density increases flow resistance and the **work of breathing**. Breathing an O_2/helium mixture partly offsets this density increase. Helium has less density than atmospheric air and, therefore, reduces the work of breathing.

3. **Turbulence:** The Poiseuille law above assumes that airflow through the lungs is streamline, but this is generally not the case. The airways consist of a series of branching tubes. Each branch point creates a local eddy current that disrupts streamline flow and increases airway resistance. In practice, the eddy currents cause flow through the airways to be proportional to ($\Delta P + \sqrt{\Delta P}$) rather than ΔP alone.

C. Factors affecting airway resistance

Airways are the primary source of resistance in the lung and, therefore, changes in airway radius can significantly impact lung function. Airway radius is governed by airway musculature and by lung volume.

1. **Smooth muscle:** Bronchioles are lined with smooth muscle cells. When the muscles contract, they decrease airway radius and increase resistance to airflow. Flow through the airways may decrease as a result. Smooth muscle relaxation and bronchiolar dilation reduces resistance and facilitates increased airflow. Airway muscles are regulated by the autonomic nervous system (ANS) and by local factors.

   a. **Autonomic control:** Airways are controlled by both parasympathetic and sympathetic (SNS) branches of the ANS.

      i. **Parasympathetic:** Parasympathetic nerve fibers from the vagus nerve release acetylcholine (ACh) from their terminals when active. ACh binds to M_3 muscarinic ACh receptors and causes bronchoconstriction, which reduces airflow.

      ii. **Sympathetic:** Sympathetic activation causes bronchioles to dilate, mainly by inhibiting ACh release rather than through direct effects on the musculature. SNS terminals release norepinephrine, which binds to a presynaptic β_2-adrenergic receptor. This receptor is particularly sensitive to epinephrine release from the adrenal medulla during SNS activation. SNS-mediated bronchodilation is important for facilitating increased airflow to the blood–gas interface during exercise, for example.

   b. **Local factors:** Local irritants and allergens constrict bronchioles and obstruct airways. Airway muscle contraction is a response to histamine and other inflammatory mediators.

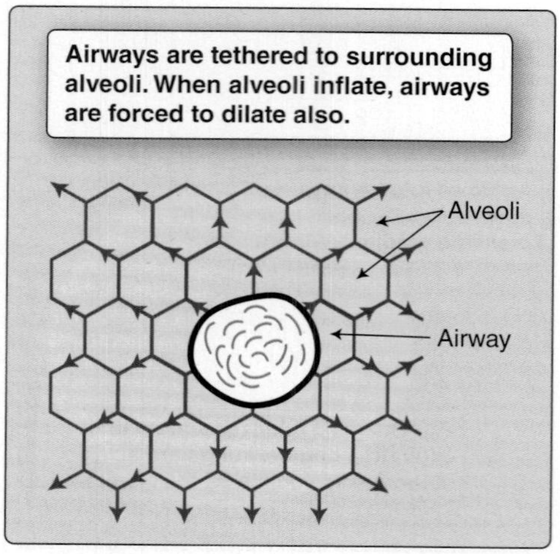

Figure 22.19
Radial traction on airways during lung inflation.

2. **Lung volume:** Airway resistance is highly dependent on lung volume. Net airway resistance is low at high lung volumes and high at low volumes.

a. **High volumes:** The decrease in P_{pl} that establishes a gradient for airflow during inspiration is transmitted to the airways as well as to alveoli. This causes airway resistance to fall during lung expansion. Airways are also dilated by **radial traction**. Traction results from a mechanical tethering between the alveoli and all surrounding structures. In practice, when the alveoli are expanded, radial traction on intervening airways increases their radius and lowers their resistance (Figure 22.19).

b. **Low volumes:** At low lung volumes, radial traction is reduced, and airway resistance increases.

D. Airway collapse during expiration

Airways tend to collapse and limit flow during expiration, an effect known as **dynamic compression of the airways**. The reasons and consequences of collapse are easiest to appreciate during a forced expiration after a deep inspiration (Figure 22.20). Forced expiration begins with contraction of the abdominal muscles and internal intercostals, which forces the chest wall downward and inward, and causes P_{pl} to become positive. The positive pressure is transferred to and compresses the alveoli, decreasing their volume and causing P_A to rise above P_B. Compression thereby establishes the pressure gradient that drives expiratory outflow. The larger airways have a relatively high resistance to flow that limits lung-emptying rates, so there is a time period during which alveoli remain filled with pressurized air. High intraalveolar pressure maintains patency, even though P_{pl} may be positive and favoring alveolar collapse. Airway pressure falls with distance from the alveoli and proximity to the main site of resistance (bronchi and trachea). Thus, whereas intraalveolar pressure may be strongly positive (relative to P_B), pressure within the larger airways may be much closer to zero (i.e., P_B) and thus more susceptible to collapse by P_{pl} (see Figure 22.20[3]). The larger airways are equipped with cartilage that helps maintain patency during forced expiration, but it may be inadequate to prevent collapse. As air leaves the lungs and P_A drops, the collapse zone moves distally and involves increasingly smaller airways. Compression and collapse of conducting airways is the self-regulating, limiting factor that determines how fast air escapes the lungs during expiration. If a subject attempts to speed outflow with a more forceful muscular contraction, the pressure gradient driving outflow is raised, but so are the forces favoring airway collapse with a net zero sum gain (Figure 22.21).

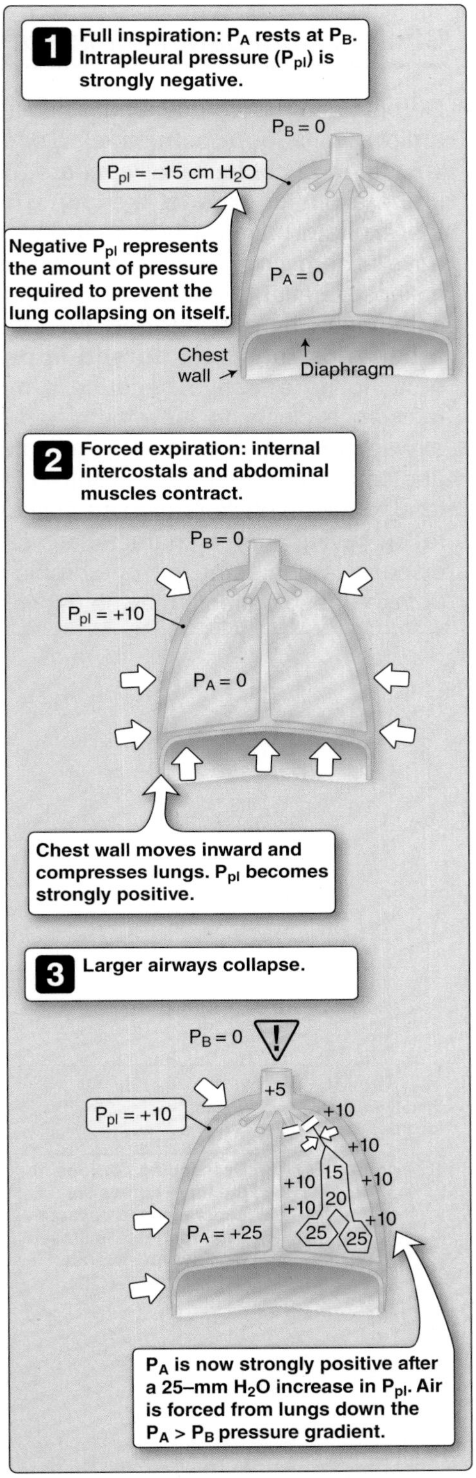

1 Full inspiration: P_A rests at P_B. Intrapleural pressure (P_pl) is strongly negative.
P_B = 0
P_pl = –15 cm H2O
Negative P_pl represents the amount of pressure required to prevent the lung collapsing on itself.
P_A = 0
Chest wall
Diaphragm
2 Forced expiration: internal intercostals and abdominal muscles contract.
P_B = 0
P_pl = +10
P_A = 0
Chest wall moves inward and compresses lungs. P_pl becomes strongly positive.
3 Larger airways collapse.
P_B = 0
+5
P_pl = +10
+10
+10
+10
15
+10
+10
20
+10
P_A = +25
25
25
P_A is now strongly positive after a 25–mm H2O increase in P_pl. Air is forced from lungs down the P_A > P_B pressure gradient.

Figure 22.20
Airway collapse during forced expiration. P_A = intraalveolar pressure; P_B = barometric pressure; P_{pl} = intrapleural pressure. All values are given in cm H_2O.

E. Work of breathing

Breathing requires that the respiratory muscles contract to expand the lungs against resistance. The **work of breathing** normally accounts for ~5% of total energy usage at rest, but it can rise to >20% of total during exercise. Such workloads are normally insignificant in a healthy individual, but some patients with pulmonary disease have difficulty expanding their lungs, and even resting breathing movements can fatigue their respiratory muscles and precipitate respiratory failure (see 40·VI).

1. **Work components:** Many factors contribute to the work of breathing. The two principal factors are elastic work and resistive work.

Clinical Application 22.2: Pursed-Lipped Breathing

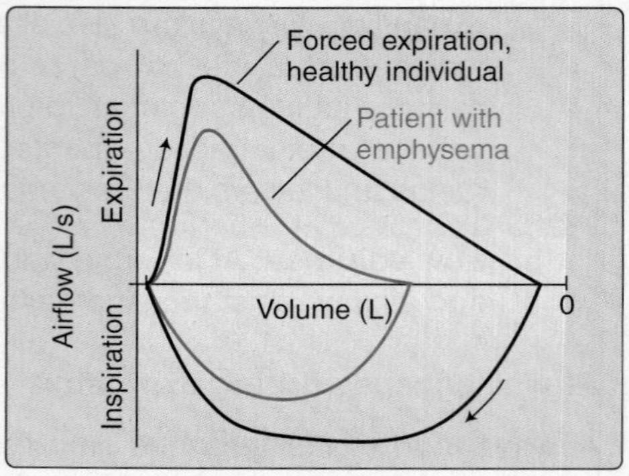

Effects of emphysema on airflow.

Chronic obstructive pulmonary disease (COPD) is characterized by airflow limitation (obstruction). Spirometry testing reveals a flow–volume loop contour that appears "scooped-out" (concave upward) in the expiratory limb of the loop. There may also be a long tail on the expiratory limb, which manifests because patients with COPD have a hard time exhaling due to loss of elastic recoil and airway collapse. Patients can partly compensate for the loss of mechanical support by pursing their lips (as if whistling) during expiration, a behavior known as **pursed-lipped breathing**, or puffing. This behavior is effective because it moves the site of main airway resistance closer to the mouth and extends the time during which airway pressure remains high and the airways patent. Patients with anatomic tissue loss (emphysema) in addition to airflow obstruction tend to hyperventilate and use accessory muscles to help with expiration, giving them a characteristic pink complexion ("pink puffers"). This contrasts with COPD patients whose disease is characterized by chronic bronchitis and excessive mucus production that interferes with oxygen uptake (these patients may be described as "blue bloaters").

Elastic work includes the work required to counter a lung's elastic recoil during inspiration, which is proportional to its compliance. Work is also required to displace the chest wall outward and the abdominal organs downward. **Resistive work** involves moving air through the airways against airway resistance.

2. **Measuring work:** Work is calculated as the amount of force required to move an object a given distance. In pulmonary terms, the work of breathing is calculated from the product of the force needed to change the transpulmonary pressure gradient and the air volume moved per unit time. Work can be represented graphically as the area to the left of the inspiratory phase of the pressure–volume loop (Figure 22.22).

3. **Pulmonary diseases:** COPD and pulmonary fibrosis both increase the work of breathing (see Figure 22.22). Patients with COPD work harder to exhale against high airway resistance (increased resistive work). Pulmonary fibrosis stiffens the lung and requires that a patient generate higher transpulmonary pressures than normal to expand the lungs during inspiration (increased elastic work).

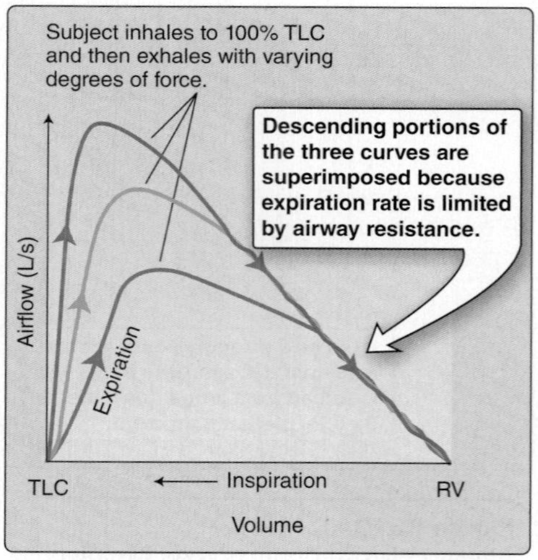

Figure 22.21
Airway resistance limits flow during forced expiration. RV = residual volume; TLC = total lung capacity.

IX. LUNG VOLUMES AND CAPACITIES

Normal quiet breathing uses less than 10% of TLC. Exercise increases this amount significantly, but there is always a small residual volume that communicates with the ventilated space but does not itself participate in ventilation, even at maximal levels of exercise. Clinically, it is important to determine the contribution of this volume to the mix of gases in the lungs and to assess how lung volume(s) may be impacted by the progression of various pulmonary diseases. In addition to airflow

measurement with **spirometry, pulmonary function tests** (**PFTs**) typically measure four primary **lung volumes**, which are then combined to derive several **lung capacities** (Figure 22.23). PFTs also assess the efficiency of the blood–gas interface ("diffusing capacity" is discussed in Chapter 23·V).

A. Volumes

The volume of air inspired or expired with each breath, typically ~500 mL in an average adult, is called the **tidal volume** (**TV**). **Inspiratory reserve volume** (**IRV**) and **expiratory reserve volume** (**ERV**) are the volumes that can be inspired or expired, respectively, over and above TV. **Residual volume** (**RV**) is the volume of air remaining in the lung after a maximal expiration (~1.2 L in a normal individual). A spirometer is unable to provide information about RV, so pulmonary function testing often includes more specialized body plethysmography or techniques that monitor intrapulmonary concentrations of gases over time (i.e., helium-dilution and nitrogen-washout assays).

B. Capacities

The sum of all four lung volumes (TLC) amounts to ~6 L in a normal individual (see Figure 22.23). **Functional residual capacity** (**FRC**) is the volume remaining in the lungs after expelling a tidal breath. **Inspiratory capacity** (**IC**) is the sum of the TV and the IRV. **Vital capacity** (**VC**) is the sum of the TV, IRV, and ERV, and is the maximal TV achievable (i.e., the biggest breath one can take). **Forced vital capacity** (**FVC**) is the volume of air that can be *forcibly* expired after a maximal inspiration.

C. Forced expiratory volume

FEV_1 is the volume of air that can be forcibly expired *in 1 second* following a maximal inspiration and is an important clinical measure of lung function (see Clinical Application 22.3).

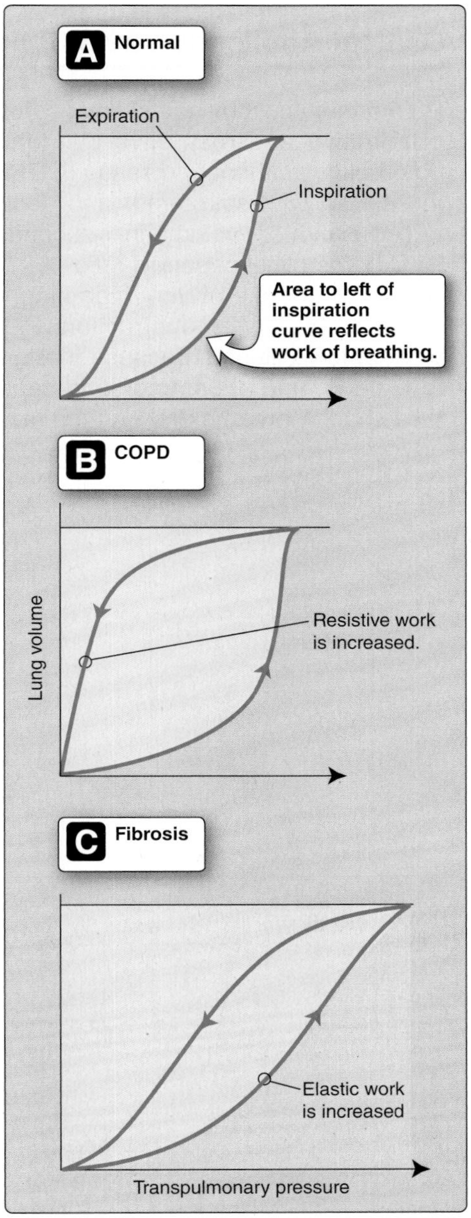

Figure 22.22
Effects of pulmonary disease on the work of breathing. COPD = chronic obstructive pulmonary disease.

X. DEAD SPACE AND VENTILATION

Gas exchange occurs at the alveolar surface. By the time that inspired air contacts the gas exchange interface, its O_2 and CO_2 concentration has been modified through mixing with gases lingering in the RV, which itself is influenced by how often the contents of the lung are refreshed (**ventilation**). Alveolar gas concentration is also influenced by the amount of inhaled air that does not participate in gas exchange because it fills **dead space**.

A. Dead space

The lung contains two types of dead space: anatomic and physiologic.

1. **Anatomic:** The pharynx, trachea, bronchi, and other conducting airways contain ~150 mL of air that is moved out during expiration without ever contacting the gas exchange interface. This represents **anatomic dead space**.

Clinical Application 22.3: Pulmonary Function Tests

Pulmonary function tests are useful in detecting the presence of obstructive and restrictive pulmonary pathophysiology. Chronic obstructive pulmonary disease (COPD) is best identified by measuring airflow with spirometry and documenting obstruction (reduced forced expiratory volume in 1 second [FEV_1] in a setting of an FEV_1/FVC [forced vital capacity] ratio of <70%). Patients with COPD typically also operate at very high lung volumes because exhalation is impaired by airway obstruction. Patients with pulmonary fibrosis work at low volumes because the lung is noncompliant and difficult to expand, and, thus, total lung capacity is reduced. They typically take shallow breaths and breathe rapidly.

	FEV_1	FVC	FEV_1: FVC (%)
Normal	~4.0	~5.0	>70
Obstructive	~1.3	~3.1	<70
Restrictive	~2.8	~3.1	>70

FEV_1 = volume of air forcibly expelled in 1 second; FVC = forced vital capacity. Values for FEV_1 and FVC are in liters. FVC is generally slightly less than vital capacity, hence the use of a distinguishing term.

2. **Physiologic:** In a diseased lung, a proportion of the alveoli may be ventilated but unable to participate in gas exchange because the blood–gas interface is damaged or pulmonary blood flow to these regions has been interrupted. These regions represent dead space. The term **physiologic dead space** includes anatomic dead space and contributions from these nonfunctional alveoli. In a healthy person, physiologic and anatomic dead spaces are approximately equal. In a diseased lung, the physiologic dead space may be increased by 1,500 mL or more.

3. **Calculating dead space:** Dead-space volume (V_D) can be calculated by measuring the amount of CO_2 contained in expired air (P_ECO_2). Dead space (by definition) does not participate in gas exchange and, thus, contains negligible CO_2. The amount of CO_2 in air originating from regions of the lung involved in gas exchange equals that of arterial blood (P_aCO_2), because blood gases equilibrate with alveolar gases during transit through the pulmonary circulation (i.e., $P_ACO_2 = P_aCO_2$). Thus, dead-space volume can be determined from the extent to which the amount of CO_2 in expired air has been decreased by CO_2-free air originating from dead space:

$$V_D = V_T \times \frac{P_aCO_2 - P_ECO_2}{P_aCO_2}$$

where V_T = total lung volume.

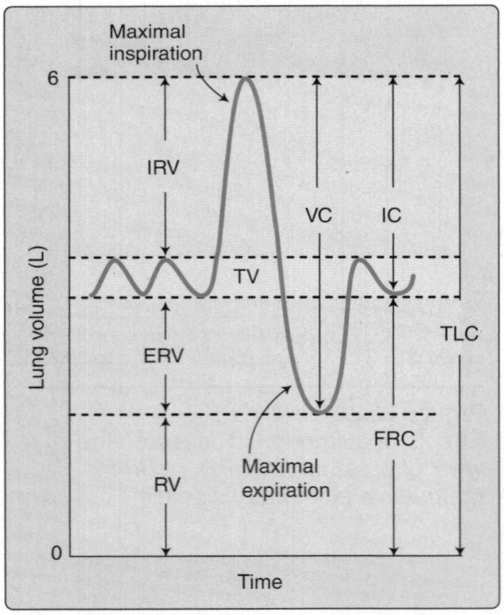

Figure 22.23
Spirometry. ERV = expiratory reserve volume; FRC = functional residual capacity; IC = inspiratory capacity; IRV = inspiratory reserve volume; RV = residual volume; TLC = total lung capacity; TV = tidal volume; VC = vital capacity.

B. Ventilation: Ventilation can be expressed as **minute ventilation** or **alveolar ventilation**. Minute ventilation (V_E) is the total volume of air inhaled and exhaled per minute:

$$\text{Minute ventilation} = TV \times \text{breaths/min}$$

Alveolar ventilation (V_A) is the volume of air per minute that enters the areas participating in gas exchange:

$$V_A = (TV - V_D) \times \text{breaths/min}$$

Where V_A represents alveolar ventilation and V_D is dead space.

Chapter Summary

- Lungs facilitate exchange of O_2 and CO_2 between blood and air. The **blood–gas interface** is located within **alveoli**, thin-walled sacs that serve to amplify interface surface area and to bring the pulmonary circulation into close proximity to inhaled air.
- Alveoli are moistened with a thin fluid film that generates surface tension. Surface tension is a force that favors lung collapse and negatively impacts lung performance. The alveolar epithelium produces **surfactant** to counter this surface tension. Surfactant is a **phospholipid** complex that helps stabilize alveolar size and increases lung compliance.
- Breathing involves repeated cycles of **inspiration** and **expiration**. Air is drawn into the lungs by contracting the **diaphragm** and other **respiratory muscles**. Contraction increases the volume of the thoracic cavity and lungs.
- The diaphragm, chest wall, and lungs move as one unit. They are linked by a thin film of **pleural fluid**, which lubricates the **visceral** and **parietal pleurae** and provides the cohesive force required to expand the lungs.
- At rest, a lung is subject to two opposing forces. Surface tension and elastic elements in lung tissue favor collapse (**elastic recoil**). Elastic elements in the chest wall favor expansion and thereby prevent collapse. Introducing air between the two pleurae (**pneumothorax**) breaks the connection between lungs and chest wall and allows a lung to collapse.
- Gravity causes significant regional differences in alveolar size in an upright lung. The base of the lung is compressed by its own mass, whereas alveoli at the apex may be expanded to 60% of their maximal volume.
- Airflow between the alveoli and the external atmosphere is driven by **pressure gradients**. Flow occurs against a **resistance** that depends largely on an airway's internal radius.
- Airway resistance is modulated by the **autonomic nervous system** but also changes passively with lung volume. During lung expansion, the airways are forced to dilate by surrounding structures acting via **mechanical tethers**, and dilation causes airway resistance to fall. When lung volumes are low, the airways are compressed by the mass of surrounding tissue, and their resistance is high.
- Airways are also sensitive to transmural pressures developed during expiration, such that their resistance becomes a pressure-dependent limiting factor on outflow.
- Air movement between lungs and atmosphere is measured using **spirometry**, one of several **pulmonary function tests** (**PFTs**) used to assess lung health. PFTs derive four lung volumes (i.e., **tidal volume**, **inspiratory reserve volume**, **expiratory reserve volume**, and **residual volume**) and capacities (i.e., **total lung capacity**, **functional residual capacity**, **inspiratory capacity**, and **vital capacity**).
- Air that is enclosed within regions of the lung that do not participate in gas exchange is known as **dead space**.

23 Gas Exchange

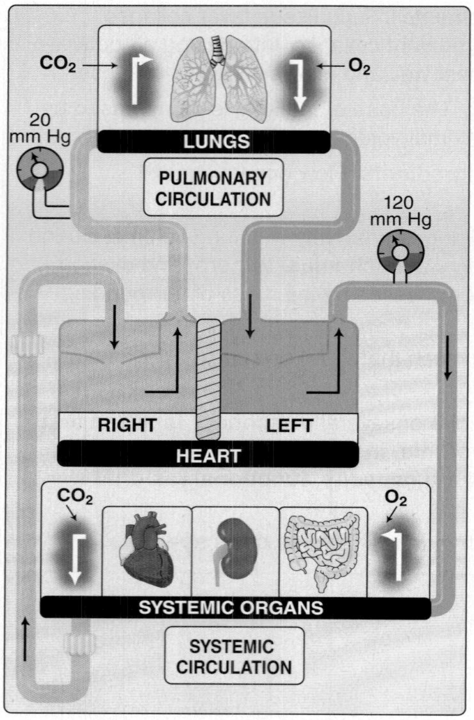

Figure 23.1
Pulmonary and systemic circulations.

I. OVERVIEW

Lungs facilitate O_2 and CO_2 exchange between blood and air. O_2 is required to help fuel adenosine triphosphate production by cells, whereas CO_2 is formed as a byproduct of aerobic metabolism. Lungs facilitate their exchange by bringing blood into close proximity to atmospheric air at a blood–gas interface. When the diaphragm and other inspiratory muscles contract, the lungs inflate. Air flows into the lungs, replenishing O_2 at the blood–gas interface and sustaining the steep O_2 and CO_2 pressure gradients required for optimal gas exchange. Exchange occurs rapidly, enhanced by the thin divide between blood and air (<1 µm) and by the large interface surface area. The efficiency of exchange is also critically dependent on the pulmonary circulation, which brings CO_2 to the lungs for disposal and carries away O_2 (Figure 23.1). Physiologic and pathologic changes in either ventilation or perfusion of the blood–gas interface can negatively impact lung performance.

II. PARTIAL PRESSURES

Gases move between air and blood by passive diffusion. The basic principles governing gas diffusion are similar to those described for solute diffusion between two fluid-filled chambers (see 1·IV). However, the issue is complicated by a need to factor in how soluble a gas might be in blood (Figure 23.2). If a gas is water insoluble, it cannot enter the circulation except under extreme, nonphysiologic circumstances. In practice, this means that we discuss the forces driving O_2 and CO_2 diffusion between blood and air in terms of **partial pressure** gradients rather than concentration gradients.

A. Gas pressures

The random motion of gas molecules exerts pressure on the walls of the vessel that contains it. The amount of pressure is directly proportional to the number of molecules within the vessel, as described by the **ideal gas law**:

$$P = \frac{nRT}{V}$$

Where P = pressure, n = number of molecules, R = universal gas constant, T is temperature, and V is container volume.

B. Partial pressure

The term "partial pressure" recognizes that atmospheric air is a mix of several different gases. The total pressure exerted by gas mixtures is equal to the sum of partial pressures of each of the individual components (**Dalton law**).

1. **Atmospheric air composition:** Atmospheric air is composed of 78.09% N_2, 20.95% O_2, 0.93% argon (Ar), 0.03% CO_2, and trace amounts of various other inert gases and pollutants. The fractional composition does not change with height above sea level or with temperature.

2. **Inspired air composition:** Air composition does change during inspiration because mucous membranes lining the nose and mouth add water vapor. By the time air reaches alveoli, it is saturated with 6.18% water. The fractional composition of the other gases is reduced correspondingly: 73.26% N_2, 19.65% O_2, 0.87% Ar, and 0.03% CO_2.

3. **Partial pressure of inspired air:** Atmospheric pressure at sea level is 760 mm Hg, reflecting the mass of air molecules stacked above. The partial pressure of the individual gases that make up inspired air reflects their fractional composition. The partial pressure of O_2 in inspired air arriving at the alveolar membrane (P_iO_2) is, thus, the product of atmospheric pressure (760 mm Hg) and fractional composition (19.7%):

$$P_iO_2 = 760 \times 0.197 = 150 \text{ mm Hg}$$

The partial pressure of CO_2 in inspired air (P_iCO_2) is 0.21 mm Hg. The latter is negligible in physiologic terms and, therefore, is usually rounded down to 0 mm Hg (Table 23.1).

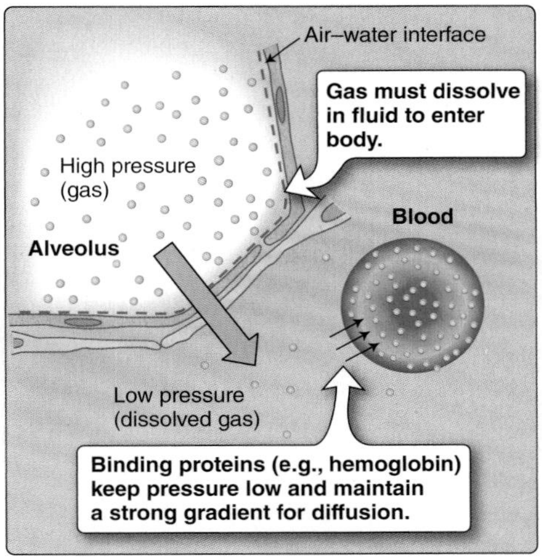

Figure 23.2
Gas diffusion between air and blood.

C. Blood gases

Alveolar ventilation brings atmospheric air to the blood–gas interface. The amount of O_2 and other air constituents that dissolve in blood is proportional to their partial pressures and their solubility in blood (**Henry law**). O_2 and CO_2 are both soluble gases that rapidly equilibrate across the blood–gas interface during inspiration. PO_2 in alveolar gas (P_AO_2) necessarily falls as O_2 molecules cross the interface and dissolve in blood: When the two compartments have equilibrated, P_AO_2 has dropped from 150 mm Hg to 100 mm Hg. At equilibrium, the concentration of O_2 dissolved in blood can be calculated from:

$$[O_2] = P_AO_2 \times s = 100 \text{ mm Hg} \times 0.0013 \text{ mmol/L/mm Hg} = 0.13 \text{ mmol/L}$$

where $[O_2]$ is dissolved O_2 concentration, and s is solubility of O_2 in blood.

The Henry law thus predicts that if blood O_2 concentration is 0.13 mmol and in equilibrium with a gas compartment, PO_2 in that compartment must be 100 mm Hg. Therefore, we consider the partial pressure of O_2 in blood to be 100 mm Hg, which allows us to discuss the pressure gradients driving gas movement between gas and liquid phases.

Table 23.1: Partial Pressures of Oxygen and Carbon Dioxide

Location	O_2 (mm Hg)	CO_2 (mm Hg)
External air	160	0
Conducting airways (during inhalation)	150	0
Alveoli	100	40
Pulmonary capillary	100	40
Systemic artery	100*	40
Pulmonary artery	40	45

*Actual value is slightly less because of physiologic shunts.

> Partial pressures reflect the amount of free gas dissolved in fluid and do not provide information about how much additional gas might be bound to hemoglobin (Hb), for example.

III. PULMONARY CIRCULATION

The pulmonary circulation, like the systemic circulation, receives 100% of cardiac output, but there the similarities end. Several features make the pulmonary vasculature unique, reflecting its location within the general circulation and a number of adaptations designed to facilitate gas exchange.

A. Overview

The pulmonary circulation has a resistance of 2–3 mm Hg • min • L^{-1}, or about fivefold less than the systemic circulation. Mean pulmonary arterial pressures are reduced correspondingly (10–17 mm Hg), as is supply-artery wall thickness. Pulmonary arterioles contain a fraction of the smooth muscle that characterizes systemic resistance vessels and makes them difficult to distinguish from veins. The paucity of muscle in the pulmonary vasculature means that the vessels readily distend in response to minor changes in filling pressure. The vasculature as a whole can accommodate up to 20% of circulating blood volume, and changes in posture routinely cause gravity-induced shifts of ~400 mL between the pulmonary and systemic circulations.

B. Blood–gas interface

Red blood cells (RBCs) are separated from atmospheric gas by the width of a capillary endothelial cell plus an alveolar epithelial cell (~0.15–0.30 μm). The density of pulmonary capillaries is so great that the alveolar surface is bathed in a near-continuous sheet of blood, which allows for highly efficient gas exchange. Pulmonary capillaries have an average length of 0.75 mm, providing ample opportunity for gas equilibration between blood and air, even at high flow rates. At rest, a single RBC traverses the length of a capillary and flows past two or three alveoli in ~0.75 s.

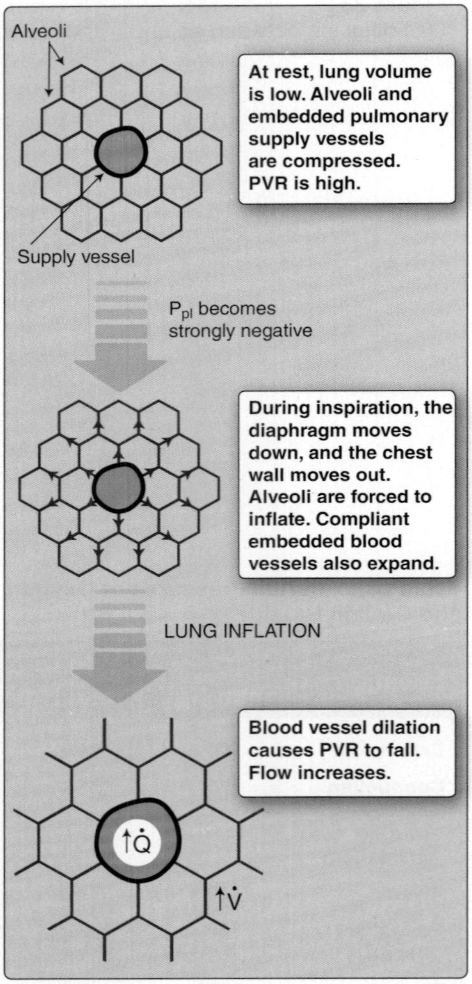

Figure 23.3
Effects of inspiration on pulmonary supply vessels. P_{pl} = pleural pressure; PVR = pulmonary vascular resistance; $\dot{Q}$ = pulmonary blood flow; $\dot{V}$ = alveolar ventilation.

C. Lung volume

The high compliance of pulmonary blood vessels means that they readily collapse when compressed by surrounding tissues. In practice, this means that changes in airway pressure during the respiratory cycle have a major impact on alveolar perfusion rates. The nature and timing of the change depends on vessel location within the bronchial tree.

1. **Supply vessels:** Flow through pulmonary supply vessels (i.e., arteries and arterioles) is very sensitive to changes in intrapleural pressure (P_{pl}). P_{pl} becomes strongly negative during inspiration, reflecting contraction and downward movement of the diaphragm and outward movement of the chest wall. The negative pressure is transmitted to the lung parenchyma, causing alveolar inflation (Figure 23.3). The negative pressure also dilates blood vessels

that are embedded in the lung parenchyma. Because vascular resistance is inversely proportional to vessel radius ($R \propto 1/r^4$), supply-vessel dilation during inspiration decreases pulmonary vascular resistance (PVR).

2. **Capillaries:** Pulmonary capillaries course through the spaces between adjacent alveoli. When alveoli expand during inspiration, their walls stretch. The embedded capillaries are stretched longitudinally, causing their internal diameter to decrease (Figure 23.4). The same effect causes skin to blanch when stretched. Stretching capillaries increases their resistance to flow and increases PVR.

3. **Pulmonary vascular resistance dependence on volume:** The differential effects of inspiration on supply-vessel and capillary resistance summate to produce a U-shaped plot of PVR against lung volume (Figure 23.5). PVR is very high at low lung volumes (supply vessels are compressed) and at total lung capacity (capillaries are stretched), but resistance is lowest during normal quiet breathing.

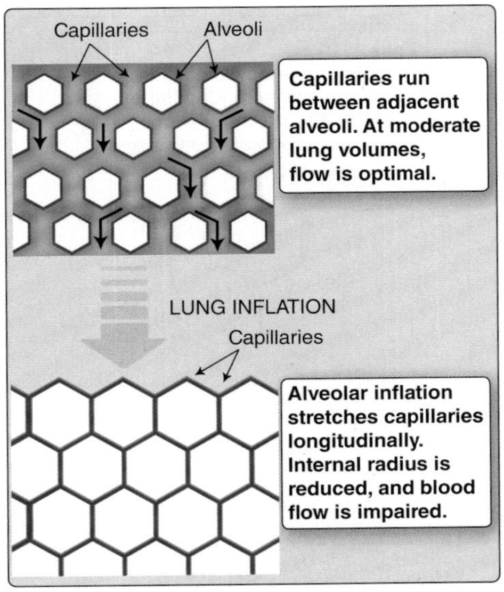

Figure 23.4
Pulmonary capillary patency during inspiration.

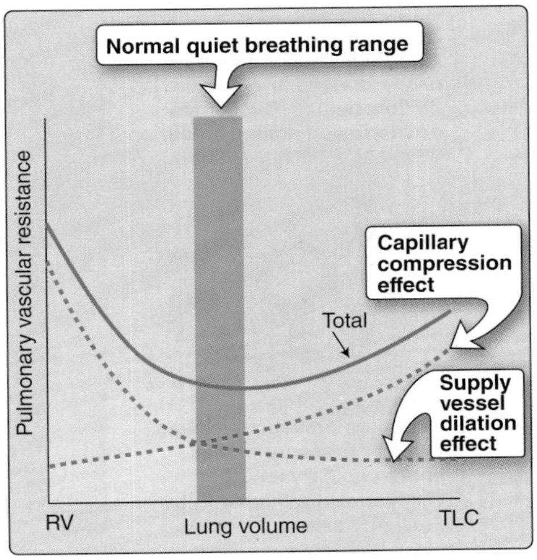

Figure 23.5
Effects of lung volume on pulmonary vascular resistance. RV = residual volume; TLC = total lung capacity.

D. Gravity

Because the pulmonary vasculature has a low resistance overall, pulmonary arterial pressures are also very low. This makes flow through the pulmonary vasculature extremely susceptible to gravitational influences.

1. **Pulmonary blood pressures:** The heart is located within the mediastinum, nestled between the right and left lungs (Figure 23.6). The pulmonary valve (where the pressure available to drive flow through the pulmonary circulation is measured) is located approximately 20 cm below the lung apex. The right ventricle generates a mean pulmonary arterial pressure (P_{pa}) of ~15 mm Hg, which approximates ~20 cm H_2O. When an individual is in a prone position, arterial pressures at the lung apex and base should both approximate 20 cm H_2O. When erect, gravity exerts a downward force that decreases arterial pressure above the heart by ~1 cm H_2O for each cm of vertical distance. Gravity increases pressures below the heart by the same amount.

2. **Regional differences:** The effects of gravity on P_{pa} mean that when a person is upright, pulmonary flow is lowest at the apex and increases progressively with decreasing height (see Figure 23.6). We can distinguish three distinct zones (1 through 3) based on flow characteristics.

   a. **Zone 1—minimal flow:** At the apex, **alveolar pressure > arterial pressure > venous pressure**. Because P_{pa} falls with height above the heart, pressure within an arteriole located ~20 cm above the ventricle is zero. Pulmonary venular pressure (P_{pv}) is less than zero at the same height (−9 cm H_2O). This creates a 9–cm H_2O pressure gradient available to drive flow through apical capillaries, but, in practice, they are collapsed. Collapse occurs because pressure within an alveolus (P_A) at rest is also 0 cm H_2O (i.e., barometric pressure), which is greater than the perfusion pressure maintaining

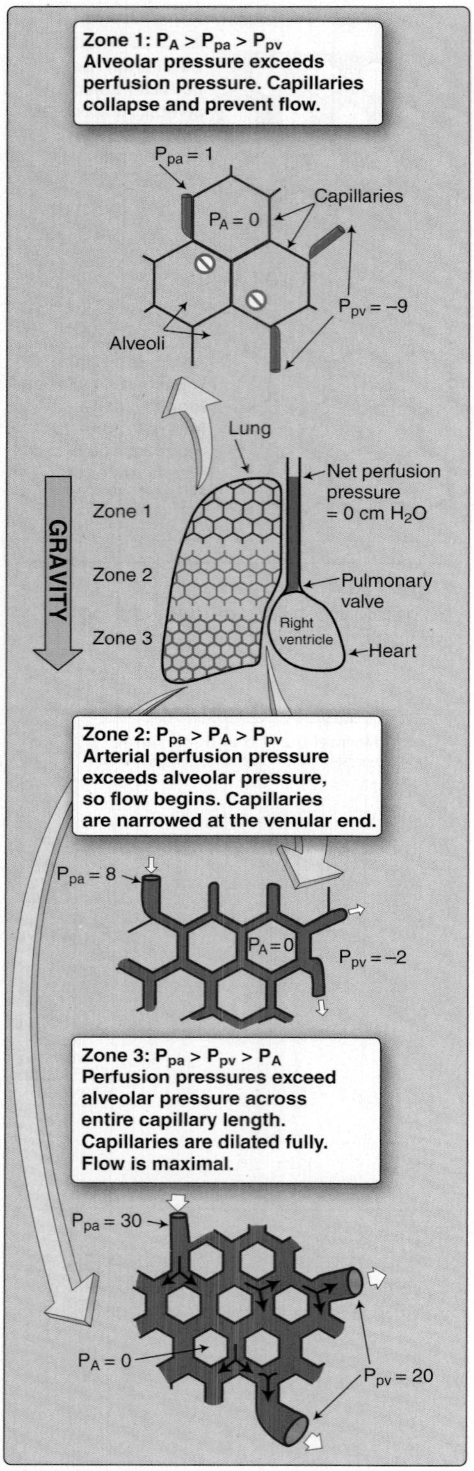

Figure 23.6
Regional perfusion and flow patterns in a static, upright lung. Values are given in cm H_2O. P_A = intraalveolar pressure; P_{pa} = pulmonary arteriolar pressure; P_{pv} = pulmonary venular pressure.

capillary patency (see Figure 23.6, upper panel). Zone 1 only exists at the very pinnacle of the lung, when pulmonary vascular pressures are critically low (e.g., during hemorrhage or other form of circulatory shock) or when alveolar pressure is raised artificially by **positive pressure ventilation**.

b. **Zone 2—moderate flow:** In zone 2, **arterial pressure > alveolar pressure > venous pressure**. Zone 2 includes the apex and the middle of the lung, regions in which P_{pa} and mean capillary pressure (P_c) are greater than P_A. P_{pv} in zone 2 is still lower than P_A, so the capillary tends to be compressed at the venular end, but flow continues. The resistance created by extravascular compression gradually decreases with lung height, reflecting the coincident rise in both P_{pa} and P_{pv} (note that P_A is insensitive to position because it is determined by barometric pressure).

c. **Zone 3—maximal flow:** The lung base is located below the pulmonary valve. Gravity enhances perfusion pressures in this region, so **arterial pressure > venous pressure > alveolar pressure**. Vascular collapse is no longer an issue here. Instead, capillaries at the lung base are typically distended by high, gravity-enhanced perfusion pressures. In the systemic circulation, resistance vessels tightly control P_c through reflex constriction and dilation of smooth muscle layers that make up the vessel walls. Pulmonary arterioles contain so little smooth muscle that they are relatively ineffective pressure regulators. Thus, P_c rises in concert with P_{pa} and P_{pv}, and the capillary swells beyond normal capacity. Flow through blood vessels is proportional to the fourth power of internal radius, and, therefore, flow is disproportionately high also (see Figure 23.6, lower panel).

E. Flow regulation

Blood flow through systemic resistance vessels is controlled by the sympathetic nervous system, bloodborne agents, rising metabolite levels, and other factors. By contrast, pulmonary resistance vessels are relatively insensitive to sympathetic activity or humoral factors. The vasculature is mildly sensitive to rising interstitial CO_2 and H^+ levels, but whereas systemic resistance vessels would dilate reflexively, pulmonary vessels *constrict* when CO_2 and H^+ levels rise. A predominant force controlling pulmonary resistance vessels and PVR is P_{AO_2}. Low O_2 levels promote **hypoxic vasoconstriction** of pulmonary resistance vessels. This reflex is again the complete opposite of how systemic resistance vessels respond to hypoxia, but it has clear advantages for optimizing pulmonary function. Hypoxic vasoconstriction steers blood away from poorly ventilated areas, redirecting it to well-ventilated regions where gas exchange can occur.

F. Venous admixture

Ideally, blood would leave the pulmonary circulation and enter the systemic circulation at 100% saturation. In practice, this never occurs because there is always some degree of **venous admixture**, or the mixing of deoxygenated (venous) and oxygenated blood prior to blood

entering the systemic arterial system. There are two main causes: **shunts** and low **ventilation/perfusion** (**$\dot{V}_A/\dot{Q}$**) **ratios**.

1. **Shunts: Shunts** allow venous blood to bypass the normal process of gas exchange. There are two types: **anatomic shunts** and **physiologic shunts** (Figure 23.7).

   a. **Anatomic:** Anatomic shunts have a structural basis, comprising fistulas or blood vessels. Examples include an atrial septal defect that allows blood from the right atrium to enter the left atrium or an anastamosis between a pulmonary artery and a pulmonary vein. These are also known as **right-to-left shunts**.

   b. **Physiologic: Physiologic shunting** occurs when atelectasis, pneumonia, or some other problem affecting ventilation of the blood–gas interface prevents gas exchange. Hypoxic vasoconstriction redirects flow, but there is always some residual perfusion of a nonfunctional interface. Blood from these regions escapes oxygenation and reduces arterial O_2 saturation levels when it enters the systemic circulation.

2. **Low ventilation/perfusion ratios:** $\dot{V}_A/\dot{Q}$ ratios are discussed in more detail below, but if the blood–gas interface is perfused at rates that exceed its diffusional limits, O_2 saturation cannot occur. Venous admixture is the result.

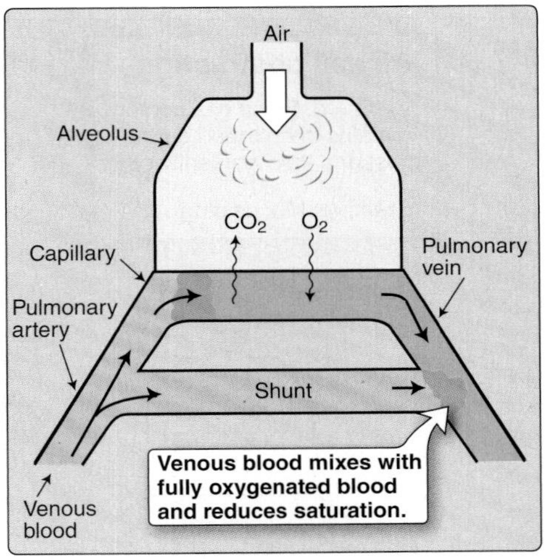

Figure 23.7
Shunts allow venous admixture.

IV. VENTILATION/PERFUSION RATIO

At rest, the pulmonary circulation is perfused with ~5 L/min of blood ($\dot{Q}$) representing the entire output of the right heart. Lung inflation maximally draws ~4 L of air into the alveoli sacs during this time (alveolar ventilation is abbreviated as $\dot{V}_A$), so the pulmonary $\dot{V}_A/\dot{Q}$, = 0.8. In an ideal lung, all alveoli would be ventilated and perfused optimally, but there are many physiologic causes of mismatch.

A. Model lung mechanics

The function of alveolar ventilation is to bring outside air into close proximity to blood so that O_2 may be loaded and CO_2 offloaded. Outside air contains 150 mm Hg O_2 and negligible CO_2 (Figure 23.8). Blood arriving at the alveolus from the pulmonary arterioles (**mixed venous blood**) is rich in CO_2 (P_{CO_2} = 45 mm Hg) but O_2-poor (P_{O_2} = 40 mm Hg). During normal quiet breathing, equilibration of both gases between air and blood completes before blood has progressed even a third of the way through the capillary, raising $P_{A}CO_2$ to 40 mm Hg and lowering $P_{A}O_2$ to 100 mm Hg. Alveoli have no means of modifying these values further, so blood exiting a pulmonary capillary also contains 40 mm Hg CO_2 and 100 mm Hg O_2. Changes in either ventilation or perfusion will affect these values, however.

1. **Airway obstruction:** If an airway is obstructed by a mucus plug, for example, the $\dot{V}_A/\dot{Q}$ ratio drops to zero. In the absence of ventilation, alveolar gas equilibrates with mixed venous blood at a $P_{A}CO_2$ of 45 mm Hg and a $P_{A}O_2$ of 40 mm Hg. Blood leaving the area of obstruction has no opportunity to exchange O_2 or CO_2

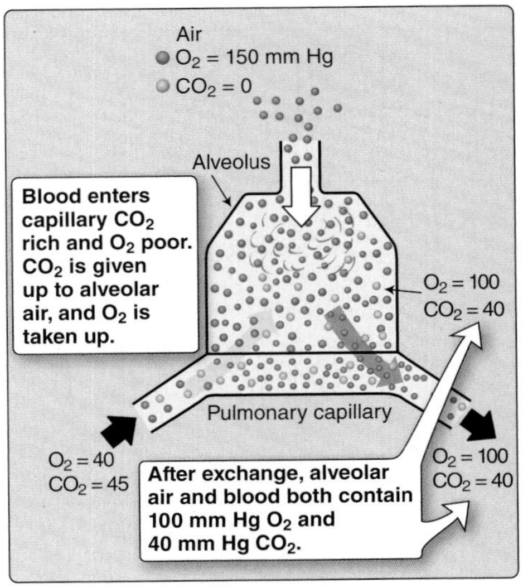

Figure 23.8
CO_2 and O_2 exchange between pulmonary blood and alveolar air. Partial pressures are given in mm Hg.

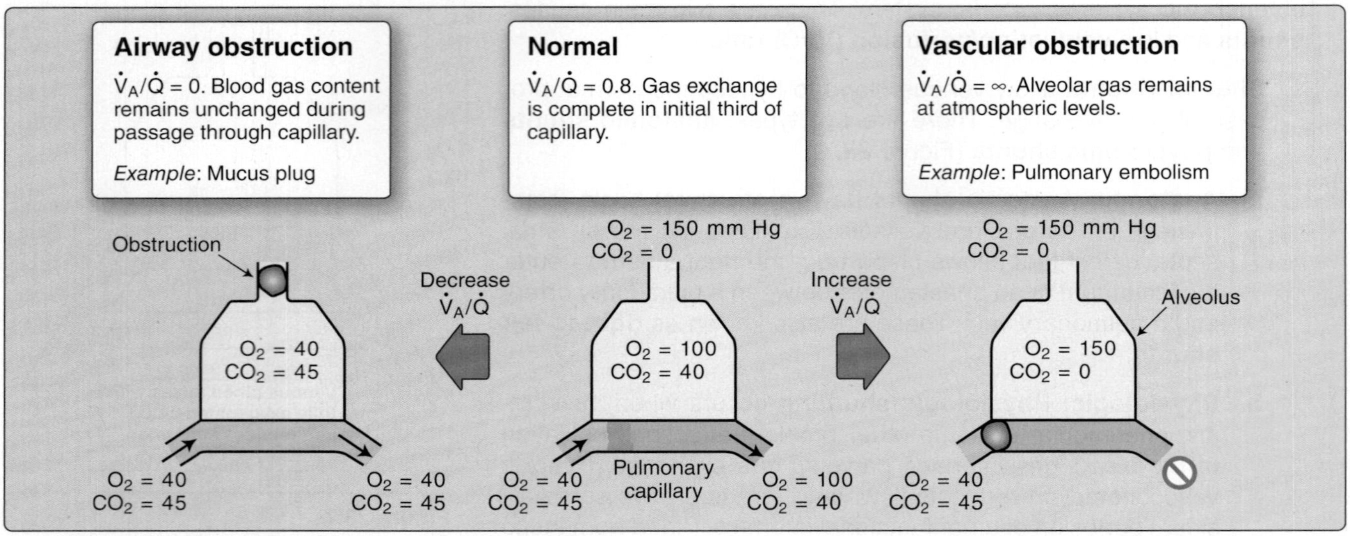

Figure 23.9
Effect of obstructing either ventilation or perfusion on Po_2 and Pco_2 in the lung. All partial pressures are given in mm Hg. $\dot{V}_A$ = alveolar ventilation; $\dot{Q}$ = alveolar perfusion.

and, thus, remains unchanged during passage across the alveolar sac (Figure 23.9, left). This creates a physiologic shunt, as discussed above.

2. **Blood flow obstruction:** If blood flow is prevented by an embolus, for example, the $\dot{V}_A/\dot{Q}$ ratio approaches infinity. Alveolar gas composition remains unchanged following inspiration because there is no blood contact (see Figure 23.9, right).

B. Ventilation/perfusion ratios in an upright lung

Gravity significantly affects alveoli ventilation and perfusion (see Figure 23.6; also see Figure 22.14). This creates a broad spectrum of $\dot{V}_A/\dot{Q}$ ratios in an upright lung (Figure 23.10).

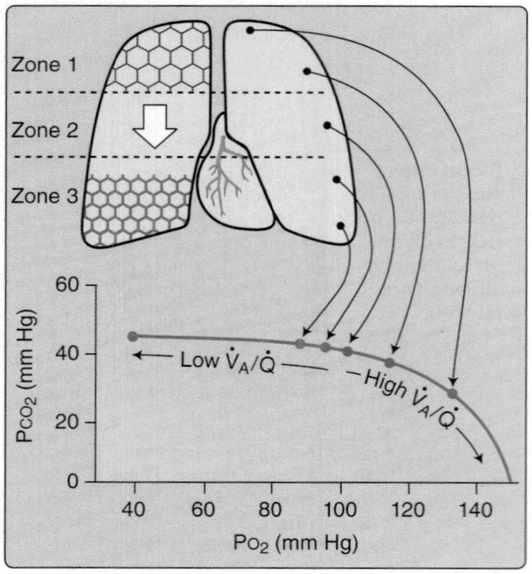

Figure 23.10
Distribution of $\dot{V}_A/\dot{Q}$ ratios in an upright lung. $\dot{V}_A$ = alveolar ventilation; $\dot{Q}$ = alveolar perfusion.

1. **Zone 1—highest ratio:** Alveoli at the lung apex ventilate poorly because they are inflated to 60% of maximal volume even at rest. Perfusion in this region is minimal because the vasculature is compressed by alveolar pressures that exceed perfusion pressures. Thus Po_2 and Pco_2 in the small volumes of blood exiting this region approaches that of inspired air ($\dot{V}_A/\dot{Q} \sim \infty$).
2. **Zone 2—moderate ratio:** Ventilation improves slowly with decreasing lung height. Perfusion increases more steeply, however, causing the $\dot{V}_A/\dot{Q}$ ratio to fall rapidly toward the base.
3. **Zone 3—lowest ratio:** Alveoli at the lung base are compressed at rest and ventilate very well upon inspiration. Pulmonary perfusion pressures are also very high in this region, so flow rates are maximal.
4. **Net effect:** The extent to which the different regions contribute to the composition of the blood leaving the lung is determined by their perfusion rates. Thus, the $\dot{V}_A/\dot{Q}$ extremes seen at the apex have minimal effect on overall saturation levels. The O_2 and CO_2

Clinical Application 23.1: Tuberculosis

The microorganism that causes tuberculosis, *Mycobacterium tuberculosis*, favors lung regions where O_2 levels are high and typically establishes itself at the apices, where alveolar gas composition most closely resembles that of atmospheric air. In advanced cases, lung tissue is destroyed, and large cavities develop. The cavities are avascular, which can make infection difficult to treat. Multiple drugs must be given together for a long period to fully eradicate tubercular organisms from the tissue.

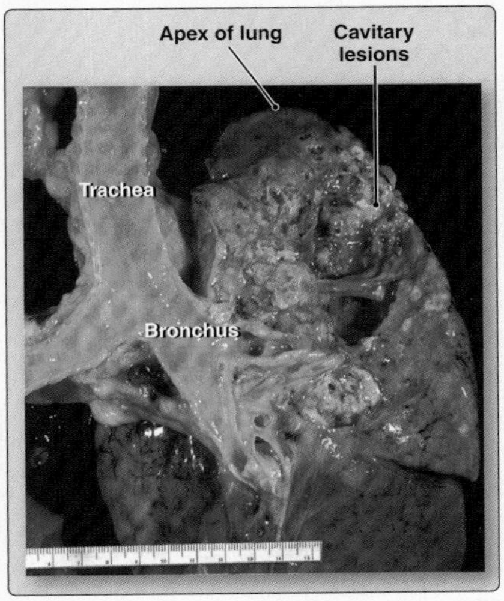

Postmortem specimen showing apical lung lesions caused by tuberculosis.

content of systemic arterial blood is determined largely by the heavily perfused regions at the base.

C. Ventilation/perfusion mismatches

Blood flow through the lung base is so high that it exceeds the ventilatory capacity of the blood–gas interface and causes a local $\dot{V}_A/\dot{Q}$ mismatch. Blood leaving the area has a P_{O_2} of around 88 mm Hg, or 12 mm Hg below optimum, whereas P_{CO_2} is higher by ~2 mm Hg. Some degree of physiologic shunting caused by $\dot{V}_A/\dot{Q}$ mismatch occurs normally even in a healthy individual, but can become severe when an airway is obstructed by, for example, aspiration of a foreign body, tumor growth, or during an asthma exacerbation. The $\dot{V}_A/\dot{Q}$ ratio is an important measure of pulmonary function and health. Both parameters can be visualized clinically using radioactive tracers, but imaging techniques are generally used only if gross deficiencies in either ventilation or perfusion are suspected such as those caused by pulmonary embolism (Figure 23.11).

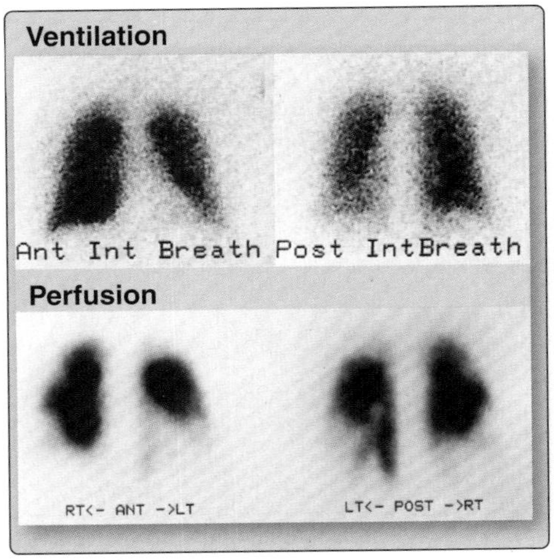

Figure 23.11
This ventilation scan (visualized radiographically using radioactive xenon gas) is normal, but the perfusion scan (visualized radiographically using radiolabeled albumin) shows many areas devoid of radioisotope, a pattern characteristic of pulmonary embolism.

D. Alveolar–arterial oxygen difference

Potential problems with either ventilation or perfusion can also be assessed fairly simply from the **alveolar–arterial difference for O_2 ($A–aDO_2$)**, which compares P_{O_2} in alveoli with that of systemic arterial blood. Ideally the two values should be the same. In practice, there is always a 5–15 mm Hg P_{O_2} difference between alveolar gas and blood, depending on age. P_{AO_2} is assessed using a simplified form of the **alveolar gas equation**:

Equation 23.1

$$P_{AO_2} = P_{iO_2} - \frac{P_{ACO_2}}{R}$$

Where P_{iO_2} is the partial pressure of O_2 in inspired air, P_{ACO_2} is alveolar P_{CO_2}, and R is the respiratory exchange ratio. P_{ACO_2} is determined

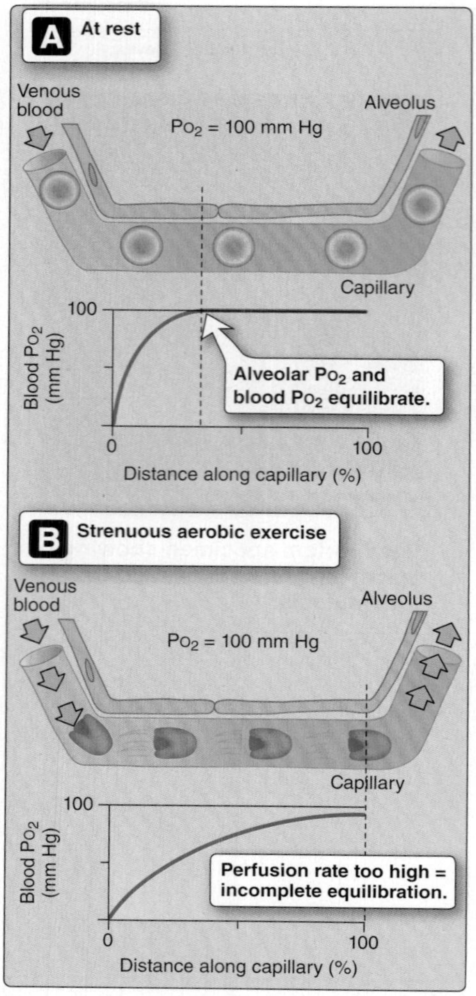

Figure 23.12
Effect of capillary perfusion rate on oxygenation saturation.

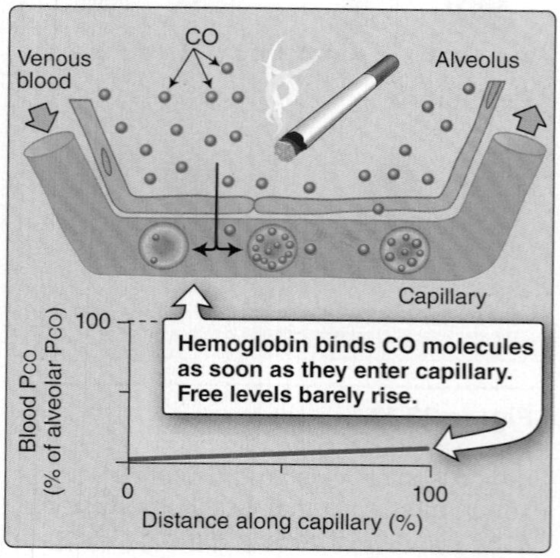

Figure 23.13
Diffusion-limited gas exchange. Pco = partial pressure of carbon monoxide.

by analyzing gas captured at the very end of expiration. "R" (normally 0.8) represents the ratio of CO_2 produced:O_2 consumed by internal respiration. P_aO_2 can be measured by arterial blood gas analysis. The difference between P_AO_2 and P_aO_2 for a healthy individual can be predicted as:

$$\text{A-a gradient} = 2.5 + 0.21 \times \text{age in years.}$$

An A-a difference that is wider than predicted indicates that O_2 uptake at the blood–gas interface is impaired (see Example 23.1).

V. GAS EXCHANGE

The rate at which gases diffuse across the blood–gas interface (i.e., gas flow, or $\dot{V}$) is determined by the pressure difference across the interface (ΔP), the surface area available for exchange (A), and barrier thickness (T):

$$\dot{V} = \frac{\Delta P \times A \times D}{T}$$

where D is a diffusion coefficient that takes into account the molecular weight and solubility of a gas. In practice, surface area, thickness, and the diffusion coefficient can be combined to yield a constant that describes the lung's **diffusing capacity** (**D_L**) for gas. Gas flow across the barrier can then be estimated from:

$$\dot{V} = \Delta P \times D_L$$

Lungs' design maximizes flow by providing a large surface area for diffusion and by restricting barrier thickness to the width of a pneumocyte plus a capillary endothelial cell. Ventilation and perfusion maintain steep partial pressure gradients across the interface.

A. Diffusion-limited exchange

Blood traverses the length of a pulmonary capillary in ~0.75 s at rest. Equilibration of O_2 between alveolar gas and blood occurs within a fraction of this time, so uptake is not normally limited by the rate at which O_2 diffuses across the exchange barrier (Figure 23.12A). During maximal exercise, however, cardiac output increases, and capillary transit time decreases to <0.4 s. Blood may exit the capillary before being fully O_2 saturated (see Figure 23.12B). O_2 uptake is now considered to be **diffusion limited** because exchange has been limited by the rate at which O_2 diffuses across the blood–gas interface. The effects of diffusion limitations can best be appreciated by studying the characteristics of carbon monoxide uptake, which is always diffusion limited (Figure 23.13).

> The "CO" abbreviation is generally used to denote cardiac output in cardiovascular and other areas of physiology, but CO indicates carbon monoxide in pulmonary physiology.

1. **Carbon monoxide uptake:** Hb binds CO with an affinity that is ~240 times greater than that for O_2. In practice, this means CO molecules bind to Hb as fast as they can diffuse across the exchange barrier, and alveolar CO never has the chance to equilibrate with plasma CO. A diffusion limitation such as this might be offset by increasing the pressure gradient driving diffusion or by increasing the D_L for CO (D_{LCO}).

2. **Changing perfusion rate:** Intuitively, one would think that slowing blood flow through the capillary would be beneficial in terms of increasing net uptake. A slower flow rate would allow more time for the gas and liquid phases to come into equilibration before blood exits the capillary. Although decreasing perfusion rate does allow for greater saturation, net uptake actually decreases because the volume of blood exiting the capillary per unit time is reduced also.

> Changing perfusion rate has no net effect on gas transport in a diffusion-limited exchange scenario.

3. **Diffusion limitations:** Emphysema and pulmonary fibrosis both limit O_2 and CO_2 diffusion by decreasing D_L. Erosion of alveolar sacs reduces total barrier surface area in patients with emphysema. Pulmonary fibrosis increases barrier thickness and, thereby, increases the distance separating blood and alveolar air.

B. Perfusion-limited exchange

Blood becomes fully O_2 saturated shortly after entering a pulmonary capillary (at rest). Because more O_2 *could* be transferred if flow were increased (even though this transfer may be in excess of body requirements) exchange is considered to be **perfusion limited**. The characteristics of perfusion-limited gas exchange can be best appreciated by studying N_2O uptake. Hb does not bind N_2O, so blood and alveolar partial pressures for N_2O equilibrate in <100 ms (Figure 23.14). Modest changes in barrier architecture have little effect on net uptake. Instead, net N_2O uptake is tied to flow.

> In a perfusion limited system, gas will saturate whatever amount of blood is presented to it over a wide range of values.

VI. OXYGEN TRANSPORT

O_2 uptake from the atmosphere and carriage to the tissues is required to support internal respiration. O_2 has a very poor water solubility compared with other gases, which limits the amount that can be transported in solution to ~3 mL of gaseous O_2 per liter of blood. An average adult consumes ~250 mL O_2/min at rest, so resting cardiac output would have

Example 23.1

A 50-year-old woman with a history of prior deep vein thrombosis presents in the Emergency Department complaining of shortness of breath. A room air arterial blood gas (ABG) sample is obtained, and the patient is placed on supplemental O_2.

Based on the results of the ABG, is the patient's A-a gradient normal or abnormal? What is her likely diagnosis?

ABG results:

P_{aO_2} = 70 mm Hg
P_{aCO_2} = 32 mm Hg
pH = 7.47

P_{iO_2} (room air, sea level) = 150 mm Hg

Based on age, the patient's A-a gradient would be predicted to be:

$2.5 + (0.21 \times 50) = 13$ mm Hg

Using ABG values above and Equation 23.1:

$$P_{AO_2} = P_{iO_2} - \frac{P_{aCO_2}}{R}$$
$$= 150 - 32/0.8$$
$$= 150 - 40$$
$$= 110 \text{ mm Hg}$$

Her observed A-a difference ($P_{AO_2} - P_{aO_2}$) is 110 − 70 = 40 mm Hg, or 27 mm Hg higher than predicted.

The abnormally wide A-a gradient indicates that there is a $\dot{V}_A/\dot{Q}$ mismatch, suggesting an impairment of O_2 uptake by the lungs.

These findings are consistent with pulmonary embolism.

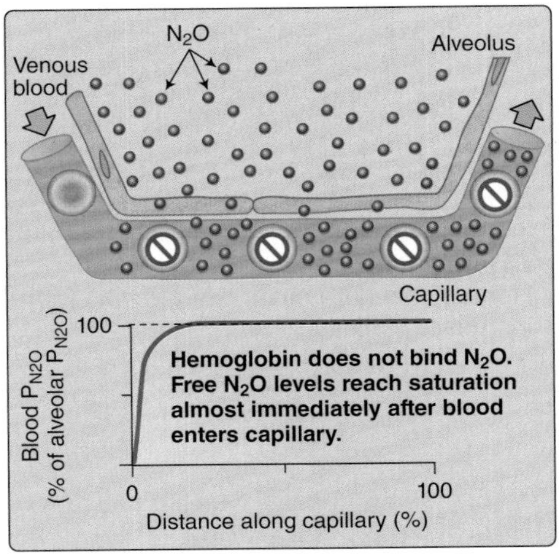

Figure 23.14
Perfusion-limited gas exchange. P_{N2O} = partial pressure of N_2O.

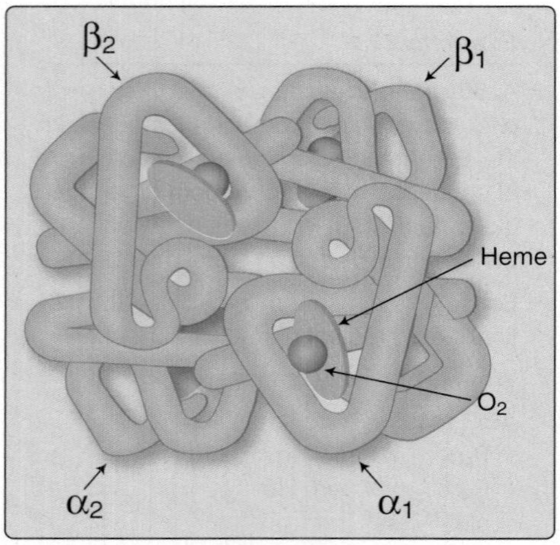

Figure 23.15
Hemoglobin structure showing location of the oxygen-binding heme group.

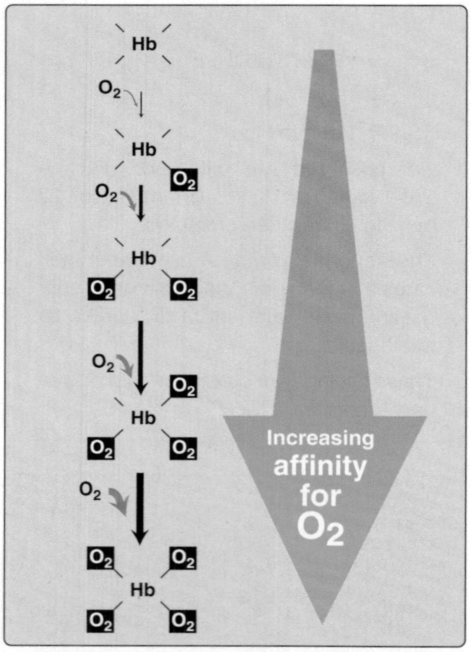

Figure 23.16
Hemoglobin (Hb) binds oxygen (O_2) with increasing affinity.

to be maintained at 83 L/min and then climb to >1,000 L/min during exercise if transport relied on O_2-solubility properties alone! Instead, blood's O_2 transport ability is greatly enhanced by the presence of Hb, a protein uniquely designed to carry O_2 through the systemic vasculature and then release it to the tissues. It then helps transport CO_2 back to the lungs for expiration.

A. Hemoglobin

Hb is a metalloprotein composed of four polypeptide chains (globins) as shown in Figure 23.15. HbA, the form found most commonly in adults, contains two alpha (α) chains and two beta (β) chains. Each globin is linked to a heme group that comprises a ferrous ion (Fe^{2+}) held within a porphyrin ring. Iron allows **deoxyhemoglobin** to bind O_2, forming **oxyhemoglobin**.

1. **Structure:** Hb comprises two dimeric subunits, each containing one α and one β chain. The chains within the subunits are stably linked by noncovalent bonds. The two subunits are linked weakly, however, and strength of association changes with O_2-binding state.

2. **Oxygen binding:** The four heme moieties give Hb the ability to bind four O_2 molecules. The interaction is reversible and is an oxygenation rather than oxidation. DeoxyHb has a relatively low affinity for O_2, but each successive O_2-binding event produces a conformational change within the protein that incrementally increases the affinity of the other sites (Figure 23.16). This binding cooperation yields a sigmoidal O_2-dissociation curve, with the curve's steepest part coinciding with the range of P_{O_2} values common to tissues (Figure 23.17). The curve approaches saturation at a P_{O_2} of 60 mm Hg. Blood **arterialization** raises P_{O_2} to 100 mm Hg but increases saturation level by only ~10%.

> Hb changes color from dark blue to bright red when O_2 binds, which makes it possible to monitor arterial O_2-saturation levels using noninvasive pulse oximetry. A light-emitting probe is attached to a finger or ear, then the relative amounts of saturated and desaturated Hb is calculated from the amount of light absorbed at 660 nm and 940 nm, respectively.

3. **Hemoglobin concentration:** The amount of O_2 that blood can carry depends on Hb concentration.

   a. **Oxygen capacity:** Blood contains ~150 g of Hb/L, or 15 g/dL (normal range is 12–16 g/dL for women and 13–18 g/dL for men). Each Hb molecule is capable of binding four O_2 molecules, which is equivalent to 1.39 mL O_2/g of Hb. Thus, blood's theoretical **O_2 capacity** is 20.8 mL/dL, a value that increases and decreases in direct proportion to blood Hb concentration.

b. **Oxygen saturation: O_2 saturation** is a measure of the number of occupied O_2-binding sites on the Hb molecule. At 100% saturation (arterial blood), all four heme groups are occupied. At 75% saturation (venous blood), three are occupied. Only two sites are occupied at 50% saturation. The degree of O_2 saturation is not dependent on Hb concentration, at least within a physiologic range.

B. Hemoglobin–oxygen dissociation curve

The dissociation curve's shape explains Hb's ability to bind O_2 in the lung and then release it on demand to the tissues.

1. **Association:** Mixed-venous blood arrives at an alveolus with a P_{O_2} of 40 mm Hg but an O_2 saturation of ~75%. The cooperative nature of O_2 binding to Hb means that the single unoccupied heme group has a very high affinity for O_2. This allows the site to capture O_2 as fast as it can diffuse across the blood–gas interface, simultaneously maintaining a steep pressure gradient for O_2 diffusion across the exchange barrier even as equilibration with alveolar gas occurs. Note that the plateau region of the O_2 dissociation curve begins at a P_{O_2} of around 60 mm Hg (see Figure 23.17). In practice, this ensures that saturation still occurs if P_{AO_2} is suboptimal (i.e., 60 mm Hg), either because ventilation is impaired or when cardiac output is increased to the point where perfusion becomes limiting.

2. **Dissociation:** Once blood arrives at a tissue, Hb must release bound O_2 and make it available to mitochondria. Transfer is facilitated by the steepness of the pressure gradient between blood and mitochondria, which maintain a local P_{O_2} of ~3 mm Hg. Hb begins releasing O_2 at a P_{O_2} of 60 mm Hg and delivers ~60% of total as P_{O_2} falls to 20 mm Hg. Each O_2 dissociation event lowers the affinity of the remaining heme groups for bound O_2, so that if a tissue's metabolic rate is very high and its need for O_2 is increased, unloading occurs with increased efficiency.

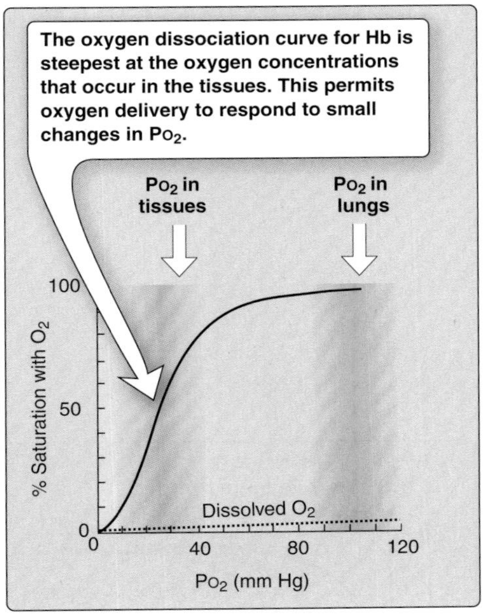

Figure 23.17
O_2 dissociation curve for hemoglobin (Hb).

C. Dissociation curve shifts

Hb is uniquely sensitive to tissue needs, allowing it to deliver increasing amounts of O_2 when metabolism increases. This is made possible through allosteric changes that decrease the protein's O_2 affinity and promote unloading. These changes manifest as a rightward shift in the Hb–O_2 dissociation curve (Figure 23.18).

1. **Rightward shifts:** Metabolism generates heat and CO_2 and acidifies the local environment. All three changes reduce Hb's O_2 affinity and cause it to unload O_2. The liberated O_2 keeps free (dissolved) O_2 levels high and maintains a steep pressure gradient between blood and mitochondria even as blood's O_2 stores are being emptied.

   a. **Temperature:** During strenuous exercise, muscle temperature rises by as much as 3°C. The Hb–O_2 dissociation curve shifts by ~5 mm Hg to the right as a result, causing more O_2 to be released to the metabolically active tissue.

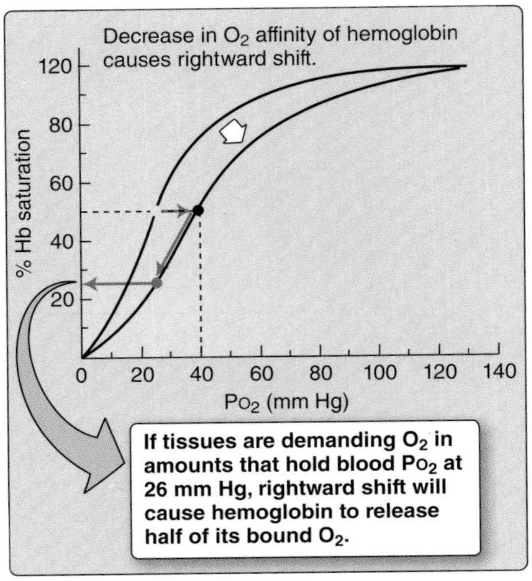

Figure 23.18
Decreasing the O_2 affinity of hemoglobin (Hb) causes O_2 unloading.

b. **Carbon dioxide:** Aerobic metabolism generates CO_2 and causes tissue P_{CO_2} to rise. CO_2 binds to terminal globin amino groups and decreases Hb's O_2 affinity. The Hb–O_2 dissociation curve shifts to the right, and O_2 is unloaded. CO_2 also dissolves in water to yield free acid, which promotes further O_2 unloading via the Bohr effect (see below).

c. **Protonation:** Protonation stabilizes the deoxy form of Hb and decreases its O_2 affinity. Metabolism generates several different acids in addition to carbonic acid, and the amount produced is proportional to metabolic activity. The Hb–O_2 dissociation curve shifts to the right, and O_2 is released (the Bohr effect).

d. **2,3-Diphosphoglycerate:** 2,3-Diphosphoglycerate (2,3-DPG) is synthesized from 1,3-DPG, which is an intermediate in the glycolytic pathway. 2,3-DPG is abundant in RBCs, its concentration rivaling that of Hb. 2,3-DPG binds preferentially to the deoxygenated form of Hb and stabilizes it, thereby reducing its O_2 affinity (Figure 23.19). The Hb–O_2 dissociation curve shifts to the right, and O_2 is unloaded. 2,3-DPG and its effects on O_2 affinity is a constant in blood, unlike the effects of temperature, CO_2, and H^+, which typically remain localized to an active tissue.

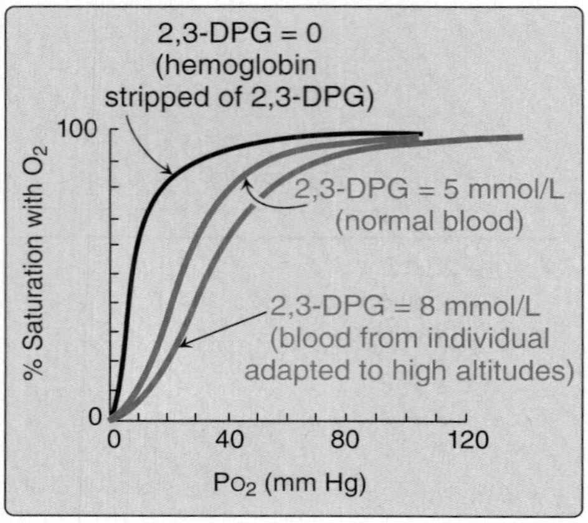

Figure 23.19
2,3-Diphosphoglycerate (2,3-DPG) decreases hemoglobin's O_2 affinity.

> Chronic hypoxemia caused by pathologic changes in lung function or living at high altitude stimulates 2,3-DPG production. Increased 2,3-DPG levels shift the Hb–O_2 dissociation curve even further to the right, which increases the tissue's accessibility to available O_2 (see Figure 23.19). Although 2,3-DPG does reduce the efficiency of O_2 loading by Hb in the lungs, the effects are minor and more than offset by the beneficial effects of assisting O_2 delivery to tissues.

2. **Leftward shifts:** Hb's O_2 affinity increases, and the Hb–O_2 dissociation curve shifts left when body temperature decreases or when CO_2, H^+, or 2,3-DPG levels decrease. All of these changes reflect decreased metabolic activity and a decreased need for O_2 delivery to tissues. A leftward shifted Hb–O_2 dissociation curve is also observed in the fetus and as a result of CO binding to Hb.

a. **Fetal hemoglobin:** Fetal Hb (HbF) contains γ chains in place of the two β chains. This causes the fetal Hb–O_2 dissociation curve to be shifted left compared with adult Hb.

i. **Mechanism:** HbF's increased O_2 affinity compared with the adult form (HbA) reflects the fact that γ-globins bind 2,3-DPG very weakly. 2,3-DPG normally stabilizes the deoxygenated form of HbA and reduces its affinity. HbF's inability to bind 2,3-DPG favors O_2 loading at low partial pressures.

> If HbA is stripped of 2,3-DPG, its O_2-dissociation curve resembles that of HbF. Storing blood causes 2,3-DPG concentrations to decline over the course of a week, causing a leftward shift in the dissociation curve (see Figure 23.19). Although RBCs replenish lost 2,3-DPG within hours to days of transfusion, giving a critically ill patient large volumes of 2,3-DPG–depleted blood presents some difficulties because such blood does not readily give up its O_2.

ii. **Benefits:** O_2 is delivered to a fetus via the placenta, which is an inefficient route for O_2 transfer compared with lungs. Fetal P_{aO_2} rarely exceeds 40 mm Hg as a result. The leftward shift in the Hb–O_2 dissociation curve brings it into closer alignment with P_{O_2} values normally encountered *in utero* and allows fetal placental blood to achieve ~80% saturation, even though P_{aO_2} is low. HbF is replaced by HbA in the months immediately following birth, although individuals with hereditary persistence of HbF may continue to express the fetal form well into adulthood (Figure 23.20).

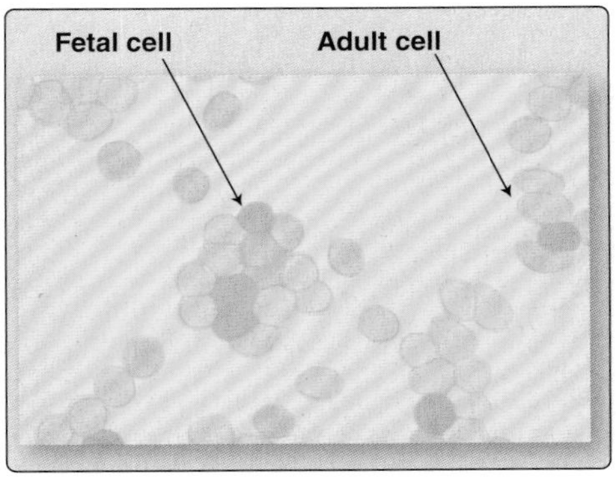

Figure 23.20
Hereditary persistence of fetal hemoglobin (HbF). Red blood cells containing HbF appear bright pink.

b. **Carbon monoxide:** Hb binds CO with high affinity to produce carboxyhemoglobin, which is bright red in color. CO occupancy of the O_2 binding sites severely reduces Hb's ability to bind and carry O_2. Inhaling the gas at a concentration of only 0.1% reduces O_2-carrying capacity by 500%. CO simultaneously stabilizes the high-affinity Hb form and shifts the Hb–O_2 dissociation curve to the left (Figure 23.21). These changes dramatically reduce Hb's ability to release O_2 to tissues and make CO an extremely deadly gas (Figure 23.22). CO poisoning is a leading cause of poisoning deaths in the United States.

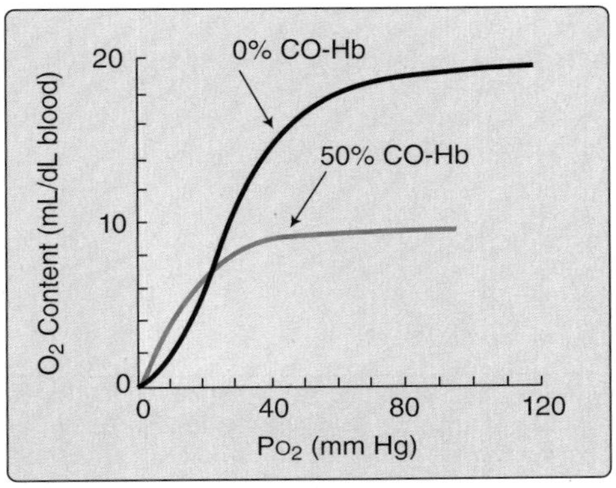

Figure 23.21
Carbon monoxide effects on hemoglobin O_2 affinity. CO-Hb = carboxyhemoglobin.

> CO is formed by combustion of hydrocarbons. Common sources of exposure include automobile exhaust, poorly ventilated heating systems, and smoke. Carboxyhemoglobin comprises up to ~3% of total Hb in nonsmokers, increasing to 10%–15% in smokers.

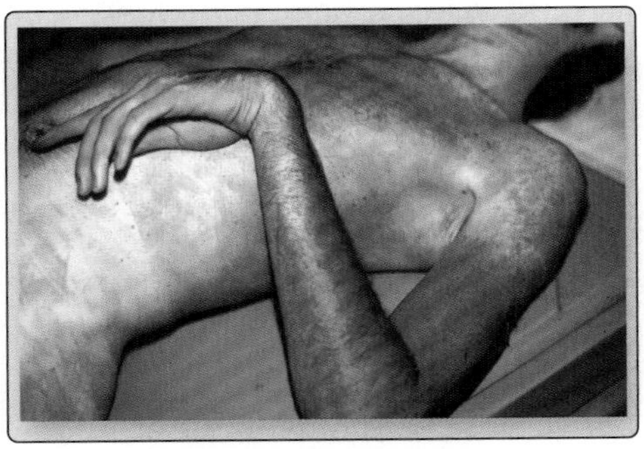

Figure 23.22
Carbon monoxide poisoning turns skin a bright cherry-red, a color that persists after death.

VII. CARBON DIOXIDE TRANSPORT

Metabolism generates ~200 mL CO_2/min in a normal person at rest. CO_2 is carried away from tissues by venous blood and then exhaled from the lungs. The body's CO_2 handling differs from the way it transports O_2 in two important respects. First, CO_2 is highly soluble in water and, therefore,

does not require a carrier protein for transport through the circulation. Secondly, CO_2 generates substantial amounts of acid when in solution, requiring the presence of a buffering system.

A. Carbon dioxide forms

CO_2 is transported through the vasculature in three principal forms: in dissolved form, as HCO_3^-, and in association with Hb.

1. **Dissolved:** CO_2 is >20 times more soluble in blood than O_2, and substantial amounts are carried in this form (~5% of total transported CO_2).
2. **Bicarbonate:** Ninety percent of CO_2 is carried as HCO_3^-. HCO_3^- forms through the spontaneous dissociation of H_2CO_3 (see reaction below), through the actions of *carbonic anhydrase* (*CA*), and through combining carbonate and H^+:

Equation 23.1

$$H_2O + CO_2 \underset{CA}{\leftrightarrows} H_2CO_3 \leftrightarrows HCO_3^- + H^+$$

3. **Carbamino compounds:** Five percent of total blood CO_2 is carried as carbamino compounds, which form by reversible reaction of CO_2 with the amine groups of proteins, principally Hb. CO_2 also binds to plasma proteins but not in significant amounts.

B. Carbon dioxide transport

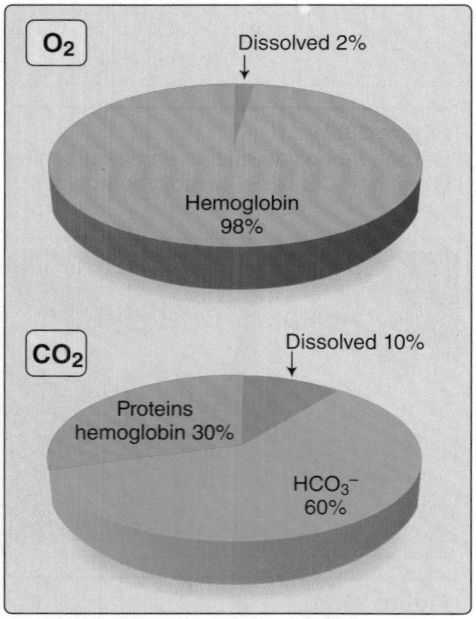

Figure 23.23
Comparison of the ways in which O_2 and CO_2 are transported between lungs and tissues.

Blood carries more than twice the amount of CO_2 than O_2 (~23 mmol/L CO_2 *versus* 9.5 mmol/L O_2). Much of this CO_2 resides in stores, and passage through the systemic capillary beds increases its total content by only 8%. CO_2 that has been newly picked up from tissues is transported to the lungs principally as HCO_3^- (~60%) as shown in Figure 23.23. The remainder is carried in dissolved form (~10%) or in association with a protein (~30%). CO_2 uptake from the tissues occurs by simple diffusion, driven by the partial pressure gradient for CO_2. Its subsequent fate can be divided into several discrete steps (Figure 23.24).

1. **Uptake by red blood cells:** RBCs contain high levels of *CA-I* that converts CO_2 to H_2CO_3 as fast as it enters cells. This helps maintains a strong partial-pressure gradient between tissues and blood that drives CO_2 diffusion. H_2CO_3 then rapidly dissociates to form HCO_3^- and H^+ (see Equation 23.1).
2. **Bicarbonate transport:** HCO_3^- is transported out of the RBC by a Cl^--HCO_3^- exchanger. The **Cl^- shift** causes a slight increase in RBC osmolarity and produces mild swelling, but this is reversed in the lungs.
3. **Hydrogen-ion buffering:** The H^+ released during HCO_3^- formation remains trapped in RBCs by the cell membrane, which is relatively impermeable to cations. This might be expected to lower intracellular pH, but H^+ accumulation occurs at the precise moment that Hb is releasing O_2 and undergoing a conformational change that favors H^+ binding. As noted above (i.e., the Bohr effect), H^+ binding actually *facilitates* O_2 unloading by shifting the

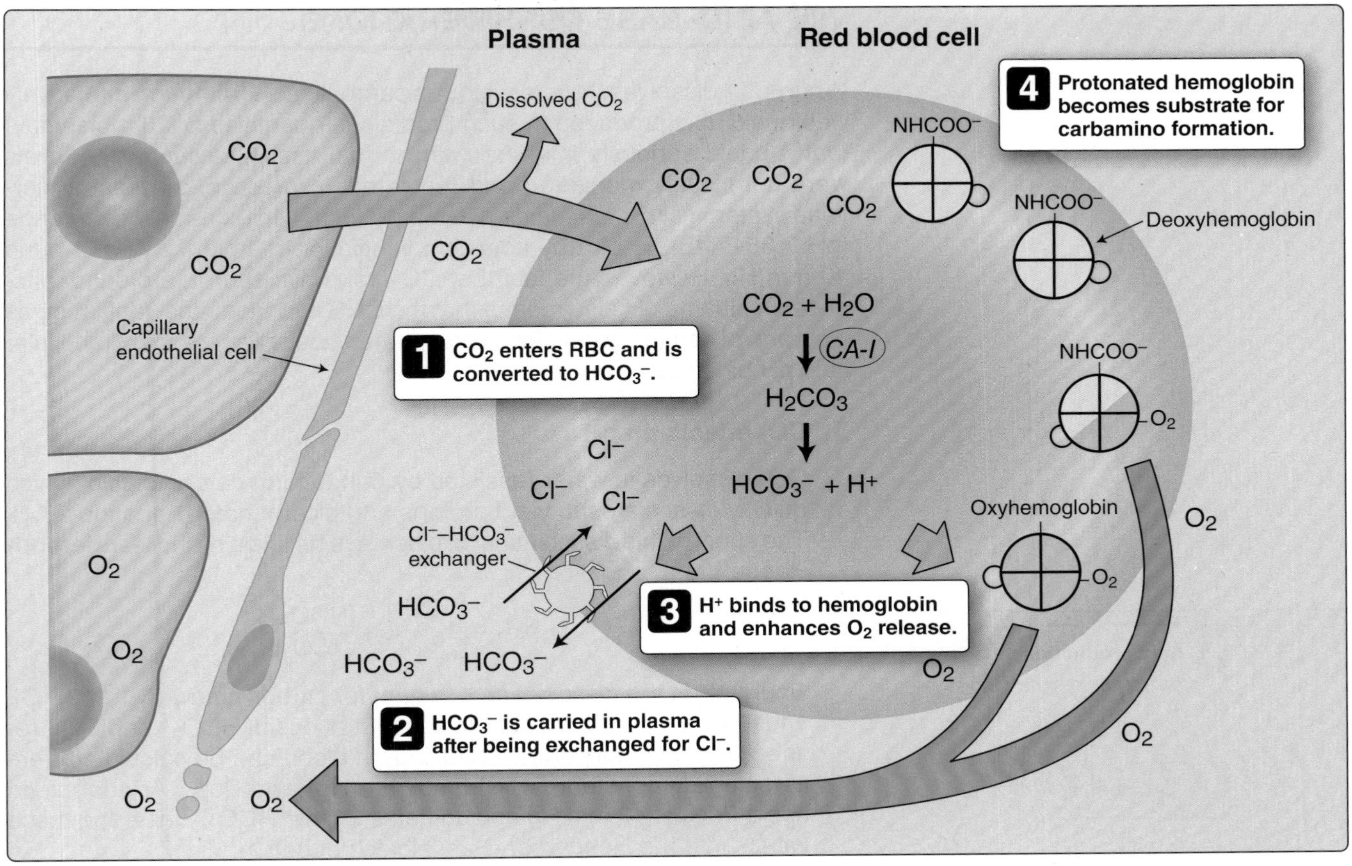

Figure 23.24
Transport of CO_2 in blood. *CA-I = carbonic anhydrase.*

Hb–O_2 dissociation curve to the right and reducing Hb's affinity for O_2. Virtually all of the acid excess caused by loss of HCO_3^- to the plasma is buffered by Hb. With intracellular H^+ kept low by Hb and the Cl^--HCO_3^- exchanger keeping HCO_3^- low, the reaction catalyzed by *CA* remains biased in favor of increased H^+ and HCO_3^- formation. The CO_2-carrying capacity of blood increases as a result.

4. **Carbamino formation:** When Hb binds H^+, it becomes a more favorable substrate for carbamino compound formation (the **Haldane effect** [Figure 23.25]). Hb carries appreciable amounts of CO_2 in carbaminohemoglobin form.

C. Unloading

When blood arrives at the lungs, the partial pressure gradients for both O_2 and CO_2 reverse compared with tissues. A high P_{O_2} causes H^+ to dissociate from Hb (the Haldane effect) and the reaction in Equation 23.1 now favors H^+ and HCO_3^- association to form H_2O and CO_2. HCO_3^- reenters RBCs in exchange for Cl^- and combines with H^+ to form H_2CO_3, which dissociates to release CO_2 and H_2O. CO_2 then diffuses out of the blood, driven by the partial pressure gradient for CO_2 between blood and the alveolar lumen.

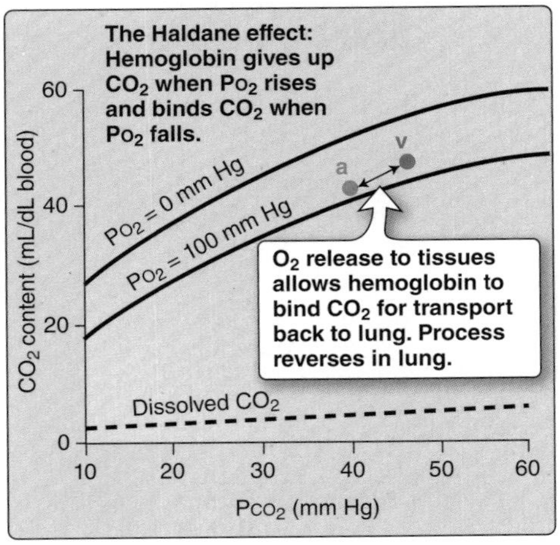

Figure 23.25
Effect of P_{O_2} on the CO_2 dissociation curve (the Haldane effect). a = arterial blood; v = venous blood.

VIII. ACID–BASE CONSIDERATIONS

When CO_2 dissolves in water, it forms carbonic acid. Although a relatively weak acid, it is produced in such prodigious quantities (>20 moles/day) that it could seriously interfere with normal tissue function if its levels were not closely monitored and regulated. In practice, the central nervous system (CNS) maintains plasma pH within an extremely tight range (pH 7.35–7.45), in part by adjusting ventilation to hold $P_{a}CO_2$ at around 40 mm Hg. However, the fact that CO_2 *can* have such a profound influence on plasma pH also means that the CNS can modulate ventilation as a means of compensating for nonrespiratory disturbances in extracellular fluid pH balance.

A. CO_2 effects on pH

CO_2 dissolves in water (assisted by *CA*) to form carbonic acid, which quickly dissociates to yield protons and bicarbonate (Equation 23.1). The effect of this dissociation on plasma pH is given by the **Henderson-Hasselbalch** equation:

$$pH = pK + \log \frac{[HCO_3^-]}{[CO_2]}$$

where pK is the dissociation constant for carbonic acid (6.1 at 37°C), and $[HCO_3^-]$ and $[CO_2]$ denote concentrations of HCO_3^- and CO_2, respectively. The concentration of CO_2 in blood can be calculated from its solubility constant (0.03 mmol/mm Hg) and P_{CO_2}. Arterial blood has a P_{CO_2} of 40 mm Hg and contains 24 mM HCO_3^-. Inserting these values into the Henderson-Hasselbalch equation:

$$pH = 6.1 + \log \frac{24}{0.03 \times 40} = 7.4$$

Note that any increase in P_{CO_2} will cause pH to fall (**acidosis**), whereas decreases will cause pH to rise (**alkalosis**).

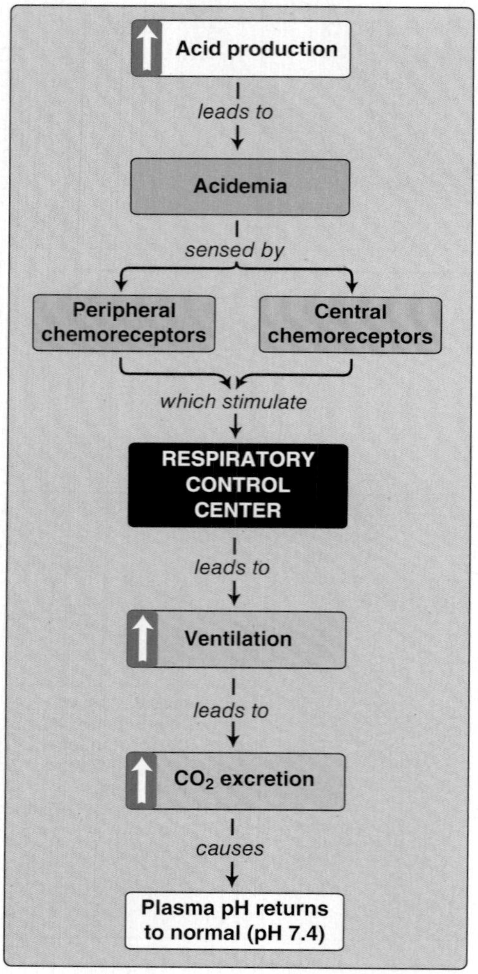

Figure 23.26
Concept map for ventilatory response to acidemia.

B. Causes of changes in extracellular fluid pH

Changes in pH caused by the lungs are referred to as **respiratory acidosis** or **respiratory alkalosis**. Nonrespiratory changes are referred to as **metabolic acidosis** or **metabolic alkalosis**.

1. **Respiratory acidosis:** Increased $P_{a}CO_2$ results from hypoventilation, $\dot{V}_A/\dot{Q}$ mismatches, or an increase in the diffusional distance between the alveolar sac and pulmonary blood supply (due to pulmonary fibrosis or edema, for example).

2. **Respiratory alkalosis:** $P_{a}CO_2$ decreases with hyperventilation, typically because of anxiety or other emotional state. It can also result from hypoxemia precipitated by ascent to high altitude.

C. Compensation

Cells are defended against excess acid accumulation in the short term by buffers, most notably the HCO_3^- buffer system and intracellular proteins such as hemoglobin (see 3·IV·B). Buffers operate on a time scale of seconds or less. Correction of an altered acid–base status ultimately requires a change in lung or kidney function. CNS

respiratory control centers continually monitor plasma pH (Figure 23.26). If pH falls, they increase pulmonary ventilation to transfer CO_2 to the atmosphere, and pH renormalizes. Conversely, a rise in plasma pH initiates a reflex ventilatory decrease, and CO_2 is retained. Ventilatory responses require several minutes to take effect and, if the underlying cause is a metabolic disturbance, may never be sufficient to compensate fully. Respiratory control pathways are considered in Chapter 24. The role of the kidneys in pH balance is detailed in Unit VI, *Urinary System*.

Chapter Summary

- **O_2 and CO_2 exchange** occurs at the **blood–gas interface** within the lung. Exchange is enhanced by the **large surface area** of the interface and the fact that the barrier between blood and air is **very thin**. Ventilation and perfusion ensure that the partial pressure gradients driving diffusion of O_2 and CO_2 across the barrier are kept high.
- The interface is perfused by blood from the **pulmonary circulation**. Pulmonary perfusion pressures are very low, and the vessels have relatively thin walls. These features mean that pulmonary vessels readily expand and collapse in response to **extravascular forces**.
- **Gravitational effects** on pulmonary blood vessels in an upright lung create three different zones of flow. **Perfusion pressures and flow** are lowest at the lung apex (zone 1). Flow is highest at the lung base (zone 3). Gravity also affects **alveolar ventilation**. Alveoli at the lung apex ventilate poorly, whereas alveoli at the lung base ventilate very well. The combined effects of gravity on perfusion and ventilation mean that most O_2 uptake occurs at the base of an upright lung.
- In an ideal lung, **alveolar ventilation and perfusion** should be perfectly matched ($\dot{V}_A/\dot{Q}$ ratio = 1.0). **Mismatches** occur because of airway obstruction or loss of perfusion, and the ratio slips toward zero or infinity, respectively.
- O_2 and CO_2 exchange occurs by diffusion, driven by partial pressure gradients for both gases. **Diffusion of gases** across the alveolar wall is influenced by barrier thickness and total surface area, both of which can become limiting in a diseased lung (**diffusion-limited exchange**). Net uptake may also be limited by inadequacy of perfusion (**perfusion-limited exchange**).
- **O_2** has limited water solubility, so an O_2-binding protein (**hemoglobin** [**Hb**]) is required to help transport it to the tissues in the quantities required for aerobic respiration. O_2 binds to four sites on Hb. The cooperative nature of O_2 binding ensures blood O_2 saturation during passage through the lungs and facilitates O_2 release as blood passes through the tissues served by the systemic circulation.
- **CO_2** is transported in dissolved form, in association with hemoglobin, and as HCO_3^-. HCO_3^- forms by dissociation of carbonic acid.
- Because CO_2 dissolves in water to form **carbonic acid**, ventilatory changes that cause CO_2 to be excreted at rates that exceed or fail to keep up with CO_2 production can result in **respiratory alkalosis or acidosis**, respectively.

24 Respiratory Regulation

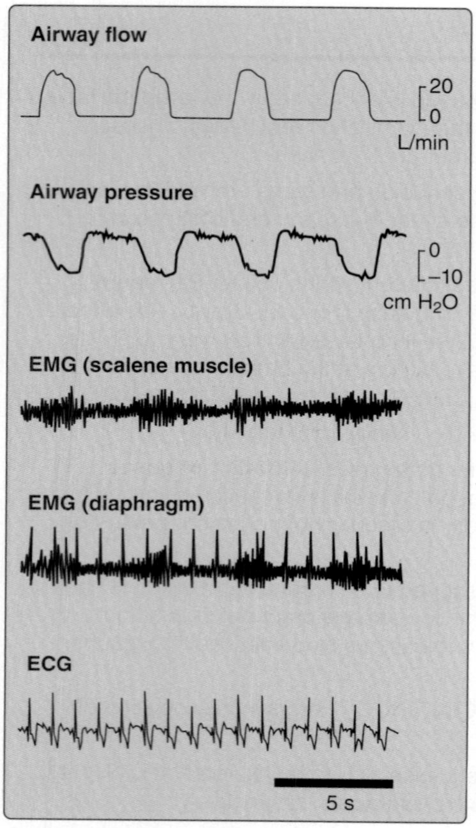

Figure 24.1
The rhythmic inspiration–expiration cycle. ECG = electrocardiogram; EMG = electromyogram.

I. OVERVIEW

Normal breathing (**eupnea**) is usually an unconscious act driven by the autonomic nervous system. The cyclical pattern is established by a respiratory control center within the brain that coordinates contraction of the diaphragm and other muscles involved in inspiration and expiration (Figure 24.1). Because the needs of internal respiration change with activity level, the pattern generator must also change its output to match prevailing needs. Sensors located within the central nervous system (CNS) and throughout the periphery continuously monitor blood and tissue P_{CO_2}, P_{O_2}, and pH levels and feed this information back to the control center for processing (Figure 24.2). The center also receives information from mechanoreceptors located in the lungs and chest wall. The control center then adjusts ventilation as needed, using motor outputs to the diaphragm, the intercostals, and the other muscles involved in breathing. Although the precise location and functions of the various neurons involved in respiratory control remain ill-defined, it is clear that the principal goal is to maintain P_{aCO_2} at a stable level while simultaneously ensuring adequacy of O_2 flow to the tissues. The dominance of CO_2 in respiratory control reflects the body's need to maintain extracellular fluid (ECF) pH within a narrow range.

II. NEURAL CONTROL CENTERS

Several regions of the brain influence breathing. The basic respiratory rhythm is established by a respiratory control center located within the brainstem medulla.

A. Medullary center

The medulla contains several discrete groups of neurons involved in respiratory control (Figure 24.3). Although functioning as a single unit, control-center neurons and their input/output pathways are mirrored on either side of the medulla. Either side is capable of generating independent respiratory rhythms if the brainstem is transected. Within the control center are two concentrations of neurons that fire in phase with the respiratory cycle and are assumed to have a key role in establishing the respiratory rhythm. These are called the **dorsal respiratory group** (**DRG**) and the **ventral respiratory group** (**VRG**).

1. **Dorsal respiratory group: DRG** neurons are active primarily during inspiration (Figure 24.4). Because inspiration is the only

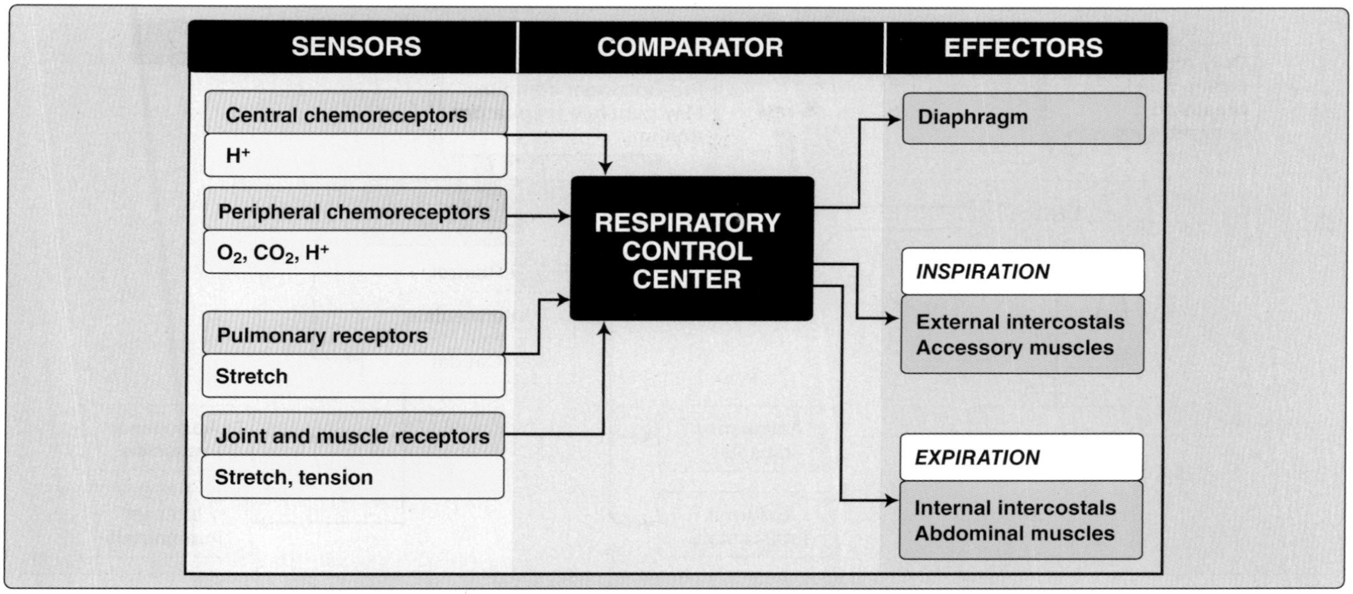

Figure 24.2
Respiratory control pathways.

active phase of quiet breathing, it is likely that the DRG implements the respiratory rhythm, although the pacemaker (a **central pattern generator**, or **CPG**) probably resides elsewhere (see below).

a. **Function:** The DRG receives sensory inputs from the thorax and abdomen and processes incoming signals using interneuron networks. If the sensors report suboptimal CO_2 or O_2 levels, the DRG formulates and executes an appropriate response.

b. **Output:** DRG neurons communicate with premotor neurons controlling the diaphragm. They also project to a control-center area involved in forced inspiration and expiration (the **VRG**).

2. **Ventral respiratory group:** The VRG can be subdivided into rostral, intermediate, and caudal regions (see Figure 24.4).

   a. **Function:** The VRG functions primarily to coordinate accessory muscles of inspiration and expiration. It is largely quiescent during quiet breathing but becomes highly active during exercise.

   b. **Output:** The VRG's intermediate region contains motor efferents to accessory muscles in the pharynx and larynx that dilate the upper airways during inspiration. The caudal region contains premotor neurons that synapse within the spinal cord to control the internal intercostals and other accessory muscles of expiration. The rostral region (the **Bötzinger complex**) communicates via interneurons with the DRG and the caudal region of the VRG. It may be involved in coordinating VRG output.

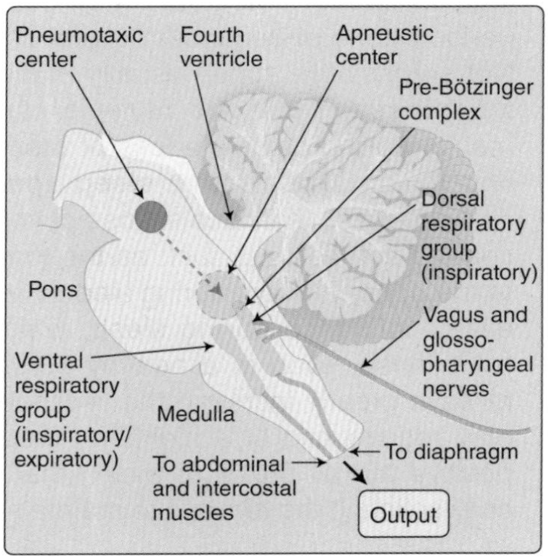

Figure 24.3
Brainstem areas involved in respiratory control.

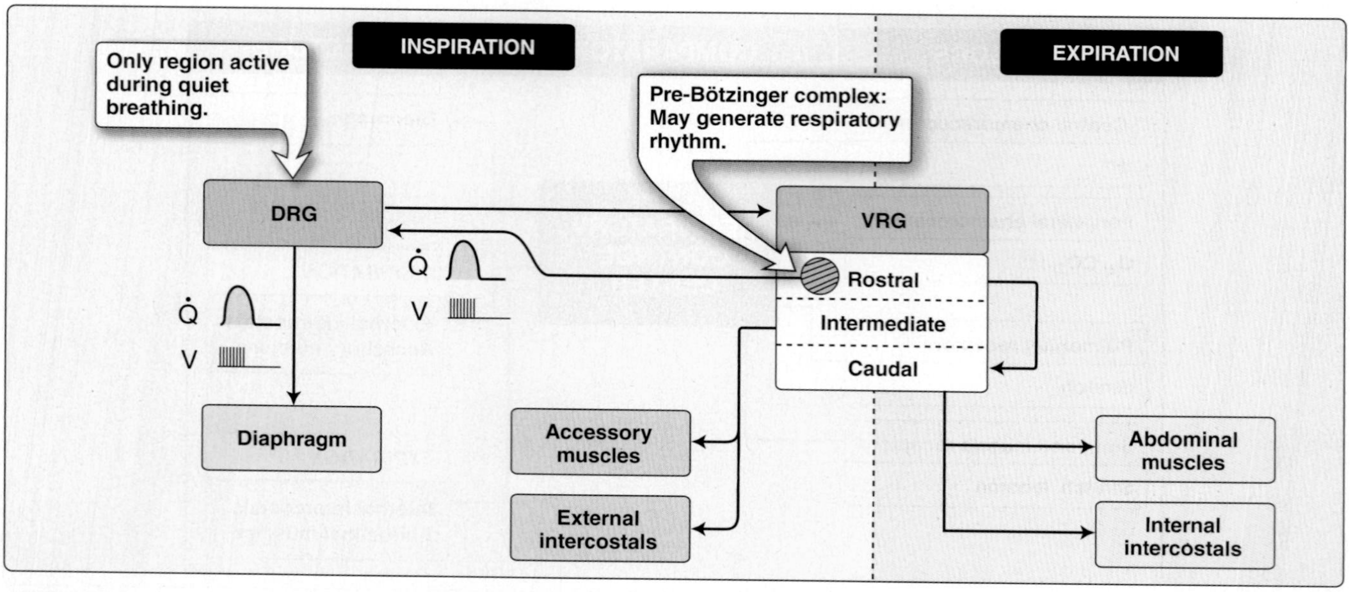

Figure 24.4
Respiratory control center organization. DRG = dorsal respiratory group; $\dot{Q}$ = airway flow; V = voltage records from respiratory motor efferent nerve fibers; VRG = ventral respiratory group.

3. **Central pattern generator:** The VRG rostral region contains a small area known as the **pre-Bötzinger complex** composed of cells that exhibit pacemaker-like activity under experimental conditions. Ablation of the complex abolishes rhythmic breathing, suggesting that it constitutes a respiratory CPG.

Clinical Application 24.1: Sleep Apnea

Sleep is usually accompanied by a pattern of normal quiet breathing, but some individuals may cease breathing for prolonged periods (tens of seconds) several times per hour (**sleep apnea**). Not surprisingly, apnea usually manifests as **excessive daytime sleepiness**. There are multiple causes. **Central sleep apnea** results from a complete loss of respiratory drive and is relatively uncommon. **Obstructive sleep apnea** is more prevalent, especially in obese individuals. Loss of ventilation results from obstruction (by soft tissues, such as the tongue and uvula) and collapse of the upper airways during sleep. Airways normally lose an active dilator influence during sleep, which makes them more prone to collapse, even in a healthy person. Fat deposits around the airways greatly increase the likelihood of collapse, however, and may require use of a nighttime breathing aid (i.e., a continuous positive airway pressure mask) that provides relief by pneumatically splinting the airways to maintain patency.

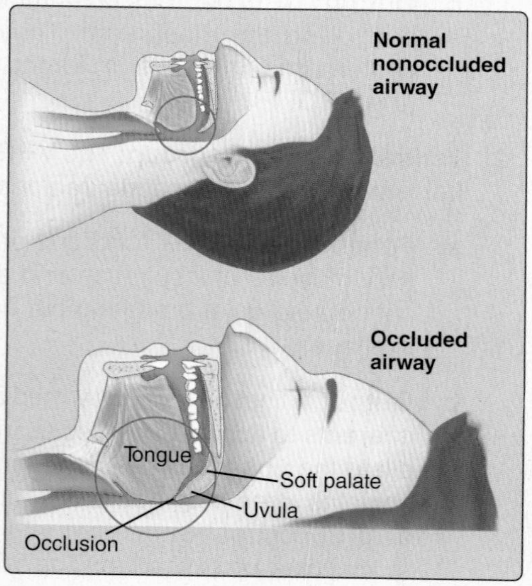

Airway obstruction during sleep causes sleep apnea.

B. Pontine centers

The pons contains two areas that influence medullary output. The **apneustic center** resides in the lower pons. Transectioning the brain above this site causes prolonged inspiratory gasping (**apneuses**), suggesting that it normally limits lung expansion. Stimulating the **pneumotaxic center** (upper pons) shortens inspiration and increases breathing rate. A role for either center in normal breathing has not been established.

C. Cerebral cortex and other brain regions

Emotions such as fear, excitement, and rage can alter respiratory rate, reflecting the ability of the hypothalamus and limbic system to modulate the CPG. The CPG also readily succumbs to the cerebral cortex to allow for talking, playing a musical wind instrument, and other activities that require fine conscious control of breathing movements. A person can also consciously override the CPG and **hyperventilate**, **hypoventilate**, or cease breathing altogether. The effects of breath-holding on blood gas concentrations are tolerated for a relatively short time before chemical control feedback pathways override voluntary control.

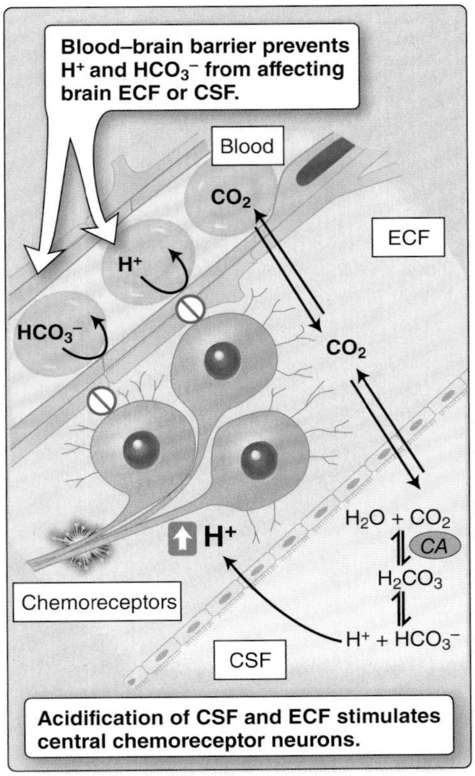

Figure 24.5
Central chemoreceptors monitor arterial P_{CO_2} through effects of CO_2 on pH of cerebrospinal fluid (CSF) and extracellular fluid (ECF). *CA = carbonic anhydrase.*

III. CHEMICAL CONTROL OF VENTILATION

A primary mission of the respiratory system is to optimize ECF pH through manipulation of P_{CO_2}. It must also maintain steep O_2 and CO_2 partial pressure gradients to maximize transfer of these two gases between tissues and the external environment. Fulfilling these roles requires that information about the chemical composition of ECF be sensed and relayed back to the respiratory control centers so that ventilation might be modified accordingly. The body employs central and peripheral chemoreceptors for this purpose.

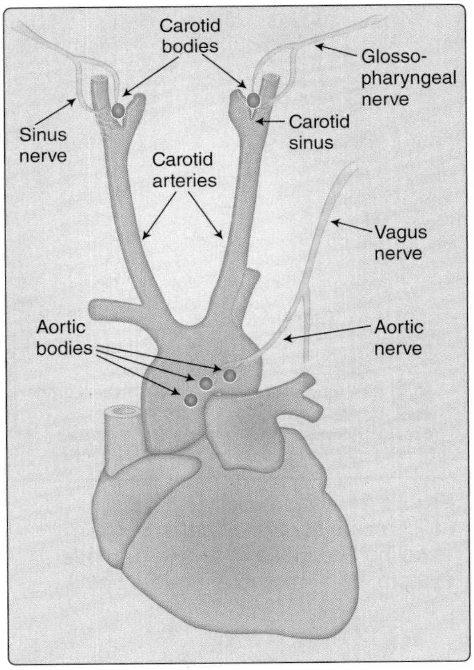

Figure 24.6
Peripheral chemoreceptors are contained within carotid bodies and aortic bodies.

A. Central chemoreceptors

Ventilatory control under resting conditions is dominated by the central chemoreceptors, which are responsive primarily to changes in P_{aCO_2}. The chemoreceptors are CNS neurons located behind the blood–brain barrier (BBB) along the medullary surface. The BBB is impermeable to virtually all blood constituents except for lipid-soluble molecules, such as O_2 and CO_2. Once inside the barrier, CO_2 dissolves to form carbonic acid, which acidifies brain ECF and cerebrospinal fluid (CSF) as shown in Figure 24.5. CSF contains minimal amounts of protein to buffer pH. The consequence is that even modest changes in P_{aCO_2} cause significant CSF and ECF acidosis. The chemoreceptor neurons respond to acid with excitatory impulses that impel the respiratory center to increase breathing rate. The BBB acts as an important information filter because, by excluding bloodborne ions such as H^+, it provides a way for the chemoreceptors to distinguish changes in P_{aCO_2} from any background changes in ECF pH.

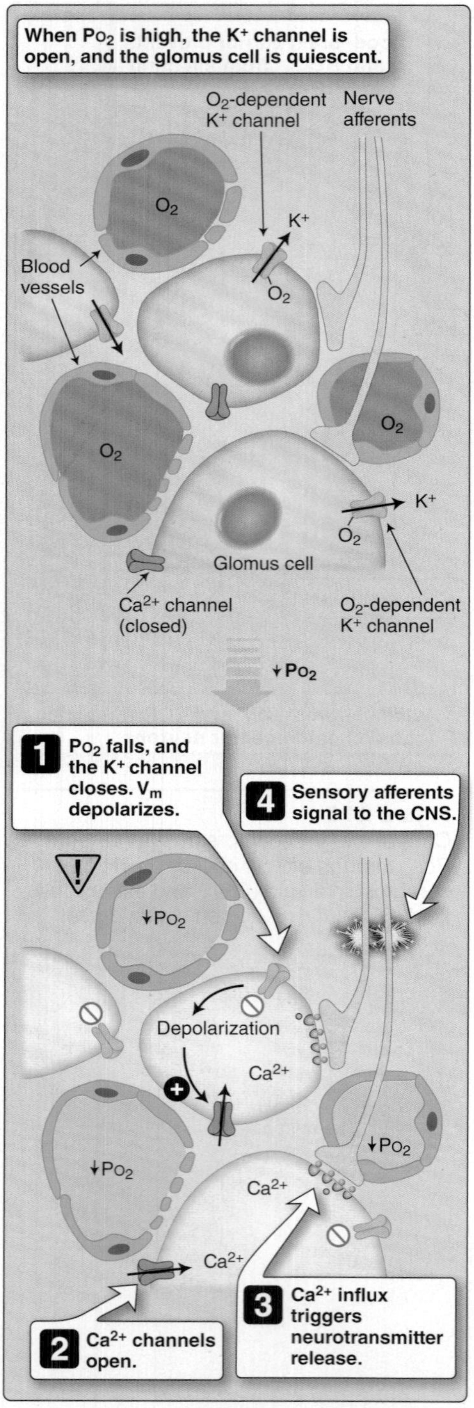

Figure 24.7
Mechanism of glomus-cell response to falling Po_2. CNS = central nervous system; V_m = membrane potential.

> Systemic changes in pH ultimately affect all tissues regardless of cause or BBB ion permeability. Central chemoreceptors play a critical role in the integrated response to changing blood pH, although reactions to metabolic acidosis or alkalosis may be slower and less intense than respiratory responses triggered by changes in P_aCO_2.

B. Peripheral chemoreceptors

Peripheral chemoreceptors were first introduced in reference to control of blood pressure (see 20·III·A·3), because they also relay information to the medullary cardiovascular control centers. The chemoreceptors with the greatest influence on respiration are located within carotid bodies located at the bifurcation of the two common carotid arteries (Figure 24.6). Chemoreceptors are also found within aortic bodies distributed along the underside of the aortic arch. Peripheral chemoreceptors monitor P_aO_2, P_aCO_2, and arterial pH.

1. **Structure:** Carotid and aortic bodies are notable for their small size (3–5 mm) and high rates of blood flow relative to their mass. The high flow rate minimizes the effect of chemoreceptor metabolism on blood gas content and, therefore, allows for a truer reading of arterial O_2 and CO_2 levels. Carotid bodies contain two types of cells arranged in clusters and in close apposition to fenestrated capillaries. **Type I**, or **glomus**, **cells** are the chemosensors (Figure 24.7). **Type II**, or **sustentacular**, **cells** play a supportive role similar to glia. Glomus cells signal the respiratory center via the carotid sinus and glossopharyngeal (cranial nerve [CN] IX) nerves (carotid bodies) or the vagus nerve (CN X; aortic bodies).

2. **Sensory mechanism:** Glomus cell membranes contain an O_2-sensitive K^+ channel whose open probability is Po_2-dependent. When P_aO_2 is high, the channel is open and allows a K^+ efflux that maintains glomus-cell membrane potential (V_m) at strongly negative levels. When P_aO_2 falls below 100 mm Hg, the channel closes, and V_m depolarizes. The change in V_m activates L-type Ca^{2+} channels and elicits a Ca^{2+} influx that stimulates neurotransmitter release onto the sensory afferents (see Figure 24.7). Glomus cells are also excited by increasing P_aCO_2 and H^+ concentrations independently of changes in P_aO_2. They help fine-tune the information that the respiratory center receives from central chemoreceptors. Changes in P_aCO_2 and pH may also act by influencing O_2-dependent K^+ channel open probability, but the mechanism(s) has not been defined.

C. Ventilatory responses

The respiratory control center processes information about P_aO_2, P_aCO_2, and arterial pH. Changes in these variables rarely occur in isolation, which forces the respiratory center to make choices about an appropriate ventilatory response (Figure 24.8). Under most circumstances, the output from the respiratory center is designed to optimize P_aCO_2, but coincident changes in P_aO_2 and pH influence system sensitivity to changes in P_aCO_2.

1. **Changing carbon dioxide levels:** The nature of control-center response to a rise in P_aCO_2 depends on whether the change is acute or chronic.

   a. **Acute:** When tissue metabolism increases, P_aCO_2 rises, and the respiratory center compensates by increasing alveolar ventilation (see Figure 24.8). CO_2 is an extremely potent ventilatory stimulus (Figure 24.9). The peripheral receptors are fast acting and trigger an immediate response. The central receptors take several minutes to activate fully, but their effects ultimately dominate the ventilatory response. In practice, ventilation increases linearly with any rise in P_aCO_2 above resting values, whereas ventilation decreases when P_aCO_2 falls below 40 mm Hg.

   b. **Chronic:** Patients with chronic pulmonary disease may not be able to ventilate at levels required to hold P_aCO_2 at 40 mm Hg. Ventilation increases initially when P_aCO_2 begins to rise because the pH of the CSF falls, but the choroid plexus responds by secreting HCO_3^- into the CSF, which, over a period of 8–24 hours, largely offsets the effect of the higher P_aCO_2 on plasma pH. Thus, although P_aCO_2 may remain high, the medullary chemoreceptors no longer register the change, and the respiratory control system adapts to a new, higher P_aCO_2.

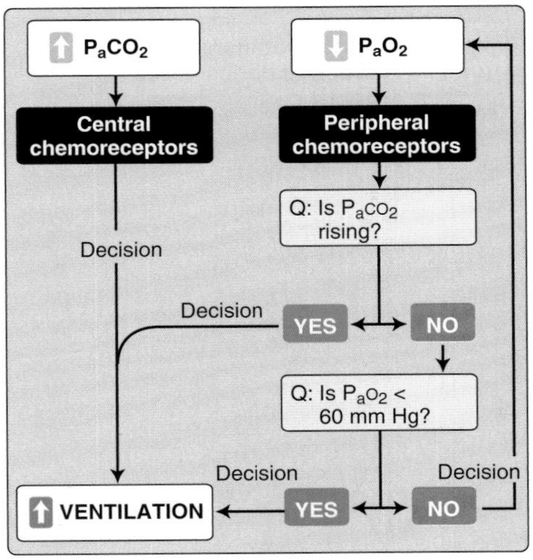

Figure 24.8
Ventilatory responses to changes in arterial PCO_2 (P_aCO_2) and PO_2 (P_aO_2).

> Patients that have adapted to hypercapnia may rely on the effects of hypoxia on the peripheral chemoreceptors to sustain their ventilatory drive. When significant hypercapnia exists, administering O_2 may help normalize P_aO_2 but it can also reduce the drive to breathe, potentially increasing P_aCO_2 further and inducing acute respiratory acidemia due to hypoventilation.

2. **Changing oxygen levels:** The peripheral chemoreceptors promote a rapid compensatory increase in ventilation if P_aO_2 falls to dangerously low levels (see Figure 24.10). More modest decreases in P_aO_2 (between 60 and 100 mm Hg) have little effect on ventilation, even though spike frequency in the peripheral chemoreceptor nerve afferents increases in direct proportion to the drop in P_aO_2. The reason is that the peripheral and central chemoreceptor work against each other for respiratory-center control, with the central chemoreceptors retaining the upper hand until P_aO_2 falls to 60 mm Hg (see Figure 24.8). Control center output in response to changes in P_aO_2 is influenced both by blood pH and PCO_2.

   a. **pH:** When P_aO_2 falls, deoxyhemoglobin concentration rises. Deoxygenation makes hemoglobin (Hb) a more favorable substrate for H^+ binding (the Bohr effect) and plasma H^+ concentrations fall as a result. The rise in pH decreases chemoreceptor sensitivity to a fall in P_aO_2.

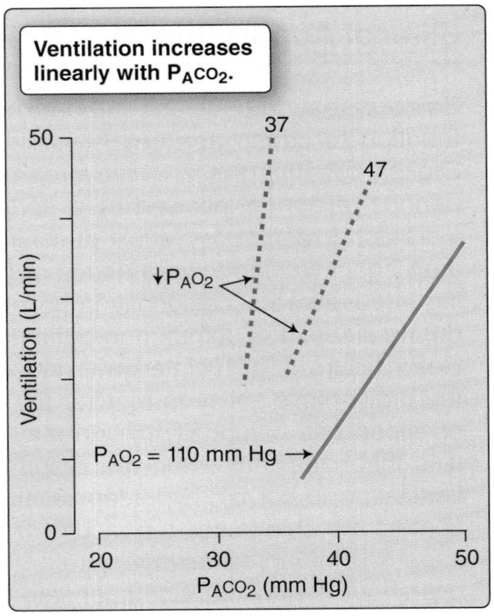

Figure 24.9
Ventilatory responses to changes in alveolar PCO_2 (P_ACO_2).

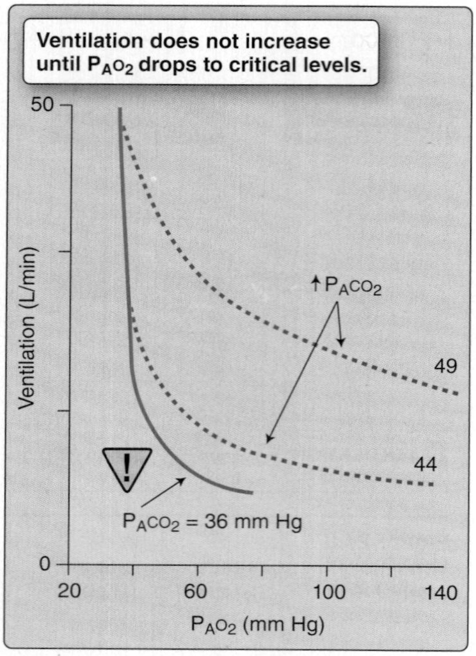

Figure 24.10
Ventilatory responses to changes in alveolar P_{O_2} (P_{AO_2}).

b. **Carbon dioxide:** When ventilation increases, P_{aO_2} increases, and P_{aCO_2} falls. Because the respiratory control centers are designed to optimize P_{aCO_2}, the response to mild hypoxia is overridden in favor of a stable CO_2 concentration.

> The chemoreceptors begin to exert their influence at a P_{aO_2} of 60 mm Hg, which, coincidentally, marks the point at which Hb begins to desaturate (i.e., the steep portion of the dissociation curve; see Figure 23.18). Thus, Hb's O_2-binding properties allow the respiratory center to adjust ventilation over a wide range to maintain a stable P_{aCO_2} with little adverse effects on O_2 delivery.

3. **Synergism:** Conditions that decrease P_{aO_2} typically cause a concomitant rise in P_{aCO_2}. Because P_{aCO_2} equates with H^+ concentration, pH falls also. Thus, it is not surprising to find that changes in P_{aO_2}, P_{aCO_2}, and arterial pH may act synergistically to elicit a ventilatory response that is greater than the sum of their individual actions.

> Hypoxia increases chemoreceptor sensitivity to hypercapnia, and rising P_{aCO_2} and H^+ concentrations sensitize the receptors to hypoxia.

Clinical Application 24.2: Cheyne-Stokes Breathing

Some disease conditions interfere with respiratory control-center function to cause abnormal resting breathing rhythms. Cheyne-Stokes breathing is a cyclic breathing pattern characterized by periods of apnea followed by a series of breaths of progressively increasing respiratory effort and airflow toward a maximum, and then again waning toward apnea. Although Cheyne-Stokes breathing is observed occasionally in normal individuals at high altitudes during sleep, it is common in patients with stroke and heart failure. Heart failure is associated with low perfusion rates, which impairs the brainstem's ability to monitor the effects of changing ventilation on blood gas composition, and may explain the abnormal rhythm. It is theorized that when the respiratory control centers increase ventilation to correct hypercapnia, there is a time delay before the centers are able to see the results of this change, during which they impel further increases in respiratory effort and cause a hypocapnic overshoot. The centers compensate by decreasing ventilation, with the same hypoperfusion delay causing respiratory effort to decline inappropriately and resulting in apnea. The cyclic breathing pattern is repeated with a variable periodicity of ~30–100 s.

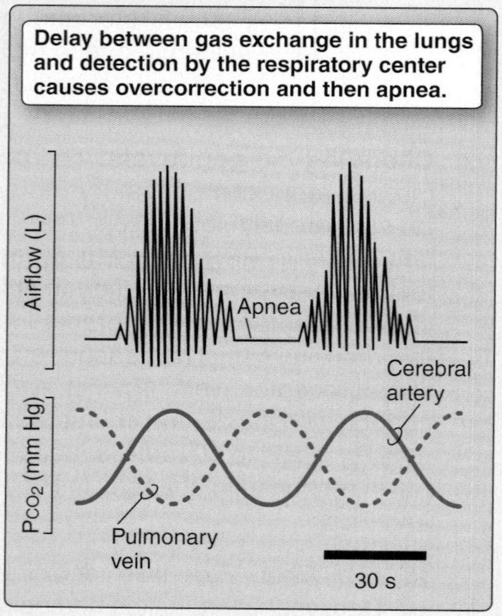

Cheyne-Stokes Breathing.

IV. PULMONARY RECEPTOR ROLE

The lung and airways contain a variety of receptors that help protect the system from foreign bodies and provide the respiratory center with feedback about lung volume (Figure 24.11). Information flow from these receptors travels primarily in the vagus nerve.

A. Irritant receptors

The epithelium of the larger conducting airways contains sensory nerve endings that respond to irritants and noxious stimuli, such as ammonia, smoke, pollen, dust, and cold air. The receptors trigger bronchoconstriction, mucus secretion, and coughing, presumably to prevent foreign materials from reaching the respiratory zone. Irritant receptors have also been implicated in the bronchoconstriction that results from histamine release during an allergic asthma attack. These receptors are also known as **rapidly adapting receptors**, a reference to their behavior during lung expansion. When the lung is inflated, the irritant receptors are stretched by transmural pressure, causing a burst of action potentials in the afferent nerve fibers. The intensity of the bursting is proportional to the rate and degree of stretch, providing information about the rate and extent of lung expansion. The bursting behavior wanes if volume holds steady at the new level, an indication that the receptors adapt rapidly to a sustained stimulus.

B. Juxtapulmonary capillary receptors

The alveolar walls contain unmyelinated nerve fibers (**C fibers**) that have similar functions and response characteristics to the irritant receptors. Also known as **juxtapulmonary capillary receptors**, or **J receptors**, they are sensitive to lung inflation, injury, pulmonary vascular congestion, and certain chemicals. When stimulated, they cause bronchoconstriction, mucus secretion, and shallow breathing.

C. Stretch receptors

Embedded within the smooth muscle layers of conducting airways are myelinated sensory fibers. They respond to stretch, with output intensity reflecting the extent of inflation. In contrast to the other two classes of receptors described above, they adapt very slowly when the stimulus is sustained. If tidal volume is sufficiently high, the stretch receptors can terminate inspiration and prolong the exhalation that follows (the **Hering-Breuer reflex)**. The function is unknown.

> The stretch receptors that mediate the Hering-Breuer reflex relay sensory signals to the respiratory control centers via the vagus nerve. Lung-transplant patients breathe normally despite having lost this pathway, showing that feedback from the receptors is not required to sustain the breathing cycle. The Hering-Breuer reflex is prominent in newborns, however, suggesting that it may have a physiologic role in preventing lung overinflation during infancy.

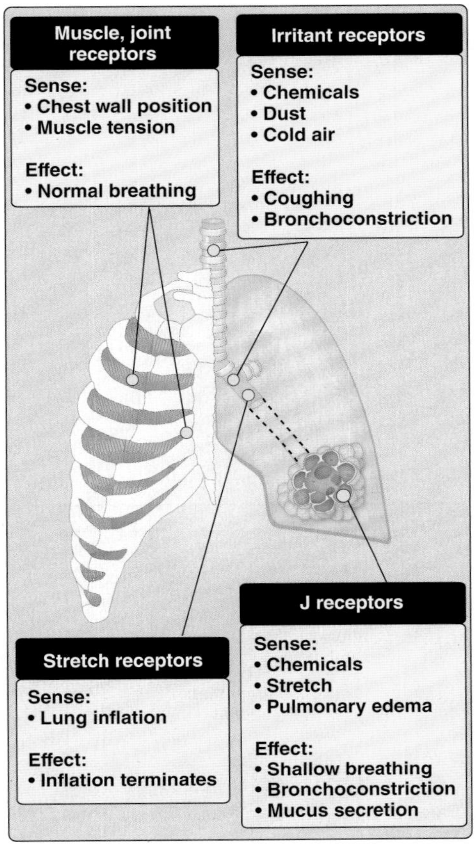

Figure 24.11
Chest wall and lung sensory receptors. J receptors = juxtapulmonary receptors.

D. Joint and muscle receptors

Stretch and tension receptors located in the chest wall sense wall movement and the amount of effort involved in breathing. Output from these receptors allows for increased force of inspiration and expiration when wall movement is impeded. Limb joints contain similar receptors that contribute to increased ventilation during exercise.

V. RESPIRATORY ADAPTATION TO ENVIRONMENT

O_2 diffusion from the atmosphere to mitochondria is driven by a partial pressure gradient. Because atmospheric pressure decreases with height above sea level, moving to altitude forces the respiratory system to adapt to a reduced partial pressure gradient. Diving dramatically increases the pressure gradient driving uptake of O_2 and other atmospheric gases, which can have severe physiologic consequences.

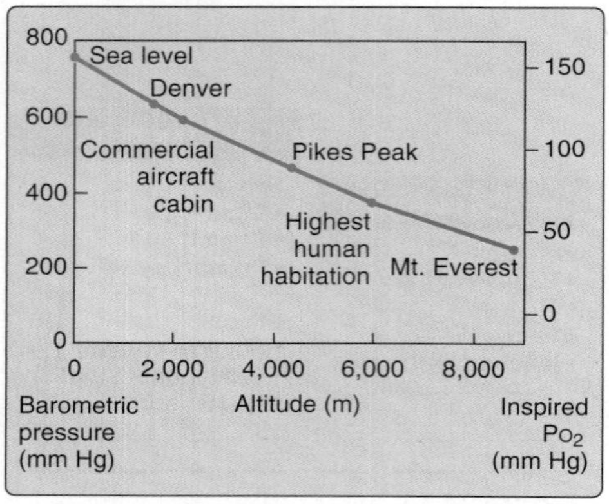

Figure 24.12
Effects of altitude on barometric pressure and P_{O_2}.

A. Altitude

The fractional composition of air does not change with altitude, but the partial pressures of its various constituents all fall with decreasing barometric pressure during ascent. The P_{O_2} of air at the peak of Mt. Everest (8,848 m), for example, is 43 mm Hg (Figure 24.12). Alveolar P_{O_2} is lower than in dry air because airways add water during inspiration, which reduces the fractional composition of the other components. P_{O_2} is reduced by altitude to a greater extent than might be predicted based on barometric pressure alone, because the rate at which lungs and airways add water and CO_2 to the lungs does not change with altitude. Falling $P_{A}O_2$ reduces the partial pressure gradient driving O_2 uptake and causes hypoxia. Sensory and cognitive functions deteriorate rapidly with altitude as a result, reflecting the acute dependence of CNS neurons on O_2 availability (Figure 24.13). The physiologic response to hypoxia can be divided into three phases: acute responses, adaptive responses, and long-term acclimation (Figure 24.14).

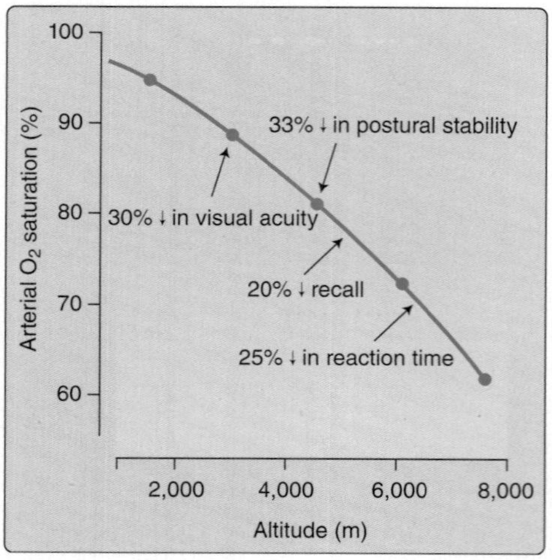

Figure 24.13
Decrease in sensory and cognitive functions caused by the decrease in arterial O_2 saturation that occurs with ascent to altitude.

1. **Acute (minutes):** Hypoxia is sensed by the peripheral chemoreceptors. The respiratory center responds by increasing ventilatory drive to ensure that $P_{A}O_2$ remains high, but $P_{a}CO_2$ falls as a result, which activates the central chemoreceptors. Ventilatory drive is blunted as a consequence. The respiratory center also suppresses the cardioinhibitory center and allows heart rate to rise (see Figure 24.14A). Resting cardiac output increases, facilitating increased O_2 uptake by increasing pulmonary perfusion. Hypoxia coincidentally causes pulmonary vasoconstriction, which increases pulmonary vascular resistance and forces the right heart to generate higher pressures to maintain output.

2. **Adaptive (days to weeks):** The central chemoreceptors adapt slowly over a period of 8–24 hours, allowing ventilation rates to climb in order to address the altitude-induced hypoxia (see Figure 24.14B). The resulting drop in $P_{a}CO_2$ causes respiratory alkalosis, but the kidneys compensate by decreasing acid excretion and blood pH renormalizes (see Figure 24.14C). Alkalosis also stimulates 2,3-diphosphoglycerate (2,3-DPG) production. 2,3-DPG decreases

Hb's O_2 affinity, causing the O_2–Hb dissociation curve to shift rightward. This change enhances O_2 unloading to the tissues.

3. **Acclimation (months to years):** Long-term acclimation to living at altitude involves changes in the properties of blood, the vasculature, and the cardiopulmonary system.

   a. **Blood:** Hypoxia stimulates erythropoietin release from kidneys and promotes red blood cell production. Hb concentration increases proportionately, from ~15 g/dL to ~20 g/dL (see Figure 24.14D). Concurrent increases in circulating blood volume can yield an overall increase in blood's O_2-carrying capacity by >50%.

   b. **Vasculature:** Hypoxia stimulates angiogenesis. Capillary density increases throughout the body, allowing for improved tissue perfusion.

   c. **Cardiopulmonary system:** The increase in pulmonary arterial pressures required to perfuse the lungs in the face of hypoxic vasoconstriction promotes vascular and ventricular remodeling. Smooth-muscle proliferation increases vascular wall thickness and the right ventricle hypertrophies to counter the increased afterload. Although the pressure increase stresses the pulmonary circulation, it is also beneficial in that it increases perfusion of the lung apex and allows apical alveoli to participate in O_2 uptake.

4. **Adverse effects:** Many individuals develop **acute mountain sickness** when they ascend to high altitude, a temporary condition characterized by headache, irritability, insomnia, dyspnea, dizziness, nausea, and vomiting. Symptoms usually dissipate over a period of several days. **Chronic mountain sickness** develops after prolonged residence at high altitude and reflects the adverse cardiovascular consequences of the adjustments noted above. Polycythemia increases blood viscosity and resistance to blood flow, forcing both ventricles to operate at higher pressures (see 19·IV·C). Decreased P_{AO_2} causes bronchoconstriction, which stresses the right side of the heart. If the hypoxia is sufficiently severe or prolonged, pulmonary veins also constrict, and the arteries become narrowed by vascular remodeling. Ultimately, this may cause pulmonary edema, right heart failure, and death.

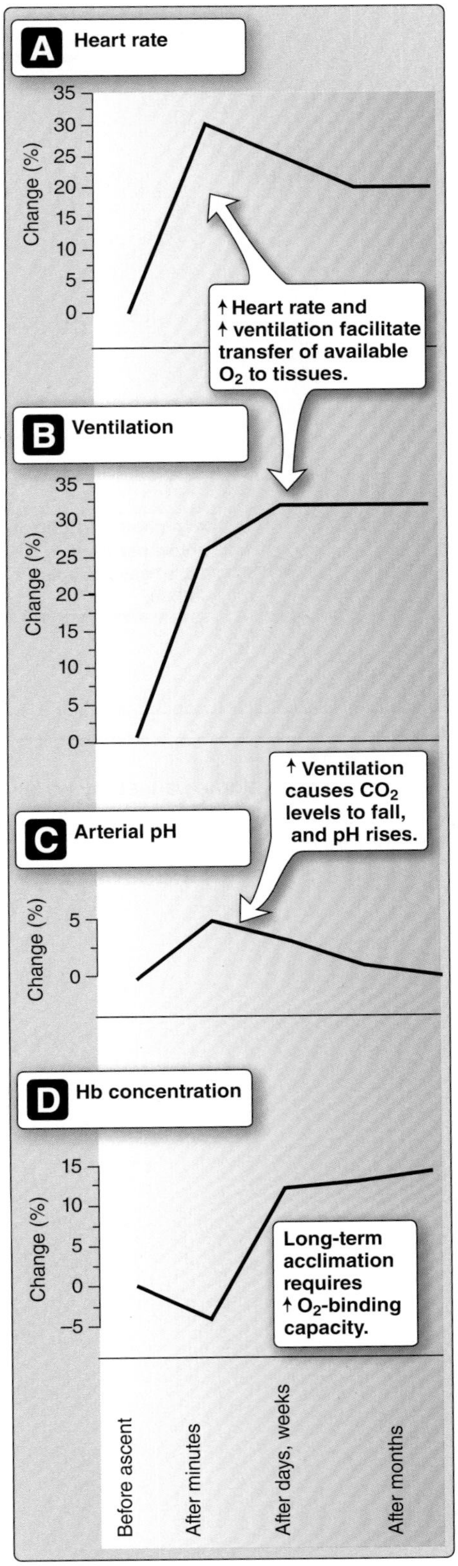

Figure 24.14
Changes in heart rate, ventilation, arterial pH, and hemoglobin (Hb) concentration after ascent and adaptation to altitude (3,000 m above sea level) shown as percentages relative to levels recorded prior to ascent.

B. Diving

Diving presents a number of challenges to the respiratory system, most of which are associated with external hydrostatic pressure at increased depth. Water is denser than air, so pressure rises quickly with depth beneath the surface. It takes a water column of only ~10 m to exert a pressure equivalent to that of the atmosphere (760 mm Hg), so a diver at ~30 m is subject to pressures approximating four atmospheres.

1. **Effects of depth:** Water squeezes and compresses a diver from all sides. It also compresses gases within the alveoli, which increases the partial pressures driving uptake of all gases and decreases alveolar volume, causing two significant challenges.

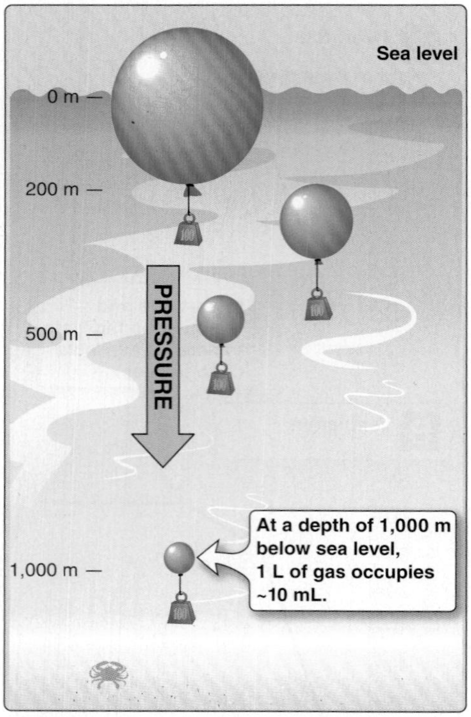

Figure 24.15
Changes in gas volume caused by water pressure at various depths below sea level (0 m).

a. **Partial pressures:** At sea level, O_2 and CO_2 are the only components of atmospheric air to dissolve in blood to any significant extent. Diving can increase the partial pressure on all constituents to such a degree that they all are forced to dissolve in potentially lethal excess.

b. **Volume:** Pressurizing a gas decreases its volume (Figure 24.15). At 30 m, 1 L of gas (sea level volume) occupies ~250 mL. Conversely, 1 L of gas expands to fill 4 L when the diver surfaces from 30 m, which potentially can cause severe damage to any tissue that contains it.

2. **Gas toxicity:** Air consists principally of N_2 (78%) and O_2 (21%), both of which become toxic when inhaled under pressure. The CO_2 composition of inspired air is insignificant and is not of concern unless the diver's breathing apparatus traps exhaled gas, allowing CO_2 to rise.

a. **Nitrogen narcosis:** N_2 has no significant effect on bodily function at sea level because it does not dissolve in tissues. However, at depths of ~40 m and below, PN_2 rises to the point where it dissolves in cell membranes in amounts sufficient to disrupt ion-channel function. Its effects are narcotic and similar to those of ethanol (**nitrogen narcosis**). The severity of its actions are related to depth and pressure, initially causing a feeling of well-being but ultimately causing loss of function at ~80 m and below.

b. **Oxygen poisoning:** O_2 is an inherently toxic molecule because of its tendency to form free radicals. At sea level, the amount of O_2 being delivered to tissues is closely regulated by Hb, which acts both as a vehicle for transporting O_2 through the circulation and also as an O_2 buffer. The delivery system is essentially saturated under normal circumstances. Breathing O_2 at high pressure causes it to dissolve in blood in amounts that exceed the buffering capacity of Hb. The tissues subsequently are exposed to a PO_2 that exceeds the normal safe range (20–60 mm Hg), causing a variety of neurologic effects, including visual disturbances, seizures, and coma.

c. **Deep-sea diving:** Divers who work at depth breathe a helium/oxygen mix (**heliox**), with the percentage of O_2 carefully tailored to yield a partial pressure that is supportive rather than harmful. Helium replaces N_2 because it dissolves in body tissues less readily, is less narcotic, and has a density that is considerably reduced compared with N_2 (14%). Inhaling the helium mix reduces airway resistance and decreases the work of breathing.

> Heliox can also be used clinically to support patients with anatomic or physiologic airway obstruction. The gas mix's decreased density allows it to slip past the obstruction site more easily than does atmospheric air and thereby helps improve patient oxygenation.

3. **Decompression sickness:** A diver breathing air at pressure for prolonged periods can accumulate significant amounts of N_2 within their tissues. The average amount of N_2 contained within the body at sea level is ~1 L. A prolonged dive at 30 m raises this amount to as much as 4 L. N_2 is taken up by diffusion across the blood–gas interface and then distributed by the circulation to all tissues, but it preferentially partitions in body fat. When a diver ascends back to the surface, N_2 is no longer subject to the pressure that forced it to dissolve at depth, so it comes out of solution and forms pure N_2 bubbles. The presence of bubbles in the blood stream causes decompression sickness, or "**the bends**." Bubbles block blood vessels, and, as small bubbles coalesce to form large bubbles, progressively larger vessels are affected. Dependent tissues become ischemic, typically manifesting as pain in joints and in limb musculature. More severe symptoms may include neurologic deficits, dyspnea, and death. Slowing the rate of ascent gives the ~3 L of excess gas dissolved in the aqueous phase more time to diffuse out of the tissues and into the circulation for transport to the lungs for exhalation (Figure 24.16). Fat is relatively avascular, however. This increases the distance over which N_2 must diffuse before it can be carried away by the circulation, thereby slowing that rate at which it can be removed. Complete renormalization of tissue N_2 levels may take several hours after ascent.

Figure 24.16
Deep sea divers carefully time their ascent to avoid "the bends."

Chapter Summary

- A cyclical pattern of inspiration and expiration is established and controlled by a **central pattern generator** (**CPG**). The CPG resides in a **brainstem respiratory control center** (medulla).
- The medulla contains two groups of cells involved in respiratory control. The **dorsal respiratory group** drives inspiration during quiet breathing. The **ventral respiratory group** coordinates accessory muscles and is believed to house the central pattern generator.
- Higher control centers are able to override unconscious breathing to allow for talking, coughing, and other voluntary acts that use the same musculature.
- Central and peripheral chemoreceptors relay information about the chemical composition of blood to the control centers. **Central chemoreceptors** monitor P_aCO_2. **Peripheral chemosensors** are sensitive to P_aO_2, P_aCO_2, and arterial pH.
- When P_aCO_2 rises, both the peripheral and central chemoreceptors are excited and cause the control center to respond with an immediate increase in ventilation. Decreases in P_aO_2 are a less effective stimulus for ventilation unless CO_2 rises simultaneously.
- Other sensors feeding information to the medulla include **irritant receptors**, **juxtapulmonary capillary receptors** that are sensitive to lung injury, **stretch receptors**, and **joint and muscle receptors**.
- Atmospheric pressure decreases with altitude above sea level. PO_2 also decreases, causing hypoxia. The respiratory and cardiovascular control centers help compensate in the short term by increasing ventilation and perfusion. Full acclimation to altitude requires months and involves an increase in lung perfusion, hematocrit, and capillary density in all tissues.
- Descent to depth in water increases the partial pressures of inspired gases. N_2 and O_2, which are poorly soluble at sea level, become toxic when forced to dissolve in tissues at depth. N_2 preferentially partitions in fat and requires many hours to extract upon returning to the surface. Premature ascent results in the formation of pure N_2 gas bubbles within the vasculature and causes severe pain known as "**the bends**."

Study Questions

Choose the ONE best answer.

V.1 A 55-year-old male with a history of interstitial pulmonary fibrosis undergoes pulmonary function testing. What parameter would most likely be decreased in this restrictive lung disease patient?

A. FVC (forced vital capacity)
B. Peak expiratory flow rate
C. FEV_1 (forced expiratory volume in 1 second)
D. FEV_1/FVC
E. Fraction of expired O_2

Best answer = A. Restrictive lung disease is associated with lung stiffening, which limits lung expansion (22·VII·B). This manifests as a decrease in forced vital capacity (FVC) in pulmonary function tests. In practice, such patients can voluntarily inhale and exhale less air volume than a healthy person of comparable age, sex, and height. Peak expiratory flow rate and forced expiratory volume in 1 second (FEV_1) may or may not be normal. The FEV_1/FVC ratio is increased because FVC is typically reduced significantly. The fraction of expired O_2 would not be changed by restrictive lung disease.

V.2 β-Adrenergic receptor agonists cause which of the following effects on pulmonary function?

A. Decreased forced vital capacity
B. Decreased total lung capacity
C. Increased diffusing capacity
D. Bronchiolar constriction
E. Bronchiolar dilation

Best answer = E. β-adrenergic receptor agonists relax airway smooth muscle, promoting bronchiolar dilation (22·VIII·C). Relaxation occurs due to inhibition of acetylcholine release from parasympathetic nerve terminals. The increased airway luminal diameter improves flow, as measured by the forced expiratory volume in 1 second (FEV_1). Total lung capacity, forced vital capacity (maximal air volume that can be forcibly expired), and diffusing capacity (a measure of blood–gas barrier exchange capacity) are not affected by β-adrenergic drugs.

V.3 A 16-year-old boy presents with shortness of breath after his family adopts a new pet. His pulmonologist suspects an underlying allergy-induced asthma and orders a pulmonary function test. Which of following is most likely to have decreased in this boy?

A. Tidal volume
B. Expiratory reserve volume
C. Forced vital capacity
D. Inspiratory capacity
E. FEV_1 (forced expiratory volume in 1 second)

Best answer = E. Allergy-induced asthma is associated with airway narrowing and obstruction (22·VII·A), which impairs the volume of air than can be forcibly expired per unit time. In most cases, forced vital capacity would be unaffected because this parameter is not time dependent. Tidal volume is similarly unaffected. Static lung volumes and capacities (expiratory reserve volume and inspiratory capacity) do not change appreciably with airflow obstruction, although residual volume may be increased by obstructive physiology when air-trapping occurs.

V.4 Pulmonary vascular resistance should be assessed when the effects of lung volume on pulmonary perfusion are minimal. When is this most likely to occur?

A. At high intrapleural pressures
B. At high alveolar pressures
C. At residual volume
D. At functional residual capacity
E. At total lung capacity

Best answer = D. Pulmonary blood vessels are thin walled, which makes then susceptible to extravascular compression (23·III). Pulmonary vascular resistance (PVR) is highest and perfusion is lowest when lung volumes are very high or very low. At total lung capacity and when alveolar pressures are high, PVR is high because the capillaries are stretched and compressed between adjacent alveoli. At residual volume and when intrapleural pressures high, arterial supply vessels are collapsed by external pressure. The nadir in the PVR-to-lung-volume curve occurs at functional residual capacity, because the combined effect of capillary and supply vessel compression is minimal.

V.5 A 58-year-old woman presents with a right-to-left shunt caused by a pulmonary arteriovenous malformation. Which of the following variables would you predict to be increased in this individual?

A. Arterial dissolved O_2 content
B. Alveolar–arterial O_2 difference
C. Venous P_{O_2}
D. Arterial P_{O_2}
E. Oxyhemoglobin levels

Best answer = B. Right-to-left shunts allow blood to pass from the right to the left heart without being oxygenated (23·III·F). The shunted blood lowers the P_{O_2} of arterial blood, thereby widening the alveolar–arterial O_2 difference. The amount of O_2 that blood carries in dissolved form is minimal normally but would be decreased further by a shunt. A right-to-left shunt would decrease venous P_{O_2} and oxyhemoglobin levels.

V.6 A hypoxemic 50-year-old man with an increased alveolar–arterial O_2 gradient is given 100% O_2 via a facemask, causing arterial P_{O_2} to increase to >500 mm Hg. Results of a lung-diffusing capacity test were normal. What is the likely cause of the hypoxemia?

A. Diffusion limitation
B. Right-to-left shunt
C. Ventilation/perfusion mismatch
D. Hypobaric ambient conditions
E. Alveolar hypoventilation

Best answer = C. A hypoxemia with an enlarged alveolar–arterial O_2 gradient ($A{-}aDO_2$) could be due to either a ventilation/perfusion mismatch or a diffusion limitation (23·IV·D). However, the lung-diffusing capacity test eliminates a diffusion limitation. Right-to-left shunts also cause an increased $A{-}aDO_2$, but 100% O_2 does not increase arterial P_{O_2} to the levels observed here. Hypobaric conditions and hypoventilation can result in hypoxemias, but they do not change $A{-}aDO_2$.

V.7 A 75-year-old man with a history of interstitial pulmonary fibrosis presents complaining of increased dyspnea on exertion. A carbon monoxide (CO) uptake test is ordered. Which of the following best describes pulmonary CO uptake during this test?

A. Perfusion is limited.
B. Diffusion is limited.
C. Ventilation is limited.
D. Solubility is limited.
E. Binding is limited.

Best answer = B. Net carbon monoxide (CO) uptake by the lungs is limited by the rate at which it diffuses across the blood–gas barrier (23·V·A). It is relatively insensitive to changes in pulmonary perfusion (unlike uptake of a perfusion-limited gas such as N_2O), which is why the test is used to assess lung-diffusing capacity. Uptake is not limited by ventilation under physiologic conditions. CO uptake binds to hemoglobin (Hb) with high affinity and, thus, uptake is not binding limited. The avidity with which Hb binds CO means that blood rarely carries appreciable amounts of gas in dissolved form.

V.8 A 25-year-old woman with normal lung function presents with anemia following childbirth (hemoglobin = 8.6 g/dL). Which of the following parameters is most likely to be reduced?

A. Arterial P_{O_2}
B. Arterial O_2 saturation
C. Arterial O_2 content
D. Right ventricular output
E. Minute ventilation

Best answer = C. Anemia, as defined by blood hemoglobin (Hb) content (normal female Hb = 12–16 g/dL) reduces the total amount of O_2 that can be carried by blood (23·VI·A). Arterial P_{O_2} is a measure of dissolved O_2 concentration and is not significantly affected by Hb concentration. Arterial O_2 saturation is a measure of Hb's O_2 binding state, which is largely independent of blood Hb concentration under physiologic conditions. A decrease in arterial O_2 content would stimulate compensatory increases in right ventricular output and minute ventilation.

V.9 A cerebrovascular accident that affects forced expirations during rest and exercise most likely damaged which neural area?

A. Apneustic center
B. Pneumotaxic center
C. Phrenic nerve center
D. Dorsal respiratory group
E. Ventral respiratory group

Best answer = E. Ventral respiratory group neurons are involved in forced expiration and coordination of labored inspiration and expiration (24·II·A). The pontine centers limit lung expansion (apneustic center) and cause rapid shallow breathing (pneumotaxic center), although the role of either center during normal breathing is uncertain. Dorsal respiratory group neurons regulate inspiration and implement the resting respiratory rhythm. The phrenic nerve contains motor neurons that control the diaphragm, which is a principal inspiratory muscle (22·V·C).

V.10 Carotid body glomus cells respond to low arterial P_{O_2} with Ca^{2+} influx, causing release of neurotransmitters that stimulate sensory nerve afferents. An increase in which of the following most likely triggers Ca^{2+} influx in glomus cells?

A. Na^+ conductance
B. Na^+ equilibrium potential
C. K^+ conductance
D. Membrane depolarization
E. Brain interstitial H^+

Best answer = D. Arterial P_{O_2} is sensed by an O_2-dependent K^+ conductance in glomus cells (24·III·B). A fall in arterial P_{O_2} allows the K^+ channel to close, decreasing K^+ efflux and causing membrane depolarization. Depolarization activates voltage-gated Ca^{2+} channels and Ca^{2+} influx. Na^+ conductances and changes in the Na^+ equilibrium potential do not have a role in this response. Changes in brain interstitial H^+ initiate ventilatory responses mediated by central chemoreceptors (24·III·A), not peripheral chemoreceptors.

V.11 A healthy 23-year-old woman reports coughing paroxysms when air temperature is below freezing. Which sensory receptors are most likely to have triggered this response?

A. Central chemoreceptors
B. Peripheral chemoreceptors
C. Irritant receptors
D. Pulmonary stretch receptors
E. Juxtapulmonary capillary receptors

Best answer = C. Irritant receptors protect the lung from noxious stimuli, such as dust, chemicals, and cold air (24·IV·A). These receptors may trigger coughing, bronchoconstriction, and mucus production when stimulated. Central and peripheral chemoreceptors respond to changes in arterial blood gas composition (24·III). Stretch receptors activate during lung inflation (24·IV·B), whereas juxtapulmonary capillary receptors respond to capillary engorgement and interstitial edema (24·IV·C).

V.12 A 29-year-old male living at sea level experiences headache and nausea after traveling to a ski resort (base = 2,500 m). Within a day, his symptoms improve, and he feels well enough to ski. Which of the following accounts for his physiologic accommodation?

A. Central chemoreceptor adaptation
B. Pulmonary stretch receptor stimulation
C. Red blood cell synthesis
D. Hemoglobin isoform alteration
E. Angiogenesis

Best answer = A. The initial response to hypobaric hypoxia at high altitudes is hyperventilation, which causes arterial P_{CO_2} to fall, suppressing the normal drive to breathe (24·V·A). The hypoxemia that results causes the symptoms associated with acute mountain sickness. Central chemoreceptors adapt slowly to lowered arterial P_{CO_2} over 8–24 hrs, allowing ventilation rate to rise and the symptoms improve. Red blood cell production and angiogenesis require weeks to months to compensate for the effects of hypoxia. There is no evidence that altitude causes a change in hemoglobin isoform or pulmonary stretch responses.

Filtration and Micturition

25

I. OVERVIEW

Metabolism generates many acids, toxins, and other waste products that can severely impair cell function if allowed to accumulate. Metabolic wastes are passed from cells to the circulation and then on to the kidneys, where they are removed by filtration and excreted in urine. **Excretion** is only one of three essential kidney functions, however. The kidney is also an **endocrine organ** that controls red blood cell production by bone marrow. The kidney also has a vital **homeostatic** role in controlling blood pressure, tissue osmolality, electrolyte and water balance, and plasma pH. Excretory and homeostatic functions both begin when blood is forced at high pressure through a filtration membrane to separate plasma from cells and proteins (Figure 25.1). The ultrafiltrate is then funneled into a tubule lined with a specialized transport epithelium. Channels and transporters in the epithelium's luminal (apical) surface then recover any *useful* components from the ultrafiltrate as it progresses toward the bladder and, ultimately, the external environment. If we were to take the same approach to domestic duties, we would carry the entire contents of our household, including laptop, MP3 player, plants, clothing, and other accoutrements of life out to the street and then carry back all items that we wished to retain. The empty pizza boxes, cans, napkins, and stray socks would be left on the curb for the city sanitation department. We would repeat this process 48 times a day! This remarkable approach to housekeeping has two major advantages. The first is speed because toxins (metabolic or ingested) can be effectively cleared from the circulation in as little as 30 min. The second is that the kidney need only be selective about what it recovers from the filtrate, not what it excretes, because anything not reabsorbed is automatically excreted. The kidney also efficiently employs osmotic gradients to recover filtered water, so that, despite the massive volume of fluid handled every day (~180 L), kidney energy use is only slightly greater than that of the heart (10% of total body energy consumption, compared with 7% for the heart).

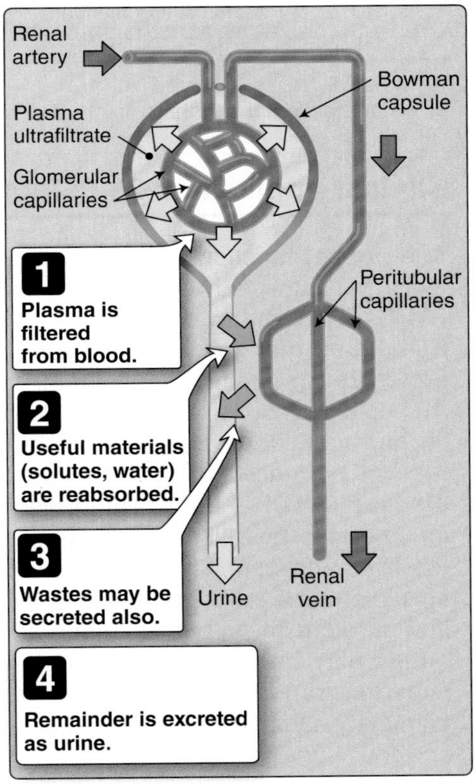
Renal artery
Bowman capsule
Plasma ultrafiltrate
Glomerular capillaries
1
Plasma is filtered from blood.
Peritubular capillaries
2
Useful materials (solutes, water) are reabsorbed.
3
Wastes may be secreted also.
Urine
Renal vein
4
Remainder is excreted as urine.

Figure 25.1
Overview of kidney function.

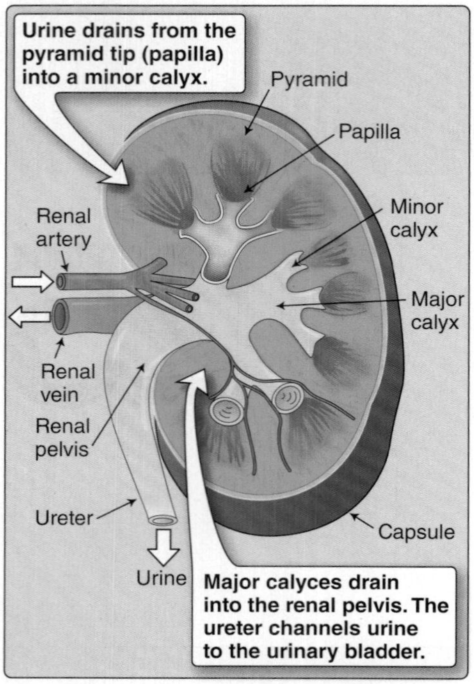

Figure 25.2
Gross anatomy of the kidney.

II. ANATOMY

The functional unit of the kidney is the **nephron**, comprising a blood filtration component (the **glomerulus**) and a filtrate recovery component (the **renal tubule**). Each kidney contains ~1,000,000 nephrons.

A. Gross

Kidneys are paired, bean-shaped organs that lie one to either side of the vertebral column close up against the abdominal wall behind the peritoneum. Each is ~11 cm long, 6 cm wide, and 4 cm deep; weighs ~115–170 g, depending on gender; and is enclosed within a resilient connective tissue **capsule** that protects its inner structures (Figure 25.2). The capsule is penetrated at the hilum by a **ureter**, a **renal artery** and **renal vein**, lymphatic vessels, and nerves. Viewed in cross section, the kidney is seen to be composed of several distinct bands. An outer band (**cortex**) lies beneath the capsule and is the site of blood filtration. The middle band (**medulla**) is divided into 8–18 conical **renal pyramids**. The pyramids contain thousands of tiny ducts that each collect urine from multiple nephrons and guide it toward the ureter. The pyramid tip (**papilla**) inserts into a collecting vessel known as a **minor calyx**. Minor calyces join to form **major calyces**, which drain into a common **renal pelvis**. The pelvis forms the head of a ureter, which propels urine to the urinary bladder for storage and voluntary release.

Clinical Application 25.1: Tubulointerstitial Disorders

The normal functioning of the renal tubule and interstitium can be negatively impacted either acutely or chronically. **Acute interstitial nephritis (AIN)** is an inflammatory condition affecting the renal interstitium. Symptoms may include an acute rise in plasma creatinine levels and proteinuria (protein in urine), both reflecting a general renal dysfunction. AIN usually results from drug exposure, β-lactam antibiotics (e.g., penicillin and methicillin) being the most common offenders.[1] Kidneys typically recover normal function after discontinuing drug use. **Polycystic kidney disease** (**PKD**) is an inherited disorder characterized by the presence of innumerable fluid-filled cysts within the kidneys and, to a lesser degree, the liver and pancreas. The cysts form within the nephron and progressively enlarge and compress the surrounding tissues, preventing fluid flow through the tubules. Although many patients remain asymptomatic, others may begin to show symptoms of impaired renal function (such as hypertension) in their fourth decade. There is no treatment, and PKD may ultimately cause complete renal failure.

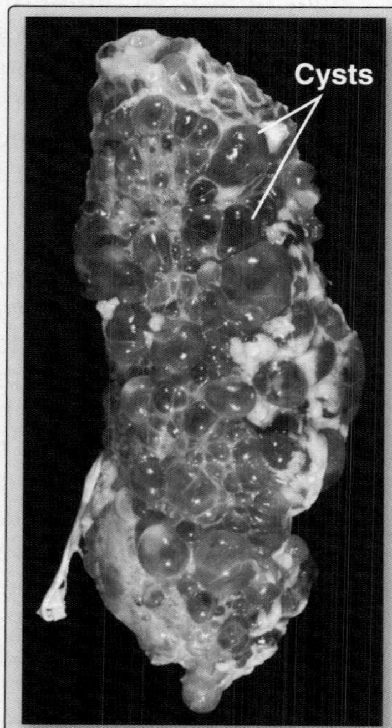

Hereditary polycystic kidney disease.

INFO LINK [1]For more information on adverse reactions to β-lactam antibiotics, see *LIR Pharmacology*, 5e, p. 386.

B. Functional

Kidneys are composed largely of fluid, as are most tissues. Although the fluid within a kidney is compartmentalized (i.e., vascular, luminal, interstitial), and flow between the compartments is limited by cellular barriers, water is still able to move relatively freely between the three compartments, driven by osmotic pressure gradients. A survey of tissue osmolality in different regions of the kidney shows gross differences between the cortex and medulla (Figure 25.3). The cortex has an osmolality that approximates that of plasma, but the osmolality of the inner medulla is increased severalfold. This osmotic gradient is essential to normal kidney function because it is used to recover virtually all of the water that is filtered from the vasculature each day (average urinary water excretion is ~1–2 L/day).

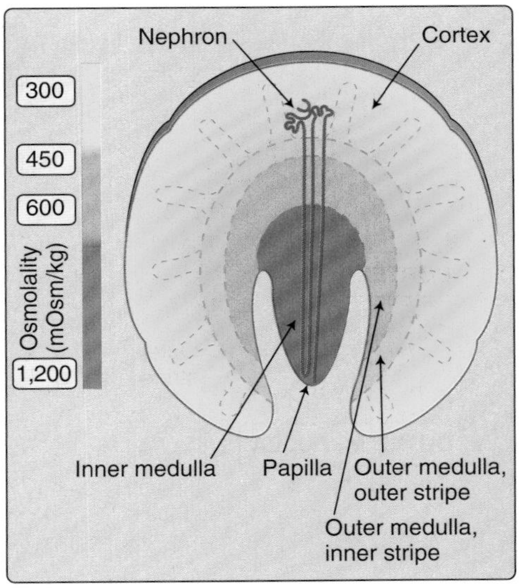

Figure 25.3
The corticopapillary osmolality gradient.

C. Vasculature

The renal nephron's *modus operandi* is to channel blood at relatively high pressure through a network of leaky blood vessels. Pressure forces plasma out of the vasculature through a filtration barrier during passage. The plasma filtrate is channeled into the renal tubule, whose function is to recover essential solutes and >99% of the fluid and return it to the vasculature. The filtration and recovery functions involve an unusual vascular arrangement. Fluid filtration is the purview of the **glomerular capillary network**. Reabsorption is the responsibility of the **peritubular capillary network**. The peritubular network is plumbed in series with and receives blood from the glomerulus (Figure 25.4).

1. **Glomerular network:** Blood enters the glomerulus from an **interlobular artery** at relatively high pressure (~60 mm Hg) via an **afferent arteriole**. Blood courses through a tuft of specialized **glomerular capillaries**. The capillaries branch and interconnect extensively via anastomoses to maximize the surface area available for filtration. Spaces between capillaries are filled with **mesangial cells**, an epithelial cell type that contracts and relaxes (a **myoepithelial cell**) as a way of modulating glomerular capillary surface area and fluid filtration rate. Blood leaves the glomerulus, not by a venule, but by an **efferent arteriole**, still at high pressure. The afferent and efferent arterioles are both resistance vessels that regulate glomerular blood flow and fluid filtration rates through vasoconstriction and vasodilation (see below).

2. **Peritubular network:** The peritubular capillary network surrounds and closely follows the renal tubule as it tracks through the kidney, sustaining the tubule with O_2 and nutrients. The network also carries away fluids and dissolved electrolytes that have been reabsorbed from the tubule lumen. Prompt removal of these materials helps maintain the concentration gradients for chemical and osmotic diffusion across the tubule epithelium that are required for normal renal function.

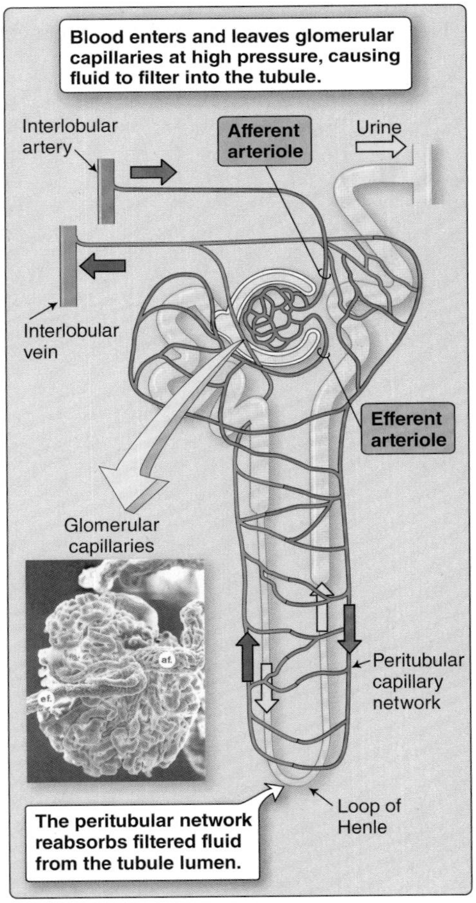

Figure 25.4
Glomerular and peritubular capillary networks.

D. Tubule

A renal tubule comprises a long, thin tube of renal epithelial cells. It can be divided into several distinct segments based on morphology and function (Figure 25.5). At its head is the **glomerular** (**Bowman**) **capsule**, which completely envelops and isolates the glomerulus from its surrounds. The capsule captures and contains fluid filtering from the glomerular

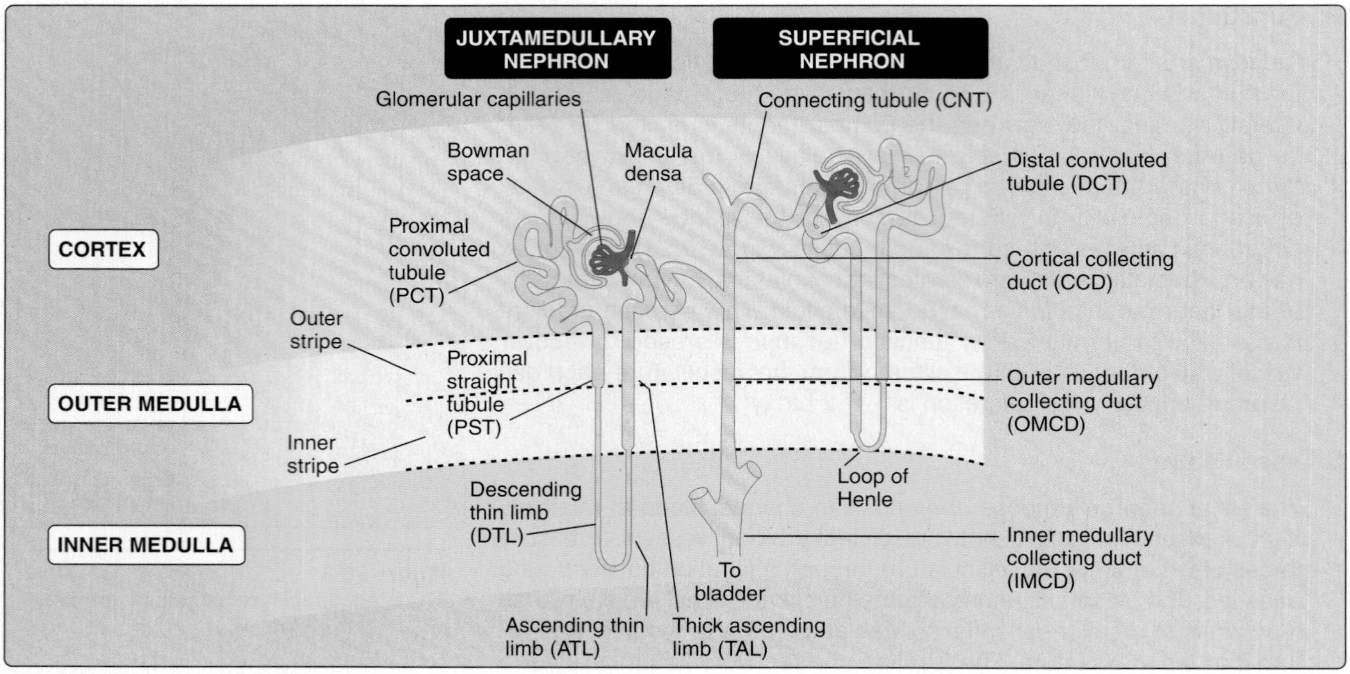

Figure 25.5
Nephron types and the collecting duct system.

capillaries. The glomerulus and capsule together comprise a **renal corpuscle**, which is located in the cortex. Plasma filtrate flows from glomerular capillaries into the **Bowman space** and then enters the **proximal tubule** (**PT**). The PT contains both a **convoluted** and a **straight** section (**PCT** and **PST,** respectively). After exiting the PT, filtrate begins a long descent into the medulla. The tubule then abruptly executes a hairpin turn and returns to the cortex. This hairpin structure is known as the **nephron loop**, or **loop of Henle**. The descending portion of the loop is known as the **descending thin limb**. The ascending portion of the loop can be divided into the **ascending thin limb** and **thick ascending limb** (**TAL**). Fluid then passes through the **distal convoluted tubule** (**DCT**), and the **connecting tubule**, before emptying into a common collecting duct. The collecting duct conveys urine to the calyx and comprises three sections: a **cortical collecting duct**, an **outer medullary collecting duct**, and an **inner medullary collecting duct**.

E. Nephron types

The kidney contains two different types of nephron: superficial nephrons and juxtamedullary nephrons.

1. **Superficial:** Superficial nephrons receive ~90% of renal blood supply, and they reabsorb a large percentage of the fluid that filters from the vasculature. Their glomeruli are located in the outer cortical regions, and their nephron loops are short. The loops dip into the outer medulla but do not enter the inner medulla (see Figure 25.5).
2. **Juxtamedullary:** Juxtamedullary nephrons receive ~10% of total renal blood supply. Their glomeruli are located within the inner cortex, and they have very long nephron loops that reach deep into the inner medulla. The peritubular network that serves

juxtamedullary nephrons is specialized. Capillaries follow the tubule down into the medulla to create a long looping vascular structure called the **vasa recta**. Juxtamedullary nephrons are designed to concentrate urine (see 27·II).

III. FILTRATION

In Chapter 19 (see 19·VII·D), we discussed the delicate balance that exists between the forces that favor fluid filtration from blood (mean glomerular **capillary hydrostatic pressure [P_{GC}]**) and the forces that favor fluid retention (glomerular capillary **colloid osmotic pressure [π_{GC}]**). In most tissues, increasing capillary hydrostatic pressure causes edema. In the kidney, increasing P_{GC} is the first step in urine formation (Figure 25.6).

A. Starling forces

The forces that control fluid movement across the glomerular capillary wall are the same as for any vascular bed. These forces are considered in the **Starling equation**:

$$GFR = K_f [(P_{GC} - P_{BS}) - (\pi_{GC} - \pi_{BS})]$$

GFR is glomerular filtration rate (i.e., net fluid flow across the capillary wall), which is measured in mL/min. K_f is a glomerular filtration coefficient, and P_{BS} and π_{BS} are the hydrostatic pressure and colloid osmotic pressure, respectively, of fluid contained within the Bowman space. Changes in any of these variables can have dramatic effects on GFR and urine formation.

1. **Filtration barrier:** K_f is a measure of glomerular permeability and surface area. The barrier comprises three distinct layers that, together, create a three-step molecular filter that produces a cell- and protein-free plasma ultrafiltrate. The barrier is composed of a **capillary endothelial cell**, a **thick glomerular basement membrane**, and a **filtration slit diaphragm** (see Figure 25.6, lower).

   a. **Layer 1:** Glomerular capillary endothelial cells are dense with fenestrations, resembling a sieve. The pores are ~70 nm in diameter, which allows free passage to water, solutes, and proteins. Cells are too large to fit through the pores, so they remain trapped in the vasculature.

   b. **Layer 2:** The glomerular basement membrane comprises three layers. An inner **lamina rara interna** is fused to the capillary endothelial cell layer. A middle layer, the **lamina densa**, is the thickest of the three. An outer **lamina rara externa** is fused to the podocytes. The basement membrane carries a net negative charge that repels proteins (which also carry a negative charge), and reflects them back into the vasculature.

   c. **Layer 3:** Glomerular capillaries are completely ensheathed in tentacle-like foot processes that project from **podocytes**. Podocytes are specialized epithelial cells. The covering is not continuous, however. The foot processes end in "toes" that interdigitate, leaving narrow slits between them. The slits are bridged by a proteinaceous **filtration slit diaphragm** that prevents proteins and other large molecules from entering the Bowman space. The fluid

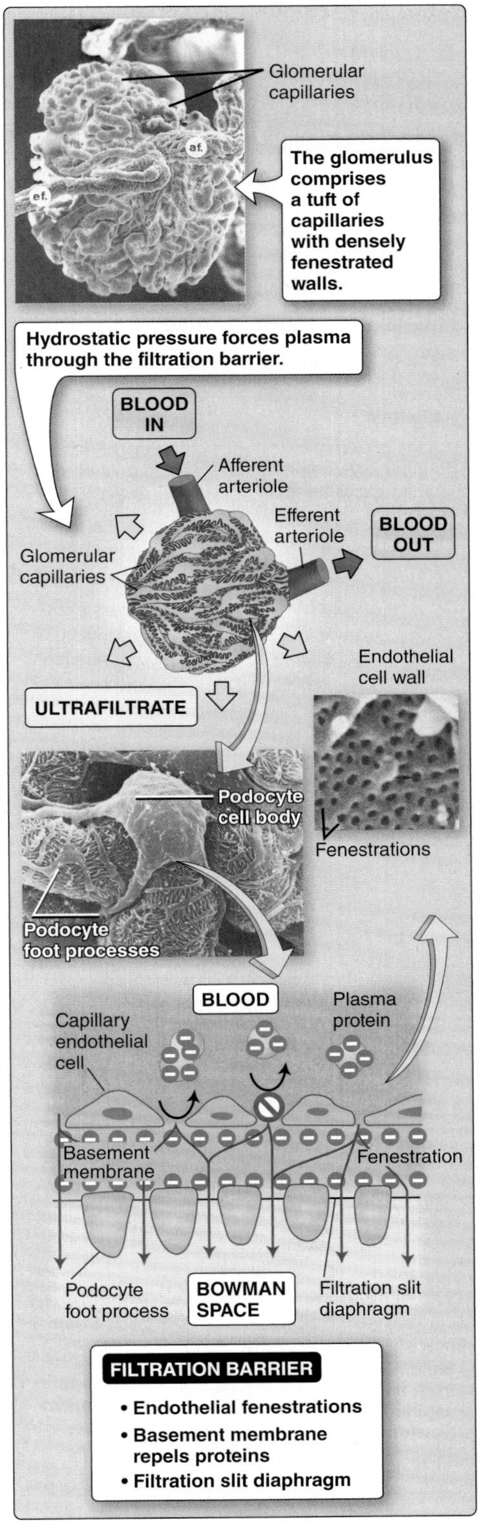

Figure 25.6
The glomerulus and its filtration barrier.

Table 25.1: Glomerular Filtration Barrier Selectivity

Substance	Molecular Weight (Da)	Molecular Radius (nm)	Permeability*
Na^+	23	0.10	1.0
Water	18	0.15	1.0
Glucose	180	0.33	1.0
Inulin	5,000	1.48	0.98
Myoglobin	17,000	1.88	0.75
Hemoglobin	68,000	3.25	0.03
Serum albumin	69,000	3.55	<0.01

*Permeability compares the plasma concentration of a substance with that in urine.

that finally enters the tubule is plasma ultrafiltrate containing electrolytes, glucose, and other small organics, but anything larger than ~5,000 Da molecular weight is excluded (Table 25.1).

2. **Hydrostatic pressure:** The glomerular capillary network is a physiologic oddity in that it is located in the midpoint of the renal arterial system rather than at its terminus. Blood enters the glomerulus via the afferent arteriole at a pressure of ~60 mm Hg, or ~25 mm Hg higher than in most capillary beds (see Figures 19.26 and 25.7). Blood leaves the glomerulus via the efferent arteriole at ~58 mm Hg, so P_c averages ~59 mm Hg.

3. **Colloid osmotic pressure:** Capillary colloid osmotic pressure is proportional to plasma protein concentration. Blood enters the glomerulus at a π_{GC} of ~25 mm Hg, as in any other circulation. Blood loses ~15%–20% of its total volume to filtrate during passage through the capillary network. The plasma proteins are prevented from leaving the vasculature by the filtration barrier, so π_{GC} rises with distance along a capillary. Blood entering the efferent arteriole has a π_{GC} of ~35 mm Hg.

4. **Bowman space:** Proteins are prevented from entering the Bowman space by the filtration barrier, so π_{BS} is 0. The sheer volume of fluid being expressed into the Bowman space causes a significant pressure to build, however (P_{BS} ~15 mm Hg). This pressure opposes filtration but is beneficial in that it creates a positive pressure gradient between the Bowman space and the renal sinus that propels fluid through the tubule.

5. **Net force:** Using values cited above, we note that there is a net positive pressure favoring ultrafiltration (P_{UF}) that gradually decreases from ~20 mm Hg to ~8 mm Hg across the length of the glomerular capillary.

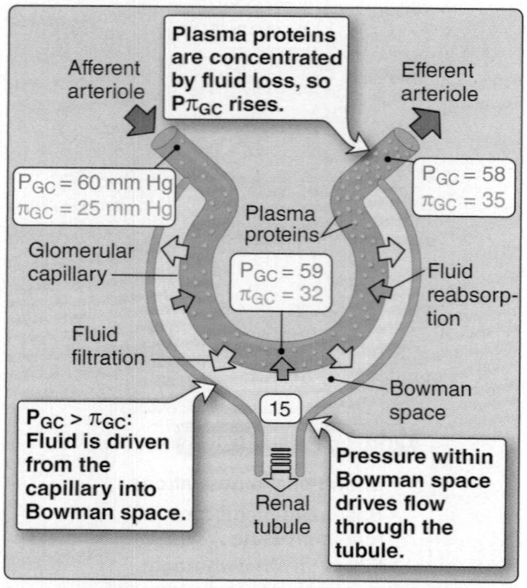

Figure 25.7
Forces favoring glomerular ultrafiltration. P_{GC} = glomerular capillary hydrostatic pressure; π_{GC} = glomerular capillary colloid oncotic pressure. All values are given in mm Hg.

B. Glomerular filtration rate

In a healthy person, P_{BS}, π_{BS}, and π_{GC} are all relatively invariant. The main factor affecting GFR is P_{GC}, which is determined by aortic pressure, renal arterial pressure, and by changes in the afferent and efferent arteriolar resistance (Figure 25.8).

> Mesangial cells may also regulate GFR through changes in glomerular capillary surface area, which affects K_f. The role of mesangial cells is minor compared with that of glomerular arterioles, however.

1. **Afferent arteriole:** Constricting the afferent arteriole decreases glomerular blood flow, just as it would in any other circulation. P_{UF} and GFR fall as a result. Afferent arteriolar dilation increases P_{UF} and GFR.

2. **Efferent arteriole:** Efferent arteriolar patency determines how easily blood can pass through the glomerular vasculature. Arteriolar constriction flattens the P_{BS} pressure gradient, and, thus, P_{UF} and GFR rise. Arteriolar dilation allows blood to flow out of the glomerular network, and P_{UF} and GRF both fall as a result.

3. **Regulation:** In practice, changes in afferent and efferent arteriolar resistance rarely occur in isolation. The two resistance vessels are regulated by a multitude of factors, as are resistance vessels in other vascular beds (see 20·II). The manner in which they are regulated is considered in Section IV below.

C. Values

GFR increases with body size and decreases with age. A normal GFR range (adjusted to reflect body surface area) averages ~100–130 mL/min/1.73 m^2. This represents a ~1,000-fold greater filtration rate than observed in skeletal musculature (for example), all due to the high P_{UF} and leakiness of the filtration barrier.

IV. REGULATION

Renal blood flow (RBF) and GFR are governed by two overriding needs that are sometimes at odds with one another. **Local** vascular autoregulatory pathways maintain RBF at rates that optimize GFR and urine formation. However, **central** homeostatic control pathways may assume control over renal function to adjust circulating blood volume and blood pressure (for example). Central control of renal function is exerted hormonally and through the autonomic nervous system (ANS).

A. Autoregulation

Autoregulation stabilizes RBF and GFR during changes in mean arterial pressure (MAP). All circulations in the body autoregulate to some degree, but the kidney's autoregulatory prowess is particularly well developed. GFR remains relatively stable over a MAP range of ~80–180 mm Hg (Figure 25.9).

B. Myogenic response

The **myogenic response** is an autoregulatory mechanism. The smooth muscle cells that line the afferent arteriole contain Ca^{2+}-permeant mechanosensory channels that activate when the vessel wall is stretched (e.g., by an increase in luminal pressure). Ca^{2+} influx initiates muscular contraction, and the arteriole constricts reflexively. The myogenic response stabilizes RBF and GFR during changes in posture, for example.

> The myogenic response is common to all vascular beds. Rising from a prone position can elicit arterial pressure pulses of >100 mm Hg, reflecting gravitational effects on the blood columns contained within the vasculature (see 20·II·A·2).

C. Tubuloglomerular feedback

Tubuloglomerular feedback (TGF) is an autoregulatory mechanism mediated by the **juxtaglomerular apparatus (JGA)** that adjusts RBF and GFR to optimize fluid flow through the renal tubule. The JGA is a functional complex that includes the renal tubule, mesangial cells, and the afferent and efferent arterioles (Figure 25.10).

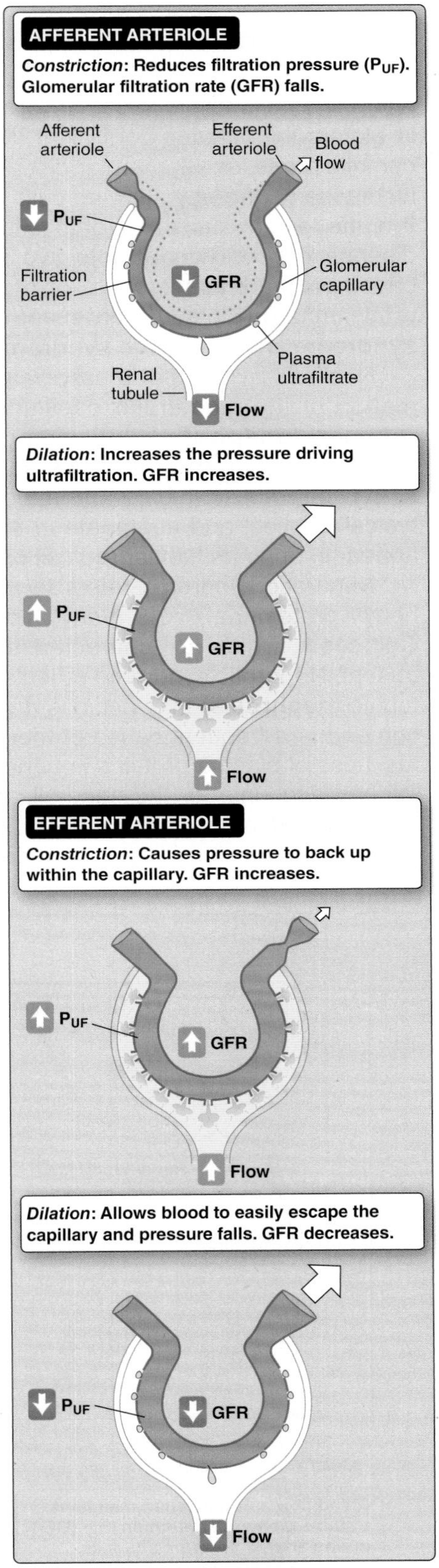
AFFERENT ARTERIOLE
Constriction: Reduces filtration pressure (P_{UF}). Glomerular filtration rate (GFR) falls.
Afferent arteriole
Efferent arteriole
Blood flow
P_{UF}
Filtration barrier
GFR
Glomerular capillary
Plasma ultrafiltrate
Renal tubule
Flow
Dilation: Increases the pressure driving ultrafiltration. GFR increases.
P_{UF}
GFR
Flow
EFFERENT ARTERIOLE
Constriction: Causes pressure to back up within the capillary. GFR increases.
P_{UF}
GFR
Flow
Dilation: Allows blood to easily escape the capillary and pressure falls. GFR decreases.
P_{UF}
GFR
Flow

Figure 25.8
Effects of changing afferent and efferent arteriolar resistance on glomerular filtration rate.

Clinical Application 25.2: Glomerular Disease

Normal renal function can be severely impacted by pathological changes in glomerular filtration coefficient (K_f), hydrostatic pressure, and colloid osmotic pressure. Glomerular disease damages the filtration barrier and increases K_f, thereby allowing cells and proteins to pass into the tubule. It is the leading cause of renal failure in the United States. Glomerular disease can be divided into two broad and overlapping syndromes based on the characteristics of proteins and cellular debris contained within urine (**urine sediments**) and the associated symptoms: **nephritic syndrome** and **nephrotic syndrome**.

Nephritic syndrome is associated with diseases that cause inflammation of the glomerular capillaries, mesangial cells, or podocytes (**glomerulonephritis**). Inflammation creates localized breaches in the filtration barrier and allows cells and modest amounts of protein to escape into the tubule and appear in urine (**proteinuria**). Red cells typically collect and aggregate in the distal convoluted tubule and then appear in urine as tubular red cell **casts**.

Nephrotic syndrome refers to a set of clinical findings that include heavy proteinuria (>3.5 g/day), lipiduria, edema, and hyperlipidemia. Cell casts, which are characteristic of an inflammatory process, are absent. Nephrotic syndrome reflects a general deterioration of the renal tubule (**nephrosis**) that includes degradation of glomerular barrier function and is a frequent cause of mortality in patients with diabetes mellitus. Loss of plasma proteins in urine causes plasma oncotic pressure to fall and accounts for the generalized edema associated with nephrotic syndrome. Hyperlipidemia reflects increased lipid synthesis that helps compensate for loss of lipids in urine.

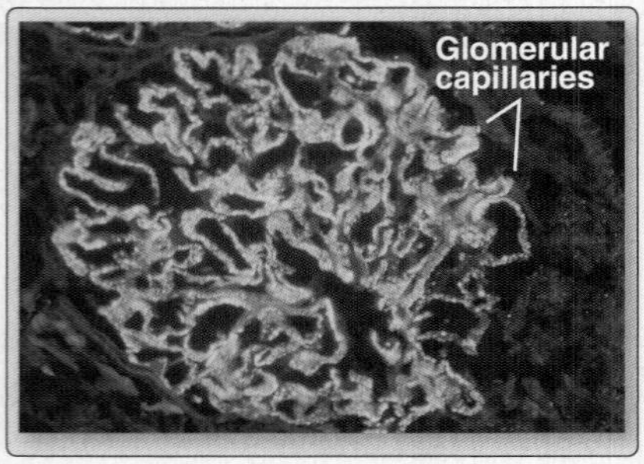

Immunoglobulin G-deposit imaging during glomerulonephritis.

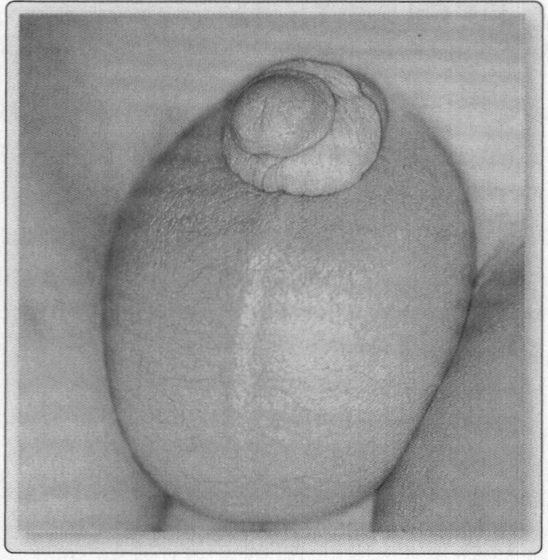

Scrotal edema in a 7-year-old boy with nephrotic syndrome.

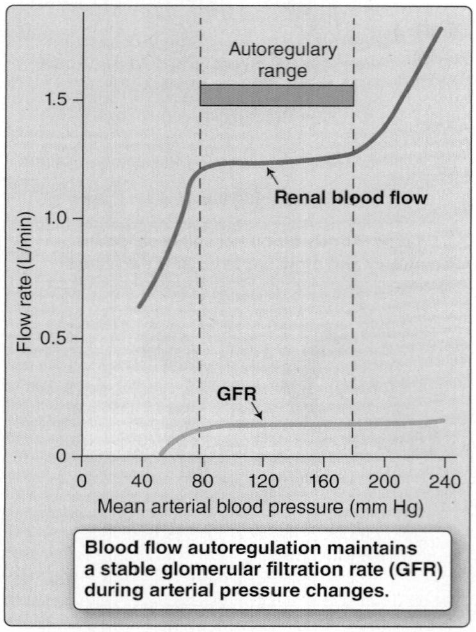

Figure 25.9
Autoregulation of renal blood flow.

1. **Tubule:** The loop of Henle comes into direct contact with the afferent and efferent arterioles after returning from the medulla. The TAL wall is modified at the contact site to form a specialized sensory region called the **macula densa** (see Figure 25.10). The macula densa monitors Na^+ and Cl^- concentrations within the tubule lumen, which, in turn, reflect RBF and GFR. Na^+ and Cl^- permeate macula densa cells via a Na^+-K^+-$2Cl^-$ cotransporter located in the apical membrane. Cl^- immediately exits via a basolateral Cl^- channel, causing a membrane depolarization whose magnitude is a direct reflection of tubule fluid NaCl concentration.

2. **Mesangial cells:** Mesangial cells provide a physical pathway for communication between the sensory (macula densa) and effector (arteriole) arms of the TGF system (see Figure 25.10). All cells in the JGA are interconnected via gap junctions (see 4·II·F), which allows for direct chemical communication between the system components.

3. **Afferent arteriole:** The afferent arteriole is notable for its adenosine receptor and for *renin*-producing **granular cells** within its walls.

   a. **Adenosine receptor:** Adenosine receptors are members of the G protein–coupled receptor superfamily that act through

the cyclic adenosine monophosphate (cAMP) signaling pathway (see 1·VII·B·2). Afferent arteriolar smooth muscle cells express a type A_1 receptor, which couples to an inhibitory G protein and decreases cAMP levels when occupied. cAMP normally inhibits smooth-muscle contractility via a *protein kinase A*–dependent pathway (see 14·III·C). Thus, when the afferent arteriole binds adenosine, it constricts.

b. **Granular cells:** Granular cells are specialized secretory cells that produce *renin*, a proteolytic enzyme that initiates the *renin*–angiotensin–aldosterone system ([RAAS] see 20·IV·C) by converting angiotensinogen to angiotensin I. Angiotensin I is subsequently converted by *angiotensin-converting enzyme* (*ACE*) to angiotensin II (Ang-II), which is vasoactive.

4. **Efferent arteriole:** The efferent arteriole expresses a type A_2 adenosine receptor, which increases intracellular cAMP levels when occupied. Adenosine binding causes efferent arteriolar dilation.

5. **Regulation:** Macula densa cell output and arteriolar response depends on whether tubule fluid flow rates are high or low (Figure 25.11).

a. **High flow rates:** When RBF and GFR are high, increased amounts of NaCl are delivered to the macula. The resulting membrane depolarization activates nonspecific cation channels in the cell membrane, causing Ca^{2+} influx. Because all cells in the JGA are coupled via gap junctions, when intracellular Ca^{2+} concentrations rise in the macula densa, they rise in the *renin*-secreting granular cells of the afferent arteriole also. Ca^{2+} strongly inhibits *renin* release. Na^+ and Ca^{2+} influx also cause macula densa cell adenosine levels to rise. Adenosine acts as a paracrine that signals to the afferent and efferent arterioles. The afferent arteriole constricts, and the efferent arteriole dilates, reducing blood flow into the glomerular capillaries and facilitating outflow. Both effects depress P_{GC}, and, thus, GFR falls (see Figure 25.11A).

b. **Low flow rates:** When GFR decreases, the amount of NaCl arriving at the macula densa falls. Macula densa cells hyperpolarize as a result, reducing Ca^{2+} influx via the nonspecific cation channel. Ca^{2+} concentrations in the granular cells fall also, thereby removing the brakes on *renin* release and allowing it to be deposited into the circulation (see Figure 25.11B). *Renin* then activates RAAS, and circulating Ang-II levels rise. Ang-II is a potent vasoconstrictor in all vascular beds, but its effects on the glomerular arterioles are not equal. The efferent arteriole is more sensitive to Ang-II. The net effect is to limit outflow from the glomerular capillaries, causing P_{GC} and GFR to rise.

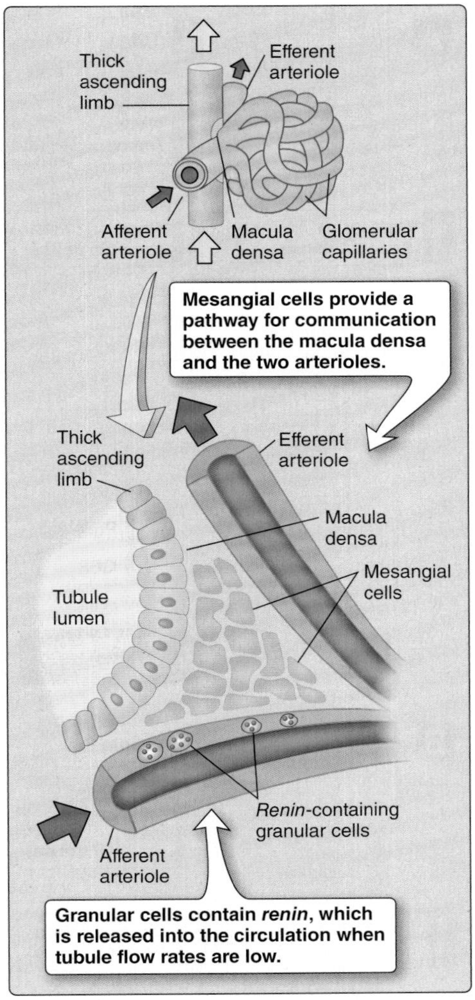

Figure 25.10
The juxtaglomerular apparatus.

D. Paracrines

Adenosine is only one of several autoregulatory paracrines produced by the kidney (Table 25.2), although their function under normal physiologic conditions is uncertain. **Prostaglandins** and **nitric oxide** both dilate glomerular arterioles and increase RBF and GFR. They may help offset intense Ang-II–mediated vasoconstriction during

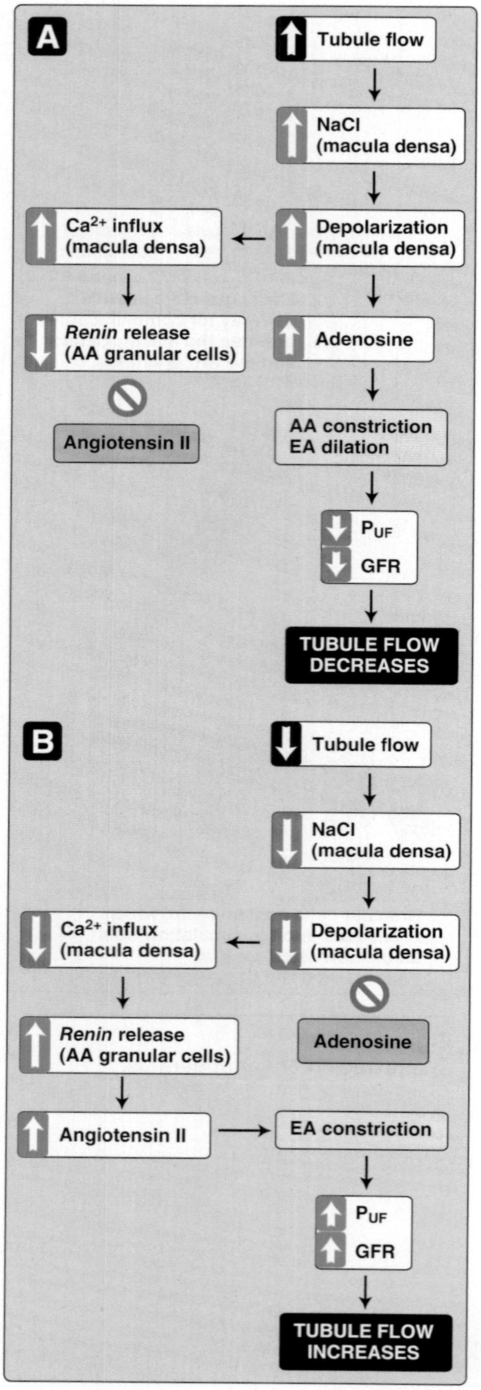

Figure 25.11
Tubuloglomerular feedback. AA = afferent arteriole; EA = efferent arteriole; P_{UF} = capillary hydrostatic ultrafiltration pressure; GFR = glomerular filtration rate.

circulatory shock, for example (see 40·IV·B·3). **Endothelins** are local vasoconstrictors released in response to Ang-II or when glomerular flow rates are damagingly high.

E. Angiotensin II

All of the components necessary for Ang-II formation and response (including *ACE*) are inherent to the kidney, suggesting that RAAS constitutes a primary renal autoregulatory system that optimizes tubule fluid flow through manipulation of RPF and GFR. However, one of the kidney's primary homeostatic functions is helping control blood pressure, and Ang-II creates an important hormonal link between blood pressure and renal function (see Chapters 20 and 28).

F. Central controls

The kidney governs total body water and Na^+ content, which, in turn, determines blood volume and MAP. The kidney also receives ~10% of cardiac output at rest, a significant blood volume that might be used to sustain more critical circulations (e.g., cerebral and coronary circulations) in the event of circulatory shock (see 40·IV·B). Renal blood flow is, thus, subject to control by the ANS, acting through neural and endocrine pathways.

1. **Neural:** Glomerular arterioles are innervated by noradrenergic sympathetic terminals that activate when MAP falls. Sympathetic activation raises systemic vascular resistance by restricting blood flow to all vascular beds, including the kidneys. Mild sympathetic stimulation preferentially constricts the efferent arteriole, which reduces RBF while simultaneously maintaining GFR at sufficiently high levels to ensure continued kidney function. Intense sympathetic stimulation severely curtails blood flow through both glomerular arterioles, and urine formation ceases. In cases of severe hemorrhage, prolonged occlusion of arteriolar supply vessels can cause renal ischemia, infarction, and failure (see 40·IV·C).

2. **Endocrine:** Hormonal regulation of RBF is mediated principally by **epinephrine** and **atrial natriuretic peptide** (**ANP**). Epinephrine is released into the circulation following sympathetic activation and stimulates the same pathways as does norepinephrine released from sympathetic nerve terminals. ANP is released from cardiac atria when they are stressed by high blood volumes. The ANP receptor has intrinsic *guanylyl cyclase* activity that dilates the afferent arteriole and increases RBF. It also relaxes mesangial cells to increase filtration barrier surface area. The net result is an increase in RBF and GFR and salt and water excretion (see Chapter 28).

V. ASSESSING RENAL FUNCTION

Patients with kidney disease can present with a diverse array of symptoms, including hypertension, edema, and bloody urine (**hematuria**), or they may be asymptomatic. It is important to be able to measure and assess the efficiency of GFR in these situations in order to narrow

the range of possible pathologies. GFR cannot be measured directly, but filtration barrier health can be assessed from studies of plasma clearance.

A. Clearance

Clearance measures the kidney's ability to "clear" plasma of any given substance and then excrete it in urine. Plasma flows through the kidneys at ~625 mL/min. If the kidney could remove every last molecule of substance X (for example) from plasma during its encounter with a nephron, substance X clearance would be ~625 mL/min. In practice, such a high degree of clearance is not possible because the glomerulus filters only a fraction of the total amount of plasma that passes through the capillary network (~20%, or ~125 mL/min). However, it is still useful to know how much of the 125 mL/min does get cleared in a healthy individual because this parameter can be used to help diagnose problems with kidney function. Clearance is calculated as:

$$C_X = \frac{U_X \times V}{P_X}$$

where C_X is clearance of substance X (mL/min), U_X and P_X are urinary and plasma concentrations of X, respectively (mmol/L or mg/mL), and V is urine flow rate (mL/min). In practice, clearance is usually measured over a 24-hr period to reduce urinary sampling errors (see below).

> Clearance is defined as the amount of plasma that is *completely* cleared of any given substance per unit time.

Table 25.2: Glomerular Regulators

Stimulus	Mediator	Effects: RBF	Effects: GFR
Tubule flow rates			
↓ NaCl	Adenosine	↑	↑
↑ NaCl	*Renin* (Ang-II)	↓	↓
Sympathetic activation			
Mild	Norepinephrine	↓	None
Intense	Norepinephrine, epinephrine, *renin* (Ang-II)	↓	↓
↑ Blood volume	Atrial natriuretic peptide	↑	↑
Uncertain	Dopamine	↑	↑
Vascular endothelium			
↓ Flow?	Prostaglandins	↑	↑
Shear stress	Nitric oxide	↑	↑
Stress, trauma, vasoconstrictors	Endothelins	↓	↓
Inflammation	Leukotrienes	↓	↓

RBF = renal blood flow; GFR = glomerular filtration rate; Ang-II = angiotensin II.

B. Glomerular filtration rate

If there were a substance that crossed the filtration barrier unhindered and then traversed the tubule without interference (i.e., no secretion, no reabsorption), we could use the rate of its appearance in urine to calculate GFR. One such substance is inulin, a fructose polymer synthesized by many plants. Inulin is physiologically inert and, thus, routinely used clinically to determine GFR. Inulin is infused intravenously to establish a known plasma concentration, and then its rate of appearance in urine measured. GFR can then be calculated from:

$$GFR = C_{in} = \frac{U_{in} \times V}{P_{in}}$$

where C_{in} is inulin clearance, U_{in} and P_{in} are urinary and plasma concentrations of inulin, respectively, and V is urine flow.

> Inulin is the gold standard for filtration markers. Alternatives include radioactive iothalamate, iohexol, diethylene triamine pentaacetic acid (DPTA), and the related ethylene diamine tetraacetic acid (EDTA).

Example 25.1

A 35-year-old woman is being evaluated for renal surgery. Her plasma creatinine concentration (P_{Cr}) is 0.8 mg/dL. A 24-hour urine collection has a creatinine concentration (U_{Cr}) of 90 mg/dL and a total volume (V) of 1,425 mL. What is her glomerular filtration rate (GFR)?

GFR can be estimated from creatinine clearance (C_{Cr}):

$$GFR = C_{Cr} = \frac{U_{Cr} \times V}{P_{Cr}}$$

Using values provided above:

$U_{Cr} = 90$ mg/dL $= 0.9$ mg/mL

$P_{Cr} = 0.8$ mg/dL $= 0.008$ mg/mL

$V = 1{,}425$ mL/24hr $= 0.99$ mL/min

$$GFR = \frac{0.9 \times 0.99}{0.008} = 111.4 \text{ mL/min}$$

Example 25.2

A healthy 22-year-old man volunteers for a research study evaluating the effects of a new drug on renal blood flow (RBF). The protocol required a urinary catheter to measure kidney output while infusing *para*-aminohippuric acid (PAH) intravenously. PAH concentration (P_{PAH}) stabilized at 0.025 mg/mL. Urine flow rate (V) was then measured at 1.2 mL/min, and urine PAH concentration (U_{PAH}) was 18 mg/mL. Hematocrit (Hct) was 48%. What was the subject's RBF?

$U_{PAH} = 18$ mg/mL

$V = 1.2$ mL/min

$P_{PAH} = 0.025$ mg/mL

RBF is calculated from RPF and Hct. RPF is calculated from PAH clearance (C_{PAH}):

$$C_{PAH} = \frac{U_{PAH} \times V}{P_{PAH}} = \frac{18 \times 1.2}{0.025}$$

$$= 864 \text{ mL/min} = RPF$$

RBF is calculated from RBF and Hct:

$$RBF = \frac{RPF}{1 - Hct} = \frac{864}{0.52}$$

$$= 1{,}661.5 \text{ mL/min}$$

C. Creatinine clearance

Inulin clearance is an expensive and cumbersome test to perform, so a preferred (albeit less accurate) alternative involves measuring creatinine clearance (Example 25.1). Creatinine is derived from creatine breakdown in skeletal muscle and is produced and excreted constantly. Creatinine filters freely from the glomerulus and is not reabsorbed by the tubule, but it is secreted. The PCT contains organic acid transporters ([OATs] see 26·IV·B) that secrete creatinine into the tubule, causing a 10%–20% overestimate of GFR. *Coincidentally*, serum creatinine measurement methods produce an equal and opposite error that cancels out the effects of secretion.

D. Renal plasma flow

Theoretically, if a substance could be identified that *was* completely cleared from plasma during a single pass (i.e., none leaves the kidney via the renal vein), similar techniques could be used to quantify **renal plasma flow** (**RPF**). There is no known substance, but ***para*-aminohippurate** (**PAH**) comes close. PAH is avidly removed from plasma and secreted into the renal tubule by OATs in the PCT epithelium. PAH clearance underestimates RPF by ~10% (i.e., 10% of the PAH that passes through the kidney escapes excretion) and is not used clinically, but it is sufficiently close to the perfect substance that it provides a convenient tool to assess the physiologic principles of kidney function:

$$RPF = C_{PAH} = \frac{U_{PAH} \times V}{P_{PAH}}$$

where C_{PAH} is PAH clearance (mL/min), U_{PAH} and P_{PAH} (mmol/L) are urinary and plasma concentrations of PAH, respectively, and V is urine flow (mL/min) as shown in Example 25.2.

E. Renal blood flow and filtration fraction

Knowing RPF makes it possible to calculate RBF and filtration fraction (see Example 25.2). RBF is calculated from:

$$RBF = \frac{RPF}{1 - \text{Hematocrit}}$$

Hematocrit is the volume of blood occupied by red blood cells.

$$\text{Filtration fraction} = \frac{GFR}{RPF}$$

Filtration fraction measures the amount of plasma that filters into the Bowman space (usually around 0.2, or 20% of RPF).

VI. MICTURITION

The fluid flowing from the collecting ducts and entering the calyces is urine in its final form. There is no further modification *en route* to or in the bladder. Urine is produced constantly and is stored in the urinary bladder until voided (**micturition**).

A. Ureters

Ureters convey urine from the kidneys to the bladder (Figure 25.12). Ureters are lined with a transitional epithelium designed to stretch without tearing to accommodate intraluminal volume increases. The ureter wall contains circular and longitudinal smooth muscle layers. The muscles are stimulated to contract by waves of depolarization that originate in pacemaker regions of the calyces and renal sinus. The waves sweep down the ureters, triggering a peristaltic contraction that increases intraluminal pressure locally and drives urine toward the bladder. Wave propagation through the musculature is facilitated by gap junctions that electrically couple adjacent smooth muscle cells. Waves propagate at ~2–6 cm/s and typically recur several times per minute.

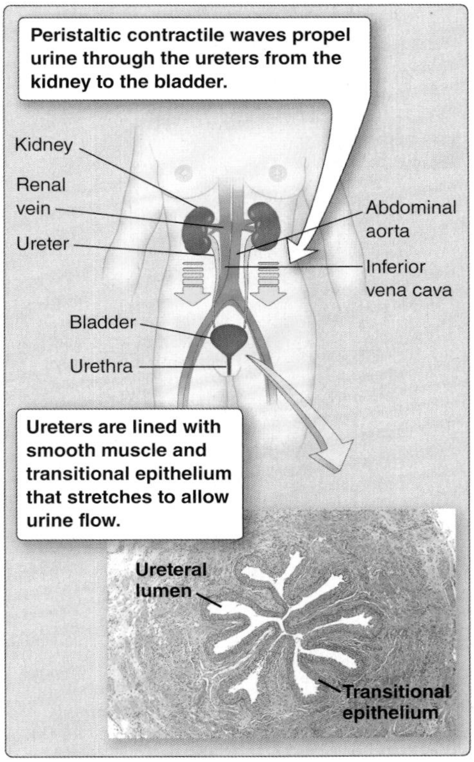

Figure 25.12
Ureters.

B. Bladder

The urinary bladder is a hollow organ comprising a large urine storage area (**body**) and a **neck** (or **posterior urethra**), which funnels urine to the **urethra** (Figure 25.13).

1. **Body:** The bladder is lined on its interior surface by transitional epithelium. When the bladder is empty, the wall is thrown up into a series of ridges called **rugae** (see Figure 25.13). The bladder wall is composed of three indistinct layers of bundled smooth muscle fibers known as **detrusor muscle**. The fibers within the layers are arranged in circular, spiral, or longitudinal fashion, so that they decrease bladder size and raise intraluminal pressure when stimulated to contract by the ANS.

2. **Valves and sphincters:** A full bladder develops considerable internal (**intravesical**) pressure, which potentially could force urine backward through the ureters. The ureters, thus, enter the bladder at an oblique angle, creating a valve that prevents ureteral reflux (Figure 25.14). The neck of the bladder comprises a mix of detrusor muscle and elastic tissue that together form an **internal sphincter** that is controlled by the ANS. The sphincter remains contracted to prevent urine from entering the urethra until micturition. A second **external sphincter** surrounds the urethra below the bladder neck. The external sphincter is composed of skeletal muscle and is under voluntary control.

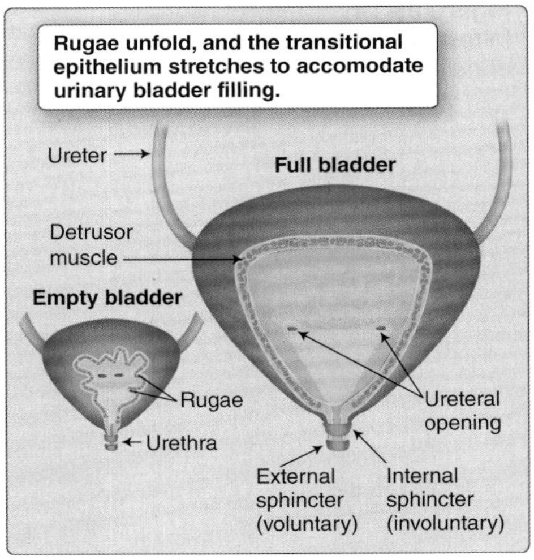

Figure 25.13
The urinary bladder.

C. Innervation

The only part of the urinary system under voluntary control is the external sphincter (see Figure 25.13). The sphincter is innervated by the **pudenal** nerve, which originates in the sacral spinal cord (S2–S4). The ureters and lower urinary tract (bladder, urethra, and internal sphincter) are all under control of the sympathetic (SNS) and parasympathetic (PSNS) nervous systems. SNS efferents originate in spinal T11–L2 segments and travel to the urinary tract via the hypogastric nerve or descend in the paravertebral chain and then travel in the pelvic nerve. SNS activity relaxes detrusor muscle and constricts the bladder neck and urethra. PSNS preganglionic fibers originate in the sacral spinal cord (S2–S4) and travel in the pelvic nerve to the pelvic plexus and bladder wall. The PSNS stimulates voiding by detrusor contraction and relaxation of the urethra and internal sphincter.

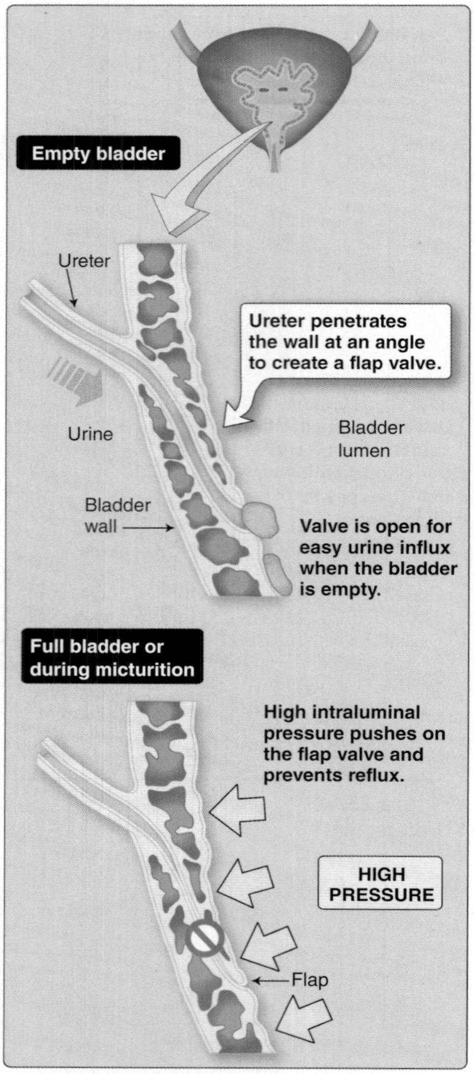

Figure 25.14
Intravesicular ureteral valve preventing urine reflux.

D. Spinal micturition reflex

Bladder capacity is ~500 mL. It fills passively, and the rugae unfold to accommodate the volume increase during this initial "**guarding**" phase (see also 14·IV·B, Smooth muscle length adaptation). Filling occurs with minimal increase in intravesical pressure. Once bladder capacity reaches ~300 mL, the bladder wall begins to stretch, activating mechanoreceptors in the detrusor layers and urothelium (Figure 25.15). Sensory afferents relay this information to the spinal cord via the pelvic and hypogastric nerves and initiate a reflex increase in PSNS efferent activity. The detrusor muscle contracts as a result, causing intravesical pressure to rise sharply and creating a sense of "urgency," with contraction frequency and discomfort level increasing with bladder volume. Sensory signals also travel rostrally to the brain. CNS control of micturition is complex and involves many different loci, including a **pontine micturition center** (**PMC**), which is believed to coordinate brain output to the lower urinary system. If voiding is inconvenient, the pontine center suppresses the presynaptic PSNS nerves that stimulate bladder contraction. Meanwhile, tonic contraction of the external urinary sphincter prevents urine flow until relaxed voluntarily.

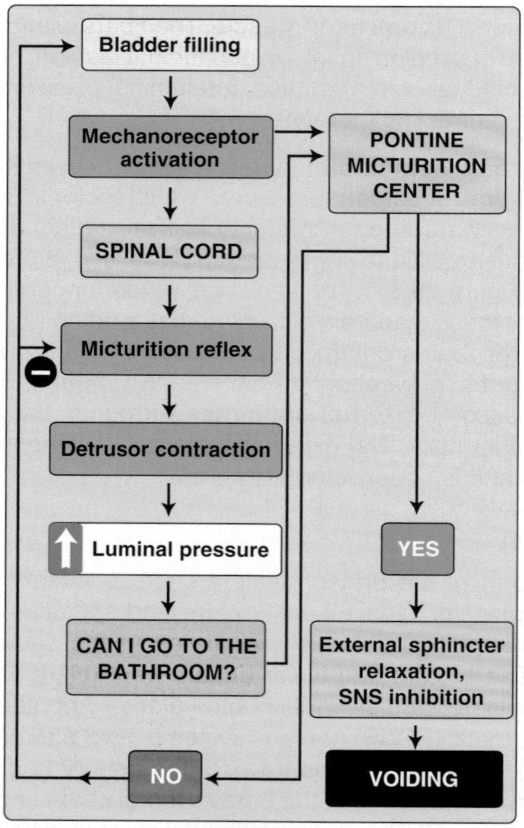

Figure 25.15
Micturition reflex initiated by bladder filling. SNS = sympathetic nervous system.

E. Voiding

Voiding begins with voluntary relaxation of the external sphincter. The PMC relaxes the internal sphincter and allows urine to enter the bladder neck and urethra. The PMC then activates PSNS outputs to the detrusor muscle and suppresses SNS outflow, and a sustained detrusor contraction begins. Mechanosensors in the urethra are stimulated by the presence of urine within the lumen, and their output to the PMC reinforces voiding. Detrusor contraction continues until the bladder is empty, although a small volume of urine (6–12 mL) typically remains after voiding is complete.

Chapter Summary

- The kidney is an **excretory** organ that cleanses blood of metabolic end products, toxins, water, and ions that may be surplus to the body's prevailing needs. It is also an **endocrine** and **homeostatic** organ controlling blood pressure, tissue osmolality, and electrolyte levels.
- The functional unit of the kidney is the **nephron**, which comprises a blood filtration module (**glomerulus**) and a filtrate recovery module (**renal tubule**). The kidney contains two types of nephrons, **superficial cortical nephrons** and **juxtamedullary nephrons**. The latter are specialized for formation of concentrated urine.
- Blood is forced under pressure (~60 mm Hg) through a **glomerular filtration barrier** to separate plasma from cells and proteins. The kidney receives ~20% of cardiac output and filters ~20% of the plasma it receives for a total of ~180 L/day (**glomerular filtration rate**).
- Glomerular filtration rate (GFR) is a function of **renal blood flow** (**RBF**). RBF is controlled by constriction and dilation of **afferent** and **efferent glomerular arterioles.** Afferent arteriolar dilation and efferent arteriolar constriction increases GFR. Afferent arteriolar constriction and efferent arteriolar dilation decreases GFR. Both arterioles are resistance vessels subject to multiple controls of local and central origin.
- The **juxtaglomerular apparatus** (**JGA**) is a functional complex comprising a section of the renal tubule and the glomerular arterioles. The JGA is a sensory system that allows renal perfusion and filtration pressures to be modulated to stabilize tubule fluid flow. Flow is sensed by the **macula densa**, a specialized region of the tubule wall, via changes in luminal Na^+ and Cl^- concentrations. If flow is too high, signals from the macula densa cause afferent arteriolar constriction and a decrease in glomerular filtration rate (GFR). If flow is too low, the afferent arteriole releases ***renin***, which activates the ***renin*–angiotensin–aldosterone system**. **Angiotensin II** constricts the efferent arteriole and raises GFR.
- Formed urine is channeled by **renal calyces** and the **renal sinus** into **ureters** and conveyed to the **urinary bladder**.
- The bladder is a hollow muscular organ that stores urine until emptying (**micturition**) is convenient. Valves prevent urinary reflux into the ureters, whereas **inner** and **outer sphincters** control outflow via the **urethra**. The outer sphincter is under voluntary control, but inner sphincter and bladder contraction is controlled by spinal reflexes and the central nervous system.
- Bladder filling stretches its muscular wall and initiates a **spinal micturition reflex**. The reflex causes parasympathetic motor efferents to stimulate bladder contraction. Emptying is prevented by the central nervous system until the outer sphincter is relaxed voluntarily.
- The efficiency of renal function can be assessed from plasma clearance of **inulin** and **creatinine**. **Clearance** refers to the amount of plasma that is completely cleared of a substance per unit time. Clearance of ***para*-aminohippurate** provides an estimate of renal plasma flow.

26 Reabsorption and Secretion

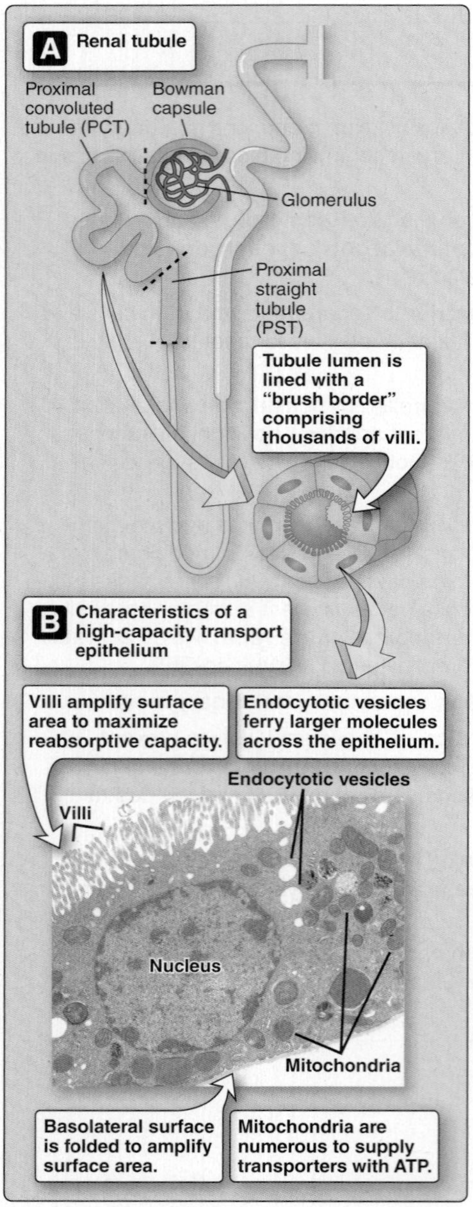

A Renal tubule
Proximal convoluted tubule (PCT)
Bowman capsule
Glomerulus
Proximal straight tubule (PST)
Tubule lumen is lined with a "brush border" comprising thousands of villi.
B Characteristics of a high-capacity transport epithelium
Villi amplify surface area to maximize reabsorptive capacity.
Endocytotic vesicles ferry larger molecules across the epithelium.
Endocytotic vesicles
Villi
Nucleus
Mitochondria
Basolateral surface is folded to amplify surface area.
Mitochondria are numerous to supply transporters with ATP.

Figure 26.1
Proximal tubule structure.
ATP = adenosine triphosphate.

I. OVERVIEW

The ultrafiltrate entering the **proximal tubule** (**PT**) from the Bowman space has a composition that is almost identical to that of plasma. It contains over 150 different components, but the major constituents are inorganic ions (Na^+, K^+, Mg^{2+}, Ca^{2+}, Cl^-, HCO_3^-, H^+, and phosphates), sugars, amino acids and peptides, creatinine, and urea. It also contains large amounts of water. The renal tubule's function is to recover >99% of the water and the majority of the solutes before they reach the bladder. Most is recovered within the first few millimeters of the PT, including virtually all organic compounds (sugars, amino acids, peptides, and organic acids) and two thirds of the filtered ions and water. Much of this material is recovered paracellularly by osmosis, made possible by the tubule wall's inherently leaky nature. The PT also actively secretes a number of organic compounds into the tubule lumen for subsequent urinary excretion. Principal sites for reabsorption, secretion, and regulation of various solutes along the nephron are summarized in Figure 27.19.

II. PRINCIPLES

The PT is a high-capacity, "leaky" transport epithelium drawn into a ~50-μm tube (Figure 26.1A). The initial portion of the tube is coiled (the proximal convoluted tubule [PCT]), and it then straightens to form the proximal straight tubule (PST). The PT's primary function is isosmotic fluid reabsorption.

> An epithelium's "leakiness" is a reflection of the ease with which solutes and water permeate the tight junctions between adjacent epithelial cells. Leaky epithelia are highly permeable, whereas intercellular junctions in tight epithelia are relatively impermeant (see 4·II·E·2).

A. Cellular structure

The PT reabsorbs ~120 L of fluid and solutes per day. The enormity of this load is reflected in the ultrastructure of the epithelial cells that make up its walls, which are packed with mitochondria, and their surface membranes are specialized to amplify surface area (see Figure 26.1B).

1. **Metabolism:** The apical and basolateral membranes of PT epithelial cells are packed with channels and transporters for retrieval and secretion of inorganic ions and other solutes. Reabsorption is driven by ion gradients generated by adenosine triphosphate (ATP)-dependent pumps, so the cytoplasm is dense with mitochondria to supply the PT's high metabolic needs.

2. **Surface area:** PT apical and basolateral membranes are extensively modified to increase their surface area. The membrane expanse is required to accommodate high numbers of channels and transporters and also to maximize area for contact between the epithelial cell and tubule contents. The numerous, densely packed microvilli that sprout from the apical surface create a **brush border** that is structurally and functionally similar to that found in the small intestine (see 31·II).

3. **Junctions:** Adjacent epithelia cells are connected at their apical surface by tight junctions that have a very loose structure. The junctions are highly permeable to solutes and water and, thus, the epithelium has very low electrical resistance.

B. Reabsorption

Reabsorption involves transferring water and solutes from the tubule lumen to the interstitium. Once in the interstitium, these materials are free to enter the peritubular capillary network by simple diffusion. The principal pathways and mechanisms involved in reabsorption were introduced in Unit I (see 4·III) and are summarized here.

1. **Pathways:** There are two paths by which materials can cross epithelia (Figure 26.2). The paracellular route lies between two adjacent epithelial cells. The permeability of the paracellular route is determined by tight junction structure. The transcellular route takes a solute through the inside of an epithelial cell and usually requires the assistance of channels or transporters to traverse the apical and basolateral membranes.

2. **Motive force:** Transepithelial transport is powered by ATP, and virtually all of the energy consumed during reabsorption is used to support Na^+-K^+ ATPase activity. The consequences of Na^+-K^+ ATPase activity can be broken down into five partly overlapping steps, all of which contribute to net reabsorption (Figure 26.3).

   a. **Step 1—Ion gradient:** The Na^+-K^+ pump is located in the basolateral membrane. It exchanges three intracellular Na^+ ions for two extracellular K^+ ions, creating an inwardly directed gradient for Na^+ diffusion across both apical and basolateral membranes.

   b. **Step 2—Voltage gradient:** Pumping Na^+ ions into the interstitium modifies the potential difference between interstitium and PT lumen. Although the difference is small (~3 mV, lumen negative), it creates a significant driving force for ion movement.

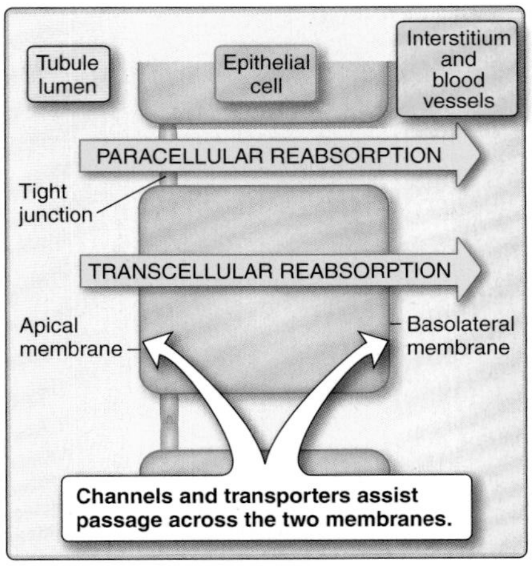

Figure 26.2
Pathways for reabsorption from the tubule lumen.

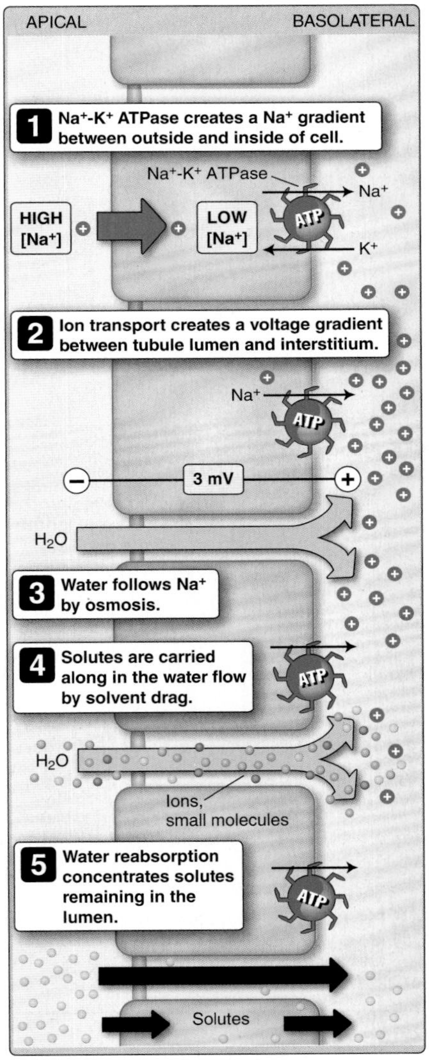

Figure 26.3
Reabsorption mechanisms.
ATP = adenosine triphosphate.

> The transepithelial voltage gradient reverses polarity, from lumen-negative to lumen-positive, in the later regions of the PT. Reversal occurs because Cl^- is reabsorbed preferentially in the later regions, leaving behind a net positive charge (see below).

c. **Step 3—Osmotic gradient:** Pumping Na^+ ions into the interstitium also creates an osmotic gradient that drives water flow from the tubule lumen across the tight junctions.

d. **Step 4—Solvent drag:** Water flowing through the intercellular junctions in response to an osmotic gradient creates solvent drag that sweeps ions and small organic molecules along with it.

e. **Step 5—Chemical gradient:** Water reabsorption concentrates the solutes that are left behind in the tubule lumen, thereby creating a chemical gradient favoring diffusional reabsorption.

C. Peritubular network

The PT's ability to reabsorb large volumes of fluid is only possible with support from the peritubular capillary network, which closely follows the tubule through the kidney (see Figure 25.4). The peritubular network sustains the tubule with O_2 and nutrients, but, just as importantly, it also clears recovered fluid from the interstitium before it has a chance to accumulate and reduce the gradients favoring reabsorption. The Starling forces governing fluid movement across the peritubular capillary wall are configured so as to promote reabsorption from the renal interstitium (Figure 26.4; see 19·VII·D). The main force favoring fluid reabsorption is plasma colloid osmotic pressure (π_{PC}). Capillary hydrostatic pressure (P_{PC}) is the principal force opposing reabsorption.

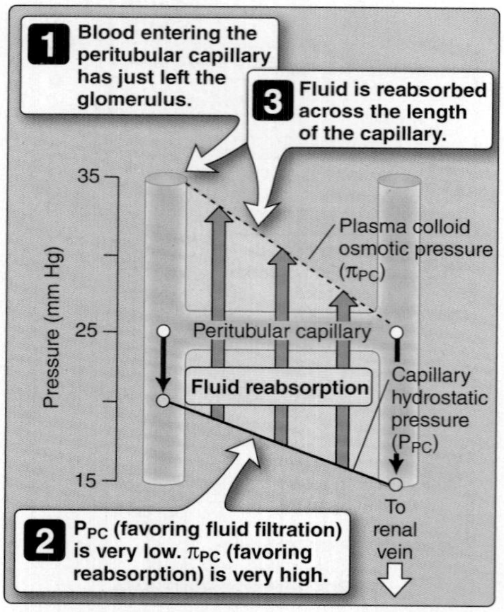

Figure 26.4
Forces controlling fluid reabsorption by peritubular capillaries.

1. **Plasma colloid osmotic pressure:** π_{PC} averages 25 mm Hg in virtually all other regions of the body, but blood entering the peritubular network has just traversed the glomerulus where ~20% of its fluid was removed by filtration. The plasma proteins are concentrated as a result, which raises π_{PC} to ~35 mm Hg.

2. **Capillary hydrostatic pressure:** Blood has to pass through an efferent arteriole before reaching peritubular capillaries. Efferent arterioles have a relatively high resistance, which decreases the pressure of blood entering the network to ~20 mm Hg. This is much lower than in other systemic capillary beds (~35 mm Hg). P_{PC} then declines over the length of the capillary. The combination of a high π_{PC} and low P_{PC} means that the driving force for fluid absorption is strongly positive across the entire length of the capillary (Flow $\propto \pi_{PC} - P_{PC}$, or ~15 mm Hg; see Figure 26.4).

III. ORGANIC SOLUTES: REABSORPTION

Plasma is laden with glucose (~4–5 mmol/L), amino acids (~2.5 mmol/L), small peptides, and organic acids (e.g., lactate, pyruvate), all of which are freely filtered into the Bowman space. These compounds represent a significant resource that must be recovered from the filtrate before it reaches the bladder. In practice, >98% of organic compounds are recovered in the early PCT and the remaining ~1%–2% are reabsorbed in the PST (Figure 26.5). Most organics are recovered by apical transporters, traverse cells by diffusion, and then are transported across the basolateral membrane to the interstitium and vasculature. Transporter involvement means that reabsorption shows saturation kinetics (Figure 26.6).

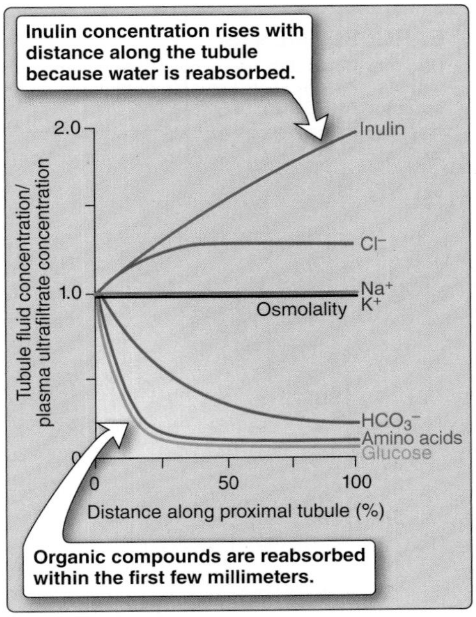

Figure 26.5
Changes in proximal tubule fluid composition with distance from the Bowman capsule.

A. Kinetics

The renal epithelium expresses a finite number of transporters, which limits solute reabsorption. If the glomerulus filters solutes in excess of maximal transporter capacity (T_m), then the transported solute will continue through the tubule and appear in urine. Plasma solute concentrations vary with intake and tissue use, but a healthy nephron is usually well equipped to recover filtered loads within a normal physiologic range. Solutes start appearing in urine in small amounts even before T_m is reached (see Figure 26.6). This region of the titration curve is said to show **splay**, reflecting transporter and nephron heterogeneity.

> "Filtered load" is the amount of any substance that filters from the glomerulus and enters the Bowman space per unit time (mg/min). Filtered loads are the products of glomerular filtration rate (GFR) and plasma concentration of the substance in question.

1. **Transporters:** Nephrons typically contain multiple transporter classes capable of transferring organic solutes across the surface membrane. The combined activity of pathways with different T_m values contributes to splay.
2. **Nephrons:** Nephrons show anatomic diversity, which causes differences in single-nephron GFR, transporter capacity, and transporter location along the tubule. These differences also contribute to splay.

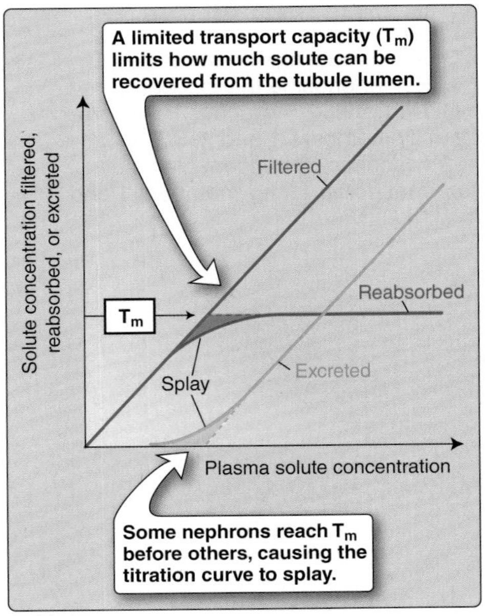

Figure 26.6
Limits to transporter-mediated solute reabsorption.

B. Glucose

Glucose plasma concentrations vary between ~3.8 and 6.1 mmol/L in a healthy person. Glucose filters freely into the tubule, and ~98% is reabsorbed in the early PT. Uptake occurs transcellularly and is transporter mediated (Figure 26.7).

1. **Apical:** Glucose is recovered using two different Na^+ cotransporters. They both harness the transmembrane Na^+ gradient to simultaneously absorb Na^+ and a glucose molecule. One of these

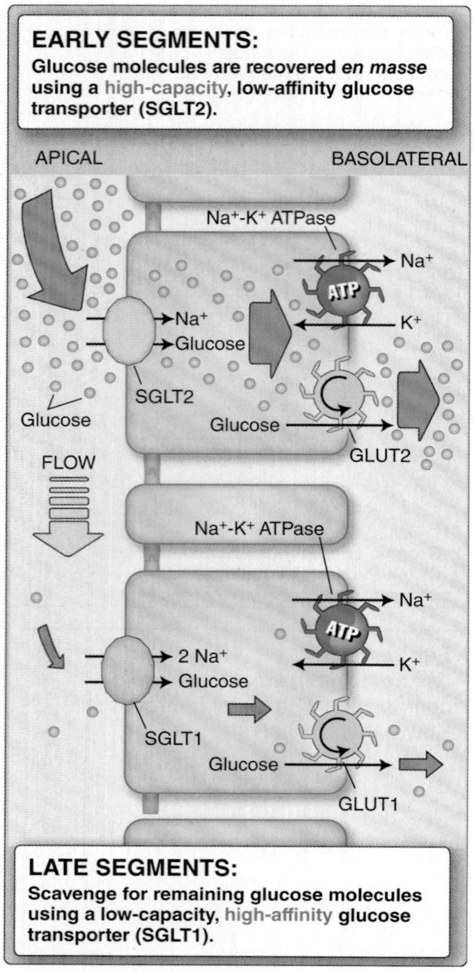

EARLY SEGMENTS:
Glucose molecules are recovered en masse using a high-capacity, low-affinity glucose transporter (SGLT2).
APICAL
BASOLATERAL
Na+-K+ ATPase
Na+
ATP
K+
Na+
Glucose
SGLT2
Glucose
Glucose
GLUT2
FLOW
Na+-K+ ATPase
Na+
ATP
K+
2 Na+
Glucose
SGLT1
Glucose
GLUT1
LATE SEGMENTS:
Scavenge for remaining glucose molecules using a low-capacity, high-affinity glucose transporter (SGLT1).

Figure 26.7
Glucose reabsorption strategies. ATP = adenosine triphosphate; GLUT1 and -2 = glucose transporter family members 1 and 2; SGLT1 and -2 = sodium-dependent glucose cotransporter family members 1 and 2.

cotransporter classes localizes primarily to the early part of the PCT, the other to the PST.

a. **Convoluted tubule:** The PCT expresses a high-capacity, low-affinity Na^+-glucose cotransporter (SGLT family, SGLT2) designed to recover the bulk of filtered glucose immediately after it enters the tubule.

b. **Straight tubule:** By the time filtrate reaches the PST, most of the glucose has been reabsorbed. The PST, thus, expresses a high-affinity, low-capacity 2 Na^+–glucose cotransporter (SGLT1), designed to recover the last of the glucose before it enters the nephron loop.

2. **Basolateral:** Glucose uptake by the epithelial cells generates a concentration gradient that drives facilitated diffusion (see 1·V·C·2) via GLUT family glucose transporters (GLUT2 and GLUT1 in the PCT and PST, respectively) across the basolateral membrane to the interstitium.

C. Amino acids

Plasma contains all of the common amino acids, and all are filtered into the renal tubule. The early PT recovers >98% of filtered amino acid load (see Figure 26.5). The amount filtered approaches T_m even under resting conditions, so urine always contains trace amounts of most amino acids. Physiologic increases in plasma amino acid levels easily overwhelm the nephron reabsorptive capacity, and significant amounts are then excreted. There are multiple pathways for amino acids to cross the apical and basolateral membranes.

1. **Apical:** There are several classes of amino acid transporter in the apical membrane. They generally have broad substrate specificity, so a single species of amino acid may have several recovery options. Anionic (acidic) amino acids are recovered by an excitatory amino acid transporter that exchanges H^+, two Na^+, and an amino acid for K^+. Cationic (basic) amino acids are taken up in exchange for a neutral amino acid. Neutral amino acids are taken up either by a Na^+ cotransporter or a H^+ cotransporter.

Clinical Application 26.1: Diabetes Mellitus

Plasma glucose concentration can rise to ~10 mmol/L before renal reabsorptive capacity is exceeded in normal, healthy individuals. Once transporter capacity is exceeded, significant quantities of glucose begin to spill over into the urine. The presence of unrecovered glucose within the renal tubule lumen causes an osmotic diuresis, manifesting as polyuria (urine output of >3 L/day). The frequent need to urinate gives rise to the term "diabetes," which is derived from a Greek verb (*diabainein*) having a similar meaning. The presence of glucose in urine gives it a sweet taste, providing a ready (albeit somewhat distasteful) means of diagnosing **diabetes mellitus** in the early days of medicine.

2. **Basolateral:** The basolateral membrane contains a different set of amino acid transporters whose substrate specificity is broader than that of the apical membrane. Cationic and many neutral amino acids are exchanged for a neutral amino acid plus Na^+. Aromatic amino acids cross to the interstitium by facilitated diffusion.

D. Peptides and proteins

The PT has three strategies for recovering peptides and proteins (Figure 26.8): uptake via small-peptide carriers, degradation and then uptake via carriers, and endocytosis.

1. **Transport:** There are many similarities in the ways the PT handles oligopeptides and glucose. The apical surface contains two peptide transporters: PepT1 and PepT2. Both are H^+-peptide cotransporters that transport di- and tripeptides in any of the >8,000 possible amino acid residue combinations. PepT1 is a low-affinity, high-capacity transporter expressed preferentially in the early part of the PT. PepT2 is a low-capacity, high-affinity transporter that scoops up remaining peptides that appear in the PST. Once inside the cell, the peptides are rapidly degraded by *proteases* and returned to the vasculature as free amino acids.

2. **Degradation:** The PT brush border resembles that of the small intestine in that it expresses many *peptidases*. These enzymes degrade large peptides (including hormones) into small peptides or their constituent amino acids, which are then reabsorbed using carriers.

3. **Endocytosis:** PT epithelial cells express endocytotic receptors (**megalin** and **cubilin**) on their apical surface that bind any proteins that might have crossed the glomerular filtration barrier and then internalizes them. Once inside the cell, proteins are digested and released on the basolateral side as free amino acids or small peptides. The PT also expresses receptors that recognize and internalize specific hormones such as somatostatin. The pharmaceutical industry has been exploring the possibility of using these receptors as vehicles for drug delivery.

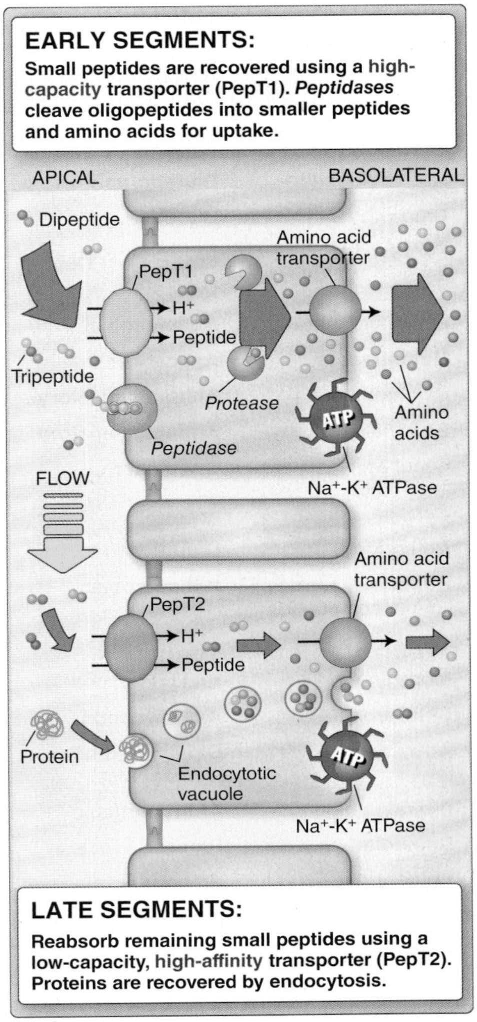

EARLY SEGMENTS:
Small peptides are recovered using a high-capacity transporter (PepT1). *Peptidases* cleave oligopeptides into smaller peptides and amino acids for uptake.
APICAL
BASOLATERAL
Dipeptide
PepT1
H+
Peptide
Tripeptide
Amino acid transporter
Protease
Peptidase
ATP
Amino acids
Na+-K+ ATPase
FLOW
Amino acid transporter
PepT2
H+
Peptide
Protein
Endocytotic vacuole
ATP
Na+-K+ ATPase
LATE SEGMENTS:
Reabsorb remaining small peptides using a low-capacity, high-affinity transporter (PepT2). Proteins are recovered by endocytosis.

Figure 26.8
Oligopeptide and protein reabsorption. ATP = adenosine triphosphate.

E. Organic acids

Plasma contains significant quantities of lactate; pyruvate; and other mono-, di-, and tricarboxylates that are freely filtered by the glomerulus and then reabsorbed by the PT using two different Na^+ cotransporters. One is specific for monocarboxylates (e.g., lactate, pyruvate), the other for di- and tricarboxylates (e.g., citrate, succinate). Monocarboxylates then exit the cell via a basolateral H^+-carboxylate cotransporter. Di- and tricarboxylates are exchanged for an organic anion by a member of the **organic anion transporter** (**OAT**) family.

IV. ORGANIC SOLUTES: SECRETION

Blood that has traversed the glomerular capillary network still contains a number of metabolic end products that are undesirable and,

Table 26.1: Drugs Secreted by the Proximal Tubule

Drug Name	Drug Class
Anions	
Acetazolamide	Diuretic, various
Chlorothiazide	Diuretic
Furosemide	Diuretic
Probenecid	Uricosuric
Penicillin	Antimicrobial
Methotrexate	Anticancer
Indomethacin	Anti-inflammatory
Salicylate	Anti-inflammatory
Saccharin	Sweetener
Cations	
Amiloride	Diuretic
Quaternary ammonium compounds	Antimicrobial
Quinine	Antimalarial
Morphine	Analgesic
Chlorpromazine	Antipsychotic
Atropine	Cholinergic antagonist
Procainamide	Antiarrhythmic
Dopamine	Pressor
Epinephrine	Pressor
Cimetidine	Antacid (H_2 blocker)
Paraquat	Herbicide

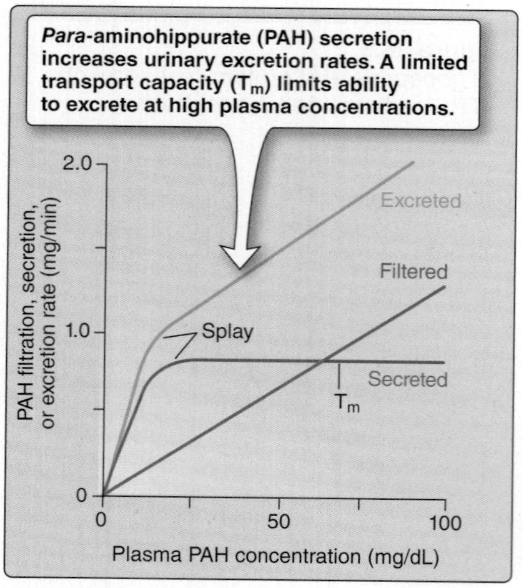

Figure 26.9
Effects of plasma *para*-aminohippurate (PAH) concentration on secretion and excretion rates.

possibly, toxic. Although these waste products would eventually be excreted during subsequent passes, the kidney supplements its passive filtration and cleansing functions with an active secretory process. Secretion occurs in the late PT and is almost 100% effective in ridding the body of a number of organic anions and cations in a single pass. Uric acid, for example, is a relatively insoluble end product of nucleotide metabolism that is actively secreted by the PT. Other secreted waste products include creatinine, oxalate, and bile salts. Secretion also helps clear exogenous toxins from the body. The secretory transporters have a very broad substrate specificity, which allows them to handle a wide array of potential chemical threats. These pathways also clear a wide range of pharmaceutical drugs from the vasculature (Table 26.1).

> The tendency for the PT to take up pharmaceuticals from the circulation puts it at grave risk because intracellular concentrations can quickly rise to toxic levels. The transporters responsible for uptake have, thus, themselves become high-priority targets for pharmaceutical intervention. Inhibiting the uptake systems not only reduces drug toxicity, but it also decreases the rate of drug elimination from the body and, thus, reduces dosing frequency.

A. Kinetics

Secretion is transporter mediated and, therefore, shows saturation kinetics, as demonstrated using *para*-aminohippurate (PAH) in Figure 26.9. PAH is a hippuric acid derivative used in studies of renal plasma flow (see 25·V·D) that is both filtered from the glomerulus and secreted from the PT via the OAT pathways mentioned above.

1. **Filtration:** PAH is freely filtered by the glomerulus in amounts that are directly proportional to GFR. Filtration removes ~20% of total plasma PAH.

2. **Secretion:** Blood entering the peritubular network still contains 80% of the original arterial PAH load. All but 10% is taken up by transporters in the basolateral membrane of the late PT and secreted into the tubule lumen. PAH excretion rises accordingly. Transporter capacity is finite, however, so the secretion curve flattens and plateaus as plasma PAH concentration approaches T_m. The secretion curve exhibits splay due to transporter and nephron heterogeneity, as discussed above in reference to glucose reabsorption.

B. Transporters

PT epithelia express a number of different broad-specificity transporters for organic anions and cations. Organic anions are taken up from the blood by several members of the OAT family. OAT1 exchanges an organic ion for a dicarboxylate such as α-ketoglutarate. A related family of **organic cation transporters** takes up amine and ammonia

Clinical Application 26.2: Gout

Organic anion transporters are one of a number of transporter families involved in reabsorption and excretion of uric acid. Most mammals metabolize urate to allantoin, but primates lost the necessary enzyme (*uricase*) during evolution. Unlike allantoin, urate is relatively insoluble, and when blood concentrations rise, it forms crystals that are often deposited in joints. The result is a painful inflammatory arthritis known as gout. Gout treatment options include drugs that inhibit the transporters that normally reabsorb urate as it passes down the tubule, thereby increasing excretion rates.

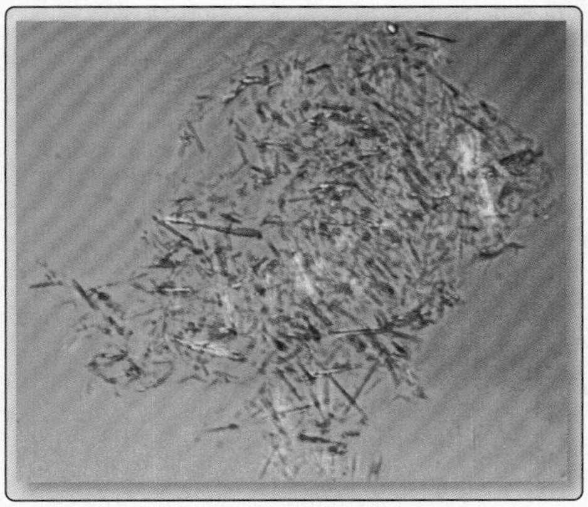

Uric acid crystals in synovial fluid from a patient with chronic gout.

compounds from blood. Anions and cations are then both extruded into the tubule lumen by one of a number of **multidrug-resistant proteins** (**MRPs**). MRPs are members of the ATP-binding cassette pump superfamily. Organic anions can also cross the apical membrane by one of a number of OATs.

V. UREA

Urea is a small organic molecule comprising two amide groups joined by a carbonyl group. It is formed in the liver[1] and excreted in urine as a way of disposing of unwanted amino acids and nitrogen (Figure 26.10). Normal plasma concentrations average 2.5–6.0 mmol/L. The PT reabsorbs ~50% of the filtered load, largely via the paracellular route. Two forces drive movement. The first is solvent drag, created by the large volumes of water being reabsorbed in the PT. Loss of water from the tubule lumen secondarily concentrates solutes in the tubule lumen, which enhances the driving force for urea diffusion across the epithelium. The kidney ultimately excretes ~20% of the filtered urea load, but it first serves an important role in helping concentrate urine. The pathways involved are discussed in Chapter 27 (see 27·V·D).

> Urea is the principal means by which nitrogenous wastes are excreted from the body, and, thus, plasma urea levels are a useful indicator of renal health and function. Clinical laboratories cite urea levels in the form of blood urea nitrogen (BUN). Normal BUN values are in the range of 7–18 mg/dL.

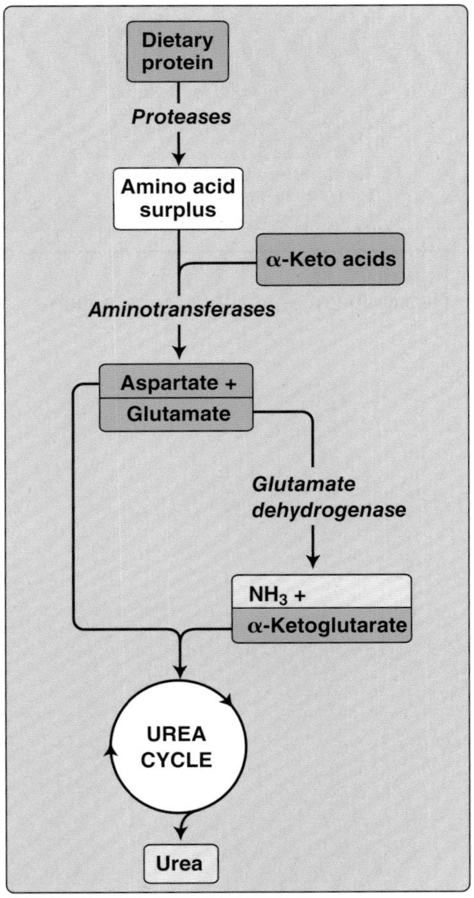

Figure 26.10
Urea formation.

INFO LINK [1]The role of urea in nitrogen excretion and details of the urea cycle are discussed at length in *LIR Biochemistry*, 5e, p. 253.

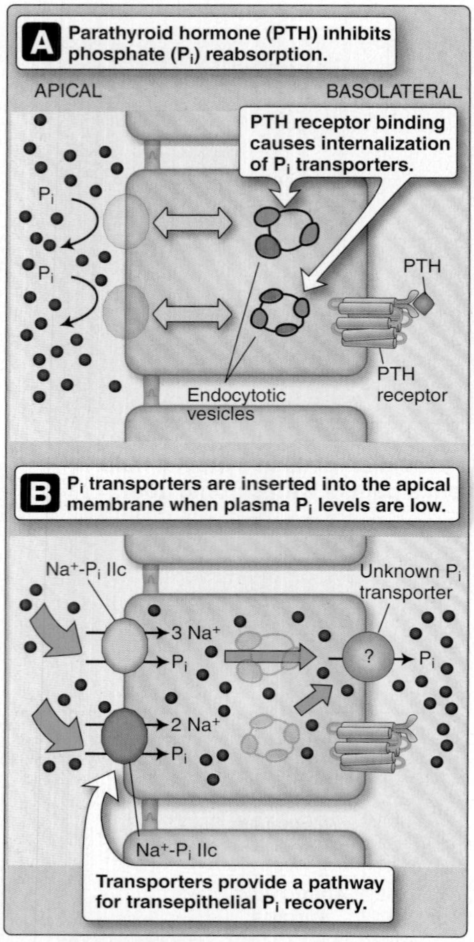
A
Parathyroid hormone (PTH) inhibits phosphate (Pi) reabsorption.
APICAL
BASOLATERAL
PTH receptor binding causes internalization of Pi transporters.
Pi
Pi
PTH
PTH receptor
Endocytotic vesicles
B
Pi transporters are inserted into the apical membrane when plasma Pi levels are low.
Na+-Pi IIc
Unknown Pi transporter
3 Na+
Pi
?
Pi
2 Na+
Pi
Na+-Pi IIc
Transporters provide a pathway for transepithelial Pi recovery.

Figure 26.11
Regulation of phosphate reabsorption.

VI. PHOSPHATE AND CALCIUM

Plasma contains a total of ~1.0–1.5 mmol/L inorganic phosphorus (P_i) and ~2.1–2.8 mmol/L Ca^{2+}. Both are critically important for normal cell function. P_i is a component of RNA and DNA, powers metabolism in the form of ATP, and is found associated with numerous lipids and proteins. Ca^{2+} is a vital second messenger that activates enzymes, initiates muscle contraction, and triggers neurotransmitter secretion. About half of total plasma phosphorus and calcium exists in ionized form (as HPO_4^{2-}, $H_2PO_4^-$, and Ca^{2+}), the remainder being complexed with proteins and other molecules. Plasma contains only a tiny fraction of total body phosphorus and calcium, however. The vast majority of phosphorus (>80%) and calcium (>99%) is locked in hydroxyapatite crystals in a mineral vault called bone. Plasma P_i and Ca^{2+} concentrations are regulated by similar mechanisms. Total body concentrations of both ions represent a precise balance between bone deposition and resorption, intestinal secretion and absorption, and renal filtration and reabsorption. All three processes are regulated by parathyroid hormone ([PTH] discussed in more detail in Chapter 35).

A. Phosphate

The kidney tubule reabsorbs ~90% of the filtered P_i load, of which ~80% is reclaimed in the PT and the remaining 10% in the distal convoluted tubule (DCT). The PT is the principal site of P_i regulation, effected through PTH and plasma P_i concentrations (see Figure 27.19).

1. **Reabsorption:** P_i is reabsorbed using two apical Na^+-P_i cotransporters (Na^+-P_i IIa and Na^+-P_i IIc) as shown in Figure 26.11. The mechanism by which P_i crosses the basolateral membrane is under investigation.

2. **Regulation:** PTH blocks P_i recovery from the tubule lumen by promoting endocytosis and subsequent degradation of the apical P_i transporters. In the absence of a recovery pathway, P_i then passes through the tubule and is excreted. Low dietary intake causes cotransporters to be inserted into the apical membrane, thereby increasing the PT's ability to reabsorb filtered P_i.

B. Calcium

Plasma free Ca^{2+} concentrations are tightly regulated in the range of 1.0–1.3 mmol/L, and virtually all filtered Ca^{2+} is reabsorbed during passage through the nephron (see Figure 27.19). The PT recovers ~65%, largely via the paracellular route. The motive force is partly solvent drag and, in the later stages of the PT where the lumen is positively charged with respect to blood, the transepithelial voltage difference. Most of the remaining 35% of filtered load is reabsorbed in the thick ascending limb ([TAL] ~25%) and the DCT (~8%). The DCT is the main site of Ca^{2+} regulation (see 27·III·C).

VII. MAGNESIUM

Mg^{2+} is a vital cofactor required for the normal function of hundreds of enzymes, its positive charge helping stabilize protein structural integrity.

It also regulates ion flow through ion channels, so physiologic decreases in plasma free concentrations cause membrane hyperexcitability, arrhythmias, and muscle tetany. The majority of total body Mg^{2+} is complexed in bone or associated with proteins and other small molecules. Plasma concentrations are normally maintained in a range of ~0.75–1.00 mmol/L, of which ~60% is in the free form. Mg^{2+} is a common ingredient in most foods, so ~2%–5% of the filtered load typically is excreted in urine to balance daily intake. The PT recovers ~15% of the filtered load. Reabsorption occurs paracellularly by solvent drag and diffusion. Reabsorption is favored by the small lumen-positive potential difference that exists across the more distal regions of the PT epithelium. The bulk of filtered Mg^{2+} (~70%) is recovered in the TAL, which is also the principal site of Mg^{2+} homeostatic regulation (see 27·III·B and Figure 27.19).

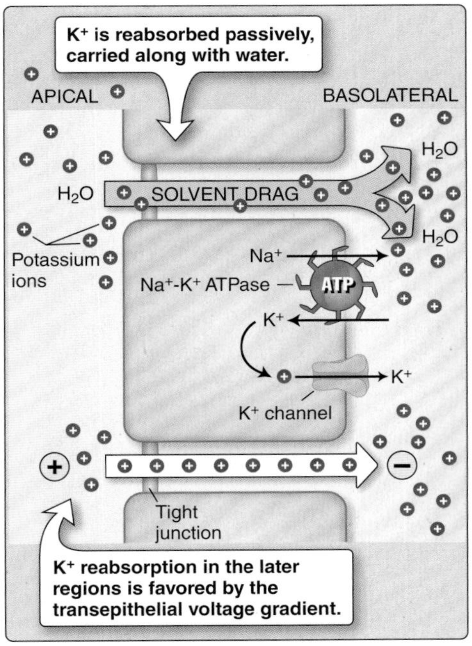

Figure 26.12
Potassium reabsorption pathways in the proximal tubule. ATP = adenosine triphosphate.

VIII. POTASSIUM

K^+ is unique among electrolytes in that even modest changes in plasma K^+ concentrations can be life threatening, causing potentially fatal cardiac dysrhythmias and arrhythmias (see Clinical Application 2.1). Plasma concentrations are tightly regulated within the range of 3.5–5.0 mmol/L. K^+ is filtered freely across the glomerulus, so the nephron handles a daily load of ~0.6–0.9 mol. The PT reabsorbs ~80% of the filtered load, primarily via the paracellular route (Figure 26.12). As is the case for Ca^{2+} and Mg^{2+}, absorption occurs as a result of solvent drag, and by diffusion that is enhanced by a transepithelial voltage gradient. Another 10% is recovered in the TAL (see 27·II·B), but regulation of K^+ reabsorption (and excretion) occurs primarily in the distal segments (see 27·IV·C and Figure 27.19).

IX. BICARBONATE AND HYDROGEN IONS

One of the kidney's most important functions is to help maintain extracellular fluid (ECF) pH at ~7.40. Metabolism generates immense quantities of volatile acid (H_2CO_3) that is expelled via the lungs and another ~50–100 mmol/day of nonvolatile acid (sulphuric, phosphoric, nitric, and other minor acids; see 3·IV·A) that must be excreted by the kidneys. Although all portions of the nephron are involved in acid–base homeostasis to some degree (see Figure 27.19), the PT is a principal site for HCO_3^- recovery and H^+ secretion.

A. Bicarbonate

Excreting HCO_3^- causes the ECF to become acidic, so the first goal of pH homeostasis is to recover 100% of the filtered HCO_3^- load. The PT recovers ~80% of total. Because HCO_3^- is anionic, it cannot diffuse freely across membranes, so the PT secretes molar amounts of H^+ into the tubule lumen to titrate the HCO_3^- and then uses *carbonic anhydrase* (*CA*) to convert the H_2CO_3 to CO_2 and H_2O. Both molecules are then recovered by simple diffusion. Reclamation is a four-step process (numbers below correspond to steps shown in Figure 26.13):

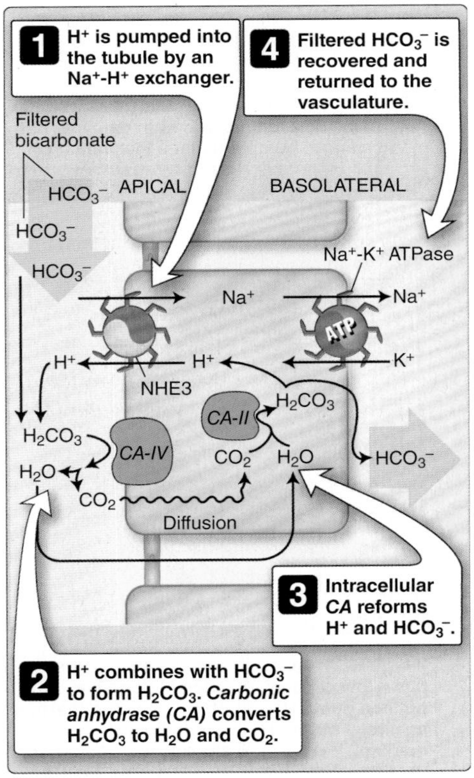

Figure 26.13
Bicarbonate reabsorption pathway in the proximal tubule. ATP = adenosine triphosphate.

1. H^+ is transported into the tubule lumen by an apical Na^+-H^+ exchanger (NHE3). The exchange is powered by the transmembrane Na^+ gradient.

2. H^+ combines with luminal HCO_3^- to form H_2CO_3, which dissociates to form H_2O and CO_2. The reaction is catalyzed by *CA-IV*, which is expressed on the epithelium's apical surface:

$$HCO_3^- + H^+ \leftrightarrows H_2CO_3 \underset{CA}{\leftrightarrows} CO_2 + H_2O$$

3. CO_2 diffuses into the cell and combines with H_2O to reform HCO_3^- and H^+. The reaction is catalyzed by intracellular *CA-II*.
4. HCO_3^- is reabsorbed across the basolateral membrane to the interstitium and then into the vasculature, although the mechanism is unclear. H^+ is pumped back into the tubule lumen to repeat the reabsorption cycle.

HCO_3^- reabsorption causes a slight acidification of the tubule contents, from pH 7.4 at the glomerulus to ~pH 6.8 in the late PT.

Acetazolamide is a *CA* inhibitor that blocks HCO_3^- and Na^+ reabsorption by the PT, causing diuresis. The drug acts on both the apical (*CA-IV*) and intracellular form (*CA-II*) of the enzyme. As a class, the *CA* inhibitors are relatively ineffective diuretics because the more distal regions of the tubule compensate for their effects on PT function.[1] The main indication for *CA* inhibitor use is in patients with metabolic alkalosis, because the drugs impair the tubule's ability to reabsorb HCO_3^- and, thereby, cause excess base to be excreted in urine.

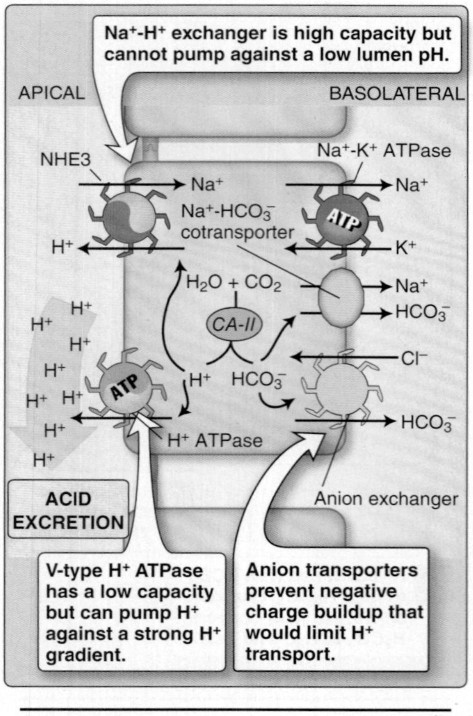

Figure 26.14
Acid secretion by the proximal tubule. ATP = adenosine triphosphate; *CA-II* = *carbonic anhydrase II.*

B. Hydrogen ions

The PT is a principal site for H^+ secretion, although final determination of urine pH and regulation of ECF pH occurs in the distal segments (see 27·V·E). H^+ is secreted by the NHE3 Na^+-H^+ exchanger mentioned above, and by a H^+ pump (Figure 26.14).

1. **Sodium–hydrogen ion exchange:** The NHE3 Na^+-H^+ exchanger uses the Na^+ gradient created by the basolateral Na^+-K^+ ATPase to power H^+ secretion. The dependence on the Na^+ gradient means that its ability to *concentrate* H^+ in the lumen is limited, but it has a very high *capacity* that accounts for ~60% of net H^+ secretion in the PT.

The NHE3 exchanger is also a principal pathway by which the PT recovers Na^+ from the tubule lumen (see below).

2. **Proton pump:** The PT also actively secretes H^+ into the tubule using a vacuolar-type H^+ pump (V-type H^+ ATPase). The H^+ pump accounts for ~40% of net secretion in the PT and is capable of

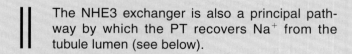

[1]For more information on acetazolamide use, see *LIR Pharmacology*, 5e, p. 287.

establishing a strong H^+-concentration gradient across the apical membrane. The pump is **electrogenic**, meaning that it causes a negative charge to build within the cell. This charge can become limiting to further transport, so H^+ secretion is balanced by HCO_3^- movement across the basolateral membrane via a Na^+-HCO_3^- cotransporter and an anion exchanger (see Figure 26.14).

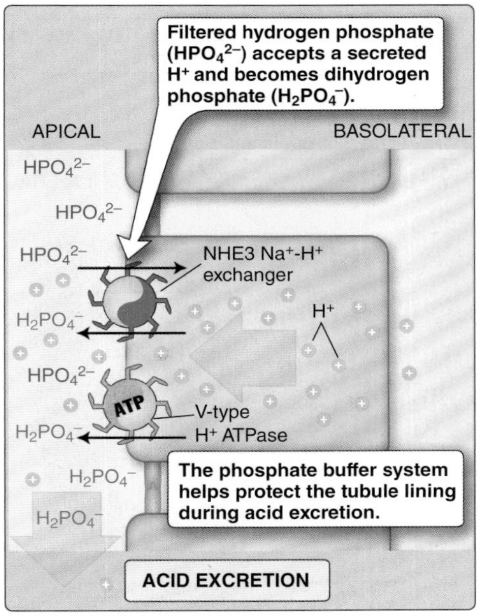

Figure 26.15
Phosphate buffer system. ATP = adenosine triphosphate.

C. Nonvolatile acid

Ideally, the H^+ excess created by nonvolatile acid formation would be transported to the kidney and then dumped into the tubule and excreted without further ado. In practice, the amount of nonvolatile acid generated is large, and the ability of available H^+ transporters to pump H^+ against a concentration gradient is limited. The V-type H^+ ATPase mentioned above can create a lumen pH of ~pH 4.0 at best (i.e., 0.1 mmol/L H^+), which is insufficient to handle the daily acid excess. Two different workarounds have evolved to allow H^+ to be excreted in the quantities required to maintain pH balance. The first is to simultaneously excrete urinary buffers (**titratable acids**) that limit a rise in free H^+ concentration even as acid is being pumped into the tubule lumen. The second is to attach H^+ to ammonia (NH_3) and excrete it as an ammonium ion (NH_4^+).

1. **Titratable acids:** The plasma filtrate contains several buffers, and the PT secretes several more. These include hydrogen phosphate (pK = 6.8), urate (pK = 5.8), creatinine (pK = 5.0), lactate (pK = 3.9), and pyruvate (pK = 2.5). Collectively, these buffers are known as "titratable acids" that complex with and, thereby, limit rises in tubule H^+ concentration. Hydrogen phosphate's pK makes it a more effective urinary buffer than the other titratable acids. Hydrogen phosphate accepts H^+ to become dihydrogen phosphate (Figure 26.15):

$$H^+ + HPO_4^{2-} \leftrightarrows H_2PO_4^-$$

The PT reabsorbs ~80% of filtered phosphate, but the remaining 20% remains to buffer lumen pH during nonvolatile H^+ excretion.

2. **Ammonia:** Plasma does not normally contain NH_3, but PT cells are able to synthesize it from glutamine, which is converted to NH_3 and α-ketoglutarate. NH_3 is lipid soluble, so it diffuses out of the cell into the tubule lumen and combines with H^+ to form NH_4^+. Some NH_4^+ is formed inside the PT cells and moved into the tubule lumen by the Na^+-H^+ exchanger, which is able to bind NH_4^+ in place of H^+ (Figure 26.16).

3. **New bicarbonate:** Excreting ~50–100 mmol of nonvolatile acid generated every day creates a sizeable deficit in the body's buffer systems. This must be matched precisely by the formation of new buffer or ECF would rapidly become acidotic. Excreted buffer is replaced by generation of "new" HCO_3^-. Some is formed *de novo*, and some is created from α-ketoglutarate after NH_3 is formed from glutamine. α-Ketoglutarate is metabolized to glucose and then to CO_2 and H_2O. *CA* then catalyzes H_2CO_3 formation, which dissociates to yield HCO_3^- and H^+. The newly formed HCO_3^- diffuses into blood and is ultimately used to buffer nonvolatile acid at its formation site within tissues.

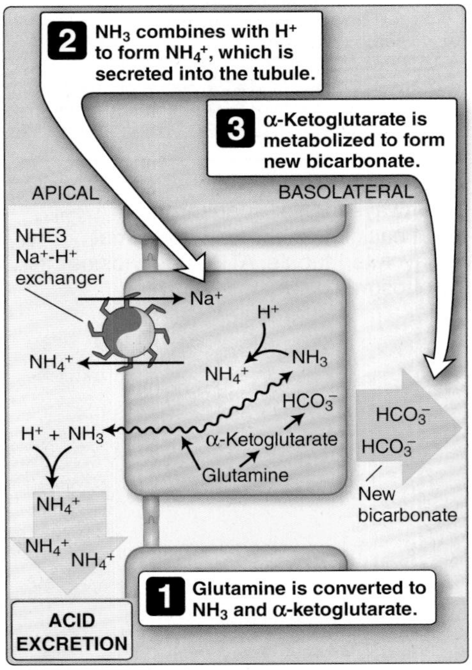

Figure 26.16
Excreting acid in the form of the ammonium ion.

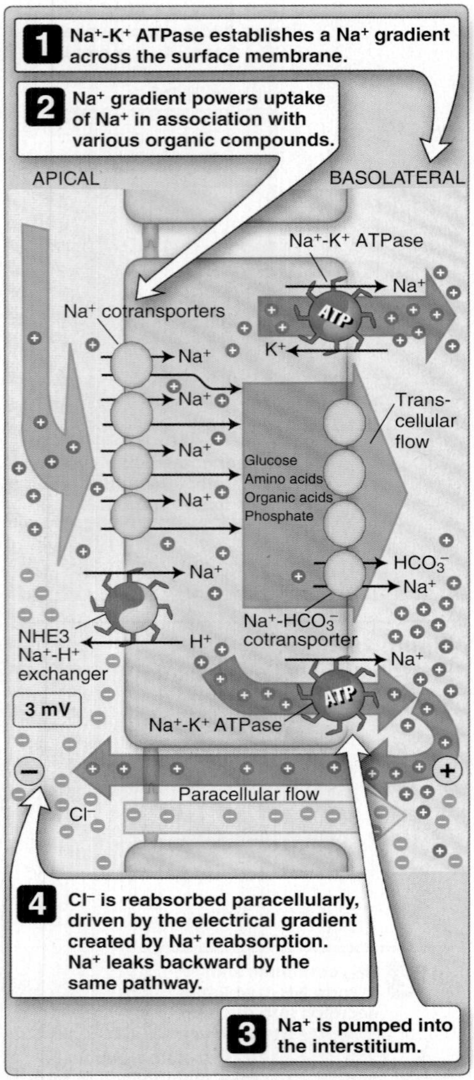
1 Na+-K+ ATPase establishes a Na+ gradient across the surface membrane.
2 Na+ gradient powers uptake of Na+ in association with various organic compounds.
APICAL
BASOLATERAL
Na+-K+ ATPase
Na+ cotransporters
Na+
K+
Na+
Na+
Na+
Na+
Trans-cellular flow
Glucose
Amino acids
Organic acids
Phosphate
HCO3−
Na+
Na+-HCO3− cotransporter
NHE3
Na+-H+ exchanger
H+
Na+
3 mV
Na+-K+ ATPase
Paracellular flow
Cl−
4 Cl− is reabsorbed paracellularly, driven by the electrical gradient created by Na+ reabsorption. Na+ leaks backward by the same pathway.
3 Na+ is pumped into the interstitium.

Figure 26.17
Pathways for Na^+ reabsorption and backflow in the early proximal convoluted tubule. ATP = adenosine triphosphate.

X. SODIUM, CHLORIDE, AND WATER

Plasma Na^+ concentration is maintained at between ~136 and 145 mmol/L, primarily as a way of controlling how water distributes between the three body compartments (intracellular, interstitial, and plasma; see 3·III·B). Na^+ moves freely across the glomerular filtration barrier, so the daily filtered load exceeds 25 mol. Approximately 99.6% of the filtered load is reabsorbed during passage through the renal tubule, the bulk (~67%) being recovered by the PT (see Figure 27.19). Cl^- follows Na^+ across the epithelium, driven inward by sodium's positive charge. Reabsorption of Na^+, Cl^-, and organic solutes creates a strong osmotic potential that also drives water from the tubule lumen toward the interstitium. The net effect of these and all of the other reabsorptive and secretory processes described in the previous sections is that the fluid reabsorbed by the PT is isosmotic and has a composition that resembles plasma. There are regional differences in the way that Na^+ and Cl^- are reabsorbed between the early and late regions of the PT, however.

A. Early proximal convoluted tubule

Early PT epithelial cells are specialized to recover virtually all useful organic solutes and HCO_3^- in association with Na^+, which leads to significant transcellular Na^+ reabsorption. Some of this Na^+ then leaks backward paracellularly (Figure 26.17).

1. **Transcellular:** The primary force driving reabsorption is the basolateral Na^+-K^+ ATPase, which establishes a Na^+ gradient that drives Na^+-coupled glucose, amino acid, organic acid, and phosphate reabsorption from the tubule. Large quantities of Na^+ also enter cells via the NHE3 Na^+-H^+ exchanger. Na^+ is then moved to the interstitium by the Na^+-K^+ ATPase and, to a lesser degree, by a basolateral Na^+-HCO_3^- cotransporter. Cotransport is driven by high intracellular HCO_3^- concentrations following reabsorption and *de novo* synthesis.

2. **Paracellular:** The cotransporters that recover organic solutes from the plasma filtrate are electrogenic, leaving an excess of negative charges in the tubule lumen. These charges create a ~3-mV difference between tubule and interstitium, which creates a significant force that drives paracellular Cl^- reabsorption. The paracellular route also permits significant amounts of reabsorbed Na^+ (~30%) to leak backward from the interstitium to tubule lumen. Movement is driven by the voltage gradient.

B. Proximal straight tubule

The fluid entering the PST has been stripped of all useful organic solutes and most HCO_3^-, but contains relatively high concentrations of Cl^-. Na^+ and Cl^- are reabsorbed via both transcellular and paracellular routes.

1. **Transcellular:** The late PT takes up Na^+ in exchange for H^+, which creates a transcellular Na^+ flux. This region of the PT also contains a Cl^--base exchanger (CFEX) that allows for significant transcellular Cl^- uptake. CFEX exchanges Cl^- for formate, oxalate, OH^-, or HCO_3^-.

2. **Paracellular:** High luminal Cl^- concentrations drive diffusion of Cl^- out of the lumen via the paracellular route. This leaves an excess of positive charge in the lumen that favors Na^+ reabsorption, so Na^+ follows Cl^- across the tight junctions and into the interstitium.

Chapter Summary

- The proximal tubule (PT) recovers ~67% of the fluid and up to 100% of some solutes that are filtered into the renal tubule by the glomerulus. PT epithelial cells possess apical **microvilli** that increase surface area, and the junctions between cells are leaky to maximize free flow of water and dissolved solutes.
- The proximal tubule reabsorbs fluid **isosmotically**. **Transcellular absorption** is powered mainly by the transmembrane Na^+ gradient established by a basolateral Na^+-K^+ ATPase. Reabsorption also occurs by diffusion via tight junctions (**paracellular absorption**) and paracellular **solvent drag**.
- Reabsorbed fluid is returned to the vasculature via the **peritubular network**. Blood reaches the peritubular capillaries by way of the glomerulus. Glomerular filtration concentrates the plasma proteins and, thereby, increases **plasma colloid osmotic pressure**. The efferent arteriole has a high resistance that lowers **capillary hydrostatic pressure**. These features together create a situation in which fluid uptake from the interstitium is strongly favored, which facilitates reabsorption.
- The proximal tubule (PT) recovers almost 100% of filtered **glucose** and **amino acids**, principally via Na^+ cotransport. The PT also recovers **small peptides** by **H^+ cotransport**. Larger peptides and proteins are degraded to small peptides and are then reabsorbed or taken up by **endocytosis**.
- The proximal tubule actively secretes a number of **organic acids**, **toxins**, and **drugs** using **organic anion** or **cation transporters** or **multidrug resistance proteins**.
- Phosphate is recovered from the proximal tubule (PT) by **Na^+-phosphate cotransporters**. Reabsorption is regulated by **parathyroid hormone**. Ca^{2+} reabsorption by the PT occurs paracellularly.
- Mg^{2+} reabsorption by the proximal tubule is minimal (~15% of filtered load) and occurs paracellularly.
- Approximately 80% of the filtered K^+ load is recovered in the proximal tubule.
- The lungs and kidneys together are responsible for maintaining the pH of extracellular fluids within a narrow range (pH 7.35–7.45). Lungs excrete the daily load of **volatile acids** (CO_2) generated during metabolism. Kidneys excrete **nonvolatile acids** (sulphuric, phosphoric, nitric, and other minor acids).
- **pH homeostasis** begins in the proximal tubule (PT) with recovery of 80% of filtered HCO_3^-, the body's primary pH buffer. Excretion of nonvolatile acid requires that buffers be excreted also to control luminal free H^+ concentration. The primary **urinary buffers** are **phosphate** and **ammonium**, the latter newly synthesized from glutamine in the PT.
- Na^+ reabsorption by the proximal tubule (PT) is driven by the basolateral Na^+-K^+ ATPase through cotransport with organic solutes and in exchange for H^+. Cl^- absorption occurs principally in the late PT by the paracellular route or by a Cl^--base exchanger. Water reabsorption occurs by osmosis, driven by influx of Na^+, Cl^-, and solutes.

27 Urine Formation

I. OVERVIEW

The fluid leaving the **proximal tubule** (**PT**) and entering the **loop of Henle** (**nephron loop**) has been stripped of almost all useful organic molecules, such as glucose, amino acids, and organic acids. The residual fluid (~60 L/day) comprises water, inorganic ions, and excretory products. The function of the loop and distal nephron segments is to recover remaining useful components (principally water and inorganic ions) before the fluid reaches the bladder and is excreted as urine. The amount of fluid and electrolytes recovered is determined by homeostatic needs and is heavily regulated (see Chapter 28; the main sites of water and solute recovery and regulation are summarized in Figure 27.19). The first step is to begin extracting water. One way to achieve this might be to pump water out of the tubule, much as one might bail a waterlogged boat. Nature has yet to devise the cellular equivalent of a bilge pump, however, so, as an alternative, the tubule contents are forced to run an osmotic gauntlet created within the renal medulla expressly for the purpose of extracting water from the tubule lumen. The tubule contents are exposed to the osmotic torments of the medulla twice before finally being deposited in the bladder. The first trip involves passage around the loop of Henle.

Note that the corticopapillary osmotic gradient is established by juxtamedullary nephrons alone (see Figure 25.5). Superficial nephrons do not participate and are not considered further in this chapter.

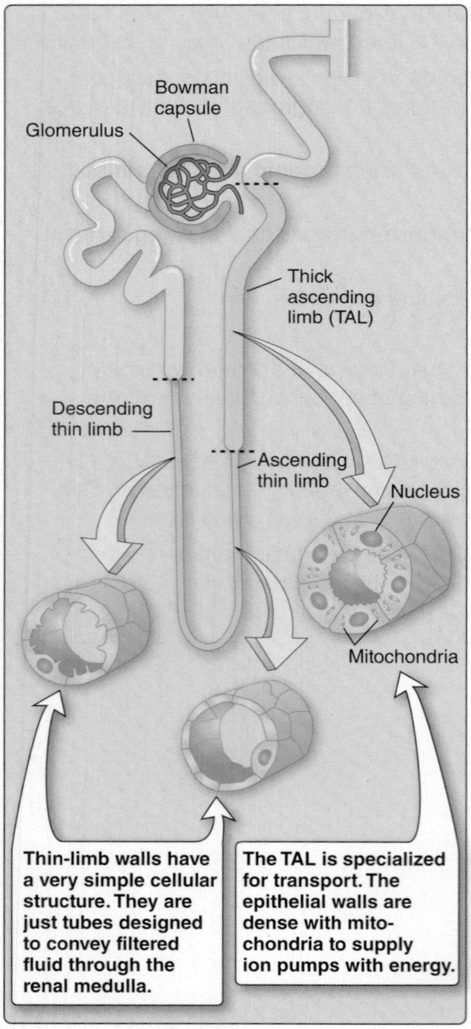

Figure 27.1
Loop of Henle tubule structure.

II. LOOP OF HENLE

The loop of Henle comprises three sections: a **descending thin limb** (**DTL**), an **ascending thin limb** (**ATL**), and a **thick ascending limb** (**TAL**) as shown in Figure 27.1. Thin limb function is very simple: It conveys fluid down through the inner reaches of the medulla and exposes it to the corticopapillary osmotic gradient (see 25·II·B). Water and solutes exit and reenter passively during the fluid's passage. The ATL transitions gradually at the inner–outer medullary junction to become the TAL. The increasing wall thickness reflects an abundance of mitochondria and other cellular machinery required to support the activity of numerous ion pumps. The TAL establishes the corticopapillary gradient.

A. Thin limbs

The proximal tubule ends abruptly at the border between outer and inner stripes of the outer medulla. The DTL and ATL are both com-

posed of thin epithelial cells with a few stubby microvilli. Adjacent cells are extensively coupled by wide tight junctions. Water and solutes move across the tubule wall (transcellularly and paracellularly) passively, driven by a pronounced interstitial corticopapillary osmotic gradient, although selectivity of passage is regulated and changes from one section to the next.

1. **Corticopapillary gradient:** The osmotic gradient is established within the medullary interstitium by a **countercurrent multiplication** mechanism, described below. Cortical osmolality approximates that of plasma (~290–300 mOsm/kg) but increases progressively with distance toward the papillary tips (Figure 27.2). The gradient's magnitude varies according to the body's need to conserve or excrete water (**diuresis**). When water conservation is necessary, papillary tip osmolality may increase to ~1,200 mOsm/kg, whereas during hypervolemic conditions, tip osmolality may be closer to 600 mOsm/kg.
2. **Descending thin limb:** The DTL is relatively impermeable to urea and Na^+, but the epithelial cell membranes contain aquaporins (AQPs) that allow free passage of water. Water exits the tubule by osmosis as the fluid is carried deeper into the medulla, causing luminal Na^+ and Cl^- to become progressively more concentrated. The DTL reabsorbs ~27 L of water per day, or 15% of glomerular filtrate.
3. **Ascending thin limb:** The tubule epithelium transitions at the turn of the loop from being water permeable to water impermeant (the ATL does not express AQPs), which prevents further water movement until the tubule contents reach the collecting ducts ([CDs] see Figure 27.2B). ATL epithelial cells *are* permeable to Cl^-, however. Cl^- leaves the tubule lumen during fluid's passage back up to the cortex, driven by the transepithelial electrochemical gradient. Na^+ follows Cl^- paracellularly.

> Forcing fluid around the loop of Henle extracts water but does not increase its osmolality because solutes are extracted also. Urine only becomes concentrated when exposed to the corticopapillary gradient a second time during passage through the CDs.

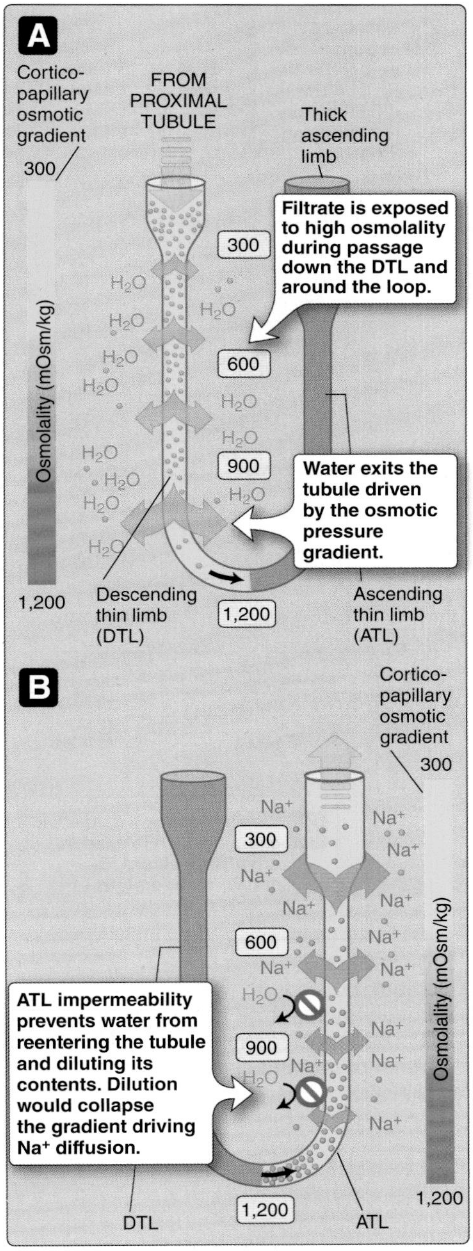

Figure 27.2
Water and Na^+ reabsorption in the loop of Henle.

B. Thick ascending limb

The TAL actively recovers significant amounts of Na^+, Cl^-, K^+, Ca^{2+}, and Mg^{2+} from the tubule lumen (summarized in Figure 27.19).

1. **Sodium, chloride, and potassium:** The TAL reabsorbs ~25% of the filtered load of Na^+ and Cl^- and 10% of the K^+ load. Reabsorption occurs via both transcellular and paracellular routes and is so efficient that it leaves the tubule contents hyposmotic relative to plasma, even though there has been no net water movement. The TAL is sometimes referred to as the **diluting segment** for this reason.
   a. **Transcellular:** Reabsorption is facilitated by the transmembrane Na^+ gradient generated by the basolateral Na^+-K^+ ATPase. Ion reabsorption can be broken down into several steps (Figure 27.3).

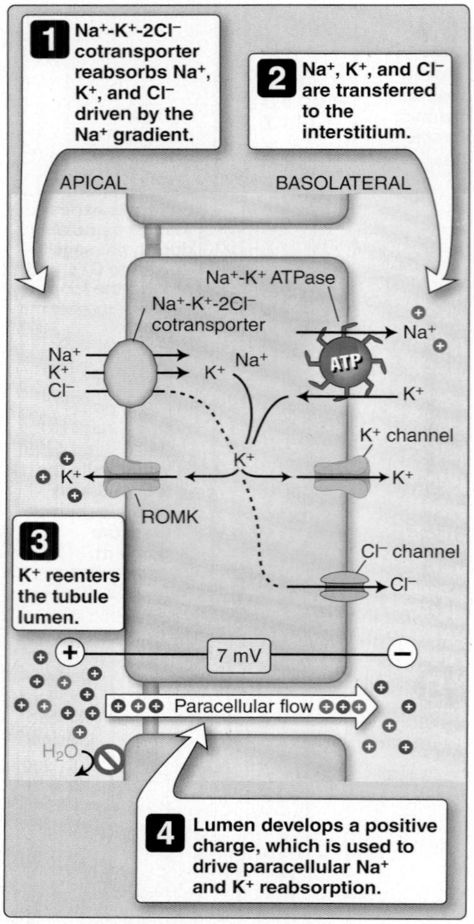

Figure 27.3
Sodium, potassium, and chloride recovery by the thick ascending limb. ATP = adenosine triphosphate; ROMK = renal outer medullary K^+ channel.

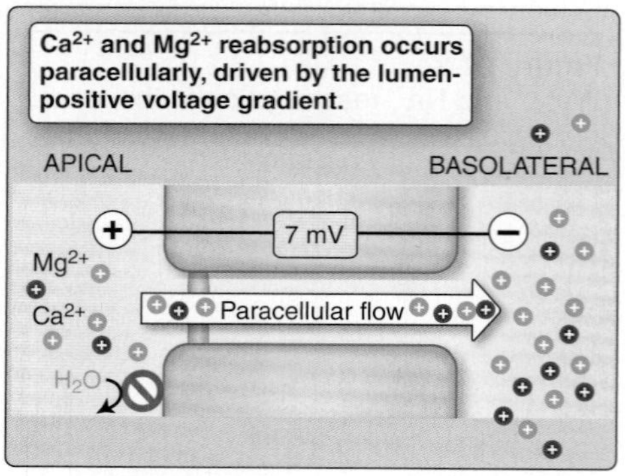

Figure 27.4
Calcium and magnesium recovery by the thick ascending limb.

Clinical Application 27.1: Loop Diuretics

Physiologic regulation and fine-tuning of urine composition occurs in segments distal to the thick ascending limb (TAL), but drugs that inhibit the Na^+-K^+-$2Cl^-$ cotransporter have proved to be very powerful clinical tools for treating edema. As a class, these drugs are known as **loop diuretics** and include furosemide (Lasix is one common brand), bumetanide, ethacrynic acid, and torsemide.[1] Inhibiting the cotransporter prevents reabsorption of Na^+, Cl^-, and K^+ directly and indirectly prevents reabsorption of water. Inhibition also prevents a positive charge developing within the tubule lumen and, thereby, reduces Ca^{2+} and Mg^{2+} reabsorption. Segments distal to the TAL do not have the ability to compensate for loss of cotransporter function, so the loop diuretics all cause copious urine formation. Although reduced salt and water retention effectively reduces circulating blood volume and helps prevent edema, the concomitant loss of K^+ and Mg^{2+} to urine can cause hypokalemia and hypomagnesemia.

i. **Sodium, potassium, and chloride reabsorption:** An apical Na^+-K^+-$2Cl^-$ cotransporter facilitates Na^+, K^+, and Cl^- reabsorption. Na^+ is pumped out of the cell by the Na^+-K^+ ATPase, whereas K^+ and Cl^- flow into the interstitium down their respective electrochemical gradients via basolateral Cl^- and K^+ channels.

ii. **Potassium secretion:** The apical membrane also contains a renal outer medullary K^+ channel (ROMK), which allows K^+ to cross back to the tubule lumen. This pathway is necessary to prevent luminal K^+ depletion, an event which would grind Na^+-K^+-$2Cl^-$ cotransport to a halt.

b. **Paracellular:** K^+ secretion creates an ~7 mV electrical gradient between tubule lumen and interstitium that drives paracellular Na^+ and K^+ reabsorption.

2. **Calcium and magnesium:** The TAL reabsorbs ~25% of the filtered Ca^{2+} load and ~65%–70% of filtered Mg^{2+}. Most of this reabsorption occurs paracellularly (see Clinical Application 4.2) and is driven by the voltage difference between the tubule lumen and the interstitium (Figure 27.4).

3. **Bicarbonate and acid:** Fluid leaving the PT still contains ~20% of the filtered HCO_3^- load. Virtually all of this is recovered, either in the TAL or the distal segments.

a. **Bicarbonate:** HCO_3^- is reabsorbed using the same strategies seen in the PT (see Figure 26.13). *Carbonic anhydrase* (*CA*) facilitates H^+ and HCO_3^- formation from H_2O and CO_2. H^+ is pumped across the apical membrane by a H^+-ATPase

[1]For more information on the mechanism of action and use of loop diuretics, see *LIR Pharmacology*, 5e, p. 284.

and a Na^+-H^+ exchanger, where it combines with filtered HCO_3^- to form CO_2 and H_2O, the reaction again catalyzed by *CA*. HCO_3^- is reabsorbed across the basolateral membrane in exchange for Cl^- and via a Na^+-HCO_3^- cotransporter.

b. **Acid:** The PT generates NH_3 as a way of excreting H^+ in the form of NH_4^+ (see Figure 26.16). The TAL reabsorbs a portion of the NH_4^+ via an apical Na^+-K^+-$2Cl^-$ cotransporter (NH_4^+ substitutes for K^+) and then transfers it to the interstitium, where, like Na^+, it aids formation of the corticopapillary osmotic gradient through countercurrent multiplication.

C. Corticopapillary osmotic gradient

The loop diuretics are effective because they collapse the corticopapillary osmotic gradient that is used to draw water from the DTL and to later concentrate urine during its passage through the CDs. The gradient is established by the TAL, but it affects all vessels traveling through the medulla.

1. **Tubule arrangement:** Textbook figures (e.g., see Figure 25.5) traditionally separate the various nephron segments across the width of a page to make labeling easier, but, in real life, the DTL, ATL, CDs, and vasa recta are all bundled together like a handful of drinking straws (Figure 27.5). The interstitial space between them is minimal, so the interstitium and tubule contents (i.e., filtrate, urine, blood) are generally in osmotic equilibrium. Changes in one compartment affect the others almost instantaneously. The fact that some tubules (e.g., DTLs) carry fluid down toward the papilla at the same time as other tubules (e.g., ATLs) in the bundle carry fluid back up to the cortex allows for amplification of an osmotic difference between tubule lumen and interstitium generated by TAL epithelial cells.

2. **Countercurrent multiplication:** The corticopapillary gradient is easiest to understand when broken down into a series of theoretical steps. Prior to multiplication, tubule contents and interstitium are all assumed to be in equilibrium at 300 mOsm/kg (Figure 27.6[1]).

   a. **Single effect:** The TAL reabsorbs Na^+ from the tubule via the Na^+-K^+-$2Cl^-$ cotransporter and transfers it to the interstitium using the basolateral Na^+-K^+ ATPase. This transfer generates a maximal 200-mOsm/kg osmolality difference between the tubule lumen and interstitium (see Figure 27.6[2]). Thus, if interstitial and tubule osmolality are both initially at 300 mOsm/kg, Na^+ reabsorption causes lumen osmolality to fall to 200 mOsm/kg and interstitial osmolality to rise to 400 mOsm/kg. The DTL, which lies next to the TAL, is filled with fluid arriving from the PT, which has an osmolality of 300 mOsm/kg. Because the DTL is highly water permeable, water is drawn from its lumen by the osmotic pressure gradient until it equilibrates with the interstitium at 400 mOsm/kg (the water is subsequently carried away by the peritubular vasculature). This same phenomenon occurs simultaneously down the length of the TAL and DTL and is known as a "**single effect**."

   b. **Fluid displacement:** Fluid continues arriving at the DTL from the PT, displacing the 400-mOsm/kg fluid downward and

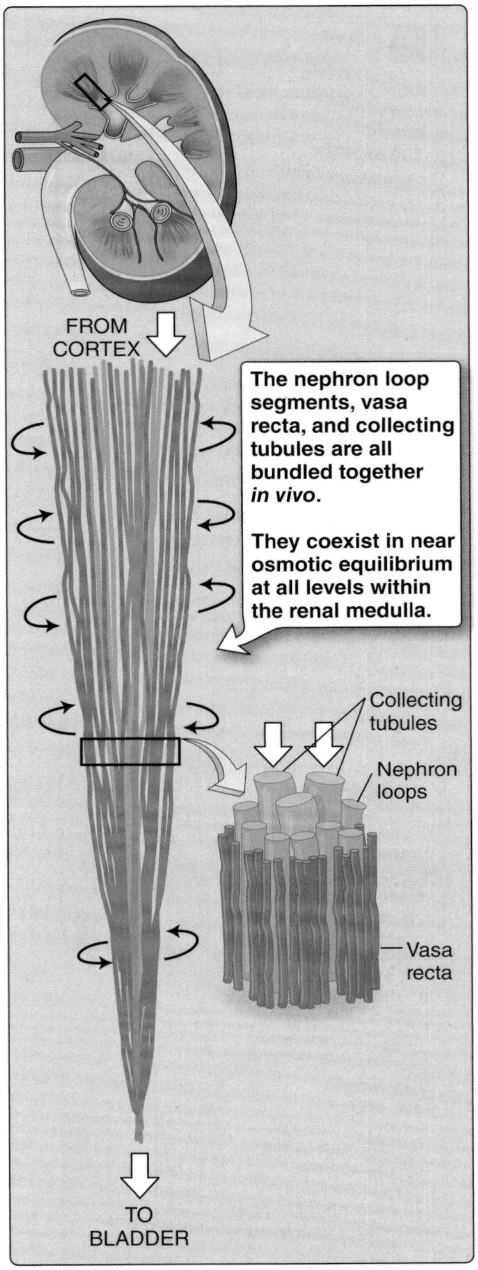
FROM
CORTEX
The nephron loop segments, vasa recta, and collecting tubules are all bundled together *in vivo*.
They coexist in near osmotic equilibrium at all levels within the renal medulla.
Collecting tubules
Nephron loops
Vasa recta
TO
BLADDER

Figure 27.5
Arrangement of tubule segments and vasa recta in renal medulla.

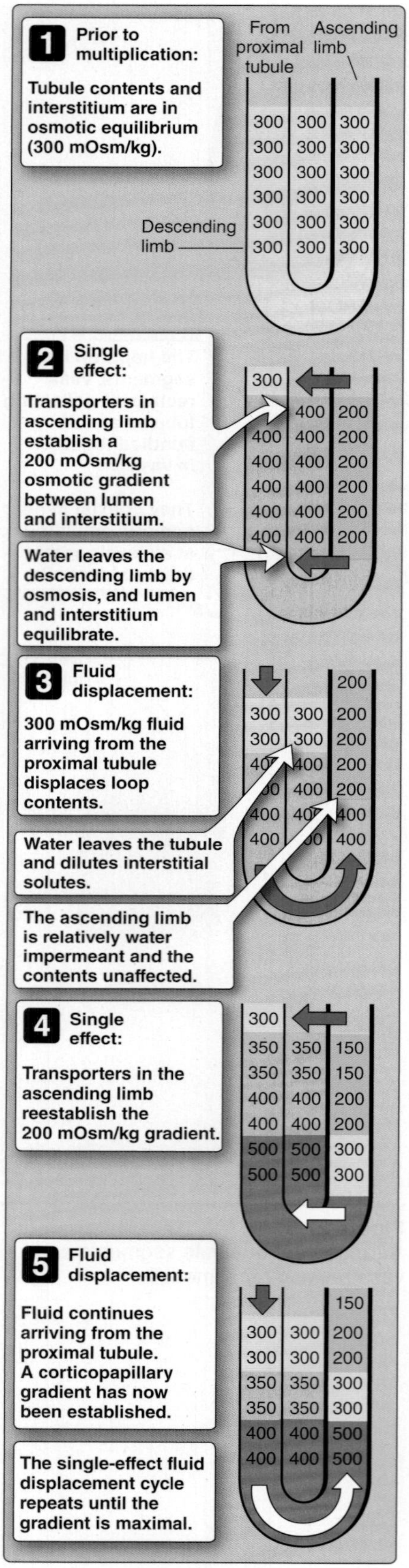

Figure 27.6
Countercurrent multiplication in the loop of Henle.

around the loop tip (see Figure 27.6[3]). The interstitium at the corticomedullary junction re-equilibrates at 300 mOsm/kg. The TAL is water impermeable, so the fluid within remains at 200 mOsm/kg. At the loop tip, the two limbs and interstitium remain equilibrated at 400 mOsm/kg.

c. **Single effect:** TAL cells continue transferring Na^+ from tubule lumen to interstitium, but lumen osmolality begins this cycle at 200 mOsm/kg (see Figure 27.6[4]). Na^+ reabsorption reestablishes the 200-mOsm/kg gradient across the tubule wall, causing lumen osmolality to fall to 150 mOsm/kg and interstitial osmolality to rise to 350 mOsm/kg. Further down the TAL toward the medulla, lumen osmolality drops from 400 mOsm/kg to 300 mOsm/kg, and interstitial osmolality rises to 500 mOsm/kg. Even after only two conceptual cycles, a corticopapillary gradient has begun to form. Each cycle multiplies the gradient further.

d. **Fluid displacement:** Fluid with an osmolality of 300 mOsm/kg continues to arrive at the DTL from the PT, decreasing the local interstitial osmolality and pushing high-osmolality fluid around the tip of the loop (see Figure 27.6[5]). The next transport cycle reduces the tubule osmolality at the top of the TAL and further increases osmolality at the tip.

3. **Urea:** Countercurrent multiplication eventually generates a papillary osmolality of 600 mOsm/kg, but this can rise to 1,200 mOsm/kg when water must be conserved. The gradient's magnitude determines how much water can be reclaimed from the filtrate and is regulated according to prevailing needs. Achieving a 1,200 mOsm/kg gradient is only possible with assistance from urea. When water conservation is necessary, the CDs allow urea to pass from the duct lumen to the medulla, further enhancing its osmolality and water reabsorptive capabilities. The pathways involved are described in Section V below.

> The corticopapillary osmotic gradient is created by juxtamedullary nephrons, which represent a relatively small proportion of total nephron number (~10%). The remaining ~90% are superficial and have short loops, which limits the maximal degree to which urine can be concentrated. Desert rodents such as the Australian hopping mouse (Genus *Notomys*; Figure 27.7) can produce ~10,000 mOsm/kg urine. Their kidneys contain a much higher proportion of juxtamedullary nephrons compared with superficial nephrons, and concentrating ability is increased accordingly. This remarkable ability to conserve fluid means that hopping mice are able to subsist on water extracted from their food (e.g., roots, leaves, and berries) and never need to drink, which has clear survival advantages in an arid environment.

D. Vasa recta

The nephron loops require extensive vascular support, not only to supply O_2 and nutrients, but also to carry away reabsorbed water and electrolytes. Because plasma has an osmolality of ~300 mOsm/kg and capillaries are inherently leaky vessels, there is a danger that blood entering the medulla could wash out the corticopapillary osmotic gradient and, thereby, prevent urine concentration. Washout is largely prevented by two important features of the vasa recta (Latin for "straight vessels"). Flow rate is slow, and the vessels form a hairpin loop that creates a **countercurrent exchange system** (Figure 27.8).

1. **Flow rate:** The medulla receives <10% of total renal blood flow. The vasa recta has a high intrinsic resistance due to its length, which keeps flow to a nutritional minimum. The slow flow rate allows for near-complete equilibration of water and solutes as the blood is carried through the medulla.
2. **Countercurrent exchange:** The vasa recta is intimately associated with the nephron loop and the CDs, closely paralleling the DTL down to the papilla and then back up to the cortex alongside the ATL and TAL (see Figures 25.4 and 27.5). The blood vessels are leaky, so water leaves, and solutes enter passively, which maintains an osmotic equilibrium between blood and interstitium (see Figure 27.8). On the way back up to the cortex, water reenters the blood vessels, and solutes leave passively. Therefore, blood flow through the vasa recta has minimal net effect on the corticopapillary gradient when perfusion rates are low.

Figure 27.7
Australian hopping mouse.

III. EARLY DISTAL TUBULE

The transition from TAL to distal convoluted tubule (DCT) is marked by a fivefold increase in tubule wall thickness. The epithelial cells are filled with platelike structures packed with mitochondria. The apical surface bears slender microvilli, and the basolateral membrane is folded, both modifications designed to increase surface area. These anatomical features all point to the early DCT as being the site of active solute reabsorption. The DCT has very low water permeability and is the principal site for homeostatic regulation of Mg^{2+} and Ca^{2+}.

A. Sodium and chloride

The early DCT reabsorbs only a small fraction of the Na^+ and Cl^- filtered load, primarily via an apical Na^+-Cl^- cotransporter. Na^+ is then pumped out of the cell to the interstitium by the basolateral Na^+-K^+ ATPase, whereas Cl^- exits via a Cl^- channel. The DCT is impermeable to water, so extracting NaCl from the tubule lumen further dilutes its contents.

B. Magnesium

By the time tubule fluid reaches the DCT, 85% of filtered Mg^{2+} load has been reabsorbed, principally in the TAL. The DCT is the only segment that recovers Mg^{2+} in a regulated fashion, the amount recovered reflecting homeostatic needs. There are no further opportunities for recovery once Mg^{2+} leaves the DCT. Mg^{2+} is reabsorbed from the tubule lumen via TRPM6 (a member of the transient receptor potential superfamily; see 2·VI·D), which is expressed in the early part of the

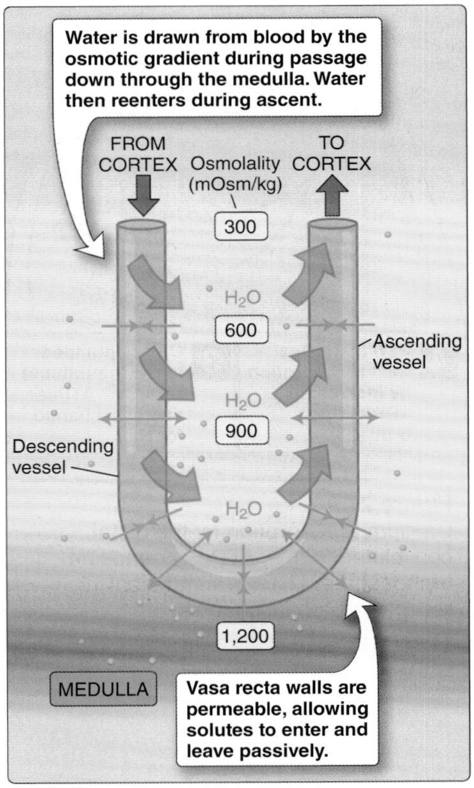

Figure 27.8
Countercurrent exchange in the vasa recta.

Clinical Application 27.2: Thiazide Diuretics

Thiazide diuretics (e.g., hydrochlorothiazide) inhibit Na^+ reabsorption by the distal convoluted tubule (DCT) Na^+-Cl^- cotransporter. The DCT reabsorbs relatively small amounts of NaCl, so thiazide diuresis is of limited help in reducing edema, although the DCT does help determine the final plasma Na^+ content, which, in turn, helps determine blood pressure. Thiazides are, therefore, useful for treating hypertension. Inhibiting Na^+ influx causes DCT epithelial cell hyperpolarization, which increases the electrochemical gradient driving Ca^{2+} reabsorption. Thiazide diuresis sometimes causes hypercalcemia for this reason.[1]

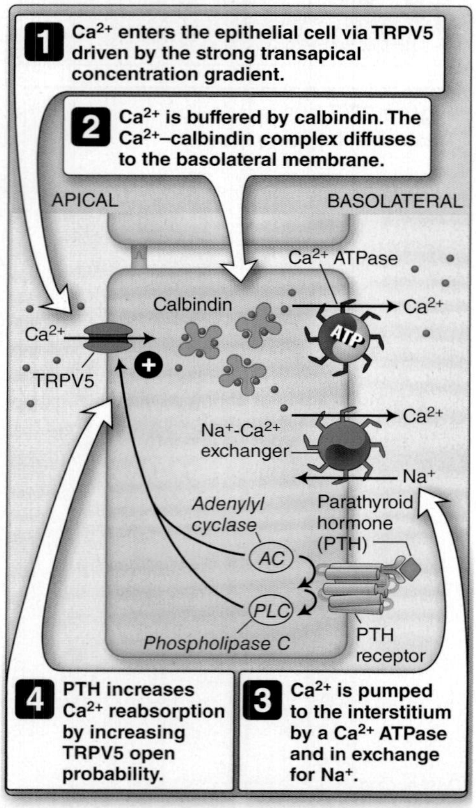

1 Ca2+ enters the epithelial cell via TRPV5 driven by the strong transapical concentration gradient.
2 Ca2+ is buffered by calbindin. The Ca2+–calbindin complex diffuses to the basolateral membrane.
APICAL
BASOLATERAL
Ca2+ ATPase
Calbindin
Ca2+
Ca2+
TRPV5
ATP
Ca2+
Na+-Ca2+ exchanger
Na+
Adenylyl cyclase
AC
PLC
Parathyroid hormone (PTH)
Phospholipase C
PTH receptor
4 PTH increases Ca2+ reabsorption by increasing TRPV5 open probability.
3 Ca2+ is pumped to the interstitium by a Ca2+ ATPase and in exchange for Na+.

Figure 27.9
Calcium reabsorption by the distal convoluted tubule. ATP = adenosine triphosphate; TRPV5 = transient receptor-potential channel.

DCT. Mg^{2+} recovery is regulated by epidermal growth factor through increased TRPM6 activity. Influx is passive, driven by the electrochemical gradient across the apical membrane. The means by which Mg^{2+} crosses the basolateral membrane is uncertain at present.

C. Calcium

The late DCT reabsorbs ~8% of the filtered Ca^{2+} load. Reabsorption occurs passively via an apical membrane channel, but net uptake is regulated by parathyroid hormone (PTH) as shown in Figure 27.9.

1. **Apical:** Ca^{2+} crosses the apical membrane via TRPV5, another TRP family channel member, reabsorption being powered by the electrochemical gradient for Ca^{2+}. All cells need to maintain a very low intracellular Ca^{2+} concentration (see 1·II) and could easily be overwhelmed by the amount of Ca^{2+} crossing the apical membrane. Therefore, DCT epithelial cells contain large amounts of high-affinity Ca^{2+}-binding protein (**calbindin**) that buffers Ca^{2+} influx until it can be pumped across the basolateral membrane. Intracellular buffering also maintains a steep electrochemical gradient favoring Ca^{2+} reabsorption from the tubule lumen.

2. **Basolateral:** Interstitial Ca^{2+} concentrations are ~10,000 times higher than intracellular concentrations, so Ca^{2+} has to be actively pumped out of the epithelial cell by a basolateral Ca^{2+} ATPase. The Ca^{2+} ATPase functions much like a sump pump. When intracellular Ca^{2+} levels are rising, its activity increases, and the excess is deposited into the interstitium. The basolateral membrane also contains a Na^+-Ca^{2+} exchanger that helps support Ca^{2+}-pump activity when intracellular Ca^{2+} concentrations are high (see Figure 27.9).

3. **Regulation:** Ca^{2+} reabsorption is regulated by PTH. When plasma Ca^{2+} concentrations are suboptimal, PTH is released into the circulation from the parathyroid glands (see 35·V·B). PTH binds to a G protein–coupled receptor (GPCR) on the basolateral membrane of DCT cells, which activates both *adenylyl cyclase* (*AC*) and *phospholipase C*–signaling pathways. Both increase

[1]For more information on the mechanism of action and use of thiazide diuretics, see *LIR Pharmacology*, 5e, p. 281.

TRPV5 open probability and Ca^{2+} reabsorption. PTH effects on Ca^{2+} reabsorption are potentiated by vitamin D, which increases the expression of most (perhaps all) of the proteins involved in Ca^{2+} transport, including calbindin.

IV. DISTAL SEGMENTS

The late DCT, connecting tubule (CNT), and cortical collecting duct (CCD) have similar structures and functions and are referred to collectively as the **distal segments** (Figure 27.10). These segments are notable for their **intercalated cells**, which comprise ~20%–30% of the tubule epithelium. Intercalated cells secrete acid. The other 70%–80% of the epithelium comprises Na^+-reabsorbing cells. In the CCD, these cells are known as **principal cells**.

A. Epithelial structure

The late DCT is the most distal portion of the renal nephron. The CNT connects the DCT to the CD system and, ultimately, the renal pelvis. Several CNTs fuse before joining a CCD, each CCD draining ~11 nephrons. In these regions, cells filled with mitochondria characterize the epithelial wall, whose basolateral membrane is amplified by extensive infoldings.

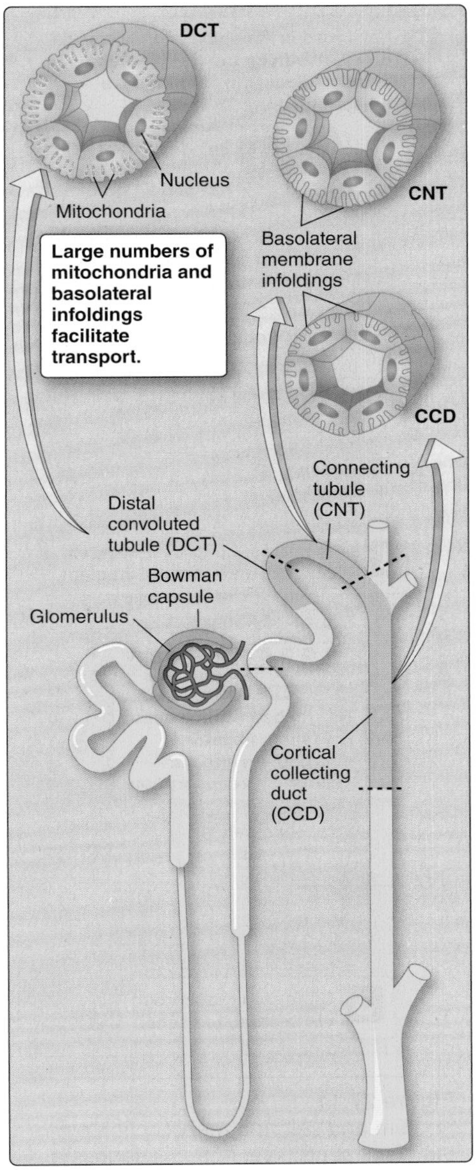

Figure 27.10
Renal tubule distal segments.

Clinical Application 27.3: Potassium-Sparing Diuretics

The distal segments are the site of action of two general classes of drugs that promote Na^+ and water excretion while simultaneously promoting K^+ retention, which is why they are referred to as "K^+-sparing" diuretics.[1] Because very little of the original Na^+ load remains by the time filtrate reaches the distal segments, these drugs have limited natriuretic effect. They are usually used in combination with loop or thiazide diuretics to limit K^+ loss. One class of K^+-sparing diuretics inhibits the epithelial Na^+ channel (ENaC), and the other inhibits aldosterone binding to the mineralocorticoid receptor (MR).

Amiloride and triamterene both inhibit ENaC and prevent Na^+ reabsorption by principal cells. Na^+ remains in the tubule and acts as an osmotic diuretic. Reducing the amount of Na^+ entering principal cells secondarily decreases Na^+-K^+ ATPase activity and, consequently, reduces K^+ uptake and subsequent secretion.

Spironolactone and eplerenone are competitive inhibitors of aldosterone binding to the MR. They act by reducing aldosterone-stimulated increases in ENaC, Na^+-K^+ ATPase, and K^+-channel expression. The net result is a decrease in Na^+ reabsorption and a decrease in K^+ secretion.

INFO LINK [1]For more information on the mechanism of action and use of K^+-sparing diuretics, see *LIR Pharmacology*, 5e, p. 286.

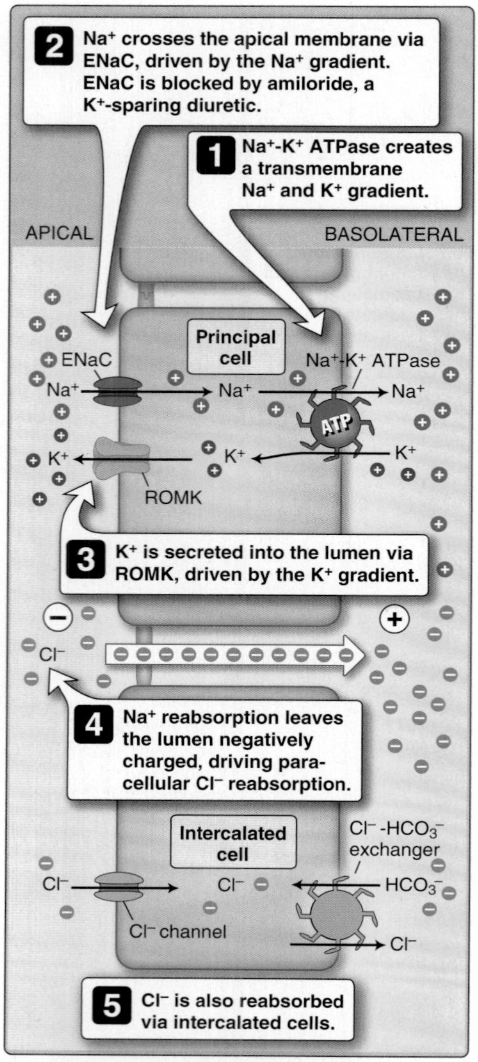

Figure 27.11
Sodium and chloride reabsorption by the distal segments. ATP = adenosine triphosphate; ENaC = epithelial Na^+ channel; ROMK = renal outer medullary K^+ channel.

B. Sodium and chloride

The fluid arriving at the late DCT is relatively dilute and contains low concentrations of Na^+ and Cl^-. The distal segments together reabsorb only ~5% of the NaCl filtered load, but this is the principal site for regulation by hormones concerned with extracellular fluid Na^+ homeostasis and, thus, is one of the more critical stages of recovery (see 28·III·C).

1. **Pathways:** Na^+ and Cl^- reabsorption occurs transcellularly, driven by the transmembrane Na^+ gradient established by the basolateral Na^+-K^+ ATPase. The late DCT expresses the same thiazide-sensitive Na^+-Cl^- cotransporter noted in Section III·A above, but the predominant pathway for Na^+ reabsorption in the distal segments is via an **epithelial Na^+ channel** (**ENaC**), which appears in the late DCT (Figure 27.11). Na^+ crossing the apical membrane via ENaC leaves the tubule lumen very negative. K^+ reentry via an apical ROMK partly offsets the charge, but, even so, the tubule lumen rests at around −40 mV compared with blood. This creates a strong driving force for paracellular Cl^- reabsorption. Cl^- is also recovered transcellularly via α-intercalated cells. An apical Cl^- channel allows influx from the tubule lumen, and the ion then crosses to the interstitium via a Cl^--HCO_3^- exchanger.

2. **Regulation:** Na^+ recovery by principal cells is regulated by aldosterone (Figure 27.12). Aldosterone is released from the adrenal cortex in response to angiotensin II (Ang-II) or an increase in plasma K^+ concentrations (hyperkalemia). Ang-II is formed during activation of the ***renin–angiotensin–aldosterone system*** (**RAAS**) when blood pressure and renal blood flow is low (see 20·IV·C). Aldosterone binds to a basolateral mineralocorticoid receptor (MR) and then is internalized and translocated to the nucleus, where it upregulates transcription and expression of numerous proteins involved in Na^+ reabsorption and K^+ secretion (see below). These proteins include ENaC, ROMK, and the Na^+-K^+ ATPase. Aldosterone also stimulates basolateral membrane elaboration to increase its surface area and facilitate an increase in Na^+-K^+ pumping capacity. Synthesis of new channel and transporter subunits is relatively slow, requiring ~6 hr to implement, but aldosterone also has short-term effects mediated by a *serum- and glucocorticoid-activated kinase* (*SGK*). *SGK* increases apical Na^+ permeability by reducing ENaC turnover rates and increasing basolateral Na^+-K^+ ATPase activity.

C. Potassium

Hyper- and hypokalemia both adversely affect cardiac excitability and function (see Clinical Application 2.1), so the kidneys must excrete K^+ when dietary intake exceeds homeostatic needs and conserve K^+ when dietary intake is limited. Plasma K^+ concentration is determined in the distal nephron segments and in the outer medullary collecting duct (OMCD).

1. **Secretion:** K^+ is secreted and excreted by principal cells using the same pathways that reabsorb Na^+ (see Figure 27.11). K^+

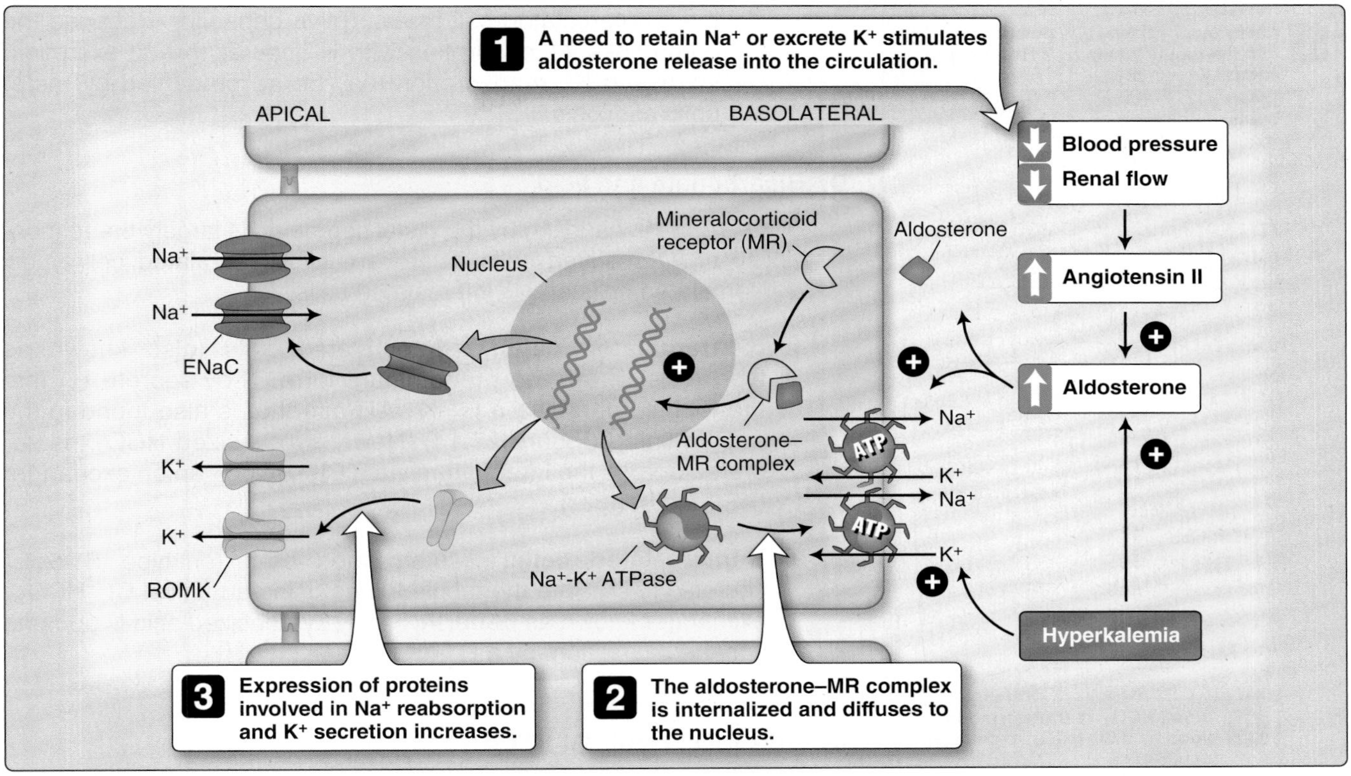

Figure 27.12
Aldosterone regulation of sodium reabsorption and potassium secretion by principal cells in the distal segments. ATP = adenosine triphosphate; ENaC = epithelial Na^+ channel; ROMK = renal outer medullary K^+ channel.

is taken up from blood by the basolateral Na^+-K^+ ATPase and transferred to the tubule lumen via apical ROMK. Secretion is favored both by a high intracellular K^+ concentration and by the net negative charge within the tubule lumen (see Figure 27.11). Hyperkalemia promotes K^+ secretion directly by increasing basolateral Na^+-K^+ ATPase activity. Hyperkalemia is also a potent stimulus for aldosterone release from the adrenal cortex. Aldosterone increases expression of proteins involved in Na^+ reabsorption and K^+ secretion, as shown in Figure 27.12.

2. **Reabsorption:** K^+ reabsorption relies on α-intercalated cells, which express a H^+-K^+ ATPase on their apical membrane (Figure 27.13). The ATPase pumps H^+ into the tubule lumen in exchange for K^+, which subsequently exits the cell via basolateral K^+ channels. K^+ reabsorption increases during hypokalemia and involves regulation of both principal cells and α-intercalated cells.

   a. **Principal cells:** Hypokalemia decreases circulating aldosterone levels, thereby reducing expression of proteins involved in K^+ secretion. Hypokalemia also reduces K^+ uptake by principal cells through direct effects on basolateral Na^+-K^+ ATPase activity (see Figure 27.12).

   b. **α-Intercalated cells:** Hypokalemia upregulates H^+-K^+ ATPase numbers in the apical membrane, which increases

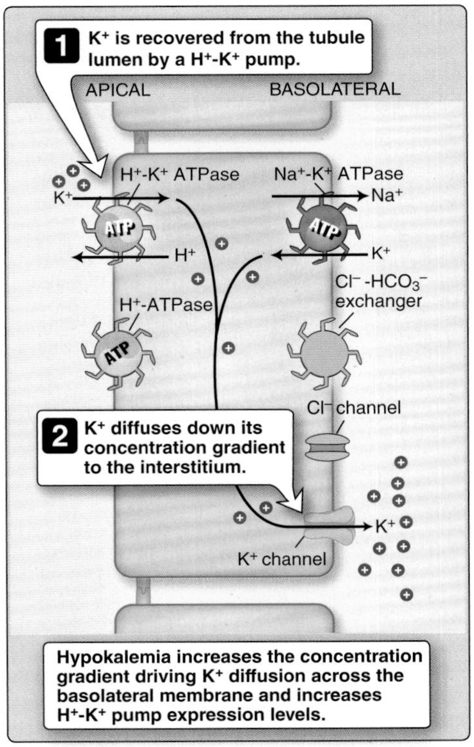

Figure 27.13
Potassium reabsorption by α-intercalated cells in the distal segments. ATP = adenosine triphosphate.

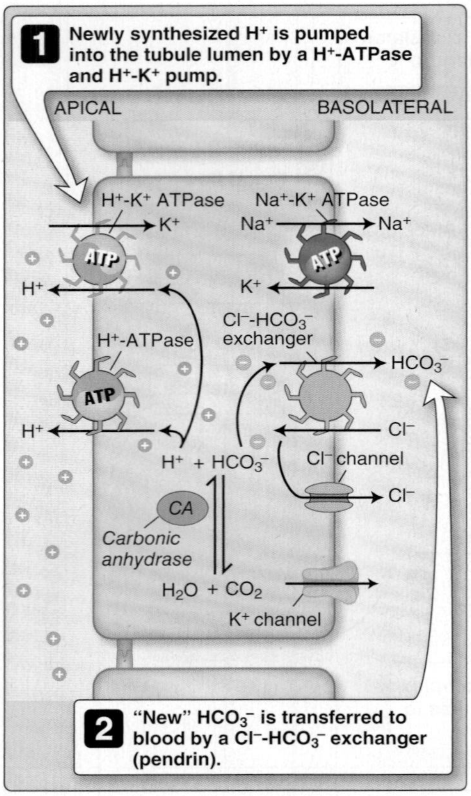

Figure 27.14
Acid secretion by α-intercalated cells in the distal segments. ATP = adenosine triphosphate.

the α-intercalated-cell reabsorptive capability. Because the pump links K^+ absorption with H^+ secretion and excretion, increased K^+ reabsorption may be accompanied by metabolic alkalosis.

D. Bicarbonate and acid

HCO_3^- and H^+ handling by the distal segments is largely the responsibility of intercalated cells. There are two intercalated cell types: α-intercalated cells and β-intercalated cells.

1. **α-Intercalated cells:** α-Intercalated cells (also known as type A cells) are the predominant form. They secrete H^+ into the tubule lumen via a H^+-K^+ ATPase that is also found in the gastric lining (Figure 27.14). Newly synthesized HCO_3^- is secreted across the interstitium via a Cl^--HCO_3^- anion exchanger (AE1 exchanger).

2. **β-Intercalated cells:** β-Intercalated cells (or type B cells) secrete HCO_3^- into the tubule lumen using an apical Cl^--HCO_3^- exchanger known as **pendrin**. Newly synthesized acid is pumped into the interstitium by a H^+-K^+ ATPase.

V. COLLECTING DUCTS

The fluid that enters the CD system has been denuded of all valuable solutes and is very dilute (~50 mOsm/kg) compared with the surrounding cortex (~300 mOsm/kg). The fluid is now poised to once again run the corticopapillary osmotic gauntlet to extract water. If body water intake exceeds homeostatic needs, the tubule fluid will flow through the CDs to the renal sinus and bladder without further water recovery, potentially at a rate of up to 20 L/day. If water intake is limited (as is usually the case), AQPs are inserted into the CD epithelium to allow water to flow out of the ducts and back to the vasculature. The driving force for movement is the osmotic potential created by the corticopapillary gradient, which becomes ever more powerful as urine flows toward the renal sinus.

A. Epithelial structure

The OMCD is a straight, unbranching tube that passes through the outer medulla (Figure 27.15). IMCDs fuse successively toward the papillary tip, gaining diameter and increasing wall thickness with each fusion. IMCD epithelial cells bear stubby microvilli on both their apical and basolateral surfaces, and their basolateral membrane is folded extensively, consistent with their high potential reabsorptive capacity.

B. Urine volume determinants

A normal, healthy individual excretes ~1–2 L of ~300–500 mOsm/kg urine every day. There can be considerable deviation from this range depending on the amount of water ingested and the amount lost to the environment through evaporation (skin, mucous membranes,

lungs) and nonurinary excretion (i.e., feces; see 28·II·A), but there are physiologic limits to output.

1. **Maximal output:** Maximal urinary output is around 20 L/day. Although excretion rates can go higher, flow volumes in excess of 20 L/day exceed the kidneys' ability to recover Na^+ and K^+ from the tubule lumen. The results are hyponatremia and hypokalemia. Hyponatremia causes nausea, headaches, confusion, and seizures (all symptoms of cerebral edema) and, like hypokalemia (see above), can be fatal.

2. **Minimal output:** The human body generates ~600 mOsm of solutes every day that must be excreted in urine. The kidney's ability to concentrate urine is limited by the corticopapillary gradient to ~1,200 mOsm/kg, so the 600 mOsm of excreted solutes are accompanied by at least 0.5 L of water per day. If additional solutes must be excreted (e.g., as the result of eating too many salty chips), then urine volume increases accordingly.

3. **Free water clearance:** Free water clearance (C_{H2O}) is a measure of the kidney's water-handling ability. For the purposes of this discussion, dilute urine (i.e., with an osmolality less than that of plasma, or ~300 mOsm/kg) can be considered to comprise two components. The first is the volume needed to dissolve excretory solutes to a final osmolality of 300 mOsm/kg. The second is **free water**, or the amount of water in urine in excess of that required to dissolve the excreted solutes. C_{H2O} cannot be calculated directly and must be determined by measuring total urine volume and then subtracting out the amount of water needed to create an isosmotic solution from the amount of excreted osmolytes contained in urine. This latter component is measured from osmolal clearance (C_{Osm}):

$$C_{Osm} = \frac{U_{Osm} \times V}{P_{Osm}}$$

where U_{Osm} is urine osmolality, V is urine flow rate, and P_{Osm} is plasma osmolality. Free water clearance is then calculated as:

$$C_{H2O} = V - C_{Osm} = V \times \frac{(1 - U_{Osm})}{P_{Osm}}$$

A negative C_{H2O} indicates that urine is concentrated (hyperosmotic). A positive value indicates that urine is dilute (hyposmotic).

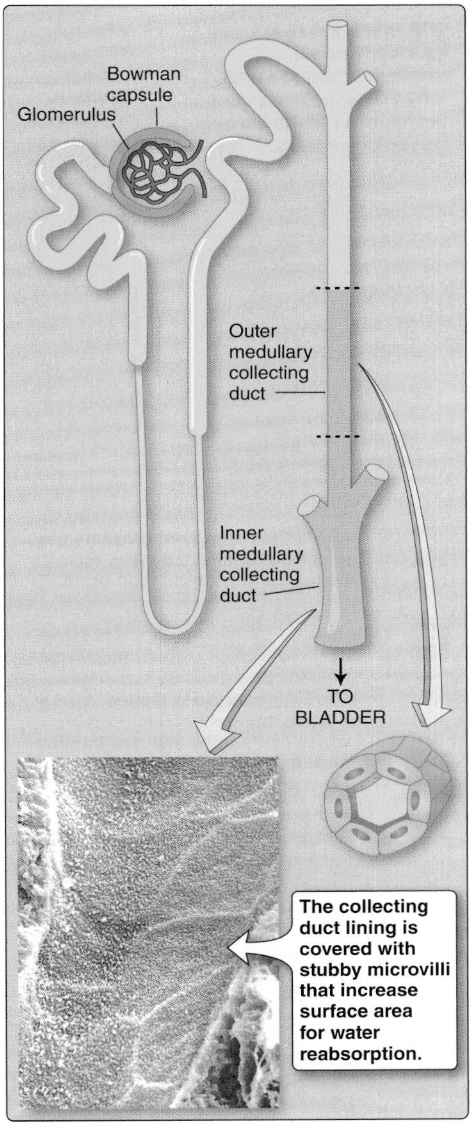

Figure 27.15
Renal medullary collecting ducts.

C. Water reabsorption

When water intake exceeds homeostatic needs, dilute urine passes through the CDs to the bladder largely unchanged, as if flowing through a cast-iron downspout. If there is a need to conserve water, virtually all of the fluid (aside from the ~0.5 L/day of obligatory loss) can be recovered. Recovery of water and final urine concentration is governed by the presence of aquaporins in the ductal epithelium and is regulated by **antidiuretic hormone** ([**ADH**] also known as **arginine vasopressin**).

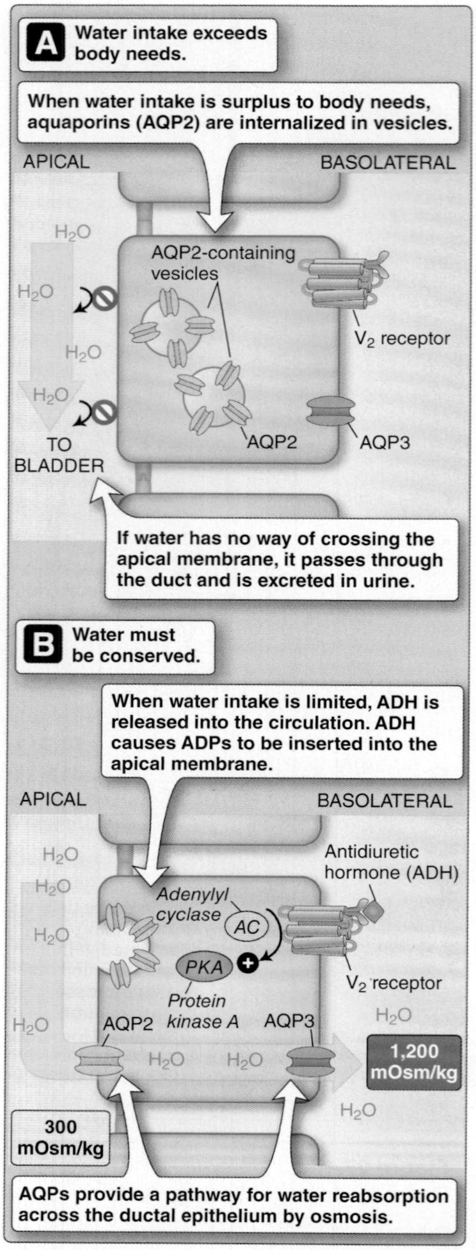

Figure 27.16
Water reabsorption by the collecting ducts. ADP = adenosine diphosphate; V_2 = vasopressin type 2.

1. **Aquaporins:** Aquaporins form pores that allow water to cross the lipid bilayer (see 1·V·A). Aquaporins are expressed constitutively in the apical and basolateral membranes of the PT and DTL, which gives these segments high water permeability. The ATL and TAL do not express aquaporins, so they are water impermeant. Principal cells in the CNT, CCD, OMCD, and IMCD all express aquaporin 2 (AQP2), but the channels are not inserted into the apical membrane until ADH binds to a basolateral vasopressin (V_2) receptor (Figure 27.16A).

2. **Antidiuretic hormone:** ADH is released from the posterior pituitary in response to an increase in plasma osmolality or a decrease in mean arterial blood pressure. The hormone is carried via the peritubular capillary network to the CD system, where it binds to ADH V_2 receptors. V_2 receptors are GPCRs, which, when occupied, activate *protein kinase A* (*PKA*) via the *AC* signaling pathway (see 1·VI·B·2). *PKA* phosphorylates intracellular trafficking proteins, causing AQP2-containing vesicles to shuttle to the cell surface and fuse with the apical membrane.

3. **Water reabsorption:** The basolateral membrane of CD principal cells also contains an aquaporin isoform (AQP3) that is not ADH dependent. AQP2 and AQP3 together provide a pathway for transcellular reabsorption of water, driven by the osmotic gradient between tubule and interstitium. Note that the fluid entering the CD system from the DCT has a lower osmolality than that of the cortex (~100 mOsm/kg, compared with ~300 mOsm/kg). This difference causes substantial amounts of water to be reabsorbed even before the fluid runs the corticopapillary osmotic gauntlet. As the tubule contents progress toward the papilla, more water is reabsorbed, and urine osmolality reaches its maximal value.

4. **Aquaporin recycling:** When water intake increases and circulating ADH levels fall, AQPs are removed from the membrane by endocytosis and returned to subapical vesicles. Principal cells then remain water impermeant until ADH release resumes, and AQPs are returned to the apical surface.

D. Urea recycling

The fluid that enters the IMCD is now close to urine in its final form. The principal excretory components are (in relative order based on molar amounts) urea, creatinine, ammonium salts, and organic acids. The final step in urine formation is reabsorption of urea, which is regulated by ADH.

1. **Reabsorption:** Urea is reabsorbed by facilitated diffusion, driven by high ductal concentrations and facilitated by urea transporters (UTs) in the IMCD (Figure 27.17). The basolateral membrane contains a UT that is constitutively active. The apical membrane contains a UT (UT-A1) that is minimally active unless ADH is circulating in the vasculature. ADH causes *PKA*-dependent phosphorylation of UT-A1, thereby creating a

pathway for urea to leave the duct and reenter the medullary interstitium. Allowing urea to equilibrate across the tubule wall also helps prevent an osmotic diuresis that might otherwise result from the presence of highly concentrated fluid within the ductal lumen. It also contributes to the corticopapillary osmotic gradient that is used to concentrate urine (Figure 27.18).

2. **Recycling:** Remembering the close anatomical arrangement between the CDs, blood vessels, and limbs of the nephron loop (see Figure 27.5), urea that reenters the interstitium from the IMCD could potentially be reabsorbed by earlier tubule segments or be carried away by the circulation. In practice, it does both.

   a. **Loop of Henle:** Urea reenters the tubule via facilitated transport across the epithelia of both the DTL and ATL (see Figure 27.18). It then recycles back through the distal segments and the CDs. From here, it either may be excreted in urine or take one more trip through the medulla.

   b. **Vasa recta:** The descending vasa recta expresses UT-B transporters, which allow urea to enter the vasculature by facilitated diffusion. Uptake by the vasa recta is beneficial because it increases the osmolality of blood during its passage through the medulla, thereby preventing washout of the osmotic gradient. Urea exits the vasa recta and reenters the interstitium during the return trip to the cortex (see Figure 27.18), so the amount that is ultimately returned to the systemic circulation is minimal (~5% of original filtered load).

3. **Excretion:** Ultimately, the amount of urea excreted in urine depends on the need to conserve water. When water intake is limited, urea is recycled through the medulla, and excretion rates are minimal. When water intake is unlimited, ADH release is suppressed, and there is no significant pathway for urea to escape the IMCD. It is excreted in urine as a result.

E. Acid handling

α-Intercalated cells continue secreting H^+ during urine's passage through the CD and can cause significant urine acidification (pH 4.4, the minimal value attainable). Creatinine (pK = 5.0) becomes a viable buffer at such low pH values, allowing it to assist in H^+ excretion, but the bulk of acid is excreted in the form of NH_4^+. NH_4^+ is excreted as a result of **"diffusion trapping"** or through direct secretion.

1. **Diffusion trapping:** NH_4^+ is formed from NH_3 as a result of glutamine metabolism in the PT (see 26·IX·C). NH_3 is lipid soluble, allowing it to diffuse out of PT epithelial cells and enter the interstitium, where it accumulates in relatively high concentrations. Some NH_3 may then diffuse into the tubule or CD lumen, where it immediately combines with H^+ to form NH_4^+. NH_4^+ is not lipid soluble and, therefore, is now trapped in the tubule or duct unless provided with a carrier that facilitates reabsorption (diffusion trapping). NH_4^+ that is trapped in the proximal segments is actively

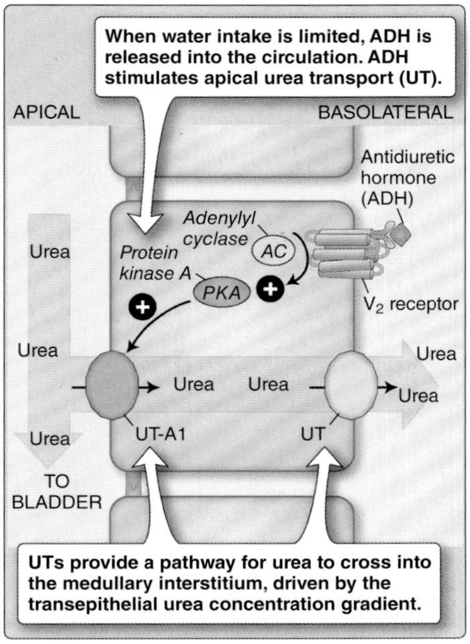

Figure 27.17
Urea reabsorption by the inner medullary collecting duct.

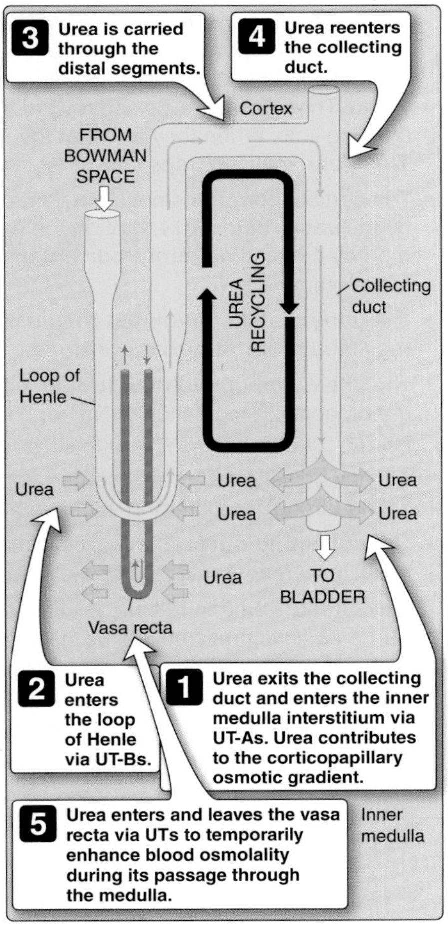

Figure 27.18
Urea recycling. UT = urea transporter.

reabsorbed by a Na^+-K^+-$2Cl^-$ cotransporter in the TAL and then transferred to the interstitium to help generate the corticopapillary osmotic gradient through countercurrent multiplication (see Section II·B·3 above).

2. **Transport:** Some of the NH_4^+ that is transferred to the interstitium by the TAL enters the vasa recta and is carried away by the circulation (washout). This portion eventually reaches the liver, where it is converted to urea. A significant proportion is also transferred to the CD lumen for excretion in urine, although the pathways involved are not well defined.

The principal sites of solute reabsorption, secretion, and regulation in the renal tubule are summarized in Figure 27.19.

Chapter Summary

- The function of the nephron segments distal to the proximal tubule is to recover inorganic ions and to concentrate urine. These segments are the primary sites of **homeostatic regulation** of Na^+, K^+, Ca^{2+}, Mg^{2+}, Cl^-, and water.
- The **loop of Henle** comprises three segments that convey the tubule contents through a **corticopapillary osmotic gradient** designed to extract water from the filtrate.
- The corticopapillary osmotic gradient forms by **countercurrent multiplication** of a transepithelial osmotic gradient created by the **thick ascending limb** (**TAL**) epithelium. The TAL pumps Na^+ and other ions (e.g., NH_4^+) into the cortical interstitium. These ions then diffuse into the **thin descending limb**, causing an increase in the osmolality of the fluid within. These ions are carried toward the papilla, around the tip of the loop, and back up toward the TAL via the thin ascending limb. When they arrive at the TAL, they are pumped back out into the interstitium for a return trip to the medulla. Therefore, the loop traps ions in the medulla and causes this region to develop a high osmolality.
- The corticopapillary osmotic gradient extracts water and ions from the tubule lumen. Reabsorbed fluid is carried away by the **vasa recta**. Flow through the descending and ascending limbs of the vasa recta occurs in opposite directions, thereby creating a **countercurrent exchange system** that prevents incoming arterial blood from washing out the osmotic gradient.
- The **early distal convoluted tubule** is the primary site of regulated Ca^{2+} and Mg^{2+} reabsorption. Ca^{2+} reabsorption is regulated by **parathyroid hormone**.
- The **late distal convoluted tubule**, **connecting tubule**, and **collecting duct** (the **distal segments**) are the primary site of homeostatic regulation of Na^+ and K^+. Na^+ reabsorption is regulated by **aldosterone**. Aldosterone increases epithelial Na^+ permeability by upregulating expression of Na^+ channels and pumps in the apical and basolateral membranes.
- K^+ may be secreted or reabsorbed depending on plasma K^+ concentrations. Secretion is stimulated by hyperkalemia acting through aldosterone. Aldosterone increases epithelial K^+ permeability and Na^+-K^+ ATPase activity.
- The **collecting ducts** (cortical collecting duct and inner and outer medullary collecting ducts) determine final urine osmolality by reabsorbing water. Reabsorption is regulated by **antidiuretic hormone**.
- Antidiuretic hormone release is stimulated by increased plasma osmolality or decreased blood pressure. The hormone exerts its antidiuretic effects by stimulating insertion of **aquaporins** into the apical membranes of ductal epithelial cells. Water is reabsorbed by osmosis, driven by the corticopapillary osmotic gradient.
- The outer medullary collecting duct is also the site of urea reabsorption. Urea is recovered by facilitated diffusion. When water intake is limited, the tubule fluid becomes highly concentrated, and there is a strong driving force for urea movement into the medullary interstitium. The presence of urea in the medulla contributes to the corticopapillary osmotic gradient.

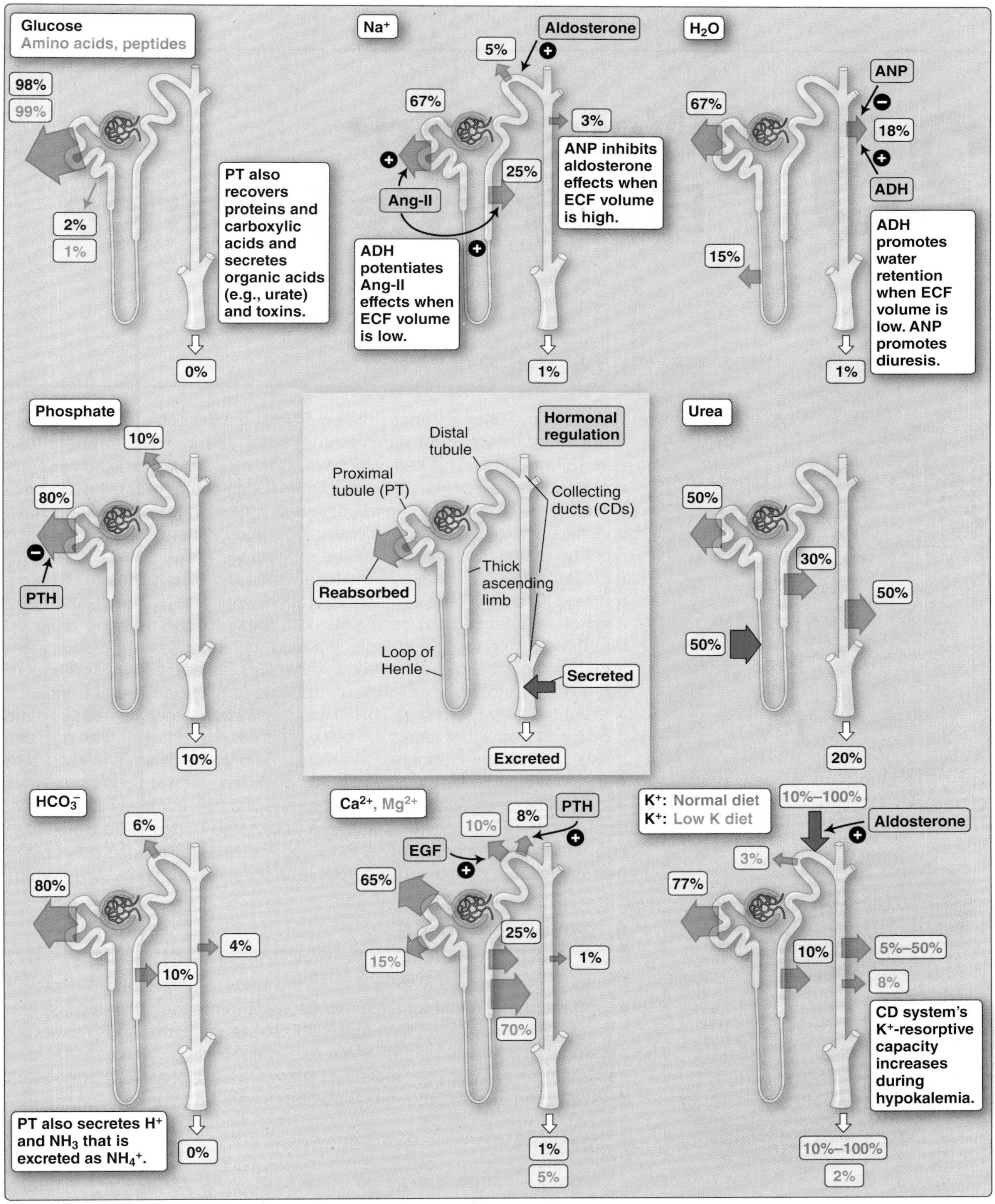

Figure 27.19
Principal sites of solute and water recovery and secretion by the renal nephron. ADH = antidiuretic hormone; Ang-II = angiotensin II; ANP = atrial natriuretic peptide; ECF = extracellular fluid; EGF = epidermal growth factor; PTH = parathyroid hormone.

28 Water and Electrolyte Balance

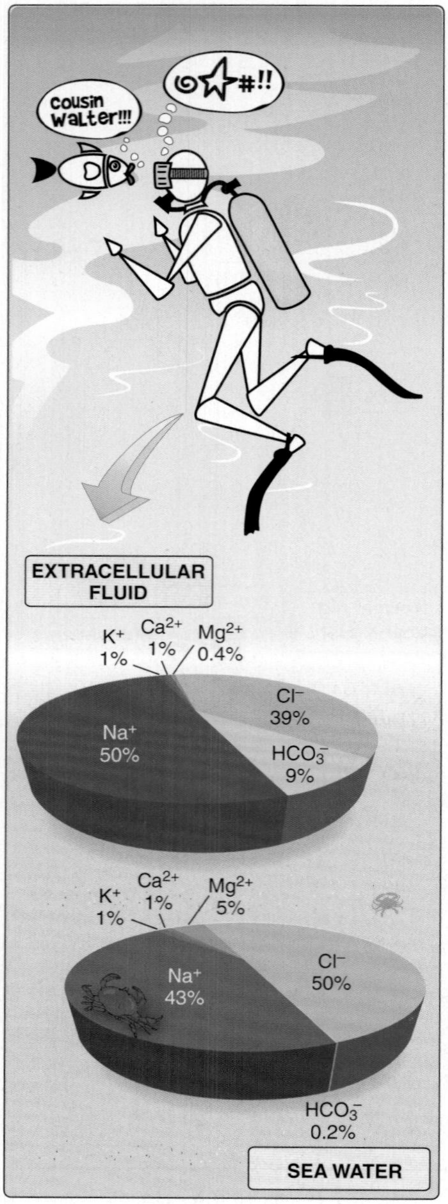

Figure 28.1
Electrolyte composition of the internal and external seas.

I. OVERVIEW

When our evolutionary ancestors emerged from the oceans and laid claim to the land, they carried within them a small sea in which to bathe their cells (Figure 28.1). This sea, which we know as extracellular fluid (ECF), is comprised largely of Na^+, Cl^-, and water. It also contains smaller amounts of HCO_3^-, K^+, Ca^{2+}, Mg^{2+}, and phosphates. All of these constituents have specific roles to play in human physiology, and the concentrations of each must be maintained within a limited range if we are to survive and thrive (see Table 1.1). We continuously lose water and electrolytes to the environment as a result of secretion, excretion, and evaporation. If water and electrolyte balance is to be maintained, these losses must be replaced through drinking fluids and ingesting food, but ingestion and subsequent absorption of salts by the gastrointestinal (GI) system is largely unregulated, being tied to the absorption of nutrients (e.g., glucose and peptides). The central nervous system (CNS) can modify behavior to increase intake if ECF salt and water levels fall below optimal (through salt cravings and thirst), but the main regulated step in salt and water balance is excretion, which is mediated by the kidneys. Although the body includes pathways that maintain stable plasma concentrations of all the common electrolytes, discussions of ECF homeostasis are dominated by Na^+ and water. Na^+ and water together determine ECF volume, which, in turn, determines plasma volume, cardiac output (CO), and mean arterial pressure (MAP).

II. WATER BALANCE

Maintaining water balance is one of the body's most fundamental and important homeostatic functions. Because water is the universal solvent, when total body water (TBW) levels fall, solute concentrations rise to the detriment of bodily function. TBW's role in supporting CO and MAP (discussed in more detail in Section III below) means that TBW regulatory pathways are layered and influence both intake and output.

A. Tally sheet

Individuals ingest and lose ~2.5 L of water per day on average. Actual water requirements are less (1.6 L/day) as shown in Table 28.1, dictated by the amount of **insensible** water loss (evaporation) and obligatory water loss (water needed for urine formation; see 27·V·B).

1. **Intake:** The tally sheet for intake includes water formed through metabolism ($C_6H_{12}O_6 + 6\ O_2 \rightarrow 6\ CO_2 + 6\ H_2O$), ingested with

food, and imbibed by drinking fluids. Drinking is the primary regulated intake step, and net input can vary considerably.

2. **Output:** Output includes water evaporation from respiratory and cutaneous epithelia (insensible losses), sweat, fecal water content, and urine. Water loss from the respiratory epithelia is dependent on respiratory rate and air humidity, but such losses stress water balance only under extreme conditions (e.g., climbing at high altitude). Cutaneous evaporation remains relatively constant under normal conditions. Sweat formation reaches 1.5–2.0 L/hr during heat stress. Vomiting and diarrhea (see Clinical Application 4.4) can greatly accelerate water loss from the GI tract, but normal fecal water loss is modest. Urine formation is the primary regulated output mechanism.

B. Sensory mechanism

TBW is sensed through changes in ECF osmolality (see 23·II·B for a discussion of osmolarity and osmolality), which is normally 275–295 mOsm/kg. Osmolality is sensed by osmoreceptors located in two CNS **circumventricular organs**, the **organum vasculosum of the lamina terminalis** (**OVLT**) and **subfornical organ** ([**SFO**] see Figure 7.10). Osmoreceptors are neurons that are responsive to changes in cell volume, which is dependent on ECF osmolality (see 3·II·E; Figure 28.2). When TBW falls and ECF osmolality rises, water is drawn osmotically from the osmoreceptors, and they shrink (see Figure 28.2B). Shrinkage is transduced by a mechanosensitive transient receptor potential channel ([TRPV4] see 2·VII·D), which opens to allow cation influx and osmoreceptor depolarization. When TBW rises, the osmoreceptor neurons swell, and TRPV4 open probability is reduced. The membrane hyperpolarizes, suppressing signaling.

C. Regulation

Osmoreceptor neurons project to the nearby **hypothalamus**, which functions as an **osmostat** (osmolality regulation center). Responses to changes in osmolality are effected by neurosecretory cells located in the **supraoptic nucleus**. The neurosecretory cells are themselves osmosensitive, which creates an additional layer of osmoregulatory control.

Water balance is achieved by modulating water intake and urinary output.

> Some individuals develop **reset osmostat**, a rare disorder in which osmoreceptors are strongly excited even when ECF osmolality is within a normal range. Reset osmostat is one cause of the syndrome of inappropriate antidiuretic hormone release ([SIADH] see Clinical Application 28.1).

1. **Intake:** A need to drink water is perceived as thirst, which causes an individual to seek a thirst-quenching beverage. The sensation is mediated by higher cortical areas, including the **anterior cingulate cortex** and **insular cortex**. Thirst is sated well before tissue osmolality changes, probably reflecting sensory input from oropharyngeal and GI osmoreceptors.

Table 28.1: Water Intake and Output Routes

Pathway	mL/day
Intake	
Metabolism	300
Food	800
Beverages	500*
Total	1,600
Output	
Feces	200*
Skin	500
Lungs	400
Urine	500*
Total	1,600

*Regulated steps.

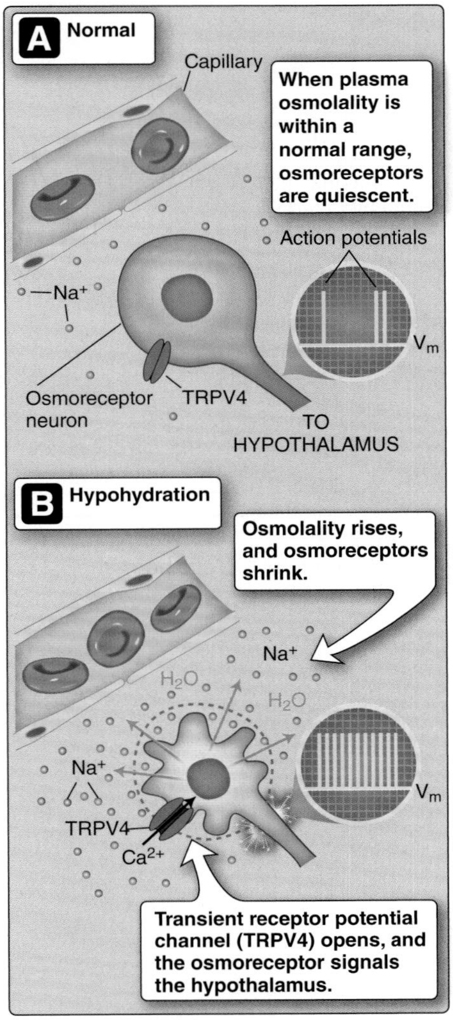

Figure 28.2
Effect of increasing osmolality on osmoreceptor output. V_m = membrane potential.

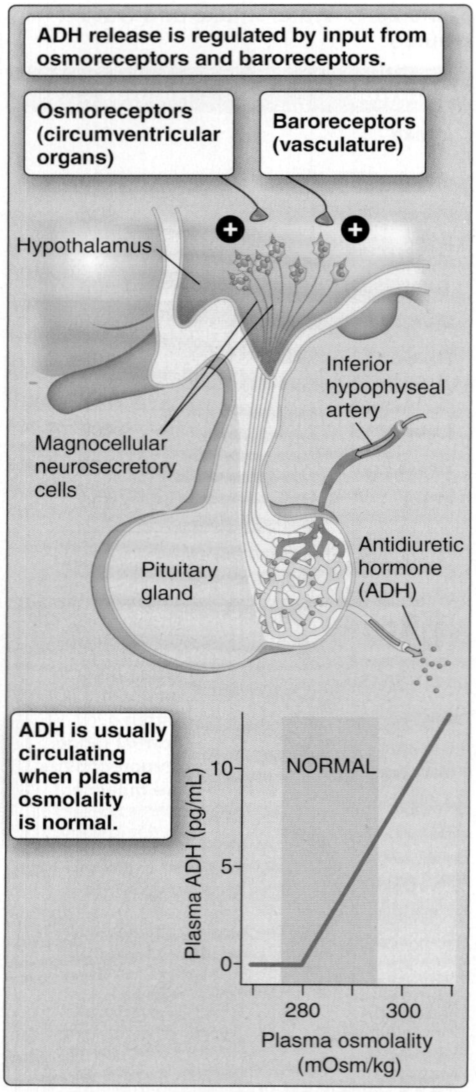

Figure 28.3
Regulation of antidiuretic hormone release.

2. **Output:** The glomerulus filters water into the renal tubule at a rate of ~125 mL/min. Approximately 67% of the filtrate is immediately reabsorbed by the proximal tubule (PT), another 15% is recovered in the nephron loop's descending thin limb (DTL), and most of the remaining 18% is reabsorbed in the outer and inner medullary collecting duct system (OMCD and IMCD, respectively). Output regulation occurs in these distal segments via **antidiuretic hormone (ADH)**. ADH is a small peptide hormone produced by hypothalamic neurosecretory neurons and moved by fast axonal transport for release from terminals in the posterior pituitary as shown in Figure 28.3 (also see 7·VII·D).

   a. **Release:** A rise in ECF osmolality stimulates dose-dependent ADH release into the circulation. The release threshold is 280 mOsm/kg, so small amounts of ADH circulate even when plasma osmolality is within a normal range. ADH has a half-life of 15–20 minutes before being metabolized by kidney and liver *proteases*.

> Because MAP is critically dependent on ECF volume, ADH release thresholds are modulated so as to optimize MAP. Thus, when MAP is low, ADH release continues even though ECF osmolality may have renormalized (Figure 28.4).

Clinical Application 28.1: Antidiuretic Hormone–Release Disorders

The **syndrome of inappropriate antidiuretic hormone** secretion (**SIADH**) is a relatively common disorder characterized by increased circulating antidiuretic hormone (ADH) levels and water retention. Patients typically develop hyponatremia as a result. Although some cases are idiopathic, common causes of SIADH include central nervous system disorders (e.g., stroke, infection, or trauma), drugs (anticonvulsants, such as carbamazepine and oxcarbazepine, and cyclophosphamide, which is used to treat certain cancers), and some pulmonary diseases and carcinomas. For example, small-cell lung malignancies may secrete ADH in an unregulated manner, thereby causing SIADH.

Central diabetes insipidus (**CDI**) describes a polyuria caused by decreased circulating ADH levels. Patients may also present with nocturia and polydipsia. CDI, most commonly idiopathic in etiology, is characterized by degeneration of hypothalamic ADH-secreting cells. CDI may also be caused by trauma and surgery. Familial CDI is a dominant hereditary form caused by ADH gene mutation. The most common familial form causes an ADH processing defect and accumulation of misfolded hormone. The secretory cells degenerate as a result of these accumulations, although the cause is still under investigation.

b. **Actions:** ADH provides pathways for water to flow out of the renal collecting ducts and rejoin the ECF (see 27·V·C). In the absence of such pathways, water is channeled to the bladder and excreted.

c. **Negative feedback:** ADH-mediated ECF volume expansion is limited by **atrial natriuretic peptide** (**ANP**). ANP is released from atrial myocytes when ECF and blood volumes are high. ANP has many actions (detailed below), including antagonizing ADH release and ADH-mediated water retention (Figure 28.5).

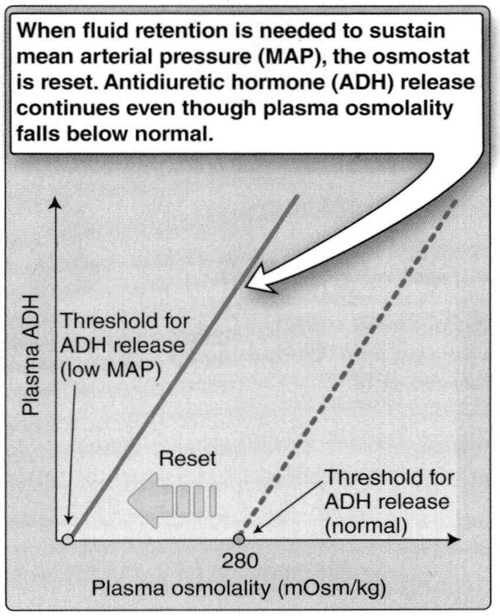

Figure 28.4
Effect of arterial pressure on antidiuretic hormone release.

III. SODIUM BALANCE

Although it is possible to identify mechanisms and sites involved in Na^+ balance, the pathways involved are so closely interwoven with those controlling water balance and MAP that Na^+ balance cannot be discussed in isolation.

A. Tally sheet

The body contains ~75 g of Na^+ on average, almost half of which is immobilized in bone osteoid. Na^+ is obtained from dietary sources and leaves the body via feces, sweat, and urine.

1. **Intake:** The average diet contains far more Na^+ than is required to offset losses. The U.S. recommended dietary allowance (RDA) is 1.5–2.3 g/day, but per capita consumption worldwide is typically much greater (up to 7 g/day). A low-Na^+ diet triggers salt craving, which manifests as a need to seek out and ingest salty foods. The pathways involved are not well defined.

2. **Output:** A small amount of ingested Na^+ is lost in feces. Sweat is also a minor pathway for Na^+ loss unless sweating is prolonged and profuse (see 16·VI·C). Urine is the primary route for Na^+ output. Because there is no obligatory Na^+ loss as there is for water, urinary Na^+ excretion typically balances the amount ingested. When Na^+ intake is limited, however, the renal tubule can recover 100% of the filtered load and generate Na^+-free urine for several weeks. The principal sites for recovery are the PT (67% of filtered load) and the thick ascending limb ([TAL] ~25%). The distal segments and collecting ducts recover the remaining 8% and are the principal sites of output regulation (see 27·VI·B).

B. Sodium and blood pressure relationship

When Na^+ is ingested, most of it ends up in the ECF (~85%) because, although Na^+ exchanges freely between ECF and intracellular fluid (ICF), all cells actively eliminate Na^+ from the ICF via the Na^+-K^+ ATPase (see 3·III·B). Ingesting Na^+ thus increases ECF osmolality, creating an urge to drink water and stimulating water retention (see Figure 28.5). Because plasma is an ECF component, Na^+ ingestion also increases circulating blood volume, which raises central venous pressure ([CVP] see Figure 20.24). A rise in CVP increases left ventricular (LV) preload. The LV responds with an increase in stroke volume (SV) and CO via the Frank-Starling mechanism, which raises MAP (see 18·III·D). Increasing MAP has immediate and wide-ranging

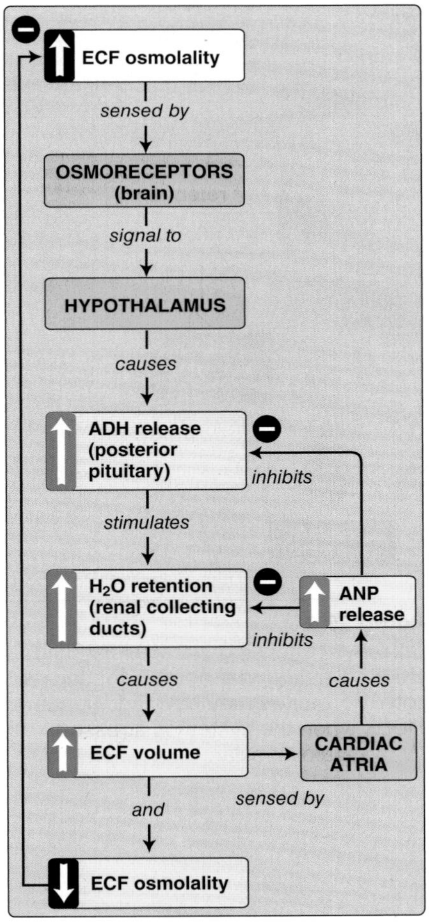

Figure 28.5
Regulation of antidiuretic hormone (ADH) release. ANP = atrial natriuretic peptide; ECF = extracellular fluid.

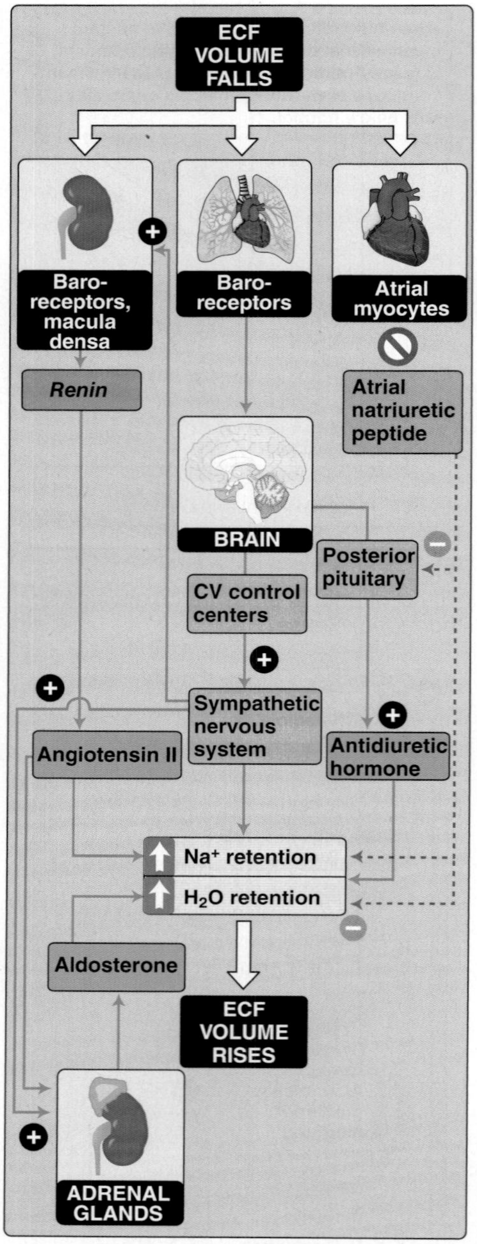

Figure 28.6
Pathways regulating extracellular fluid (ECF) volume under normal conditions. Note: Hormones may have additional effects when the system is stressed. CV = cardiovascular.

repercussions for both the cardiovascular system and the kidneys, all aimed at excreting the Na^+ and water excess to reduce ECF volume and renormalize MAP.

C. Regulation

ECF volume is determined by four different but interdependent pathways that regulate Na^+ balance, water balance, and MAP (Figure 28.6). They include the ***renin*–angiotensin–aldosterone system** ([**RAAS**] see Figure 20.17), the sympathetic nervous system (SNS), ADH, and ANP. Three of the four pathways activate when ECF volume and MAP are low, such as might occur when a marathon runner has become hypohydrated due to inadequate replenishment of salt and water loss. When ECF volume and MAP are high, these pathways are inhibited.

1. ***Renin*–angiotensin–aldosterone system:** RAAS activates following a decrease in glomerular afferent arteriolar perfusion pressure (sensed by renal baroreceptors), a decrease in fluid flow past the macula densa (see 25·IV·C), and an increase in sympathetic activity triggered by a decrease in MAP (see 20·IV·C). RAAS effects are mediated by Ang-II, whose actions are all geared toward retaining Na^+ and water and raising MAP (Table 28.2).

2. **Sympathetic nervous system:** The SNS activates when the brainstem cardiovascular control centers detect a need to raise MAP, sensed by arterial baroreceptors (see 20·III·A). The SNS innervates most tissues in the body. It has wide-ranging effects when activated and includes many of the same targets as Ang-II.

3. **Antidiuretic hormone:** ADH's primary role is in maintaining water balance, but severe cardiovascular stress (e.g., hemorrhage) can cause circulating ADH levels to rise to the point that they constrict resistance vessels. ADH effects on glomerular blood flow are similar to those of Ang-II. ADH can also stimulate Na^+ reabsorption from the TAL and cortical collecting duct (CCD), which further enhances fluid retention.

Clinical Application 28.2: Sodium Reabsorption and Blood Pressure

The relationship between Na^+, extracellular fluid volume, and blood pressure means that mutation in any of a number of key proteins involved in renal Na^+ absorption may potentially cause hypo- or hypertension.

Liddle syndrome is a very rare congenital disorder that increases epithelial Na^+ channel (ENaC) expression by the distal segments, resulting in increased Na^+ reabsorption. The syndrome is characterized by hypertension and may be associated with hypokalemia and metabolic alkalosis.

Pseudohypoaldosteronism type 1 mutations cause hyponatremia, hypotension, and hyperkalemia. The dominant form prevents mineralocorticoid receptor expression, causing aldosterone resistance. Recessive forms inhibit ENaC activity, thereby preventing Na^+ reabsorption in the distal segments.

4. **Atrial natriuretic peptide:** ANP is released when cardiac atria are stretched by high blood volumes and provides a negative feedback pathway that limits ECF volume expansion. Its principal effects are to antagonize Ang-II's and ADH's actions and to stimulate natriuresis through direct effects on the glomerulus.

   a. **Angiotensin II:** ANP inhibits Na^+-H^+ exchange in the PT, Na^+-Cl^- cotransport in the distal tubule, and Na^+ channels in the collecting ducts, all of which promote natriuresis.

   b. **Antidiuretic hormone:** ANP suppresses ADH release and prevents ADH-stimulated insertion of aquaporins into the apical membranes of the collecting ducts. These actions prevent water reabsorption and promote diuresis.

   c. **Vasculature:** ANP vasodilates to increase flow through the glomerulus and peritubular system. Glomerular filtration rate increases markedly as a result, causing a pronounced diuresis.

> Brain natriuretic peptide (BNP) is a related peptide that is released from the atria and ventricles when filling volumes are high. Although ANP is rapidly metabolized by the liver, circulating BNP is more stable and provides an early, sensitive indicator of heart failure. BNP measurements are rapid and inexpensive and used clinically to determine the presence and severity of failure and to help exclude congestive heart failure as a possible cause of dyspnea.

Table 28.2: Angiotensin II Effects

Target	Action	Effects
Adrenal Cortex		
	Aldosterone release	↑ Na^+ reabsorption, renal distal segments ↑ ENaC ↑ ROMK ↑ Na^+-K^+ ATPase
Kidney		
	Proximal tubule, thick ascending limb	↑ Na^+ reabsorption ↑ Na^+-H^+ exchanger
	Distal segments	↑ ENaC
Vasculature		
	Vasoconstriction (resistance vessels)	↑ Systemic vascular resistance
Central Nervous System		
Hypothalamus	Antidiuretic hormone release (posterior pituitary)	↑ H_2O reabsorption, renal distal segments
Cortex	Thirst and salt craving	↑ H_2O and NaCl intake

ENaC = epithelial Na^+ channel; ROMK = renal outer medullary K^+ channel.

D. Glomerulotubular balance

Na^+ balance is maintained, in part, by a phenomenon known as **glomerulotubular (GT) balance**. GT balance refers to the PT's tendency to reabsorb a constant fraction of the filtered Na^+ load regardless of glomerular filtration rate (GFR). Normally, fractional reabsorption is ~67%, although this value may change during ECF volume contraction and expansion. Thus, when GFR increases (e.g., due to a rise in glomerular filtration pressure), the PT increases net Na^+ reabsorption to compensate for the increased amounts of Na^+ appearing in the tubule lumen. GT balance helps ensure that Na^+ is not excreted inappropriately when GFR rises. GT balance relies on changes in peritubular and tubule function.

1. **Peritubular:** When GFR increases due to an increase in filtration fraction (filtration fraction = GFR ÷ RPF), blood leaving the glomerulus via the efferent arteriole has a higher colloid osmotic pressure (π_c) compared with previously because the plasma proteins have been concentrated by glomerular filtration to a greater degree. A higher π_c enhances fluid reabsorption by the peritubular network serving the PT. Conversely, when GFR falls due to a decrease in filtration fraction, the osmotic potential favoring fluid

Table 28.3: Conditions Affecting Internal Potassium Balance

Causal Event	Mechanism
Shifts from ECF to ICF (hypokalemia)	
↑ Insulin	↑ Na^+-K^+ ATPase
↑ Epinephrine	↑ Na^+-K^+ ATPase
Alkalosis	↑ Na^+-K^+ ATPase to compensate for cation (H^+) efflux
Shifts from ICF to ECF (hyperkalemia)	
Exercise	↑ Excitation and K^+ channel opening
Acidosis	H^+ displaces K^+ and inhibits uptake pathways
↑ ECF osmolality	Cells lose water and K^+ follows by diffusion
Cell trauma, necrosis	Loss of cellular K^+ containment

ECF = extracellular fluid; ICF = intracellular fluid.

reabsorption by peritubular blood is reduced, thereby facilitating GT balance.

2. **Tubule:** The PTs reabsorptive capacity typically exceeds the normal filtered load for most organic and inorganic solutes, including Na^+. When GFR and filtered load increases, the reabsorptive reserve allows the PT to compensate by increasing net uptake, which helps maintain GT balance.

E. Pressure-induced natriuresis

Hypertension produces a **pressure natriuresis** that occurs independently of the pathways delineated above. Pressure natriuresis acts as a safety valve to reduce ECF through Na^+ and water excretion, thereby bringing MAP back down to normotensive levels. Natriuresis results primarily from a hypertension-induced removal of Na^+-H^+ exchangers from PT villi, which reduces the segment's Na^+ reabsorptive capacity.

> Under resting conditions, most adjustments to blood volume are effected through the kidney's autoregulatory mechanisms in concert with osmoreceptor-mediated release of ADH. Other pathways are only called into action when the cardiovascular system is stressed.

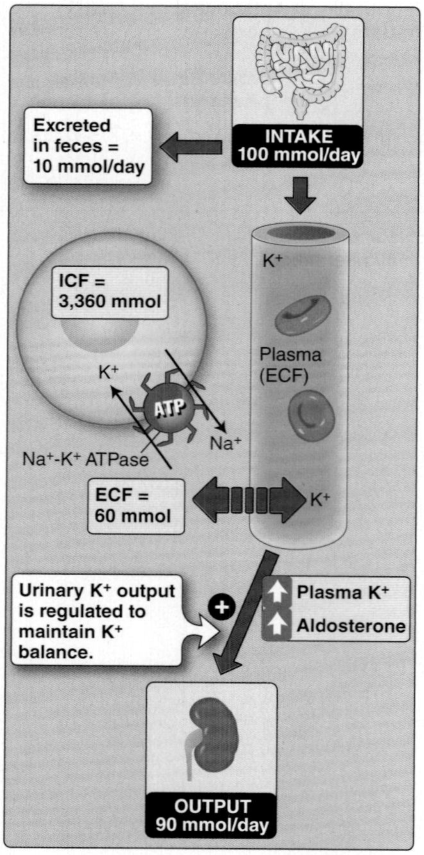

Figure 28.7
Potassium balance. ATP = adenosine triphosphate; ECF = extracellular fluid; ICF = intracellular fluid.

IV. POTASSIUM BALANCE

The body contains ~3.6 mol (~140 g) of K^+, ~98% of which is concentrated within cells by the plasma membrane Na^+-K^+ ATPase. However, all cells express K^+ channels and K^+ transporters on their surface membranes that allow K^+ to move relatively freely between the ICF and ECF. These pathways make significant shifts in K^+ localization possible (e.g., during changes in pH balance; see below), causing an **internal K^+ balance** disturbance (Table 28.3). Despite these challenges, kidneys are able to maintain plasma K^+ concentrations within a fairly narrow range (3.5–5.0 mmol/L).

A. Tally sheet

Maintaining stable plasma K^+ concentrations involves a simple balance between ingestion and urinary excretion (Figure 28.7).

1. **Intake:** The U.S. RDA for potassium is 4.7 g (~120 mmol/day). Net intake varies widely with diet. Fruits and vegetables are particularly rich in K^+ and provide more than adequate K^+ to meet bodily needs under normal circumstances. Most K^+ ingested is subsequently absorbed during transit through the GI tract.

2. **Output:** Kidneys are the only significant route for K^+ output. 80% of filtered K^+ is reabsorbed isosmotically in the PT. Another 10% is reabsorbed in the TAL. Regulation of K^+ balance occurs in the distal segments (see 27·IV·C). When dietary K^+ intake exceeds need (which is generally the rule), the distal segments secrete K^+ for urinary excretion. When the body is severely K^+ depleted, the tubule can reabsorb >99% of the filtered load.

B. Regulation

K^+ balance is effected primarily by the distal segments. The responsibility for maintaining K^+ balance shifts from principal cells to α-intercalated cells depending on whether K^+ intake is high, and the excess must be secreted, or K^+ intake is restricted, requiring reabsorption.

1. **Potassium secretion:** The GI tract can transfer several tens of millimoles of K^+ to the vasculature during a typical meal. Processing such a significant K^+ load while maintaining plasma K^+ levels within a safe range requires that the excess be stored temporarily to give the kidneys time to excrete the surplus.

    a. **Interim storage:** Ingesting a meal causes circulating insulin levels to rise. Insulin has many effects on cell metabolism (see 34·IV), including stimulation of Na^+-K^+ ATPase activity. Ingested K^+ moves temporarily from ECF to ICF as a result.

    b. **Decreased reabsorption:** When plasma K^+ rises, so does the concentration of K^+ entering the PT, which reduces reabsorption of all cations, including K^+ and Na^+.

    c. **Increased secretion:** Hyperkalemia stimulates aldosterone release from the adrenal cortex. Aldosterone targets principal cells in the distal segments, promoting an increase in basolateral Na^+-K^+ ATPase and apical renal outer medullary K^+ channel (ROMK) expression. The Na^+-K^+ pump creates the driving force, and ROMK provides a pathway for increased K^+ secretion into the tubule lumen.

2. **Potassium reabsorption:** Hypokalemia suppresses aldosterone release, inhibiting K^+ secretion. Hypokalemia simultaneously stimulates H^+-K^+ ATPase activity in collecting duct α-intercalated cells, promoting K^+ reabsorption from the ductal lumen.

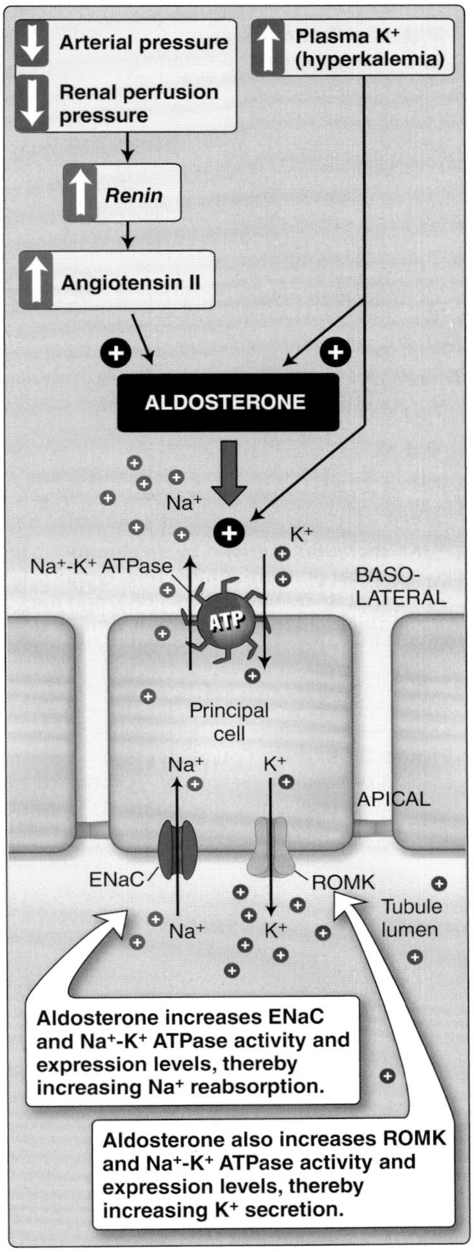

Figure 28.8
Convergence of pathways regulating Na^+ reabsorption and K^+ secretion in the distal segments. ATP = adenosine triphosphate; ENaC = epithelial Na^+ channel; ROMK = renal outer medullary K^+ channel.

C. Sodium and potassium balance relationship

The pathways regulating Na^+ balance and K^+ balance converge on basolateral Na^+-K^+ ATPase activity in the distal segments (Figure 28.8). The Na^+-K^+ ATPase exchanges Na^+ for K^+, simultaneously enhancing Na^+ reabsorption and K^+ secretion. Because there are situations in which secreting K^+ during Na^+ reabsorption (or *vice versa*) would be deleterious, the two processes must be functionally uncoupled. Uncoupling is achieved through the potent effects of tubular flow rate on K^+ excretion.

1. **Flow:** K^+ secretion by principal cells is powered by the transepithelial K^+ concentration gradient. When tubule flow rates are low, K^+ diffusing from principal cells causes luminal K^+ concentrations to rise significantly, which blunts the driving force for further diffusion and secretion (Figure 28.9A). When tubule flow rates are high, K^+ is flushed out of the distal segments at an accelerated rate, and the concentration gradient favoring K^+ secretion remains high (see Figure 28.9B).

2. **Diuresis:** When ECF volume is too high, the kidney excretes Na^+ and water at increased rates. The first steps in excretion are to increase GFR and reduce reabsorption from the PT, which raises

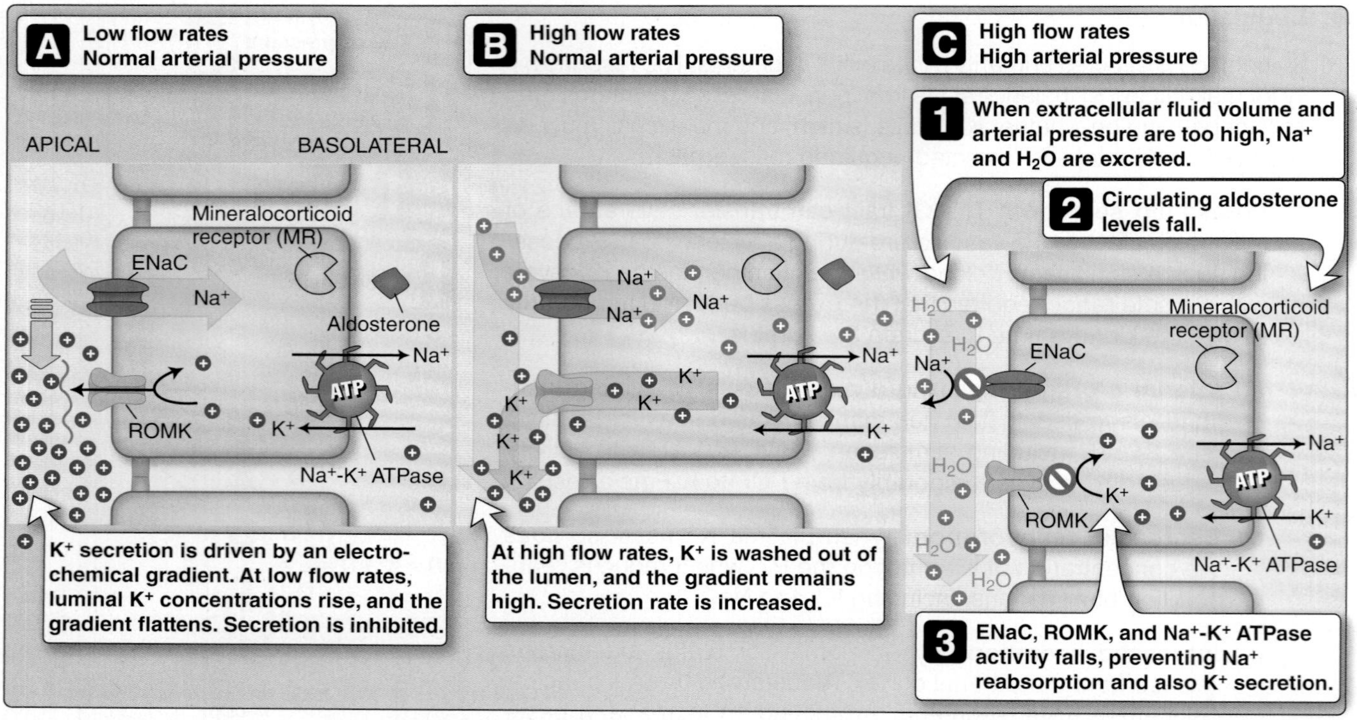

Figure 28.9
Effects of tubule flow rate on potassium excretion. ATP = adenosine triphosphate; ENaC = epithelial Na^+ channels; ROMK = renal outer medullary K^+ channel.

flow rates through the distal tubule. High flow rates would be expected to cause excessive K^+ secretion, but when ECF volume and MAP are high, RAAS is suppressed. In the absence of aldosterone, K^+ secretion by the distal segments is attenuated, thereby preventing excessive flow-induced K^+ loss (see Figure 28.9C).

3. **Volume expansion:** When ECF volume and MAP are low, RAAS is activated, and the tubule's Na^+ reabsorptive capacity increases. This allows for Na^+ and water retention, but it simultaneously upregulates the pathway that mediates K^+ secretion by principal cells. Although this might be expected to increase K^+ excretion, it occurs in the context of low flow rates through the tubule, which blunts the driving force for K^+ secretion.

D. pH and potassium balance relationship

K^+ balance is very sensitive to changes in pH balance. Acidosis causes hyperkalemia, whereas alkalosis causes hypokalemia. These disturbances reflect combined effects of H^+ on ICF and renal function.

1. **Intracellular fluid:** H^+ has a variety of ways of crossing cell membranes, so, when plasma H^+ concentrations rise, ICF concentration rises also. Because H^+ carries a positive charge, H^+ influx would be expected to depolarize membrane potential (V_m), but the cell responds with a counterbalancing K^+ efflux to maintain V_m at resting levels. ECF K^+ concentration rises as a result (hyperkalemia). Alkalosis has the opposite effect. When plasma H^+ concentrations fall, H^+ diffuses out of the cell, and K^+ is taken up from the ECF to redress the charge imbalance. The result is hypokalemia.

2. **Renal function:** Although the kidneys might be expected to correct such K^+ imbalances, in practice, H^+ has simultaneous effects on tubule function that cause the imbalance to worsen. Acidosis inhibits K^+ secretion by the distal segments by inhibiting principal cell Na^+-K^+ ATPase activity. Inhibition reduces K^+ uptake from blood and reduces the concentration gradient driving K^+ efflux across the apical membrane into the tubule lumen. H^+ also inhibits apical K^+ channels in principal cells, reducing K^+ secretion directly and further potentiating hyperkalemia. Alkalosis has the opposite effect, promoting K^+ secretion and hypokalemia. Factors affecting K^+ excretion are summarized in Table 28.4.

Table 28.4: Urinary Potassium Excretion Determinants

Variable	Net Effect on K^+ Excretion
Hypokalemia	Inhibit
Hyperkalemia	Increase
Alkalosis	Increase
Acidosis	Decrease
↓ Tubule flow	Decrease
↑ Tubule flow	Increase
Aldosterone	Increase

E. Sodium, potassium, and pH balance relationship

Na^+ reabsorption by the PT occurs, in part, via an apical Na^+-H^+ exchanger (NHE3), which uses the transmembrane Na^+ gradient to power H^+ secretion. This Na^+-H^+ coupling means that pathways modulating Na^+ reabsorption can also affect pH balance. When MAP or ECF volume falls, Na^+-H^+ exchanger activity increases due to Ang-II release, and the ensuing H^+ secretion increase results in **contraction alkalosis**. Aldosterone further potentiates the alkalosis by increasing PT Na^+-H^+ exchanger expression levels and by stimulating H^+-K^+ ATPase activity in the distal segments. Conversely, when MAP or ECF volume rises, Na^+ reabsorption and H^+ secretion are attenuated, causing acidosis. Alterations in K^+ balance also affect pH balance. Hypokalemia causes alkalosis by stimulating Na^+-H^+ exchange and NH_3 production in the PT and by stimulating H^+-K^+ pump activity in the distal segment. Conversely, hyperkalemia causes acidosis.

V. pH BALANCE

pH balance is achieved through the combined actions of the lungs and kidneys. The lungs excrete volatile acid (H_2CO_3, which is expired as CO_2). The kidneys excrete nonvolatile acid (Figure 28.10).

A. Tally sheet

The tally sheet for acid is unusual insofar as most acid is generated internally by metabolism rather than being ingested.

1. **Intake:** The majority of daily acid "intake" (~15–22 mol) is formed as a result of carbohydrate metabolism. An additional 70–100 mmol/day of nonvolatile acid (nitric, sulphuric, and phosphoric acid) is generated through breakdown of amino acids and phosphate compounds.

2. **Output:** Most of the CO_2 generated during metabolism and converted to H^+ and HCO_3^- for blood transport is subsequently excreted by the lungs. A small amount of volatile acid remains trapped in the body when HCO_3^- is lost in feces and must be excreted by the kidneys as nonvolatile acid. Nonvolatile acid is excreted primarily as titratable acid and ammonium in the PT (see 26·IX·C).

B. Regulation

Volatile acid is sensed by CNS chemoreceptors in the brainstem (see 24·III·A) and regulated by adjusting ventilation. All nephron seg-

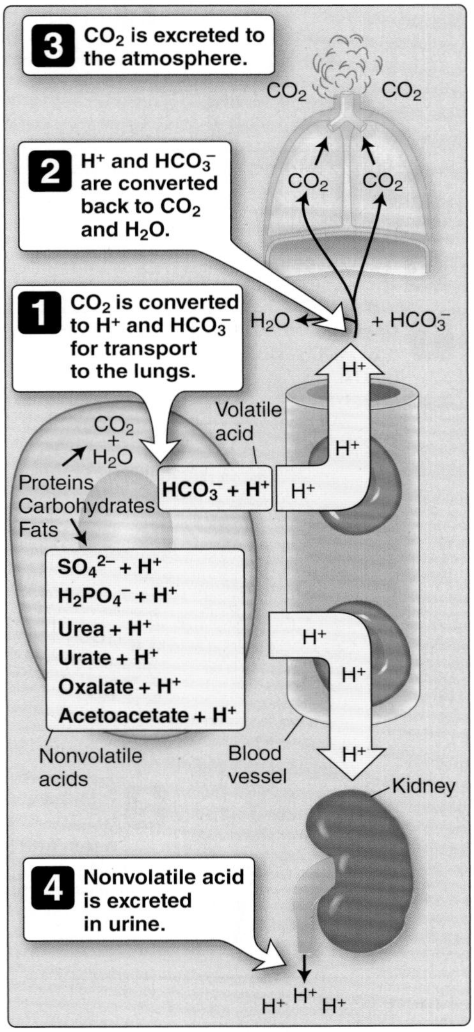

Figure 28.10
Volatile and nonvolatile acid excretion.

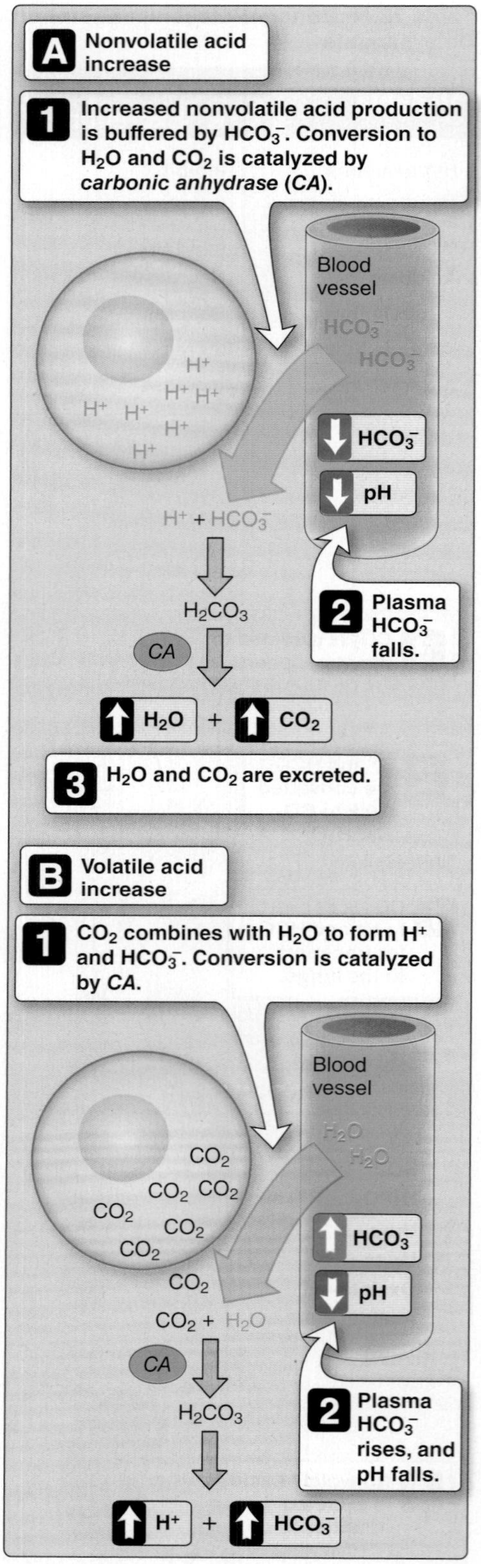

Figure 28.11
Effects of nonvolatile and volatile acids on plasma bicarbonate and pH.

ments are involved in excreting nonvolatile acid, but the PT and intercalated cells in the distal segments play prominent roles.

1. **Volatile acid:** Brainstem respiratory control centers monitor plasma P_aCO_2 via changes in cerebrospinal fluid pH. If either parameter is higher than optimal, the control centers increase ventilation to transfer additional volatile acid to the atmosphere. If P_aCO_2 or H^+ levels are lower than normal, ventilation rates and CO_2 transfer decreases.

2. **Nonvolatile acid:** The PT secretes the bulk of the daily nonvolatile acid load. Acidosis increases H^+ secretion and upregulates NH_3 synthesis by the PT, whereas alkalosis decreases expression of these pathways. The primary effectors of pH balance are intercalated cells in the distal segments (see 27·V·E). Chronic metabolic acidosis increases the proportion of acid-secreting α-intercalated cells, whereas metabolic alkalosis reverses this change and increases β-intercalated cell density.

VI. ACID–BASE DISORDERS

pH balance can be upset by numerous changes in pulmonary, GI, and renal function and can be triggered through altered regulation of acid or base production. In practice, this means that acid–base disorders are encountered frequently in clinical medicine.

A. Types and compensation

There are four basic types of "simple" acid–base disorders. Respiratory acidosis and alkalosis are primary disorders of CO_2 handling by the lungs. Metabolic acidosis and alkalosis manifest as a primary disorder in plasma HCO_3^- levels, although there may be many underlying causes (discussed below).

> When more than one type of simple acid–base disturbance is present, a "mixed" acid–base disorder is said to exist. The number of identifiable disorders never exceeds three, because a body cannot simultaneously over- and underexcrete CO_2. A "triple" disorder is, therefore, two metabolic disorders plus one respiratory disorder.

Cells are protected from acid–base changes by three primary defense mechanisms with varying time courses and efficacy: buffers (immediate), the lungs (minutes), and the kidneys (days; see also 3·IV).

1. **Buffers:** Buffers limit the effects of acid–base changes until compensation can occur. The principal intracellular buffers include proteins (including hemoglobin in red blood cells) and phosphates. The principal buffer in ECF is HCO_3^-, which combines with H^+ to form H_2O and CO_2 via a reaction catalyzed by *carbonic anhydrase* ([*CA*] Figure 28.11).

Equation 28.1: $$H^+ + HCO_3^- \leftrightarrows H_2CO_3 \underset{CA}{\leftrightarrows} CO_2 + H_2O$$

An increase in nonvolatile acid production is buffered by HCO_3^-, causing ECF (including plasma) HCO_3^- levels to fall. Conversely, increased CO_2 production (volatile acid) increases plasma HCO_3^- levels even as plasma pH falls (see Figure 28.11).

2. **Lungs:** Respiratory control centers located in the brainstem adjust ventilation to increase or decrease CO_2 (volatile acid) transfer to the atmosphere. Because respiratory rate is normally 12–15 breaths/min, compensation occurs rapidly.

3. **Kidneys:** Kidneys are the third and final line of acid defense, adjusting the amount of H^+ they secrete to maintain strict control over pH balance. Upregulation of the necessary enzymatic pathways takes hours to implement, making renal compensation much slower than respiratory compensation (up to 3 days).

B. Clinical assessment

A physician evaluating a patient with an acid–base disorder typically reviews arterial pH, P_{CO_2}, and HCO_3^- values, which provide a starting point from which the nature of the acid–base disturbance (e.g., simple *versus* mixed, respiratory *versus* metabolic) may be ascertained.

> Assessment of a patient's acid–base status requires data from an arterial blood gas sample and a basic metabolic panel. The arterial gas analysis provides data on pH, P_aCO_2, P_aO_2, and HCO_3^-. The metabolic panel provides contiguous data that helps interpret the metabolic origin of an acid–base disturbance.

1. **Davenport diagram:** Davenport diagrams are typically not used clinically, but they are helpful in understanding how acid–base disorders manifest as changes in arterial pH, P_{CO_2}, and HCO_3^-. The diagram is a pictorial representation of the Henderson-Hasselbalch equation (Figure 28.12):

$$pH = pK + \log \frac{[HCO_3^-]}{[CO_2]}$$

where $[HCO_3^-]$ and $[CO_2]$ represent plasma HCO_3^- and CO_2 concentrations, respectively (the latter calculated from the product of P_{CO_2} and CO_2 solubility). At a plasma pH of 7.4 and P_aCO_2 of 40 mm Hg, $[HCO_3^-]$ is ~26 mmol/L. Changes in plasma H^+ and P_{CO_2} cause HCO_3^- concentrations to shift in a predictable manner (see Equation 28.1 and Figure 28.12).

2. **Anion gap:** The anion gap is an important clinical determination that helps identify and distinguish among types of metabolic acidosis (see Section E below). An anion gap is calculated by comparing total serum cation and anion concentrations, which, according to the principle of bulk electroneutrality, must always be equal. The principal plasma cation is Na^+ (see Figure 28.1). The principal anions are Cl^- and HCO_3^-. Typical serum values for these ions are 140 mmol/L Na^+, 100 mmol/L Cl^-, and 25 mmol/L

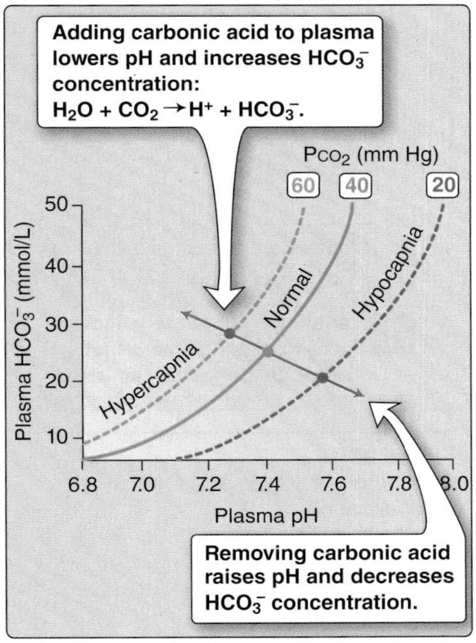

Figure 28.12
Effects of P_{CO_2} on plasma bicarbonate concentration and pH.

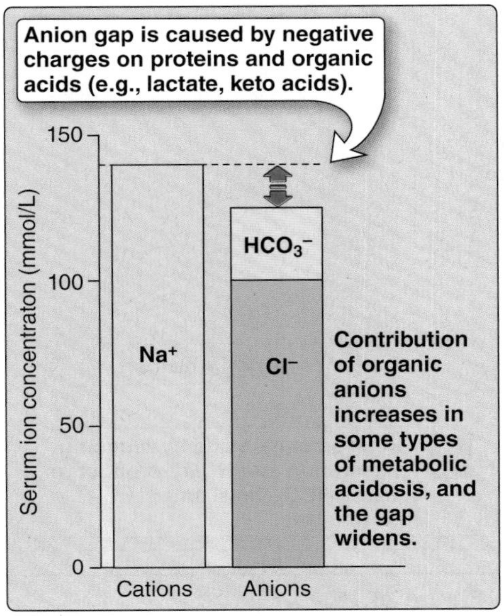

Figure 28.13
Serum anion gap.

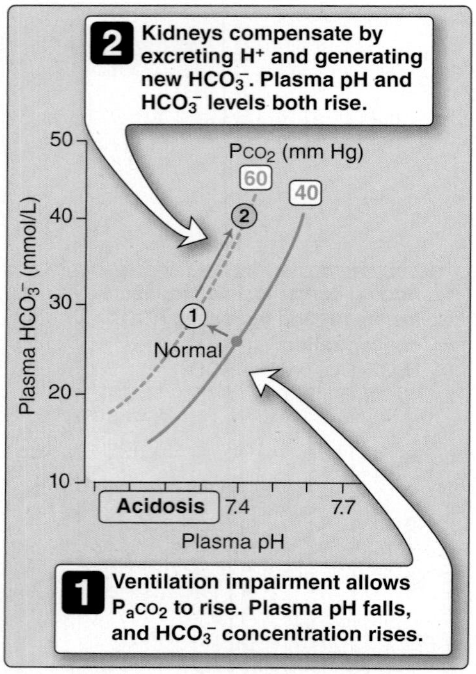

Figure 28.14
Respiratory acidosis and renal compensation.

HCO_3^-. The difference between these values is the anion gap (Figure 28.13):

$$\text{Normal anion gap} = [Na^+] - ([Cl^-] + [HCO_3^-])$$
$$= 140 - (100 + 25) = 15 \text{ mmol/L}$$

The anion gap is normally in a range of 8–16 mmol/L. The gap represents the sum of all the minor serum anions, including proteins and organic ions such as phosphate, citrate, and lactate. Some forms of metabolic acidosis are caused by accumulation of lactate, keto acids, or other such anions, which causes the gap to widen.

C. Respiratory acidosis

Respiratory acidosis is usually caused by hypoventilation but can result from any condition that allows P_aco_2 to rise. Respiratory acidosis is characterized by an elevated P_aco_2 and a low arterial pH.

1. **Causes:** Causes of respiratory acidosis include decreased ventilatory drive, an air pump disorder, and processes that interfere with gas exchange.

   a. **Ventilatory drive:** Because ventilatory drive and the respiratory rhythm originates in the brainstem, congenital CNS disorders or tumors affecting brainstem function can potentially cause respiratory acidosis. For example, **Ondine curse** is a rare form of congenital central hypoventilation syndrome in which ventilatory drive and respiratory reflexes are absent. Drugs that suppress CNS function (e.g., opiates, barbiturates, and benzodiazepines[1]) can also cause respiratory depression and increase P_aco_2.

> In medicine, the term "Ondine curse" is synonymous with hypoventilation associated with loss of autonomic respiratory drive. The term has its origins in European mythology, referring to a water nymph (Undine), who became mortal in order to marry a man that she had fallen in love with. When she aged, her husband fell into the arms of a younger woman. Ondine punished him with a curse that forced him to have to remember to breathe. Once he finally fell asleep, he died. In reality, there is no record of any such curse: The myth was misquoted in the medical literature.

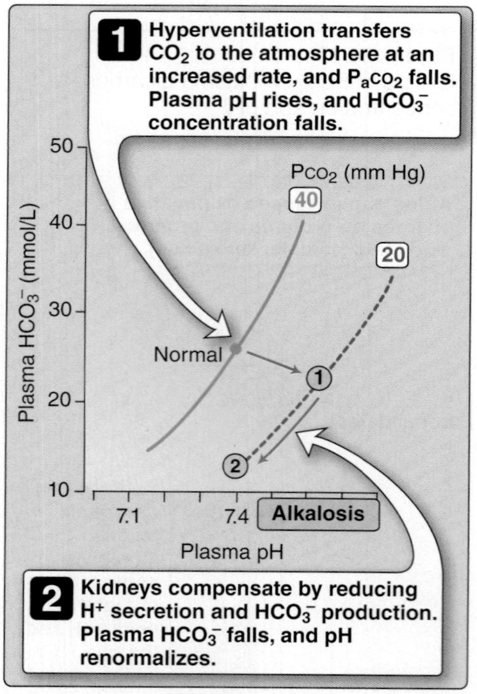

Figure 28.15
Respiratory alkalosis and renal compensation.

   b. **Air pump:** Inspiration is effected through contraction of inspiratory muscles (diaphragm and external intercostals) that expand the lungs and create the pressure gradient that drives air into the alveoli. Any disease process that affects these muscles or

INFO LINK [1]For more information on drugs that cause central nervous system and respiratory depression, see *LIR Pharmacology*.

their motor command pathways potentially may cause respiratory acidosis. Common examples include amyotrophic lateral sclerosis, myasthenia gravis (see Clinical Application 12.2), muscular dystrophy (see Clinical Application 12.1), and infectious diseases such as polio (see Clinical Application 5.1).

c. **Gas exchange:** Airway obstruction can prevent normal alveolar ventilation and cause P_aCO_2 to rise. Causes include aspiration of a foreign body, bronchospasm, chronic obstructive lung diseases, and obstructive sleep apnea (see Clinical Application 24.1). Conditions that cause the alveoli to fill with fluid (pulmonary edema), pus (pneumonia), or other infiltrates (e.g., acute respiratory distress syndrome; see 40·VI·D) may also increase P_aCO_2.

2. **Compensation:** The effects of acidosis caused by an acute rise in P_aCO_2 are limited by the HCO_3^- buffer system. Hypercapnia biases Equation 28.1 in favor of HCO_3^- formation, so plasma HCO_3^- rises even as pH falls (Figure 28.14). Compensation occurs over a period of several days and involves an increase in renal H^+ secretion and NH_3 production. The "new" HCO_3^- generated during H^+ secretion and NH_3 synthesis is transferred to ECF, so plasma HCO_3^- rises further during compensation.

D. Respiratory alkalosis

Respiratory alkalosis is *always* caused by hyperventilation and is characterized by a low P_aCO_2 and an elevated arterial pH.

1. **Causes:** There are fewer primary causes of respiratory alkalosis compared with respiratory acidosis. They include increased ventilatory drive and hypoxemia.

a. **Ventilatory drive:** Hyperventilation is a common response to anxiety, such as might be induced by fear or pain, panic attacks, and hysteria. A mild respiratory alkalosis may also occur during pregnancy (see 37·IV·E). Aspirin poisoning also causes respiratory alkalosis by stimulating the respiratory control centers directly.[1]

b. **Hypoxemia:** Hypoxemia raises respiratory rate and can cause respiratory alkalosis in some circumstances. Ascent to high altitude stimulates hyperventilation to compensate for reduced O_2 availability and can precipitate respiratory alkalosis (see 24·V·A). Respiratory alkalosis may also occur when O_2 uptake is impaired due to pulmonary embolism or severe anemia.

2. **Compensation:** An acute fall in P_aCO_2 is accompanied by a decrease in plasma HCO_3^- levels (Figure 28.15). Compensation involves reduced H^+ secretion and decreased NH_3 synthesis by the kidneys. Because less "new" HCO_3^- is formed, plasma HCO_3^- level falls further during compensation.

[1]For more information on the side effects of salicylates, see *LIR Pharmacology*, 5e, p. 530.

Table 28.5: Renal Tubular Acidosis (RTA)

Type 1 RTA (Distal RTA)	
Characteristics	**Impaired H^+ secretion by the distal segments** • **Urine pH >5.3** • **Plasma HCO_3^- variable**
Renal defect	**↓ H^+-K^+ ATPase** **↑ Tubule permeability, allowing H^+ backflow** **↓ Na^+ reabsorption**
Etiology	**Familial autoimmune disorders** • **Sjögren syndrome** • **Rheumatoid arthritis** **Drugs, toxins**
Type 2 RTA (Proximal RTA)	
Characteristics	**Impaired proximal HCO_3^- reabsorption** • **Urine pH variable** • **Plasma HCO_3^- 12–20 mmol/L**
Renal defect	**Nonspecific tubule dysfunction or mutations in genes involved in HCO_3^- reabsorption**
Etiology	**Familial** **Fanconi syndrome** **Drugs, toxins** ***Carbonic anhydrase* inhibitors**
Type 4 RTA (Hypoaldosteronism)	
Characteristics	**Impaired aldosterone release or response** • **Urine pH <5.3** • **Plasma HCO_3^- >17 mmol/L** • **Hyperkalemia**
Renal defect	**Impaired Na^+ reabsorption via epithelial Na^+ channel**
Etiology	**Congenital hypoaldosteronism (Addison disease)** • **Aldosterone resistance** • **Diabetic nephropathy** • **Drugs** • **Diuretics**

E. Metabolic acidosis

Metabolic acidosis is caused by increased nonvolatile acid accumulation. It may also result from excessive HCO_3^- loss from the body. Metabolic acidosis is characterized by a low plasma HCO_3^- and a low arterial pH.

1. **Causes:** Metabolic acidosis can result from a number of different endogenous and exogenous mechanisms, including excess nonvolatile acid production, poisoning, HCO_3^- loss, and an impaired ability to excrete H^+.

    a. **Acid production:** The body normally generates ~1.5 mol of lactic acid per day, almost all of which is metabolized, mostly by the liver. Strenuous muscle activity can increase lactate production temporarily, but the liver has a high metabolic capacity, and lactate levels typically renormalize within 30 min. When the liver is damaged, lactate is allowed to accumulate and causes a lactic acidosis. Ketoacidosis is a metabolic acidosis resulting from ketone body production (i.e., acetone, acetoacetic acid, β-hydroxybutyrate) and metabolism. Ketoacidosis usually is associated with an insulin deficiency (**diabetic ketoacidosis**; see Clinical Application 33.1). Lactic acidosis and ketoacidosis both cause a high anion gap metabolic acidosis.

    b. **Drugs and poisons:** Aspirin is an acid that can produce a mixed disorder with a high anionic gap when ingested at toxic levels. Other common causes of **toxic acidosis** include methanol and ethylene glycol ingestion. Methanol is often consumed as a cheap ethanol substitute, whereas ethylene glycol is an antifreeze typically ingested accidentally. Neither poison is toxic until metabolized. Methanol is converted to formaldehyde and formic acid, whereas ethylene glycol is metabolized to glycoaldehyde and glycolic and oxalic acids. Both toxins cause a high anionic gap metabolic acidosis.

    c. **Bicarbonate loss:** The small and large intestines secrete HCO_3^- which may be excreted inappropriately during episodes of severe diarrhea, causing metabolic acidosis. HCO_3^- loss may also result from congenital or acquired disorders that impair HCO_3^- reabsorption by the PT. The resulting acidosis is known as **type 2 renal tubular acidosis** ([**RTA**] Table 28.5). Diuretics, especially *CA* inhibitors, (see 26·IX·A) can also cause HCO_3^- loss via urine.

    d. **Impaired acid excretion:** Type 1 and type 4 RTA are both characterized by a decreased ability to excrete H^+. Type 1 RTA is usually due to a congenital inability to acidify urine in the distal segments. Type 4 RTA results from hypoaldosteronism or an impairment of the renal tubule's ability to respond to aldosterone (see Table 28.5).

2. **Compensation:** Nonvolatile acid excesses are buffered by plasma HCO_3^-, causing plasma concentrations to fall (Figure 28.16). Acidosis initiates a reflex increase in ventilation to transfer volatile acid to the atmosphere, and plasma HCO_3^- falls further.

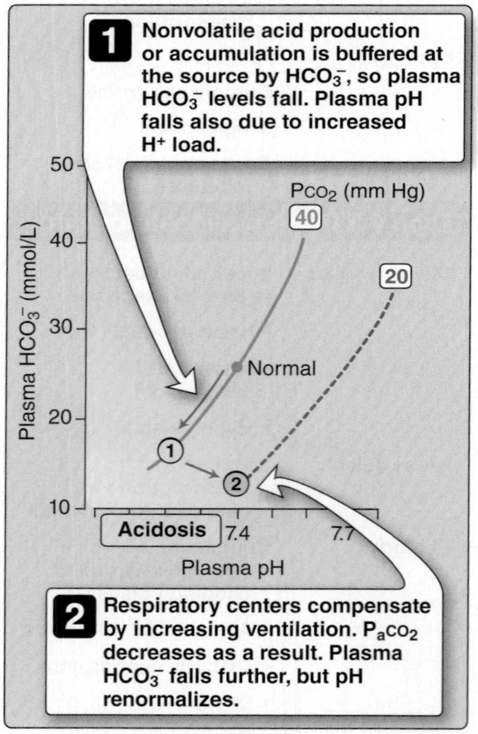
1 Nonvolatile acid production or accumulation is buffered at the source by HCO3⁻, so plasma HCO3⁻ levels fall. Plasma pH falls also due to increased H⁺ load.
50
40
30
20
10
Plasma HCO3⁻ (mmol/L)
PCO2 (mm Hg)
40
20
Normal
1
2
Acidosis
7.4
7.7
Plasma pH
2 Respiratory centers compensate by increasing ventilation. PaCO2 decreases as a result. Plasma HCO3⁻ falls further, but pH renormalizes.

Figure 28.16
Metabolic acidosis and respiratory compensation.

F. Metabolic alkalosis

Metabolic alkalosis results when the body takes in HCO_3^- or loses H^+ and is characterized by an elevated plasma HCO_3^- and arterial pH.

1. **Causes:** Although metabolic alkalosis can be caused by excessive $NaHCO_3$ intake ($NaHCO_3$ is used as an antacid), the most common causes include diuretics, vomiting, and nasogastric (NG) suctioning. Vomiting and NG suctioning cause stomach acid to be lost to the body's exterior, leaving a HCO_3^- excess that manifests as alkalosis (Figure 28.17).

2. **Compensation:** Metabolic alkalosis acutely increases plasma HCO_3^- levels and raises pH, but the respiratory system soon compensates by decreasing ventilation and allowing P_aCO_2 to rise. The kidneys may also assist compensation by reducing H^+ secretion and allowing filtered HCO_3^- to pass through the tubule to the bladder. Fluid loss during prolonged vomiting may also result in ECF volume contraction, which favors HCO_3^- reabsorption and manifests as a contraction alkalosis.

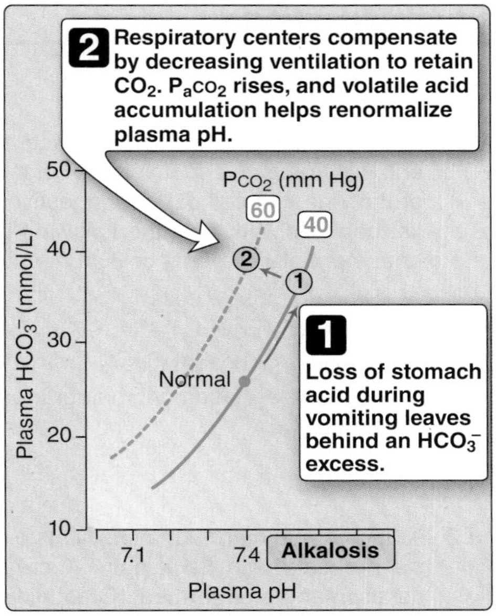

Figure 28.17
Metabolic alkalosis and respiratory compensation.

Chapter Summary

- **Total body water** is sensed through changes in the **osmolality** of extracellular fluid. **Osmosensors** are located within **circumventricular organs** within the brain, in close proximity to the **hypothalamus**.
- A decrease in total body water increases extracellular fluid osmolality. The osmoreceptors respond by stimulating **antidiuretic hormone** (**ADH**) release from the **posterior pituitary**. ADH causes **aquaporins** to be inserted into the epithelial lining of the **collecting ducts**, which permits water reabsorption. Osmoreceptor activation also increases **thirst**.
- Extracellular fluid osmolality also depends on plasma Na^+ levels. Na^+ balance is controlled principally by **aldosterone**-induced increases in renal Na^+ retention. Aldosterone is released during ***renin*–angiotensin–aldosterone system** activation.
- Water and Na^+ balance are dominated by the need to optimize **mean arterial pressure** (**MAP**). MAP is determined, in part, by extracellular fluid (ECF) volume. When ECF volume is low, MAP falls, and the *renin*–aldosterone–angiotensin system (RAAS) activates. **Angiotensin II** (Ang-II) is the primary RAAS effector hormone. Ang-II stimulates aldosterone release, modulates glomerular filtration rate, stimulates Na^+ reabsorption from the renal tubule, and promotes water retention via antidiuretic hormone release.
- **Atrial natriuretic peptide** (**ANP**) provides a negative feedback pathway that limits extracellular fluid (ECF) volume expansion. ANP is released from cardiac atria when ECF volume is high. ANP antagonizes the actions of angiotensin II and promotes **natriuresis** and **diuresis**.
- K^+ balance is controlled by aldosterone. Aldosterone is released as a direct response to **hyperkalemia** and stimulates K^+ secretion by the distal tubule. **Hypokalemia** stimulates K^+ reabsorption, primarily in the distal tubule segments.
- The kidneys and lungs together maintain pH balance. The lungs excrete **volatile acid** (H_2CO_3). The kidneys excrete **nonvolatile acid** and can help compensate for changes in pH balance caused by respiratory disorders.
- Plasma pH is normally maintained within a narrow range (7.35–7.45). A rise in P_aCO_2 causes a **respiratory acidosis** and **acidemia** (pH $<$7.35). The kidneys compensate by excreting additional H^+. **Hyperventilation** causes a **respiratory alkalosis** and **alkalemia** (pH $>$7.45). The kidneys compensate by reducing H^+ secretion.
- Accumulation of nonvolatile acids (e.g., lactic acid and ketone bodies), toxins, and renal disturbances in H^+ secretion or HCO_3^- reabsorption can cause **metabolic acidosis**. The lungs compensate by increasing ventilation and transferring CO_2 to the atmosphere. Loss of stomach H^+ as a result of prolonged vomiting causes a **metabolic alkalosis**. The lungs compensate by retaining CO_2, and the kidneys decrease H^+ secretion.

Study Questions

Choose the ONE best answer.

VI.1 A patient taking penicillin for a bacterial infection presents with nausea and vomiting. Urinalysis reveals mild proteinuria and cell casts, suggestive of acute interstitial nephritis. Which of the following glomerular structures normally prevents cells from entering the tubule?

A. Smooth muscle cells
B. Mesangial cells
C. Capillary endothelial cells
D. Glomerular basement membrane
E. Podocytes

Best answer = C. The glomerular filtration barrier comprises capillary endothelial cells, a basement membrane, and a filtration slit diaphragm located between podocyte foot processes (25·III·A). Capillary walls are fenestrated to enhance plasma filtration, but the pores are small (~70 nm), effectively trapping the cells in the vasculature. Smooth muscle cells are located in glomerular arterioles, whereas mesangial cells are located between the glomerular capillaries. Although the latter regulate barrier surface area, they are not directly involved in fluid filtration.

VI.2 A 65-year-old man with a family history of nephrolithiasis presents with flank pain. A creatinine clearance assessment is performed. "Creatinine clearance" best equates with which of the following?

A. Renal blood flow
B. Renal plasma flow
C. Amount of creatinine traversing the glomerulus per minute
D. Amount of creatinine entering the urinary bladder per minute
E. Plasma volume completely cleared of creatinine per minute

Best answer = E. "Clearance" defines the kidneys' ability to completely clear a known volume of plasma of a given substance during passage through the renal vasculature (25·V·A). Creatine clearance is used clinically to estimate glomerular filtration rate (25·V·C). Clearance of other substances (e.g., *para*-aminohippuric acid) can be used to estimate renal plasma flow and, if hematocrit is known, renal blood flow (25·V·D). A change in clearance might affect how much creatinine enters the bladder, but excretion rate does not equate with clearance. Clearance is unrelated to the amount of a substance traversing the glomerular network per unit time.

VI.3 A 17-year-old male presents with urethral burning following urination. He is asked to provide a urine sample and swabbed to test for a possible bacterial infection. Which of the following is responsible for initiating micturition when providing a urine sample?

A. Pontine micturition center
B. Uroepithelial mechanoreceptors
C. Spontaneous detrusor contractions
D. Rising intravesical pressure
E. Internal urethral sphincter relaxation

Best answer = A. Voiding is initiated and coordinated by the pontine micturition center, which relaxes the internal (involuntary) urethral sphincter and facilitates detrusor muscle contraction once voluntary relaxation of the external urethral sphincter has occurred (25·VI·D). Although internal sphincter relaxation is required for urine flow, it does not initiate voiding. Uroepithelial mechanoreceptors trigger spontaneous detrusor contractions when intravesical pressure rises during bladder filling, but bladder emptying is suppressed by the pontine micturition center until voiding is convenient.

VI.4 A 31-year-old male with a body mass index of 35 is found to have glycosuria during a routine physical. Elevated urinary glucose levels correlate with unmanaged type 2 diabetes mellitus. Why does glucose appear in the urine of patients with untreated diabetes?

A. Tubule glucose levels exceed transport capacity.
B. Glucose causes an osmotic diuresis that increases glucose excretion.
C. Hyperglycemia downregulates glucose transporters.
D. High plasma insulin levels are nephrotoxic.
E. High plasma insulin inhibits Na^+-K^+ ATPases.

Best answer = A. Transporters exhibit saturation kinetics, which limits the tubule's ability to reabsorb solutes (26·III·A). Although glucose transport maximum is seldom reached in a healthy individual, the plasma ultrafiltrate of patients with untreated diabetes may contain glucose levels that exceed the tubule's reabsorptive capability, causing it to appear in urine. Glucose can cause an osmotic diuresis, but such an event would be a consequence of exceeding transporter maximum, not the cause. Possible effects of hyperglycemia on transporter numbers and insulin-induced nephrotoxicity is not a significant physiologic concern. Insulin does modulate the Na^+-K^+ ATPase, but it increases pump activity rather than inhibiting it.

VI.5 Fanconi syndrome is associated with proximal tubule (PT) dysfunction, the symptoms including polyuria, glycosuria, hypocalcemia, hypomagnesemia, and hypophosphatemia. A healthy PT normally recovers ~100% of which of the following filtered solutes?

A. Peptides
B. Uric acid
C. Ca^{2+}
D. PO_4^{3-}
E. Na^+

Best answer = A. The proximal tubule (PT) reabsorbs a high percentage of most materials filtering from blood, including Ca^{2+}, PO_4^{3-}, and Na^+, but it is the principal site for reabsorption of 100% of proteins, peptides, amino acids, and glucose (26·III). The PT recovers 65% of Ca^{2+}, the remainder being recovered in the thick ascending limb and distal segments. The PT recovers 80% of the PO_4^{3-} filtered load, with the remainder being recovered distally. The PT recovers 67% of Na^+, although this amount can increase in the presence of angiotensin II. The PT secretes uric acid, oxalate, and other wastes (26·IV).

VI.6 A 66-year-old woman receiving cisplatin therapy for metastatic ovarian cancer develops proximal tubule (PT) nephrotoxicity and symptoms associated with renal impairment. Which of the following best describes PT function in a healthy person?

A. Antidiuretic hormone is a primary regulator.
B. Aldosterone is a primary regulator.
C. It accomplishes isosmotic fluid reabsorption.
D. It creates the corticopapillary gradient.
E. The tubule has a high electrical resistance.

Best answer = C. The proximal tubule (PT) epithelium actively takes up many organic solutes (including drugs such as cisplatin) from blood and excretes them into the tubule (26·IV). Concentrating such materials through uptake can cause them to rise to toxic levels. The PT is also specialized for isosmotic fluid reabsorption, which gives the epithelium a low electrical resistance (26·II·A). Antidiuretic hormone acts principally on the collecting ducts (27·V·C), whereas aldosterone targets the distal tubule segments (27·IV). The corticopapillary osmotic gradient is established by the loop of Henle (27·II·C).

VI.7 Increasing which of the following variables would decrease the magnitude of the renal corticopapillary osmotic gradient that allows for urine concentration?

A. *Renin* release from the afferent arteriole
B. Thick ascending limb Na^+-K^+-$2Cl^-$ cotransport
C. Urea reabsorption by the collecting ducts
D. Blood flow through the vasa recta
E. Sympathetic nervous system activation

Best answer = D. The corticopapillary osmotic gradient is established by countercurrent multiplication in the loop of Henle (27·II·C). The countercurrent multiplier relies on Na^+-K^+-$2Cl^-$ cotransport by the thick ascending limb, so the gradient collapses when the cotransporter is inhibited by loop diuretics. Increasing flow through the vasa recta washes ions out of the medulla, thereby diminishing the osmotic gradient. *Renin* is released when arterial pressure falls or when the sympathetic nervous system activates, conditions that signal a likely need to conserve water. Gradient magnitude increases as a result, in part through increased urea reabsorption from the collecting ducts.

VI.8 Genetic evaluation of a 6-year-old boy with growth and mental retardation identified alleles associated with Bartter syndrome. Bartter syndrome mimics loop diuretics by causing thick ascending limb (TAL) dysfunction. Which of the following best describes the TAL in healthy individuals?

A. Fluid leaves the thick ascending limb at ~600 mOsm/kg.
B. It is known as the "concentrating segment."
C. It has a high water permeability.
D. It is the primary site of Ca^{2+} reabsorption.
E. It extracts Na^+, K^+, and Cl^- from the lumen.

Best answer = E. The thick ascending limb (TAL) reabsorbs Na^+, K^+, and Cl^- from the tubule lumen via Na^+-K^+-$2Cl^-$ cotransport and transfers these ions to the interstitium, where they help form the corticopapillary osmotic gradient (27·II·B). The TAL has a low water permeability that prevents H_2O from following ions into the interstitium, so the tubule fluid becomes relatively dilute (<300 mOsm/kg). The TAL may be referred to as the "diluting (not "concentrating") segment" for this reason. Ca^{2+} is reabsorbed primarily in the proximal tubule, with regulated reabsorption occurring in the distal tubule (27·III·C).

VI.9 A 77-year-old woman is taking a thiazide diuretic to treat hypertension but has become hypercalcemic. Thiazides inhibit Na^+-Cl^- reabsorption by the distal convoluted tubule. Why do thiazide diuretics also cause hypercalcemia?

A. Thiazide diuretics also inhibit Ca^{2+} ATPases.
B. The Na^+-Cl^- cotransporter also carries Ca^{2+}.
C. Apical Na^+-Ca^{2+} exchange increases.
D. The gradient driving Ca^{2+} uptake steepens.
E. Paracellular Ca^{2+} uptake increases.

Best answer = D. Ca^{2+} reabsorption by the distal convoluted tubule (DCT) is mediated by a Ca^{2+} channel (a transient receptor-potential channel, TRPV5) and driven by the electrochemical gradient across the tubule epithelium's apical membrane (27·III·C). Inhibiting the Na^+-Cl^- cotransporter reduces Na^+ influx into the epithelial cell, so the interior becomes more negative. This negativity increases the driving force for Ca^{2+} reabsorption and causes hypercalcemia. Thiazides have no significant effect on Ca^{2+} ATPases. The DCT does not reabsorb significant amounts of Ca^{2+} via the Na^+-Cl^- cotransporter, an apical Na^+-Ca^{2+} exchanger, or paracellularly.

VI.10 A researcher observes a consistent 75% decrease in renal blood flow in subjects performing maximal exercise. Which of the following best accounts for the decreased flow?

A. Decrease in mean arterial pressure
B. Decrease in renal arterial pressure
C. Sweat-induced hypovolemia
D. Increased renal sympathetic nerve activity
E. Antidiuretic hormone release

Best answer = D. The sympathetic nervous system (SNS) increases cardiac output and decreases flow to nonessential organs (such as the kidney) to sustain mean arterial pressure (MAP) during skeletal muscle vasodilation (28·III·C; 39·V). The SNS reduces renal blood flow by constricting resistance vessels (arterioles, including glomerular arterioles, and small arteries). Renal arterial pressure, which is closely tied to MAP, should not be affected to a significant degree. Although antidiuretic hormone can vasoconstrict under some circumstances, these effects are secondary to SNS activation. Hypovolemia may potentiate SNS effects on renal flow during exercise, but, again, this is secondary to SNS effects.

VI.11 A physician notes that an underweight teenage girl's tooth enamel is eroded. A basic metabolic panel reveals hypokalemia and metabolic alkalosis, suggestive of an eating disorder and repeated purging. Which of the following would also be consistent with such a diagnosis?

A. Renal tubular acidosis
B. Decreased β-intercalated cell activity
C. Hypoventilation
D. Increased NH_4^+ excretion
E. High plasma aldosterone levels

Best answer = C. Loss of stomach acid during repeated vomiting leaves an HCO_3^- excess that manifests as metabolic alkalosis (28·VI·F). The respiratory centers help compensate by decreasing volatile acid (H_2CO_3) transfer to the environment by decreasing ventilation (hypoventilation). Renal tubular acidosis is a metabolic acidosis that may have a number of underlying causes. NH_4^+ excretion helps dispose of nonvolatile acid, so excretion rates would fall during alkalosis. β-Intercalated cells secrete HCO_3^- into the tubule lumen, and, thus, their activity would be increased during alkalosis. Aldosterone is involved in Na^+ balance, not pH balance.

VI.12 A 25-year-old patient with recurrent flash (rapid onset) pulmonary edema is evaluated for renal hypertension using Doppler ultrasonography. Tests confirm renal artery stenosis. An *angiotensin-converting enzyme* inhibitor might have which of the following effects in this patient?

A. Unchanged hypertension
B. Increased plasma creatinine
C. Decreased plasma *renin*
D. Increased glomerular filtration rate
E. Increased systemic vascular resistance

Best answer = B. Renal artery stenosis impairs glomerular perfusion and decreases ultrafiltration pressure (P_{UF}). The afferent arteriole responds by releasing *renin* (28·IV·C). Plasma angiotensin II (Ang-II) levels rise as a result, causing systemic vasoconstriction and a rise in mean arterial pressure (MAP). The MAP increase helps restore glomerular flow and P_{UF} rises. Ang-II also constricts the efferent arteriole to potentiate a rise in P_{UF} (25·IV·C). *Angiotensin-converting enzyme* (*ACE*) inhibition would, thus, decrease P_{UF} and glomerular filtration rate, which would allow plasma creatinine levels to rise. The afferent arteriole would respond with increased *renin* release. An *ACE* inhibitor would also attenuate Ang-II effects on systemic vessels, decreasing systemic vascular resistance, thereby reducing MAP.

Principles and Signaling

29

I. OVERVIEW

The gastrointestinal (GI) system is a complex tube bounded by the mouth at one end and the anus at the other. Food enters the mouth; travels through the esophagus, stomach, small intestine (duodenum, jejunum, and ileum), large intestine (ascending, transverse, and descending colon), and rectum; and then exits via the anus (Figure 29.1). This tube's primary function is the **absorption** of dietary nutrients. To maximize nutrient absorption, **secretions** are added to food from the salivary glands, stomach, liver, gallbladder, and pancreas to convert complex molecules into simpler ones. This conversion, called **digestion**, is effected by enzymes and H^+. Dietary contents and secretions are mixed and propelled along the tube (**motility**) from one specialized compartment to another by coordinated peristaltic contractions and relaxations of the tube walls (Figure 29.2). Two other important GI functions include **storage** (e.g., food is stored in the stomach and fecal matter in the colon) and **excretion** of undigested materials and biliary waste products.

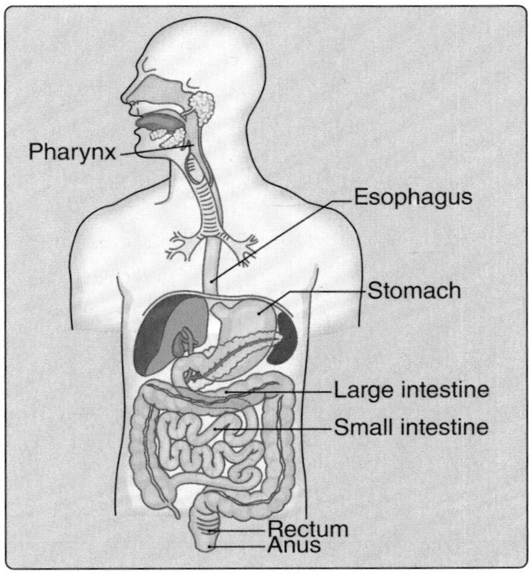

Figure 29.1
Gastrointestinal tract.

II. GASTROINTESTINAL LAYERS

The intestinal tract is composed of multiple layers, each possessing a distinct function. Depending on structure–function relations, a particular layer's prominence changes along the length of the tube. Moving from the lumen to the outside of the tube, the layers include epithelium, lamina propria, muscularis mucosa, submucosa, submucosal plexus, circular muscle, myenteric plexus, longitudinal muscle, and serosa (Figure 29.3).

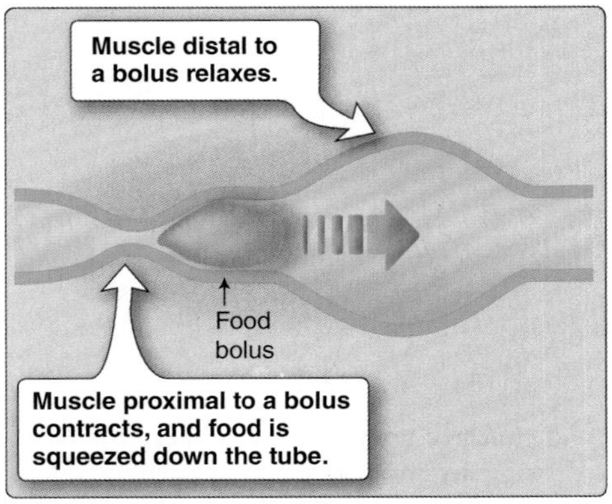

Figure 29.2
Peristalsis.

A. Mucosa

The epithelium, lamina propria, and muscularis mucosa together form the **mucosa**. The **epithelium** is a single cell layer forming a continuous lining of the GI tract. GI epithelial cells are shed and replaced every 2–3 days. The apical side of the epithelium faces the GI lumen, and the basolateral side faces the interstitium and vasculature. Apical surfaces may be enhanced with **villi** (thumblike projections)

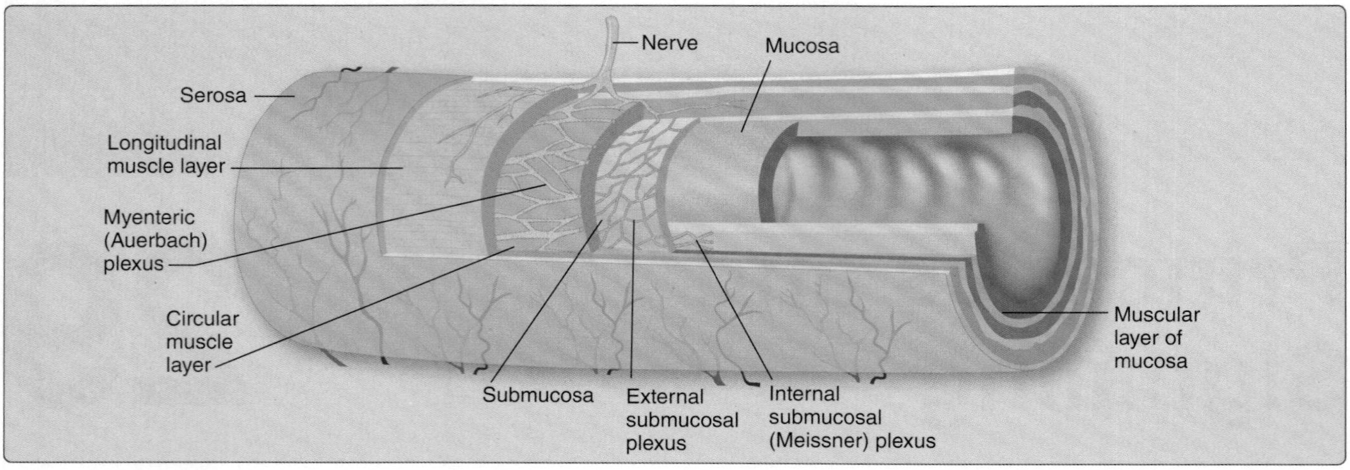

Figure 29.3
Gastrointestinal tract layers.

and **crypts** (invaginations) to increase surface area and maximize contact between epithelium and intestinal contents (Figure 29.4). Absorptive areas (e.g., small intestine) contain numerous apical enhancements. Areas primarily involved with motility (e.g., esophagus) do not. The **lamina propria** is a loose connective tissue composed of elastin and collagen fibers that contains sensory nerves, blood and lymph vessels, and some secretory glands. The **muscularis mucosa** is a thin layer of smooth muscle that further increases surface area by creating mucosal ridges and folds.

B. Submucosa

The **submucosa** is a thicker layer with a similar composition to the lamina propria. It incorporates blood vessels and bundles of nerves that collectively form a **submucosal plexus** (Meissner plexus), which is an integral part of the enteric nervous system (ENS). The ENS is described in more detail below.

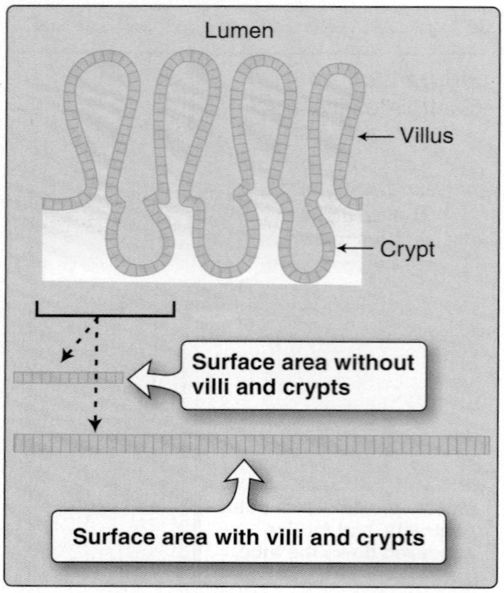

Figure 29.4
Surface area enhancements.

C. Muscularis externa

The **muscularis externa** comprises **circular muscle**, the **myenteric plexus**, and **longitudinal muscle** layers. The two smooth muscle layers are named based on their orientation. The circular muscle layer is arranged in rings and pinches the tube when it contracts. The longitudinal layer is arranged in parallel and shortens the tube when it contracts. The ENS coordinates circular and longitudinal muscle contraction to mix intestinal contents and move them between compartments. Circular muscle also forms sphincters, which regulate flow of food from one compartment to the next by modulating lumen diameter. The myenteric plexus (Auerbach plexus), also part of the ENS, is located between the circular and longitudinal muscle layers.

D. Serosa

The **serosa** comprises an outermost layer of connective tissue and a layer of squamous epithelial cells. Some portions of the GI tract (e.g., the esophagus) do not have a serosal layer but rather connect

directly to the adventitia, which is connective tissue that blends into the abdominal or pelvic wall.

III. INNERVATION AND NEUROTRANSMITTERS

GI function is regulated by three divisions of the autonomic nervous system (ANS): the parasympathetic nervous system (PSNS), sympathetic nervous system (SNS), and the ENS.

A. Parasympathetic nervous system

Parasympathetic innervation is derived from the vagus (medulla oblongata) and pelvic–splanchnic nerves (S2–S4) and has both motor and sensory components (Figure 29.5). The sensory components respond to stretch, pressure, temperature, and osmolarity and participate in **vagovagal reflexes**. Vagovagal reflexes occur when the vagus nerve (cranial nerve X) participates in both afferent sensation and efferent responses without central nervous system involvement. Primary neurotransmitters used directly or indirectly by the PSNS are **acetylcholine (ACh)**, **gastrin-releasing peptide**, and **substance P** (Table 29.1). In general, signals from the PSNS stimulate GI secretions and motility, which facilitates digestion and absorption of nutrients.

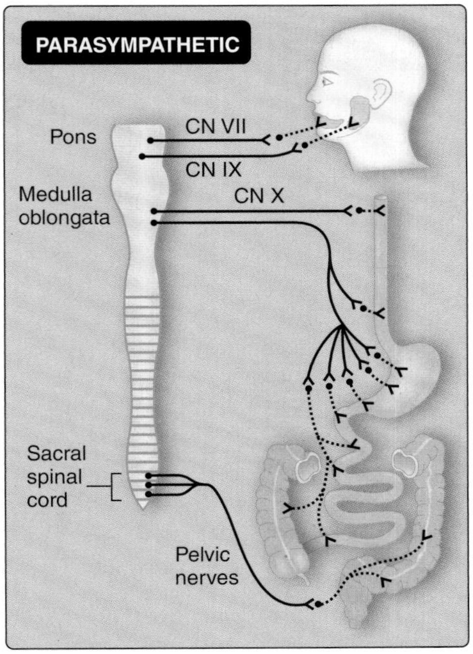

Figure 29.5
Parasympathetic innervation.
CN = cranial nerve.

B. Sympathetic nervous system

Sympathetic nerves originate in the thoracic (T5–T12) and lumbar (L1–L3) regions and synapse in one of three ganglia: celiac, superior mesenteric, or inferior mesenteric for the lower GI system (Figure 29.6). The upper GI tract (e.g., salivary glands) is innervated by SNS nerves that synapse within the superior cervical ganglion (see Figure 29.6). Unlike the PSNS, the SNS component does not contain a direct sensory arm and generally decreases GI secretions and motility when active. The primary SNS neurotransmitters are **norepinephrine** and **neuropeptide Y** (see Table 29.1).

Table 29.1: Gastrointestinal Neurotransmitters and Neuromodulators

Neurotransmitter	Releasing Nerves	Structures	Function
Acetylcholine	Parasympathetic, cholinergic	Smooth muscle, glands	Contracts wall muscle; relaxes sphincters; increases salivary, gastric, and pancreatic secretion
Vasoactive intestinal peptide	Parasympathetic, cholinergic, enteric	Smooth muscle, glands	Relaxes sphincters; increases pancreatic and intestinal secretion
Norepinephrine	Sympathetic, adrenergic	Smooth muscle, glands	Relaxes wall muscle; contracts sphincters; decreases salivary secretions
Neuropeptide Y	Sympathetic, adrenergic, enteric	Smooth muscle, glands	Relaxes wall muscle; decreases intestinal secretions
Gastrin-releasing peptide	Parasympathetic, cholinergic, enteric	Glands	Increases gastrin secretion
Substance P	Parasympathetic, cholinergic, enteric	Smooth muscle, glands	Contracts wall muscle; increases salivary secretions
Enkephalins	Enteric	Smooth muscle, glands	Constrict sphincters; decrease intestinal secretions

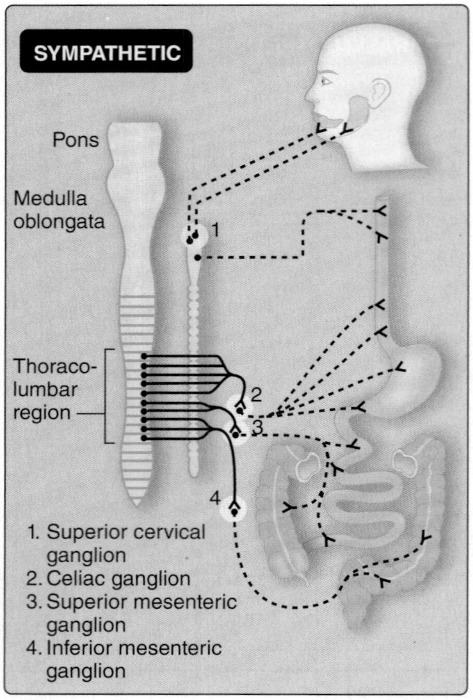

Figure 29.6
Sympathetic innervation.

C. Enteric nervous system

PSNS and SNS nerves usually synapse with components of the ENS. Although the ENS is modulated by these extrinsic neural inputs, it can operate autonomously via intrinsic regulation and sensory reflexes. ENS nerves are organized into myenteric and submucosal plexuses.

1. **Plexuses:** The myenteric plexus forms a dense parallel neuronal configuration that primarily regulates intestinal smooth muscle and participates in tonic and rhythmic contractions. Some myenteric neurons also synapse with neurons in the submucosal plexus or directly on secretory cells. The submucosal plexus primarily regulates intestinal secretions and the local absorptive environment but also can synapse on blood vessels, circular and longitudinal muscle, and the muscularis mucosa. ENS neurons are supported by **enteric glial cells**, which structurally and functionally resemble astrocytes in the brain.

2. **Reflexes:** Many GI reflex actions are regulated solely by neural circuits in which a mechanoreceptor or chemoreceptor is stimulated in the mucosa and transmits the signal back to neurons in the submucosal plexus, which stimulates other neurons in the submucosal or myenteric plexus that regulate endocrine or secretory cells.

3. **Neurotransmitters:** There are a number of neurotransmitters and regulatory molecules used in ENS communication (see Table 29.1). **Enkephalins** constrict circular muscle around sphincters. In the submucosal plexus, secretory neurons primarily use **vasoactive intestinal peptide** (**VIP**) and ACh as neurotransmitters, whereas sensory nerves use substance P. In the myenteric plexus, motor neurons use ACh and nitric oxide, sensory neurons use substance P, and the interneurons use ACh and **serotonin** (**5-hydroxytryptamine**). These enteric neurotransmitters are also used elsewhere in the body and are important pharmacologically. For example, a person on serotonin reuptake inhibitors may experience decreased GI motility as a side effect because these drugs alter serotonin levels.

Clinical Application 29.1: Chagas Disease

Submucosal and myenteric plexus neuropathy can impair motility. For example, a protozoan infestation (*Trypanosoma cruzi*, often delivered by kissing bugs) of these plexus neurons can lead to **Chagas disease**. Among other pathologies, Chagas disease causes distention and structural enlargements of the esophagus and colon because regions with neuropathy can constrict but not relax muscular layers. The asymptomatic sections continue to deliver food, which is retained just proximal to the constricted area. Such retention stretches these areas and, over time, enlarges and contorts them.

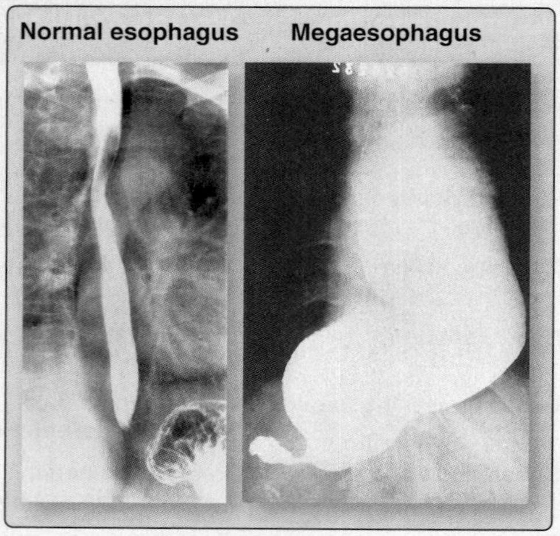

Megaesophagus.

Table 29.2: Gastrointestinal Hormones

Hormone	Releasing Cells	Structures	Function
Cholecystokinin	I cells	Pancreas, gallbladder, stomach	Increases enzyme secretion; contracts gallbladder; increases gastric empting
Glucose-dependent insulinotropic peptide	K cells	Pancreas, stomach	Releases insulin; inhibits acid secretion
Gastrin	G cells	Stomach	Increases gastric acid secretion
Motilin	M cells	Gastrointestinal smooth muscle	Increases contractions and migrating motor complexes
Secretin	S cells	Pancreas, stomach	Releases HCO_3^- and pepsin

IV. NONNEURAL SIGNALING MOLECULES

In addition to neurotransmitters, **hormones** and **paracrine** signaling molecules also regulate and control GI function.

A. Hormones

GI peptide hormones include **cholecystokinin** (**CCK**), **gastrin**, glucose-dependent insulinotropic peptide ([GIP] formerly known as gastric inhibitory peptide), **motilin**, and **secretin** (Table 29.2). Endocrine cell types are located in different densities in various locations throughout the stomach and intestines (Figure 29.7). Gastrin is secreted in the stomach antrum and then tapers off in the small intestine. CCK, secretin, GIP, and motilin are primarily secreted in the duodenum and jejunum, and CCK and secretin continue to be secreted in the ileum, albeit to a lesser degree.

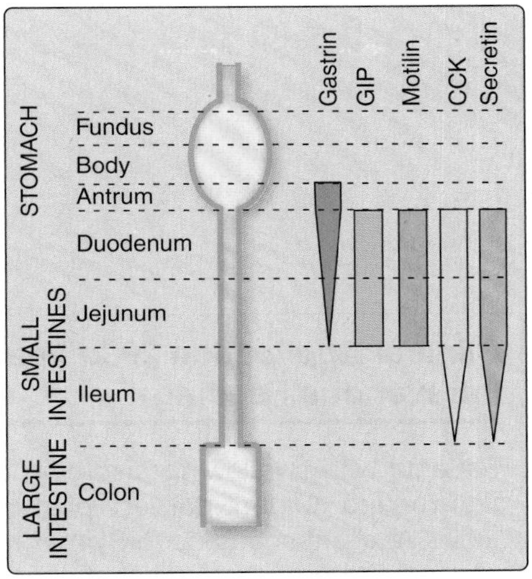

Figure 29.7
Principal sites of gastrointestinal hormone release. CCK = cholecystokinin; GIP = glucose-dependent insulinotropic peptide.

B. Paracrines

GI paracrines are both released and act locally. The primary paracrine GI signaling molecules are **histamine**, **prostaglandins**, and **somatostatin** (Table 29.3). Of these, only somatostatin is a peptide. Histamine is classified as a monoamine, and prostaglandins are eicosanoid-signaling molecules. Histamine is released in the stomach, whereas both prostaglandins and somatostatin are more widespread in their release and actions.

Table 29.3: Gastrointestinal Paracrines

Paracrine	Releasing Cells	Structures	Function
Histamine	Enterochromaffin-like cells, mast cells	Stomach	Increases gastric acid secretion
Prostaglandins	Cells lining gastrointestinal tract	Mucosa	Increase blood flow and mucus and HCO_3^- secretion
Somatostatin	D cells	Stomach and pancreas	Inhibits peptide hormones and gastric acid secretion

> Prostaglandins are *cyclooxygenase* products derived from arachidonic acid. Prostaglandins have an important role in maintaining mucosal integrity and, thus, *cyclooxygenase* inhibitors (i.e., aspirin and other **nonsteroidal anti-inflammatory drugs**) can cause stomach irritation.[1]

V. DIGESTIVE PHASES

Stomach and duodenal function can be divided into three discrete phases: cephalic, gastric, and intestinal.

A. Cephalic phase

The **cephalic phase** is triggered by the thought of food or conditions suggestive of previous food intake (e.g., classical conditioning to eat after hearing a dinner bell). Chemoreceptors and mechanoreceptors in the oral and nasal cavities and throat that are stimulated by tasting, chewing, swallowing, and smelling food also contribute. The cephalic phase is primarily neural and causes ACh and VIP release. ACh and VIP stimulate secretion by the salivary glands, stomach, pancreas, and intestines.

Clinical Application 29.2: Feeding Tubes and Intravenous Feeding

Patients with swallowing disorders or on mechanical ventilation require nutrient delivery past obstructed areas. Feeding tubes (e.g., nasogastric [NG] and nasoduodenal [ND] tubes) are used to provide nutritional support for these patients. NG tubes deliver food directly to the stomach, whereas ND tubes deliver food directly to the duodenum. Feeding tubes thereby bypass the majority of the digestive phase initiation cues. This requires the feeding tube formula to be prepared in a manner that will not require upper gastrointestinal (GI) processing of food. Nutrients can also be directly infused intravenously, which bypasses the entire GI system. Care must be taken to include all required nutrients, although less total kcals are necessary insofar as ~7% of energy consumed by mouth is used to digest and absorb nutrients.

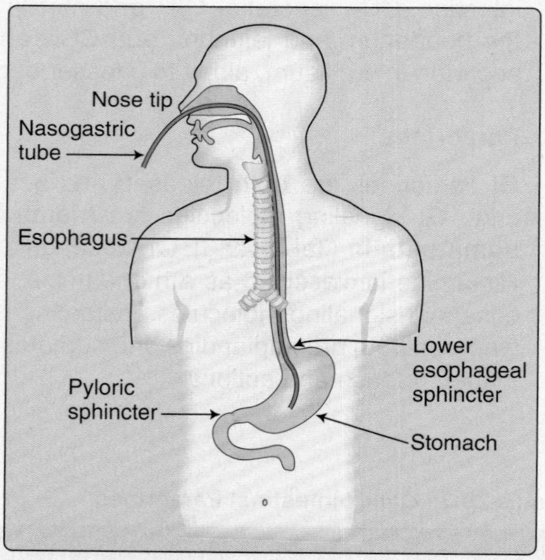

Nasogastric tube.

[1]For a discussion of the gastrointestinal effects of nonsteroidal anti-inflammatory drugs, see *LIR Pharmacology*, 5e, p. 531.

B. Gastric phase

The **gastric phase** begins when food and oral secretions enter the stomach. It coincides with distention and stomach contents (amino acids and peptides) and elicits neural, hormonal, and paracrine GI responses. A good example of this combination of signaling molecules is in gastric acid secretion, which includes ACh (neural), gastrin (hormonal), and histamine (paracrine).

C. Intestinal phase

The **intestinal phase** begins when stomach contents enter the duodenum. It is linked to digested constituents of proteins and fats as well as H^+ and initiates primarily hormonal but also paracrine and neural responses. CCK, gastrin, secretin, and GIP are all secreted during this phase.

Chapter Summary

- **Absorption** is the process of transporting dietary contents across the gastrointestinal barrier into the body.
- To prepare nutrients for absorption, the body mechanically and chemically breaks down food into smaller, simpler particles. The chemical breakdown of food is **digestion**, and the mechanical breakdown of food involves smooth (i.e., as in mixing) or skeletal (i.e., as in chewing) muscle contractions.
- **Secretion** is the act of transporting molecules or fluid from the body to the gastrointestinal lumen. Secretion facilitates digestion by delivering enzymes and water and protects the endothelial surface by secreting **HCO_3^-** and **mucus**.
- The autonomic nervous system innervates the entire gastrointestinal (GI) system. The parasympathetic nervous system most often facilitates secretion and motility, whereas the sympathetic nervous system decreases these functions. The **enteric nervous system** can operate independently and is involved with reflexes and the majority of GI functions.
- Gastrointestinal hormones include **cholecystokinin**, which is released from I cells and participates in pancreatic and biliary secretions; **gastrin**, which is released from G cells and primarily functions in H^+ secretion; **glucose-dependent insulinotropic peptide**, which is released from K cells and primarily functions to increase **insulin** release and decrease H^+ secretion; motilin, which primarily functions to increase motility; and secretin, which is released from S cells and primarily functions to increase water and HCO_3^- secretion and decrease H^+.
- Gastrointestinal (GI) paracrines include **histamines**, which are derived from enterochromaffin-like cells and mast cells and have many functions such as increasing H^+ production; **prostaglandins**, which have many functions including decreasing H^+ production and maintaining GI barrier properties; and **somatostatin**, which decreases GI secretions.
- The phases of digestion (**cephalic**, **gastric**, and **intestinal**) allow for preparation and timing and regulation feedback. The cephalic is primarily a feedforward regulation, and the gastric and intestinal phases are feedback mechanisms.

30 Mouth, Esophagus, and Stomach

I. OVERVIEW

The upper half of the gastrointestinal (GI) tract (mouth, esophagus, and stomach) plays a minimal role in nutrient absorption but contributes in the transportation and preparation of food to be absorbed in the small intestine. This preparation involves mechanically breaking down food into smaller pieces to increase its surface area. Preparation also involves chemical actions, such as secreting enzymes and acid to break down food and hydrating it to improve the local aqueous environment for enzymatic action.

II. MOUTH

The mouth serves as the first site of mechanical and chemical digestion of food. **Mastication** (chewing) breaks food down into smaller pieces to increase the surface area available to digestive enzymes and to ease swallowing. Saliva provides the majority of the oral hydration and lubrication and performs some protective and digestive functions.

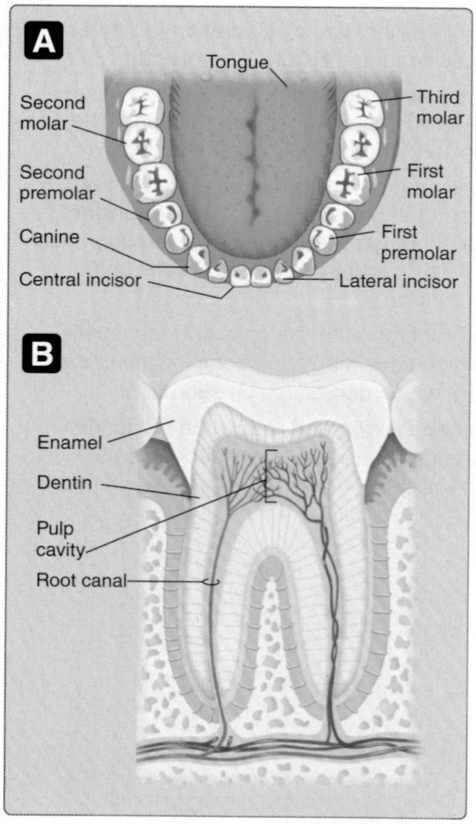

Figure 30.1
Teeth classification and anatomy.

A. Teeth

Teeth aid in cutting (incisors), tearing and piercing (canines), and grinding and crushing (premolars and molars) food (Figure 30.1). The crown portion of teeth is coated with enamel, which is >95% calcium hydroxyapatite (see 15·II·A). This extremely hard shell allows for mastication functions and, along with dentin (hard but less mineralized connective tissue), protects the pulp cavity (containing nerves and blood vessels) and root canal (see Figure 30.1). Jaw muscles provide mechanical force and movement for the teeth to perform their functions.

B. Tongue

The tongue grips and repositions food during mastication. The tongue contains intrinsic skeletal muscles (fibers running longitudinally, vertically, and in a transverse plane of the tongue), which allow the tongue to change shape, and extrinsic skeletal muscles that the tongue uses to change position, such as to protrude and to move from side to side. The tongue also contains taste buds (see 10·II·A) and serous and mucous glands. However, these glands do not secrete solutions in sufficient quantity to adequately hydrate food without saliva.

C. Salivary glands

Salivary glands produce a watery fluid that lubricates the mouth, begins food digestion, and is protective. Individuals normally produce1.0–1.5

liters of saliva daily, the majority of which is secreted by the **sublingual**, **submandibular**, and **parotid glands**.

1. **Anatomy:** Salivary glands comprise numerous **lobules**. Each lobule contains several **acini**, each lined with epithelial (**acinar**) cells that are specialized for synthesis and secretion of protein and a serous fluid. The fluid has an ionic composition that resembles plasma. The **primary secretion** drains from an acinus via an **intercalated duct** into a larger **striated duct**. Striated ducts, in turn, drain fluid into an **interlobular duct**. Cells lining these ducts modify the ionic composition of the primary secretion during its passage, which is assisted by **myoepithelial cell** contraction. Acinar and ductal epithelia also contain mucous cells that secrete mucin, a glycoprotein that gives mucus its lubrication properties. Sublingual and submandibular glands secrete a mixed serous and mucous solution, whereas the parotid gland primarily secretes serous fluid.

2. **Serous secretion:** Saliva is always hypotonic with respect to plasma (Figure 30.2), but the primary acinar secretion is close to being isosmotic. Salivary secretion is facilitated by the ion gradients established by the basolateral Na^+-K^+ ATPase (Figure 30.3). Na^+, K^+, and Cl^- are taken up from plasma via the interstitium and a basolateral Na^+-K^+-$2Cl^-$ cotransporter, with K^+ then passing across the apical membrane via a K^+ channel. Cl^- is secreted into the acinar lumen by a Cl^--HCO_3^- cotransporter. HCO_3^- is generated from CO_2 and H_2O in a reaction catalyzed by *carbonic anyhydrase*. The H^+ generated during HCO_3^- formation then leaves the cell via a basolateral Na^+-H^+ exchanger. Cl^- and HCO_3^- secretion creates a trans-epithelial potential difference that favors paracellular Na^+ movement into the acinar lumen. H_2O follows transcellularly and paracellularly, driven by the osmotic gradients created by ion secretion and facilitated by aquaporins and leaky tight junctions between acinar cells.

3. **Ductal modification:** Intercalated and striated duct cells modify the composition of the primary secretion by reabsorbing Na^+ and Cl^-, while simultaneously secreting K^+ and HCO_3^-. The effects of this modification are most obvious at low salivary secretion rates (see Figure 30.2). Ductal cell transport capacity is limited, however, so salivary composition increasingly resembles the primary secretion as secretion rates rise. Na^+ is reabsorbed from the duct lumen via an epithelial Na^+ channel (ENaC) and a Na^+-H^+ exchanger located in the apical membrane and then is pumped across the basolateral membrane by the Na^+-K^+ ATPase (Figure 30.4). Cl^- reabsorption and HCO_3^- secretion is effected by an apical Cl^--HCO_3^- exchanger. Cl^- is then transferred to the interstitium via the cystic fibrosis transmembrane conductance regulator (CFTR) Cl^- channel. Secreted HCO_3^- is derived from plasma, entering the ductal cells via a Na^+-HCO_3^- cotransporter. Apical K^+ secretion may involve an apical H^+-K^+ exchanger. Ductal epithelia are relatively water impermeable due to lack of aquaporins, and, thus, saliva becomes hypotonic.

4. **Proteins:** Saliva also contains low concentrations of protective proteins and enzymes that are secreted by acinar, mucous, and ductal cells.

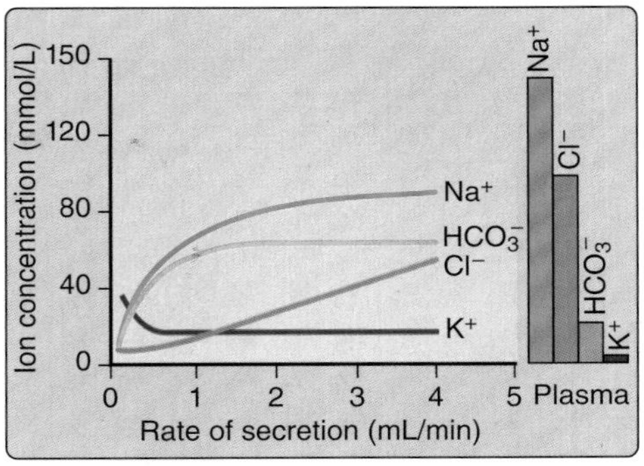

Figure 30.2
Salivary flow and ion concentration.

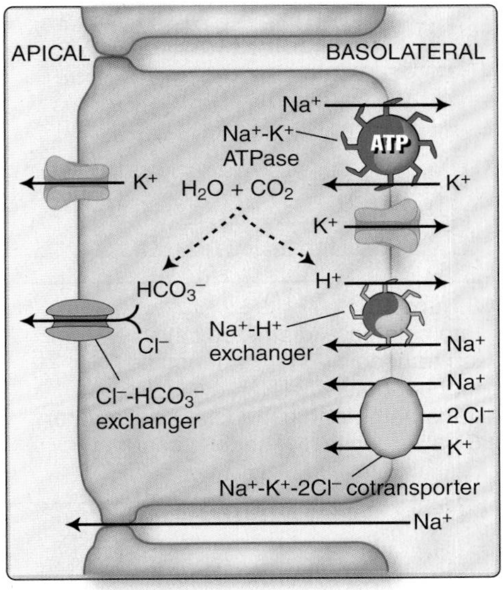

Figure 30.3
Acinar cell ion transport.
ATP = adenosine triphosphate.

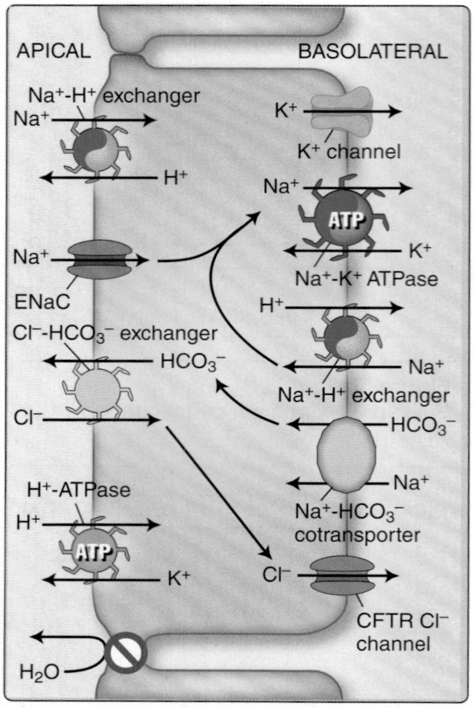

Figure 30.4
Intercalated and striated cell ion transport. ATP = adenosine triphosphate; CFTR = cystic fibrosis transmembrane conductance regulator; ENaC = epithelial sodium channel.

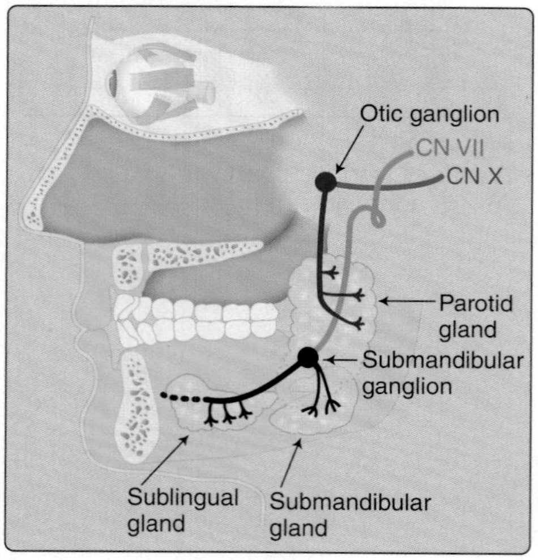

Figure 30.5
Neural innervation of salivary glands. CN = cranial nerve.

a. ***Lysozyme***: Secreted *lysozyme* has the potential to disrupt bacterial cell walls.

b. **Lactoferrin:** Lactoferrin is an iron-binding protein that can inhibit microbial growth.

c. **Immunoglobulin A:** Constituents for immunoglobulin A are secreted in the saliva and are active against both bacteria and viruses.

d. **Proline-rich proteins:** Proline-rich proteins aid in tooth enamel formation and also possess antimicrobial properties.

e. ***Salivary amylase***: *Salivary amylase* (also known as *α-amylase*, or *ptyalin*) begins the process of carbohydrate digestion but is denatured by low pH in the stomach. *Amylase* is then reintroduced into the GI tract from the pancreas.

f. ***Lingual lipase***: *Lingual lipase* hydrolyses lipids and remains active throughout the GI tract.

5. **Regulation:** Salivary flow is controlled by both the parasympathetic and sympathetic nervous systems. Although stimulation of either increases secretions, the sympathetic component is transient and produces lower volume secretions than does the parasympathetic system, which is mediated through the salivary nuclei located in the medulla. Salivary flow is increased by smell, taste, mechanical pressure in the mouth, and various reflexes (e.g., classical conditioning), whereas it is decreased by stress, dehydration, and during sleep. In addition to salivary flow, neural stimulation increases blood flow, myoepithelial cell contraction, and gland growth and development. The submandibular and sublingual glands are innervated by the facial (cranial nerve [CN] VII) and lingual nerves through the submandibular ganglion (Figure 30.5), whereas the sympathetic nerves emanating from T1–T3 synapse through the superior cervical ganglion. The parotid gland has similar sympathetic innervation, but, in terms of parasympathetic innervation, the glossopharyngeal nerve (CN X) synapses through the otic ganglion and travels along the auriculotemporal nerve rather than involving CN VII (see Figure 30.5). Parasympathetic nerves release acetylcholine (ACh), binding to muscarinic (M_3) receptors acting via the inositol trisphosphate (IP_3) signaling pathway (see 1·VII·B·3). The sympathetic nerves release norepinephrine, binding to α- and β-adrenergic receptors acting via IP_3 and cyclic adenosine monophosphate (cAMP) signaling pathways, respectively (see 1·VII·B·2).

III. ESOPHAGUS

The oropharynx and esophagus convey dietary contents and oral secretions from the back of the oral cavity to the stomach.

A. Swallowing

The act of swallowing is a coordinated act involving many structures. Swallowing is largely initiated voluntarily but becomes involuntary once initiated.

1. **Regulation:** Swallowing control is a parasympathetic process involving afferent feedback to the **swallowing center** followed by

Clinical Application 30.1: Sjögren Syndrome

Certain disorders cause oral cavity dryness (xerostomia), one of which is the autoimmune disorder **Sjögren syndrome**. In addition to xerostomia, these patients present with a burning feeling in the mouth and throat, difficulty swallowing food, increased incidence of dental caries, and some speech issues. The pathophysiology of the disorder revolves around the inability to transport ions and water via Cl^--HCO_3^- exchangers and aquaporin-5 pores after cholinergic stimulation in order to produce adequate saliva. In patients with less severe involvement, salivary stimulants can be used to increase salivary secretions and function to prevent drying and cracking of mucosal membranes.

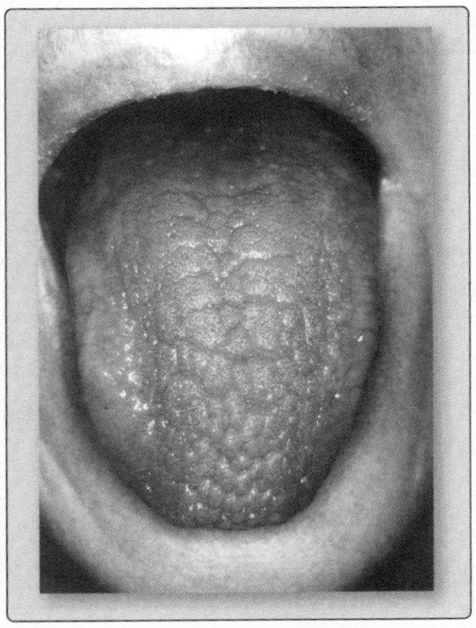

Xerostomia in Sjögren syndrome.

efferent responses through other nuclei, including the **nucleus ambiguus** and **dorsal motor nucleus**. This control system allows muscle to be contracted in a proximal-to-distal manner and coordinates with other physiologic functions, such as respiration and speech, which cannot occur concurrently. The involuntary components of swallowing are shown in Figure 30.6.

2. **Pathophysiology: Dysphagia** is difficulty swallowing. Problems with swallowing can be classified into two broad origins: 1) mechanical such as from stomach protrusion past the diaphragm (**hiatal hernia**) and 2) functional such as the inability to coordinate the event sequence during swallowing seen after a **cerebrovascular accident** (stroke). A person with dysphagia not only has difficulty swallowing solid food but also may have difficulty swallowing liquids. An esophagoscopy or barium swallow study can be used to assess swallowing and the extent of dysphagia present. Dysphagia may require an alternate diet and proper head positioning when eating or drinking, whereas severe dysphagia may require a nasogastric tube to deliver nutrients directly to the stomach.

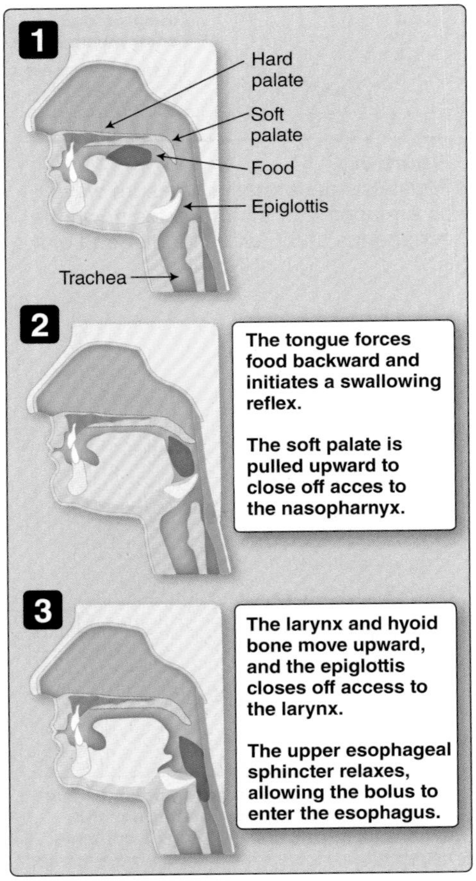

1
Hard palate
Soft palate
Food
Epiglottis
Trachea
2
The tongue forces food backward and initiates a swallowing reflex.
The soft palate is pulled upward to close off acces to the nasopharnyx.
3
The larynx and hyoid bone move upward, and the epiglottis closes off access to the larynx.
The upper esophageal sphincter relaxes, allowing the bolus to enter the esophagus.

Figure 30.6
Swallowing reflex.

B. Esophageal peristalsis

Once food has moved past the **upper esophageal sphincter** (**UES**), further movement is accomplished by a series of coordinated muscle contractions and relaxations known as **peristalsis**. Think of this as providing a single "wave" that is sustained via these contractions that a surfer (or food) rides in to the shore (or stomach). These coordinated movements send the positive pressure wave down the esophagus until it reaches the **lower esophageal sphincter** (**LES**) and stomach. The LES is tonically constricted, but, as the peristaltic wave reaches the sphincter, it relaxes and allows food to enter the stomach. Changes in sphincter tone are mediated by ACh, **nitric oxide** (**NO**), and **vasoactive intestinal peptide** (**VIP**). Food traverses the esophagus in about 6–10 seconds. If food is not cleared by the first pressure wave (**primary peristalsis**), repetitive waves (**secondary peristal-**

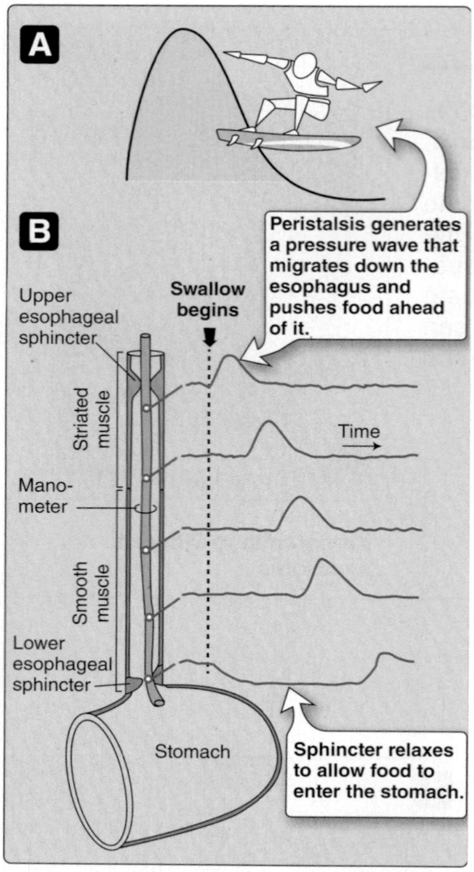

Figure 30.7
Peristalsis generates a pressure wave on a manometer that migrates down the esophagus and pushes food ahead of it.

sis) may be initiated. These secondary peristaltic waves involve the esophageal smooth muscle. The esophagus musculature is unique in that the first third of the esophagus is largely skeletal muscle, and the final two thirds is largely smooth muscle (Figure 30.7). The difference in muscle type extends to the sphincters as well, with the UES being skeletal muscle, and the LES being smooth muscle.

IV. STOMACH

The stomach serves a number of important physiologic functions: accepting and storing food, mixing food with secretions, digesting food, and delivering food to the small intestine in timed increments. The fundus, body, and antrum comprise the three anatomic areas of the stomach (Figure 30.8). In terms of motility, the upper half accepts food from the esophagus, and the lower half mixes and delivers food to the small intestine.

A. Accommodation

The primary function of the upper half of the stomach is to accommodate food from the esophagus. During swallowing, the LES relaxes, allowing food to move from an area of higher pressure in the esophagus to an area of lower pressure in the stomach. The stomach must be prepared for this bolus. This is accomplished by relaxing the upper portion of the stomach, which normally is contracted. This relaxation is termed **receptive relaxation** and is mediated by NO and VIP. This relaxation is coordinated by the vagus nerve in response to vagus afferent stimulation and, thus, is referred to as a **vagovagal reflex**. The average stomach can accommodate ~1.5 L of food.

B. Mixing

The mechanical mixing and grinding of food occurs in the lower half of the stomach. Mechanical stomach contraction occurs in phases mediated by **slow waves** (or basal electrical rhythms).

1. **Slow waves:** Slow waves are generated by **interstitial cells of Cajal** (**ICCs**) at a rate of 3–5 cycles/min and propagate toward the

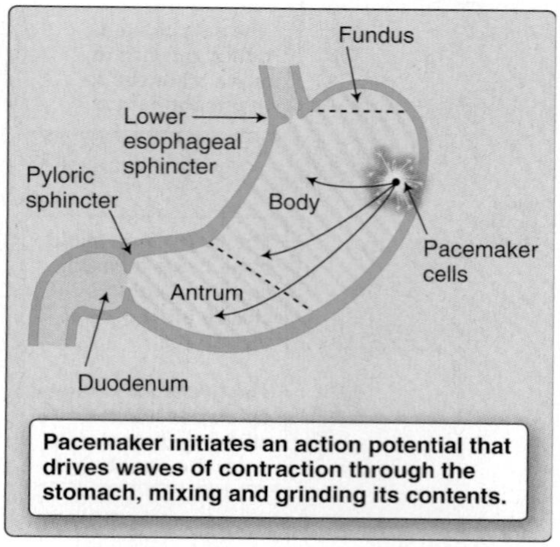

Figure 30.8
Stomach structure and pacemaker cells.

Clinical Application 30.2: Achalasia and Gastroesophageal Reflux Disease

Lower esophageal sphincter (LES) pathology can involve contraction without relaxation or incomplete contraction. In **achalasia**, the LES does not relax because of a loss of neurons that contain nitric oxide and vasoactive intestinal peptide. Other neurons can also be involved as the disease progresses. Thus, dietary contents are retained just before the LES and can lead to dilatation of the esophagus. In **gastroesophageal reflux disease**, gastric contents enter the esophagus through the LES because the sphincter provides an incomplete barrier. This is exacerbated during times when the LES briefly relaxes such as during belching. Although some gastric reflux is normal and is resolved by secondary peristalsis, H^+ and enzymes from gastric fluid can damage the esophagus, causing heartburn. If exposure is chronic, endothelial damage and esophageal remodeling can occur.

pylorus. These electrical signals do not necessarily elicit a corresponding muscular contraction when inhibited by norepinephrine. A threshold related to voltage amplitude and duration is reached in normal conditions and a greater frequency when stimulated by ACh. This occurs as membrane potential (V_m) depolarizes, opening voltage-gated Ca^{2+} channels and causing spikes. As Ca^{2+} concentration increases, this opens Ca^{2+}-dependent K^+ channels, which, in turn, hyperpolarizes V_m due to K^+ efflux. The decrease in V_m eventually closes the voltage-gated Ca^{2+} channels. This attenuates K^+ efflux, and V_m once again begins to rise (Figure 30.9). These spikes can eclipse threshold to induce muscle contractions. These spikes and overall V_m are inhibited by norepinephrine released by sympathetic neurons and are stimulated by mechanical stretch and the ACh released by parasympathetic and enteric neurons.

2. **Muscle contractions:** Slow-wave induced muscle contractions propagate from the ICCs to the pylorus. Interestingly, this contraction-induced pressure wave eventually overtakes the food bolus (i.e., the pressure wave moves faster than the food is moving) and, thus, begins to push food in both directions. This results in a small amount of food entering the duodenum and the majority of the food being pushed back toward the middle of the stomach. This brief backward movement, termed **retropulsion**, allows for better food mixing and mechanical breakup.

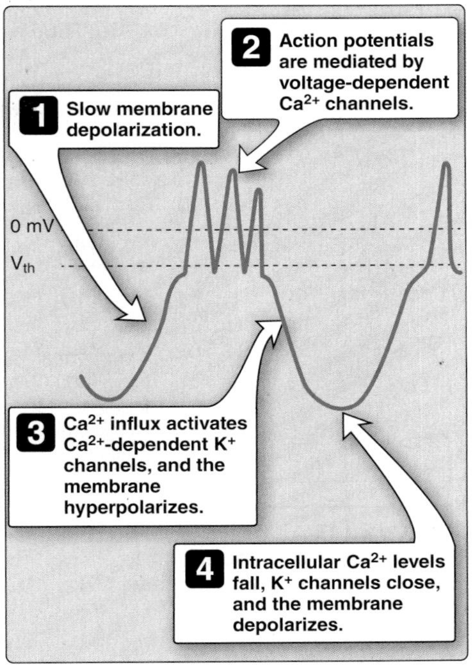

Figure 30.9
Mechanism of slow-wave development. V_{th} = voltage threshold for action potential formation.

C. Secretions

Gastric secretions are derived from gastric invaginations called **pits**. Pits are lined with many different secretory cell types (Figure 30.10). Mucous neck cells within gastric pits secrete mucus, which is vital to the gastric lining's barrier function in protecting the stomach from gastric acid and *pepsin*. Chief cells produce *gastric lipase* and pepsinogen, which is the inactive form of *pepsin*, and parietal cells (also known as **oxyntic cells**) secrete H^+ and intrinsic factor (both of these cell types are described in more detail below). G and D cells are endocrine cells that secrete gastrin and somatostatin, respectively (see Table 29.2). Regional differences exist in the number of cell types lining a gastric pit. Pits near the lower esophageal and pyloric sphincters contain more cells that produce more of the protective secretions, such as mucus and HCO_3^-, whereas pits in the rest of the stomach contain more of the secretory cells that produce more of the digestive secretions, such as H^+ and pepsinogen.

1. **Hydrogen ion secretion mechanism:** Acidification of the gastric lumen is accomplished by the transport of H^+ across the apical membrane by H^+-K^+ ATPase (H^+ pump). Cl^--HCO_3^- exchangers on the basolateral membrane exchange Cl^- for HCO_3^-. Cl^- then exits the parietal cell across the apical membrane via Cl^- channels (Figure 30.11). This leaves H^+ and Cl^- in the luminal space of the gastric pit, and these ions can combine to form hydrochloric acid (HCl). The interstitial space becomes slightly basic in the process of secreting H^+ because of the addition of the HCO_3^-, referred to as an "alkaline tide." This alkaline tide may help protect adjacent cells against the large pH change mediated by parietal cells.

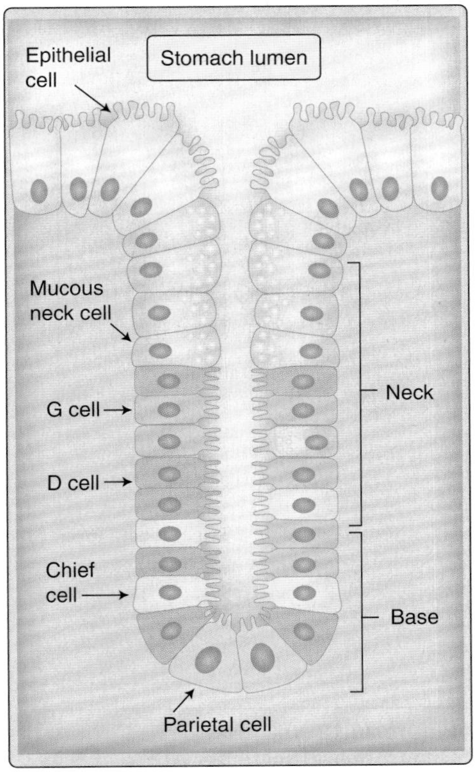

Figure 30.10
Gastric pit cells.

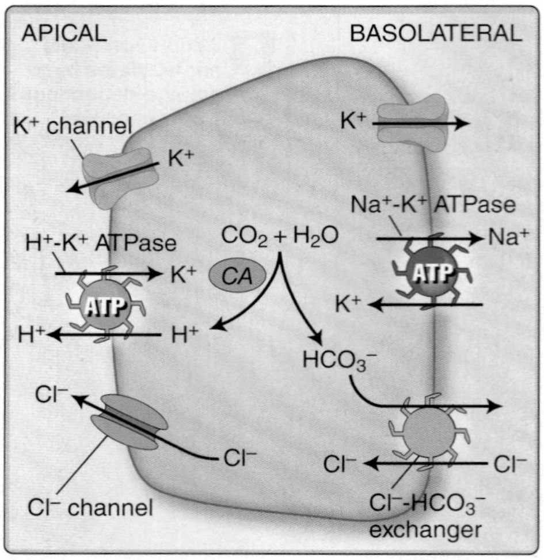

Figure 30.11
Mechanisms of H^+ secretion. ATP = adenosine triphosphate; *CA* = *carbonic anhydrase*.

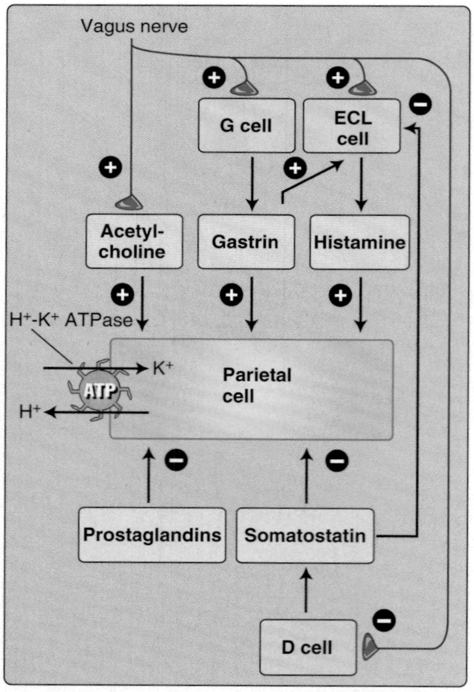

Figure 30.12
Control of acid secretion. ATP = adenosine triphosphate; ECL = enterochromaffin-like cell.

2. **Regulation:** The control of gastric secretions is dependent on the phase of digestion (see 29·V).

   a. **Cephalic phase:** The **cephalic phase** accounts for ~40% of gastric secretions via the vagus nerve's actions on **parietal cells, enterochromaffin-like cells (ECL cells),** and **G cells**. Vagally derived ACh stimulates H^+ production by parietal cells directly. The vagus nerve also releases ACh to initiate H^+ production by the stimulation of ECL cells to produce histamine and of G cells to produce gastrin. Gastrin-releasing peptide is also involved in stimulating both parietal and ECL cells (Figure 30.12). ACh also stimulates mucus secretion to protect the stomach lining from H^+. This neural input allows for the gastric secretions to be activated in preparation for the food to enter the stomach.

   b. **Gastric phase:** The gastric phase accounts for ~50% of gastric secretions. This primarily occurs via a two-pronged feedback: 1) directly through vagus afferents, which then allows for the vagus to mediate a response (vagovagal reflex) and 2) via local enteric reflexes. Distention appears to be the primary stimulant acting via vagal afferents and local reflexes. Besides distention, proteins, peptides, and, especially, amino acids additionally stimulate G cells to release gastrin. The negative feedback of this phase is provided by H^+ stimulation of D cells to produce somatostatin, which inhibits both parietal and G cells and, therefore, decreases H^+ and gastrin secretions.

   c. **Intestinal phase:** The intestinal phase accounts for ~10% of gastric secretions. The digestion of **chyme** (post-stomach food and secretion mix), in particular the digestion of proteins, continues to directly stimulate intestinal G cells as well as to stimulate gastric G cells via proteins and amino acids in the portal circulation. The negative feedback in the intestinal phase is provided by intestinal distension, which releases glucose-dependent insulinotropic peptide, which, in turn, inhibits parietal cells.

3. **Control of hydrogen ion secretions:** H^+ production control involves both direct and indirect neural pathways. The direct parietal pathway involves the vagus nerve releasing ACh to stimulate M_3 receptors and gastrin binding to cholecystokinin type-B (CCK_B) receptors. In contrast, indirect stimulation involves histamine release from ECL cells, binding to H_2 receptors on the parietal cell. The ECL cells are stimulated by both ACh and gastrin. Both somatostatin and prostaglandins decrease H^+ production via binding to their own cell surface receptors on the parietal cell. Despite these multiple agonists and antagonists, there are two common signaling pathways to regulate the H^+-K^+ ATPase. First, gastrin and ACh act via the IP_3 signaling pathway (see 1·VII·B·3). Second, histamine, somatostatin, and prostaglandins act via cAMP signaling pathway (see 1·VII·B·2). Histamine increases cAMP, and somatostatin and prostaglandins decrease cAMP. These multiple pathways and indirect-versus-direct stimulation allow for fine-tuning of the regulation of the H^+-K^+ ATPase and thus the amount of H^+ in the gastric lumen. The augmentation of these pathways to secrete greater H^+ than any one pathway alone is termed **potentiation**.

Clinical Application 30.3: Peptic Ulcer Disease

In **peptic ulcer disease,** ulcers, or small breaks in the mucosal surface of the stomach or duodenum, are formed. The ulcers can rupture one layer of the stomach lining or perforate the entire lining as in the figure. The bacterium *Helicobacter pylori* and nonsteroidal anti-inflammatory drugs are responsible for most ulcers.[1] Treatment involves eradicating *H. pylori* and decreasing H^+ secretions (using a H^+-K^+ ATPase inhibitor) to allow healing of the ulcerated area.

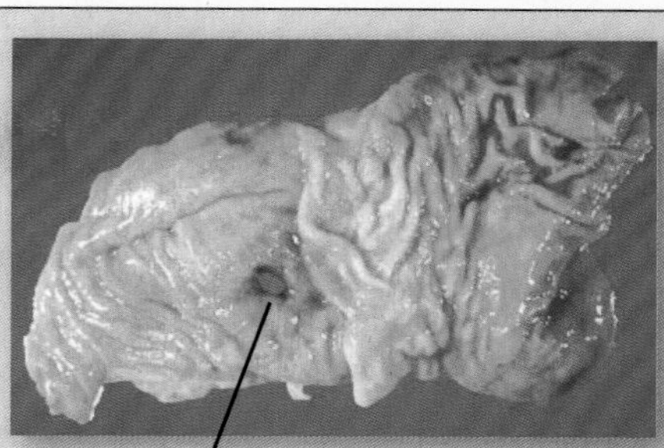

Peptic ulcer.

4. **Effect of secretion rate:** Similar to salivary gland secretion, the concentration of gastric secretion constituents is a function of secretory rate (Figure 30.13). As flow rate increases, concentration of H^+, K^+, and Cl^- increase, and Na^+ decreases in parietal secretions.

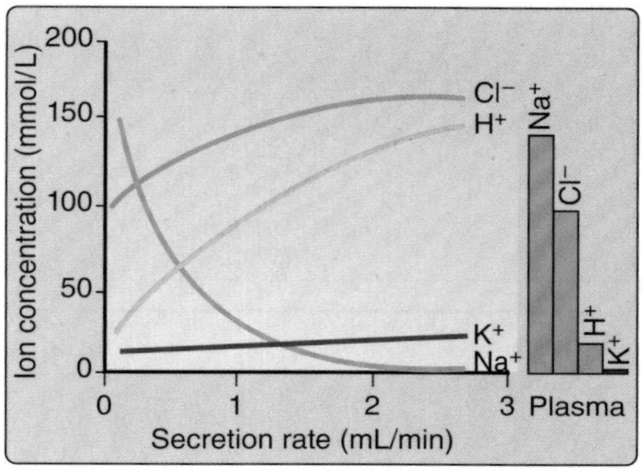

Figure 30.13
Gastric secretion rate and ion concentration.

D. Digestion

The low pH of the gastric contents helps denature and break down proteins. Breakdown is assisted by the proteolytic enzyme *pepsin*, which is secreted in inactive form (pepsinogen) by chief cells and converted to *pepsin* by the low pH. *Pepsin* is an *endopeptidase* that cleaves aromatic amino acids and has an optimal pH between 1 and 3. *Pepsin* will be deactivated in the duodenum once pH increases toward the neutral range. *Gastric lipase* also has a low optimal pH (3–6), and acts primarily on ester bonds to form fatty acid and diglyceride products.

INFO LINK [1]For further discussion of the pharmacological treatment of peptic ulcer disease, see *LIR Pharmacology*, 5e, pp. 329–335.

Chapter Summary

- The mouth reduces food to an optimal size and mixes food with secretions from three salivary glands: **sublingual**, **submandibular**, and **parotid**.
- Salivary secretions are controlled by both the sympathetic and parasympathetic nervous systems and involve a two-step process: First, Cl^-, Na^+, and water are transported into the duct lumen. Second, ductal cells modify this fluid by reabsorbing Na^+ and Cl^- and secreting K^+ and HCO_3^-.
- The esophagus transports food from the mouth to the stomach. **Swallowing** is a conscious act that moves food from the mouth through the upper esophageal sphincter. **Esophageal peristalsis** then propels food down the esophagus ahead of a contraction-induced pressure wave.
- The stomach has three primary motility functions: **accommodation** via receptive relaxation, **mixing** via slow wave–initiated contractions and retropulsion, and **gastric emptying**.
- Gastric secretions include ions and water, mucus from mucous **neck cells**, pepsinogen from **chief cells**, and intrinsic factor and H^+ from **parietal cells**.
- Regulation of H^+ secretion occurs at the level of the H^+-K^+ ATPase. **Acetylcholine** from nerves, **gastrin** from G cells, and **histamine** from enterochromaffin-like cells increase secretion, and **somatostatin** from D cells and **prostaglandins** decrease secretion.

31 Small and Large Intestines

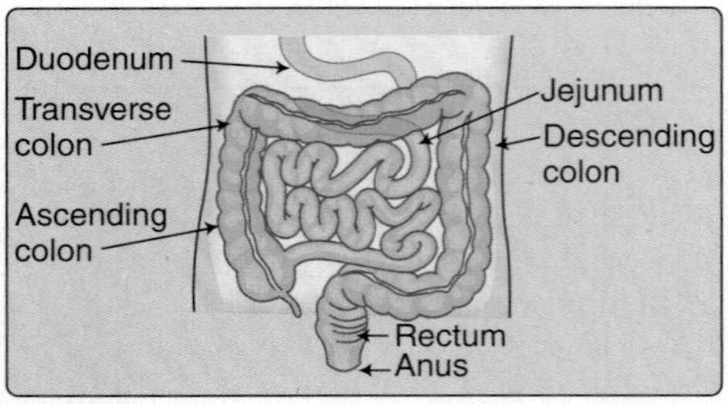

Figure 31.1
Intestines.

I. OVERVIEW

In the upper portion of the gastrointestinal (GI) tube, food was liquefied and reduced in size but was not absorbed. The small intestine is where nutrients start to be removed in earnest. To facilitate this absorption, complex carbohydrates, lipids, and proteins are chemically digested into simpler forms for transport. Absorption requires transport across both the apical and basolateral membranes of enterocytes (absorptive cells of the intestinal epithelium). Not only macronutrients (carbohydrates, lipids, proteins, and water) are absorbed, but also ions, vitamins, and minerals. Once the necessary items are extracted, the remainder needs to be eliminated from the body. The distal colon, rectum, and anus participate in bowel movements, which involve both voluntary and involuntary components, to eliminate the remnants (Figure 31.1).

II. SMALL INTESTINE

The small intestine is the longest section of the GI tract, at about 6 m. It is divided into three functional segments: **duodenum** (first ~0.3 m), **jejunum** (next ~2.3 m), and **ileum** (final ~3.4 m). The majority of macronutrient, vitamin, and mineral absorption occurs in the small intestine. Absorption of the nutrients liberated by the digestive process is facilitated by the increase in epithelial surface area by **villi** (10-fold) and **microvilli** (20-fold) (Figure 31.2).

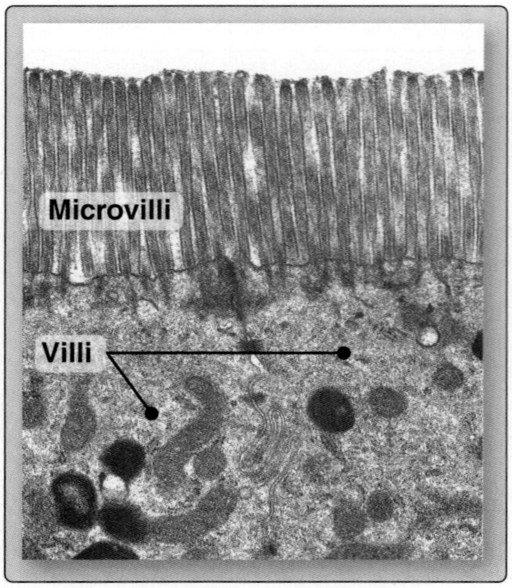

Figure 31.2
Villi and microvilli.

A. Motility and mixing

Intestinal motility not only propels chyme along the intestines via peristalsis but also allows for mixing of enzymes and other secretions from the pancreas and gallbladder (control and regulation of these secretions is discussed in Chapter 32). **Segmentation** is the back-and-forth mixing movement in the small intestine between adjacent segments. As in the stomach, smooth muscle contractions are initiated via slow waves. Slow waves in the small intestine are more frequent (~12 waves/min) than in the stomach, with the parasympathetic nervous system increasing this rate, and the sympathetic nervous system decreasing it. To aid in clearing residual contents during the fasting state, there are additional contractions regulated by motilin and known as **migrating motor complexes**

(**MMCs**), which are initiated in the stomach and continue on through the small intestine at 60–120-minute intervals. MMCs sweep the small intestine clean. Feeding disrupts these complexes in favor of peristalsis and segmentation.

B. Intestinal secretions

Intestinal secretions include both aqueous solutions and mucus. Mucus is important for the lubrication of the chyme for intestinal protection and so that peristaltic contractions can better propel the chyme. In addition, a number of endocrine cells in the intestines secrete the hormones **cholecystokinin** (**CCK**), **secretin**, and **glucose-dependent insulinotropic peptide**.

C. Carbohydrate digestion and absorption

Carbohydrates provide a substantial energy substrate for metabolism (~4 kcals/g). Carbohydrates come in many forms (e.g., starch, dietary fiber, disaccharides, and monosaccharides), but they must be broken down into monosaccharides before they can be transported across the intestinal lumen.

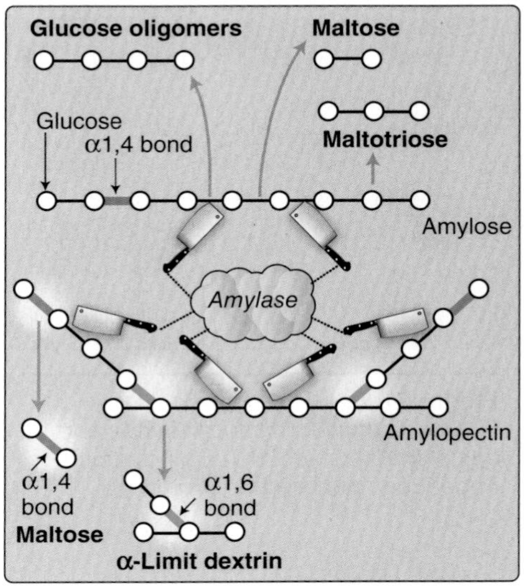

Figure 31.3
Carbohydrate digestion.

1. **Starch:** Starch is classified as linear chained (**amylose**) or branch chained (**amylopectin**). The glucose bonds that form in linear configuration are α1,4 bonds, whereas those in the branched configuration are α1,6 bonds. ***Pancreatic amylase*** breaks α1,4 bonds. The products of the *amylase* reaction are maltose, maltotriose, glucose oligomers, and α-limit dextrin (Figure 31.3). These products are a substantial size reduction of starch and are further digested by *di-* and *oligosaccharidases*.

2. **Dietary fiber:** Dietary fiber can be divided into **soluble** (e.g., pectin) and **insoluble** (e.g., cellulose) varieties (Table 31.1). Dietary fiber contains bonds that human enzymes cannot break down in the small intestine. For example, cellulose contains linear β1,4 glucose bonds, whereas both *salivary* and *pancreatic amylases* only break α1,4 glucose bonds. Because dietary fiber cannot be adequately digested, these carbohydrates cannot be absorbed and serve to increase fecal bulk. Increased fecal bulk provides some beneficial effects, such as increased intestinal motility and increased frequency of defecation.

Table 31.1: Dietary Fiber Classifications

Types	Solubility
Cellulose	Insoluble
Hemicellulose	Insoluble
Lignin	Insoluble
Gums	Soluble
Pectins	Soluble

3. **Disaccharides and oligosaccharides:** Disaccharides are derived from the breakdown of starch and from direct dietary sources (e.g., sucrose and lactose). *Amylase* activity occurs in the intestinal lumen, whereas disaccharides and oligosaccharides are broken down into monosaccharides by membrane-bound ***disaccharidases***. The *disaccharidases* can be specific for a substrate, such as *lactase*, or work on multiple substrates, such as *sucrase* and *isomaltase* to produce monosaccharide products (Table 31.2). Being membrane bound allows for a close association between enzyme products and absorption transporters. For example, lactose breakdown into glucose and galactose is facilitated by *lactase*, which is located close to cotransporters (**SGLT1**) for the products' absorption (Figure 31.4).

Table 31.2: Membrane-bound *Disaccharidases*

Enzyme	Substrate(s)	Product(s)
Glucoamylase	Maltose and maltotriose	Glucose
Isomaltase	α-Limit dextrins, maltose, and maltotriose	Glucose
Lactase	Lactose	Glucose and galactose
Sucrase	Maltose, maltotriose, and sucrose	Glucose and fructose

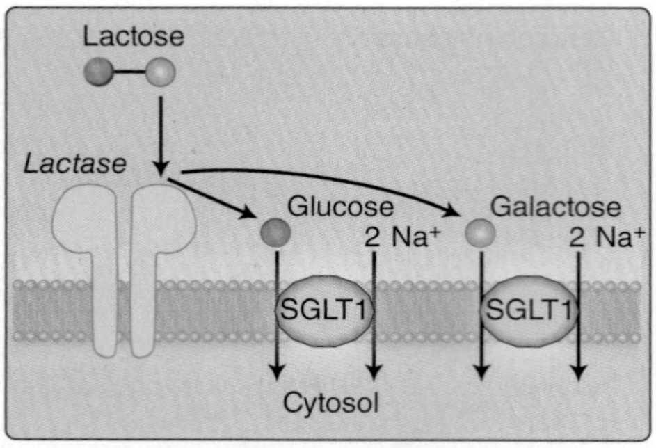

Figure 31.4
Relation between *disaccharidase* and apical transporters. SGLT1 = sodium-dependent glucose transporter 1.

Clinical Application 31.1: Lactose Intolerance

Dietary lactose can be consumed in excess of *lactase* capacity in the small intestine. Excess nonabsorbed lactose is then broken down by intestinal microbes further down the gastrointestinal tract, which can then cause symptoms such as diarrhea, bloating, and cramping. Some persons are born with lower concentrations of *lactase*, but there is a progressive decrease in *lactase* expression across the lifespan, such that a person has higher concentrations early in life but lower concentrations later in life. Individuals with conditions such as inflammatory bowel disease are especially prone to lactose intolerance because of the associated small intestine inflammation. The ability to digest and absorb lactose can be measured by giving 100 g of oral lactose, followed by a blood draw every 30 min for 2 hr. Those with lactose intolerance exhibit an attenuated increase in blood glucose (<20 mg/dL) because glucose and galactose are not formed from *lactase* in sufficient quantities to be absorbed.

4. **Monosaccharides:** Monosaccharides, such as glucose, fructose, and galactose, are transported across the apical and basolateral membranes of small intestine enterocytes. Because monosaccharides are hydrophilic, transporters are needed to move these nutrients across these membranes.

   a. **Apical membrane transport:** Glucose and galactose are transported across the apical membrane by SGLT1, a Na^+-glucose cotransporter. The Na^+-K^+ ATPase provides a low Na^+ environment within the enterocyte to allow for Na^+ to be used as a driving force for the movement of glucose across the apical membrane. Fructose is transported by GLUT5 (glucose transporter) as shown in Figure 31.5.

   b. **Basolateral membrane transport:** Transport of monosaccharides across the basolateral membrane from the inside of the enterocyte to the interstitium is facilitated by GLUT2 and GLUT5 transporters. GLUT2 transports both glucose and galactose, and GLUT5 transports fructose across the basolateral membrane. These nutrients can then diffuse into the portal circulation to be carried to the liver.

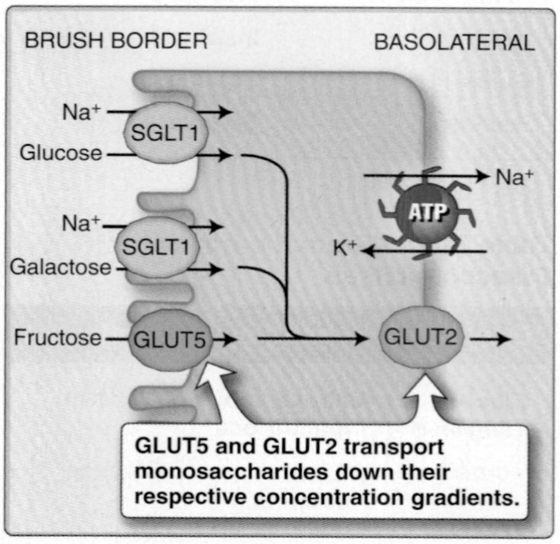

Figure 31.5
Apical and basolateral monosaccharide transport. ATP = adenosine triphosphate; GLUT2 and 5 = members of the glucose transporter family; SGLT1 = sodium-dependent glucose transporter 1.

D. Protein digestion and absorption

Proteins also can be used for energy production (~4 kcal/g), but, in a fed state, proteins are primarily used as building blocks for reassembly into other proteins. Protein digestion begun in the stomach through the action of *pepsin* is then continued by several other *proteases* secreted in the small intestine.

1. **Luminal *proteases*:** A small amount of proteins and peptides are absorbed via phagocytosis across the apical membrane of enterocytes and specialized mucosal immune, or M, cells. However, the majority of proteins are broken down into amino acids and

oligopeptides to facilitate absorption. End products of *endopeptidases* (***trypsin***, ***chymotrypsin***, and ***elastase***) are oligopeptides, peptides normally 6 or fewer amino acids in length. *Exopeptidases* (***carboxypeptidase A*** and ***B***) cleave off single amino acids from oligopeptides (Figure 31.6). It is thought that these luminal actions convert about ~70% of proteins to oligopeptides and ~30% to amino acids.

2. ***Apical peptidases***: *Apical peptidases* (also termed brush border *peptidases*) break down small peptides and oligopeptides into individual amino acids.

3. **Apical dipeptide, tripeptide, and amino acid transport:** Amino acids are transported across the apical membrane via different classes of amino acid cotransporters. Di- and tripeptides are transported across the apical membrane by an **H^+-oligopeptide cotransporter** (**PepT1**). Di- and tripeptides are then broken down by cytosolic *peptidases* into individual amino acids (Figure 31.7).

4. **Basolateral amino acid transport:** Individual amino acids are transported across the basolateral membrane without the need for cotransport. Many different amino acid transporters are located on the basolateral membrane and provide specificity (see Figure 31.6).

E. Lipid digestion and absorption

Fats are calorically denser (~9 kcal/g) than carbohydrates and proteins and are a substantial energy substrate for metabolism. Lipid absorption does not require the same transporter machinery because lipids are hydrophobic and can diffuse across the apical membrane. Lipids must be solubilized to ensure adequate mixing with enzymes. Lipid digestion begins in the mouth and stomach with *lingual* and *gastric lipase*, although the vast majority occurs in the small intestine. Assisting with lipid digestion, liver-derived **bile salts** surround and emulsify lipids so that *lipase* and *colipase* can interact with the lipid. ***Pancreatic lipase*** is the active enzyme that digests triglycerides into fatty acids and monoacylglycerols. ***Colipase*** functions

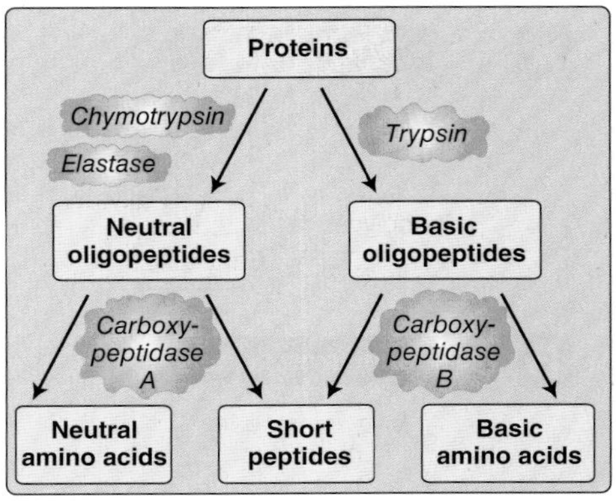

Figure 31.6
Protein and peptide digestion.

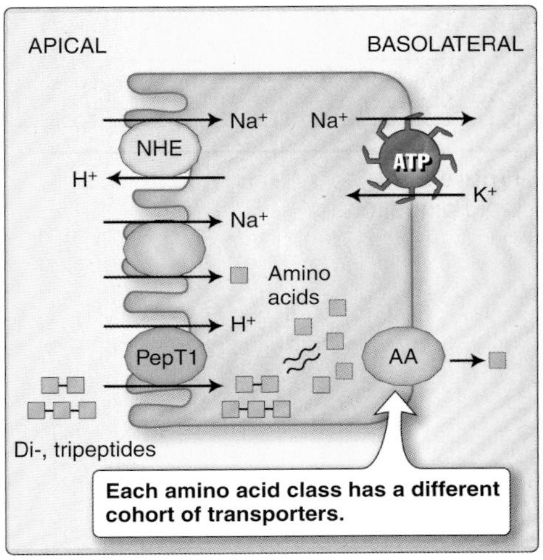

Figure 31.7
Amino acid (AA) and di- and tripeptide transporters. ATP = adenosine triphosphate; NHE = Na^+-H^+ exchanger; PepT1 = H^+-oligopeptide cotransporter.

Clinical Application 31.2: Hartnup Disease

Hartnup disease is an autosomal recessive disorder of neutral amino acid transport in gastrointestinal and renal systems. The specific transporter affected is an apical membrane Na^+-amino acid cotransporter (*SLC6A19* gene). These inherent disorders can lead to amino acid deficiencies, but it is possible that the other modes of protein absorption (i.e., PepT1 and phagocytosis) can partially accommodate this transport defect because some neutral amino acids can be absorbed by these routes. Deficiencies in neutral amino acids like tryptophan can lead to niacin availability issues because niacin is derived from tryptophan metabolism. This results in symptoms such as skin lesions and neurologic manifestations.

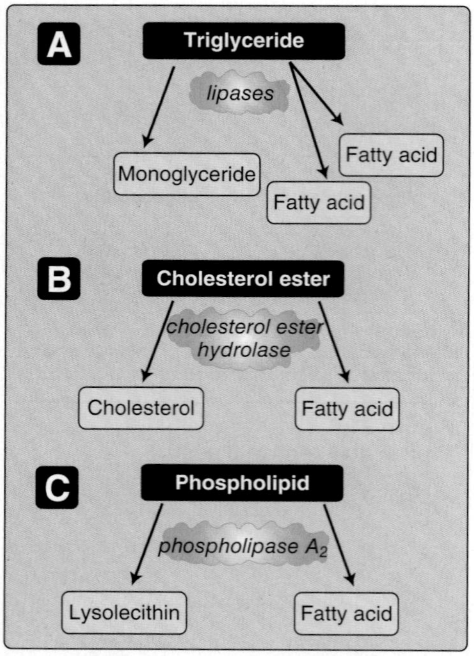

Figure 31.8
Lipid digestion.

to position and stabilize *pancreatic lipase*. *Lipases* do not digest phospholipids and cholesterol, requiring other pancreatic enzymes. Dietary cholesterol esters are digested into cholesterol and fatty acids by ***cholesterol esterase*** (*carboxyl ester hydrolase*) as shown in Figure 31.8. ***Phospholipase A_2*** breaks down phospholipids into fatty acids and lysolecithin.

1. **Free fatty acids:** Fatty acid length (long, medium, or short) determines rate of absorption and assimilation. This length distinction is partly related to solubility: The longer the fatty acid, the less soluble it is in an aqueous environment.

   a. **Long-chain fatty acids:** Long-chain fatty acids are concentrated into micelles in the small intestine lumen. Lipids often form micelles, in which hydrophilic portions face outward toward the water, and hydrophobic portions face the center. This is a stable conformation in aqueous environments and allows lipids to enter the unstirred layer surrounding the intestinal lumen in order to come in contact with the apical membranes of enterocytes. Close to the surface of the apical membrane, the micelle begins to disperse, possibly due to a pH change. Long-chain fatty acids are freed and can either diffuse directly across the apical membrane or be transported by **fatty acid–binding proteins**. These binding proteins speed absorption across the apical membrane. In the cytosol, long-chain fatty acids are attached to monoacylglycerols and diacylglycerols to form triglycerides within the enterocyte. Triglycerides are packaged in apoprotein vesicles called **chylomicrons** and, to a lesser degree, in **very-low-density lipoproteins** (**VLDLs**). Chylomicrons are then exocytosed through the basolateral membrane into the interstitial space. From the interstitial space, chylomicrons do not enter the circulation because of capillary fenestration size restrictions but, rather, move into the lymphatic system for transportation (Figure 31.9).

   b. **Medium-chain fatty acids:** Medium-chain fatty acids (6–12 carbons long) are more soluble in water than are long-chain fatty acids. This allows them to cross the apical membrane by moving through the cytosol without the need to be repacked into a chylomicron. Medium-chain fatty acids cross the basolateral membrane into the interstitial space and then into the portal circulation. This is in contrast to long-chain fatty acids, which enter the lymphatic circulation (see Figure 31.9).

> Medium-chain fatty acids can be used as dietary supplements to increase total absorbed kilocalories (energy). Due to both the solubility and transport method of these supplements, individuals with diseases such as bile duct obstruction are able to absorb these fats without the need for bile salts.

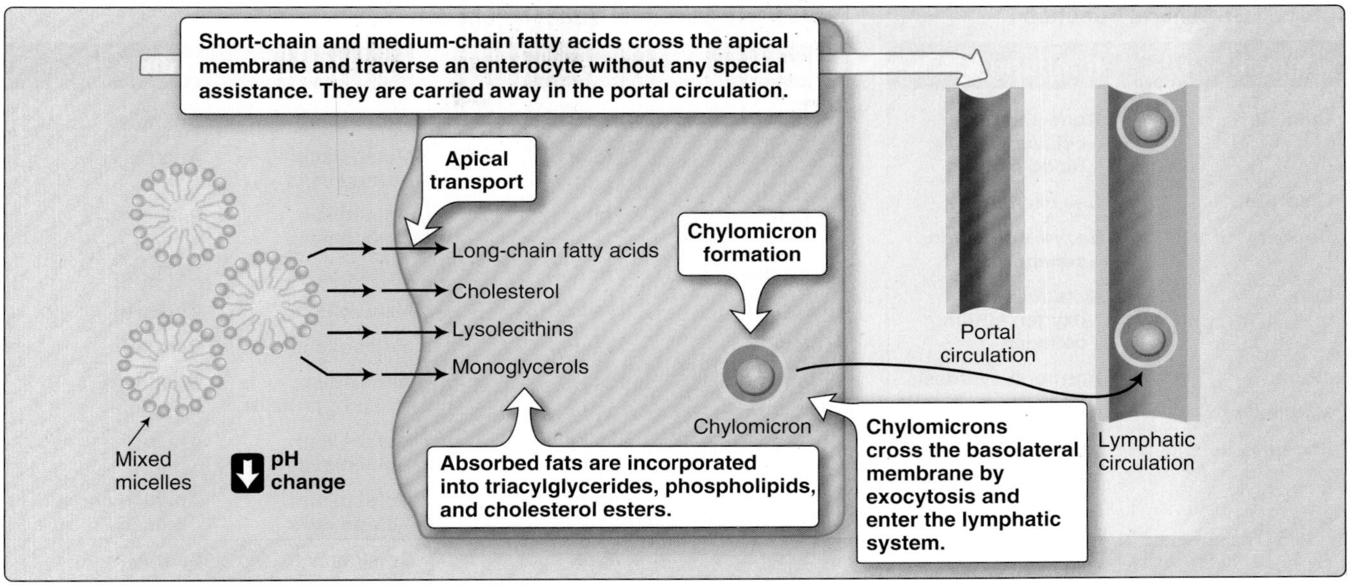

Figure 31.9
Lipid apical and basolateral transport.

c. **Short-chain fatty acids:** Short-chain fatty acids are less than 6 carbons in length. These fatty acids are absorbed and assimilated in a manner similar to that of medium-chain fatty acids.

2. **Monoacylglycerols and glycerols:** Monoacylglycerol is packaged in micelles (if there is a heterogeneous group of lipids, it is called a mixed micelle), released just prior to the enterocyte, and moves across the apical membrane through passive diffusion. In the enterocyte, monoacylglycerols are combined with long-chain fatty acids to form triglycerides and are secreted in chylomicrons (see Figure 31.9). Glycerol is absorbed directly across the enterocyte and is not repackaged. After glycerol exits the basolateral membrane of the enterocyte into the interstitial space, it can then diffuse directly into the portal circulation.

3. **Cholesterols:** Cholesterol esters are also packaged in micelles and released just prior to the enterocyte. Cholesterol esters appear to both diffuse through and be transported across the apical membrane. One of these transporters is NPC1L1 (Niemann-Pick C1 like 1), the pharmacologic blockade of which decreases cholesterol uptake and lowers circulating levels of cholesterol in some patients. In the enterocyte, cholesterol esters are esterified, packaged into chylomicrons, and secreted (see Figure 31.9).

4. **Lysolecithins:** Phospholipids are also packaged in micelles, released just prior to the enterocyte, and move across the apical membrane through passive diffusion. In the enterocyte, phospholipids are esterified into lysolecithin, packaged into chylomicrons, and secreted into the interstitial space to be picked up by the lymphatic system (see Figure 31.9).

Table 31.3: Mineral Function

Mineral	Functions
Calcium	Bone and teeth; cell excitability; blood clotting
Chloride	Cell excitability
Copper	Enzyme cofactor; collagen
Iron	Metabolism; oxygen binding; collagen
Iodine	Hormone synthesis
Magnesium	Metabolism
Phosphorus	Bone and teeth; energy storage; cell signaling
Potassium	Cell excitability
Sodium	Cell excitability
Zinc	Enzyme cofactor

Table 31.4: Essential Vitamin Function

Vitamin	Solubility	Function	Deficiency
Biotin	Water	Metabolism	Unknown
Folate	Water	Metabolism; blood cells	Anemia
Niacin	Water	Metabolism; blood cells	Pellagra
Pantothenic acid	Water	Metabolism	Unknown
Riboflavin	Water	Metabolism	Cheilosis
Thiamin	Water	Metabolism	Beriberi
Vitamin A	Fat	Antioxidant; vision; proteins	Blindness
Vitamin B6	Water	Metabolism; blood cells	Anemia
Vitamin B12	Water	Metabolism; blood cells	Anemia; nerve damage
Vitamin C	Water	Collagen; antioxidant	Scurvy
Vitamin D	Fat	Proteins	Rickets; osteomalacia
Vitamin E	Fat	Antioxidant	Anemia
Vitamin K	Fat	Blood cells	Poor clotting

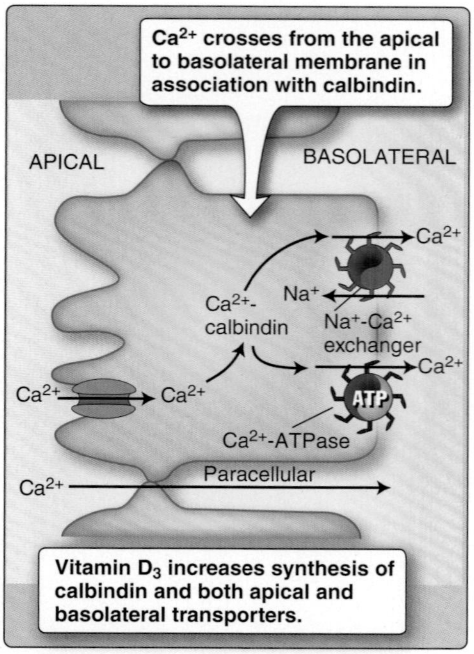

Figure 31.10
Calcium absorption. ATP = adenosine triphosphate.

F. Vitamin and mineral absorption

In addition to macronutrients, small amounts of vitamins and minerals must be in the diet to directly prevent disease (Tables 31.3 and 31.4).

1. **Vitamins:** Fat-soluble vitamins are incorporated in micelles and absorbed similar to long-chain fatty acids and packaged in chylomicrons. Water-soluble vitamins, with the exception of vitamin B12, are absorbed by Na^+ cotransport. Vitamin B12 is absorbed in a four-step process. First, vitamin B12 is liberated from dietary proteins. Second, vitamin B12 binds to haptocorrin released from G cells. Third, pancreatic secretions cause the release of haptocorrin, which is how intrinsic factor binds vitamin B12. Intrinsic factor is released from parietal cells. Fourth, the intrinsic factor/vitamin B12 complex is absorbed by phagocytosis in the ileum.

2. **Minerals:** Monovalent ions and electrolytes will be discussed with the large intestine, later in this chapter. Divalent ions (Ca^{2+}, Mg^{2+}, Fe^{2+}, Cu^{2+}, and Zn^{2+}) are absorbed in the small intestine. A good example of ion transport regulation can be seen with Ca^{2+}, insofar as it can be absorbed through either a paracellular route (throughout the small intestine) or transcellular route (in the duodenum). The transcellular route involves an apical Ca^{2+} channel, cytosolic binding by calbindin, and basolateral Ca^{2+} ATPase and Ca^{2+}-Na^+ exchanger (Figure 31.10). Vitamin D_3 stimulates the expression of these four proteins, which allows for greater Ca^{2+} absorption through the transcellular route.

G. Water absorption

The small intestine is the site of the majority of water absorption, nearly 80%. This fluid includes both what is eaten and drunk as well as secretions from the salivary glands, gastric, liver, pancreas, and intestinal lining. The majority of this absorption occurs via osmosis because of the apical transport of NaCl from the intestinal lumen.

III. LARGE INTESTINE

The large intestine comprises the cecum; ascending, transverse, descending, and sigmoid colon; rectum; and anus (Figure 31.11). The large intestine plays a lesser role in digestion compared to the small intestine but is intricately involved in ion and water absorption.

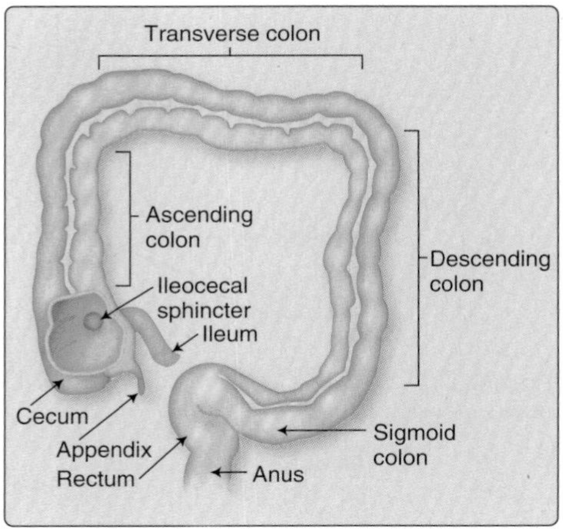

Figure 31.11
Large intestine.

A. Motility

Motility is one of the prime functions of the large intestine. There are three main movement patterns in the large intestine: segmentation, peristalsis, and mass movement contractions. Besides the anatomic divisions, the large intestine can contract into smaller segments called **haustra**, which are seen as the beadlike appearance of the large intestine (Figure 31.12). Segmentation contractions increase the opportunity for contact between the luminal contents and intestinal epithelium, thereby allowing ion and water removal. Segmentation contractions do not propel chyme forward, but both peristalsis and mass movement contractions perform this function. Mass movement contractions occur a few times per day and involve a massive peristaltic wave that results in a significant chyme movement along the large intestine.

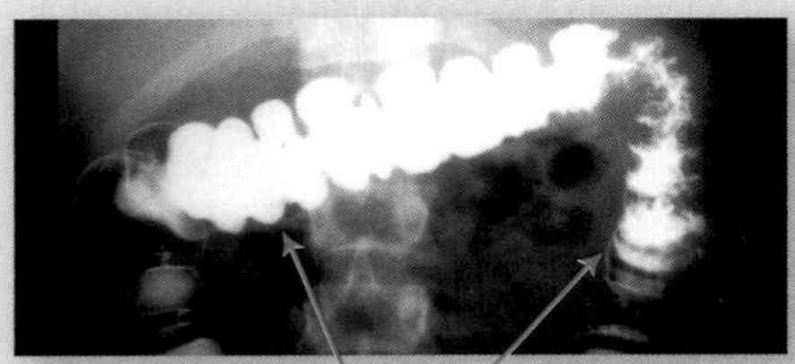

Figure 31.12
Haustra.

1. **Ileocecal sphincter:** The ileocecal sphincter prevents backflow from the large to small intestine (see Figure 31.11). Ileum distention and irritation (stimulation of chemical afferents) initiates ileum peristalsis and relaxes the sphincter, whereas cecum distention and irritation inhibits peristalsis and contracts the sphincter. Immediately following meal ingestion, the ileocecal sphincter relaxes and the ileum contracts. This response is known as the **gastroileal reflex** and is likely controlled by gastrin and CCK.

2. **Other reflexes:** The **gastrocolic reflex** is the urge to defecate shortly after food intake. It is thought both to have a neural component and be mediated by both mechanical- and chemical-sensitive neurons and functions to clear the colon and ready it for the remnants of the new meal. The **orthocolic reflex** is an urge to defecate after standing. This reflex is thought to be mediated by mechanosensitive neurons and the enteric nervous system via gravity-induced distention. For those on medical bed rest, this reflex should be periodically elicited to prevent constipation.

3. **Anal sphincters:** The anus comprises two sphincters: one internal, the other external. The internal sphincter is composed of

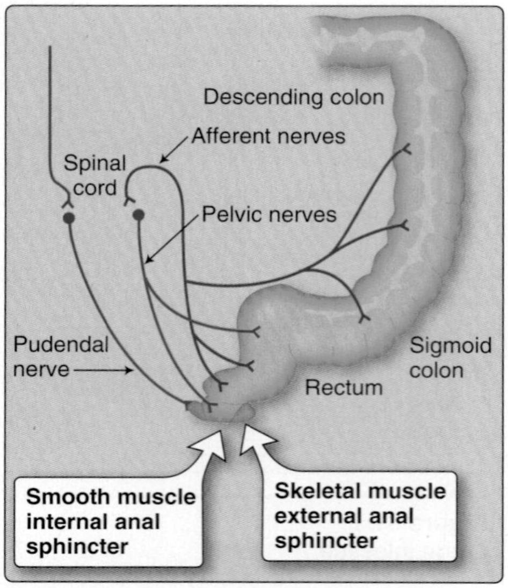

Figure 31.13
Innervation of the colon, rectum, and anus.

smooth muscle. The external sphincter is composed of skeletal muscle that is under somatic control innervated by the pudendal nerve (Figure 31.13). **Defecation** is a multistep process involving both sphincters as well as both enteric and somatic regulation. A peristaltic wave from the large intestine forces feces from the rectum toward the anus. The internal sphincter then relaxes by inhibiting contraction of the smooth muscle within this area (termed the **rectosphincteric reflex**). If the external sphincter is voluntarily relaxed, then defecation occurs. If the external sphincter remains contracted, then defecation is delayed, and feces is retained. Peristaltic waves may cause a sense of urgency, which may or may not lead to defecation depending on the external sphincter. Alternatively, a person may voluntarily increase thoracic and abdominal pressure by a straining maneuver (to push feces downward to start peristaltic waves, which involuntarily relax the internal sphincter) and then voluntarily relax the external sphincter in order to pass the feces.

B. Transport

The large intestine both absorbs and secretes ions. This ion transport also allows for water absorption and regulation during periods of water deprivation and dehydration. Finally, some fats are transported across the apical membrane.

1. **Electrolytes:** Na^+ and water are absorbed via endothelial Na^+ channels (ENaCs) in the distal colon (Figure 31.14). Cl^- is passively absorbed paracellularly. In the proximal colon, Cl^- crosses the apical membrane by Cl^--HCO_3^- exchangers. K^+ is passively secreted in the distal colon (see Figure 31.14). Active secretion can also occur by the insertion of apical K^+ channels in the large intestine with increased concentration of aldosterone or certain second messengers.

2. **Short-chain fatty acids:** Short-chain fatty acids are transported across the apical membrane to be used by colonic epithelial cells as an energy substrate (see Figure 31.14).

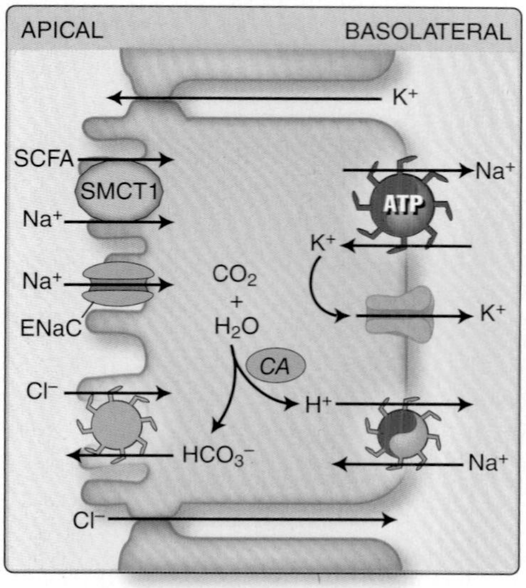

Figure 31.14
Ion and fatty acid transport. ATP = adenosine triphosphate; ENaC = endothelial sodium channel; SCFA = short-chain fatty acid; SMCT1 = short-chain fatty-acid transporter 1.

Clinical Application 31.3: Fecal Incontinence

Fecal incontinence is involuntary defecation. The severity can range from a partial ability to control defecation (except when there are increases in abdominal or thoracic pressure, such as during a cough or when straining to lift an object) to little or no voluntary control. The pathophysiology is often related to trauma, injury to the pelvic floor such as during childbirth or surgery, or a prolapsed rectum. Patients' rectosphincteric reflexes are typically normal, but the pathophysiology is associated with the external sphincter. Treatments for fecal incontinence are dependent on the cause and severity and include garments to collect feces, fecal bulking agents (because liquid stool is more difficult to contain), strengthening pelvic floor and sphincter muscles, and surgical procedures.

Clinical Application 31.4: Diarrhea and Metabolic Acidosis

Chronic (>4 weeks), persistent (2–4 weeks), or acute (<2 weeks) diarrhea, if severe enough, can result in excretion of large amounts of HCO_3^- and other ions. Diarrhea is a frequent semisolid or fluid stool that can have a number of causes (infection, toxins, etc.), but pathophysiology involves osmotic pressure being developed in the gastrointestinal lumen, which favors fluid being retained in the lumen or even dehydrating the surrounding large intestine interstitial spaces. This results in not only water loss but also a decrease in plasma HCO_3^-, thereby decreasing plasma pH. Because Na^+ and Cl^- are lost along with HCO_3^-, the anion gap does not appreciably change. Thus, this type of acid–base disturbance can be classified as **metabolic acidosis** with a normal anion gap.

3. **Water:** The large intestine also plays an important role in water absorption (Figure 31.15). Only 1% of fluid that is delivered (including that from both the diet and GI secretions) to the GI tract is excreted. The large intestine is responsible for ~20% of the fluid absorption. The capacity for water absorption can be increased twofold during hypohydrated states (e.g., when there is an aldosterone-mediated increase in Na^+ transport that allows for more water to be osmotically absorbed).

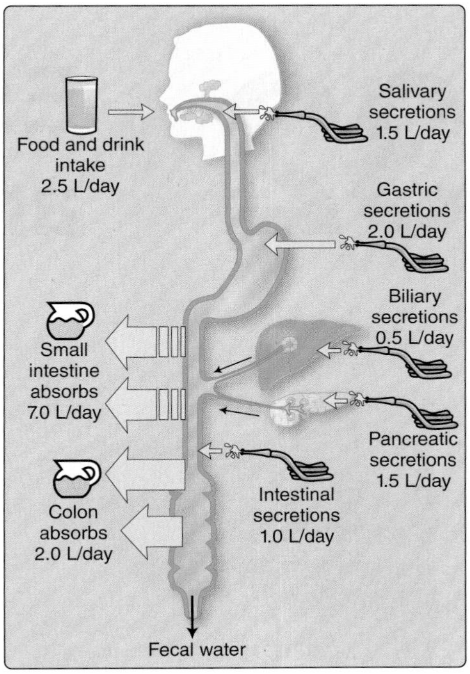

Figure 31.15
Fluid intake, secretion, and absorption.

Chapter Summary

- **Motility** in the small intestine involves both mixing via **segmentation** and **propulsion** via peristalsis. Migrating motor complexes sweep the intestinal lumen free of residual particles between meals.
- ***Pancreatic amylase*** begins starch digestion by cleaving **α1,4 glucose bonds**, and apical membrane-bound ***disaccharidases*** convert the starch remnants to **monosaccharides** (glucose, galactose, and fructose) for absorption.
- Monosaccharide absorption involves cotransport of glucose and Na^+ across the apical membrane, whereas fructose moves through without the aid of cotransport. Basolateral transport also does not involve cotransportation.
- ***Proteases*** secreted by the pancreas (***trypsin***, ***chymotrypsin***, ***elastase***, and ***carboxypeptidases***) cleave amino acid bonds to form smaller peptides. These peptides are further digested by membrane-bound ***peptidases*** forming amino acids, dipeptides, and tripeptides.
- Amino acids are transported across the apical membrane with Na^+, and small peptides are transported with H^+. Within the cytosol, small peptides are broken down into amino acids. Basolateral amino acid transport occurs by specific amino acid–class transporters.
- **Bile acids** emulsify lipids so that *pancreatic lipase* can cleave fatty acids from triglycerides. Dietary cholesterol esters are digested into cholesterol and fatty acids by ***carboxyl ester hydrolase***. These products are then formed into **micelles**.
- Long-chain fatty acids and cholesterol both diffuse across the apical membrane. They are then reconstituted and repackaged into chylomicrons within the enterocyte. Chylomicrons are then secreted and enter the lymphatic circulation.
- The **ileocecal sphincter** regulates the amount of chyme entering the large intestine, and the internal and external anal sphincters regulate the feces exiting the gastrointestinal system. Motility in the large intestine consists of **segmentation**, **peristalsis**, and **mass movement** as well as a number of reflexes that control sphincter contraction and relaxation.
- The large intestine absorbs Na^+, Cl^-, and water and secretes K^+ and HCO_3^-.

32 Exocrine Pancreas and Liver

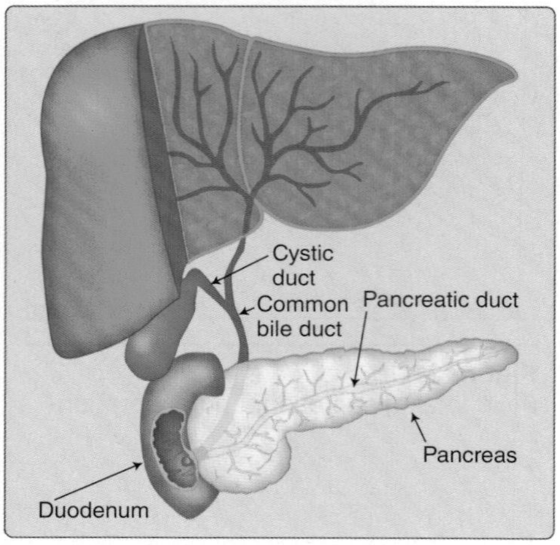

Figure 32.1
Gastrointestinal accessory organs.

Table 32.1: Zymogen Granule Enzymes and Enzyme Precursors

Enzyme	Class/Action
Amylase	Carbohydrate enzyme
Chymotrypsinogen	Protein enzyme precursor
Deoxyribonuclease	Nucleic acid enzyme
Lipase	Lipid enzyme
Procarboxypeptidase A and B	Protein enzyme precursor
Proelastase	Protein enzyme precursor
Prophospholipase A2	Lipid enzyme precursor
Procolipase	Lipid enzyme precursor
Ribonuclease	Nucleic acid enzyme
Trypsinogen	Protein enzyme precursor

I. OVERVIEW

The pancreas, gallbladder, and liver serve as accessory organs for the intestines (Figure 32.1), providing specialized secretions to digest carbohydrates, proteins, and lipids in the small intestine. Pancreatic secretions are highly regulated by neural and hormonal means both in anticipation of eating and in response to food in the gastrointestinal (GI) tract. Hepatobiliary secretions are steadily produced but then stored in the gallbladder for regulated secretion into the small intestine. Once digested and absorbed, most nutrients travel via the **portal circulation** to the liver to either be extracted and processed or to pass through to the systemic circulation (see Figure 21.11).

II. EXOCRINE PANCREAS

The primary functions of the exocrine pancreas are to neutralize acid and deliver enzymes for macronutrient digestion within the duodenum. The **acinar cells** are the primary secretory cells. Small clusters of acinar cells are connected by **intercalated ducts**, which converge on the **collecting duct** (Figure 32.2). The cells lining the intercalated duct add ions and serous secretions to the enzyme and ion secretions of the acinar cells.

A. Regulation

Regulation of pancreatic secretions is dependent on the phase of digestion: cephalic, gastric, or intestinal.

1. **Cephalic phase:** During the cephalic phase, the vagus nerve stimulates pancreatic secretions by releasing **acetylcholine** (**ACh**) and **vasoactive intestinal peptide** (**VIP**) and is thought to account for about 25% of pancreatic secretions.

2. **Gastric phase:** The gastric phase accounts for about 10% of pancreatic secretions and is mediated by **vagovagal reflexes** stimulated by stomach distension.

3. **Intestinal phase:** The intestinal phase accounts for the majority of pancreatic secretions (~65%) and is controlled hormonally control via **secretin** and **cholecystokinin** (**CCK**) Secretin is released in response to H^+, and CCK is released in response

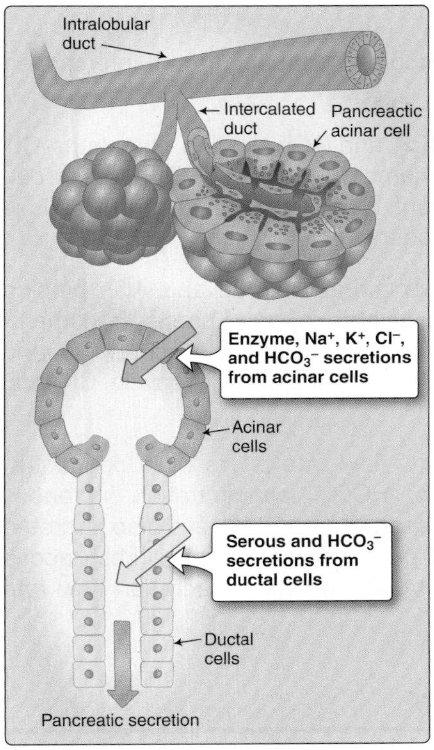

Figure 32.2
Pancreatic acinar and intercalated cells.

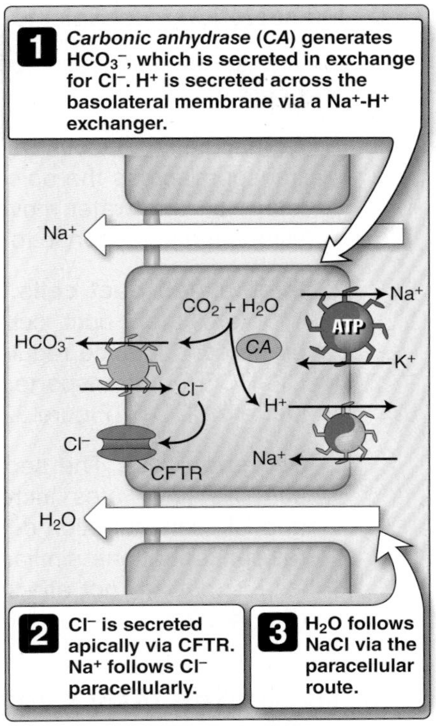

Figure 32.3
Intercalated cell ion secretion. ATP = adenosine triphosphate; CFTR = cystic fibrosis transmembrane conductance regulator.

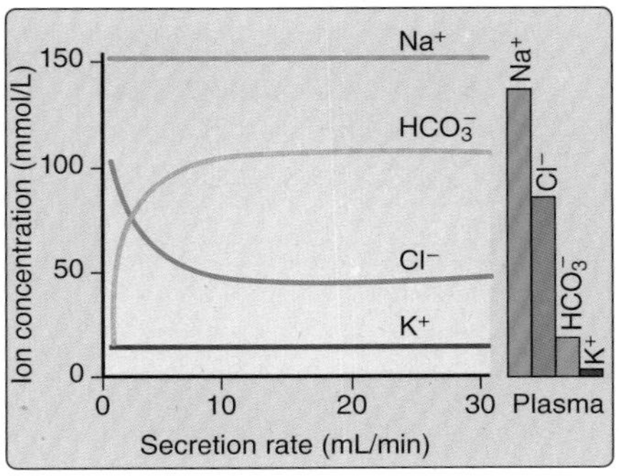

Figure 32.4
Effect of pancreatic secretion rate.

to amino acids, fatty acids, and monoacylglycerols. The primary inhibitors of pancreatic secretions are **somatostatin** and a decrease in chyme macronutrients.

B. Enzyme secretion mechanisms

CCK is released from I cells in the small intestine. CCK and, to a lesser extent, VIP and **gastrin-releasing peptide**, are the primary signals responsible for pancreatic enzyme secretion from acinar cells. Acinar cells contain **zymogen granules** that house some active but mostly inactive digestive enzymes (Table 32.1). When stimulated, acinar cells exocytose zymogen granules into the luminal space. Enzyme packaging occurs in the Golgi apparatus, and large vacuoles are condensed into zymogen granules prior to docking and fusing with the apical membrane. Exocytosis is regulated hormonally and neurally.

1. **Classical hormonal signaling:** CCK is released into the interstitial space and enters the bloodstream. It then travels in the circulation to pancreatic acinar cells where it binds to CCK_A receptors.

2. **Vagal afferent stimulation:** CCK also binds to CCK_A receptors on vagal afferents. This binding stimulates the afferents, eliciting the efferent stimulation of pancreatic acinar cells via VIP.

Clinical Application 32.1: Cystic Fibrosis

Cystic fibrosis leads to pancreatic insufficiency because of a gene mutation encoding for the cystic fibrosis transmembrane conductance regulator. By not having a functional version of this epithelial transporter, secretions are thickened, which can eventually partially block ducts and lead to pancreatic tissue damage. This inhibits the release of pancreatic enzymes, leading to malabsorption of proteins, fats, and fat-soluble vitamins.

C. Ion secretion mechanisms

Ion and serous fluid secretion occurs in both acinar and intercalated duct cells.

1. **Acinar cells:** Basolateral CCK and ACh binding stimulates Cl^- transport across the apical membrane, which facilitates paracellular Na^+ and water movement. Secretin release from S cells is stimulated in response to duodenal acidification.

2. **Intercalated duct cells:** Basolateral secretin and ACh binding in intercalated duct cells activates cystic fibrosis transmembrane conductance regulators (CFTRs), other Cl^- channels, and Cl^--HCO_3^- cotransporters. These transporters recycle Cl^- and secrete HCO_3^- (Figure 32.3).

3. **Secretion rate:** The secretion flow rate alters ionic concentration. As flow rates increase, HCO_3^- concentration increases and Cl^- concentration decreases. Na^+ and K^+ are also secreted (at concentrations similar to plasma for Na^+ and slightly above for K^+) but are not affected by alterations in secretion flow rate (Figure 32.4).

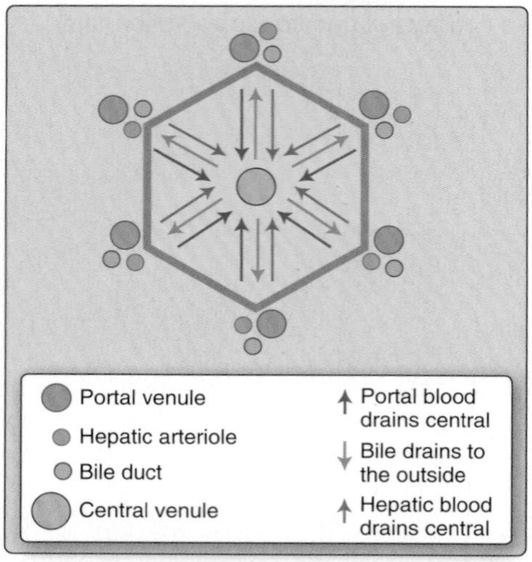

Figure 32.5
Hepatocytes and blood and bile flow.

III. HEPATOBILIARY SYSTEM

The liver produces and secretes bile (termed **hepatic bile** to distinguish it from bile from the gallbladder). Bile is secreted by **hepatocytes** into canaliculi, then traverses a series of bile ducts, which become less numerous but progressively larger in diameter until they form the common hepatic duct. The flow from hepatocytes is in the opposite direction (peripherally) of the blood from the hepatic artery and portal vein, which flow centrally (Figure 32.5). From this junction, bile can move through either the common bile duct into the duodenum or the cystic duct to the **gallbladder**. The **sphincter of Oddi** controls the path. When the sphincter is contracted, the common bile duct has high resistance to bile flow, and, thus, bile travels to the gallbladder. When the sphincter is relaxed, bile flows from the common hepatic duct and often from the gallbladder into the duodenum (Figure 32.6). Sphincter relaxation is regulated primarily by CCK.

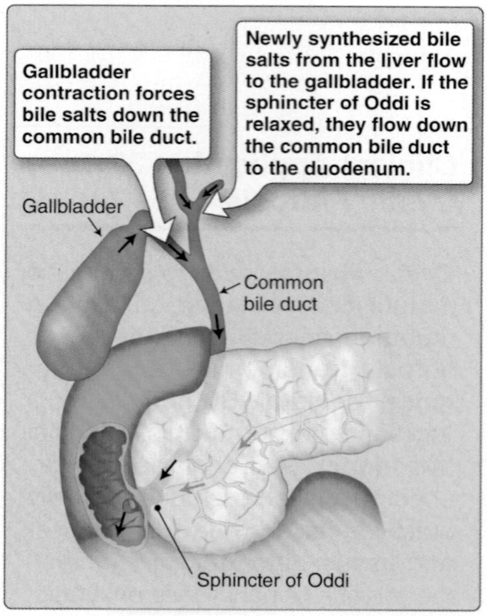

Gallbladder contraction forces bile salts down the common bile duct.
Newly synthesized bile salts from the liver flow to the gallbladder. If the sphincter of Oddi is relaxed, they flow down the common bile duct to the duodenum.
Gallbladder
Common bile duct
Sphincter of Oddi

Figure 32.6
Bile storage and secretion.

A. Bile components

Bile components include bile acids, electrolytes, cholesterol, phospholipids, and bilirubin. Gallbladder bile is significantly more concentrated than hepatic bile, with the exception of osmotic ions, such as Na^+ and Cl^- (Table 32.2).

1. **Bile acids:** Bile acids emulsify lipids to aid in their digestion by *pancreatic lipase* in association with *colipase*. Without bile acids, lipid digestion occurs slowly and is often incomplete because of a dramatic decrease in the surface area available for enzymes. Bile acids are formed from cholesterol, and there are two general forms of bile acids: primary and secondary.

   a. **Primary bile acids:** Cholic and chenodeoxycholic acids are synthesized in hepatocytes via ***7α-hydroxylase***. Thus, the

formation of these two bile acids is a principal route of cholesterol metabolism.

b. **Secondary bile acids:** Deoxycholic and lithocholic acids are not synthesized in hepatocytes. Instead, bacteria in the large intestine and terminal ileum contain *7α-dehydroxylase*, which converts cholic acid to deoxycholic acid and chenodeoxycholic acid to lithocholic acid. These bile acids are then passively reabsorbed and transported back to the liver via the **enterohepatic circulation**. These bile acids can be conjugated or unconjugated, where conjugation simply refers to a salt attachment. Bile acid type affects the specific intestinal or hepatocyte membrane transporter utilized.

2. **Water and electrolytes:** Ions including Na^+, K^+, Ca^{2+}, Cl^-, and HCO_3^- are secreted from hepatocytes isotonically. Some additional water and HCO_3^- are secreted by duct cells. Bile concentration is completed in the gallbladder, and this concentration can be quite dramatic (up to tenfold). This occurs via Na^+ and Cl^- reabsorption, which causes isosmotic water reabsorption that occurs paracellularly and cellularly via aquaporins (AQPs) 1 and 8. In the process of Cl^- reabsorption, HCO_3^- is secreted (Figure 32.7).

3. **Cholesterol and phospholipids:** Besides the cholesterol converted to primary bile acids, small amounts of cholesterol are secreted in the bile. Phospholipids, primarily lecithin, are also secreted and help solubilize some of the bile constituents.

4. **Pigments and organic molecules:** The major pigment in bile is bilirubin. Bilirubin is formed from the catabolism of hemoglobin and is transported in the circulation in a complex with albumin. Hepatocytes secrete this bile pigment, which is either ultimately excreted directly in the intestines or temporarily reabsorbed and then excreted in the urine. Organic ions are also bile components, which serve as a method for the liver to excrete toxins, drugs, and related compounds.

B. Gallbladder

The liver constantly produces bile but not in sufficient quantity to properly emulsify lipids in the small intestine. The gallbladder serves as the bile storage and distribution center. Therefore, when needed, a large quantity can be released. The stored bile in the gallbladder is concentrated. (Think of it as concentrated dishwashing detergent: A little bit can go a long way.) The gallbladder can contract to propel out the bile with CCK stimulation. CCK is the same substance that causes the sphincter of Oddi to relax. This combined effect allows for sufficient quantities of bile acids to be secreted (Figure 32.8). Vagal stimulation also can cause weak gallbladder contraction. Both somatostatin and norepinephrine inhibit bile acid secretion.

IV. NONBILIARY LIVER FUNCTIONS

There are a number of integrative physiology processes of the liver related to metabolism, detoxification, and immune system function. One of

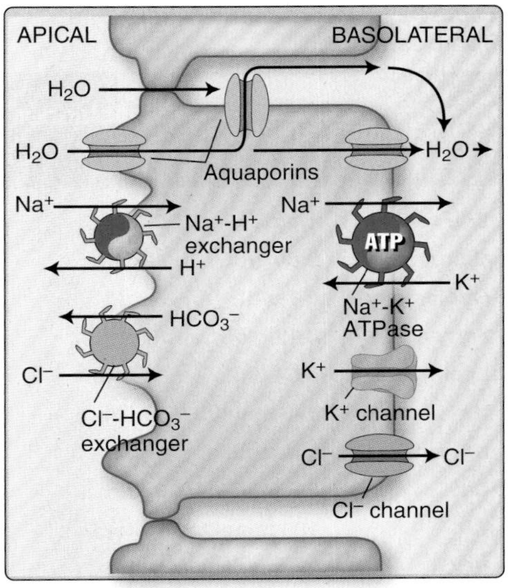

Figure 32.7
Gallbladder concentrations. ATP = adenosine triphosphate.

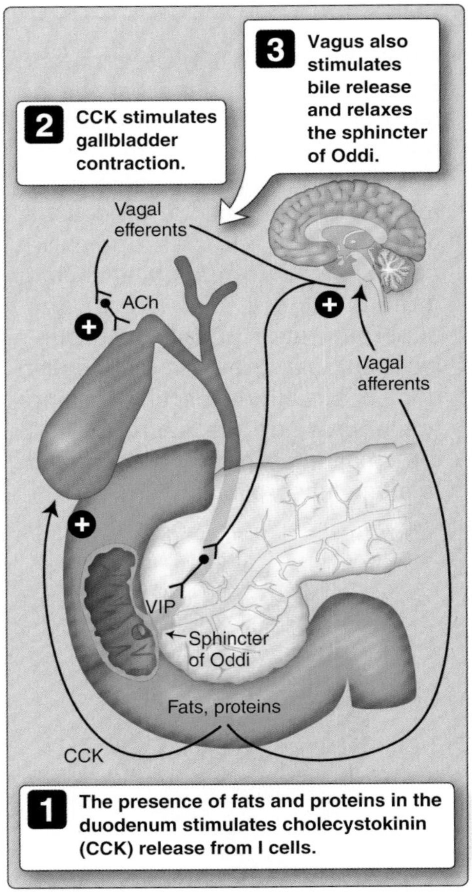

Figure 32.8
Neural and endocrine control of bile secretion. ACh = acetylcholine; VIP = vasoactive intestinal peptide.

Table 32.2: Hepatic and Biliary Bile Composition

Substance	Liver Bile	Gallbladder Bile
Bile salts	1 g/dL	5-fold ↑
Bilirubin	0.04 g/dL	10-fold ↑
Cholesterol	0.1 g/dL	5-fold ↑
Fatty acids	0.12 g/dL	6-fold ↑
Lecithin	0.04 g/dL	10-fold ↑
Na^+	145 mmol/L	Slight ↓
K^+	5 mmol/L	3-fold ↑
Ca^{2+}	2.5 mmol/L	5-fold ↑
Cl^-	100 mmol/L	10-fold ↓
HCO_3^-	28 mmol/L	3-fold ↓

Clinical Application 32.2: Cholelithiasis

Cholelithiasis is the presence of gallstones. The stones can be of two primary types: calcium bilirubinate stones and cholesterol stones. Cholesterol stones are more common, and a number of processes contribute to the pathophysiology of stone formation involving genetic factors, bile stasis, and supersaturation of bile with cholesterol. Gallstones can lead to bile duct obstruction, thereby limiting the amount of bile secreted into the small intestine, which can lead to fat malabsorption.

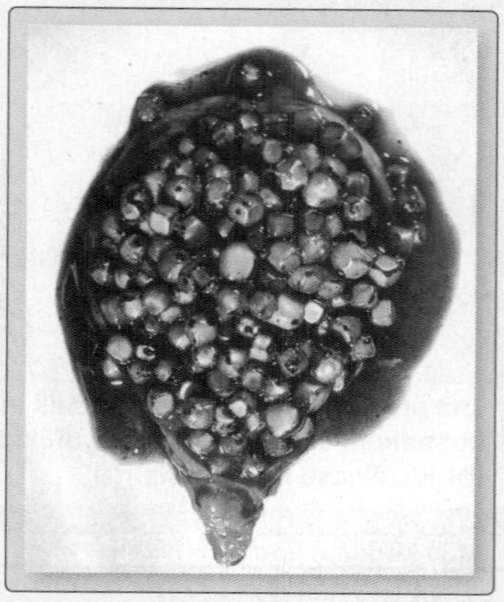

Gallstones.

the main functions is to provide energy substrates to other cells in the body, especially in times when food is scarce.

A. Metabolism

The liver participates in carbohydrate, fat, and protein metabolism. The liver can also store and subsequently release large quantities of carbohydrates in the form of glycogen and certain vitamins and minerals.

1. **Carbohydrate:** The liver plays a major role in the storage and subsequent breakdown of **glycogen**. The average liver can store ~100 g of glycogen. Glycogen breakdown is called **glycogenolysis**, which liberates glucose for released into the systemic circulation. Besides releasing glucose, the liver can convert fructose and galactose into glucose as well as convert amino acids and triglycerides into glucose through a process known as **gluconeogenesis**.
2. **Lipid:** The liver contains the enzymes to undergo large amounts of lipid metabolism. Here the liver can mobilize fatty acids, through a process called **lipolysis**, to be released into the systemic circulation. The liver also produces lipoproteins, phospholipids, ketone bodies, and cholesterol and has the ability to convert amino acids and carbohydrates into new lipids.
3. **Protein:** The liver is involved in protein synthesis and amino acid uptake and metabolism. Proteins synthesised include plasma proteins, prohormones, clotting factors, apoproteins, and transport-binding proteins. The liver also has the capacity to deaminate amino acids.
4. **Vitamins and minerals:** Many vitamins and minerals are delivered to the liver by the portal circulation. The liver has the capacity to store lipid-soluble vitamins such as vitamins A, D, E, and K. This storage of fat-soluble vitamins allows for a short-term reserve for when dietary sources are not available. The liver also stores certain minerals, such as iron and copper.

B. Detoxification

The liver participates in a number of detoxification and removal reactions. Two of the most important of these processes are removal of ammonia and ethanol, and it also mediates various biotransformations.

1. **Ammonia:** The intestines (primarily the large intestine) are responsible for about ~50% of the ammonia produced. The liver receives the majority of this ammonia via the portal circulation. The liver removes most of the circulating ammonia through a series of reactions, which comprise the urea cycle. Urea is released into the systemic circulation, where the majority can be excreted by the kidney.
2. **Ethanol:** The liver contains *alcohol dehydrogenase*, which facilitates the conversion of ethanol into acetaldehyde and reduced nicotinamide adenine dinucleotide. These two products can then be converted into acetyl coenzyme A by peripheral tissues such as skeletal muscles.

3. **Drug biotransformations:** Biotransformation reactions involve two phases. These phases can be described as **phase I**, or **oxidation**, and **phase II**, or **conjugation** and **elimination**.

   a. **Phase I reactions:** Phase I reactions utilize cytochrome P450 enzymes to oxidize organic molecules. These reactions metabolize the majority of drug classes, and there are only a few phase I cytochrome P450–independent reactions of amine-containing compounds. Phase I reactions can also be used to activate some drugs.

   b. **Phase II reactions:** Phase II reactions conjugate the products to aid in solubility for release into the systemic circulation to be filtered and excreted in the kidney or to be secreted into the small intestine with bile for ultimate excretion.

> The liver is involved in the first pass metabolism of oral pharmaceuticals via biotransformation reactions. Certain drugs are almost entirely metabolized in this first pass through the liver via the portal circulation. This is the reason why some drugs must be dosed and delivered in a topical, inhaled, or injectable form.

C. Immune functions

The portal circulation delivers nutrients from the intestines, but often bacteria are also among the sampling of portal blood. However, in the healthy person, there are no intestinal bacteria in the systemic circulation. **Kupffer cells** are specialized phagocytic macrophages located in the liver, which engulf and digest these intestinal bacteria (Figure 32.9). The liver is also the major site of both lymph production and immunoglobulin A release.

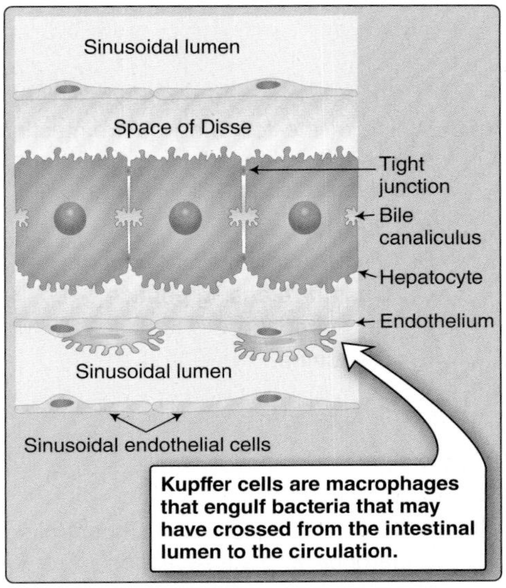

Figure 32.9
Kupffer cells.

Chapter Summary

- Exocrine pancreas regulation occurs via the stimulatory effects of **secretin** and **cholecystokinin** as well as the inhibitory effects of **somatostatin**. The exocrine pancreas secretes enzymes, ions, and serous solutions. Enzymes are secreted in their inactive forms to be activated in the small intestine, and HCO_3^- is secreted to aid in the neutralization of stomach acid.
- Bile components are bile acids (both primary and secondary), electrolytes, cholesterol, phospholipids, and bilirubin. The gallbladder is the primary bile storage site. Bile in the gallbladder is concentrated compared to the liver. **Cholecystokinin** causes gallbladder contractions to move bile toward the small intestine.
- The **sphincter of Oddi** is the narrowing that regulates bile release into the small intestine. Cholecystokinin causes the sphincter of Oddi to relax, thereby permitting bile to enter the small intestine.
- The liver participates in carbohydrate, lipid, and protein metabolism. The liver can either store or release these substrates depending on the fed *versus* fasted state. The liver also stores lipid-soluble vitamins and certain minerals.
- The liver participates in both the **detoxification** and **removal of drugs**, hormones, and ammonia. In addition, the liver produces large amounts of lymph and is involved in many immune-related functions.

Study Questions

Choose the ONE best answer.

VII.1 A 35-year-old woman with a recent breast cancer diagnosis reports gastrointestinal (GI) distress following chemotherapy treatments targeting fast-replicating cells. Which GI layer or signaling molecule is most likely affected by the treatment?

A. Longitudinal muscle
B. Submucosal plexus
C. Epithelium
D. Motilin
E. Vasoactive intestinal peptide

Best answer = C. Gastrointestinal (GI) epithelium has a very high turnover rate (29·II·A) and, thus, is most affected by chemotherapy. Longitudinal muscle and the submucosal plexus are important GI layers for gastric emptying and intestinal motility but have slower turnover rates. Motilin derived from M cells stimulates gastric and intestinal motility (29·III·C), and vasoactive intestinal peptide from parasympathetic nerves relaxes GI smooth muscle (29·IV·A), but such hormones and neurotransmitters do not originate from fast-replicating epithelial cells.

VII.2 Which of the following gastrointestinal signaling substances is released by sympathetic nerve terminals and decreases intestinal secretions?

A. Substance P
B. Vasoactive intestinal peptide
C. Gastrin-releasing peptide
D. Neuropeptide Y
E. Histamine

Best answer = D. Neuropeptide Y relaxes wall muscle and decreases intestinal secretions (29·III·B). Vasoactive intestinal peptide increases pancreatic and intestinal secretion but is released by parasympathetic and enteric neurons (29·III·A). Substance P increases secretions, partially in the salivary glands. Gastrin-releasing peptide increases gastrin secretion. Histamine increases gastric secretions and is released by enterochromaffin-like cells (29·IV·B).

VII.3 A 40-year-old man with uncontrolled Crohn disease undergoes an ileal resection to remove damaged tissue. Synthesis and release of which gastrointestinal hormone would be the most affected by this surgery?

A. Gastrin
B. Motilin
C. Glucose-dependent insulinotropic peptide
D. Prostaglandins
E. Cholecystokinin

Best answer = E. Cholecystokinin (CCK) is secreted by I cells throughout the small intestine, including the ileum (29·IV·A). CCK acts on the stomach, pancreas, and gallbladder to promote secretion and gastric emptying. Motilin and glucose-dependent insulinotropic peptide are secreted by M and K cells, respectively, in the duodenum and jejunum, but not the ileum. Gastrin is secreted both in the stomach and the small intestine. Prostaglandins are not considered hormones but rather are classified as gastrointestinal paracrines.

VII.4 A 52-year-old woman taking scopolamine (a cholinergic antagonist) for motion sickness during airplane travel also develops symptoms consistent with xerostomia as a side effect. Which of the following changes is most consistent with xerostomia?

A. Increased parotid cell inositol trisphosphate
B. Stimulated parotid cell *adenylyl cyclase*
C. Increased mucus production
D. Decreased salivary Cl^- concentration
E. Decreased salivary K^+ concentration

Best answer = D. Salivary secretion is controlled primarily by the parasympathetic nervous system (30·II·C). When active, acetylcholine release increases salivary secretion via the inositol trisphosphate (IP_3) signaling pathway. Blocking cholinergic signaling reduces IP_3 levels and decreases salivary flow. Salivary ionic composition is dependent on flow rates. When flow rate decreases, Cl^- content decreases, whereas K^+ concentration increases. Scopolamine does not stimulate adrenergic receptors, and, therefore, no change in *adenylyl cyclase* would be expected. Anticholinergics decrease mucus production by salivary glands.

VII.5 Which of the following best describes the gastrointestinal pacemaker cells known as interstitial cells of Cajal?

A. They generate 15–20 cycles/min.
B. They require voltage-gated Na^+ channels.
C. They require voltage-gated Ca^{2+} channels.
D. Depolarization is initiated in the antrum.
E. Depolarization is initiated in the fundus.

Best answer = C. Interstitial cells of Cajal (ICC) are pacemakers located in the body of the stomach, not the antrum or fundus (30·IV·B). They generate waves of depolarization (slow waves) at a frequency of ~3–5 cycles/min. The slow waves generate action potentials and initiate waves of contraction that are mediated by voltage-gated Ca^{2+} channels rather than voltage-gated Na^+ channels. The contractile waves are responsible for mixing and grinding stomach contents to help break down food prior to its delivery to the small intestine.

VII.6 If gastric D-cell function were impaired by immune or inflammatory mediators, acid secretion would increase through which of the following mechanisms?

A. Reduced potentiation
B. Increased acetylcholine release
C. Increased prostaglandin E_2 synthesis
D. Decreased G cell secretion
E. Loss of parietal cell inhibition

Best answer = E. Gastric D cells secrete somatostatin, which normally inhibits H^+ secretion from parietal cells (30·IV·C). Reducing somatostatin levels would set up the potential for increased H^+ secretion. Prostaglandins also decrease H^+ secretion normally, but through pathways that do not require D cells. G cells secrete gastrin, which stimulates H^+ secretion from parietal cells. Acetylcholine (ACh) also increases H^+ secretion by a number of direct and indirect routes. Potentiation refers to the observation that H^+ secretion increases to a greater extent when two stimulatory factors bind simultaneously (e.g., gastrin plus ACh) than might be expected from the sum of their individual actions.

VII.7 A 35-year-old woman complains of heartburn and stomach pains, which frequently wake her at night. She is subsequently discovered to have a peptic ulcer. Which of the following best explains how the duodenum normally protects itself against ulcer formation?

A. It has a thick layer of viscous mucus.
B. It has a thick apical membrane.
C. S cells release secretin.
D. Enterochromaffin-like cells release histamine.
E. *Peptidases* are released in inactive form.

Best answer = C. Peptic ulcers occur in the stomach and duodenum (Clinical Application 30.3). They are often precipitated by *Helicobacter pylori*, but intestinal wall erosion is due to acid and enzymes. The duodenum's main defense against acid is secretin released from S cells when stimulated by acid. Secretin triggers HCO_3^- release from the pancreas (32·II·A). Unlike the stomach, the duodenum does not have a thick protective mucus layer, which makes it vulnerable to acid. It does not possess a thick apical membrane, which would impair nutrient absorption. Pancreatic *peptidases* are released in inactive form but activate immediately in the intestinal lumen. Histamine is a local gastric parietal cell control factor.

VII.8 Na^+ is required for absorption of which of the following substances by the small intestinal epithelium?

A. Apical fructose uptake
B. Basolateral glucose transport
C. Apical dipeptide uptake
D. Basolateral amino acid transport
E. Apical glycerol uptake

Best answer = C. Apical absorption of dipeptides occurs via PEPT1, which is a cotransporter powered by an inward Na^+ gradient (31·II·D). Apical fructose transport occurs via GLUT5, and basolateral glucose transport occurs via a GLUT2 transporter (31·II·C). GLUT family transporters facilitate diffusional uptake of substrates down their own concentration gradients and independently of Na^+. Basolateral amino acid transport also occurs via either individual or group transporters independently of ion gradients. Apical glycerol uptake does not require the assistance of any ion or specialized transport protein. Glycerol uptake occurs by diffusion across the epithelial cell membrane.

VII.9 A 65-year-old woman on a strict 1,500-kcal meal plan presents with visceral pain and excessive fullness. A blood sample identifies elevated serum bilirubin, and an ultrasound of the right upper quadrant reveals gallstones obstructing the common bile duct. This obstruction would most affect the digestion and absorption of which of the following meal plans?

A. 55% carbohydrate, 15% protein, 30% fat
B. 20% carbohydrate, 30% protein, 50% fat
C. 70% carbohydrate, 10% protein, 20% fat
D. 40% carbohydrate, 40% protein, 20% fat
E. 50% carbohydrate, 20% protein, 30% fat

Best answer = B. The meal plan consisting of 20% carbohydrate, 30% protein, 50% fat contains the highest fat content and, therefore, would be the most difficult for this individual to digest and absorb. Fats require the emulsification properties of bile acids (31·II·E). Without this emulsification, lipid digestion is compromised, and steatorrhea (fat in the feces), pain, and bloating can result. The protein and carbohydrate portions of the meals will not be directly influenced by a reduction in bile acids.

VII.10 A 28-year-old woman recently gave birth to her second child by cesarean delivery. She is now experiencing both urinary and fecal incontinence during straining maneuvers. A pudendal nerve conduction test indicates that the pudendal nerve is the cause of the fecal incontinence. Which sphincter is most likely affected?

A. Pyloric
B. Ileocecal
C. Rectosigmoid
D. Internal anal
E. External anal

Best answer = E. The pudendal nerve innervates the external anal sphincter, which is a skeletal muscle under voluntary, somatic motor control (31·III·A). The internal anal sphincter is composed of smooth muscle and is innervated by the pelvic nerves and under involuntary control. The pyloric sphincter regulates gastric emptying into the duodenum. The rectosigmoid is a junction rather than a sphincter. The ileocecal valve controls movement of waste materials between the small and large intestines but is not directly involved in defecation.

VII.11 During a hepatic surgery, bile is sampled from the liver and then from the gallbladder. Compared with liver bile, how might the composition of the gallbladder contents differ?

A. Lower bile salt concentration
B. Lower fatty acid concentration
C. Lower cholesterol concentration
D. Higher bilirubin concentration
E. Higher Cl^- concentration

Best answer = D. Bile is produced by the liver and stored by the gallbladder until needed to aid fat digestion (32·III·A). The gallbladder concentrates and modifies bile composition during storage, causing bilirubin levels to rise 10-fold. Bile salts, fatty acids, and cholesterol all increase concentration also. Cl^- is reabsorbed along with some other ions during biliary concentration, its levels falling 10-fold.

VII.12 To evaluate possible cholecystitis, cholecystokinin (CCK) is given during a cholescintigraphy procedure in which bile constituents are radioactively labeled, and biliary secretions tracked. What is the primary function of CCK in this test?

A. To decrease primary bile salt formation
B. To decrease secondary bile salt formation
C. To stimulate local sympathetic efferents
D. To inhibit bicarbonate secretion
E. To contract the gallbladder

Best answer = E. Cholecystokinin (CCK) has several roles in gastrointestinal function, including to facilitate bile release into the intestinal lumen. Release is effected by relaxing the sphincter of Oddi and contracting the gallbladder (32·III·B). CCK also increases HCO_3^- secretion. Bile release is also stimulated by parasympathetic nervous system acetylcholine release. The sympathetic nervous system does not contribute to bile release, and norepinephrine is classified as an inhibitor of bile secretion. CCK does not regulate bile salt formation.

Unit VIII
Endocrine System

Endocrine Pancreas and Liver 33

I. OVERVIEW

All cells need energy to survive and grow. The endocrine pancreas and liver regulate bloodborne energy substrate availability, namely, glucose, fatty acids, ketone bodies, and amino acids. Of these, glucose forms the backbone of cellular energetics (i.e., glycolysis, citric acid cycle, and oxidative phosphorylation). Blood glucose levels are regulated by the pancreatic hormones **insulin**, which promotes glucose entry into cells, and **glucagon**, which increases blood glucose levels primarily through effects on the liver. Besides addressing the cellular need for immediate energy, these hormones are also involved in both short- and long-term energy storage. Energy can be stored in the form of **glycogen** or as lipids in the liver and peripheral tissues, such as adipose and muscle. Energy is not only necessary for immediate cellular needs but also for growth, division, and repair. **Growth hormone** (**GH**) and **insulin-like growth factor** (**IGF**) from the **hypothalamic–pituitary–liver axis** mediate many of these actions. Because GH and IGF-1 are so heavily involved in anabolism (building up the body from smaller compounds), it is not surprising that these two hormones also influence energy delivery and interact with the main pancreatic hormones (i.e., insulin and glucagon). Insulin, glucagon, GH, and IGF are peptide hormones that are first produced in a "prepro" form, which is modified into a "pro" form and, finally, in the Golgi apparatus, converted into an "active" form. This active form is often secreted with additional cleaved sequences. Insulin, for example, is secreted along with C-peptide (Figure 33.1).

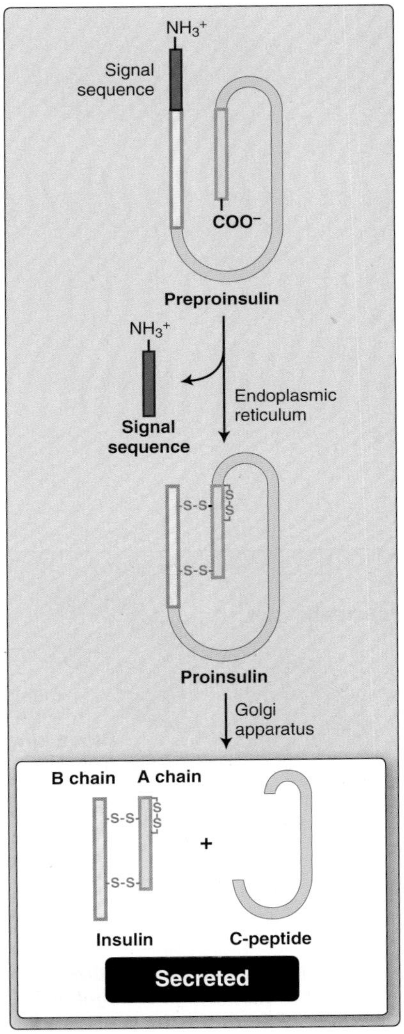

Figure 33.1
Insulin processing steps.

II. ENDOCRINE PANCREAS

The pancreas contains two types of glands. Exocrine glands secrete digestive enzymes and HCO_3^- into the intestinal lumen, as discussed in Chapter 32. Endocrine glands are highly vascularized clusters of hormone-producing cells known as the **pancreatic islets** (**islets of Langerhans**). Exocrine gland products help digest food to liberate energy substrates for absorption, whereas the endocrine gland secretions control availability and use of these energy substrates following absorption.

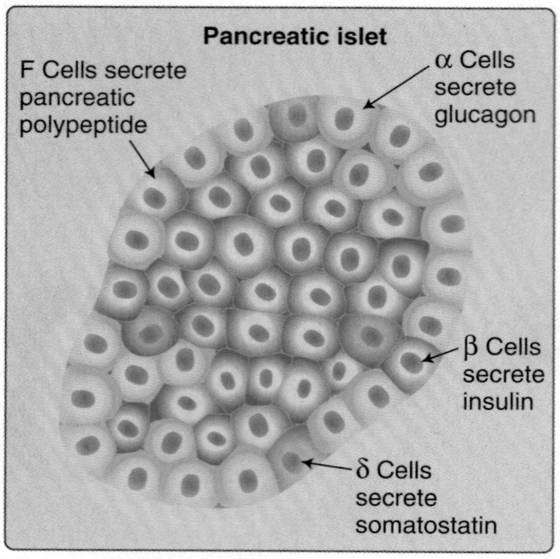

Figure 33.2
Pancreatic islet cellular composition.

A. Islet structure

Islets contain four principal endocrine cell types, each of which produces a specific hormone. **α Cells** secrete glucagon, **β cells** secrete insulin, **δ cells** secrete **somatostatin**, and **F cells** secrete **pancreatic polypeptide** (Figure 33.2). Insulin-secreting cells are numerous and are located centrally, whereas glucagon-secreting cells are located more peripherally. Adjacent cells within the islet are connected via gap junctions, allowing for direct cell-to-cell communication (see 4·II·F).

B. Blood flow

Arterial blood enters islets at their center and then flows toward the edge (Figure 33.3), much as the water in a fountain bubbles up from the center and flows outward. This flow pattern allows for local hormonal signaling (paracrine) within an islet. For example, δ cells release somatostatin that acts locally to decrease glucagon and insulin secretion from α and β cells, respectively. Secreted pancreatic hormones drain with the blood into the portal circulation and are carried to the liver (see Figure 21.11). Hepatocytes are key targets for many of the pancreatic hormones, which is not surprising given the liver's significant role in energy substrate storage and metabolism.

C. Innervation

Islet cells are innervated by both the sympathetic and parasympathetic nervous systems. Postsynaptic muscarinic receptors (cholinergic) mediate parasympathetic effects, and both α- and β-adrenergic receptors mediate the sympathetic effects (see 7·IV·B·2). In general, sympathetic stimulation increases energy substrate release into blood for cellular use, whereas parasympathetic stimulation causes cells to take up and store energy substrates.

III. GLUCAGON

Glucagon is a small (29 amino acid) peptide hormone synthesized by islet α cells. It is formed by proteolysis of proglucagon, liberating glucagon and two inactive protein fragments. Glucagon's half-life is 5–10 minutes after release into the circulation. It is degraded and removed from the circulation by the liver.

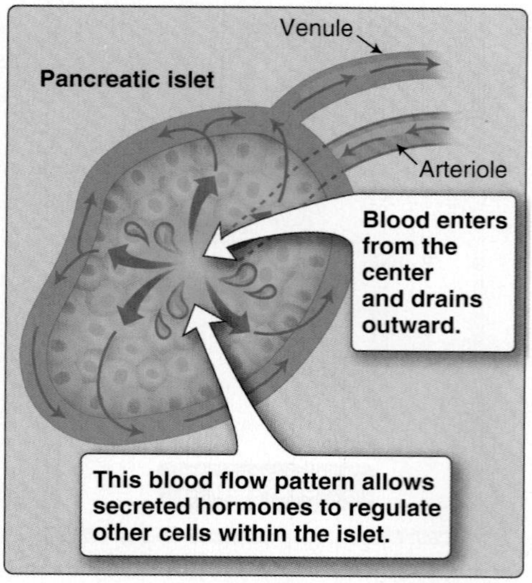

Figure 33.3
Islet blood flow.

A. Function

Glucagon's main function is to mobilize and make energy substrates available for use by tissues during times of stress or between meals. Glucagon's primary target is the liver, but it has secondary targets that include striated myocytes and adipocytes. **Glucagon receptors** are part of the G protein–coupled receptor (GPCR) superfamily and mediate a number of cellular effects, including increases in blood glucose, fatty acid, and ketone body concentration by **glycogenolysis**, **gluconeogenesis**, **lipolysis**, and **ketogenesis**.

1. **Glycogenolysis:** Glucagon stimulates hepatic glycogen breakdown by ***glycogen phosphorylase*** and *glucose 6-phosphatase*, liberating glucose for release into the circulation. *Glycogen phosphorylase* is activated by *protein kinase A* (*PKA*)-dependent phosphorylation following glycogen-receptor binding. *PKA* simultaneously phosphorylates and inhibits glycogen synthesis

by *glycogen synthase*, thereby facilitating glucose mobilization.[1] Glucose then exits the cell via GLUT2, a member of the glucose transporter family (Figure 33.4). Glucagon also stimulates glycogenolysis in muscle to support an increase in contractile activity.

2. **Gluconeogenesis:** Glucagon also stimulates glucose synthesis from noncarbohydrate sources, such as lipids and proteins. Gluconeogenesis is mediated by pathways that include *glucose 6-phosphatase* and *fructose 2,6-bisphosphatase* (Figure 33.5).[2] Glucagon simultaneously inhibits enzymes involved in glucose breakdown, including *glucose kinase, phosphofructokinase,* and *pyruvate kinase*.

3. **Lipolysis:** Glucagon also targets adipocytes, causing them to break down triglycerides into glycerol and free fatty acids. Lipolysis is mediated by *hormone-sensitive lipase* (*HSL*), thereby increasing plasma free fatty acids and fatty acid utilization as both direct (lipid metabolism) and indirect (converted back to glucose, then metabolized) substrates.

4. **Ketogenesis:** Ketone bodies (i.e., acetoacetate, β-hydroxybutyrate, and acetone) form in hepatocytes from incomplete oxidation of free fatty acids. Fatty acids are absorbed and produced by hepatocytes. These fatty acids are then transported into the mitochondria via a carnitine shuttle system for processing. Ketone bodies are released from the hepatocyte into the circulation. They are soluble in aqueous solutions and easily absorbed by extrahepatic tissues, where ketone bodies are converted back into acetyl coenzyme A for use in aerobic metabolism, liberating their stored energy.

> Ketone bodies such as acetone (familiar as nail-polish remover) are organic volatiles that have a characteristic fruity aroma. They are easily detected on the breath of individuals metabolizing them. Excess ketone production can also cause ketoacidosis, a high anion gap metabolic acidosis (see 28·VI·E).

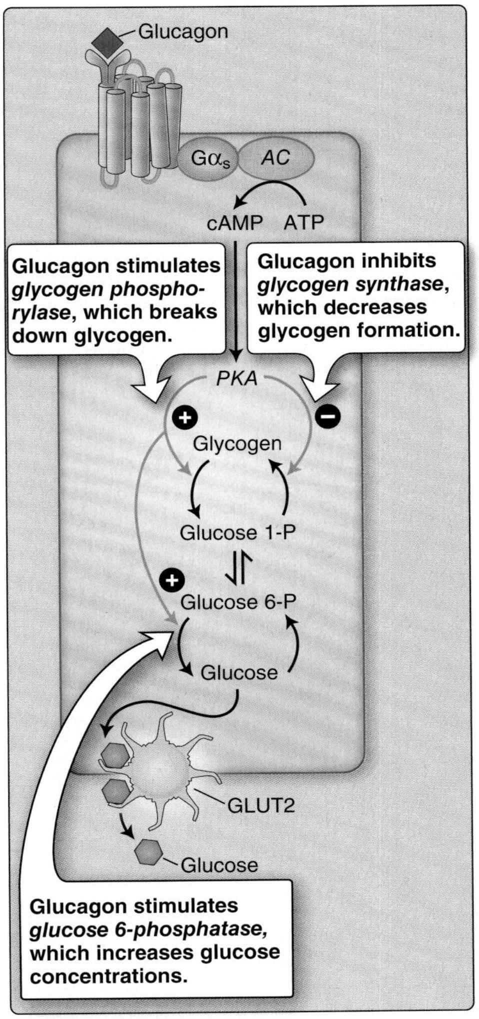

Figure 33.4
Glycogenolysis. *AC* = *adenylyl cyclase*; ATP = adenosine triphosphate; cAMP = cyclic adenosine monophosphate; Glucose 1-P = glucose 1-phosphate; Glucose 6-P = glucose 6-phosphate; GLUT2 = member of the glucose transporter family; *PKA* = *protein kinase A*.

B. Secretion

Glucagon release is regulated by circulating substrates (amino acids, ketone bodies, and glucose) and by neural and hormonal mechanisms.

1. **Increased secretion:** Cholecystokinin (CCK) and higher blood concentrations of amino acids (such as result from consuming protein) stimulate glucagon secretion. Glucagon is also stimulated by decreases in blood glucose via negative feedback. The sympathetic nervous system (SNS) increases glucagon secretion during stressful events to increase energy substrate availability in the form of blood glucose, fatty acid, and ketone bodies to working tissues.

[1]For more information on glycogen breakdown, see *LIR Biochemistry*, 5e, p. 125.

[2]For more information on gluconeogenesis, see *LIR Biochemistry*, 5e, p. 117.

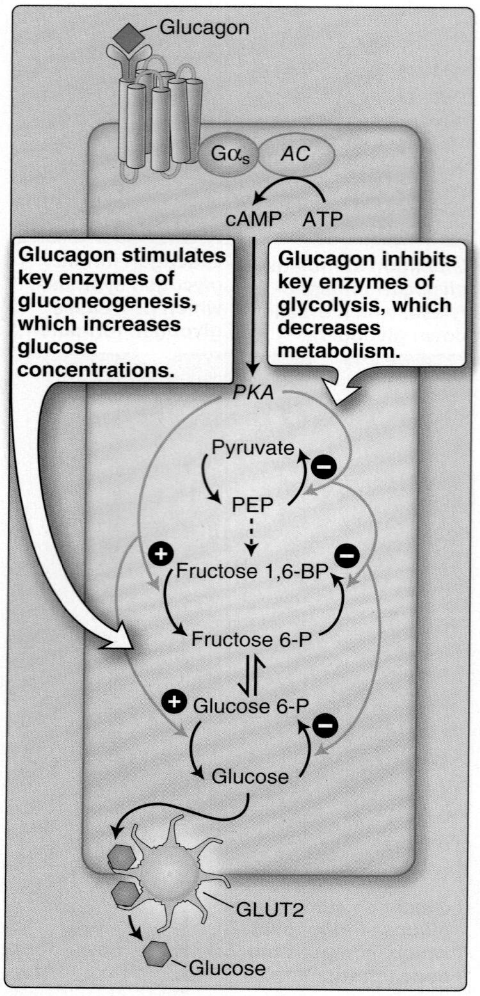

Figure 33.5
Gluconeogenesis. *AC* = *adenylyl cyclase*; ATP = adenosine triphosphate; cAMP = cyclic adenosine monophosphate; Fructose 1,6-BP = fructose 1,6-bisphosphate; Fructose 6-P = fructose 6-phosphate; Glucose 6-P = glucose 6-phosphate; GLUT2 = member of the glucose transporter family; *PKA* = *protein kinase A*; PEP = phosphoenolpyruvate.

2. **Decreased secretion:** Insulin and somatostatin decrease glucagon secretion by the islet. Increases in blood glucose, fatty acids, and ketone bodies also decrease glucagon secretion by negative feedback. Circulating **glucagon-like peptide 1** (**GLP-1**) secreted from intestinal L cells also suppresses glucagon secretion.

C. Glucagon-like peptides

In the pancreas, there are two inactive protein fragments that are secreted with glucagon. These fragments are **GLP-1** and **glucagon-like peptide 2** (**GLP-2**). They are inactive because of extra amino sequences attached to each. In L cells, the same gene is transcribed, yielding active forms of GLP-1 (involved in insulin secretion, see below) and GLP-2 (involved in stabilizing the intestinal mucosa) and an inactive form of glucagon. This process indicates the importance of not only having the gene transcribed but also the correct enzymes and posttranslational modifications within a tissue to create active peptide hormones.

IV. INSULIN

In islet β cells, proinsulin is broken down into insulin and C-peptide (see Figure 33.1). Insulin is a protein hormone consisting of two peptide chains. Insulin's half-life is about 3–8 minutes. Insulin is degraded by the liver during its first pass, which removes more than 50% with additional degradation occurring in the kidneys and other peripheral tissues.

> C-peptide (31 amino acids) is biologically inert but secreted at the same ratio as insulin. It is not removed by the first pass through the liver and has a longer half-life than that of insulin. Thus, clinically it can be used to monitor β-cell function.

A. Function

Insulin's major function is to lower blood glucose levels. The primary insulin targets are the liver, skeletal muscle, and adipose tissue, which, when stimulated, allow for the uptake of glucose, fatty acids, glycerol, ketone bodies, and amino acids from the blood. Think of insulin as the key to a crowd control gate where there are people (glucose) waiting on the sidewalk (blood) to get into a show (the cell). Once some of the people are let in, fewer remain outside on the sidewalk. The cellular effects of insulin are transduced by a ***tyrosine kinase receptor*** (see 1·VII·C) and important docking proteins, insulin receptor substrates (IRSs). Both the *tyrosine kinase* portion of the insulin receptor and IRS activate other proteins by phosphorylation to mediate myriad cellular effects.

1. **Glucose uptake:** Insulin increases glucose uptake by upregulating and inserting GLUT4 transporters into muscle (Figure 33.6) and adipose tissue. Cell membrane glucose transporters can be insulin sensitive, like GLUT4, or insensitive like GLUT2 (e.g., liver). Thus, muscle can dramatically decrease blood glucose levels when insulin is high (e.g., after meals) but does not appreciably affect blood glucose levels during periods of low insulin (e.g., between meals).

2. **Glycogenesis:** Insulin stimulates the formation of glycogen via stimulating *glycogen synthase* and inhibiting *glycogen phosphorylase* in muscle and liver (Figure 33.7). Glycogen formation is also increased by facilitating the conversion of glucose to glucose 6-phosphate. This facilitation is mediated by the insulin-sensitive *glucokinase.*

3. **Glycolysis:** Glycolysis is stimulated by insulin-induced activation of *pyruvate dehydrogenase* and *phosphofructokinase* in muscle and liver. The liver additionally activates *pyruvate kinase*. This, coupled with the aforementioned increase in *glucokinase*, facilitates the utilization of glucose. At the same time that glycolysis is being stimulated, the reverse pathway (gluconeogenesis) is being repressed in hepatocytes to prevent competition for substrates and products of these pathways.

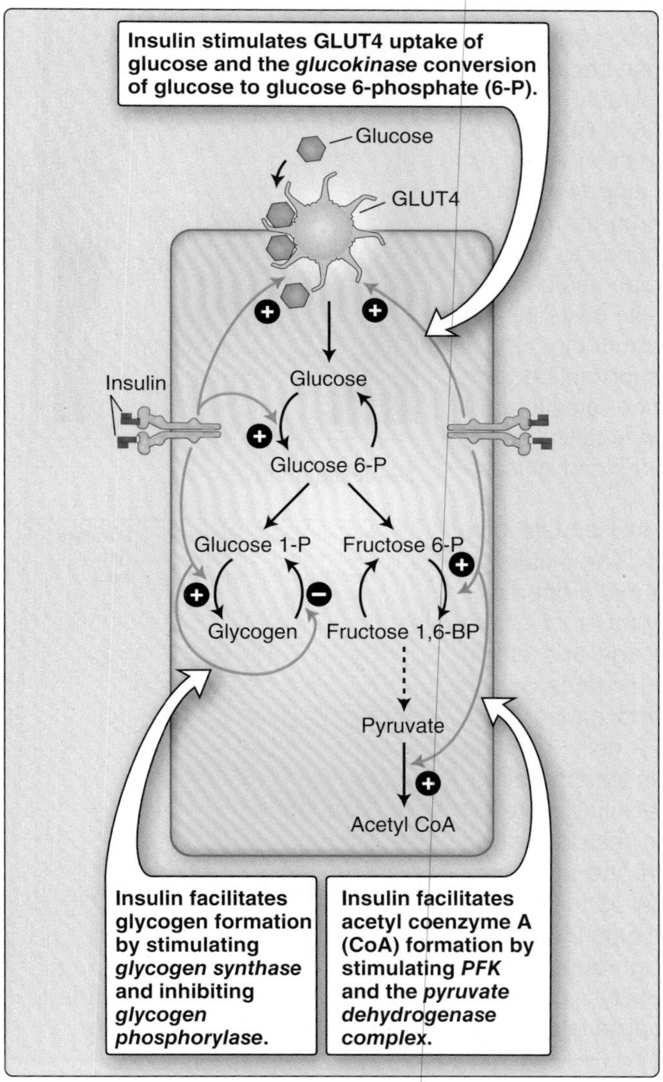

Figure 33.6
Skeletal myocyte. Fructose 1,6-BP = fructose 1,6-bisphosphate; Fructose 6-P = fructose 6-phosphate; Glucose 1-P = glucose 1-phosphate; GLUT4 = member of the glucose transporter family; *PFK* = *phosphofructokinase*.

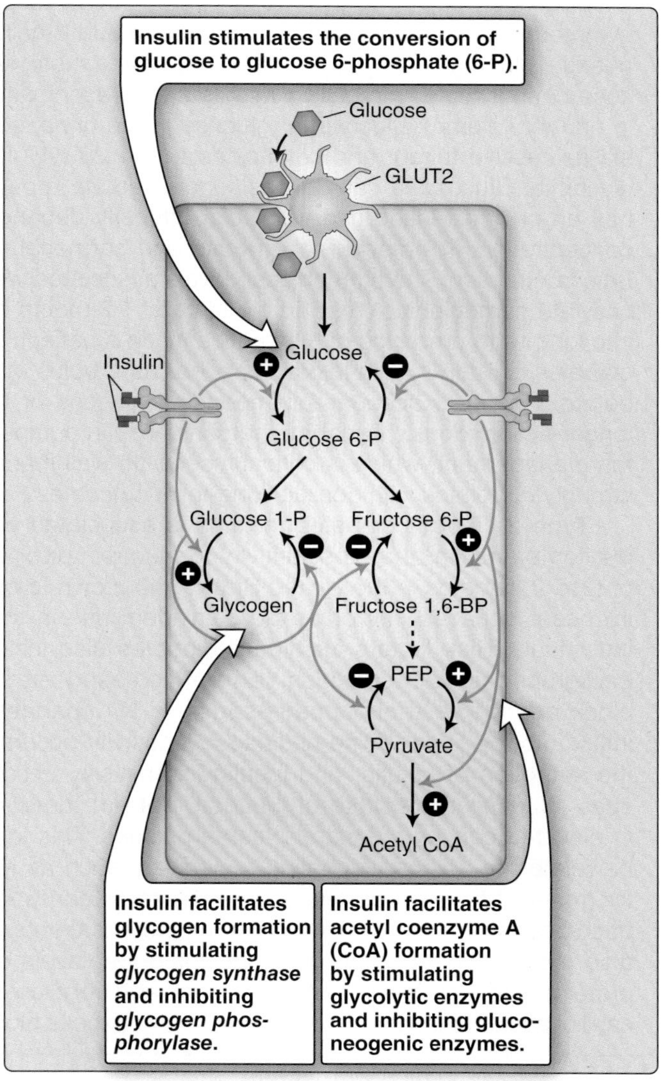

Figure 33.7
Hepatocyte. Fructose 1,6-BP = fructose 1,6-bisphosphate; Fructose 6-P = fructose 6-phosphate; glucose 1-P = glucose 1-phosphate; GLUT2 = member of the glucose transporter family; PEP = phosphoenolpyruvate.

Clinical Application 33.1: Diabetes Mellitus

The regulation of blood glucose is essential for normal tissue function. High blood sugar (hyperglycemia) can lead to diabetes mellitus. Low blood sugar (hypoglycemia) can be the result of diet; a symptom of another medical condition, such as sepsis; or the result of a medication. Diabetes mellitus is a worldwide health epidemic and affects ~8.3% of the U.S. population. Complications of diabetes include cardiovascular (endothelial dysfunction, hypertension, heart disease, and stroke), renal, and ocular problems as well as peripheral neuropathy. Part of this widespread damage is because hyperglycemia can damage fundamental structures, such as basement membranes and endothelial tissue. These complications can also negatively combine, such as when peripheral vascular damage and peripheral neuropathy lead to foot ulcers. Hyperglycemia in diabetes is defined as fasting plasma glucose ≥126 mg/dL or ≥200 mg/dL after an oral glucose tolerance test (rapid ingestion of 75 g glucose in 300 mL water and glucose monitoring over the following 120 min). There are two broad classifications of the disease.

Type 1: Type 1 diabetes mellitus results from destruction of the islet β cells, often due to a virus or an autoimmune response. The onset of symptoms is often quite rapid. Without adequate β-cell function and insulin release, blood glucose levels, especially after a meal, will rise because glucose cannot move into cells. Patients also present with polyuria (increase in urinary volume), glucosuria (glucose in the urine, see 26·III·B), polydipsia (excessive thirst), and ketoacidosis (see 28·V·E) in addition to hyperglycemia. Glucosuria can be easily detected via a paper strip test, which has an enzyme-linked dye with color intensity dependent on the glucose concentration. Polydipsia is the neural and hormonal response to the volume contraction. The high glucose levels associated with diabetes can glycosylate hemoglobin A1c. Red blood cells' ~2-month half-life gives insight into long-term presence of hyperglycemia in a patient. Treatment for type 1 diabetes is often insulin injections either as a bolus of a short-acting form before a meal to account for impending glucose or routine injections of longer-acting forms.[1] For some persons, insulin pumps can simplify diabetes management, which is critical insofar as well-managed blood glucose strongly correlates with positive long-term outcomes.

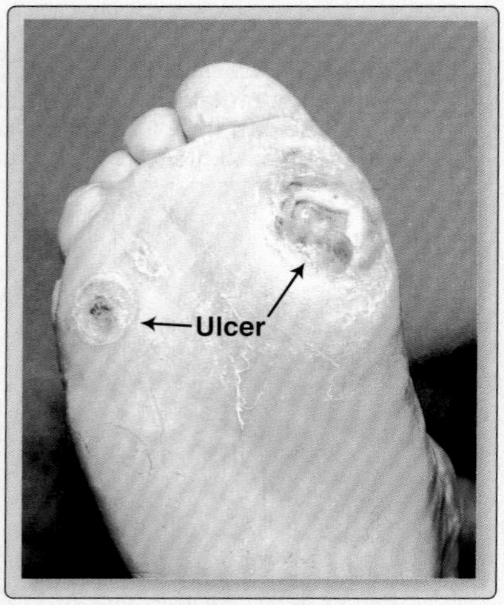

Diabetes ulcer.

Type 2: Type 2 diabetes mellitus is associated with inadequate production of insulin or an insensitivity to glucose and insulin. The incidence of type 2 diabetes in the United States has grown to epidemic proportions in recent decades, mirroring the coincident rise in the number of obese individuals. Risk factors for type 2 diabetes also include age and ethnic background. Insulin secretion can be increased via sulfonylureas, which block adenosine triphosphate–sensitive K^+ channels and cause Ca^{2+} influx into the β cell.[2] The glucose insensitivity occurs in β cells, blunting the secretion of insulin, and insulin insensitivity occurs in peripheral tissues, blunting the uptake of glucose. Both of these insensitivities result in elevated concentrations of plasma glucose. This insulin resistance can be related to adipose signaling molecules, such as leptin and adiponectin (more of these substances are released when a fat cell is fuller), and decreases with weight loss. Similar to type 1, patients with type 2 diabetes also present with dyslipidemia, including low levels of high-density lipoproteins and high levels of chylomicrons and very low–density lipoproteins in the blood, which can contribute to cardiovascular disease. This is likely due to excess bloodborne fats that are not moved into cells via insulin.

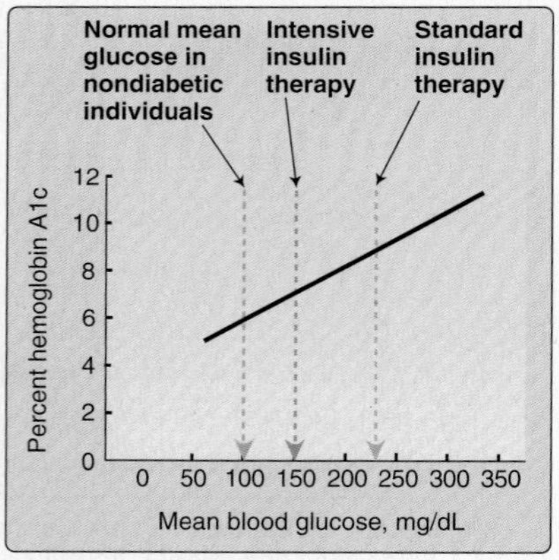

Hemoglobin A1c.

[1]For more information on exogenous insulin preparations, see *LIR Pharmacology*, 5e, pp. 304–306 and 309.

[2]For more information on insulin secretagogues, see *LIR Pharmacology*, 5e, p. 307.

4. **Lipogenesis:** Insulin increases ***lipoprotein lipase*** (***LPL***) and decreases *HSL* activity in adipocytes. *LPL* facilitates the breakdown of chylomicrons and other low-density lipoproteins into free fatty acids, which then can be absorbed. The increase in cellular free fatty acids increases triglycerides and the formation of lipid droplets.

5. **Ketone bodies:** In hepatocytes, ketone body formation and secretion are inhibited in the presence of insulin because of insulin's inhibition of the rate-limiting carnitine shuttle. The carnitine shuttle consists of *transferase* and *translocase* enzymes that move fatty acyl coenzyme A into the mitochondria for processing.[1]

6. **Protein synthesis:** In skeletal muscle and hepatocytes, insulin promotes protein synthesis and inhibits protein catabolism. The anabolic effect of insulin involves both the mTOR (mammalian target of rapamycin) pathway and the cellular increase in amino acid uptake. The mTOR pathway decreases proteolysis and increases ribosomal production and assembly.

B. Secretion

Insulin secretion is regulated by neural, hormonal, and circulating substrate mechanisms (Figure 33.8).

1. **Increased secretion:** Increases in blood glucose, fatty acids, and amino acids stimulate insulin secretion by inhibiting **ATP-sensitive K^+ channels**. Inhibition of K^+ efflux depolarizes membrane potential, which opens voltage-gated Ca^{2+} channels. The subsequent increase in cytosolic Ca^{2+} facilitates docking and fusion of the insulin-containing vesicles with the cell membrane to allow for insulin secretion. Glucagon, **glucose-dependent insulinotropic peptide** (**GIP**), GLP-1, CCK, acetylcholine, and β-adrenergic stimulation increase cytosolic Ca^{2+} or activate *PKA* to increase insulin secretion (see Figure 33.8). GIP and GLP-1 (collectively termed incretin hormones) are secreted by the intestines in response to increased glucose levels in the gut. It is thought that this signal of impending glucose (think of this as a news flash, indicating that there is glucose in the gut that will soon be in the blood) could account for up to half of the insulin response to a carbohydrate meal.

2. **Decreased secretion:** Decreases in blood glucose provide negative feedback to decrease insulin secretion. Somatostatins from adjacent islet cells suppress insulin secretion, as does α-adrenergic stimulation by the SNS. These latter two effects occur via inhibition of *adenylyl cyclase* and *PKA* (see Figure 33.8).

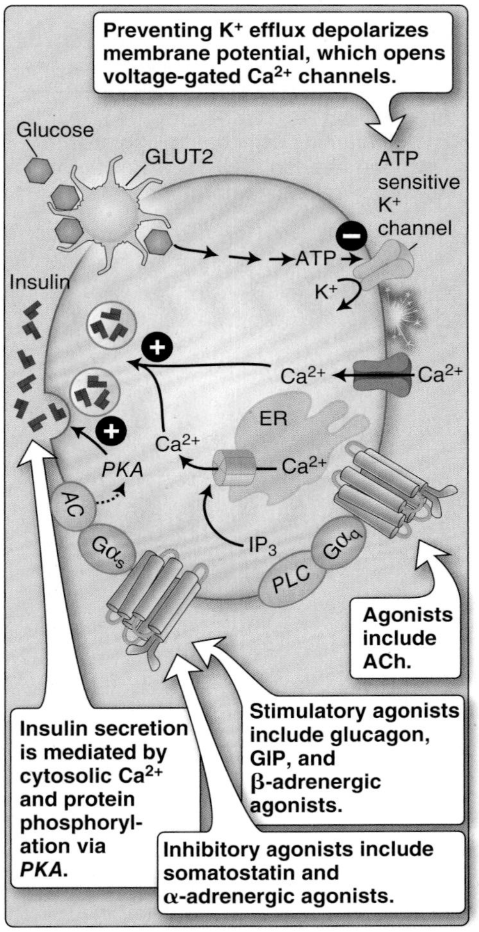

Figure 33.8
Regulation of insulin secretion by pancreatic β cells. *AC* = *adenylyl cyclase*; ACh = acetylcholine; ATP = adenosine triphosphate; ER = endoplasmic reticulum; GIP = glucose-dependent insulinotropic peptide; GLUT2 = member of the glucose transporter family; IP_3 = inositol trisphosphate; *PKA* = *protein kinase A*; *PLC* = *phospholipase C*.

V. HYPOTHALAMIC–PITUITARY–LIVER AXIS

Hormone secretion is often regulated by a multitiered axis system. The benefits of axis control are similar to a microscope that has adjustment knobs for both gross (large) and fine (small) focus. These multiple tiers, however, also mean that a number of locations could be disease sites. For example, pathology could result in a decrease in pituitary release (pituitary

INFO LINK [1]For more information on the carnitine shuttle, see *LIR Biochemistry*, 5e, p. 191.

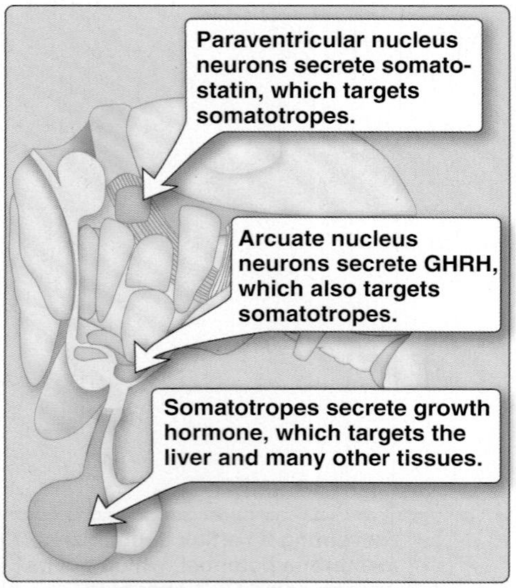

Figure 33.9
Hypothalamic nuclei controlling growth hormone release. GHRH = growth hormone–releasing hormone.

hypofunction), such as in craniopharyngioma (hypothalamic–pituitary tumor), despite having normal endocrine function downstream. The hypothalamic–pituitary–liver axis is unique in that both the second (GH) and third (IGF-1) secretions are hormones with widespread biological effects. In the other endocrine axes (discussed in Chapters 34, 35, and 36), only the third hormone in the axis is biologically active in tissues outside of the axis.

A. Hypothalamus

The hypothalamus contains two nuclei that are important for control of GH and IGF-1. The paraventricular nucleus secretes **somatostatin**, which inhibits GH release, whereas the arcuate nucleus secretes **GH-releasing hormone** (**GHRH**) into the hypophyseal portal circulation. These hormones target **somatotropes** in the anterior pituitary (Figure 33.9).

B. Pituitary gland

Somatotropes are the most numerous (~50%) cell type in the anterior pituitary (see Table 7.3). GHRH and somatostatin receptors on these cells are part of the GPCR superfamily, with GHRH increasing and somatostatin decreasing cAMP. cAMP activates *PKA*, which facilitates Ca^{2+} influx to allow GH-containing vesicles to dock and release their contents into the circulation.

C. Liver

The liver is a key target organ site of the hypothalamic–pituitary axis and produces IGF-1. IGF-1 is not solely produced in the liver, but, on average, hepatocytes contain 100-fold more IGF mRNA than do other tissues. It is thought that these extrahepatic tissues use IGF in autocrine or paracrine rather than endocrine signaling. **GH receptors** in the liver use a *tyrosine kinase* from the *JAK*/STAT pathway. This pathway is named from *Janus kinase* (*JAK*) and signal transducers and activators of transcription (STAT), which involves both protein phosphorylation and the regulation of gene transcription once activated. Receptor activation increases IGF-1 production and release into the circulation.

VI. GROWTH HORMONE

GH is a peptide hormone that occurs in a 20-kDa form and a more abundant 22-kDa form. Preprohormone is produced in the rough endoplasmic reticulum (ER) of somatotropes and then converted to prohormone in the smooth ER and Golgi apparatus, with final processing occurring in the Golgi apparatus and secretory granules. Once secreted, a portion of GH binds weakly to **GH-binding protein** and other plasma proteins before being ultimately broken down by the liver. The half-life of GH in the circulation is about 20 min.

A. Function

GH has a number of targets: liver, cartilage and bone, muscle, and adipose tissue. In cartilage and muscle, GH stimulates amino acid uptake and protein synthesis. Collagen formation and chondrocyte size and number increase in the presence of GH. In adipose tissue, GH increases the breakdown of triglycerides and decreases glucose uptake. This decrease in glucose uptake is sometimes referred to as an "anti-insulin effect."

Clinical Application 33.2: Acromegaly

Acromegaly is associated with excess growth hormone (GH) secretion, often caused by a GH-producing tumor. This causes excessive growth of long bones if epiphyseal plates have not closed, leading to gigantism. Bone growth also leads to a prominent brow and mandible as well as soft tissue overgrowth leading to large hands, feet, and nose. Treatment is often to remove the adenoma or pharmacologically suppress (via somatostatin analogues) the anterior pituitary to control excess GH and insulin-like growth factor 1 secretion.

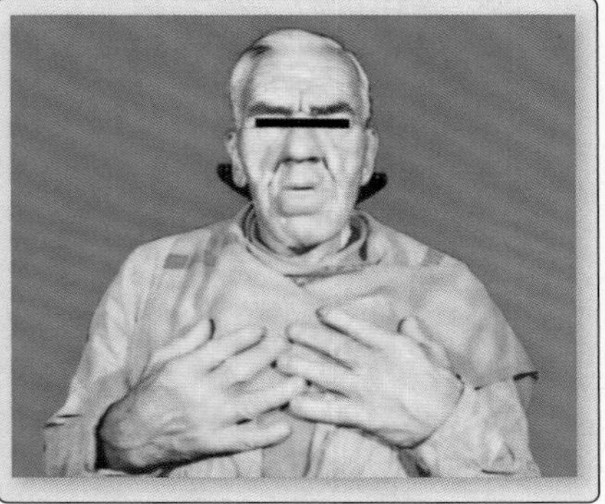

Gigantism.

B. Secretion

GH secretion is regulated by a number of circulating substrates (discussed below) as well as behavioral states, such as sleep and stress, both of which increase its secretion. GH is released in pulses and is cyclic throughout the day. Why is pulsatile secretion helpful in an endocrine axis? Endocrine signaling can often be sustained for long periods, but target tissues tend to "tune out" a constant signal (as do students listening to a humdrum faculty lecture). One method to abate the need for sustained high signaling is to briefly increase (pulse) signal intensity. Oscillatory pulsing may be maintained for a day or a month or be more frequent in certain life stages.

1. **Increased secretion:** Decreases in blood glucose concentration and fatty acids stimulate GH release. Perceived or actual physical or biochemical (e.g., hypoglycemia) stress increases release. This stress can be traumatic in nature or occur via normal stressful activities such as during exercise. Nighttime and deep levels of sleep also stimulate GH release and are thought to be related to growth and repair functions.
2. **Decreased secretion:** Increasing blood glucose and fatty acids inhibits GH release. Conditions such as obesity as well as aging decrease release through less understood mechanisms. Direct negative feedback is provided by GH and IGF-1 on both the anterior pituitary and the hypothalamus (Figure 33.10).

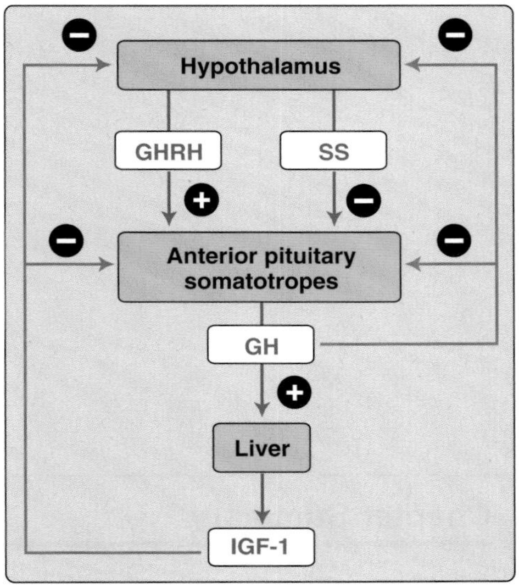

Figure 33.10
Hypothalamic–pituitary–liver hormone axis regulation. GH = growth hormone; GHRH = growth hormone–releasing hormone; IGF-1 = insulin-like growth factor 1; SS = somatostatin.

VII. INSULIN-LIKE GROWTH FACTOR 1

IGF-1 (somatomedin C) is produced and secreted from hepatocytes. IGF-1 is a peptide hormone with some structural similarity to insulin (hence, "insulin-like"). Unlike GH, IGF-1 tightly binds plasma proteins. This results in ~90% circulating in the bound form and a half-life of ~20 hr.

Because of the pulsatile nature and short half-life of GH, measurement of the more stable IGF-1 in the plasma can provide a better way of assessing hypothalamic–pituitary–liver axis status.

Clinical Application 33.3: Insulin-like Growth Factor 1 Deficiencies

Insulin-like growth factor 1 (IGF-1) deficiencies are seen in high prevalence in certain ethnic groups such as the Bayaka of central Africa (one of the traditional pygmy peoples). In this ethnic group, many persons are proportioned normally but have very short stature. Adult men are often <5 ft (150 cm) tall. In these individuals, growth hormone levels are normal to high, whereas IGF-1 concentration is very low.

A. Function

IGF-1 functions very similar to GH. The majority of sustained actions of the hypothalamic–pituitary–liver axis are mediated by IGF-1. IGF-1 effects focus more on the musculoskeletal system, increasing amino acid and glucose uptake and protein synthesis. Increased IGF-1 correlates to growth spurts such as during adolescence.

B. Secretion

IGF-1 secretion is mediated by GH levels. If GH is increased, IGF-1 increases, and *vice versa*. Thus, the factors that alter GH secretion indirectly alter IGF-1 levels. IGF-1 participates in negative feedback regulation of the hypothalamic–pituitary–liver axis at the hypothalamus (see Figure 33.10).

Chapter Summary

- **Glucagon** is secreted by **pancreatic α cells**. The primary function of glucagon is to increase circulating energy substrate levels. This occurs by breaking down glycogen, triglycerides, and proteins as well as forming new glucose from noncarbohydrate sources.
- **Insulin** is secreted by **pancreatic β cells**. The primary function of insulin is to facilitate uptake of energy substrates from the blood. **GLUT4** glucose transporters are insulin sensitive and are inserted into the membrane of skeletal muscle and adipose tissue. **GLUT2** transporters are insulin insensitive and are constituently active in tissues such as the liver.
- **Insulin receptors** stimulate the production of glycogen, fat, and protein within the cell.
- **Diabetes mellitus** is a major health epidemic that involves an inability to adequately regulate blood glucose. Uncontrolled hyperglycemia damages tissues in the cardiovascular, renal, and nervous systems. **Type 1** diabetes mellitus occurs with the inability to secrete insulin, whereas **type 2** diabetes mellitus results from glucose or insulin insensitivity.
- **Somatostatin** is secreted by **pancreatic δ cells**. Somatostatin inhibits the secretion of both glucagon and insulin.
- **Growth hormone** (**GH**) is secreted from the anterior pituitary in response to hypothalamic release of **GH-releasing hormone** (**GHRH**). GH can acutely affect growth and uptake of glucose and amino acids, but the majority of the functions of this axis are performed by insulin-like growth factor 1.
- **Insulin-like growth factor 1** (**IGF-1**) is secreted from the liver in response to increases in growth hormone (GH). IGF-1 sustains many of the same functions as GH, including amino acid uptake and *protein synthase* in cartilage and muscle and increase in the breakdown of triglycerides in adipocytes.

Adrenal Glands

34

I. OVERVIEW

The adrenal (suprarenal) glands provide the bloodborne signals of stress, **epinephrine**, and **cortisol**. The sounding of the body's alarms and defenses helps an individual survive physical threats, endure pain, and tap the body's physical and metabolic reserves. In modern humans, stress is often more mental and social in nature, but such events elicit very similar stress responses as does climbing a tree to escape a pack of wolves. Besides stress, adrenal glands regulate plasma Na^+ via **aldosterone** and certain secondary sex characteristics by the **adrenal androgens**. Stress, salt, and sex are heavy responsibilities for this small (~1.5 by 7.5 cm and weighing ~8–10 g) set of glands located just above each kidney. Each gland can be divided into two main sections: the **cortex** (~90% of gland weight) and the **medulla** (~10%) as shown in Figure 34.1. The cortex is controlled and regulated, in part, by the hypothalamic–pituitary axis and is further divided into the **zona glomerulosa**, **zona fasciculata**, and **zona reticularis** (see Figure 34.1). The zona glomerulosa produces and secretes aldosterone, which regulates plasma volume by controlling how much Na^+ is retained by the kidney. Cortisol is primarily produced and secreted by the zona fasciculata and increases metabolism and catabolism as well as suppresses inflammation and immunity. The adrenal androgens, which are **dehydroepiandrosterone** (**DHEA**), **dehydroepiandrosterone sulfate** (**DHEAS**), and **androstenedione**, are primarily produced and secreted by the zona reticularis and participate in secondary sex characteristics (e.g., hair growth) during puberty and adolescence. The adrenal medulla is controlled and regulated by the sympathetic nervous system (SNS), and its major hormonal product is epinephrine (adrenaline). Similar to the SNS "fight-or-flight" response, epinephrine provides a rapid stress signal but delivered via the circulation rather than the nervous system.

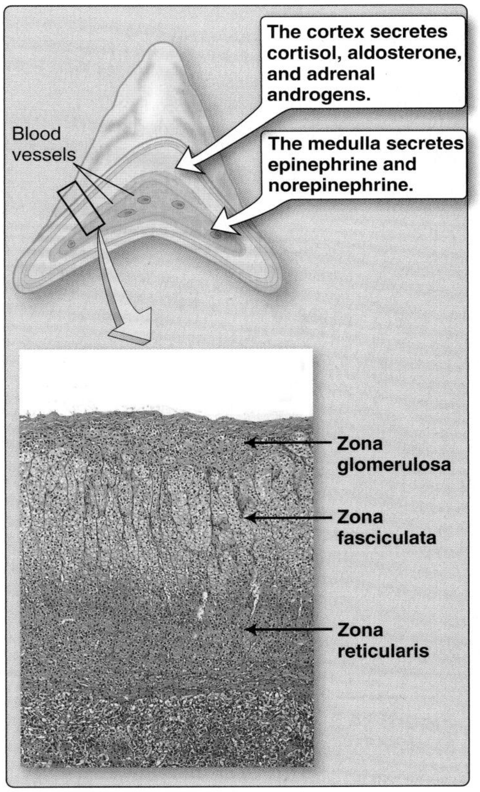

The cortex secretes cortisol, aldosterone, and adrenal androgens.
Blood vessels
The medulla secretes epinephrine and norepinephrine.
Zona glomerulosa
Zona fasciculata
Zona reticularis

Figure 34.1
Adrenal gland structure.

II. HYPOTHALAMIC–PITUITARY–ADRENAL AXIS

The adrenal cortex is controlled and regulated by an endocrine axis, providing a multitiered response that allows for both gross and fine hormonal adjustment. Axis control is directed primarily at the zona fasciculata and reticularis. The zona glomerulosa is regulated primarily by other hormones (**angiotensin II** [**Ang-II**]) and ions (K^+).

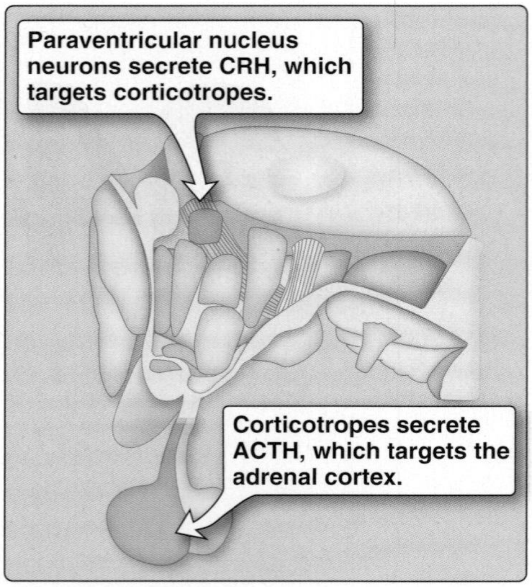

Figure 34.2
Hypothalamus and pituitary gland.
ACTH = adrenocorticotropic hormone;
CRH = corticotropin-releasing hormone.

A. Hypothalamus

Corticotropin-releasing hormone (**CRH**) is synthesized (see Figure 7.11) in the **paraventricular nucleus** and released into **hypophyseal portal circulation** for carriage to the anterior pituitary (Figure 34.2). A number of higher brain centers stimulate CRH release during physical, biochemical (e.g., low blood glucose), and emotional stress. CRH release follows a circadian rhythm, peaking just before waking and then pulsing throughout the day, based on the above stressors. The paraventricular nucleus also produces antidiuretic hormone (ADH), which can further regulate CRH release and stimulate **corticotropes**.

B. Pituitary gland

CRH binds to corticotrope type 1 corticotropin-releasing hormone receptor (CRH-R1), which is part of the G protein–coupled receptor (GPCR) superfamily that acts primarily through the *adenylyl cyclase* (*AC*) second-messenger system. CRH-R1 binding activates transcription factors to express the ***preproopiomelanocortin*** (***POMC***) **gene**, which encodes **adrenocorticotropic hormone** (**ACTH**), which is released into the bloodstream. ACTH's target is the adrenal cortex.

C. Adrenal cortex

Adrenocortical hormone (i.e., aldosterone, cortisol, DHEA, DHEAS, and androstenedione) synthesis begins with cholesterol. A small amount of cholesterol is synthesized by the cortex, but the majority is taken up from blood and then stored in a cytosolic pool. Cortical activity is stimulated by ACTH from the pituitary, acting via **melanocortin 2 receptors**, which are part of the GPCR superfamily. These receptors act primarily through the *AC* second messenger system (Figure 34.3) to activate enzymes that aid cholesterol uptake as well as a specialized ***side-chain cleavage enzyme complex*** (sometimes termed *cholesterol desmolase*, or *cytochrome P450 SCC*). *Side-chain cleavage enzyme complex* is one of the key rate-limiting steps for adrenal cortex hormone production. There are a number of common enzymes and intermediates in the synthesis of cortex hormones (Figure 34.4). The activation or inhibition or even the presence of one enzyme but not another can preferentially shunt the production to cortisol rather than an adrenal androgen, or *vice versa*.

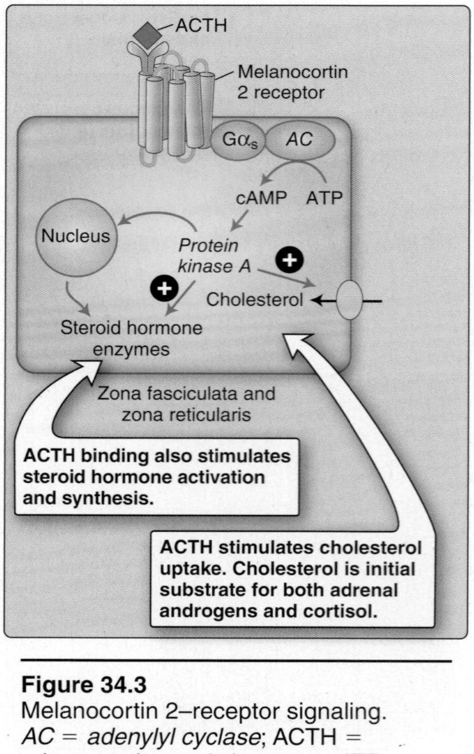

Figure 34.3
Melanocortin 2–receptor signaling.
AC = *adenylyl cyclase*; ACTH = adrenocorticotropic hormone; ATP = adenosine triphosphate; cAMP = cyclic adenosine monophosphate;
CRH = corticotropin-releasing hormone.

III. ALDOSTERONE

Aldosterone is synthesized in the zona glomerulosa. This is the only cortical region to express ***aldosterone synthase*** (and other *CYP11B2* gene product enzymes), which facilitates the final step in the conversion of cholesterol into aldosterone. Once released into the circulation, aldosterone binds with low affinity to corticosteroid-binding protein and albumin. The hormone has a half-life of ~20 min.

A. Function

Aldosterone increases Na^+ and water reabsorption as well as K^+ secretion from renal tubules (see 27·IV). Aldosterone also increases Na^+

Clinical Application 34.1: Addison Disease

Primary adrenal insufficiency (**Addison disease**) commonly results from an autoimmune response that destroys the adrenal cortex. Symptoms include fatigue, dehydration, hyponatremia, and hypotension associated with loss of glucocorticoids and mineralocorticoids. Adrenal hormone deficiency stimulates corticotropin-releasing hormone release and *preproopiomelanocortin* gene expression through a negative feedback pathway, which increases the circulating levels of adrenocorticotropic hormone (ACTH). Hyperpigmentation of hands, feet, nipples, axillae, and the oral cavity occurs because of the elevated ACTH. Treatment involves fluid replacement and exogenous glucocorticoids such as hydrocortisone. Once symptoms have stabilized, mineralocorticoid replacement therapy can be implemented until the postural drop in blood pressure can be adequately controlled.

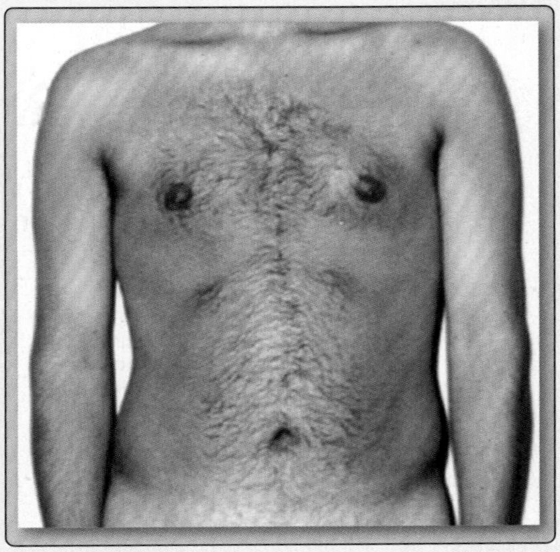

Bronze skin and nipple hyperpigmentation.

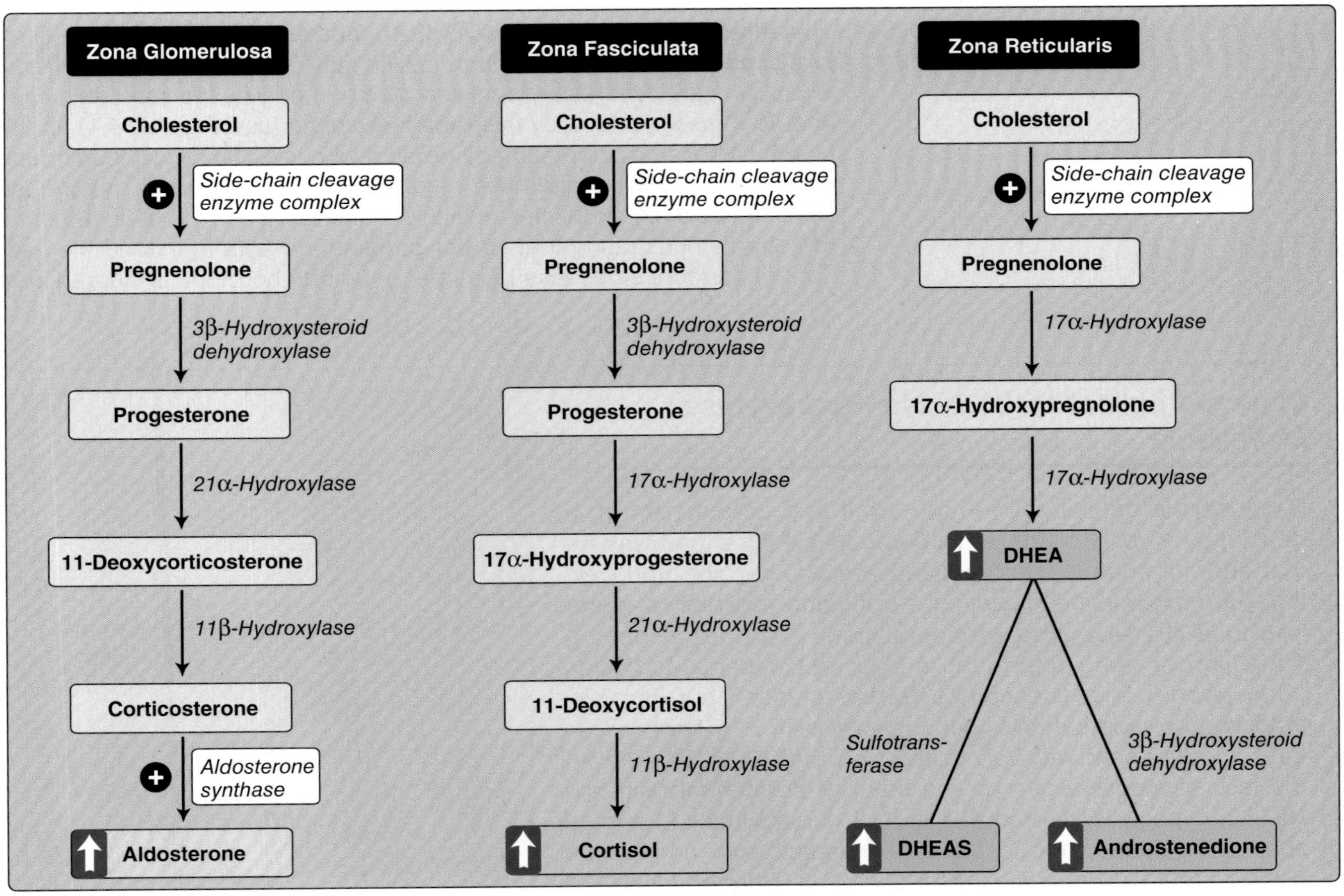

Figure 34.4
Cortex hormone biosynthesis. DHEA = dehydroepiandrosterone; DHEAS = dehydroepiandrosterone sulfate.

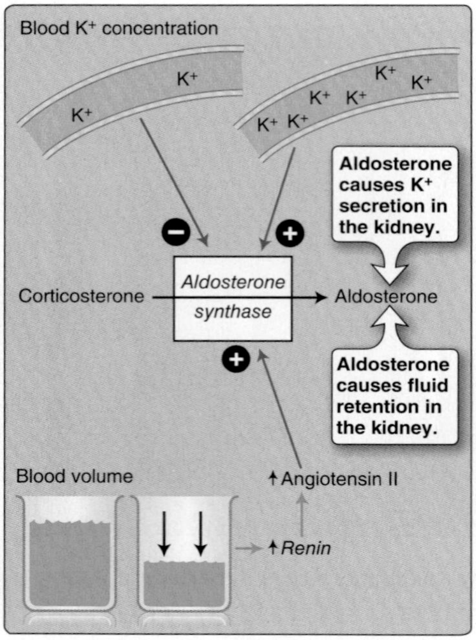

Figure 34.5
Aldosterone regulation.

reabsorption by intestinal enterocytes, which increases the body's Na^+ stores. The effect of aldosterone on ions (minerals) is reflected in its class name, mineralocorticoid. Aldosterone acts through cytosolic **mineralocorticoid receptors** in target cells to facilitate Na^+ and water reabsorption in the kidney and absorption in the gastrointestinal (GI) system (see Figure 27.12).

B. Secretion

Aldosterone synthase produces aldosterone from corticosterone. *Aldosterone synthase* is the gatekeeper of aldosterone production and is regulated by Ang-II and plasma K^+ levels. Ang-II, a hormone within the *renin*–angiotensin–aldosterone system, is stimulated by low circulating fluid volume, low pressure in the glomerulus, and increases in SNS activity (see 28·III·C). An increase in ACTH, which is vital for regulation of other renal cortex hormones, must be present but is less of a stimulator for the final step in aldosterone synthesis. Feedback for aldosterone secretion is not aldosterone itself, but rather comes in the form of its effects of decreasing fluid volume and plasma K^+ levels (Figure 34.5).

IV. ANDROGENS

The adrenal androgens (DHEA, DHEAS, and androstenedione) are often produced as a cohort rather than individually (see Figure 34.4). Adrenal androgens are synthesized and secreted primarily by the zona reticularis and, to a lesser extent, by the zona fasciculata (see Figure 34.1). In the blood, DHEA and androstenedione bind with low affinity to albumin and other blood globulins and have a half-life of 15–30 minutes. In contrast, DHEAS has a higher affinity for albumin and has a half-life of 8–10 hours, thereby demonstrating that carrier proteins are able to extend the half-lives of hormones because less free (unbound) hormone is cleared from

Clinical Application 34.2: *21α-Hydroxylase Deficiency*

Because the pathways in Figure 34.4 are interconnected, a deficiency in one of the enzymes can bias the pathway so that one hormone is overproduced and another not produced. *21α-Hydroxylase deficiency* is a condition in which a mutation in *CYP21A2* gene products results in nonfunctioning *21α-hydroxylase*. Thus, there is a lack of mineralocorticoids (aldosterone) and glucocorticoids (like cortisol) but an overproduction of adrenal androgens. Infants with *21α-hydroxylase* deficiency present with 1) hypotension and dehydration from the lack of aldosterone and the inability to adequately retain Na^+, 2) hypoglycemia from the lack of cortisol-induced energy substrate release, and 3) excess virilization and ambiguous genitalia (in females) is a result of androgen overproduction.

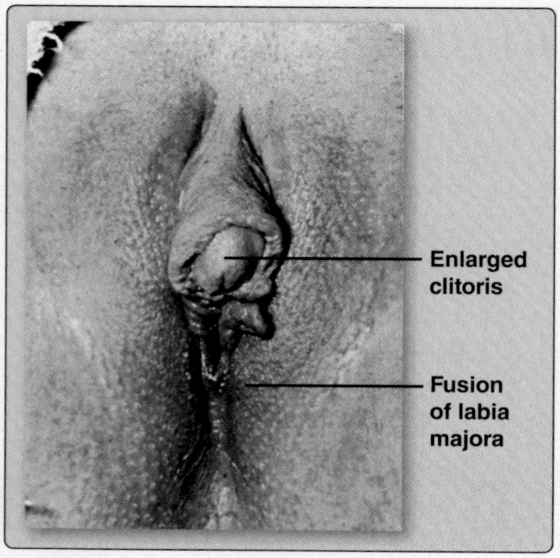

Ambiguous genitalia.

the blood and can serve as a small temporary storage facility for a hormone. Why do some hormones need a carrier protein? Think of a carrier protein as an additive that binds oil, so that it does not separate from water, allowing it to be transported anywhere water is.

A. Function

DHEA and DHEAS are less potent than androgens produced by the gonads but do have functional effects on secondary sex characteristics and are involved in development during childhood and adolescence. The beginning of androgen release (adrenarche) during development stimulates axillary and pubic hair growth. DHEA can be converted into androstenedione, which can then be converted to more potent androgens, such as **testosterone** and **estrogens**, in peripheral tissues. *17-Ketosteroid reductase* is a key enzyme in facilitating the conversion of androstenedione to testosterone. This androgen conversion is an important source of testosterone in women.

B. Secretion

DHEA, DHEAS, and androstenedione are controlled by the negative feedback loops of CRH and ACTH (Figure 34.6). These multiple feedback loops provide a finer regulation of hormone production than does a single feedback loop. Events that trigger the release of ACTH facilitate synthesis and release of adrenal androgens. The input rhythms associated with growth and development during puberty and across the lifespan affect ACTH production and release.

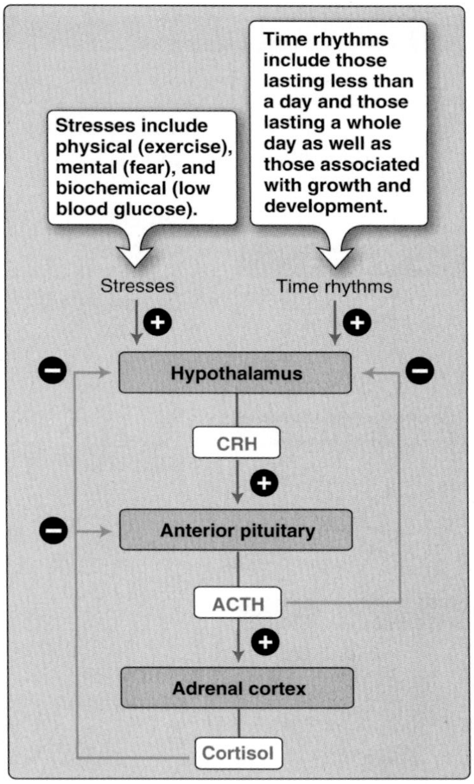

Figure 34.6
Hypothalamic-pituitary-adrenal regulation. ACTH = adrenocorticotropic hormone; CRH = corticotropin-releasing hormone.

V. CORTISOL

Cortisol and corticosterone are synthesized and secreted primarily by the zona fasciculata and, to a lesser extent, by the zona reticularis (see Figure 34.1). Cortisol synthesis, in contrast to adrenal androgen synthesis, requires two *hydroxylases* (*21α-hydroxylase* and *11β-hydroxylase*) to eventually convert progesterone and 17-hydroxyprogesterone into their final products (see Figure 34.4). In the blood, cortisol binds **corticosteroid-binding protein** with a high affinity and has a half-life of ~60 minutes.

A. Function

Cortisol and corticosterone prepare the body for stress. Cortisol diffuses across the cell membrane and binds to a cytosolic **glucocorticoid receptor**. The hormone–receptor complex translocates to the nucleus and binds a **glucocorticoid response element** on DNA. Cortisol also binds with low affinity to mineralocorticoid receptors and, thus, induces some minor collateral aldosterone-like responses. Cortisol causes a number of physiologic effects (Figure 34.7).

1. **Metabolic:** Cortisol increases plasma glucose and free fatty acid concentration in order to provide energy substrates to body tissues for their response to the stressful event that stimulated cortisol production.

   a. **Increased catabolism:** Cortisol increases skeletal muscle protein catabolism, liberating amino acids that are then converted

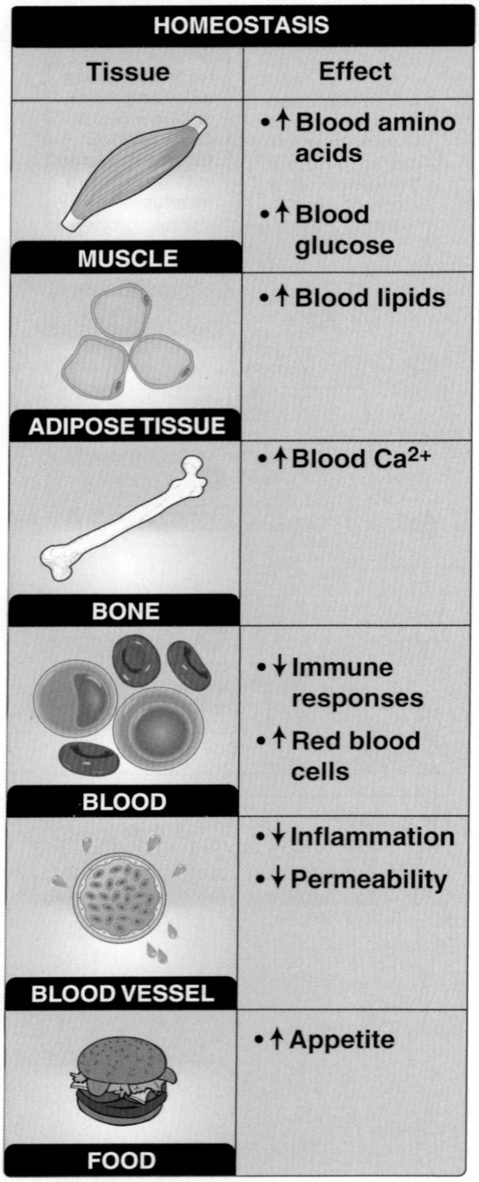

HOMEOSTASIS
Tissue
Effect
MUSCLE
• ↑ Blood amino acids
• ↑ Blood glucose
ADIPOSE TISSUE
• ↑ Blood lipids
BONE
• ↑ Blood Ca2+
BLOOD
• ↓ Immune responses
• ↑ Red blood cells
BLOOD VESSEL
• ↓ Inflammation
• ↓ Permeability
FOOD
• ↑ Appetite

Figure 34.7
Glucocorticoid effects.

to glucose via gluconeogenesis in the liver (see Chapters 32·IV·A and 33·III·A·2). This glucose response is part of the origin of the glucocorticoid classification of cortisol.

b. **Increased lipolysis:** Cortisol stimulates white adipose tissue to undergo lipolysis to liberate free fatty acids and triglycerides. The fatty acids and triglycerides are then transported in the blood for use as an energy source by other tissues.

c. **Increased intake:** Cortisol stimulates appetite. Acutely, this is beneficial to provide energy substrates to respond to the stressful event. However, if the stressful event does not involve physical work, then this increased appetite can lead to weight gain.

2. **Immune:** Cortisol suppresses both immune responses and inflammation.[1] Although this response may seem counterproductive in stressful conditions, when the life of the organism is in danger, fighting illness with the immune system becomes less important than immediate survival. The mechanisms by which this immunosuppression is accomplished are via decreased production of lymphocytes and interleukins 1 and 6 (IL-1 and IL-6) and T-cell suppression. The anti-inflammatory effects of cortisol are due to decreases in capillary permeability as well as reductions in both prostaglandin and leukotriene synthesis that mediate increases in local blood flow.

3. **Musculoskeletal:** Cortisol increases bone resorption and decreases Ca^{2+} absorption from the GI tract and reabsorption from the kidney. Chronic high levels of cortisol can lead to osteoporosis. Cortisol decreases collagen formation throughout the body. Protein catabolism to increase plasma glucose levels can eventually lead to muscle weakness and early fatigue onset during physical activity.

4. **Cardiovascular:** Cortisol increases erythropoietin release, which stimulates red blood cell production. Cortisol potentiates vasoconstrictor responses by blocking local vasodilators, such as nitric oxide and prostaglandins, and through glucocorticoid receptors in vascular smooth muscle by altering Ca^{2+} homeostasis within these cells. Glucocorticoids increase the effectiveness of catecholamine actions, such as inotropy and vasoconstriction, through the upregulation of adrenergic receptors.

B. Secretion

Cortisol and corticosterone release are controlled by the negative feedback loops of CRH and ACTH (see Figure 34.6). Section II, Hypothalamic–Pituitary–Adrenal Axis, described how physical, emotional, and biochemical stress stimulate the release of CRH, ACTH,

[1]Immune suppressive and anti-inflammatory effects of glucocorticoids can be exploited pharmacologically. Drugs like prednisone, which is structurally similar to cortisol, can be used as immune suppressants for autoimmune diseases. For more information, see *LIR Pharmacology*, 5e, p. 334.

Clinical Application 34.3: Cushing Syndrome

Patients with **Cushing syndrome** may present with muscle weakness, osteoporosis, hypertension, diabetes, and weight gain with fat redistribution. These symptoms reflect chronic elevations in glucocorticoid levels. Muscle weakness results from skeletal muscle protein catabolism, osteoporosis by Ca^{2+} resorption in bone, hypertension by cortisol's mineralocorticoid effects on Na^{+} retention, diabetes by increases in plasma glucose, weight gain by increases in appetite, and fat redistribution by unutilized fatty acid release. For less-understood reasons, unutilized fatty acids are redeposited in the face and upper back, causing a "moon face" appearance and the development of "buffalo hump." The cause of the excess cortisol secretion is often a pituitary adenoma, which causes an excess secretion of adrenocorticotropic hormone.

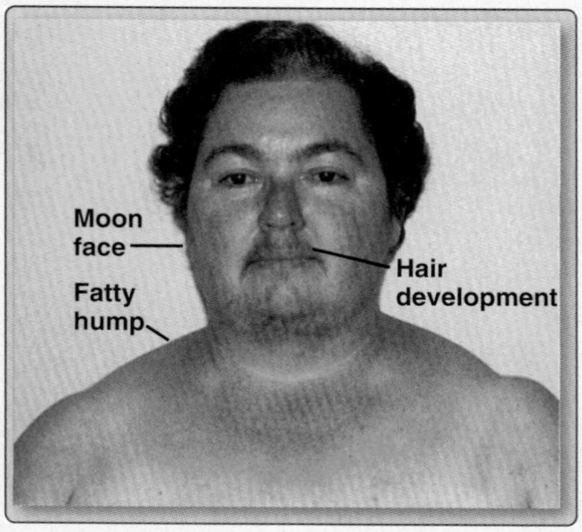

Female with Cushing syndrome.

and cortisol. The control of CRH is primarily for regulation of cortisol and less so for adrenal androgens or aldosterone.

VI. CATECHOLAMINES

The adrenal medulla is derived from the neural crest rather than the mesodermal mesenchyme, which forms the cortex. In practice, this means that the medulla functions as an extension of the SNS. The medulla is composed of small clusters of **chromaffin cells** (medullary cells), which synthesize catecholamines from the amino acid tyrosine (see Figure 5.7). Dopamine is synthesized in the cytosol, and a **catecholamine-H^{+} exchanger** (**VMAT1**, for vesicular monoamine transporter 1) transports it into secretion vesicles. Dopamine is then converted to norepinephrine via ***dopamine β-hydroxylase***. Unlike postganglionic adrenergic nerves of the SNS, chromaffin cells contain ***phenylethanolamine N-methyltransferase***. This enzyme is located in the cytosol and facilitates the conversion of norepinephrine to epinephrine. Therefore, norepinephrine must be transported back into the cytosol to be converted to epinephrine, which is in turn transported back into the secretion vesicle. Epinephrine and norepinephrine are then stored with **chromogranin** (binding protein) in preparation for vesicle exocytosis and hormonal release (Figure 34.8). Chromaffin cells secrete norepinephrine and epinephrine in an approximate 1:4 ratio into the fenestrated medullary capillary network for delivery to various body tissues. Catecholamine half-lives range from 10–90 s. Although seemingly short, they are longer than the SNS release and clearance of norepinephrine in the synaptic cleft. This allows for a more sustained SNS response.

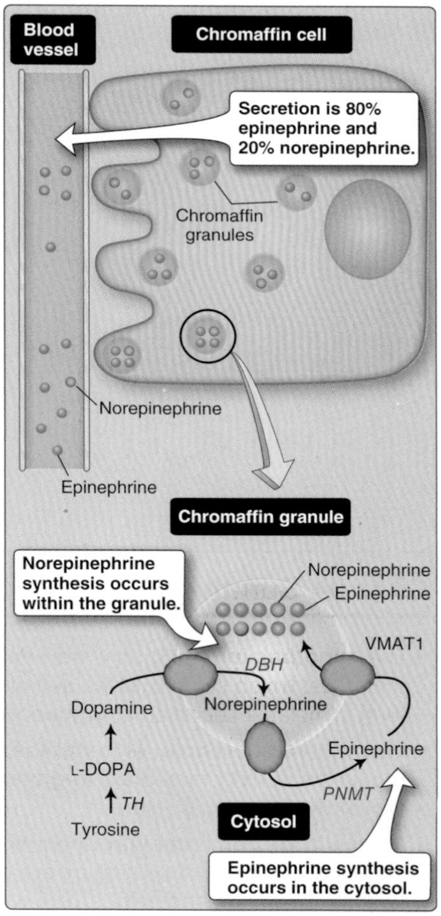

Figure 34.8
Chromaffin cell. *DBH* = *dopamine β-hydroxylase*; L-DOPA = L-3,4-dihydroxyphenylalanine; *PNMT* = *phenylethanolamine N-methyltransferase*; *TH* = *tyrosine hydroxylase*; VMAT1 = vesicular monoamine transporter 1.

A. Function

Epinephrine (adrenaline), produces classic fight-or-flight responses, or the "adrenaline rush." Thus, the key actions of epinephrine and norepinephrine are similar to those of the SNS (see Figure 7.4). Delivery via the circulation means that responses to hormones, although typically slower, are wider ranging because they can reach receptor populations that are not specifically located within a SNS synaptic

Clinical Application 34.4: Pheochromocytoma

Pheochromocytomas are catecholamine-producing tumors located in the adrenal medulla or preganglionic neurons. The classic triad of symptoms includes headaches, palpitations (tachycardia), and profuse sweating. Palpitations and sweating are caused by high circulating epinephrine and norepinephrine levels. Headaches can be caused by direct vasoconstriction of cerebral blood vessels or the high blood pressure (hypertension) induced by peripheral vasoconstriction. These symptoms can be episodic or sustained, depending on the nature of the catecholamine release caused by the tumor.

cleft. The functional effects are related to the amount secreted and the tissue responsiveness.

B. Secretion

Catecholamine release is regulated by the SNS rather than the hypothalamic–pituitary–adrenal axis. Thus, secretion is increased during stresses to homeostasis; strong emotions, such as anger and fear; and exercise. Cholinergic preganglionic SNS neurons stimulate secretion from chromaffin cells via **nicotinic type 2 acetylcholine receptors** to increase chromaffin granule secretion (see Figure 7.5).

C. Regulation

Adrenergic receptor expression is dynamic. With high levels of circulating catecholamines, such as during continual stress, membrane receptors can be internalized, thereby reducing responsiveness to subsequent catecholamine stimulation. Conversely, tissue catecholamine responses may be increased by cortisol and triiodothyronine (for example), by increased receptor synthesis or increased receptor trafficking to the cell membrane.

Chapter Summary

- The **hypothalamic–pituitary–adrenal axis** involves the secretion of **corticotropin-releasing hormone** from the hypothalamus, which stimulates the secretion of **adrenocorticotropic hormone** (**ACTH**) from the anterior pituitary. ACTH then stimulates secretion of **glucocorticoids** and **adrenal androgens** from the adrenal cortex.
- The **mineralocorticoid**, **aldosterone**, is only under minor control of the hypothalamic–pituitary–adrenal axis. The major regulators of aldosterone are **angiotensin II** and plasma K^+. Aldosterone increases Na^+ and water reabsorption to preserve circulating fluid volume.
- Adrenal androgens (**dehydroepiandrosterone**, **dehydroepiandrosterone sulfate**, and **androstenedione**) participate in the development of secondary sex characteristics and serve as substrates in peripheral conversion of androgens to testosterone and estrogens.
- Glucocorticoids (**cortisol** and **corticosterone**) increase blood glucose and suppress immunity and inflammation, among other physiologic responses.
- **Catecholamines** (**epinephrine** and **norepinephrine**) are produced and secreted from the adrenal medulla by **chromaffin cells**, which are regulated by the sympathetic nervous system. Catecholamines prepare the body to face stressful events by increasing heart rate and inotropy and by converting stored energy sources to usable metabolic substrates.

Thyroid and Parathyroid Hormones

35

I. OVERVIEW

The cells that make up the human body vary widely with respect to their metabolic and development rates. Thyroid hormones provide the brain with a global method to govern these processes outside of conditions of acute stress. The **thyroid gland** is located in the neck just below the larynx (Figure 35.1) and is regulated by the hypothalamus and the pituitary gland. The active hormones of the hypothalamic–pituitary–thyroid axis are **triiodothyronine (T_3)** and **thyroxine (T_4)**. T_3 and T_4 induce transcription, translation, and synthesis of pumps, transporters, enzymes, and cellular scaffolding and contractile elements, thereby increasing the metabolic rate of peripheral tissues. The thyroid gland also has a minor role in Ca^{2+} homeostasis via **parafollicular C cells**. These cells release **calcitonin**, which decreases Ca^{2+} and PO_4^{3-} levels through increased urinary excretion. The major regulators of Ca^{2+} homeostasis are the **parathyroid glands** located in the inferior and superior margins of the thyroid gland (see Figure 35.1) and **vitamin D**. **Parathyroid hormone** (**PTH**) increases circulating Ca^{2+} by stimulating bone resorption and increasing renal Ca^{2+} reabsorption. Vitamin D also increases circulating Ca^{2+} and PO_4^{3-} levels by increasing absorption of these ions in the gastrointestinal (GI) tract. Ca^{2+} homeostasis is important for a host of cellular events, such as cell signaling (e.g., initiating muscle contraction and vesicle secretion) and sustaining action potentials (e.g., in cardiac myocytes) as well as maintaining bone mineral density.

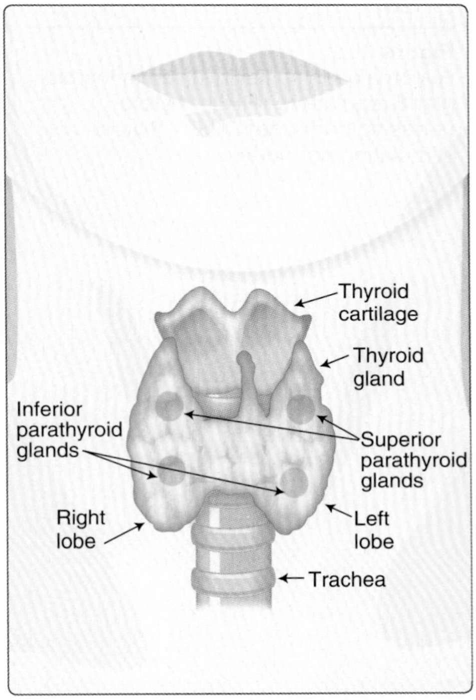

Figure 35.1
Thyroid and parathyroid glands.

II. HYPOTHALAMIC–PITUITARY–THYROID AXIS

Thyroid hormone secretion is regulated by a multitiered endocrine axis involving the hypothalamus and pituitary. **Thyroid-stimulating hormone** (**TSH**) released from the anterior pituitary increases production and secretion of T_3 and T_4 from the thyroid gland.

A. Hypothalamus

Thyroid hormone secretion is regulated by three hypothalamic nuclei: the **paraventricular nucleus**, the **arcuate nucleus**, and the **median eminence** (Figure 35.2). Parvocellular neurons in these areas project to

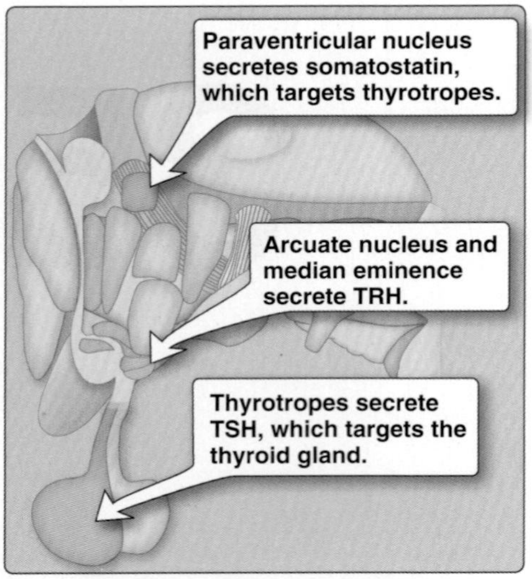
Paraventricular nucleus secretes somatostatin, which targets thyrotropes.
Arcuate nucleus and median eminence secrete TRH.
Thyrotropes secrete TSH, which targets the thyroid gland.

Figure 35.2
Hypothalamic nuclei involved in thyroid gland regulation. TRH = thyroid-releasing hormone; TSH = thyroid-stimulating hormone.

and secrete **thyroid-releasing hormone** (**TRH**) and **somatostatin** into the hypophyseal portal circulation. Both hormones target **thyrotropes** in the pituitary gland.

B. Pituitary gland

TRH binds to thyrotrope TRH receptors, which are members of the G protein–coupled receptor (GPCR) superfamily that act through the *phospholipase C* second-messenger system (see 1·VII·B·3). TRH receptor occupancy stimulates TSH synthesis and release from secretory granules. In contrast, **somatostatin** (sometimes termed "somatotroph release–inhibiting factor") decreases TSH production and release. Somatostatin binds to a different GPCR that inhibits *adenylyl cyclase* (*AC*) and cyclic adenosine monophosphate (cAMP) signaling (see 1·VII·B·2). Neurons within hypothalamic nuclei that initiate TSH release also express TSH receptors, which provides a negative feedback pathway by which high circulating TSH levels can inhibit further release.

C. Thyroid gland

TSH regulates the thyroid gland, which is an assemblage of numerous, 200 to 300-μm diameter hollow spheres (**follicles**) filled with a protein-rich fluid matrix known as **colloid** (Figure 35.3). Follicles are the site of thyroid hormone synthesis and secretion. Parafollicular C cells, which synthesize calcitonin, are randomly distributed between follicles throughout the gland. The TSH receptor is a GPCR that stimulates cAMP formation when occupied. cAMP, in turn, regulates most aspects of thyroid hormone secretion by regulating expression of most of the proteins involved in thyroid hormone synthesis (see Steps 1–8 in Section III·A).

1. **Follicles:** Follicles comprise a specialized epithelium composed of follicular cells resting on a basement membrane. The epithelium is supported by an extensive basolateral vasculature. The epithelium's apical side faces the colloid-filled follicular lumen. Follicular epithelial cells synthesize and secrete thyroid hormones when stimulated by TSH. TSH also increases expression of cellular components needed for thyroid hormone synthesis.

Clinical Application 35.1: Goiter

A normal thyroid gland weighs ~20 g, but its mass can increase quite dramatically under pathologic conditions. Enlarged thyroid glands (goiters) can be smooth or nodular depending on etiology. Thyroid cancers and infiltrates can cause nodular goiters. Goiters occur more commonly due to an excess production of thyroid-stimulating hormone (TSH) or activation of the TSH receptor. TSH receptor stimulation either by TSH or an autoimmune response causes gland growth (hypertrophy).

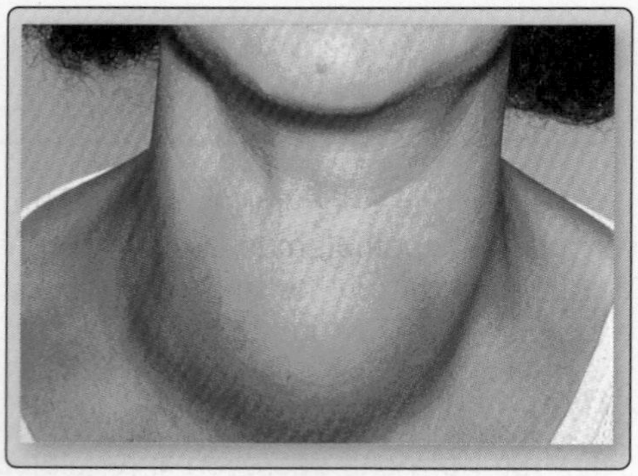

Goiter.

2. **Colloid:** The oxidative chemistry involved in thyroid hormone synthesis can be very harmful to cells, so it is performed extracellularly within colloid (see Figure 35.3). This is a similar concept to walling off hydrogen peroxide reactions within cytosolic peroxisomes. TSH increases the production of colloid.

III. THYROID HORMONES

The thyroid gland produces and secretes T_3 and T_4 in a ratio of ~1:10. T_4 has a longer half-life than T_3 but has a relatively low biologic activity. Conversion of T_4 to T_3 occurs primarily in target tissues. The thyroid hormones target virtually all cells in the body and exert their effects via cytosolic receptors that modulate gene expression.

A. Synthesis

Thyroid hormone synthesis and secretion is a multistep process involving iodination and conjugation of adjacent tyrosine residues (amino acid projections to which iodine can bind) on **thyroglobulin**. Thyroglobulin's primary purpose is to hold thyroid hormone precursors in close proximity to enable the synthesis steps that occur in the colloid. Thyroid hormone synthesis begins with iodide uptake from the blood and can be broken into eight sequential steps (Figure 35.4).

1. **Iodide uptake:** Thyroid hormone synthesis begins with I^- uptake (also called I^- trapping) from the vasculature by follicular cells. I^- is transported across the basolateral membrane from blood by a Na^+-I^- cotransporter (or, sodium/iodide symporter **[NIS]**), powered by the Na^+ gradient established by the basolateral Na^+-K^+ ATPase.

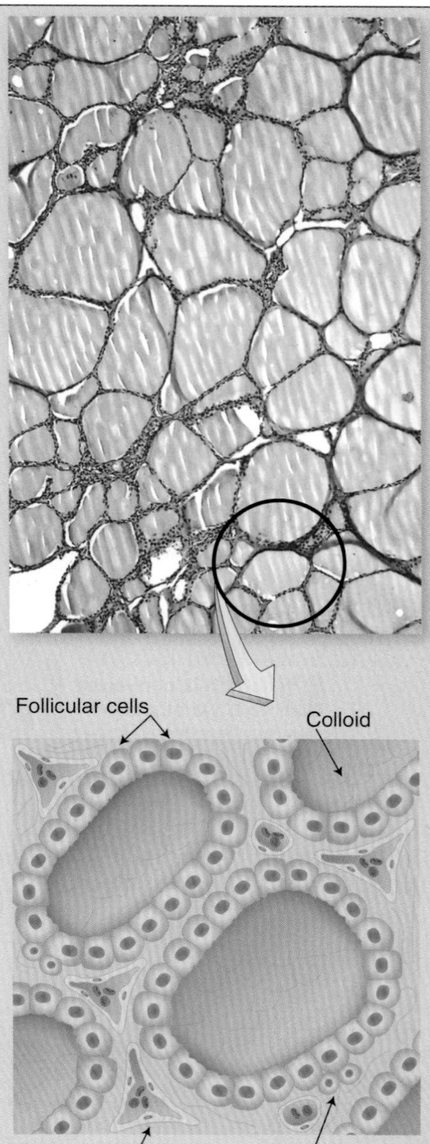

Figure 35.3
Cellular organization of the thyroid gland.

> Thyroid hormones cannot be produced without I^-, which must be obtained from dietary sources (U.S. recommended dietary allowance is 150 g). Iodine deficiency results in hypothyroidism and presents as goiter (see Clinical Application 35.1).

2. **Apical secretion:** I^- is then transported across the apical membrane primarily by a specialized Cl^--I^- cotransporter known as a **pendrin channel**. Other mechanisms also exist for apical iodide transport, but, when this channel is defective, such as in Pendred syndrome, the patient presents with low circulating thyroid hormones. Thyroglobulin is synthesized in the follicular cells and is exocytosed across the apical membrane into the colloid.

3. **Oxidation:** The thyroglobulin-laden secretory vesicles express ***thyroid peroxidase*** (***TPO***), a heme-containing enzyme, on their inner surfaces. When the vesicles fuse with the apical membrane, *TPO* is presented to the colloid lumen and immediately catalyzes an oxidation reaction in which iodide is combined with H_2O_2 to

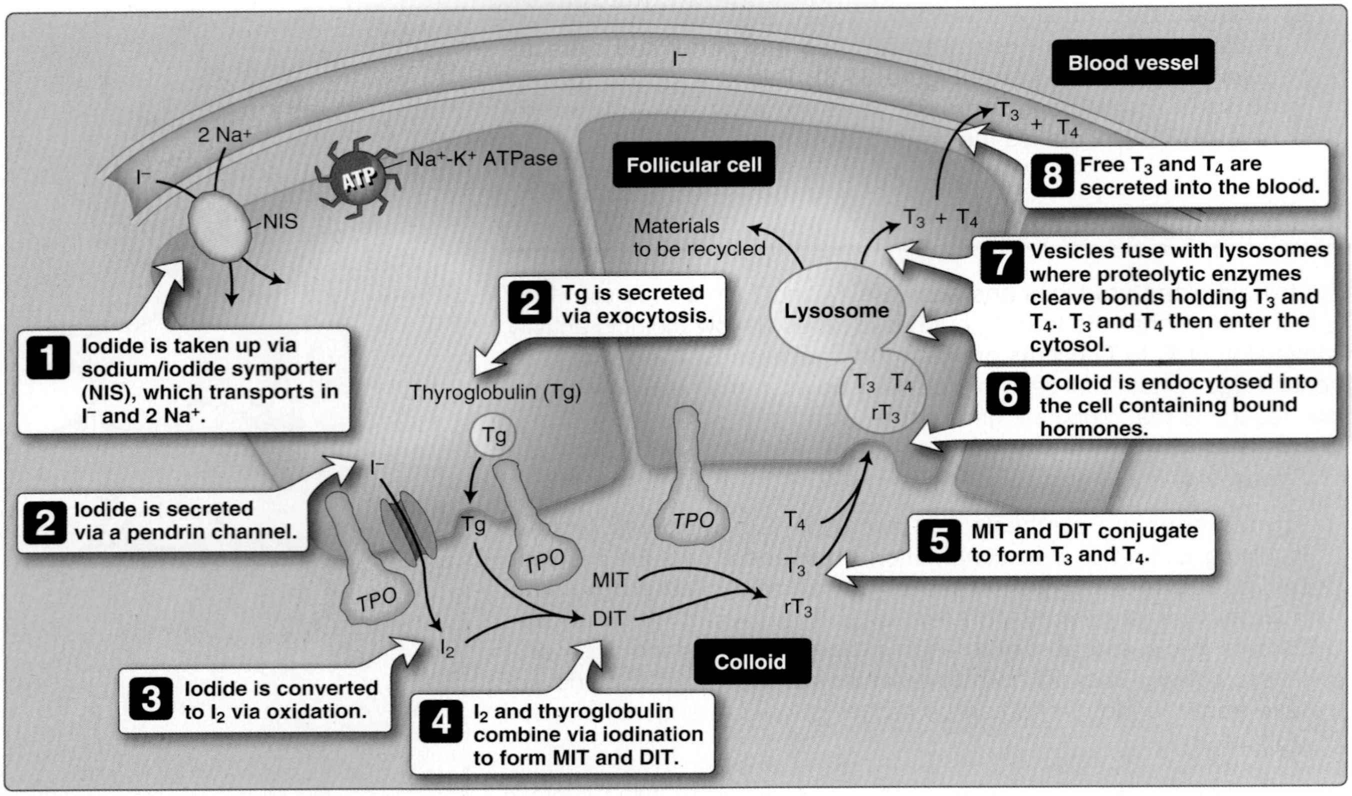

Figure 35.4
Thyroid hormone biosynthesis. ATP = adenosine triphosphate; DIT = diiodotyrosine; I^- = iodide; I_2 = iodine; MIT = monoiodotyrosine; rT_3 = reversed T_3; T_3 = triiodothyronine; T_4 = thyroxine; *TPO = thyroid peroxidase*.

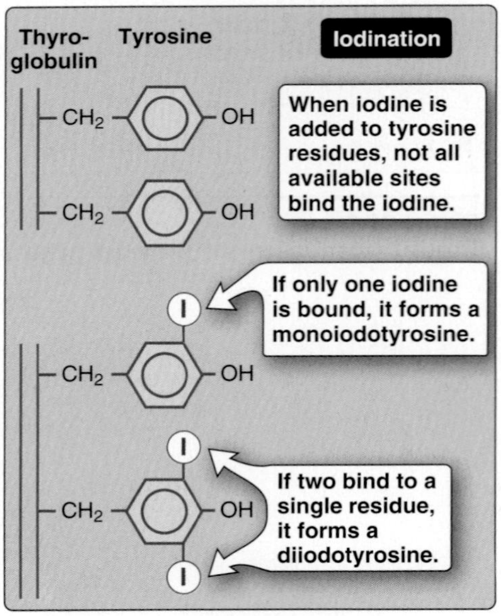

Figure 35.5
Thyroglobulin iodination.

form iodine (I_2) and H_2O. The H_2O_2 is derived from apical *dual oxidase 2* (*DUOX2*) transporters that combine intermediates from the pentose phosphate pathway[1] with O_2 in the follicular cytosol to form H_2O_2 in the colloid.

4. **Iodination:** *TPO* also facilitates iodination (or **organification**) of thyroglobulin tyrosine residues to form **monoiodotyrosine** (**MIT**) and **diiodotyrosine** (**DIT**) as shown in Figure 35.5. The precise reason for whether one or two iodines will bind to a particular residue is not clearly understood.

5. **Conjugation:** MIT and DIT combine (a process called **conjugation**) to form $\mathbf{T_3}$ and **reverse** $\mathbf{T_3}$ (**r**$\mathbf{T_3}$), whereas two DIT residues combine to form T_4. The hormones remain attached to thyroglobulin until internalized by the follicular cells. Conjugation is also facilitated by *TPO*.[2]

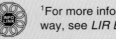

INFO LINK

[1]For more information on the pentose phosphate pathway, see *LIR Biochemistry,* 5e, p. 145.

[2]*Thyroid peroxidase* can be inhibited pharmacologically by methimazole and propylthiouracil, which are used to treat a hyperfunctioning thyroid gland. See *LIR Pharmacology,* 5e, p. 298.

6. **Endocytosis:** The iodinated and conjugated thyroglobulin is then endocytosed back into follicular cells, initiated by megalin receptors. TSH regulates megalin receptor expression and, thereby, indirectly controls the amount of colloid endocytosed.

7. **Proteolysis:** The endocytosed vesicle containing colloid then fuses with a lysosome, and the iodine-containing molecules are cleaved from thyroglobulin. T_3 and T_4 are released into the follicular cytosol near the basolateral membrane, and the remaining molecules and colloid material are recycled.

8. **Secretion:** The final step is secretion of T_3 and T_4 from the follicular cell into the blood. Cytosolic thyroid hormones diffuse through the basolateral cell membrane to the interstitial space, where they enter the capillary network and blood vessels of the highly vascularized thyroid gland.

B. Transport and regulation

Approximately 99% of the T_3 and T_4 released into the circulation binds to thyroid hormone-binding globulin and, to a lesser degree, albumin and transthyretin. Binding increases the half-life to as long as a week and serves as a thyroid hormone reserve for short periods. Both free (unbound) T_3 and T_4 participate in negative feedback control at the level of the hypothalamus and thyrotropes. In addition, thyroid hormones also increase somatostatin, which further decreases TSH release from thyrotropes (Figure 35.6).

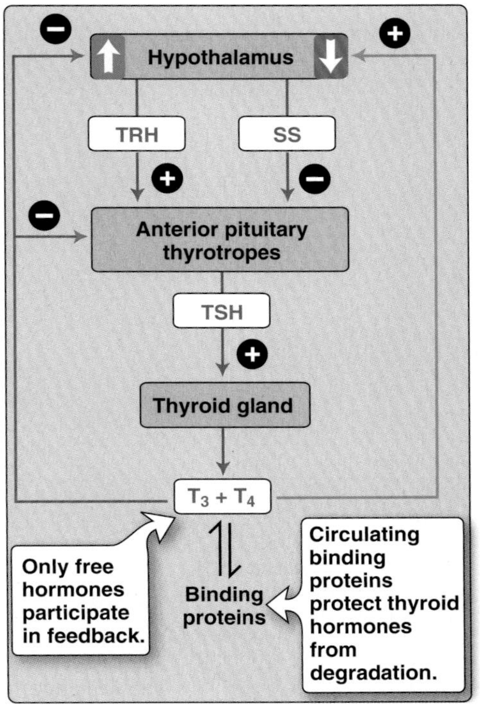

Figure 35.6
Feedback regulation of thyroid hormone release. SS = somatostatin; T_3 = triiodothyronine; T_4 = thyroxine; TRH = thyroid-releasing hormone; TSH = thyroid-stimulating hormone.

C. Effects

T_4 and T_3 diffuse across the target cell membrane, and T_4 is converted to T_3 (the more biologically active form) by *5′-deiodinase* (also called *5′/3′-monodeiodinase* or *deiodinase I*). T_3 then binds to a nuclear thyroid receptor that complexes with a retinoid X receptor (RXR). This receptor complex then binds to the thyroid-response element of DNA which, through both the addition of a coactivator and release of a corepressor, begins transcription (Figure 35.7). Proteins synthesized mediate a wide range of cellular responses, including increases in growth and development, glucose and fatty acid availability, and metabolic rate.

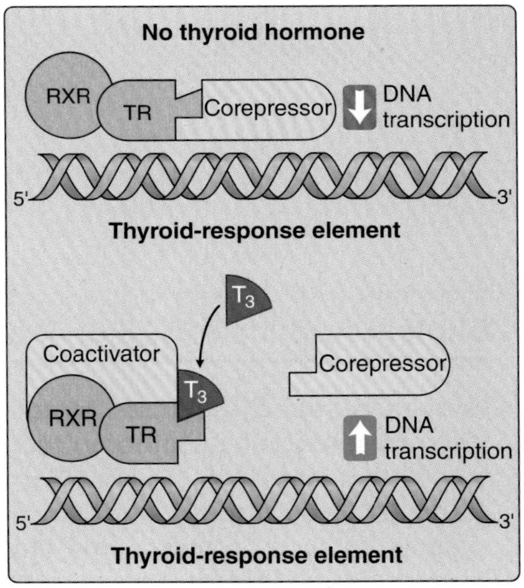

Figure 35.7
Thyroid hormone–receptor binding. RXR = retinoid X receptor; T_3 = triiodothyronine; TR = thyroid hormone receptor.

1. **Growth and development:** Growth and development of nervous tissue and bone are dependent on thyroid axis hormone synthesis and release. In nervous tissue, T_3 and T_4 aid in the timing and rate of development, which affects, for example, the development of stretch reflexes (see 11·III·B). In bone, thyroid hormone increases ossification and linear growth in children and adolescents. Thyroid hormone deficiencies can, thus, result in mental impairments and short stature in children.

2. **Macronutrient metabolism:** Thyroid hormones not only alter the rate of metabolism but also affect energy substrates. T_3 and T_4 increase both the breakdown of glycogen (glycogenolysis) and the formation of glucose (gluconeogenesis). T_3 and T_4 also increase the formation of lipids (lipogenesis) followed by promoting lipolytic enzymes, which break down the stored lipids into free fatty

Clinical Application 35.2: Hashimoto Disease

Hashimoto disease is the most common form of **hypothyroidism**, characterized by an autoimmune response that targets the thyroid gland. Inflammation and destruction of the gland decreases triiodothyronine and thyroxine production and release. A goiter may also form but is the result of the lymph and T-cell infiltrates rather than glandular hypertrophy. Patients lack energy, and are easily fatigued. They gain weight due to metabolic rate decline. If low levels of thyroid hormone occur during growth, growth can be stunted and development slowed.

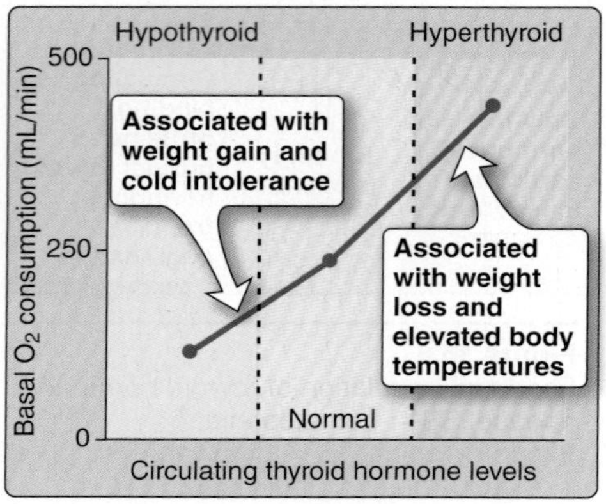

Figure 35.8
Thyroid hormone effects on metabolism.

acids to be used as an energy substrate. Having a low thyroid axis hormone level has the opposite effect on the breakdown of carbohydrates, proteins, and lipids.

3. **Basal metabolic rate:** Metabolism, through its inherent reaction inefficiencies, produces heat. Na^+-K^+ ATPase expression provides a good example of this phenomenon, in which thyroid hormone induces expression of the "always-on" pump uses more energy and produces more heat. The opposite occurs if pump expression is reduced by hypothyroid conditions (Figure 35.8).

4. **Thyroid hormone and catecholamine synergy:** When T_3 and T_4 and norepinephrine (from the sympathetic nervous system) are released in concert (e.g., during severe cold stress), the physiological functions of both are heightened. Thyroid hormones also upregulate β-adrenergic receptors, which potentiates these synergistic effects.

IV. CALCITONIN

Calcitonin is a small (32 amino acid) peptide hormone produced by thyroid parafollicular C cells. Calcitonin is released in response to high

Clinical Application 35.3: Graves Disease

Graves disease produces **hyperthyroidism** via an antibody (**thyroid-stimulating antibody**) that mimics the effects of thyroid-stimulating hormone (TSH). TSH receptor stimulation causes glandular hypertrophy, which, in turn, causes goiter and other symptoms, such as myxedema and ophthalmopathy. Myxedema involves dermal edema and increases in connective tissue, which is caused by an overproduction of mucopolysaccharides (modified carbohydrates). Ophthalmopathy involves swelling around the eyes and also includes an excess production of mucopolysaccharides, which causes the eyes to protrude.

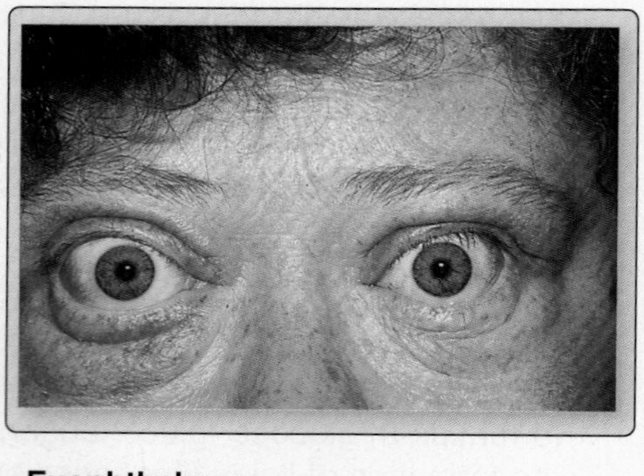

Exophthalmos.

plasma Ca^{2+} concentrations, and its primary effect is to block osteoclast-mediated bone resorption and Ca^{2+} mobilization (see 15·IV). Calcitonin is not thought to play any significant role in Ca^{2+} homeostasis in humans, but it can be used as a biomarker for thyroid cancer, and its physiologic effects can be exploited therapeutically.[1] Plasma Ca^{2+} and PO_4^{3-} levels are regulated primarily by PTH and vitamin D derivatives, as discussed below.

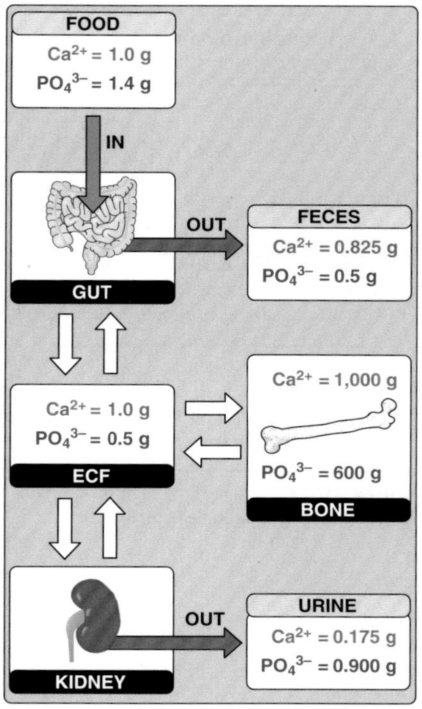

Figure 35.9
Calcium and phosphate balance. ECF = extracellular fluid.

V. PARATHYROID HORMONE AND VITAMIN D

PTH and vitamin D together regulate Ca^{2+} and PO_4^{3-}. The major targets of regulation are the GI system (absorption), kidneys (reabsorption), and bone (deposition and resorption).

A. Calcium and phosphate pools

Large quantities of Ca^{2+} and PO_4^{3-} are stored in bone (think of bone as a Ca^{2+} bank depository). This large Ca^{2+} source can be mobilized (withdrawn) during bone resorption in a process mediated by osteoclasts (see 15·IV). Alternately, Ca^{2+} can be actively stored (deposited) within bone during bone deposition in a process mediated by osteoblasts. Besides storage and release, Ca^{2+} balance is maintained by Ca^{2+} excretion from the kidney and Ca^{2+} intake by the GI system (Figure 35.9). PO_4^{3-} intake, excretion, and storage sites are similar to those of Ca^{2+}.

B. Parathyroid hormone

PTH is released from parathyroid glands in response to a decline in circulating Ca^{2+} and Mg^{2+} levels. PTH actions are geared toward increasing Ca^{2+} availability.

1. **Regulation:** Parathyroid cells express a specialized GPCR that functions as a Ca^{2+} sensor. When plasma Ca^{2+} levels are high, the receptor tonically inhibits PTH secretion (Figure 35.10). The relationship between free Ca^{2+} and PTH release is sigmoidal, with the steepest portion of the curve being in the physiologic range of plasma Ca^{2+}. PTH release is similarly dependent on plasma free Mg^{2+}.

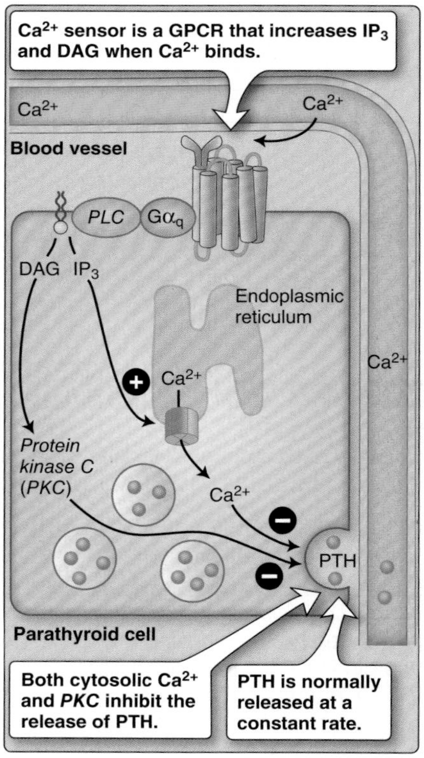

Figure 35.10
Calcium sensing. DAG = diacylglycerol; GPCR = G protein–coupled receptor; IP_3 = inositol trisphosphate; *PLC* = *phospholipase C*; PTH = parathyroid hormone.

> Lithium sensitizes the Ca^{2+} receptor to changes in plasma Ca^{2+}, causing increased PTH release in response to a given Ca^{2+} stimulus. This is the reason why bipolar patients on lithium salts for manic episodes can also have hypercalcemia.

[1]For a discussion of the use of calcitonin to treat osteoporosis, see *LIR Pharmacology,* 5e, p. 366.

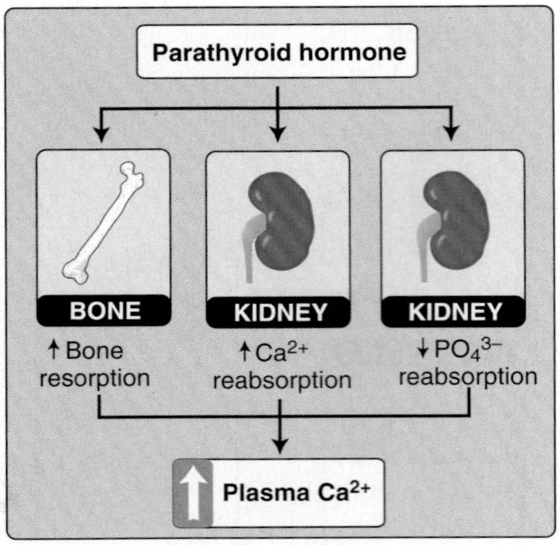

Figure 35.11
Parathyroid hormone effects.

2. **Function:** PTH increases circulating Ca^{2+} levels by two principal mechanisms. First, it stimulates bone resorption by binding to receptors on the surface of osteoblasts, which then recruit osteoclasts precursors to a bone resorption site. Second, it stimulates Ca^{2+} reabsorption by the renal tubule (Figure 35.11). Proximal tubule reabsorption of PO_4^{3-} is decreased by PTH. Although this increases PO_4^{3-} excretion, it does not significantly alter circulating plasma PO_4^{3-} levels because PTH also increases PO_4^{3-} liberation from the bone. PTH also indirectly increases circulating Ca^{2+} by its stimulatory effects on **1α,25-dihydroxyvitamin D_3 [1,25-$(OH)_2D_3$]** synthesis.

C. Vitamin D derivatives

Vitamin D is functionally related to PTH, but, structurally, it is very different. Vitamin D and its derivatives are hydrophobic and transported in the blood primarily by vitamin D-binding protein. The primary active vitamin D derivative is 1,25-$(OH)_2D_3$, which increases Ca^{2+} and PO_4^{3-} absorption by the small intestine. The synthesis of 1,25-$(OH)_2D_3$ from vitamin D2 or D3 involves a multistep process that includes both the liver and the kidney. The major regulatory enzyme for 1,25-$(OH)_2D_3$ synthesis is ***25(OH)D1-α-hydroxylase***. Vitamin D3 can be synthesized in the skin by keratinocytes (see 16·III·A·1) via interaction with UV light with 7-dehydrocholesterol, and vitamin D2 and D3 can also come from dietary sources.

1. **Regulation:** The primary factor that regulates 1,25-$(OH)_2D_3$ is PTH. PTH increases the activity of *25(OH)D1-α-hydroxylase* and decreases the activity of *25(OH)D24-hydroxylase* in the kidney. This shifts the reactions toward the production of 1,25-$(OH)_2D_3$. Low levels of both Ca^{2+} and PO_4^{3-} as well as hormones, such as growth hormone, prolactin, and estrogen, also increase 1,25-$(OH)_2D_3$ levels.

2. **Function:** 1,25-$(OH)_2D_3$ is very effective in aiding the absorption of Ca^{2+} and PO_4^{3-} from the GI tract and reabsorption of these ions from the renal tubule. Ca^{2+} absorption is improved by 1,25-$(OH)_2D_3$ via increased synthesis of apical, basolateral, and cytosolic transport proteins. 1,25-$(OH)_2D_3$ binds to a nuclear vitamin D receptor that complexes with an RXR and induces transcription from the vitamin D response element. Thus, most of the effects of 1,25-$(OH)_2D_3$ are genomic, but there are some faster effects mediated by a cell membrane vitamin D receptor. These faster cell membrane responses are observed primarily in the GI system. PO_4^{3-} absorption is also improved by 1,25-$(OH)_2D_3$ effects on apical Na^+- PO_4^- cotransporter synthesis. 1,25-$(OH)_2D_3$ also aids in the maturation of osteoclasts, which potentially allows for more bone resorption. Finally, 1,25-$(OH)_2D_3$ decreases PTH through negative feedback in addition to the negative feedback provided by increases in circulating Ca^{2+} due to increased intestinal absorption (Figure 35.12).

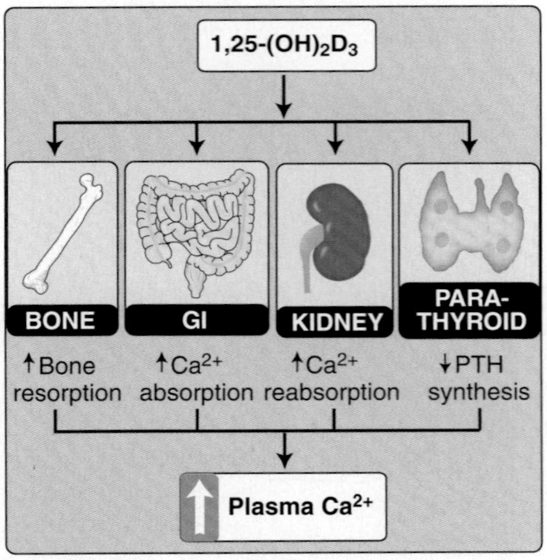

Figure 35.12
Vitamin D3. GI = gastrointestinal; PTH = parathyroid hormone.

Clinical Application 35.4: Rickets

Rickets is caused by a lack of vitamin D, Ca^{2+}, or PO_4^{3-}. Without adequate 1α,25-dihydroxyvitamin D_3 [1,25-$(OH)_2D_3$], Ca^{2+} and PO_4^{3-} cannot be absorbed from the gastrointestinal tract, causing plasma levels to fall. Parathyroid hormone is released as a result, promoting bone resorption to raise serum Ca^{2+} levels. Over time, bone minerals are depleted, causing them to weaken and bow. The lack of 1,25-$(OH)_2D_3$ can result from low exposure to sunlight, lack of access to dietary sources of vitamin D or a genetic disorder producing a low functioning *25(OH)D1-α-hydroxylase*.

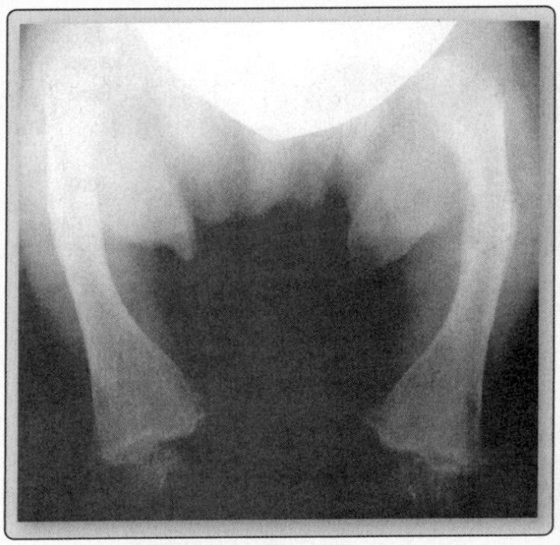

Bowed bone.

Chapter Summary

- Thyroid axis hormones consist of **thyroxine (T_4)**, **triiodothyronine (T_3)**, and **reverse triiodothyronine**. Thyroid intermediates consist of **monoiodotyrosine** (**MIT**) and **diiodotyrosine** (**DIT**). T_4, which is formed from two DITs, is the most prevalent blood variant. T_3 is the most biologically active in peripheral tissues and is formed from one DIT and one MIT.
- The steps of thyroid hormone synthesis are: **uptake**, **colloid secretion**, **oxidation**, **iodination**, **conjugation**, **endocytosis**, **proteolysis**, and **glandular secretion**. **Thyroid-stimulating hormone** (**TSH**) regulates the secretion of triiodothyronine and thyroxine via its actions on the Na^+-I^- symporter, pendrin channels, ***thyroid peroxidase***, megalin receptors, and proteolytic enzymes. TSH also causes enlargement of the thyroid gland.
- The majority of triiodothyronine (T_3) and thyroxine (T_4) in blood is bound to **thyroid hormone-binding globulin** rendering them inactive, but some T_3 and T_4 circulates in the active free form.
- Once triiodothyronine (T_3) and thyroxine (T_4) enter a cell, the majority of the T_4 is converted to T_3, and T_3 has a higher binding affinity for thyroid receptors. Thyroid hormones act via nuclear receptors, to modulate transcription and translation and, thereby, increase protein synthesis.
- Depending on which proteins are synthesized, thyroid hormones increase metabolic rate, heat production, and glycogen and fat breakdown and utilization. Thyroid hormones are also imperative for normal growth and development.
- **Parathyroid hormone** (**PTH**) increases circulating Ca^{2+} levels by increasing Ca^{2+} reabsorption in the kidney and by increasing bone resorption. PTH also activates **vitamin D3**.
- Vitamin D3 is converted to its active form via ultraviolet light exposure or enzymatic reaction in the kidney. The active form of vitamin D is **1α,25-dihydroxyvitamin D_3 [1,25-$(OH)_2D_3$]**, and its primary function is to increase circulating Ca^{2+} levels via absorption from the intestines, reabsorption from the kidney, and bone resorption.

36 Female and Male Gonads

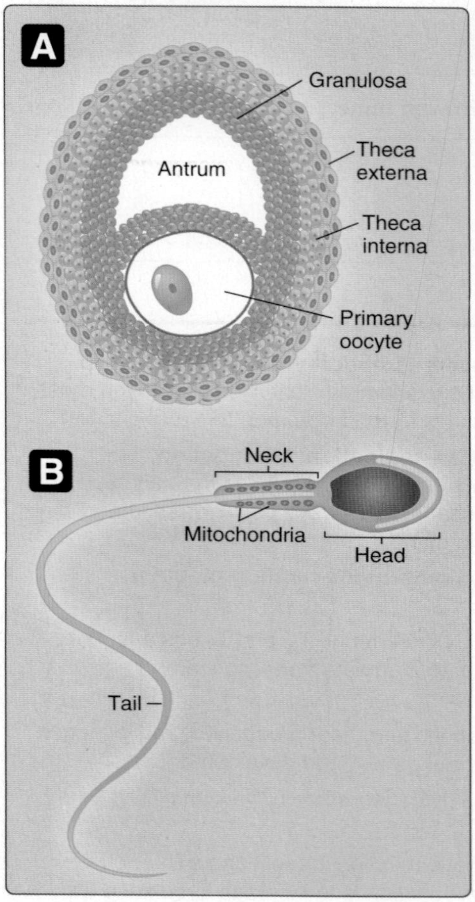

Figure 36.1
Oocyte and sperm.

I. OVERVIEW

"Gonad" is derived from the Greek word *gonos*, which translates to "seed" or "family." Gonads produce seed cells (gametes) that can divide and replicate into an organism, and reproduction forms a family lineage. In a basic sense, the primary purpose of any species is to pass along a unique set of deoxyribonucleic acids (DNA) to the next generation (although teaching them to eat and avoid danger helps ensure that the lineage survives). The gonads form oocytes in females and sperm in males (Figure 36.1), and the means to signal the body to allow for (in the broadest sense) reproduction. Gonadal hormones are involved in a wide array of functions in sexual maturation: **oogenesis** (formation and development of an oocyte) and **spermatogenesis** (formation and development of sperm) as well as supporting conception, pregnancy, and lactation. These hormone effects are not limited to the reproductive organs but also affect bone, muscle, and blood vessels. Gonadal hormones are under hypothalamic–pituitary axis control, with both males and females using the same hypothalamic signaling hormone (**gonadotropin-releasing hormone** [**GnRH**]) and pituitary hormones, including **luteinizing hormone** (**LH**) and **follicle-stimulating hormone** (**FSH**). In females, LH and FSH target the ovaries, which secrete two main hormones: **progesterone** from **granulosa cells** and **estradiol** in cooperative fashion (sharing of precursors) between **theca** and granulosa cells. In males, the pituitary hormones target the testis, which primarily secretes **testosterone** from **Leydig cells**.

II. SEX AND GENDER

Chromosomes determine sex: XX are female, and XY are male. **Gonadal sex** is defined by the type of gonads. Gonadal females have ovaries, and gonadal males have testes. **Phenotypic sex** is determined by the characteristics of the genital tract and external genitalia. Finally, **gender** is a psychosocial term used to identify a person as a man or woman based on a set of characteristics, attributes, or social norms. These definitions become blurred in genetic variants like Klinefelter syndrome, which produces an extra X chromosome, 47, XXY ("47" denotes an extra chromosome, as normally there are 23 pairs, or 46). Alternately, a person could have his or her gonads removed by castration surgery or have external genitalia added or removed. For clarity, in this chapter, we will use gonadal sex denotations and introduce genetic and phenotypic components that align with the gonads.

III. HYPOTHALAMIC–PITUITARY–OVARIAN AXIS

The female gonads, the ovaries, secrete estrogens and progestins. The ovaries are subject to multitiered endocrine axis feedback, which allows for precise regulation of function.

A. Hypothalamus

The principal hypothalamic areas involved in ovarian control are the preoptic and supraoptic nuclei (Figure 36.2). Parvocellular neurons in these areas synthesize and secrete GnRH. GnRH is a peptide hormone that is produced in the soma as a prohormone and then modified into its active form and secreted into the hypophyseal portal system. The release of GnRH is pulsatile, meaning that there is not a constant release from the hypothalamus. Pulsatile secretion has the advantage of using less energy and does not desensitize target tissue receptors. The perception of stress and other inputs from the higher brain centers as well as from the brain's rhythm centers help influence the pulsatile secretion of GnRH.

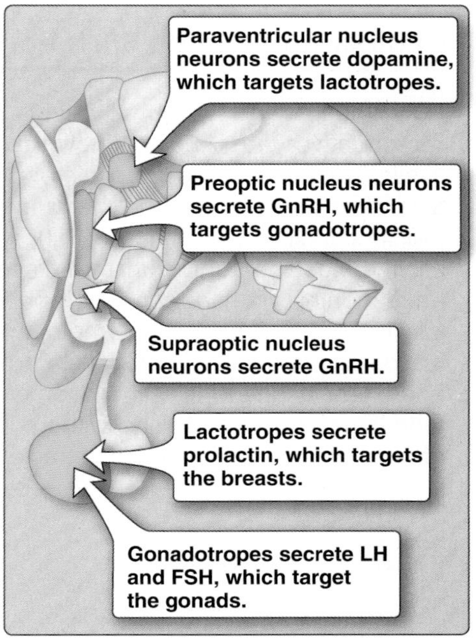

Figure 36.2
Hypothalamic nuclei involved in ovarian control. FSH = follicle-stimulating hormone; GnRH = gonadotropin-releasing hormone; LH = luteinizing hormone.

B. Pituitary gland

The hypophyseal portal circulation delivers GnRH to anterior pituitary **gonadotropes** (see Figure 7.11), which subsequently secrete LH and FSH. GnRH receptors are part of the G protein–coupled receptor (GPCR) superfamily and function primarily through the *phospholipase C* (*PLC*)-induced diacylglycerol and inositol trisphosphate second messenger system (see 1·VII·B·3).

C. Ovaries

The ovaries house female germ cells (**oocytes**) containing follicles in various stages of development: primordial, primary, secondary, tertiary, and Graafian follicles. The endocrine portion of the ovaries is primarily related to the latter follicles and involves theca and granulosa cells. These cells work cooperatively to synthesize and secrete estradiol.

1. **Theca cells:** Theca cells are a superficial layer of the follicle that transports **low-density lipoprotein** (**LDL**) into the cells via cell membrane LDL receptors in clathrin-coated pits. Once the LDL receptor binds ligand, LDL is endocytosed, and cholesterol is liberated. Cholesterol is the initial substrate of the first reaction of steroid hormone synthesis as it is in the adrenal gland (see Figure 34.4). LH receptors are in the GPCR family. They primarily work through an *adenylyl cyclase* (*AC*)-induced cyclic adenosine monophosphate (cAMP) second messenger process (see 1·VII·B·2), which activates a ***side-chain cleavage enzyme complex*** that facilitates the conversion of cholesterol into pregnenolone (Figure 36.3). Thus, LH works in a similar manner to adrenocorticotropic hormone in the adrenal glands (see 34·II·C). Steroid hormone synthesis continues in the theca cell, producing **androstenedione** and testosterone. The majority of these androgens exits the theca cell and enters nearby granulosa cells, because there are insufficient quantities of ***aromatase*** in theca cells to facilitate the

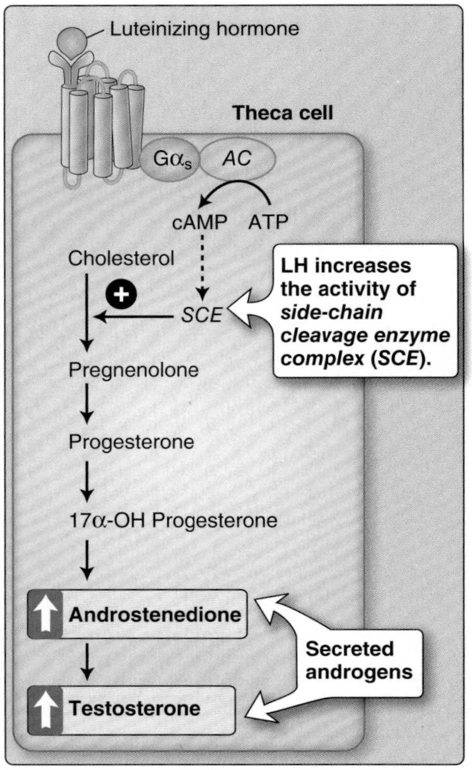

Figure 36.3
Theca cell. *AC* = *adenylyl cyclase*; ATP = adenosine triphosphate; cAMP = cyclic adenosine monophosphate.

Clinical Application 36.1: Polycystic Ovary Syndrome

Polycystic ovary syndrome affects 5%–10% of females in their reproductive years. For unknown reasons, ovaries become polycystic (have a thickened capsule and prominent subcapsular cysts), and there are elevations in the ratio of luteinizing hormone (LH) to follicle-stimulating hormone (FSH), weight gain, and insulin insensitivity. Most persons with this disease are amenorrheic (do not have monthly menses) or have abnormal uterine bleeding and are infertile. Other presenting symptoms are hirsutism, acne, and virilization due to excess androgens. Part of the reason for these excess androgens is due to LH stimulation of theca cells, producing androstenedione and testosterone without corresponding FSH stimulation of granulosa cells to produce estradiol from these theca androgens.

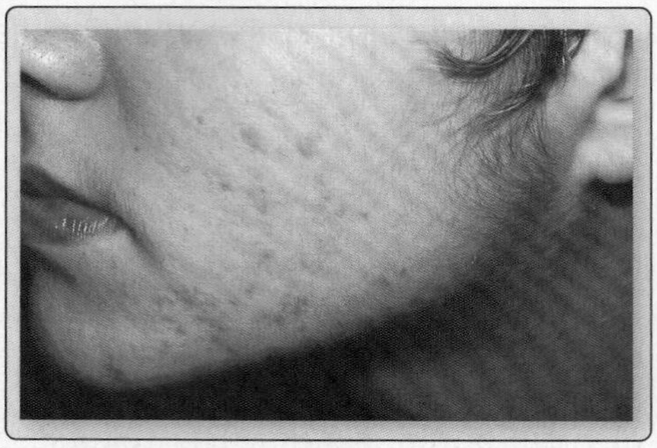

Hirsutism.

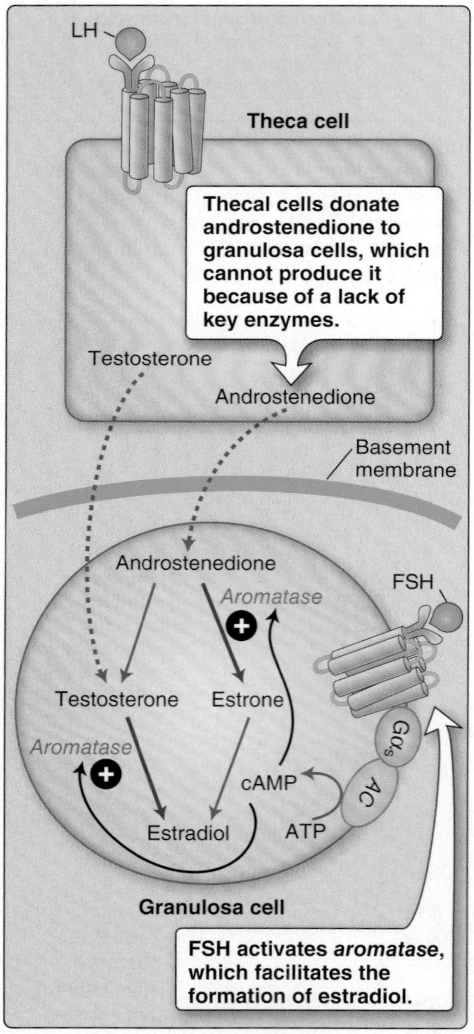

Figure 36.4
Theca and granulosa cells. *AC* = *adenylyl cyclase*; ATP = adenosine triphosphate; cAMP = cyclic adenosine monophosphate; FSH = follicle-stimulating hormone; LH = luteinizing hormone.

conversion of either androstenedione or testosterone into estrogens (Figure 36.4).

2. **Granulosa cells:** The granulosa cell is deep compared to theca cells in follicles (see Figure 36.1). The granulosa cell layer dramatically increases during the development from primary to secondary follicles. Granulosa cells express both LH and FSH receptors. Therefore, not only is the conversion of cholesterol to pregnenolone facilitated as in theca cells, but, in addition, the enzyme *aromatase* is also activated. FSH receptors are in the GPCR superfamily and primarily work through the *AC*-induced cAMP second messenger system to activate *aromatase*. Thus, products such as estradiol can be synthesized and then secreted into the bloodstream. Granulosa cells do not contain sufficient quantities of *17α-hydroxylase* or *17,20-desmolase* and, thus, rely on androstenedione and testosterone to be secreted from the theca cell to complete sex steroid synthesis (see Figure 36.4).

3. **Oogenesis:** Oogenesis begins during the fetal stage. Early in development, primordial germ cells (oogonia) dramatically increase in number. A portion of these oogonia mature into oocytes. By about 20 weeks of gestational age, the number of oocytes is at its maximum. Thereafter, the number of oocytes continually decreases until exhausted.

IV. ESTROGENS

There are three main estrogens. The most potent is estradiol, although **estrone**, which is also formed in peripheral tissues, and **estriol**, which is secreted in higher concentrations during pregnancy (see 37·III·C·3), also have functional effects. As mentioned in the previous chapter, adrenal gland–derived androstenedione can also be converted peripherally into estrogens (see 34·IV·A). Estradiol has a high binding affinity to **sex steroid–binding globulin** (**SSBG**) and a moderate binding affinity to albumin, which keeps the amount in the active free form low in the blood. The liver processes estrogens, and these products are secreted in the urine.

A. Function

Estrogens have a number of functional effects, both genomic and nongenomic. The nongenomic effects are mediated by cell membrane receptors and do not directly induce transcription, translation, and protein synthesis. Most of estrogens' effects are genomic and utilize a similar mechanism to that of other steroid hormones. There are two classes of **estrogen receptors** (**ERs**): **ERα** and **ERβ**. ERα is primarily expressed in the reproductive organs, whereas ERβ is primarily expressed in granulosa cells and in the nonreproductive organs. ERα and ERβ are cytosolic and nuclear. Once estrogens bind the receptor, the receptor homodimerizes and binds to an estrogen-response element on DNA, inducing specific gene expression. The nature of the proteins synthesised and their effect is dependent on the tissue (Table 36.1).

Table 36.1: Effect of Estrogens

Tissue	Effect
Bone	↑ Growth via osteoblasts
Endocrine	↑ Progesterone responses
Liver	↑ Clotting factors
	↑ Steroid-binding proteins
	↓ Total and LDL
	↑ HDL
Reproductive organs	↑ Uterine growth
	↑ Vaginal and fallopian tube growth
	↑ Breast growth
	↑ Cervical mucus secretion
	↑ LH receptors on granulosa cells

LDL = low-density lipoprotein; HDL = high-density lipoprotein; LH = luteinizing hormone.

B. Secretion

The regulation of estrogens comprises an interrelated set of feedback loops at each level of the hypothalamic–pituitary–ovarian axis. Estrogens, progestins, inhibins, and activins provide axis feedback (Figure 36.5). These multiple layers of control allow for precise timing of hormonal signaling, despite the two main hormone classes (estrogens and progestins) using the same control system axis.

1. **Estrogens:** Estrogens secreted from the granulosa cells negatively feed back to both the anterior pituitary and hypothalamus, and some evidence suggests that there may be additional feedback to higher brain centers that can stimulate or inhibit the axis. Estrogens normally exert negative feedback, but this feedback shifts to positive feedback midcycle. This shift is caused by the upregulation of receptors, such as GnRH in the anterior pituitary, when circulating estrogen levels are elevated. The functional result of the shift is a surge in LH and FSH just prior to ovulation.

2. **Progestins:** Progestins (discussed more in the subsequent section) also provide negative feedback to the anterior pituitary and hypothalamus.

3. **Inhibins:** Granulosa cells synthesize and secrete peptide hormones called inhibins that feed back to the anterior pituitary. There are two inhibins, A and B, both of which appear to be functional in females. Inhibins decrease secretion of FSH. FSH is the primary stimulus for the production of inhibins, and, thus, inhibin increase lags slightly behind FSH in the menstrual cycle but does provide negative feedback for FSH regulation.

4. **Activins:** Activins are peptide hormones secreted by the granulosa cells that stimulate secretion of FSH from the anterior pituitary as well as local FSH receptor upregulation. Activin levels are highest during follicle development.

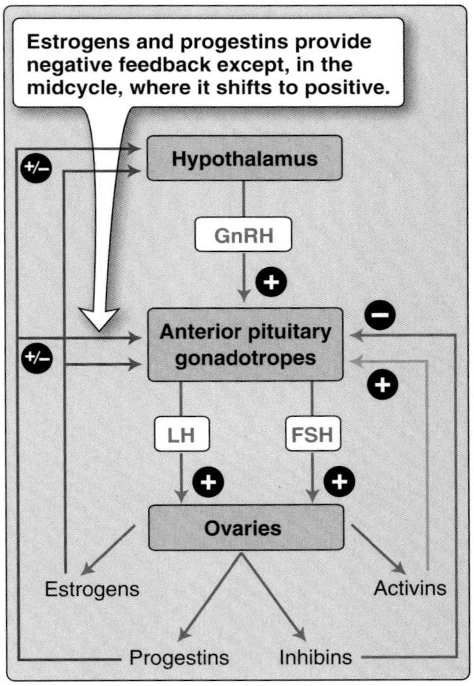

Figure 36.5
Female feedback regulation.
FSH = follicle-stimulating hormone; GnRH = gonadotropin-releasing hormone; LH = luteinizing hormone.

V. PROGESTINS

Progesterone is the most common and biologically active progestin. A second, less potent but measureable, progestin is **17α-hydroxyprogesterone**. Progesterone is produced in both theca and granulosa cells. Progesterone

Table 36.2: Effect of Progestins

Tissue	Effect
Breast	↑ Lobular development ↓ Milk production
Reproductive organs	↓ Endometrial growth ↑ Endometrial secretions Mucosal secretions become thicker
Temperature	↑ Internal temperature

binds to albumin with low affinity, and therefore has a fairly short half-life of about 5 min in the circulation. The liver processes progesterone similarly to other steroid hormones, and these products are secreted in the urine.

A. Function

Progesterone functions are more limited compared to those of estrogens, primarily initiating and maintaining pregnancy (see Figure 37.5). The effects of progestins are mediated by progesterone receptors that have A and B half-sites and form a homodimer, which is then bound to a progesterone-response element to transcribe specific genes to cause the functional effects in various tissues (Table 36.2).

B. Secretion

The control of progestin secretion is intricately linked to those of estrogens and thus was discussed above.

VI. OVARIAN AND ENDOMETRIAL CYCLES

The menstrual cycle is actually two distinct cycles: the **ovarian cycle** and the **endometrial cycle**. The ovarian cycle deals with follicle development and the endometrial cycle with changes associated in the endometrial lining. Both are controlled and regulated by the hypothalamic–pituitary–ovarian axis. The mean duration of these cycles is approximately 28 days, but normal menstrual cycles can vary by a number of days. The most variability in cycle duration occurs earlier and later in the reproductive years.

A. Ovarian cycle

The ovarian cycle is divided into follicular and luteal phases. Each phase lasts for half of the duration of the cycle (Figure 36.6A). The events that divide these phases are ovulation and the beginning of menses.

1. **Follicular phase:** The primary result of the follicular phase is the development of a mature Graafian follicle and secondary oocyte. Follicular phase duration is variable. Estrogens gradually increase, causing FSH and LH to peak, whereas progesterone remains low throughout.

2. **Luteal phase:** The luteal phase is dominated by the actions of the corpus luteum (residual theca and granulosa cells of the follicle after oocyte release), which synthesizes and secretes estrogen and progesterone. These hormones are necessary for implantation and maintenance of any fertilized oocytes. If fertilization does not occur, the corpus luteum regresses and eventually forms a nonfunctional scarlike structure (corpus albicans). The corpus albicans slowly migrates deeper into the ovary and is slowly degraded. The regression of the corpus luteum occurs about 10–12 days after ovulation in the absence of **human chorionic gonadotropin** (**hCG**). Thus, the 14 days of the luteal phase is fairly constant. Progestins gradually rise, and estrogens first fall but then increase again. Body temperature increases.

B. Endometrial cycle

The uterine inner wall lining (endometrium) undergoes many changes during a typical month in a woman in her childbearing years.

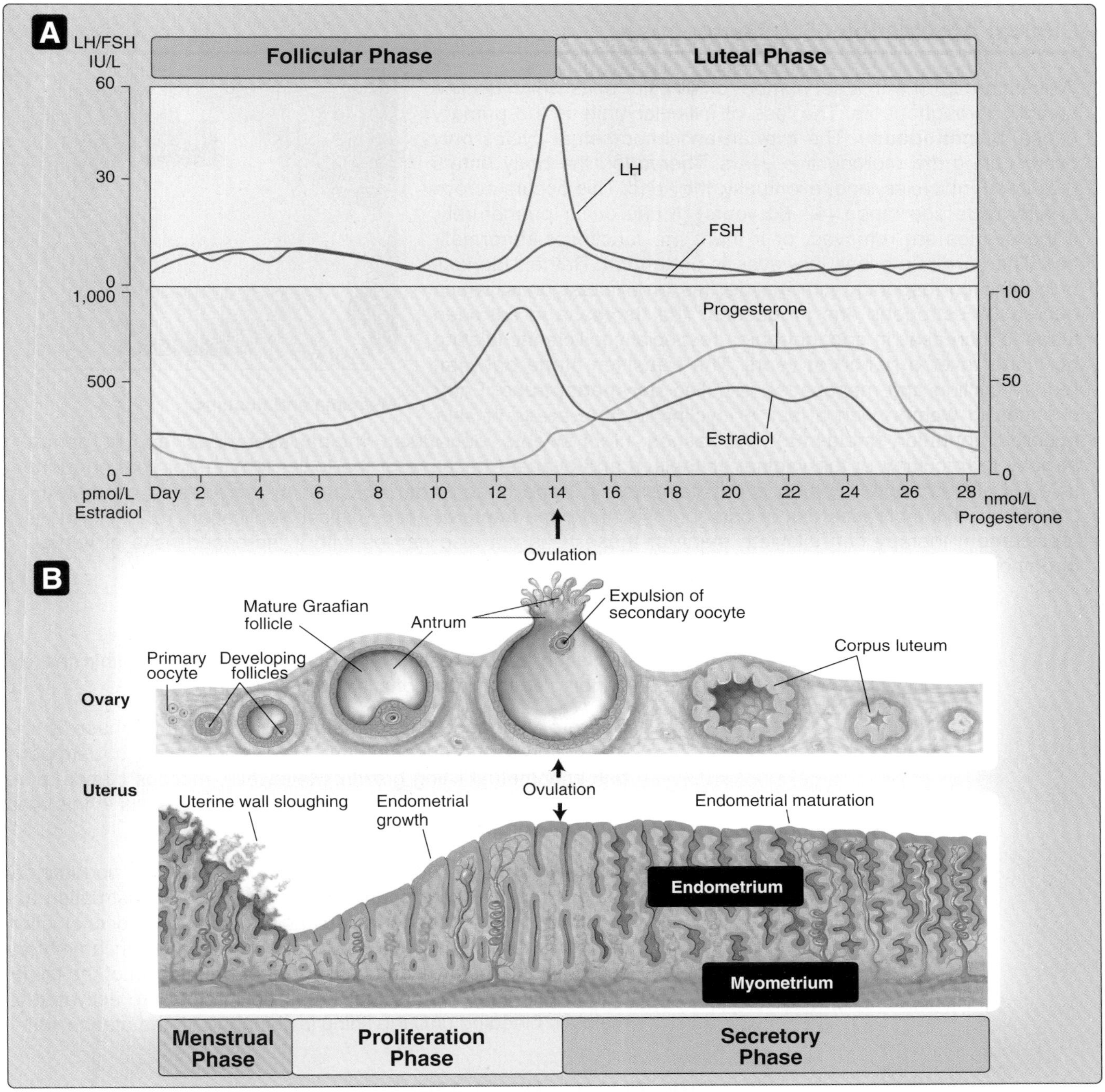

Figure 36.6
Ovarian and endometrial cycle. FSH = follicle-stimulating hormone; LH = luteinizing hormone.

The endometrial cycle is divided into a proliferative phase, a secretory phase, and menstruation (see Figure 36.6B).

1. **Proliferative phase:** Endometrial growth is the primary outcome of this phase and is mediated by increases in estrogens. Growth is pronounced, with endometrial thickness increasing from 1–2 mm to 8–10 mm by the end of the phase, which is marked by

Clinical Application 36.2: Menopause

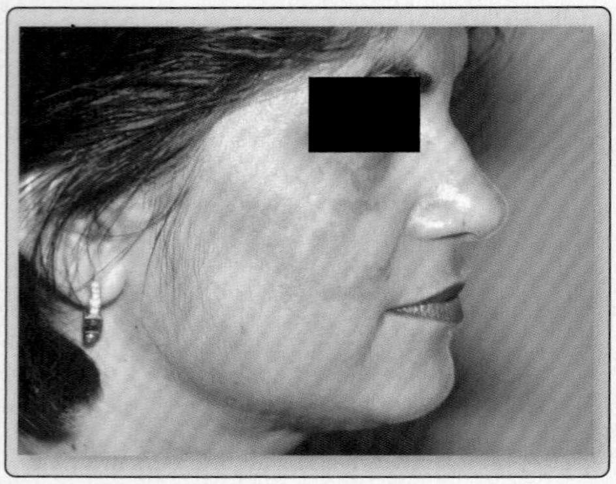

Melasma and flushing.

Women are born with a set number of follicular units which decline steadily throughout life. The loss of follicular units is the primary cause of **menopause**. The ovarian and endometrial cycles only occur during the reproductive years. Thereafter, the body enters less frequent cycles, and, eventually, they end. This occurs across a fairly wide age range (42–60 years). It can occur prematurely, if the ovaries are removed, or if there are functional abnormalities. The normal process, however, is not abrupt. Rather, the loss of cycling corresponds with the decline in follicular units, which decreases estrogens and progestins. The decrease in estrogen leads to increases in gonadotropins (especially follicle-stimulating hormone) over a period of years. The transition stage between regular cycling and menopause is called **perimenopause**. Early changes in women with a loss of cycling include areas of skin hyperpigmentation (melasma), hot flashes, night sweats, decreased vaginal secretions, and urogenital atrophy (particularly of the vaginal epithelium and ovaries) because of the lower female gonadal hormone levels. Later changes include a net decrease in bone mineral density and an increase in cholesterol. Combined, this increases risk of osteoporosis and bone fractures (see Clinical Application 15.3). Hormone replacement therapy can alleviate many of these risks but also carries with it increased risks of venous thrombosis and certain cancers.[1]

ovulation. Blood vessels and gland growth occur within the expanding stratum functionalis of the endometrium.

2. **Secretory phase:** The primary outcome of this phase is the maturation of the endometrium. Decreasing levels of estrogens halt endometrial lining growth. Meanwhile, mucous glands more fully develop, and both the glands and blood vessels in this area increase surface area and coil.

3. **Menstruation:** If conception does not occur, the endometrial lining is replaced to prepare for the next cycle. Menstruation begins with a pronounced prostaglandin-mediated vasoconstriction of spiral arteries, which causes local ischemic injury. Inflammatory cells infiltrate the area and cause further breakdown of the lining. During this time, factors that break down clots are activated to maintain bleeding until the lining is sloughed off the uterine wall.

VII. MAMMARY GLANDS

The breast and mammary glands provide optimal nourishment for infants. Although lactation (the period during which milk is produced and secreted) occurs just after birth, the development of breast tissue and preparation for this act occurs during puberty. This growth and de-

[1]For additional information on hormone replacement therapy, see *LIR Pharmacology*, 5e, p. 319.

velopment is mediated by female gonadal hormones as part of secondary sex characteristics. In direct preparation for lactation, the breast develops more fully via high levels of estrogens, progestins, hCG from the fetus, and **prolactin**. Besides beginning milk production and sustaining it, which is primarily mediated by prolactin, milk must be "let down" and ejected to allow for suckling, a process mediated by the posterior pituitary hormone **oxytocin** (Figure 36.7).

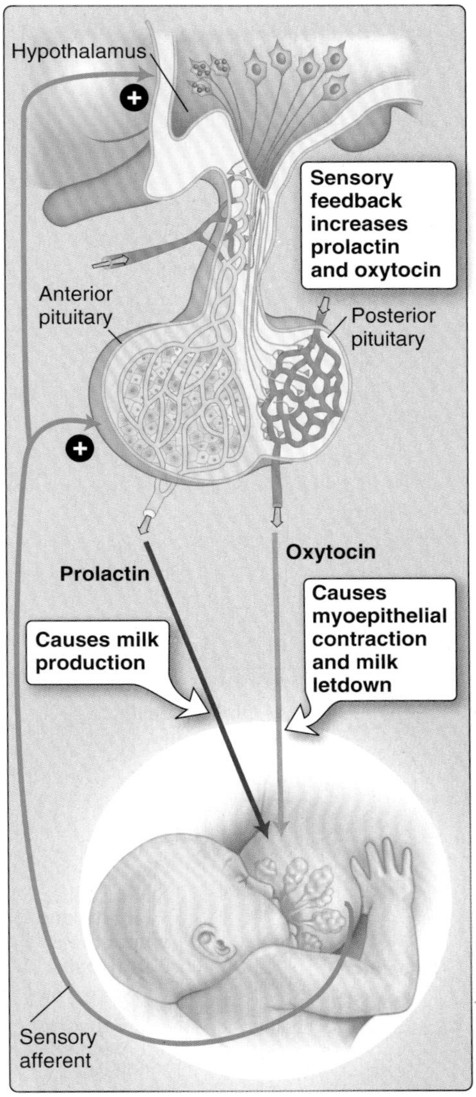

Figure 36.7
Hormonal regulation of lactation.

A. Prolactin

Prolactin is a peptide hormone produced and secreted from lactotropes in the anterior pituitary gland. Unlike the other anterior pituitary hormones, it is not associated with a hormone axis and is produced and secreted in both males and females. In females, lactotropes hypertrophy, and prolactin secretion increases during pregnancy. Prolactin is not associated with a hormone-binding protein and has a half-life of about 20 min.

1. **Function:** Prolactin causes mammary glandular tissue growth and development, ductal proliferation, synthesis of breast milk, and preparation of the breast for lactation. The effects of prolactin on the breast and mammary gland are mediated by a cytokine cell membrane receptor that stimulates the *Janus kinase* (*JAK*) and signal transducer and activator of transcription (STAT) pathway, commonly known as the *JAK*/STAT signaling pathway (see 33·V·C). The function of prolactin in males is not well understood.

2. **Secretion:** Prolactin secretion by lactotropes is normally suppressed by tonic **dopamine** secretion from the paraventricular and arcuate nuclei of the hypothalamus. Prolactin has a negative feedback loop to the hypothalamus to adjust the release of dopamine. Nursing and breast manipulation (see Figure 36.7) as well as estrogen, oxytocin, thyroid-releasing hormone, sleep, and stress all increase prolactin secretion. Somatostatin and growth hormone decrease prolactin secretion. These alterations in prolactin secretion occur either directly at the level of the lactotrope or via inhibition of the hypothalamic dopaminergic neurons.

B. Oxytocin

Oxytocin is a small peptide hormone produced in magnocellular neurons of the paraventricular and supraoptic nuclei of the hypothalamus and secreted from the posterior pituitary (see 7·VII·D·4). Once in the blood, oxytocin has a very short (3–5 min) half-life.

1. **Function:** Oxytocin has two primary functions in females. First, it stimulates contraction of myoepithelial cells in breast tissue. This allows for milk ejection during lactation (milk letdown). Next, it stimulates contraction of the uterine muscle during parturition (see 37·VI·B·3). Secondary functions in females include inducing maternal behavior (such as caring for an infant), stimulating prolactin release, and decreasing nociception. The effects of oxytocin on the breast, uterus, and central nervous system are mediated by cell membrane oxytocin receptors (GPCR superfamily) and work primarily through the *PLC*-induced second messenger system.

2. **Secretion:** Cervical stretch and breast suckling stimulate oxytocin secretion through neuroendocrine reflexes (see Figure 36.7). Strong or intense emotional stimuli, such as fear and pain, can decrease oxytocin levels in the blood. Other methods that control functional effects of oxytocin are up- or down-regulating oxytocin receptors. Estrogen dramatically increases protein synthesis of oxytocin receptors during pregnancy, which, in turn, potentiates oxytocin effects during pregnancy.

VIII. HYPOTHALAMIC–PITUITARY–TESTICULAR AXIS

The male gonads are the testes. The testes are under endocrine axis control in a manner similar to that of the female gonads.

A. Hypothalamus and pituitary gland

The hypothalamic areas involved in testicular control are identical to those that regulate ovaries. GnRH is also secreted into the hypophyseal portal system, binding to anterior pituitary gonadotropes. There, GnRH stimulates GnRH receptors to secrete the peptide hormones LH and FSH, as occurs in females. Gender differences are confined to the target organ (i.e., testis).

B. Testis

The testis contains Leydig cells that produce testosterone; blood vessels; and seminiferous tubules, which produce sperm and house Sertoli cells. The testicular endocrine functions reside in Leydig cells and Sertoli cells, which work cooperatively in a similar manner to theca and granulosa cells, to synthesize testosterone and, to a lesser extent, estradiol.

1. **Leydig cells:** Leydig cells transport cholesterol as do theca cells, and cholesterol provides the initial structure for steroid-hormone synthesis. LH binds to surface membrane LH receptors, which activate and produce steroid hormone synthetic enzymes such as *side-chain cleavage enzyme complex*. The end product of this pathway is testosterone. Testosterone diffuses out of the Leydig cell with a portion entering the circulation, and a portion migrating to nearby Sertoli cells (Figure 36.8).

2. **Sertoli cells:** Adjacent Sertoli cells form tight junctions to create a functional **blood–testis barrier**. This barrier is selectively permeable for substances like testosterone but inhibits passage of many other substances. Sertoli cells primarily express FSH cell surface receptors. FSH receptors are in the GPCR superfamily and primarily work through the *AC*-induced cAMP second messenger system to stimulate synthesis of enzymes such as *aromatase*, inhibins for FSH negative feedback, and various growth factors. Sertoli cells rely on Leydig cells for testosterone. *Aromatase* activation facilitates the conversion of testosterone into estradiol, which regulates much of the protein synthesis in both Sertoli and Leydig cells. Sertoli cells secrete **androgen-binding protein** (**ABP**) along with testosterone into the seminiferous tubular lumen.

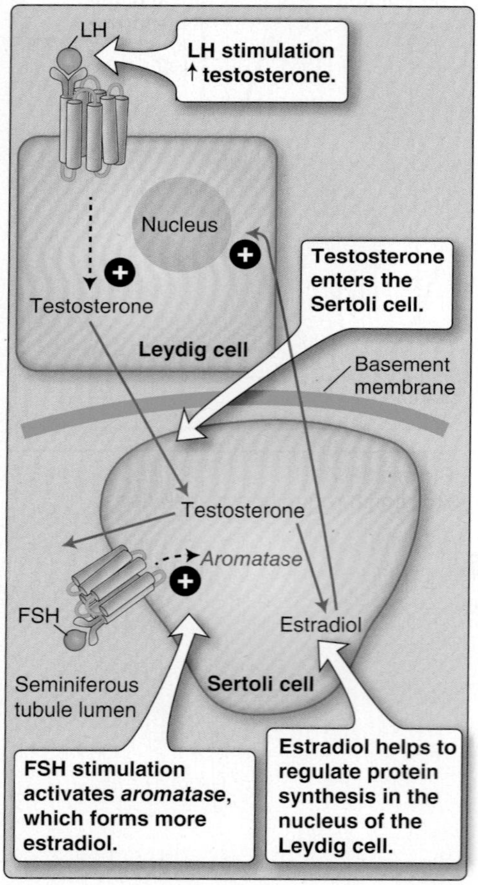

Figure 36.8
Leydig and Sertoli cells. FSH = follicle-stimulating hormone; LH = luteinizing hormone.

3. **Spermatogenesis:** The development of spermatogonia into primary spermatocytes; secondary spermocytes; spermatids; and, finally, into spermatozoa is a process called spermatogenesis (Figure 36.9). The final stage (spermiogenesis) consists of cell elongation, cytoplasmic removal (forming a second structure, a residual body), and organelle reorientation. Notably, in this stage, the flagellum is formed, which will allow for spermatozoan motility. Spermatogenesis is regulated by testosterone, and testosterone level is maintained by ABP.

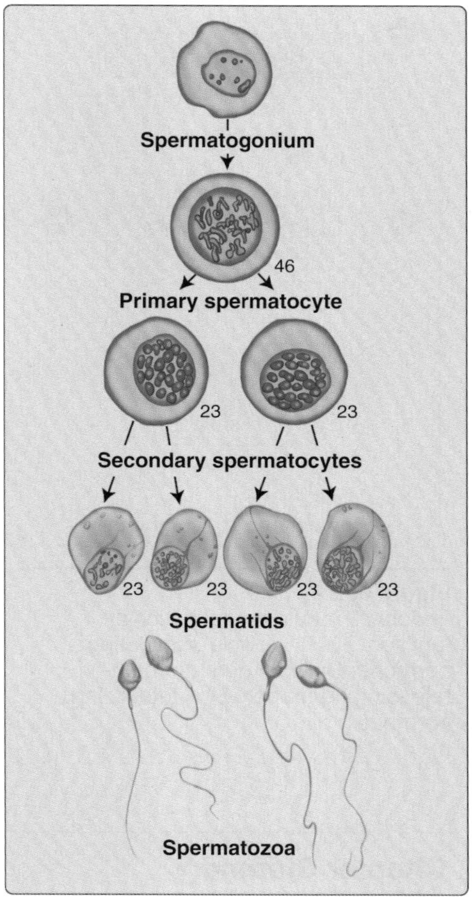

Figure 36.9
Spermatogenesis.

IX. TESTOSTERONE

Testosterone is produced from the conversion of androstenedione via *17β-hydroxysteroid dehydrogenase*. In addition to direct usage of testosterone, some target tissues convert testosterone into **dihydrotestosterone** (**DHT**) via *5α-reductase*. Similar to estrogens, testosterone has a high binding affinity to SSBG and, to a lesser extent, albumin. Approximately 2% circulates in the biologically active free form. The liver produces SSBG and also inactivates and processes testosterone. Testosterone and byproducts are secreted in the urine and feces.

Clinical Application 36.3: Male Hypogonadism

Male patients with primary gonadal failure typically present with an underdeveloped penis and scrotum; overdeveloped mammary glands (gynecomastia); infertility; low libido; and little, if any, facial hair because of an impaired ability to synthesize and secrete testosterone. Circulating luteinizing hormone and follicle-stimulating hormone levels are typically elevated, reflecting loss of negative feedback control by testosterone on the anterior pituitary gonadotropes. Treatment often involves androgen therapy.[1]

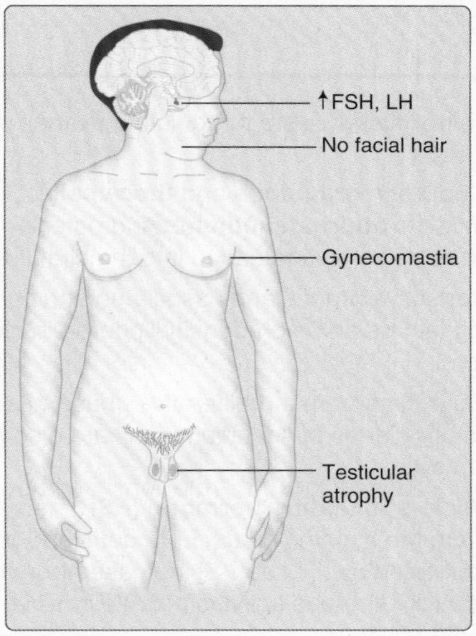

Hypogonadism. FSH = follicle-stimulating hormone; LH = luteinizing hormone.

[1]For additional information on hormone replacement therapy, see *LIR Pharmacology*, 5e, p. 319.

Table 36.3: Effect of Testosterone

Tissue	Effect
Bone	↑ Growth of bone and connective tissue
Muscle	↑ Growth of muscle and connective tissue
Reproductive organs	↑ Growth and development of testes, prostate, seminal vesicles, and penis ↑ Growth of facial, axillary, and pubic hair ↑ Growth of larynx ↑ Spermatogenesis
Skin	↑ Sebaceous gland size and secretions

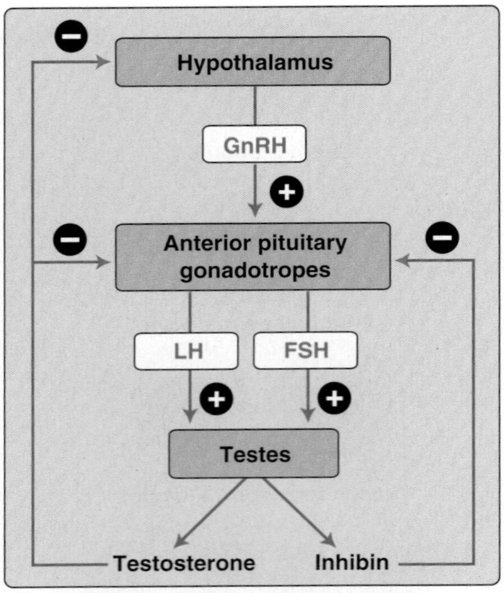

Figure 36.10
Feedback regulation of testicular function. FSH = follicle-stimulating hormone; GnRH= gonadotropin-releasing hormone; LH = luteinizing hormone.

A. Function

Testosterone and DHT have a number of androgen effects (for adrenal androgens, see 34·IV·A). Once testosterone binds the receptor, a homodimer is formed that translocates to the testosterone-response element. The functional effects are dependent on the target tissue (Table 36.3).

B. Secretion

The pathways regulating testosterone secretion are similar to those regulating the ovaries, with two exceptions: There is only one primary hormone (testosterone) instead of two (estrogens and progestins), and no role for activins has been functionally determined.

1. **Testosterone:** Testosterone secreted from Leydig cells exerts negative feedback on both the hypothalamus to decrease GnRH and the anterior pituitary to decrease both LH and FSH. There is even some evidence to suggest that there may be additional feedback to higher brain centers that can affect axis control. Testosterone also stimulates Sertoli cells to release inhibins.

2. **Inhibins:** Inhibin B provides negative feedback to the anterior pituitary in males, where it decreases secretion of FSH (Figure 36.10).

Chapter Summary

- The hypothalamic–pituitary–ovarian axis is the multitiered system regulating ovarian synthesis and release of **estrogens** and **progestins** from **granulosa cells**.
- Production of estrogens involves cooperation between **granulosa** and **theca cells**. Both are capable of producing progesterone, but only theca cells can process this into **androstenedione**. Androstenedione and testosterone migrate to granulosa cells, where they are converted into estradiol and released into the circulation.
- Estrogens stimulate growth of the female reproductive organs and associated structures. Bone growth is also stimulated by estrogens. Estrogens increase clotting factors and high-density lipoprotein cholesterol, while lowering glucose tolerance and low-density lipoprotein cholesterol.
- Progestins aid in the conversion of endometrial cycling from a proliferative phase to a secretory phase. During pregnancy, progestins are involved in preparing mammary glands for lactation. They are also responsible for body temperature elevation during the luteal phase of the ovarian cycle.
- In women, **luteinizing hormone** (**LH**) and **follicle-stimulating hormone** (**FSH**) secretion are stimulated by **gonadotropin-releasing hormone** release from the hypothalamus. Estrogens and progestins both provide negative feedback. Estrogens and progestins are activated by LH and FSH released from the anterior pituitary. Estrogens, progestins, and inhibins provide negative feedback, whereas activins provide positive feedback at the level of the anterior pituitary.
- The hypothalamic–pituitary–testicular axis is the multitiered system regulating synthesis and release of **testosterone** from testicular **Leydig cells**. Some peripheral cells convert testosterone to **dihydrotestosterone** to mediate androgen effects.
- Testosterone and dihydrotestosterone cause embryonic development of the male genitals and accessory organs, growth of genitals and hair, spermatogenesis, and anabolic effects on the musculoskeletal system.
- In men, luteinizing hormone (LH) and follicle-stimulating hormone (FSH) are activated by gonadotropin-releasing hormone release from the hypothalamus to which testosterone provides negative feedback to the hypothalamus. Testosterone and inhibin B provide negative feedback at the level of the anterior pituitary.

Study Questions

Choose the ONE best answer.

VIII.1 A 22-year-old woman is participating in a drug study affecting pancreatic hormones. If the test drug greatly elevates glucagon levels while having no effect on insulin release, which of the following processes are most likely to increase?

A. Lipolysis in adipocytes
B. Glycolysis in skeletal and cardiac muscle
C. Gluconeogenesis in neurons
D. Glucose uptake in hepatocytes
E. Ketone uptake in neurons

Best answer = A. Glucagon's principal function is to mobilize energy substrates and release them to the circulation for use by cells. Its actions include stimulating *hormone-sensitive lipase* in adipocytes (33·III·A). The *lipase* breaks down triglycerides into free fatty acids and glycerol (lipolysis), which are then released into the circulation. Glycolysis (glucose breakdown) is inhibited by glucagon. Glucagon increases gluconeogenesis (glucose synthesis) and ketone release in hepatocytes, but not neurons. Glucagon simulates glucose release from hepatocytes, not uptake.

VIII.2 A 14-year-old boy with an autoimmune disease that destroyed his pancreatic β cells is most likely to exhibit which of the following signs and symptoms?

A. Hyperglycemia and diuresis
B. Hyperkalemia
C. Enhanced protein storage in muscle
D. Decreased circulating fatty acid levels
E. Enhanced glucose uptake by adipocytes

Best answer = A. Selective loss of pancreatic β cells results in type 1 diabetes mellitus (Clinical Application 33.1). Symptoms include high blood glucose (hyperglycemia) that can spill over into the urine and cause an increase in urinary water loss (diuresis). Insulin does normally help regulate K^+ balance through its effects on the Na^+-K^+ ATPase, but hyperkalemia is associated with kidney disease, not pancreatic disease. Insulin's actions also normally include stimulating glucose uptake by adipocytes and protein storage in muscle, so both actions would be decreased in type 1 diabetes, not increased. Patients with type 1 diabetes have increased levels of circulating fatty acids and triglycerides.

VIII.3 Blood tests on a 34-year-old man identified high levels of circulating adrenocorticotropic hormone (ACTH). Levels of which of the following adrenal cortex hormones would be least likely to be affected by high ACTH?

A. Androstenedione
B. Dehydroepiandrosterone sulfate
C. Cortisol
D. Corticosterone
E. Aldosterone

Best answer = E. Adrenocorticotropic hormone (ACTH) is released by the anterior pituitary and targets the adrenal cortex. Its principal actions are to regulate corticosteroid production and release (34·II·C). The primary regulators of aldosterone release are angiotensin II and low plasma levels of K^+, whereas ACTH has only minimal effects. In contrast, adrenal androgens (androstenedione and dehydroepiandrosterone sulfate; 34·IV·B) and glucocorticoids (cortisol and corticosterone; 34·V·B) are all directly controlled by ACTH. ACTH stimulates *side-chain cleaving enzyme complex*, which is one of the key rate-limiting steps for adrenal cortex hormone production.

VIII.4 A 32-year-old male with suspected adrenocortical insufficiency is being treated with a synthetic cortisol (hydrocortisone). High doses improve his symptoms. If this dosing regimen is continued for a prolonged period of time, what is the most likely result?

A. Muscle weakness
B. Bone deposition and collagen formation
C. Virilization
D. β-Adrenergic receptor desensitization
E. Adrenal gland hypertrophy

Best answer = A. Corticosteroids normally prepare the body for stress by mobilizing substrates, such as glucose and free fatty acids (34·V·A). This is accomplished in part through skeletal muscle protein catabolism, so prolonged cortisol administration can cause muscle weakness. Cortisol stimulates Ca^{2+} resorption from bone, not deposition. Glucocorticoids can stimulate mineralocorticoid receptors at high levels, but the androgen receptors involved in virilization are relatively insensitive. β-Adrenergic receptor desensitization occurs in response to chronically high catecholamine levels (34·VI·C), not glucocorticoids. Cortisol suppresses adrenal gland growth due to feedback inhibition of adrenocorticotropic hormone.

VIII.5 A 45-year-old female suffers from symptoms associated with hypothyroidism. A blood sample reveals that thyroid-stimulating hormone levels are above normal. Which of the below statements best describes its action on thyroid follicular cells?

A. Inhibits pendrin insertion
B. Inhibits growth
C. Increases iodide uptake
D. Increases blood flow
E. Increases thyroxine-binding globulin synthesis

Best answer = C. Thyroid follicular cells are epithelial cells specialized for thyroid hormone synthesis and release (35·II·C). Thyroid-stimulating hormone (TSH) regulates many steps in the synthetic pathway, including iodide uptake by Na^+-I^- symporters ([NIS] or cotransporters). TSH upregulates NIS. Pendrin is an apical Cl^--I^- cotransporter required to move I^- across the apical membrane and into the follicular lumen, and it is not regulated by TSH. TSH stimulates thyroid tissue growth, rather than inhibiting it. Neither the acute control of blood flow nor liver production of thyroxine-binding globulin is directly regulated by TSH.

VIII.6 Triiodothyronine (T_3) and thyroxine (T_4) have a multitude of peripheral effects. In what form are T_3 and T_4 most biologically active?

A. Bound to albumin
B. Bound to transthyretin
C. Bound to thyroglobulin
D. Bound to thyroxine-binding globulin
E. Unbound

Best answer = E. Blood binding proteins, such as albumin and thyroxine-binding globulin, are important in maintaining a circulating "pool" of triiodothyronine (T_3) and thyroxine ([T_4] 35·III·B). However, while bound, hormones such as T_3 and T_4 are not biologically active. Only free hormones can exert peripheral effects, and this is one reason why both the free and bound states of thyroid hormones are measured in the blood during a thyroid panel. Thyroglobulin is a protein involved in thyroid hormone synthesis by the thyroid gland. Transthyretin (also known as prealbumin) is an albumin precursor protein that binds T_3 and T_4 in the circulation.

VIII.7 A 20-year-old woman who was administered a gonadotrope-stimulating drug responded with an increase in plasma luteinizing hormone levels but follicle stimulating hormone levels remained low. Levels of which of the following hormones would also be expected to remain unaffected by such a drug?

A. Estradiol
B. Progesterone
C. Androstenedione
D. Testosterone
E. Dehydroepiandrosterone

Best answer = A. Gonadotropes, which are located in the anterior pituitary, normally respond to a stimulating hormone by releasing both follicle-stimulating hormone (FSH) and luteinizing hormone ([LH] 36·II·B). FSH stimulates *aromatase* within granulosa cells, which converts androgens from surrounding theca cells into estrogens, including estradiol (36·II·C). If FSH release does not occur, then estradiol levels would remain low also. LH stimulates *side-chain cleaving enzyme complex* activity and increases production of progesterone, androstenedione, and testosterone. Dehydroepiandrosterone is primarily an adrenal androgen that is not directly stimulated by LH or FSH, although small increases could occur in the gonads.

VIII.8 A 16-year-old suffers from a *5α-reductase* deficiency. This individual was raised as a girl, but, at puberty, male secondary sex characteristics emerged, and male genital growth occurred. Levels of which of the following steroids are most likely to be reduced as a result of this deficiency until overwhelmed during puberty?

A. Estradiol
B. Estrone
C. Progesterone
D. Androstenedione
E. Dihydrotestosterone

Best answer = E. *5α-Reductase* is an enzyme normally found in several tissues that converts testosterone to dihydrotestosterone ([DHT] 36·IX). A *5α-reductase* deficiency would, thus, result in reduced DHT levels. DHT mediates many androgen effects, so individuals with a *5α-reductase* deficiency do not express many male secondary sex characteristics until puberty, when testosterone levels dramatically increase. Androstenedione is a substrate for *17-ketosteroid reductase*, a key enzyme involved in testosterone synthesis (34·IV·A). This reaction, however, is located upstream of *5α-reductase*, which means that androstenedione levels would not be actively reduced. Estradiol, estrone, and progesterone are all gonadal hormones, but none are directly associated with male secondary sex characteristics.

UNIT IX
Living and Dying

Pregnancy and Birth

37

I. OVERVIEW

Pregnancy and birth are exceptional phenomena that place extreme demands on both mother and fetus. Although the likelihood of success seems improbable once the complexity of the underlying physiology is appreciated, the current global population of ~7 billion shows it to be a highly reliable way of perpetuating the species. A successful pregnancy requires that several challenges be met. After fertilization, the developing embryo must implant in the uterine endometrium. The placenta must assume hormonal control of uterine growth to create an environment that allows the fetus to develop undisturbed for the next several months. An interface must be established between maternal and fetal circulations that allows for exchange of nutrients and waste products. The mother's body must adapt to meet the needs of the growing fetus. Finally, at term, the link between mother and fetus must be broken in a manner that allows both individuals to survive and thrive. Pregnancy begins with fertilization of the ovum and ends at **parturition** (childbirth). Pinpointing the moment of fertilization is typically difficult, so the progression of pregnancy is usually measured with reference to the first day of a woman's last menstrual period. By this measure, pregnancy lasts approximately 40 weeks.

II. IMPLANTATION

Fertilization typically occurs within the **fallopian tube** ampulla (Figure 37.1). The fallopian tube is lined with motile cilia that sweep the newly fertilized ovum downward toward the uterine cavity. The embryo remains free in the mother's reproductive tract for 6 or 7 days, during which time it undergoes a series of rapid divisions to form a **blastocyst** with a fluid-filled cavity (**blastocoele**) at its center. A thin layer of **trophoblast** cells around the central cavity's outer edge ultimately becomes the **placenta** and the membranes that enclose and protect the developing embryo. The embryo develops from an inner cell mass. By the time the blastocyst is ready to attach to and invade the uterine wall, the endometrium (decidua) has been readied for implantation (**decidualization**) under the influence

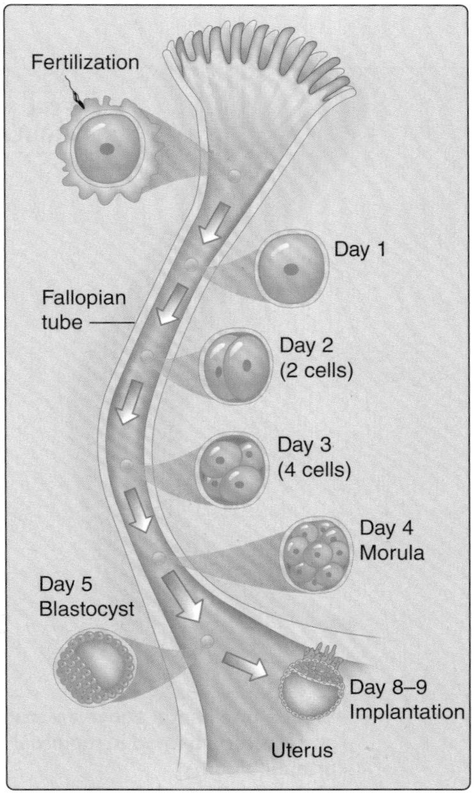

Figure 37.1
Embryo development and implantation.

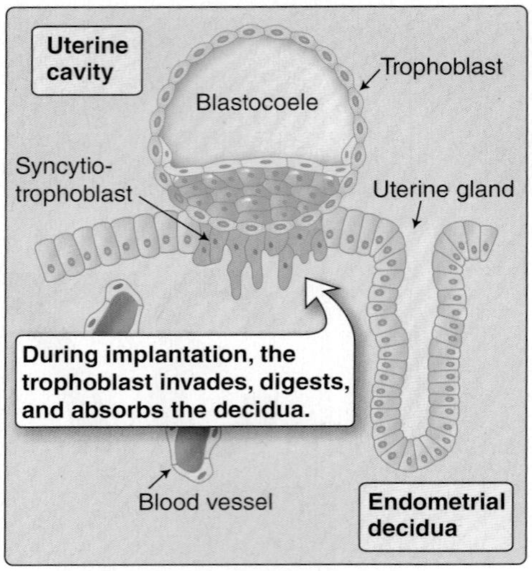

Figure 37.2
Implantation.

of progesterone from the **corpus luteum**. During implantation, the trophoblast layer enzymatically digests and invades the maternal uterine endometrium (Figure 37.2). **Decidual cells** within the endometrium sustain the embryo with glycogen and other nutrients until the placenta is formed and functional. Erosion of capillaries by the invading trophoblast allows blood to escape the maternal vasculature. Small pools (**lacunae**) ultimately coalesce to form a lake of maternal blood that fills the space between maternal and fetal placenta.

III. PLACENTA

The placenta is a disk-shaped organ that represents an interface between a fetus and its mother (Figure 37.3). The placenta has three important functions. First, it anchors the fetus to the uterus. Second, it brings fetal and maternal blood into close apposition to facilitate exchange of materials between the two circulations. Third, it is an endocrine organ that manipulates maternal reproductive physiology in order to sustain the pregnancy.

A. Structure

The placenta comprises a **fetal placenta** and a **maternal placenta**.

1. **Fetal:** The fetal placenta is attached to the fetus by the **umbilical cord**, a ropelike, muscular tether containing two **umbilical arteries** and an umbilical vein (see Figure 37.3). O_2-poor blood

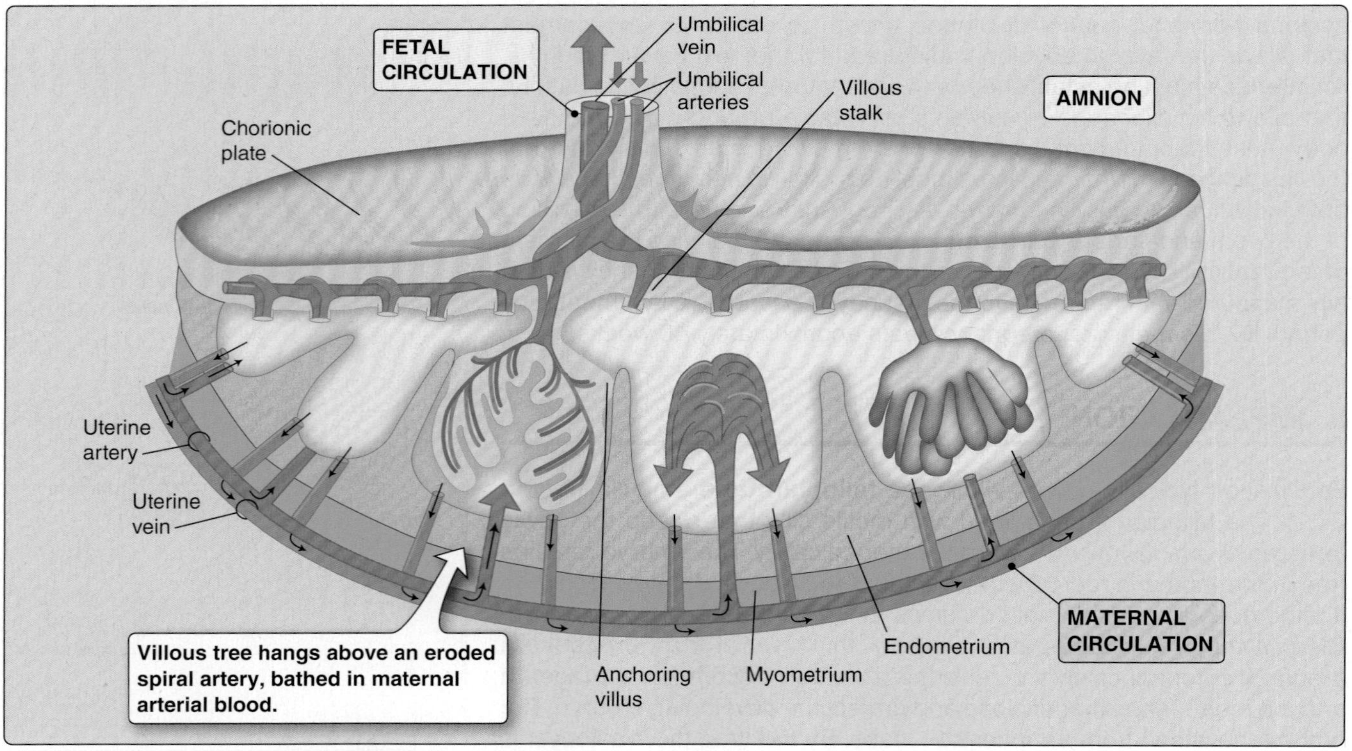

Figure 37.3
Placental anatomy and blood flow.

is carried from the fetus to the placenta by umbilical arteries. They penetrate the placental **chorionic plate** and then branch and distribute arterial blood to 60–70 **villous trees** that are assembled in groups called **cotyledons** (15–20 per placenta). The villous trees are branching structures covered with **chorionic villi**. Villi contain fetoplacental capillaries and are the primary site of exchange between maternal and fetal circulations. O_2- and nutrient-rich blood is carried from the villi to the fetus by the umbilical vein. Some villi are structural. They are physically attached to the maternal placenta and serve as placental anchors.

2. **Maternal:** The maternal **placental bed** (the area below the fetal placenta) resembles an egg carton. The endometrium is hollowed out to form an array of blood-filled sinuses in which the villous trees hang (see Figure 37.3). The intervillous space between the fetal placenta and endometrium is filled with ~500 mL of maternal blood. Blood enters this space via **uteroplacental** vessels, which are the remnants of spiral arteries that have been eroded by the fetal trophoblast layer during implantation and placental development. Blood drains from the space via uterine veins located in the floor of the maternal placental bed.

> Although fetal and maternal circulations are brought into close proximity with each other to facilitate exchange, the vascular contents do not mix appreciably under normal circumstances.

B. Exchange

Everything the fetus needs to develop and grow must cross the fetoplacental barrier separating maternal and fetal circulations. Most materials cross by simple or facilitated diffusion, driven by concentration gradients. Small amounts of material may cross by pinocytosis. Three features of placental design optimize transfer: the minimal nature of the barrier, its large surface area, and positioning of villous trees above maternal blood vessels (Figure 37.4).

1. **Barrier:** Blood flowing through the maternal placenta is outside the usual confines of the maternal vasculature. In practice, this means that the barrier between maternal and fetal blood comprises a single endothelial cell layer (fetal capillary wall), a thin layer of connective tissue (basal lamina), and a thin layer of **syncytiotrophoblast**. By the end of gestation ("**term**"), this barrier has thinned to <5 μm and represents a minimal barrier to diffusion.

2. **Surface area:** Fetal villi hang from the villous trees like bunches of bananas. The syncytiotrophoblast's apical (maternal) surface is densely packed with **microvilli**, which greatly amplifies the surface area available for diffusion. By term, total villous surface area amounts to ~10–12 m^2.

3. **Villi:** Villi develop directly above and around plumes of blood spewing from an eroded maternal artery (see Figure 37.3). In practice, this means that they are washed continually by arterial

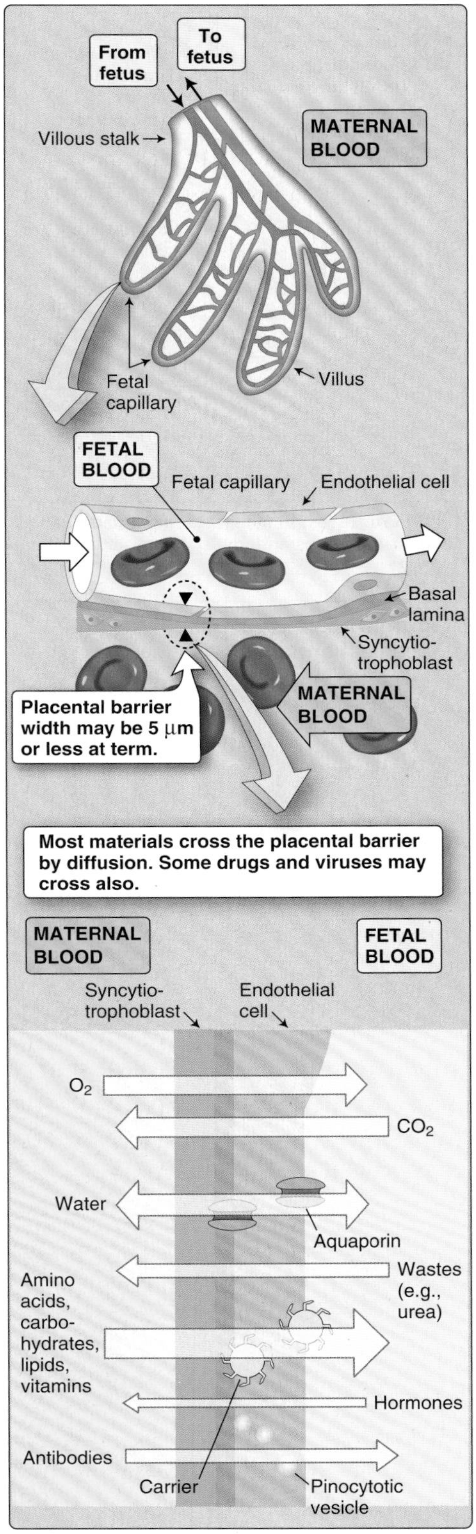

Figure 37.4
Exchange across the placental barrier.

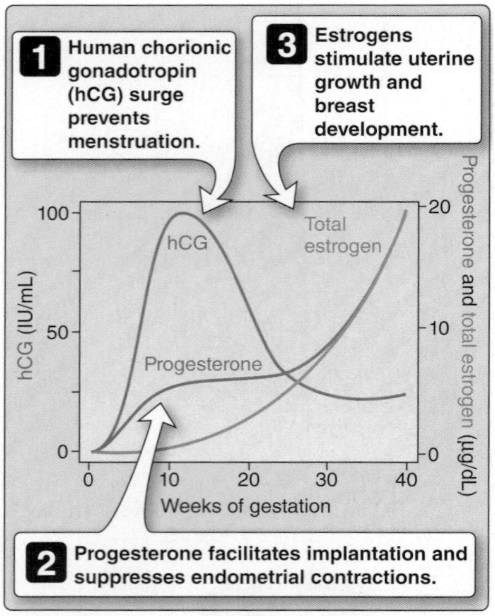

Figure 37.5
Placental hormones.

blood. Constancy of flow is important for maintaining the steep concentration gradients that drive diffusional exchange of nutrients and waste products between maternal and fetal blood.

C. Endocrine functions

A nonpregnant female's uterus sheds its lining (the outer endometrial layer) every 4 weeks and then begins the **menstrual cycle** anew, the cycle's timing being controlled by female reproductive hormones (see 36·VI·B). A successful pregnancy requires that the menstrual cycle be interrupted and the fetus left undisturbed for ~9 months. Cycle interruption is accomplished by the fetal placenta, which secretes several key hormones that manipulate maternal reproductive physiology, including **human chorionic gonadotropin** (**hCG**), **progesterone**, and **estrogens** (Figure 37.5).

1. **Human chorionic gonadotropin:** The syncytiotrophoblast (the fetoplacental precursor) begins secreting hCG within days of fertilization. hCG signals the corpus luteum that fertilization has occurred and impels it to sustain progesterone and estrogen production. These hormones prevent the uterus from shedding its lining and prepare it for implantation. The corpus luteum continues releasing progesterone and estrogen at ever-increasing levels in response to placental hCG, until the placenta takes over hormonal control around week 10 (see Figure 37.5).

> The rapid rise in hCG that follows fertilization provides the basis for home pregnancy tests, which detect the presence of hCG in maternal urine. Test strips typically have a detection threshold of 25–50 mIU/mL, levels that are not achieved until after implantation has occurred (up to 10 days after ovulation).

2. **Progesterone:** The placental syncytiotrophoblast produces large amounts of progesterone. The hormone initially helps prepare the endometrium for implantation. Progesterone also reduces myometrial excitability, thereby preventing contractions that might expel the developing embryo. Progesterone also stimulates breast development.

3. **Estrogens:** The placenta produces several estrogens, the main one being estriol. Estrogens stimulate growth and development of the mother's uterus and breasts. The placenta does not have all of the necessary substrates (e.g., cholesterol) and enzymes (e.g., *17α-hydroxylase*) to synthesize steroids (see 34·II·C), relying on both the fetus and its mother to provide pathway intermediates.

IV. MATERNAL PHYSIOLOGY

The fetus relies on its mother to supply it with O_2 and nutrients and to dispose of CO_2, heat, and other metabolic waste products. Meeting the

fetal demands taxes all maternal organ systems and puts the maternal cardiovascular system under considerable stress. By term, maternal cardiac output (CO) and circulating blood volume have risen by 40%–50% (2 L). Much of this capacity increase is needed to perfuse the maternal placenta, but flow to the skin, kidneys, liver, and gastrointestinal (GI) tract increases substantially also.

A. Uterine blood flow

A nonpregnant female's uterus receives <5% of total CO. The main source of uterine vascular resistance is the highly muscular **spiral arteries** (Figure 37.6), which are resistance vessels that contract and relax to modulate blood flow in response to changing uterine metabolic needs (i.e., **autoregulation**; see 20·II·B·1). The developing fetal placenta erodes and invades the spiral arteries (see Figure 37.6B). Arterial walls are remodeled, the smooth muscle layers being replaced with fibrous material to create wide tortuous vessels with very high flow rates. The functional advantages of this remodeling to the fetus are obvious. Blood now pulses directly from the uterine supply arteries at a pressure of >70 mm Hg and washes over the fetoplacental villi, bringing with it needed O_2 and nutrients (see Figure 37.6C). The cardiovascular consequences for the mother are profound, both in terms of flow increase and inability to control flow through these vessels.

1. **Flow:** Resistance vessels are flow regulators that limit the amount of blood that a tissue receives to its prevailing metabolic needs (see 20·II). Spiral artery erosion and widening allows blood to flow unhindered into the placental lake, and, hence, overall uterine blood flow increases dramatically during pregnancy. Flow rises in direct proportion to falling resistance, increasing from ~50 mL/min at 10 weeks' gestation to >500 mL/min at term.

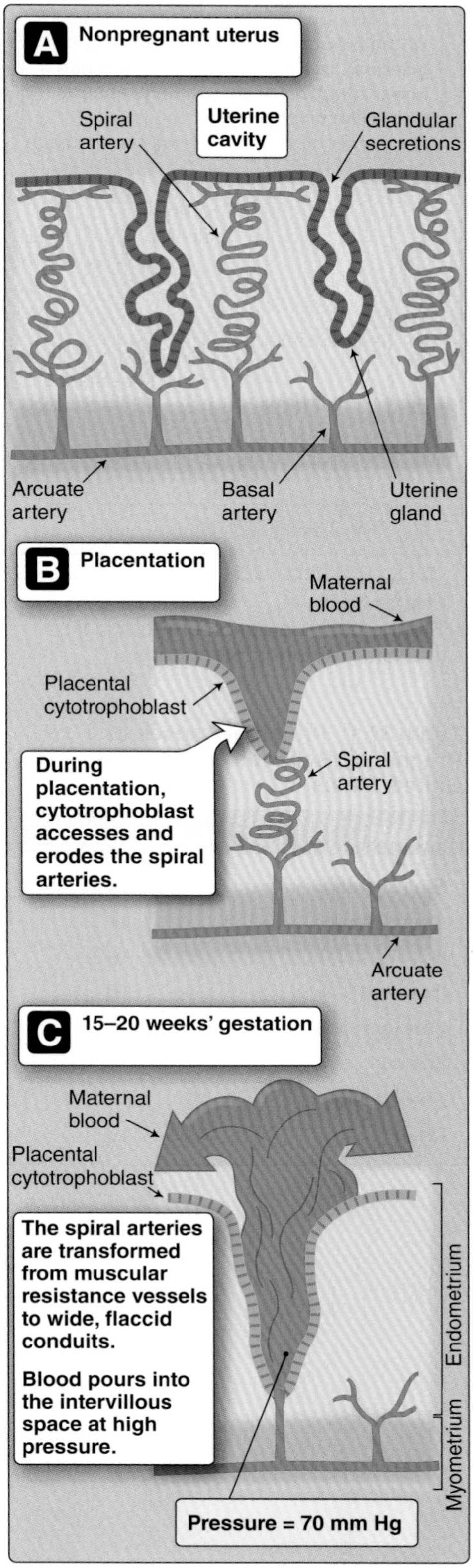

Figure 37.6
Erosion and invasion of spiral arteries during placentation.

Clinical Application 37.1: Preeclampsia

Preeclampsia is a syndrome characterized by hypertension (systolic blood pressure [SBP] of ≥140 mm Hg or diastolic blood pressure of ≥90 mm Hg) and proteinuria (≥0.3 g/24 hr) that develops after 20 weeks' gestation. Other symptoms may include severe headaches, visual disturbances, epigastric pain, and abnormal liver function. These symptoms all reflect a generalized endothelial dysfunction that causes increased vascular tone; increased vascular permeability; and a coagulopathy affecting all organs, including the brain, kidneys, liver, and placenta. Although the underlying molecular mechanisms are not yet known, preeclampsia is believed to result from incomplete remodeling of the spiral arteries during placental development. Maternal placental vessels are narrowed as a result, causing placental hypoperfusion and impaired nutrient delivery to the fetus. Hypoperfusion causes the placenta to release factors that inhibit angiogenesis and disrupt normal maternal endothelial function. **Severe preeclampsia** (SBP ≥160 mm Hg) carries a significant risk of maternal stroke and death. Immediate delivery regardless of gestational age is generally indicated.

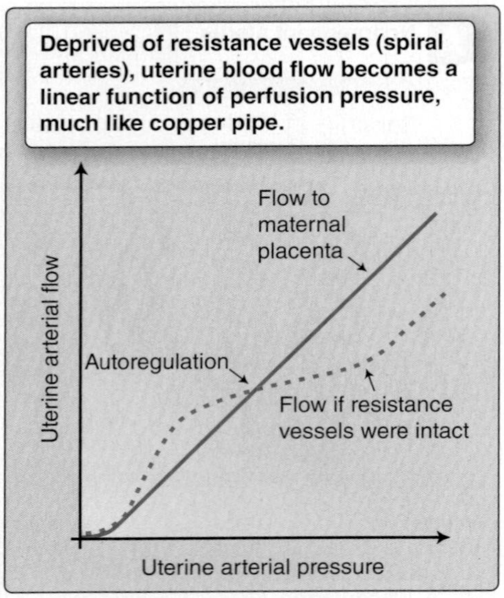
Deprived of resistance vessels (spiral arteries), uterine blood flow becomes a linear function of perfusion pressure, much like copper pipe.
Flow to maternal placenta
Autoregulation
Flow if resistance vessels were intact
Uterine arterial flow
Uterine arterial pressure

Figure 37.7
Uterine perfusion pressure–blood flow relationship during pregnancy.

Table 37.1: Principal Causes of Pregnancy-Related Deaths in the United States

Cause	% of Total
Embolism	20
Hemorrhage	17
Hypertension	16
Infection	13
Cardiomyopathy	8
Stroke	5
Anesthesia	2
Other	19

2. **Regulation:** Eradication of maternal resistance vessels maximizes flow to the placental site but simultaneously limits the uterine vascular control system's ability to regulate blood flow. Thus, flow becomes a direct function of uterine arterial pressure, as predicted by the hemodynamic equivalent of the Ohm law (Flow = Pressure ÷ Resistance) as shown in Figure 37.7 (also see 19·IV).

> Loss of uterine blood flow control mechanisms puts the mother in grave risk of massive blood loss in the event of premature placental detachment. Hemorrhage is a leading cause of pregnancy-related death in the United States (Table 37.1).

B. Hemodynamic profile

The uterus is a systemic vascular bed, so when uterine vascular resistance falls, systemic vascular resistance (SVR) falls along with it (Figure 37.8). A fall in SVR causes CO to rise to maintain mean arterial pressure (MAP): MAP = CO × SVR (see 18·III).

1. **Systemic vascular resistance:** SVR falls steadily over the first 20 weeks of gestation. The continuing erosion of maternal resistance vessels by the fetal placenta is the primary cause, but the growing need to dissipate heat and eliminate fetal waste products causes vascular resistance to fall in the cutaneous and renal vascular beds also.

2. **Cardiac output:** The growing need for increased CO is effected through increases in stroke volume (SV) and heart rate (HR). HR rises slowly during pregnancy, averaging 15–20 beats/min higher compared with nonpregnant values by 32 weeks. SV begins to rise very early in pregnancy, mediated by an increase in preload and contractility.

   a. **Preload:** The body responds to a sustained or repeated need for increased CO by increasing circulating blood volume through Na^+ and water retention. Placental hormones potentiate this effect by stimulating thirst and activating the *renin*–angiotensin–aldosterone system ([RAAS] see 20·IV).

   b. **Contractility:** Sustained increases in CO also stimulate ventricular hypertrophy. The heart enlarges to accommodate increased end-diastolic volumes (preload), and the ventricular wall thickens to increase contractility.

3. **Mean arterial pressure:** MAP must be maintained at prepregnancy levels to ensure adequate perfusion of all vascular beds, but the introduction of a low-resistance pathway into the maternal vascular circuit (i.e., the placenta) means that blood escapes the arterial system more easily during diastole (increased **diastolic runoff**; see 19·V·C·2) compared with the nonpregnant state. Thus, diastolic blood pressure falls during pregnancy, and pulse pressure widens.

C. Physiologic anemia

Increased Na^+ and water retention during pregnancy causes maternal plasma volume to increase by 40%–50%. Red blood cell (RBC) production does not keep pace with the rapid expansion of blood volume, increasing by only 25%–35%. The gap between volume expansion and RBC production results in a **physiologic anemia of pregnancy** (Figure 37.9). Although anemia reduces total O_2-carrying capacity, there are clear physiologic benefits because it reduces blood viscosity, which, in turn, reduces shear stress. It can also cause benign murmurs.

1. **Shear stress:** Blood has to move through the maternal arteries and veins at high velocity to support the sustained increases in CO that accompany pregnancy. High-velocity flow increases **shear stress** on the vascular lining, to the point where it could become damaging. Shear stress is proportional to both blood velocity and viscosity (**Reynolds equation**; see 19·V·A). Because hematocrit is the primary determinant of blood viscosity, anemia reduces stress levels and lessens the risk of vascular endothelial damage.

2. **Murmurs:** One benign consequence of decreased blood viscosity is an increased tendency for turbulent blood flow. The Reynolds equation predicts that turbulence is most likely to occur in regions of the cardiovascular system where flow velocities are highest. In practice, this means that mothers often develop **functional** (i.e., innocent) **murmurs** associated with blood ejection through the aortic and pulmonary valves. Mothers can also develop a **venous hum**, a sound associated with high-velocity, turbulent blood flow through the larger veins.

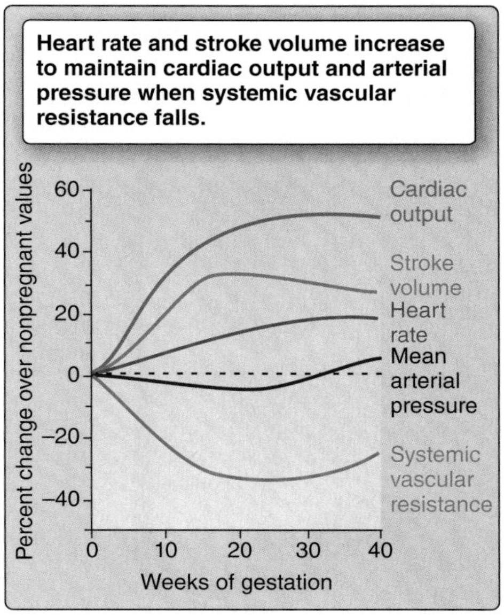

Figure 37.8
Changes in maternal hemodynamic profile during pregnancy.

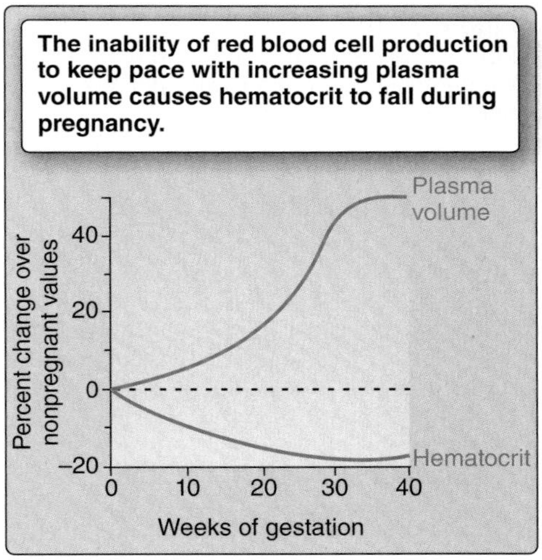

Figure 37.9
Physiologic anemia of pregnancy.

D. Edema

The combined weight of the uterus and its contents (fetus, placenta, and amniotic fluid = ~8–10 kg total at term) compresses and retards flow through the inferior vena cava and other smaller veins returning blood from the lower extremities. Compression causes venous pressures in the lower extremities to rise, which increases mean capillary pressure and increases net fluid filtration from blood to the interstitium (see 19·VII·D). The result is edema, and swelling of the feet (**pedal edema**) and ankles is common in pregnant women. The tendency for edema formation is increased by a fall in **colloid osmotic pressure** (i.e., plasma protein concentrations) by 30%–40% during pregnancy (from ~25 mm Hg prior to pregnancy to ~15 mm Hg postpartum).

E. Respiratory system

The O_2 demands of the mother and growing fetus increase rapidly during pregnancy; O_2 consumption at term is increased ~30% over nonpregnant values. These increased needs are met by a progressive increase in minute ventilation to ~50% over nonpregnant values during the second trimester. The ventilation increase is effected largely by an increase in tidal volume and only a small rise in respiratory rate (2–3 breaths/min). The net effect is that P_aO_2 rises by ~10 mm Hg, and P_aCO_2 falls by ~8 mm Hg, causing a slight respiratory alkalosis

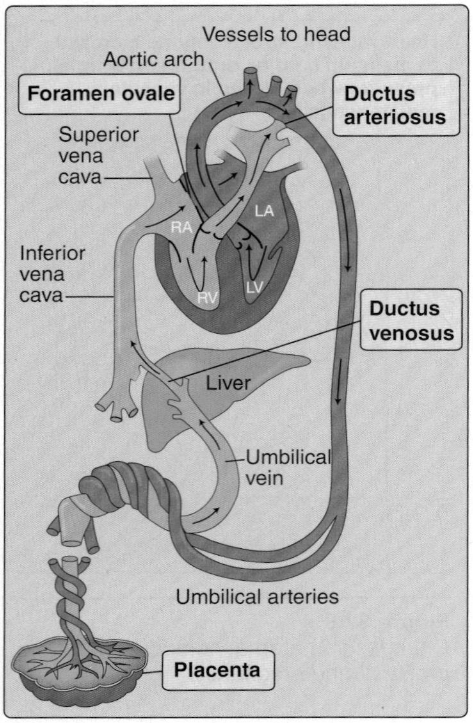

Figure 37.10
Fetal circulation. LA = left atrium; LV = left ventricle; RA = right atrium; RV = right ventricle.

(<0.1 pH units). Other significant respiratory changes include a 20% decrease in functional reserve capacity, expiratory reserve capacity, and residual volume (see 22·IX·A) caused by a rise in the diaphragm, which may limit the mother's ability to compensate for increased O_2 demand during exercise, for example.

F. Renal

Glomerular filtration rate rises steadily to ~50% above normal values at 16 weeks' gestation and remains elevated until parturition. The increase reflects a mother's need to excrete fetal wastes, including urea and nonvolatile acid.

V. FETAL PHYSIOLOGY

Because a fetus receives everything it needs for successful development from the maternal circulation via the placenta, few fetal organ systems are *required* to support normal growth, although most gain some degree of functionality before birth. The principal exception is the cardiovascular system, which becomes functional very early in pregnancy.

A. Vasculature

During initial development, the embryo relies on simple diffusion to obtain nutrients from fallopian and other maternal secretions. Once the embryo attains a size that exceeds the ability of O_2 and other nutrients to reach the innermost cells by diffusion alone, a functional cardiovascular system is required to sustain further growth. A rudimentary single-chambered heart begins pumping blood resembling interstitial fluid during the fourth week after conception. In the adult circulation, the path that blood follows is dictated by a need to pick up O_2 from the lungs and nutrients from the GI tract. The placenta provides all of the fetus' nutritional requirements, and, thus, the vascular circuitry is modified accordingly. There are four adaptations to the adult vascular circuit in the fetus: the **placenta**, **ductus venosus**, **foramen ovale**, and **ductus arteriosus** (Figure 37.10).

1. **Placenta:** The fetal placenta functions as fetal lungs, kidneys, GI tract, and liver and, thus, forms a major low-resistance circuit that receives ~40% of fetal CO (Table 37.2).

2. **Ductus venosus:** Blood coursing from the fetal placenta is shunted past the liver by the ductus venosus. In the adult, the liver filters and processes nutrient-rich blood from the GI system. In the fetus, the GI tract is largely nonfunctional, and, therefore, both GI organs are bypassed. Both receive sufficient blood to meet their nutritional needs via lesser vascular circuits.

3. **Foramen ovale:** Blood entering the fetal right heart from the inferior vena cava is O_2 rich after passing through the placenta (80% saturation). The fetal lungs do not participate in gas exchange, so passage through the pulmonary circulation would serve no purpose. Pulmonary vasculature resistance (PVR) is high also, which makes the lungs difficult to perfuse (see below). Thus, O_2-rich blood is shunted past the lungs from the right atrium directly into the left atrium via the foramen ovale.

Table 37.2: Fetal and Adult Cardiac Output Distribution*

Organ System	Fetus	Adult
Lungs	6	100
Heart	5	5
Kidneys	2	20
Brain	20	20
Musculoskeletal	20	20
Splanchnic	7	30
Placenta	40	–

*All values are approximate and given as a percentage of total cardiac output.

4. **Ductus arteriosus:** Blood entering the right heart via the superior vena cava is O_2 poor (25% saturation; see below) after having traversed the fetal systemic vascular beds. It is pumped through the right heart and then through the ductus arteriosus to the descending aorta, thereby bypassing the lungs. The foramen ovale and ductus arteriosus together create a vascular circuit in which left and right hearts are arranged in parallel with each other.

B. Vascular resistance

In the adult circulation, SVR > PVR. The adult circulation is dominated by the left heart. In the fetal circulation, PVR > SVR. Fetal lungs are fluid filled, and the air spaces are collapsed. The pulmonary vasculature is tonically constricted as a response to low O_2 levels (**hypoxic vasoconstriction**; see 23·III·E), which makes the pulmonary circuit difficult to perfuse, and, thus, fetal PVR is high. By contrast, the fetal *systemic* circulation includes the placenta, which is a very low–resistance pathway for blood flow, and, therefore, fetal SVR is low.

C. Oxygen transfer

Maternal uterine blood has an O_2 saturation of ~80%–100%. Although the barrier separating fetal and maternal circulations is minimal, the placental route is a relatively inefficient means of gas exchange compared with the lungs, and fetal blood can only achieve P_{O_2} levels of 30–35 mm Hg at best (compare to a P_{aO_2} of 98–100 mm Hg in an adult pulmonary vein). Despite the inherent limitations of the transfer route, fetal blood carries similar amounts of O_2 as does an adult circulation. This is made possible by hemoglobin F (HbF), a fetal Hb isoform that has a leftward-shifted O_2 dissociation curve (see 23·VI·C·2). HbF's high O_2 affinity is well adapted to take up O_2 at partial pressures common to the maternal placenta, meaning that blood traveling from the placenta to the fetus in the umbilical veins typically has an O_2 saturation of 80%–90% (Figure 37.11). Fetal blood also contains ~20% more Hb than adult blood, which increases overall O_2-carrying capacity.

D. Oxygen distribution

Blood traveling from the placenta to the fetus via the umbilical vein is O_2 rich. It streams around the liver via the ductus venosus, but then encounters O_2-poor blood returning from the lower regions of the body in the inferior vena cava (see Figure 37.11). A filmy membrane ensures that little mixing occurs at the point where the two bloodstreams merge, and the O_2-rich stream is preserved all the way to the right atrium (streamline flow; see 19·V·A). Here, the two streams are cleaved by the interatrial septum (**crista dividens**). The O_2-rich portion passes preferentially into the left heart and then into the aorta. The first arteries to branch off the aorta feed the myocardium and the brain, so flow streamlining ensures that these two critical circulations receive highly oxygenated blood.

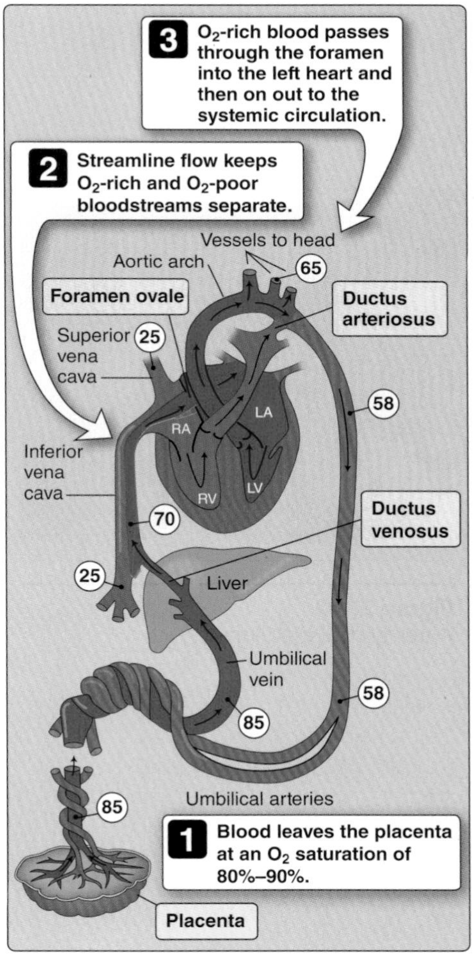

Figure 37.11
O_2 distribution by the fetal circulatory system. Circled numbers represent O_2 saturations. LA = left atrium; LV = left ventricle; RA = right atrium; RV = right ventricle.

E. Renal function

Fetal kidneys start producing urine within 9–10 weeks of conception. The ability to concentrate urine is gained around 4 weeks later, but the fetus remains dependent on the placenta for fluid and electrolyte balance throughout gestation. By 18 weeks, the kidneys are producing

over 10 mL of urine per hour, and fetal urine has become the primary source of amniotic fluid.

VI. PARTURITION

Human gestation lasts 40 weeks on average. At the end of this time, the fetus is forcibly expelled from the uterus, and the physical link with the mother is broken (**parturition**). The birthing process requires careful coordination if both mother and newborn are to survive.

A. Staging

Parturition can be divided into three stages of variable duration: **dilation**, **fetal expulsion**, and **placental**.

1. **Dilation:** Stage one begins with labor and ends when the cervix is dilated fully. The fetus is enclosed within the amniotic sac, but the main barrier preventing it from exiting the uterus is the cervix. During the first stage of parturition, the myometrium begins contracting rhythmically and with increasing intensity. Contraction begins in the uterine fundus and spreads caudally, which pushes the fetus against the cervix and causes the latter to thin and dilate. Stage one usually lasts for ~8–15 hours.

2. **Fetal expulsion:** The fetus is forcibly expelled from the uterus through the cervix and vaginal canal by frequent and intense waves of contraction. Stage two is complete within 45–100 min (Figure 37.12). The umbilical cord is traditionally clamped shortly after birth, although preterm infants may benefit from delayed clamping and milking blood from the cord toward the newborn to increase the infant's hematocrit.

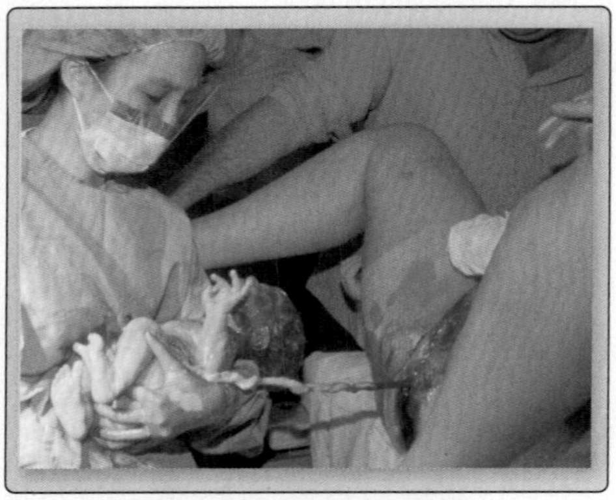

Figure 37.12
A newly delivered infant.

3. **Placental:** The uterus continues contracting after the fetus has been expelled, which causes it to shrink in size (**involution**). Shrinkage shears the placenta from the uterine wall. The placenta and associated membranes are later expelled as afterbirth. Stage three is usually completed within minutes of fetal expulsion.

B. Hormones

Irregular waves of weak uterine contraction occur throughout pregnancy (**Braxton Hicks contractions**). The reasons why contractions abruptly transition to the forceful contraction of parturition are not understood, although several hormones have been implicated. The fetus stretches the myometrium and increases its overall excitability as it grows and may be a contributing factor. The major hormones driving parturition include estrogen and progesterone, prostaglandins, oxytocin, and cortisol.

1. **Estrogen/progesterone ratio:** Progesterone suppresses uterine contraction during pregnancy. Estrogens promote excitability by increased expression of Na^+ channels, Ca^{2+} channels, and gap junctions between adjacent smooth muscle cells within the myometrium. The gap junctions allow developing waves of excitation to sweep through the uterine wall, manifesting as a wave of contraction. At parturition, the estrogen/progesterone ratio increases, and the uterus becomes excitable.

2. **Prostaglandins:** The uterus, placenta, and fetus all produce prostaglandins (PGE_2 and $PGF_{2}\alpha$), which stimulate uterine contractions. Increasing estrogen levels likewise increase prostaglandin production.

3. **Oxytocin:** Oxytocin is a powerful stimulant of uterine contractions. It is released from the posterior pituitary in response to cervical distension (see 36·VII·B), providing a positive feedback mechanism that couples fetal expulsion with the motive force required for expulsion.

4. **Cortisol:** The fetal hypothalamic–pituitary–adrenal axis is activated to release cortisol (see 34·V·B). Cortisol increases the estrogen/progesterone ratio.

C. Circulatory transition from fetal to adult

Parturition breaks the link between mother and fetus and forces the fetal vasculature to adopt the serial circulatory pattern that is common to the adult. Transition follows a rapid sequence of coincident events: an SVR increase; lung inflation; decrease in PVR; closure of the ductus arteriosus, foramen ovale, and ductus venosus; and, finally, a shift from right- to left-sided circulatory dominance (Figure 37.13).

1. **Systemic vascular resistance:** The umbilicus is a highly muscular structure that contracts spontaneously in response to the trauma of birth. Contraction occludes the umbilical arteries and vein and terminates flow to the placenta. Fetal SVR increases when this low-resistance pathway is removed from the systemic vascular circuit.

> Remnants of the umbilical arteries and vein can be observed in the adult as the **medial umbilical ligaments** and **ligamentum teres**, respectively.

2. **Lung inflation:** Compression and occlusion of umbilical vessels halts blood flow and deprives the fetus of O_2, causing asphyxia. This, together with the sudden cooling experienced by the infant at birth, stimulates respiratory control centers in the fetal brainstem, causing the neonate to gasp and take several breaths. Intra-alveolar pressure drops below atmospheric pressure, creating a pressure gradient that drives air inflow, and the lungs inflate.

3. **Pulmonary vascular resistance:** During development, PVR is high because the lungs are collapsed, and the pulmonary arteries are compressed and constricted in response to low P_{O_2}. The first breaths cause alveolar and pulmonary arterial P_{O_2} levels to rise dramatically, promoting vasodilation. Lung inflation also stretches the pulmonary vessels, thinning their walls and increasing their internal diameter. PVR drops dramatically as a result, and there is a coincident increase in pulmonary blood flow.

4. **Ductus arteriosus:** The fall in PVR and loss of flow from the umbilical vein causes right atrial pressure to fall. SVR simultaneously

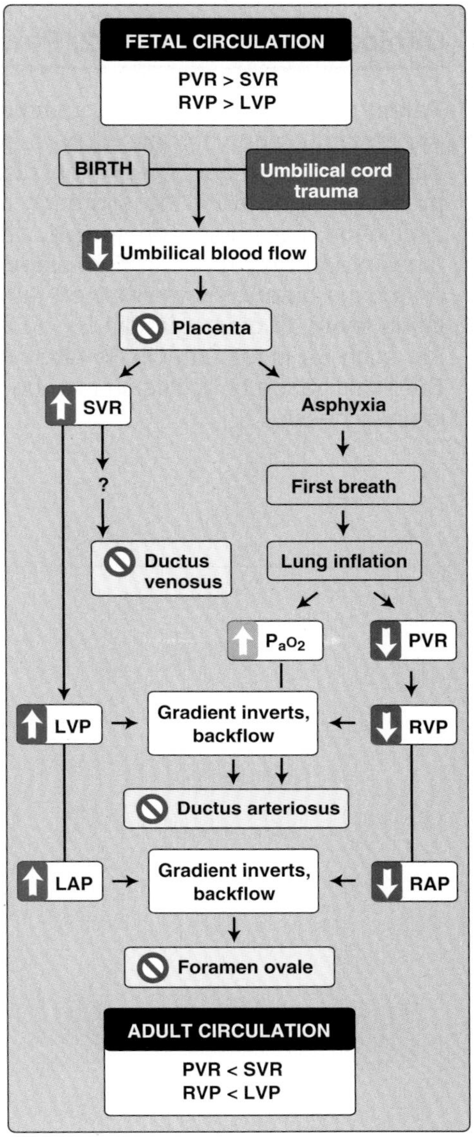

Figure 37.13
Changes in fetal vascular circuitry during parturition. LAP = left atrial pressure; LVP = left ventricular pressure; P_aO_2 = partial pressure of oxygen (arterial); PVR = pulmonary vascular resistance; RAP = right atrial pressure; RVP = right ventricular pressure; SVR = systemic vascular resistance.

Clinical Application 37.2: Patent Ductus Arteriosus

Patent ductus arteriosus is a common congenital heart defect, particularly in premature and very low–birth-weight infants among whom the incidence may be as high as 30%. If the ductus arteriosus remains open, high-pressure blood from the systemic circulation shunts into the pulmonary circulation. Depending on severity, the shunt can cause pulmonary hypertension and may result in right-sided heart failure if not addressed. Ductus arteriosus patency is maintained during development in part by high circulating levels of prostaglandin E_2, so administering a *cyclooxygenase* inhibitor, such as indomethacin, is often sufficient to prompt complete closure.[1] Surgical ligation or occlusion may be required if pharmaceutical intervention is unsuccessful.

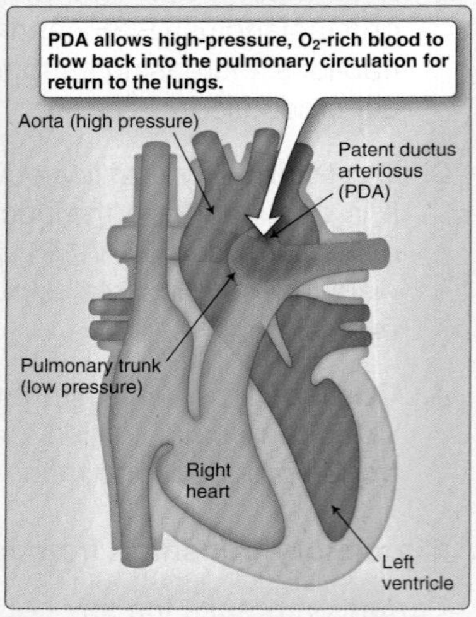
PDA allows high-pressure, O_2-rich blood to flow back into the pulmonary circulation for return to the lungs.
Aorta (high pressure)
Patent ductus arteriosus (PDA)
Pulmonary trunk (low pressure)
Right heart
Left ventricle

Patent ductus arteriosus.

increases due to loss of the placental circuit, so left ventricular and aortic pressure rises. The sudden inversion of the fetal right-to-left pressure gradient causes a reversal of blood flow in the ductus arteriosus, and it constricts, probably in response to rising P_aO_2 and falling circulating prostaglandin levels (see Clinical Application 37.2). Complete anatomic closure takes several months, and vestiges of the fetal shunt persist even in the adult as the **ligamentum arteriosum**.

5. **Foramen ovale:** The right–left blood pressure inversion pushes on a valvelike flap that then covers the foramen ovale. Gradually escalating left atrial pressures hold the flap closed to isolate the left and right sides of the heart. In time, the flap fuses with the interatrial septum to seal off the foramen permanently (seen as the **fossa ovale** in an adult heart).

Patent foramen ovale is a congenital heart lesion that affects 25%–30% of the general population. Although the pathway between right and left atria remains intact, left atrial pressure is usually higher than right atrial pressure, and, thus, the foramen remains occluded by the one-way valve. Healthy individuals usually remain asymptomatic.

6. **Ductus venosus:** The ductus venosus closes by a sphincter-like mechanism, persisting at the **ligamentum venosum** in the adult. The mechanism of closure is unknown.

[1]For more information on the actions and uses of *cyclooxygenase* inhibitors, see *LIR Pharmacology*, 5e, p. 525.

7. **Circulatory dominance:** Over ensuing weeks, the left ventricle slowly hypertrophies as a response to a rising SVR. Meanwhile, the right heart pumps against a lower PVR than it did during gestation, so its muscle mass slowly decreases relative to the left heart.

D. Maternal blood loss

A mother typically loses ~500 mL of blood from the placental site during a normal delivery. Although this represents a substantial hemorrhage, the mother has been well prepared for the loss by the massive expansion of blood volume that occurs during the first few weeks of pregnancy. Further loss is prevented by intense uterine contractions, which compress the uterine vasculature and allow hemostasis to occur. The contractions are stimulated by oxytocin during the third stage of parturition.

Chapter Summary

- Pregnancy begins with fertilization. The developing embryo divides rapidly over subsequent days to form a **blastocyst** and then **implants** in the maternal uterine wall. Implantation is effected by an outer **trophoblast** cell layer, which digests and invades the maternal **endometrium** and develops to create an interface between fetal and maternal circulations (the **placenta**).
- The placenta exchanges nutrients, hormones, and waste products between fetal and maternal circulations.
- The fetal placenta comprises 60–70 **villous trees** that serve to increase interface surface area. The maternal placenta comprises 15–20 blood-filled, eroded endometrial sinuses sculpted by fetal trophoblast during placentation. The space between fetal and maternal placenta is filled with ~500 mL of maternal blood. Blood flows in an unregulated manner at relatively high pressure (~70 mm Hg) from eroded **spiral arteries** and washes over the fetal villous trees.
- The placenta is also an endocrine organ that secretes **human chorionic gonadotropin**, **progesterone**, and **estrogens**.
- Supporting the needs of a developing fetus involves most of a mother's organ systems, including the **cardiovascular system**, **kidneys** (increased disposal of waste products), **lungs** (~30% increase in O_2 demand), **gastrointestinal tract**, **liver**, and **skin** (thermoregulation).
- Maternal **cardiac output** increases ~50% during gestation, accomplished through increases in **heart rate** and **stroke volume**. Stroke volume increases as a result of fluid retention and increased **circulating blood volume**.
- Blood volume increases faster than red blood cell production, causing a **physiologic anemia of pregnancy**. The resulting decrease in blood **viscosity** reduces **shear stress** on the heart and vasculature lining.
- The uterus and its contents gain significant weight during pregnancy, which, depending on posture, compresses and impairs blood flow from the mother's lower extremities. The result is **pedal edema**.
- The fetal circulation includes three **shunts** that allow blood from the umbilical vein to bypass the liver (**ductus venosus**) and lungs (**foramen ovale** and **ductus arteriosus**) for distribution to the developing organs.
- Fetal blood contains a **fetal hemoglobin** (**HbF**) isoform that has a high affinity for O_2. HbF helps compensate for the fact that the placenta is a less efficient route for O_2 transfer than the lungs and allows fetal blood to carry near-adult levels of O_2.
- **Parturition** is initiated and sustained by changing levels of hormones produced by both the mother and the fetal placenta. Uterine contractions expel the fetus and placenta then compress and collapse the uterine vasculature. Compression limits maternal blood loss during delivery.
- At birth, the neonate's **pulmonary vascular resistance** decreases, and **systemic vascular resistance** increases, thereby establishing a left side–dominated circulatory system common to an adult.

38 Thermal Stress and Fever

I. OVERVIEW

The ability to dissipate and retain heat, combined with the ability to adapt behaviorally to temperature extremes, has allowed humans to occupy most regions of the Earth's surface, including Plateau Station, Antarctica (average temperature = −55°C) and Dallol, Ethiopia (average temperature = 35°C). The body's internal temperature can rise to 39°–40°C without causing irreversible loss of cellular function, but body temperature is typically regulated within a much tighter range (36.5°–37.5°C) as shown in Figure 38.1. Control of internal temperature is one of the body's fundamental homeostatic functions. Internal temperatures are sensed through thermoreceptors located in the brain, spinal cord, and viscera. External temperatures are sensed through cutaneous thermoreceptors. The sympathetic nervous system (SNS), in conjunction with hypothalamic control centers, mediates thermoregulatory responses to environmental stressors. During heat stress (e.g., sitting in a Finnish sauna at 80°–90°C), there is a simultaneous increase in skin blood flow and sweat gland stimulation that mediates evaporative cooling. During cold stress (e.g., as experienced by an officer directing traffic on a cold, wet day), skin blood flow decreases to reduce cutaneous heat loss, while shivering generates heat. The hypothalamus can also actively increase internal temperature as a way of slowing infection by a pathogen. Fever is one of the oldest recognized symptoms of illness.

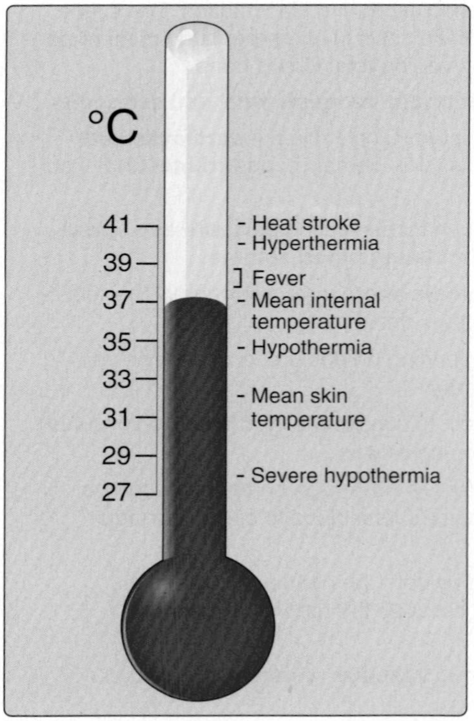

Figure 38.1
Body temperatures.

II. THERMOREGULATION

External and internal temperatures are sensed by thermoreceptors, which relay sensory information to control centers located in the hypothalamus. Thermoregulatory effector organs include the skin, brown adipose tissue, and skeletal musculature.

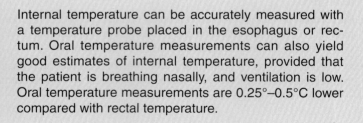

> Internal temperature can be accurately measured with a temperature probe placed in the esophagus or rectum. Oral temperature measurements can also yield good estimates of internal temperature, provided that the patient is breathing nasally, and ventilation is low. Oral temperature measurements are 0.25°–0.5°C lower compared with rectal temperature.

A. Sensors

The body possesses two diverse groups of thermosensors. Central thermoreceptors monitor internal body temperature, whereas skin thermoreceptors provide information about the external thermal environment.

1. **Central:** Thermoreceptors that monitor internal temperature are located in the hypothalamus, spinal cord, and viscera, but the sensors that have the greatest influence on thermoregulatory control center output are in the **preoptic area of the hypothalamus** (Figure 38.2). Warmth-sensitive preoptic neurons are tonically active at normal body temperature. A rise in internal temperature (as reflected by the temperature of blood bathing the preoptic area), increases their firing rate, whereas cooling decreases firing rate.

2. **Skin:** There are four primary skin thermoreceptor types. Two mediate nociceptive responses to painfully cold or hot stimuli and are discussed elsewhere (see 16·VII·B). The other two, comprising distinct populations of cold and warmth receptors, are involved in thermoregulation.

   a. **Cold: Cold receptors** mediate neutral, cool, and cold sensations (5°–45°C). Cold temperatures are believed to be sensed by TRPM8, one of the transient receptor-potential (TRP) channel family (see 2·VI·D) that mediates a depolarizing receptor potential when active. Afferent firing rate increases as a consequence.

   b. **Warmth:** Heat sensation involves **warmth receptors** that are activated from 30°–50°C. Heat reception also involves TRP channels (TRPV3 and TRPV4) that are active at neutral and warm temperatures (Figure 38.3).

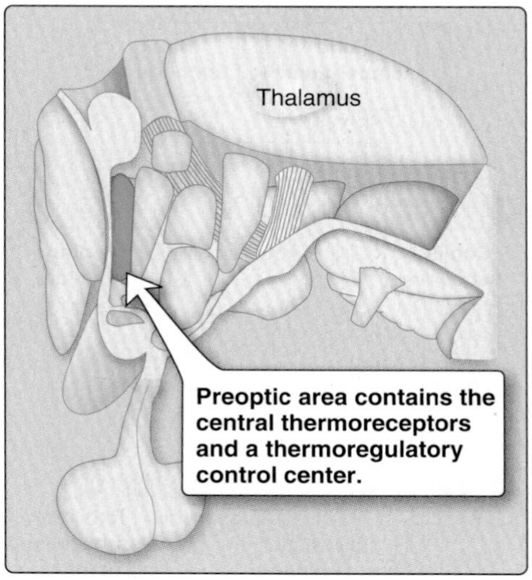

Figure 38.2
Preoptic area of the hypothalamus.

B. Control center

Body temperature is normally held at 37°C, with a circadian variance of 1°C (i.e., 36.5°–37.5°C). The temperature nadir occurs in the early morning, and the peak occurs in the late afternoon. The internal temperature that the body is trying to maintain is known as the **set point**. The preoptic area of the hypothalamus contains the thermoregulatory integration and control center. Cooling the preoptic area elicits heating responses and behaviors (e.g., putting on more clothing), whereas heating this area activates cooling responses and behaviors (e.g., seeking shade). If the preoptic area is damaged by ischemia (i.e., stroke), demyelination (e.g., multiple sclerosis), or ablation, internal temperature fluctuates over an exaggerated range, and responses to thermal stress are impaired. Control center output is governed primarily by central thermoreceptors, but the preoptic area integrates signals from many other areas also. These include skin thermoreceptors, the immune system (see Section IV below), and areas of the central nervous system that regulate other systemic variables, such as blood pressure, plasma glucose concentration, and plasma osmolality.

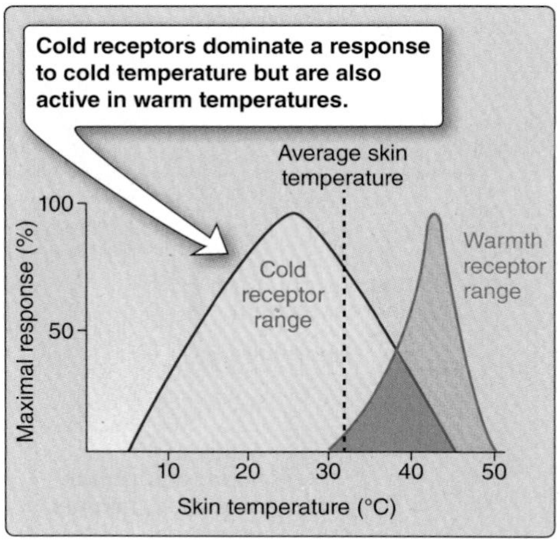

Figure 38.3
Skin thermoreceptor sensitivity.

C. Effector pathways

The hypothalamus effects most thermoregulatory responses via the SNS. Sympathetic signals travel via T1–L3 spinal nerves and synapse

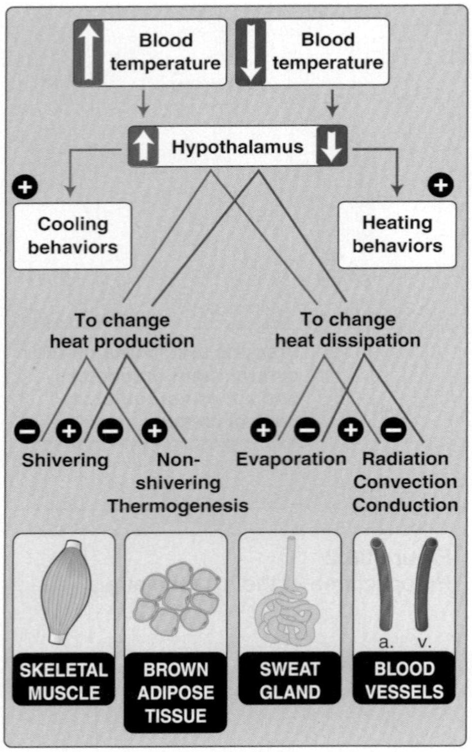

Figure 38.4
Thermoregulatory response mediators. a. = artery; v. = vein.

within sympathetic chain ganglia (see 7·III·B). Efferents project from the ganglia to skin blood vessels and to sweat glands.

D. Response

Cold stress initiates pathways that increase tissue insulation and increase metabolic rate via shivering and nonshivering thermogenesis (heat production). Conversely, heat stress reduces tissue insulation and initiates sweating. The primary thermoregulatory effectors are the skin blood vessels, sweat glands, skeletal muscles, and brown adipose tissue (Figure 38.4).

1. **Skin blood vessels:** Heat loss or gain is most effectively regulated at the level of the skin. To dissipate heat, blood flow is brought in close proximity to the body's surface, whereas to conserve heat, blood flow is shunted away from the body's surface. Glabrous skin contains deep arteriovenous anastomoses that allow blood to bypass surface capillary beds. Hairy (nonglabrous) skin does not have arteriovenous anastomoses but does have deep and superficial capillaries. The most efficient heat transfer with the environment occurs when blood is shunted through these surface capillaries. Postganglionic adrenergic nerves constrict cutaneous arteries, veins, and anastomoses, acting via α-adrenergic receptors. Removing the constrictor influence and then actively dilating vessels increases blood flow through the cutaneous vasculature (Figure 38.5). The vasodilation mechanism is less understood but involves nitric oxide and cholinergic sympathetic nerves. These vasomotor changes allow skin blood flow to change from <6 mL/min to 8 L/min. During heat stress, venous volume increases also to provide additional time for heat transfer. Heart rate and CO increase, and other vascular beds (e.g., renal and splanchnic), vasoconstrict to facilitate the skin blood flow increase.

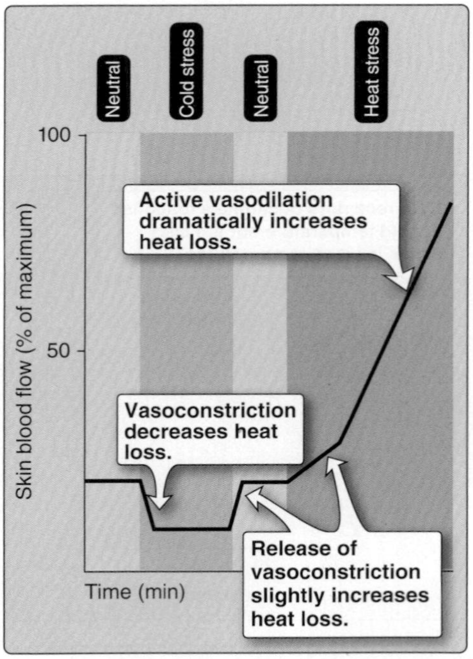

Figure 38.5
Skin blood flow during heat stress.

2. **Sweat glands:** There are three sweat gland types (see 16·VI·C), but only **eccrine sweat glands** produce sweat that mediates evaporative cooling during heat stress. Sweating is initiated by SNS cholinergic nerves, but the glands are stimulated by adrenergic compounds (e.g., epinephrine, norepinephrine) also. Sweating can dehydrate the body and cause a hypertonic volume contraction. Even small fluid loss (2% body weight) can decrease work performance and allow internal temperatures to rise during heat stress.

3. **Muscles:** Shivering is a rapid, cyclical contraction of skeletal muscles that liberates heat but produces minimal force. Muscle contractions always produce large amounts of heat because force production is only 20% efficient. The remaining 80% of expended energy is liberated as heat. Shivering muscles do not perform meaningful work, and, thus, almost all of the energy used is liberated as heat. Shivering is unique in that it is mediated by the somatic motor pathways rather than the SNS, but the response is initiated by the preoptic area.

4. **Nonshivering thermogenesis:** Nonshivering thermogenesis is a SNS-mediated increase in metabolic rate in muscle and other tissues designed to liberate heat. In brown fat, SNS stimulation activates an uncoupling protein (**thermogenin**) in the inner mitochondrial

membrane (Figure 38.6). Thermogenin is a pore-forming protein that allows H^+ to cross the inner mitochondrial membrane without generating adenosine triphosphate. Thus, oxidative phosphorylation becomes uncoupled. Infants rely on brown adipose tissue for heat production, but this pathway is less important in adults.

5. **Behavior:** Thermoregulatory behaviors can decrease or even eliminate thermal stress. These behaviors are driven by the preoptic area but can be overridden or modified by other areas of the brain. Conscious behavioral responses to cold stress involve increasing insulation (e.g., putting on a coat), increasing physical activity to increase metabolic rate, or seeking an external heat source. Heat stress–related behaviors include drinking fluids to facilitate sweating, removing clothing, seeking shade, or turning on a fan.

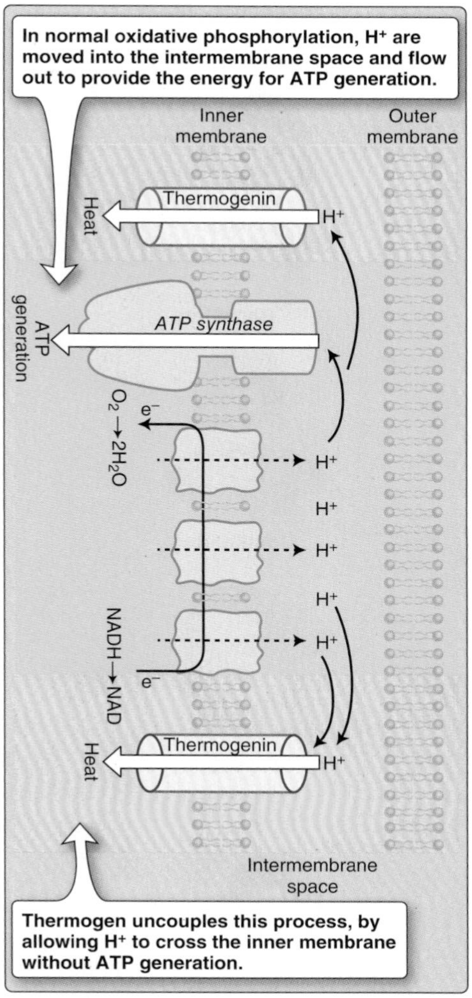
In normal oxidative phosphorylation, H+ are moved into the intermembrane space and flow out to provide the energy for ATP generation.
Inner membrane
Outer membrane
Heat
Thermogenin
H+
ATP generation
ATP synthase
$O_2 \rightarrow 2H_2O$
e−
H+
H+
H+
H+
H+
NADH→NAD
e−
Heat
Thermogenin
H+
Intermembrane space
Thermogen uncouples this process, by allowing H+ to cross the inner membrane without ATP generation.

Figure 38.6
Brown adipose tissue. ATP = adenosine triphosphate; NAD = nicotinamide adenine dinucleotide; NADH = NAD hydrogen.

III. HEAT PRODUCTION AND TRANSFER

The amount of heat stored within the body reflects a balance between the amount of heat produced and the amount transferred to the external environment. Heat storage can be quantified theoretically using the heat balance equation:

$$S = (M - Wk) \pm (R + K + C) - E$$

where S is heat storage; M is metabolism; Wk is external work; and R, K, C, and E describe radiative, conductive, convective, and evaporative heat transfer, respectively.

A. Production

Heat is a metabolic byproduct reflecting the inefficiency of the chemical pathways involved. The amount of heat produced at rest is related to **basal metabolic rate** (**BMR**), which, in turn, is related to body mass (e.g., two individuals with a mass of 50 kg and 90 kg may be expected to have BMRs of ~1,315 kcal/day and ~2,045 kcal/day, respectively). Any increase in tissue metabolism increases heat production. Food digestion and assimilation increase energy expenditure (the amount of energy used is known as the **thermic effect of food**), as do spontaneous movements and exercise.

Clinical Application 38.1: Malignant Hyperthermia

Malignant hyperthermia is a syndrome triggered by anesthetics (e.g., halothane) and muscle relaxants (e.g., succinylcholine).[1] Metabolic rate increases at a rate that far outpaces heat dissipation, due to excess sarcoplasmic Ca^{2+}, which stimulates exaggerated and prolonged excitation–contraction coupling in skeletal muscle. This heat production mechanism appears to be a genetic abnormality in the Ca^{2+}-release channels (ryanodine receptors) in the sarcoplasmic reticulum (see 12·III·A).

INFO LINK [1]For more information on the use of anesthetics, see *LIR Pharmacology*, 5e, p. 135.

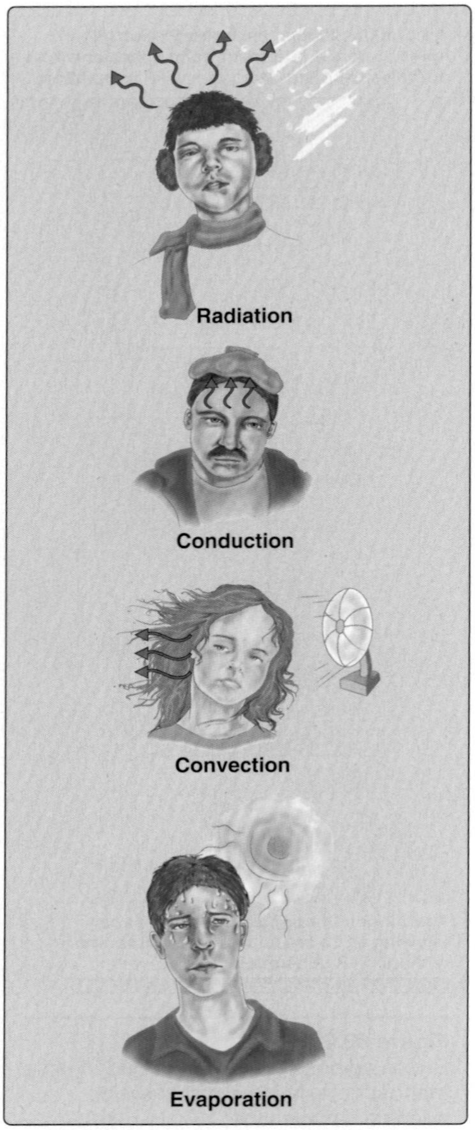

Figure 38.7
Heat transfer.

Table 38.1: Heat Dissipation Pathways

Heat Transfer	Sitting Indoors in 25°C	Walking Outside in 30°C
Radiation	60%	Minimal*
Convection	15%	10%
Conduction	5%	Minimal
Evaporation	20%	90%

*Radiation involves both heat gain and loss, but the net change is minimal.

B. Heat transfer

Heat produced by metabolism must be transferred to the external environment, principally via the skin, although a small amount of heat is transferred via the respiratory tract. The term "heat transfer" refers to a mechanism whereby heat is transferred from an area of higher temperature to an area of lower temperature. Human skin temperature is ~32°C in normothermic (i.e., temperatures that support a normal body temperature) environments. Because outside temperatures are usually lower, body heat can be transferred to air or other objects. When the external temperature is greater than skin temperature, the body gains heat. There are four primary mechanisms by which heat is transferred to the environment: **radiation**, **conduction**, **convection**, and **evaporation** (Figure 38.7).

1. **Radiation:** Radiation refers to the thermal energy that is transferred to objects in the external environment. Heat energy is carried in the infrared spectrum, and the amount transferred depends on the temperature difference and the emissivity (ability to absorb energy) of the object's surface. The majority of heat transfer at rest occurs by radiation (Table 38.1).

2. **Conduction:** Conduction of thermal energy from one body to another occurs when they are in close physical contact. The high kinetic energy of molecules in a warm region dissipates by collisions with adjacent molecules in a cool region. Solids differ enormously in their ability to conduct heat. Substances with low thermal conductivity are called thermal insulators.

3. **Convection:** Convection occurs when heat is transferred to the environment by a moving fluid (i.e., air or water). Generally, heating reduces the density of air and water, and gravity creates a "natural" fluid convection current near the skin as the warmer, low-density fluid rises. Forced convection results when an alternative energy source propels the fluid past the skin (e.g., fan, wind, water current). Convective heat loss or gain is proportional to the specific heat of the fluid, the temperature gradient, and the square root of fluid or air velocity.

4. **Evaporation:** Evaporation dissipates heat by using thermal energy to convert water from a liquid to a gaseous phase, the primary sites of evaporative heat loss being the respiratory tract and the skin. Evaporation is a very effective mode of heat dissipation, such that 1 L of sweat can remove ~580 kcal from the skin surface. The amount of evaporation is dependent on the relative humidity of ambient air: Humid air attenuates and dry air facilitates sweating. During exercise or when ambient air temperature is above skin temperature, sweat evaporation provides the primary and, often, only mode of heat dissipation (see Table 38.1).

IV. CLINICAL ASPECTS

Although internal temperature is normally maintained within a narrow range, the hypothalamus may allow it to increase in an attempt to thwart a pathogen, manifesting as **fever**. Deviation from normal may also occur

when the body's thermoregulatory systems are overwhelmed, resulting in **hypothermia** or **hyperthermia**.

A. Fever

Fever has long been recognized as a symptom of illness and is caused by **exogenous** and **endogenous pyrogens**. Exogenous pyrogens include microorganisms, such as *Staphylococcus aureus*, and their byproducts or toxins. Most often, fever is a response to endogenous pyrogens released during macrophage and monocyte activation,[1] which is related to an infection, even though the microorganism is not directly involved. Endogenous pyrogens are interferons and cytokines, including interleukins (e.g., IL-1 and IL-6) and tumor necrosis factor. Although the pathways involved are as yet unelucidated, circulating pyrogens are sensed by the circumventricular organs (see 7·VII·C), which signal their presence to the preoptic hypothalamus via prostaglandin release. Neurons in the preoptic area express a prostaglandin receptor, EP_3, that mediates the fever response. When these receptors are stimulated, the hypothalamic set point is reset, and the body begins to regulate internal temperature at a higher value (Figure 38.8). The symptoms associated with a febrile state reflect thermoregulatory effector organs trying to attain the new set point. This elevated temperature is thought to exert both a beneficial effect on the host's immune system and to decrease pathogen growth and proliferation. The febrile state is distinct from the body temperature increase associated with muscle contraction and exercise or ambient heat exposure. In both cases, the body attempts to dissipate heat with the goal of returning internal temperature to 37°C.

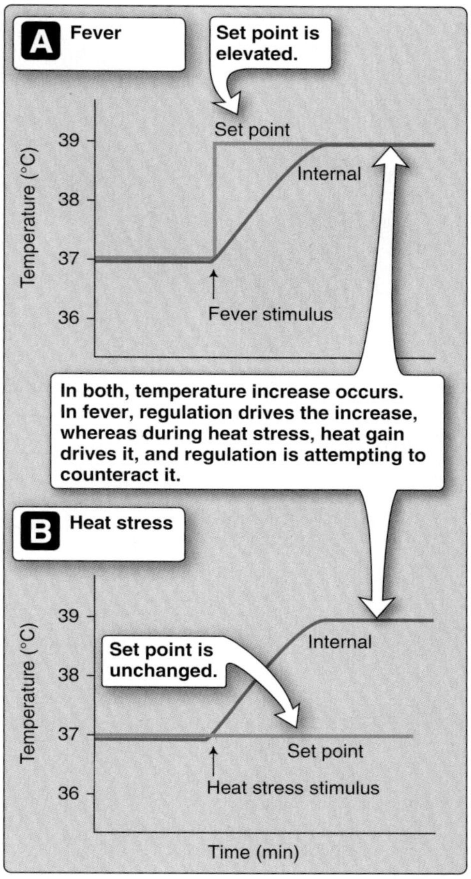

Figure 38.8
Fever *versus* heat stress.

> The hypothalamic set point can be returned toward 37°C and the symptoms of fever reduced by administering nonsteroidal anti-inflammatory drugs (NSAIDs), such as aspirin, ibuprofen, and acetaminophen.[2] NSAIDs are *cyclooxygenase* inhibitors that block prostaglandin synthesis and, thus, have antipyretic effects.

B. Hypothermia and hyperthermia

Hypothermia and hyperthermia are deviations from normal body temperature that occur due to thermoregulatory system failure. These conditions occur most commonly during weather extremes and in patients with an impaired ability to respond to thermal stress such as those with a genetic inability to secrete sweat (**congenital anhidrosis**).

INFO LINK

[1]For cellular defense mechanisms and associated cytokines, see *LIR Immunology*, 2e, p. 49.

[2]For more information on the use and actions of NSAIDs, see *LIR Pharmacology*, 5e, p. 529.

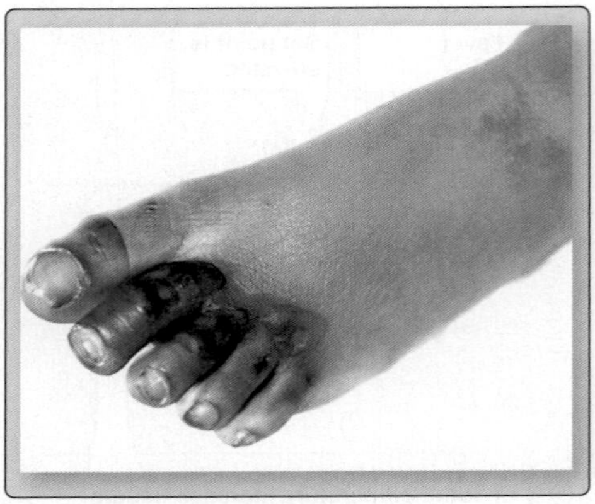

Figure 38.9
Frostbite.

1. **Hypothermia:** Hypothermia (internal temperature of <35°C) commonly results from immersion in cold water because water transfers heat 25 times faster than does air. Heat loss occurs via normal heat transfer mechanisms, but heat production cannot increase sufficiently to compensate for heat loss. Hypothermia causes symptoms associated with cold-induced decreases in neuronal metabolic rate, including drowsiness, slurred speech, bradycardia, and hypoventilation. Severe hypothermia (internal temperature of <28°C) can cause coma, hypotension, oliguria, and fatal cardiac arrhythmias (ventricular fibrillation). Peripheral tissues can also be injured by the cold. Frostbite is a condition in which fluid in skin and subcutaneous areas crystallizes (freezes), disrupting cell membranes and causing tissue necrosis (Figure 38.9). Necrosed areas often require amputation.

2. **Hyperthermia:** Precise internal temperature definitions of hyperthermia are not possible without assessing the cause. For example, internal temperatures above 40°C can be achieved during exercise, without developing a heat illness. Heat illnesses form a continuum, from a milder **heat exhaustion** to the more severe **heat stroke**. The etiology of heat exhaustion is related to a decrease in circulating blood volume caused by skin vasodilation and a sweating-induced decrease in central venous pressure (CVP). The decrease in CVP can allow blood to pool in the limbs when an individual is in the upright position, causing syncope (fainting). Heat stroke, in the classic sense, refers to failure of the heat dissipation mechanisms due to the continuing increase in internal temperature. Unfortunately, failure of these mechanisms only leads to more rapid gains in temperature. In heat stroke, internal temperatures can climb to over 41°C, which can lead to neural death and organ system failure.

Chapter Summary

- Internal temperature is normally maintained at 37.0°C ± 0.5°C.
- Internal temperature is sensed and controlled by the **preoptic area of the hypothalamus**. Environmental temperature is sensed by **warmth** and **cold skin thermoreceptors**.
- Heat stress induces skin vasodilation and increases cardiac output and sweating to aid in body heat offloading. Cold-seeking behavioral strategies are also stimulated.
- Cold stress induces skin vasoconstriction and increases shivering to decrease heat loss and increase heat production. Heat-seeking behavioral strategies are also stimulated.
- Heat balance is achieved by matching heat production with heat loss. Heat production includes the amount generated by metabolism, the thermic effect of food, spontaneous movements, and exercise. Heat transfer occurs via **radiation**, **convection**, **conduction**, and **evaporation**.
- **Fever** is the external manifestation of temperature set point resetting to a higher value. Both **exogenous pyrogens** (e.g., microbial toxins) and **endogenous pyrogens** (e.g., interferons, interleukins, and tumor necrosis factor) can increase the set point via **prostaglandin** production. The body then defends this higher value by normal means, such as shivering to raise temperature and sweating to lower temperature.
- **Hypothermia** is a low internal temperature and is clinically associated with processes that slow body metabolism. **Frostbite** results in tissue necrosis from fluid crystallizing within and between cells.
- **Hyperthermia** is an elevated internal temperature. The most serious form of heat illness is **heat stroke**, which classically involves complete thermoregulatory system failure.

Exercise

39

I. OVERVIEW

Historically, humans have engaged in physical activity to fetch water, to forage, and to journey to hunting sites in order to secure food. In modernday society, we enjoy the luxury of fresh, clean water delivered to our homes; food readily available from local stores and restaurants; and multiple transportation options to facilitate travel. When we engage in physical activity, it is usually in the form of structured exercise designed to improve fitness and health. Physical activity can take many forms, and the physiologic requirements required to execute them may be very different. For example, traveling to a distant hunting site involves repeated, rhythmic cycles of (isotonic) muscular activity, whereas carrying a pitcher of water or basket of food requires sustained (isometric) contractions. Although the tasks required of muscles may vary considerably, engaging in physical activity in any form involves the same basic pathways and principles. Muscular activity is planned and executed by the central nervous system (CNS). The energy used to develop contractile force is supplied by adenosine triphosphate (ATP), which is generated through aerobic respiration. The O_2 required to sustain aerobic metabolism is taken up from the atmosphere by the lungs and delivered to muscle, along with glucose and other necessary nutrients, by the cardiovascular system. Aerobic metabolism produces ATP relatively slowly but can be sustained for prolonged periods (e.g., during a long walk to hunting grounds). ATP can also be produced more rapidly (e.g., when initially sighting and pursuing prey) by anaerobic metabolism via ATP–creatine phosphate (CP) and lactic acid systems, but this rapid delivery cannot be sustained. Maximal aerobic exercise can place extreme demands on all of the body's homeostatic systems and push the pulmonary and cardiovascular systems to the maximum (Figure 39.1).

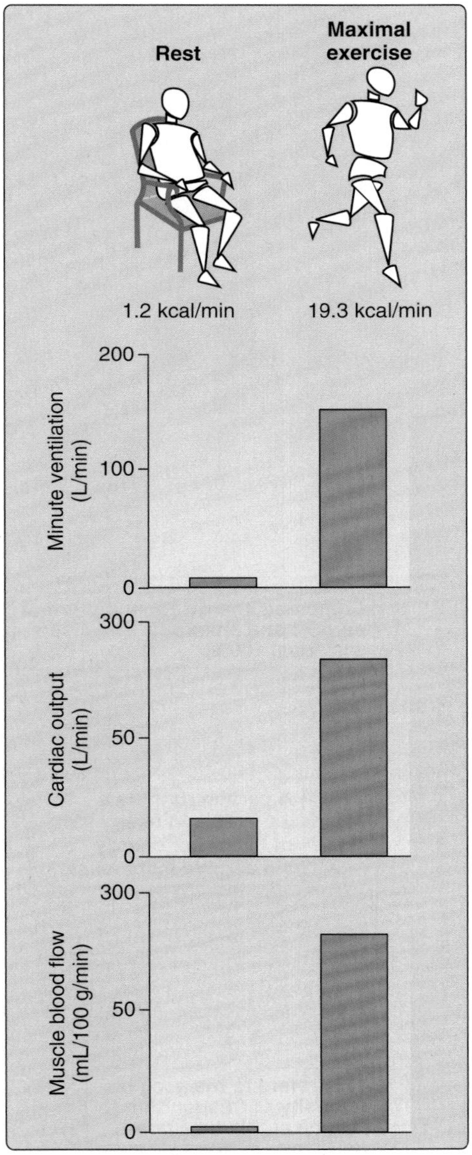

Figure 39.1
Rest *versus* maximal exercise.

II. DEFINITIONS

Physical activity and its structured component (exercise) can be classified in several different ways related to metabolism and movement and by acute responses *versus* training adaptations.

A. Aerobic *versus* anaerobic

Exercise can be classified by the predominant energy system used (Table 39.1). The two primary classifications are aerobic (exercise

Table 39.1: Exercise Classification

Exercise	Type
400-m sprint	Anaerobic
10-km run	Aerobic
Track cycling (1 km)	Anaerobic
Road cycling (40 km)	Aerobic
100-m freestyle swim	Anaerobic
1,500-m freestyle swim	Aerobic

that uses O_2) and anaerobic (exercise that does not directly involve O_2 usage).

B. Isometric *versus* isotonic

Physical activities can also be classified based on whether or not the force applied imparts external movement. The aerobic activities in Table 39.1 all employ rhythmic isotonic ("same force") muscle contractions that allow for external movement. During isometric ("same length") contractions, muscles may shorten during force development, but no joint movement occurs. Examples include muscle contractions used to maintain posture, hold a bag of groceries, and grip bicycle handlebars.

C. Acute responses *versus* exercise training

If exercise bouts occur regularly (e.g., during an exercise training program), then the body adapts to make the subsequent physical stress easier. Adaptation begins as soon as the exercise training program is initiated but may require months to years to manifest fully. Many of these adaptations can have prophylactic health benefits and can be used in physical rehabilitation to improve work capacity after injury or disease.

III. SKELETAL MUSCLE

Exercise uses skeletal muscles to generate force, which is imparted through tendons to bone. Bones then move along a force vector within a joint-specific range of motion to transfer that force to a pedal to move a bicycle or to launch a basketball toward a hoop, for example.

A. Adenosine triphosphate synthetic pathways

Muscle contractions are fueled by ATP. Myocytes store ATP in very limited quantities (~4 mmol/L), so sustained activity must be supported by ATP synthesis. The ATP-CP and lactic acid systems deliver ATP on a timescale that supports rapid activities of limited duration (seconds). Oxidative phosphorylation is slower but can sustain activities lasting hours (Figure 39.2).

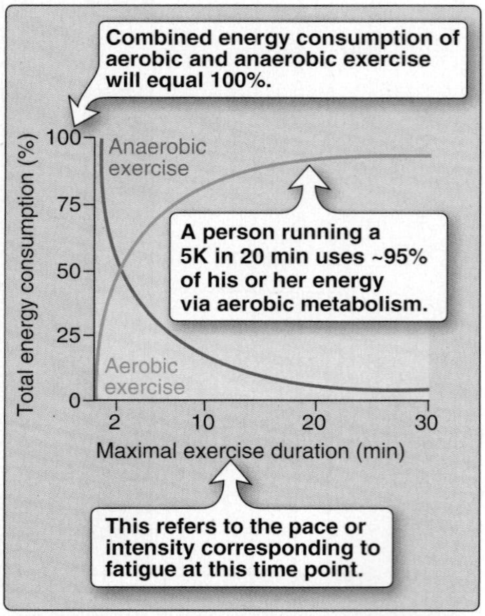

Figure 39.2
Energy system time course.

1. **Adenosine triphosphate–creatine phosphate system:** CP (also known as phosphocreatine) contains a high-energy phosphate bond that can be used to rapidly regenerate ATP from adenosine diphosphate (ADP). The conversion is catalyzed by *creatine kinase*, a sarcoplasmic enzyme. Muscles contain sufficient CP reserves to support contractions lasting 8–10 s.

2. **Lactic acid system:** The lactic acid system generates ATP at about half the rate of the ATP-CP system and uses glucose (either absorbed or from glycogen metabolism) as a substrate. Glycolysis uses two ATP molecules to produce four more, for a net gain of two ATP per glucose molecule, with the final product being pyruvic acid. Pyruvic-acid-to-lactic-acid conversion does not produce additional ATP but rather regenerates reducing

Clinical Application 39.1: *Creatine Kinase*

Creatine kinase (*CK*) can be used as an index of muscle damage because circulating levels increase when the sarcolemma has been breached. There are three different *CK* isoforms. Skeletal muscle contains a *CK-MM* isoform, *CK-MB* is specific for heart muscle, and the *CK-BB* isoform is found in nervous tissue. The *CK-MM* isoform may be released into the circulation of healthy individuals after long-duration aerobic exercise (e.g., running a marathon) that causes minor muscle damage. Individuals with Duchenne and Becker muscular dystrophies may also show 25- to 200-fold increases in blood *CK-MM* levels during muscle breakdown associated with atrophy (see Clinical Application 12.1).

equivalents.[1] Muscles continue to shuffle pyruvic acid to lactic acid to extend the maximal contraction time to 0.5–2.5 min.

3. **Oxidative phosphorylation:** Oxidative phosphorylation generates ATP at about half the rate of the lactic acid system. Oxidative phosphorylation also involves glycolysis (Figure 39.3), but pyruvic acid then enters the citric acid cycle via the *pyruvate dehydrogenase complex*. The citric acid cycle produces ATP and CO_2. Its main potential energy products are reducing equivalents that enter the electron transport chain, a process that uses O_2 as the final electron acceptor and regenerates ATP from ADP.[2] Oxidative phosphorylation nets ~30 ATP molecules per glucose molecule and can continue for hours, depending on exercise intensity and substrate availability (i.e., glucose, fatty acids, ketone bodies, and amino acids).

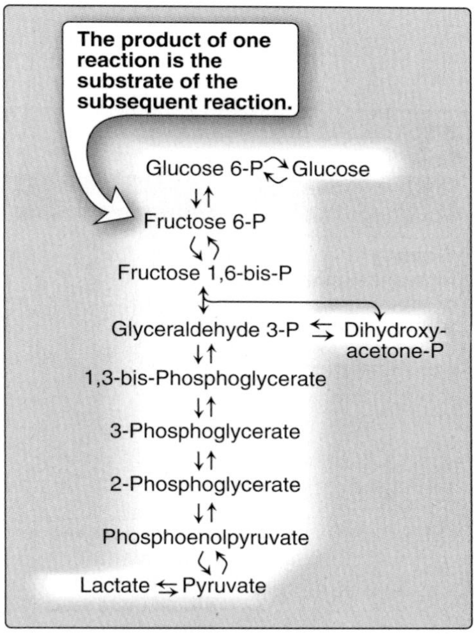

Figure 39.3
Glycolysis, an example of a metabolic pathway.

B. Aerobic exercise training

Aerobic exercise training promotes cellular adaptations that increase muscles' ability to store and then process energy substrates aerobically.

1. **Energy stores:** Aerobic exercise training increases myocyte glycogen stores. Glycogen provides a readily available carbohydrate energy source to supplement plasma glucose uptake during exercise. Once muscle glycogen reserves are depleted, a person fatigues and must slow work rate, corresponding to "hitting the wall" in marathon running.

2. **Metabolism:** Exercise training increases aerobic metabolic capacity in a number of ways. Training increases mitochondrial size and numbers and increases myoglobin content, thereby enhancing O_2

INFO LINK

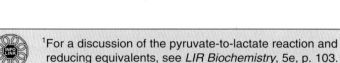

[1] For a discussion of the pyruvate-to-lactate reaction and reducing equivalents, see *LIR Biochemistry*, 5e, p. 103.

[2] For a discussion of the electron transport chain and oxidative phosphorylation, see *LIR Biochemistry*, 5e, pp. 73 and 77.

Table 39.2: Enzymes Upregulated as a Result of Aerobic Exercise Training

Pathway	Enzyme
Glycolysis	*Glucokinase*
Glycolysis	*Phosphofructokinase*
Citric acid cycle	*Citrate synthase*
Citric acid cycle	*Succinate dehydrogenase*
ETC	*Cytochrome c*
β oxidation	*Carnitine palmitoyltransferase*

ETC = Electron transport chain.

storage and transport between the sarcoplasm and mitochondria, and it upregulates oxidative enzymes involved in the citric acid cycle and in oxidative phosphorylation as well as enzymes that break down glycogen and those involved in β-oxidation (Table 39.2).

C. Anaerobic exercise training

Anaerobic exercise training (e.g., weight lifting and team sports, such as ice hockey, that involve bursts of intense activity) increases force production through muscle hypertrophy, improvements in neural recruitment, and increases muscle fatigue resistance through metabolic changes.

1. **Hypertrophy:** Anaerobic exercise training increases the cross-sectional area of type IIa and IIx muscle fibers by adding new myofibrils to the myocytes. Type II fibers are specialized for speed and force production, but they rely primarily on glycolytic pathways that make them prone to fatigue. The additional myofibrils increase muscle force–production capacity (Figure 39.4).

2. **Neural recruitment:** Muscle hypertrophy is preceded by neural adaptations that increase the efficiency of motor unit activation, which increases in muscle force production.

3. **Metabolism and energy stores:** The capacity to extend anaerobic energy production increases during training by upregulating enzymes associated with glycolysis for ATP generation in the lactic acid system and *creatine kinase* in the ATP-CP system. Training also increases intramuscular glycogen stores, as seen during aerobic exercise training.

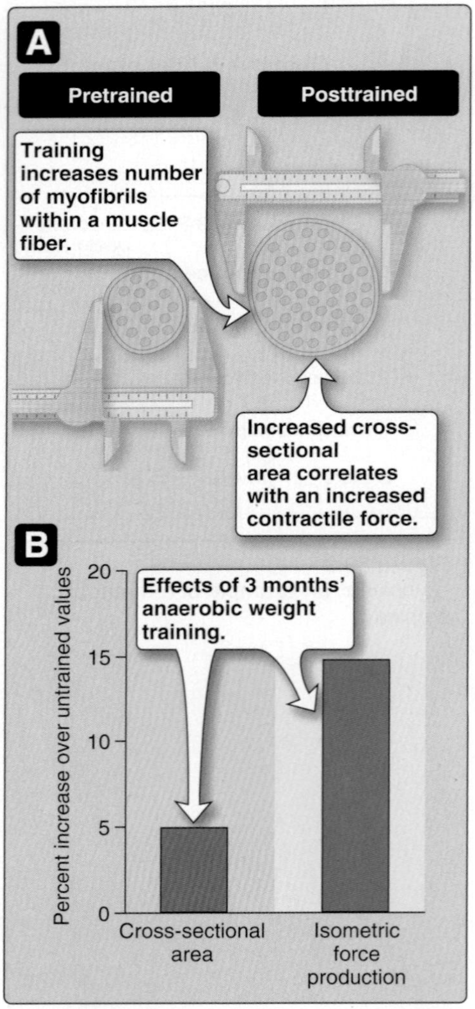

Figure 39.4
Effects of training on muscle mass and performance.

IV. MOTOR AND AUTONOMIC CONTROL

Physical activity is planned and initiated by the motor cortex, but exercise itself requires constant feedback and adjustments to motor and visceral functions involving all nervous system divisions. The way that these systems are coordinated during exercise can be illustrated by considering the pathways required to take a bike ride along a winding track.

A. Peripheral nervous system

The peripheral nervous system (PNS) conveys motor commands via motor neurons from the CNS to the various muscles required to ride a bike. The PNS also conveys sensory information from **Golgi tendon organs**, **muscle spindles**, and other muscle sensors to the CNS.

1. **Motor neurons:** Motor neurons are stimulated at the level of the spinal cord from the descending corticospinal tract. α-Motor neurons mediate force production by contacting extrafusal muscle fibers. γ-Motor neurons, which innervate intrafusal fibers, contract simultaneously to maintain the stretch sensitivity of sensory muscle spindles within the working muscle (see 11·II·A).

2. **Motor units:** Cycling on level terrain typically requires submaximal force to be applied to the pedals. Skeletal muscle fibers produce force in an all-or-none manner, but force production can be graded

by activating subsets of motor units (see 12·IV·D). Alternating active motor units ensures against individual motor unit fatigue. If bike wheel resistance increases (e.g., when ascending a small hill), more motor units are recruited to supply the necessary force to maintain forward motion (Figure 39.5).

3. **Muscle sensors:** Three types of muscle sensor provide the CNS with feedback regarding muscle and joint position during exercise. Muscle spindles convey information about limb position through changes in muscle stretch. Golgi tendon organs located in the musculotendinous junction sense muscle tension (see 11·II·B). **Muscle afferents** are free nerve endings woven throughout the muscle fiber body that monitor the local mechanical and chemical environment. They relay information back to autonomic nervous system (ANS) control centers that coordinate cardiorespiratory responses to exercise. Muscle afferents mediate an **exercise pressor reflex**, or a reflex increase in blood pressure observed during exercise. Although some sensory overlap exists, there are two main classes: class III and class IV.

   a. **Class III:** Class III afferents are thinly myelinated nerve endings located near collagen structures that respond primarily to mechanical stimuli, such as stretch, compression, and pressure. They are activated as soon as exercise (e.g., pedaling) begins.

   b. **Class IV:** Class IV afferents are unmyelinated fibers located near muscle blood and lymphatic vessels that respond primarily to metabolic byproducts, such as lactate, H^+, bradykinin, K^+, arachidonic acid, and adenosine. They activate shortly after the exercise begins and metabolite levels begin to build.

4. **Cardiovascular receptors:** Arterial and cardiopulmonary baroreceptors monitor blood pressure and allow ANS control centers to maintain arterial pressure at levels sufficient to ensure flow to active muscles and other vital systems during exercise (see 20·III). Peripheral chemoreceptors located in carotid bodies monitor arterial P_{CO_2} and H^+ levels and allow ANS respiratory centers (dorsal respiratory group neurons) in the medulla to adjust ventilation as necessary (see 24·III·B).

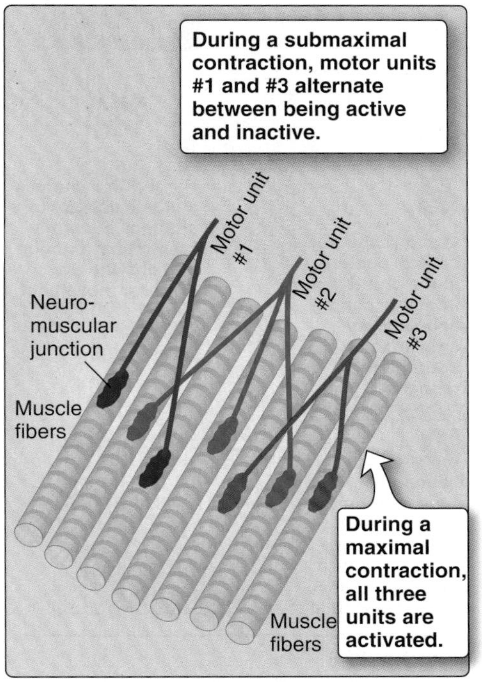
During a submaximal contraction, motor units #1 and #3 alternate between being active and inactive.
Motor unit #1
Motor unit #2
Motor unit #3
Neuro-muscular junction
Muscle fibers
Muscle fibers
During a maximal contraction, all three units are activated.

Figure 39.5
Motor unit recruitment.

B. Central nervous system

Higher brain centers provide the motivation to go on the bike ride, and they also regulate the muscle contractions required to pedal. The ANS ensures blood flow and O_2 delivery to leg muscles and those involved in maintaining stability and posture (e.g., back, arms, shoulders). Information flow from the primary senses helps one balance the bike, stay on the path, and avoid being unsaddled by the nearest tree limb.

1. **Somatic:** The premotor cortex, supplemental motor cortex, and basal ganglia aid in motor program development (see 11·IV·A), which coordinates basic motor patterns including input from sensory cues and information about where the pedals are located and

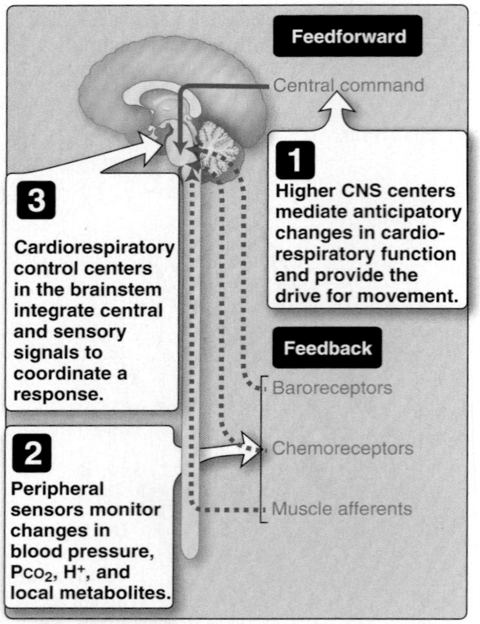

Figure 39.6
Feedforward and feedback signals. CNS = central nervous system.

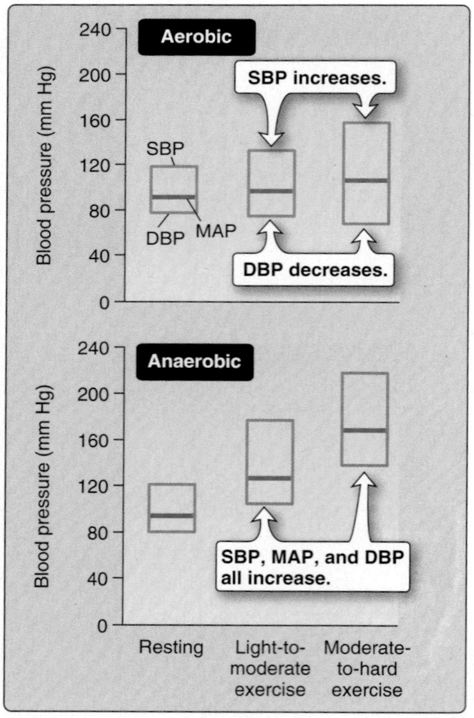

Figure 39.7
Blood pressure responses during exercise. DBP = diastolic blood pressure; MAP = mean arterial blood pressure; SBP = systolic blood pressure.

if the foot is firmly on the pedal, for example. This motor program is then executed by the primary motor cortex and signaled via the corticospinal tract. The cerebellum coordinates leg and foot movements during exercise by integrating sensory feedback with motor input.

2. **Autonomic:** Autonomic systems are required to redistribute flow among the various vascular beds to maintain arterial pressure at levels that ensure adequate flow and O_2 delivery to the active muscles. This is accomplished via sympathetic feedforward and feedback pathways (Figure 39.6).

    a. **Feedforward:** Feedforward mechanisms are mediated by central commands. This **central command** concept describes a process that increases cardiorespiratory system function immediately prior to or at the beginning of exercise. These pathways prepare for needed increases in blood flow and O_2 uptake delivery to soon-to-be working muscles. In our example, such increases would occur in anticipation of the upcoming bike ride or while donning the helmet.

    b. **Feedback:** Feedback occurs via class III and IV muscle afferents and through other autonomic afferents (baroreceptors and chemoreceptors) discussed above.

3. **Senses:** Vision plays a major role in the mountain bike ride, providing information about potential obstacles and the nature of the terrain. Hearing plays a lesser role, but it does help provide clues about the location of other riders, gearing, and under-the-tire terrain. The vestibular system provides information regarding linear acceleration (otolith organs) and head position (semicircular canals; see 9·V·A) when scanning the path ahead and looking for items on either side of the path.

V. CARDIOVASCULAR SYSTEM

A contracting skeletal muscle requires increased blood flow, both to deliver nutrients and to remove metabolic byproducts, including heat. Skeletal muscle receives ~1 L/min at rest, but strenuous exercise can increase demand to >21 L/min. Such dramatic increases in flow cannot occur without changes in both heart and vasculature function.

A. Arterial pressure

Sympathetic nervous system (SNS) activation in anticipation of exercise causes mean arterial pressure (MAP) to rise, mediated by increases in heart rate (HR), myocardial inotropy, venoconstriction, and systemic vascular resistance (SVR). During exercise, MAP increases, but the extent of the increase depends on the particular physical activity (Figure 39.7).

1. **Aerobic:** MAP increases slightly during aerobic exercise because of SNS-mediated increases in systolic blood pressure (SBP). Once exercise begins, metabolite levels within active muscles build, causing local vasodilation. Dilation facilitates increased flow

and O_2 delivery to the active muscles. Active skeletal muscles create a low-resistance circuit within the systemic vasculature that facilitates diastolic runoff. Diastolic blood pressure (DBP) stays the same or decreases slightly.

2. **Anaerobic:** Anaerobic exercise involving isometric contractions causes SBP, DBP, and MAP to rise dramatically. MAP values of >275 mm Hg have been recorded during two-leg presses (knee and hip extension), for example. The reason is that contracting muscles compress and occlude arterial supply vessels, which greatly increases muscle vascular resistance and SVR during high sympathetic drive. Similar responses occur when shoveling heavy snow, which is why older adults and patients with hypertension and coronary artery disease are advised against clearing their own walks and driveways.

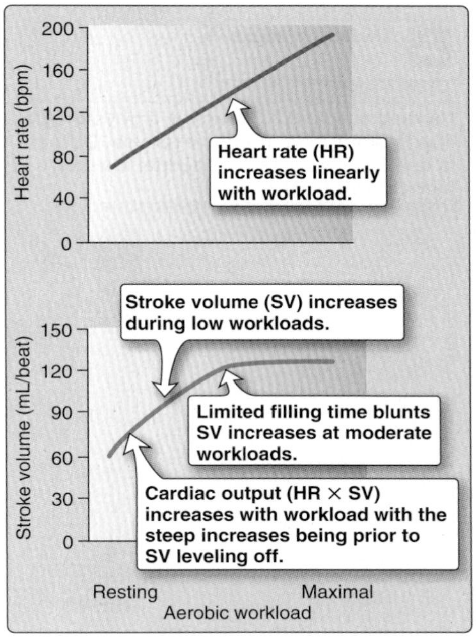
Heart rate (bpm)
200
160
120
80
40
0
Heart rate (HR) increases linearly with workload.
Stroke volume (SV) increases during low workloads.
Stroke volume (mL/beat)
150
120
90
60
30
0
Limited filling time blunts SV increases at moderate workloads.
Cardiac output (HR × SV) increases with workload with the steep increases being prior to SV leveling off.
Resting
Maximal
Aerobic workload

Figure 39.8
Heart rate and stroke volume.

B. Cardiac output

Increased flow through the skeletal musculature requires a corresponding increase in cardiac output (CO). CO during aerobic exercise is workload dependent, rising from 5 L/min at rest to >25 L/min during maximal aerobic exercise. CO increases are effected through increases in HR and stroke volume (Figure 39.8).

1. **Heart rate:** HR increases linearly with workload during aerobic exercise, which is why HR can be used as a rough estimate of how hard the body is working. The HR increase is mediated by the ANS, with coincident withdrawal of parasympathetic outflow and increase in sympathetic outflow to cardiac nodal cells causing HR to rise from ~65 beats/min at rest to a maximum of ~195 beats/min, depending on age (see 17·III·A and 40·II·A). The rise in HR also reflects direct effects of exercise-induced internal temperature rises on nodal cell automaticity (~8 beats/min/°C).

2. **Stroke volume:** Left ventricular (LV) SV is determined by myocardial inotropic state and LV end-diastolic volume (EDV), both of which are regulated by the SNS. SV increases linearly, mediated by SNS-mediated increases in inotropy and EDV (see below). At higher exercise levels, coincident increases in HR begin to limit, then decrease, the time available for ventricular filling during diastole, which causes EDV to fall. SNS-mediated decreases in atrioventricular nodal conduction time plus increases in the rate of myocardial relaxation help offset this limitation, but SV reaches a peak of 120–140 mL and may subsequently decrease at moderate-to-high levels of exercise.

C. Venous return

When CO rises to 25 L/min to support intense exercise, venous return (VR) must necessarily increase to 25 L/min also to provide blood for LV preload and continued output. Increased VR is mediated in part by the SNS, which decreases venous capacity through venoconstriction. Venoconstriction both increases effective circulating blood volume and speeds the rate at which blood traverses the system (see 20·V·B). Increased ventilation also assists VR, by increasing the pressure gradient driving flow between skeletal muscle veins and the right

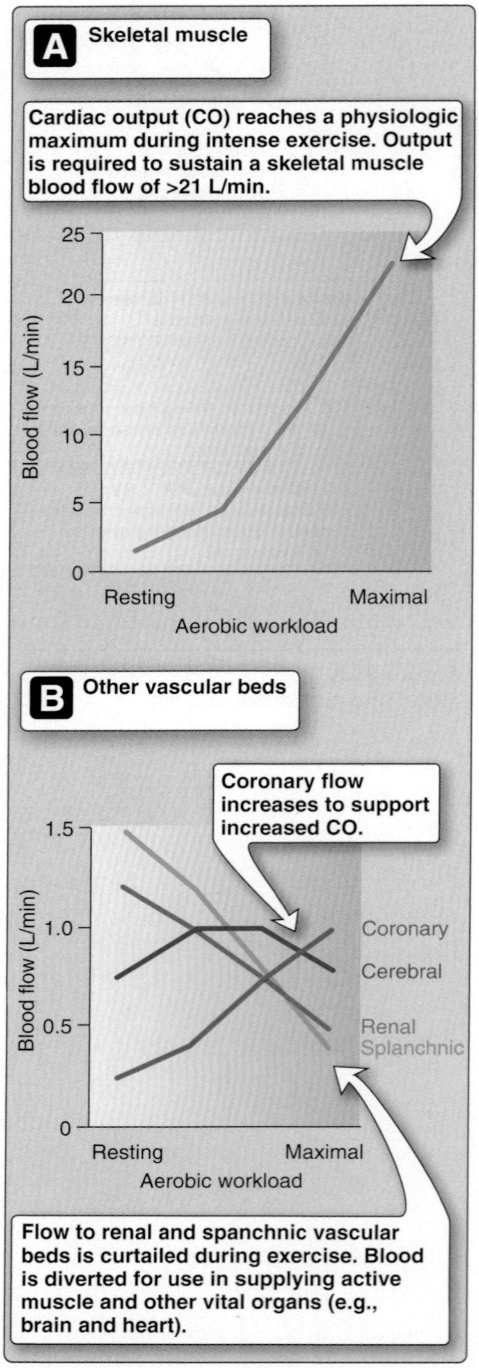

Figure 39.9
Blood flow distribution. CO = cardiac output.

atrium. The pressure gradient is enhanced during the deep inspirations that typically accompany aerobic exercise. However, the main force driving VR is a venous (or muscle) pump (see Figure 20.21). Rhythmic muscular contractions of active muscle compress the veins within, forcing blood back to the heart. The muscle pump causes central venous pressure to rise slightly during exercise, which assists ventricular preloading.

D. Flow redistribution

SNS-mediated vasoconstriction in the vascular beds supplying inactive muscles and other organs not directly involved in exercise (e.g., gastrointestinal system, kidneys) diverts blood flow temporarily to perfuse active muscles (Figure 39.9). The vasoconstrictor signal is not limited to inactive muscle but is also sent to active muscle. However, it is overridden in active muscle by local metabolic and mechanical factors that maintain vasodilation.

E. Training

The cardiovascular adaptations to long-term aerobic exercise primarily involve heart and blood vessels.

1. **Cardiac:** Aerobic exercise causes a volume-induced cardiac hypertrophy. This type of hypertrophy increases both the LV chamber diameter and LV wall mass and likely is caused by the high venous return and preload accompanying exercise. This adaptation increases resting EDV and SV, which is why trained athletes typically have a slower resting HR than untrained individuals. (Resting demand for CO is ~5 L/min in both cases, and CO = HR × SV). Maximal HRs do not change with long-term aerobic exercise.

> Anaerobic exercise training, which involves repeatedly forcing the LV to eject against an elevated MAP, stimulates a LV hypertrophy reminiscent of that seen in patients with aortic stenosis and untreated hypertension (see Clinical Application 18.2). Such hypertrophy is characterized by an increase in LV wall thickness but a decrease in lumen diameter.

2. **Vascular:** Training increases the ability of skeletal and cardiac muscle to vasodilate, probably through increased nitric oxide production. Over time, angiogenesis increases capillary density and, thereby, decreases the distance for diffusional exchange of O_2 and nutrients between blood and myocytes.

VI. RESPIRATORY SYSTEM

Aerobic exercise increases the body's O_2 requirements, from ~0.25 L/min at rest to >4.0 L/min during maximal aerobic exercise in an aerobically

trained person. These O_2 needs are met through increases in pulmonary minute ventilation (V_E) and O_2 extraction by tissues.

A. Ventilation

V_E increases from ~6 L/min at rest to ~150 L/min during maximal aerobic exercise. This increase is met through increases in both respiratory rate and tidal volume. At the beginning of exercise, there is an immediate increase in ventilation, mediated primarily by central respiratory control centers. Then, via peripheral feedback from muscles and chemoreceptors (via P_aCO_2), ventilation increases linearly throughout low-to-moderate exercise. During heavy exercise, ventilation increases to a greater extent because of added anaerobic generation of H^+, which further stimulates peripheral chemoreceptors (Figure 39.10). The point is referred to as the **ventilatory threshold**.

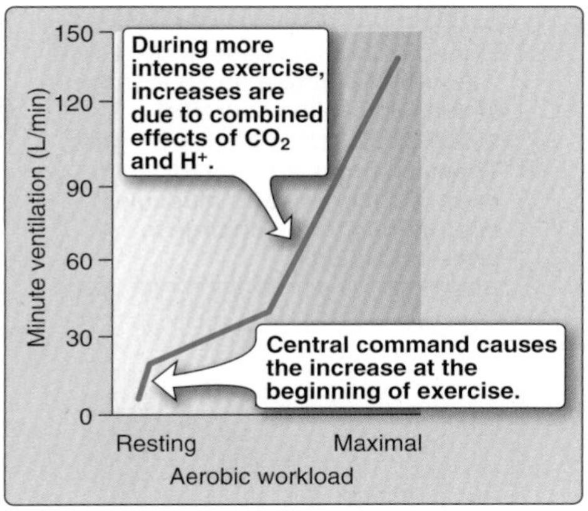

Figure 39.10
Ventilatory responses to aerobic exercise.

B. Oxygen extraction

Muscles consume O_2 at increased rates when exercised, which decreases Po_2 locally and enhances the magnitude of the gradient driving O_2 diffusion from the atmosphere to the musculature. This phenomenon manifests as a widening of the arteriovenous (a-v) O_2 difference, from ~5 mL O_2/dL at rest to ~15 mL O_2/dL during maximal aerobic exercise (Figure 39.11). O_2 delivery to the active tissues is facilitated by a decrease in hemoglobin (Hb)-O_2 binding affinity, which increases offloading. The rightward shift in the O_2-dissociation curve occurs due to increased CO_2 and H^+ production and rising local temperatures (see 23·VI·C).

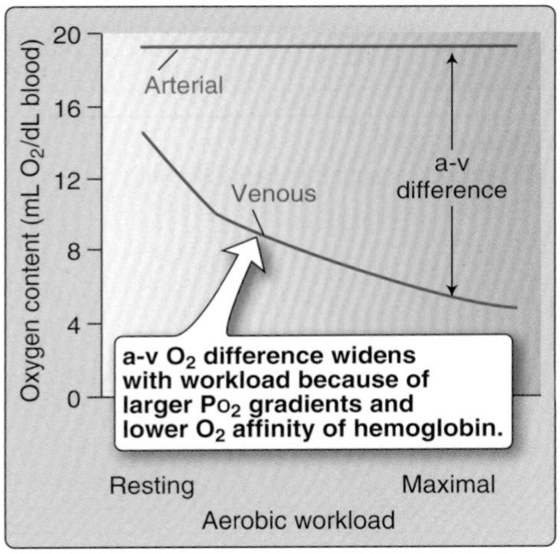

Figure 39.11
Changes in O_2 extraction during aerobic exercise. a-v = arteriovenous.

C. Excess postexercise oxygen consumption

Excess postexercise oxygen consumption (**EPOC**) describes the concept of paying back an O_2 debt incurred during initial exercise-induced increases in O_2 consumption. When exercise begins, there is a brief period during which O_2 consumption outstrips O_2 delivery, forcing the muscle to rely on high-energy phosphate groups (CP) and glycogen to generate ATP and causing accumulation of metabolic byproducts (e.g., H^+ and lactate). Upon cessation of activity, the energy stores must be regenerated and the byproducts cleared from the sarcoplasm, contributing to EPOC (Figure 39.12).

D. Training

Aerobic exercise training has no significant impact on lung volumes or capacities, but it does increase ventilation and the ability of tissues to extract O_2 from blood.

1. **Ventilation:** Maximal alveolar ventilation and V_E both increase with aerobic exercise training. This likely occurs via aerobic training adaptations in the respiratory muscles that increase fatigue resistance.

2. **Arteriovenous oxygen difference:** Aerobic exercise training increases a-v O_2 difference, reflecting an increased ability of working muscle to extract O_2 from blood. Adaptations in skeletal muscle O_2 processing, decreased diffusional distance between blood and myocytes due to increased capillary density, and increased blood

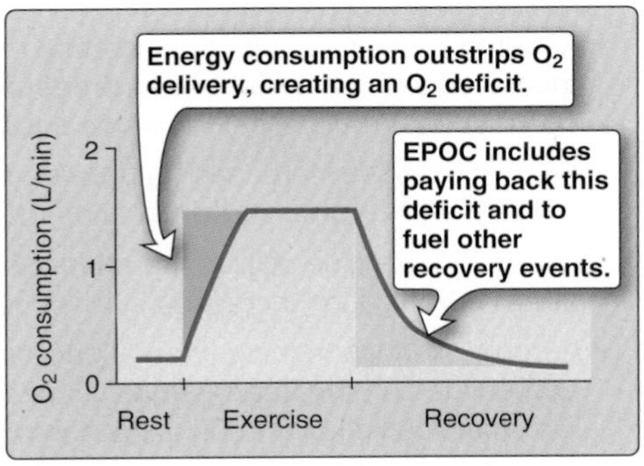

Figure 39.12
Excess postexercise O_2 consumption (EPOC).

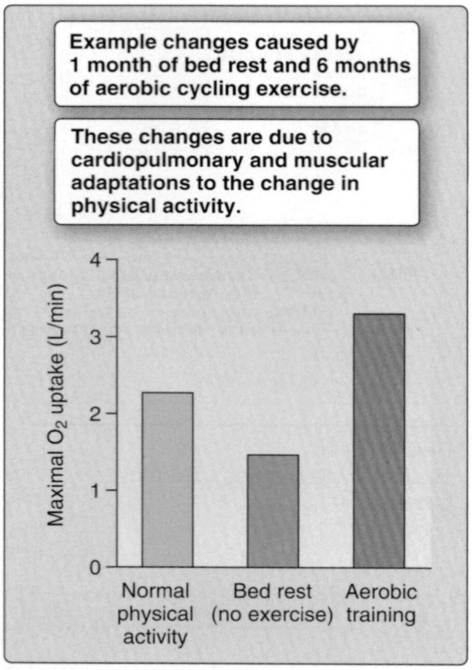
Example changes caused by 1 month of bed rest and 6 months of aerobic cycling exercise.
These changes are due to cardiopulmonary and muscular adaptations to the change in physical activity.
Maximal O2 uptake (L/min)
4
3
2
1
0
Normal physical activity
Bed rest (no exercise)
Aerobic training

Figure 39.13
Changes in aerobic capacity with training.

flow through vascular adaptations likely account for this training adaptation. Hb also increases with aerobic training, which allows more O_2 to be carried by the blood.

3. **Oxygen uptake:** The increase in CO, alveolar ventilation, and a-v O_2 difference combine to increase maximal O_2 uptake during training. During periods of physical inactivity, such as bedrest, oxygen uptake decreases (Figure 39.13).

VII. ENDOCRINE SYSTEM

There are a number of endocrine system changes associated with physical activity and exercise that occur in response to stress and the need to liberate stored energy for use by muscles. **Catecholamines**, such as epinephrine and norepinephrine, increase as part of the stress response. This increases CO and energy substrate availability for working muscles. **Cortisol** increases with strenuous aerobic exercise as part of the stress response (see 34·V·B and 34·VI·B). **Insulin** decreases with aerobic exercise, and **glucagon** increases. The stress hormones and glucagon cause increases in blood glucose, fatty acids, and amino acids by increasing glycogenolysis, gluconeogenesis, lipolysis, and proteolysis. **Antidiuretic hormone** and **aldosterone** also both increase during exercise. These hormones conserve fluid during aerobic exercise via their effects on water and sodium reabsorption by the kidney, which helps to maintain blood volume during exercise. Thyroid hormones (**triiodothyronine** [$\mathbf{T_3}$] and **thyroxine** [$\mathbf{T_4}$]) increase during exercise. T_3 and T_4 regulate metabolic rate and may participate in recovery after exercise. **Growth hormone** and **insulin-like growth factor** (**IGF-1**) increase with exercise and also contribute to recovery by stimulating tissue growth and repair by their effects on protein synthesis (see 33·VI·A).

Chapter Summary

- Exercise involves somatic control of voluntary movement and autonomic nervous system control of the cardiopulmonary system to supply working muscles with oxygenated blood.
- Aerobic exercise increases **cardiac output**, **heart rate**, **stroke volume**, and **arterial blood pressure** to ensure adequate perfusion of working muscles and other vascular beds.
- Anaerobic exercise does not increase most cardiopulmonary parameters to the same extent as aerobic exercise, with the exception of **arterial blood pressure**. Anaerobic exercise increases mean, systolic, and diastolic blood pressures in contrast to aerobic exercise, which does not increase diastolic blood pressure.
- Aerobic exercise increases $\mathbf{O_2}$ **uptake**, **ventilation**, and $\mathbf{O_2}$ **extraction** by working muscles. The **citric acid cycle** and **oxidative phosphorylation** provide the majority of the energy needed for skeletal muscle during aerobic exercise.
- Anaerobic exercise utilizes the **adenosine triphosphate** (**ATP**)**–creatine phosphate** and **lactic acid systems** to generate ATP. Ventilation increases in response to acid challenge.
- Aerobic exercise training adaptations involve many tissues. In skeletal muscle, aerobic enzymes and mitochondria are upregulated. In the heart, stroke volume and maximal cardiac output both increase. In the lungs, there are increases in maximal ventilation and O_2 extraction by peripheral tissues. Combined, these adaptations allow for increases in **maximal** $\mathbf{O_2}$ **consumption**.
- Anaerobic exercise training adaptations are focused in skeletal muscle, where there are increases in muscle size, muscle strength, anaerobic enzymes, and stored energy substrates.

Systems Failure

40

I. OVERVIEW

I like a look of Agony,
Because I know it's true –
Men do not sham Convulsion,
Nor simulate, a Throe –

The Eyes glaze once – and that is Death –
Impossible to feign
The Beads upon the Forehead
By homely Anguish strung.

Emily Dickinson

We are all destined to die.

Living depends on a delicate homeostatic balancing act. During life, we rely on our organ systems to compensate for changes in innumerable internal parameters, including P_{O_2} and P_{CO_2}, pH, electrolyte levels, and body temperature. Ultimately, however, all these compensatory systems slowly falter and then fail. At the cellular level, this process is known as senescence and apoptosis. At the organismal level, we know it as aging and death.

Figure 40.1
Emily Dickinson
(American poet, 1830–1886).

The Centers for Disease Control and Prevention periodically publishes a list of the leading causes of death in the United States (Table 40.1). The list does not include "old age" because it is based on death certificates, which require physicians to identify a specific causal event (e.g., heart failure). From a physiologic perspective, however, corporeal death usually reflects a long series of individual, aging-related cell deaths. Cell by cell, all organs age and, eventually, fail. Which organ falls off the homeostatic tightrope first may be a matter of chance or may be determined by an underlying disease or lifestyle choice. In this final chapter, we consider various causes and consequences of individual organ failure. There are many other causes of death (e.g., accidents and trauma) as shown in Table 40.1, but, regardless, death of the individual occurs when the cerebral hemispheres are O_2 deprived and the cortex dies, either because of cardiovascular failure, respiratory failure, renal failure, or multiple organ system failure. *The Eyes glaze once — and that is Death.*

Table 40.1: Leading Causes of Death in the United States in 2012

Rank	Cause of Death
1	Heart disease
2	Cancer
3	Chronic lower respiratory disease
4	Stroke
5	Accidents
6	Alzheimer disease
7	Diabetes mellitus
8	Influenza and pneumonia
9	Kidney disease
10	Suicide

II. AGING AND DEATH

Gerontology is a relatively new discipline dealing with old issues (and issues encountered by older adults). Although researchers have forwarded many ideas as to why cells and organ systems inevitably lose functionality and fail, there are no solutions to the age-old problem of why we die.

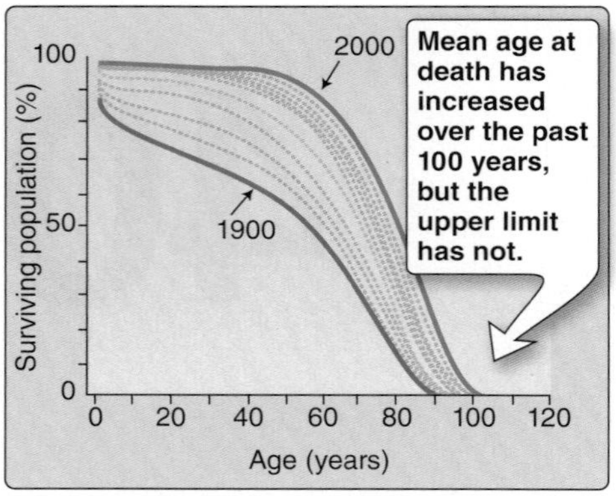

Figure 40.2
Mean life expectancy in the United States, 1900–2000.

Preprogrammed cell death (**apoptosis**) is probably just one of many contributing factors. Regardless, the human lifespan is limited to ~120 years. Medical advances over the past 100 years may have increased mean life expectancy but not the upper limit (Figure 40.2), suggesting that failure and death may be genetically predetermined.

> Apoptosis is a process whereby cells and their contents spontaneously fragment into membrane-bound **apoptotic bodies** that are rapidly engulfed by phagocytes. Apoptosis may be triggered by intrinsic factors, including genetic programming and cell damage, and by extracellular factors, such as toxins and growth factors. Apoptosis is a normal and necessary process for continued tissue health and homeostasis.

A. Physiologic aging

Individual physiologic responses to aging vary widely and can be significantly impacted by physical fitness and underlying disease, but aging is generally accompanied by progressive decreases in cell numbers and in the functionality and responsiveness of remaining cells in most organs. Coinciding with these changes are overall decreases in tissue compliance. The effects of aging on the cardiovascular system, for example, include reduced myocardial sensitivity to adrenergic agonists, which is why maximal heart rate (HR) attained during exercise is estimated from 220 minus age in years. Arteries stiffen with age due to elastin cross-link breakage (see 4·IV·B), increased collagen deposition, and calcification, causing a compensatory increase in arterial blood pressure (see 19·V·C·3). Similar changes occur throughout the body's various tissues. The deep wrinkles that develop in Caucasian skin are the most obvious external indicator of aging, but wrinkles develop largely through **photoaging** (tissue damage induced by ultraviolet light), not by the intrinsic aging process.

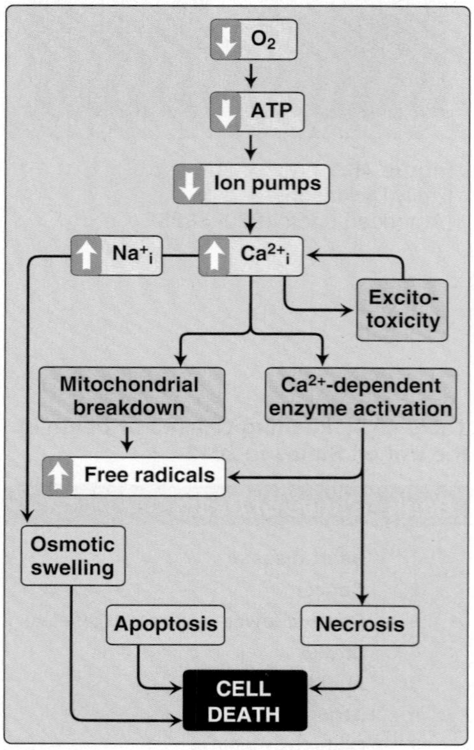

Figure 40.3
Ischemic cascade. ATP = adenosine triphosphate; Na^+_i = intracellular Na^+ concentration; Ca^{2+}_i = intracellular Ca^{2+} concentration.

B. Cell death

Death has many causes, but the final common pathway for most diseases and failing organ systems is O_2 deprivation resulting from inadequacy of perfusion (**shock**; see Sections III and IV below). All organs are dependent on O_2 for continued survival. O_2 restriction caused by interruption of local blood supply (**ischemia**) or reduced arterial O_2 levels (**hypoxemia**) initiates a sequence of biochemical events known as an **ischemic cascade**. Significant events include switchover to anaerobic metabolism, dissipation of ion gradients, Ca^{2+}-induced toxicity, mitochondrial breakdown, apoptosis, and necrosis (Figure 40.3).

1. **Anaerobic metabolism:** O_2 deprivation forces cells to convert from primarily aerobic to exclusively anaerobic metabolism to generate adenosine triphosphate (ATP). The transition is the metabolic equivalent of switching over to an emergency gas-powered generator during a domestic power outage. The generator keeps a few vital systems running, but output is limited by the size of the generator and the capacity of the gas tank (glycogen stores). Also, the exhaust fumes can be deadly in the absence of adequate

ventilation. Anaerobic "exhaust" comes in the form of lactic acid formation, which causes acidosis. Lactic acid is produced even in healthy individuals during intense muscle activity (see 39·III·A·2), but local levels remain relatively low because the circulation limits build up. If the biologic power outage reflects perfusion failure, however, lactic acid levels build rapidly, and intracellular pH falls, which further compromises cell function.

2. **Ion gradients:** Falling ATP levels limit the ability of ion pumps (e.g., Na^+-K^+ ATPase and Ca^{2+} ATPase) to maintain transmembranous ion gradients (Figure 40.4). Membrane potential depolarizes, and intracellular Ca^{2+} concentration rises as a result. In excitable cells, depolarization triggers cationic influx via voltage-dependent Na^+ and Ca^{2+} channels and K^+ efflux via K^+ channels, which effectively collapses the ion gradients within seconds. These ion movements raise intracellular fluid osmolality, causing water to enter by osmosis.

3. **Calcium toxicity:** Ca^{2+} influx and release from intracellular stores activates a number of signaling pathways that ultimately destroy the cell. These include ATPases, *lipases*, *endonucleases*, and Ca^{2+}-activated *proteases* such as *calpains*. *Calpains* are regulatory *proteases* under normal circumstances. When activated by ischemia-induced rises in intracellular Ca^{2+} concentration, *calpains* destroy the cytoskeleton and, with help from Ca^{2+}-dependent *lipases*, digest the plasma and intracellular membranes (Figure 40.5). The cell swells, lyses, and dies (**necrosis**).

> Necrosis is pathologic cell or tissue death, culminating in lysis and release of cellular contents. These materials trigger an inflammatory response that typically causes extensive cellular damage. This contrasts with apoptosis, in which damaged and dying cells stimulate phagocytosis, and their contents remain contained within membranes.

4. **Excitotoxicity: Excitotoxicity** is an aggressive positive feedback pathway that makes the brain highly vulnerable to O_2 deprivation. Ischemia-induced increases in intracellular Ca^{2+} concentration cause synaptic vesicles to fuse with the synaptic membrane, releasing their contents into the synaptic cleft (see 5·IV·C). These vesicles often contain glutamate, which is the brain's principal excitatory neurotransmitter. Postsynaptic glutamate receptors (e.g., N-methyl, D-aspartate receptors) are Ca^{2+} permeable, so intracellular Ca^{2+} levels rise even faster in neurons than they do in nonexcitable tissues (see Table 5.2). Thus, the ischemic cascade is accelerated in brain tissue.

5. **Mitochondrial breakdown:** Reduced O_2 availability impairs mitochondrial function and increases **reactive oxygen species (ROS)** accumulation. ROS include the superoxide anion ($O_2\cdot^-$), hydrogen peroxide (H_2O_2), and the hydroxyl radical (OH·), all produced by the mitochondrial electron transport chain (Figure 40.6). ROS are extremely damaging to cells because they react with and break molecular bonds in lipids, proteins, and DNA. Cells normally aggressively defend themselves against ROS using enzymes

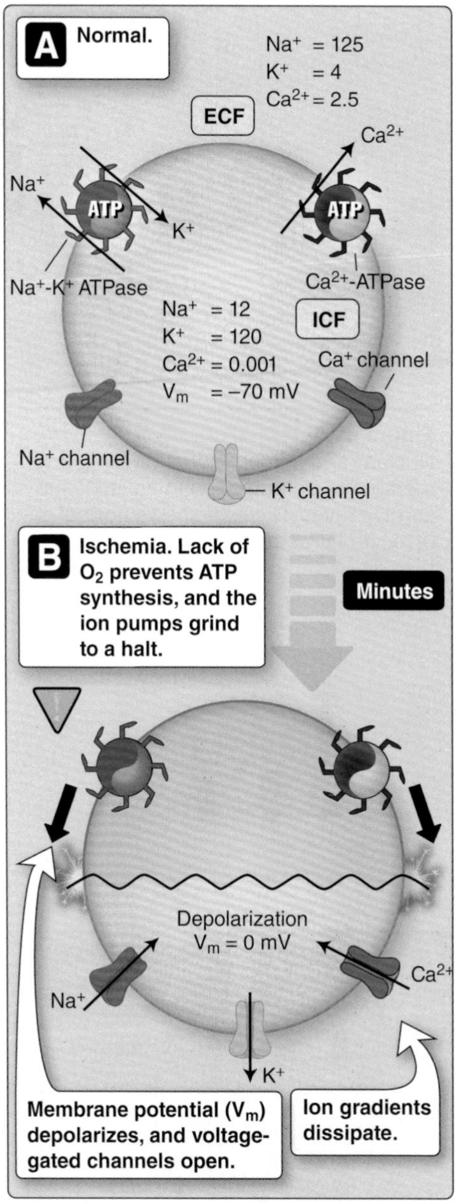

Figure 40.4
Dissipation of transmembrane ion gradient during ischemia. All ion concentrations are given in mmol/L. ECF = extracellular fluid; ICF = intracellular fluid; ATP = adenosine triphosphate.

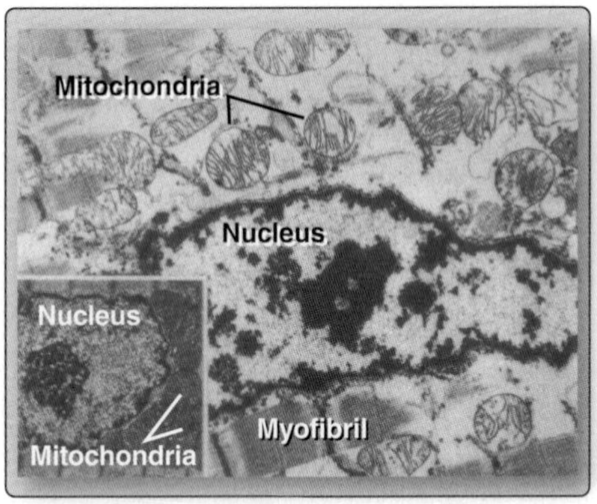

Figure 40.5
Nuclear and mitochondrial swelling and membrane deterioration in an ischemic cardiac myocyte. *Inset* shows normal myocyte ultrastructure.

Clinical Application 40.1: Therapeutic Hypothermia

Most patients (95%) who suffer a cardiac arrest outside of a hospital do not survive, even with attempted resuscitation. Death occurs largely due to neurologic damage sustained during ischemic cascade progression and exacerbated by distribution of inflammatory mediators when the circulation is restored (**reperfusion injury**). The chances of surviving myocardial infarction and avoiding neurological damage have improved significantly in the past decade through use of **therapeutic hypothermia** (**TH**), during which body temperature is reduced to 32°C–33°C for 12–24 hr following the ischemic event. Target temperatures are achieved by infusing a patient with chilled intravenous fluids, often combined with surface cooling. TH is beneficial because it reduces the extent of mitochondrial breakdown and limits inflammatory mediator release during an ischemic cascade.

(e.g., *superoxide dismutase* and *peroxidase*) and ROS scavengers (e.g., vitamins C and E). During ischemia, however, rising ROS levels increase mitochondrial membrane permeability, causing organellar swelling and release of electron-chain constituents that initiate apoptosis. If cell necrosis does not occur within the first few minutes, apoptotic pathways impel cellular suicide over a prolonged timescale, but the end result is the same nevertheless.

C. Brain death

Brain death *is* death. Although our bodily tissues can be sustained artificially following brain death, every trait we associate with being human, including personality, intellect, and awareness of self and others, is a function of the brain. Therefore, when the brain dies, we die. Verifying brain death clinically requires that a set of neurologic tests be performed. The tests are designed to establish a complete and irreversible loss of critical brain functions and reflexes, even though spinal reflexes may persist. Assessing brain function includes testing for the absence of a pupillary light reflex (see 8·II·C) or caloric reflex (response to irrigating the ear canal with warm or cold water; see Clinical Application 9.3). Both assess brainstem function. Establishing brain death also requires that a patient be provided with 100% O_2 and then disconnected from a ventilator and observed for 8–10 minutes to confirm the complete absence of spontaneous respiration even as arterial P_{CO_2} climbs >60 mm Hg (**apnea test**). Reflex increases in respiratory effort induced by hypercapnia are one of the most basic and essential brain functions (see 24·III·C). Some patients may survive a severe ischemic event and progress to a **persistent vegetative state** (**PVS**). PVS patients retain sufficient autonomic brainstem function to preserve basic cardiovascular and pulmonary reflexes, yet show no signs of awareness or comprehension. PVS patients typically die from multiorgan failure, infection, or other causes within 2 to 5 years.

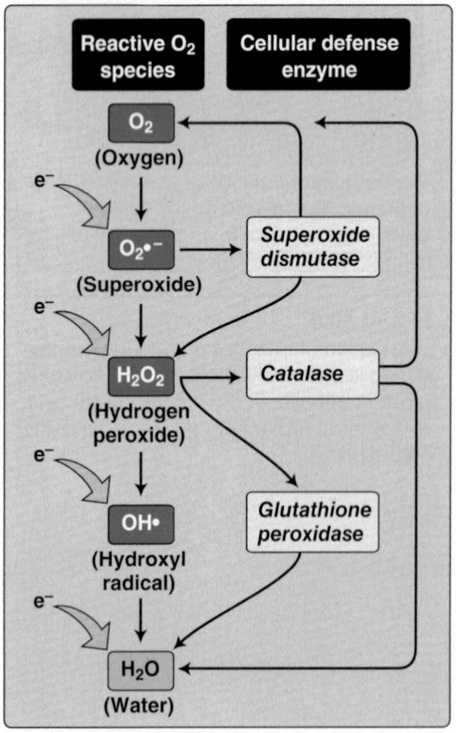

Figure 40.6
Reactive O_2 species produced by the mitochondrial electron (e^-) chain.

III. SHOCK CLASSIFICATIONS

All tissues in the body, including the heart and vasculature, are dependent on the cardiovascular system to deliver O_2 in amounts sufficient to

meet their metabolic needs. The brain has a high dependence on O_2, and loss of consciousness occurs within seconds of interrupting blood flow. Tissues with low O_2 demands can tolerate ischemia for longer periods, but ultimately all tissues die when O_2 deprived. Inadequacy of flow and O_2 delivery results in **shock**. There are three main types of shock: **hypovolemic**, **cardiogenic**, and **distributive**.

A. Hypovolemic

Hypovolemic shock is caused by a decrease in circulating blood volume. When blood volume decreases, the extent to which the left ventricle (LV) is filled during diastole (i.e., LV preload; see 18·III·D) decreases also, which compromises cardiac output (CO) as shown in Figure 40.7. When CO falls, mean arterial pressure (MAP) falls also, which reduces the amount of oxygenated blood reaching tissues. Hypovolemic shock can be further divided into two categories: hemorrhagic shock and shock caused by loss of extracellular fluid (ECF).

1. **Hemorrhagic:** Hemorrhagic shock results from loss of whole blood from the vasculature (**extravasation**). Blood loss to the external environment typically occurs as a result of trauma (see Figure 40.7) but can also occur upon rupture of esophageal or stomach varices. A bone fracture or ruptured abdominal aortic aneurism can also cause significant blood loss to internal compartments.

2. **Fluid loss:** Hypovolemic shock can also result from ECF volume contraction, due either to fluid loss to the external environment or to the interstitium and abdominal cavities (**"third spacing"**). Fluid is lost to the environment during sweating, vomiting, episodes of diarrhea, and following significant skin burns (see 16·III·B). Third spacing occurs when plasma protein concentrations fall, either as a result of liver failure and impaired ability to synthesize plasma proteins or increased capillary permeability to proteins.

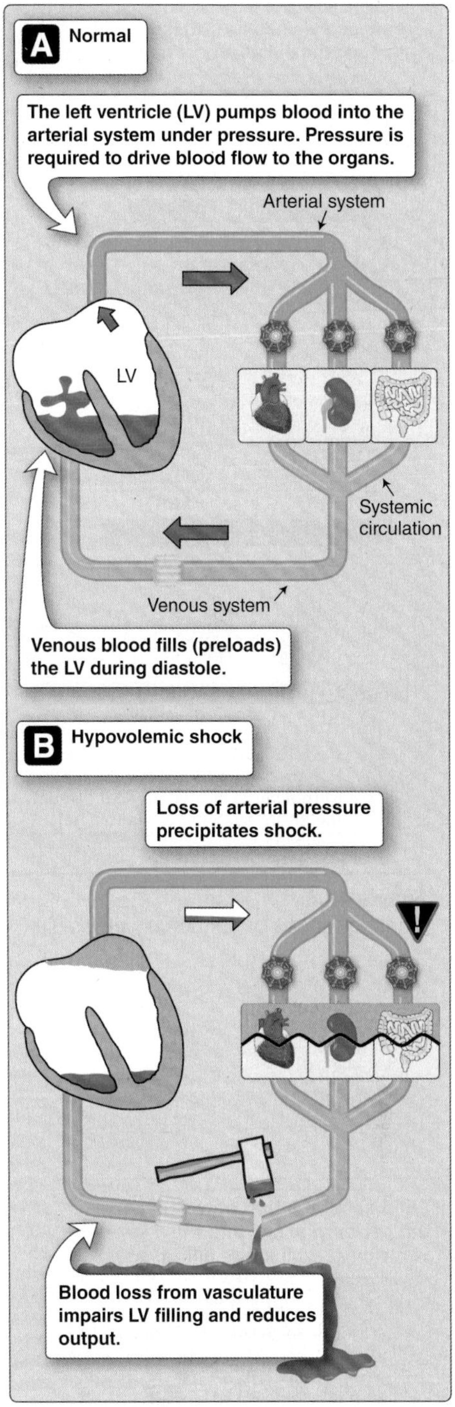

Figure 40.7
Hypovolemic shock.

> Plasma proteins create an osmotic potential (π_c) that is the main force holding fluid in the vasculature, as defined by the **Starling law of the capillary**:
>
> $$Q = K_f [(P_c - P_{if}) - (\pi_c - \pi_{if})]$$
>
> where Q is net fluid flow across the capillary wall, K_f is a filtration coefficient, P_c is capillary hydrostatic pressure, P_{if} is interstitial fluid pressure, and π_{if} is interstitial colloid osmotic pressure (see 19·VII·D).

B. Cardiogenic

Cardiogenic shock is caused by cardiac pump failure. There are four general causes: **dysrhythmia**, **mechanical issues**, **cardiomyopathies**, and **extracardiac issues**.

1. **Dysrhythmia:** Cardiogenic shock can result from atrial or ventricular dysrhythmias. Dysrhythmias prevent or impair coordinated

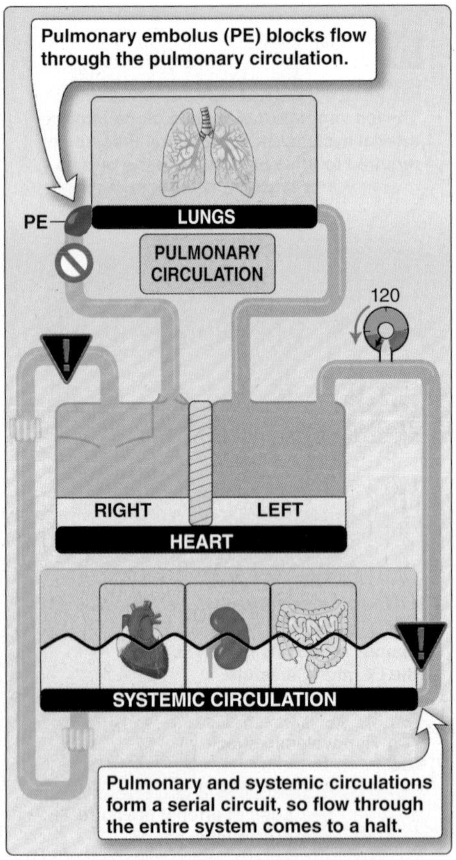

Pulmonary embolus (PE) blocks flow through the pulmonary circulation.
PE
LUNGS
PULMONARY CIRCULATION
120
RIGHT
LEFT
HEART
SYSTEMIC CIRCULATION
Pulmonary and systemic circulations form a serial circuit, so flow through the entire system comes to a halt.

Figure 40.8
Consequences of pulmonary embolism.

Clinical Application 40.2: Sepsis

Sepsis is a clinical syndrome reflecting a systemic inflammatory response to infection. It is characterized by bacteremia, fever, tachycardia, and increased respiratory rate. Although sepsis can be caused by a variety of organisms, it is often seen in association with Gram-negative infections, in which a bacterial cell wall component (lipopolysaccharide [LPS]) triggers an **inflammatory cascade.**[1] LPS binds to and is recognized by a receptor on the surface of phagocytes, which respond by releasing cytokines and initiating an inflammatory response and fever. Vascular endothelial cells respond to bloodborne cytokines by releasing more cytokines and chemokines, thereby further amplifying the inflammatory response. They also initiate blood coagulation. This inflammatory cascade also includes neutrophil activation and release of reactive O_2 species, causing extensive and widespread vascular damage. Resistance vessels and veins lose their resting tone, thereby increasing vascular capacity. Capillary permeability may be increased also, allowing plasma proteins and fluids to leak into the interstitium. Sepsis mortality rates can be as high as 50%, increasing to 90% when shock develops. Treatment options include antibiotics to address the underlying infection, intravenous fluids to help maintain effective circulating blood volume, and vasopressors to increase vascular tone.

contraction of one or more cardiac chambers, which reduces CO. Ventricular tachycardia and fibrillation cause complete loss of CO and prove rapidly fatal unless the arrhythmia is corrected by cardioversion using an external electrical defibrillator (see 17·V·D).

2. **Mechanical:** Incompetent and stenotic heart valves both reduce cardiac efficiency and challenge the ability of the myocardium to maintain a basal CO. Septal defects that allow left-to-right ventricular backflow can also precipitate cardiogenic shock.

3. **Cardiomyopathy:** The causes and consequences of heart disease are considered in more detail below. Myocardial infarction (MI) that damages >40% of the LV wall is one of the most common causes of cardiogenic shock and death.

4. **Extracardiac:** Extracardiac causes of shock include **pulmonary embolism** (**PE**), advanced **pulmonary hypertension**, **tamponade**, and **pericarditis**. PE and pulmonary hypertension limit right ventricular (RV) output, which limits LV preload. Massive PE can effectively bring the circulation to a halt and result in instant death (Figure 40.8). Tamponade is caused by fluid accumulation (e.g., blood or a pericardial effusion) between the pericardium and heart wall. The presence of fluid prevents normal ventricular filling (Figure 40.9). Inflammation-induced pericardial thickening can similarly limit ventricular filling.

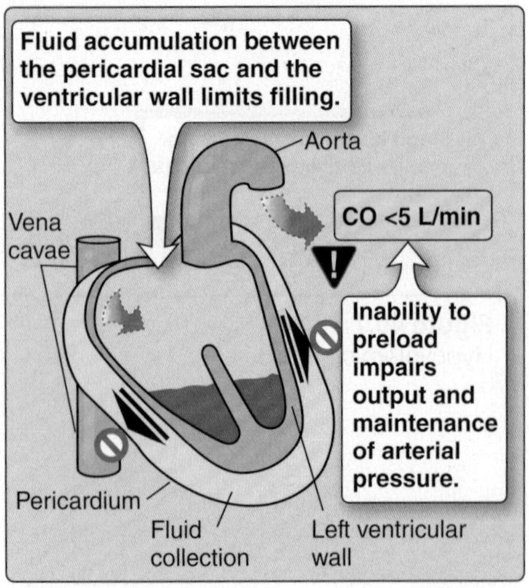

Figure 40.9
Tamponade effects on ventricular filling.

[1]For more information on differences between Gram-positive and Gram-negative bacteria, see *LIR Microbiology*, 3e, p. 51.

C. Distributive

Most arterial and venous vessels have a resting tone that is controlled by the sympathetic nervous system (SNS) as a way of limiting cardiovascular capacity to ~5 L (see 20·V). Distributive, or **vasodilatory**, shock occurs when the SNS loses control over the vasculature, and its capacity increases exponentially through vasodilation. MAP dissipates rapidly as blood flows into dilated resistance vessels and becomes trapped in capillary beds and veins (Figure 40.10). Distributive shock has many causes. The most common include **sepsis** (see Clinical Application 40.2), **systemic inflammatory response syndrome**, and **anaphylaxis**.

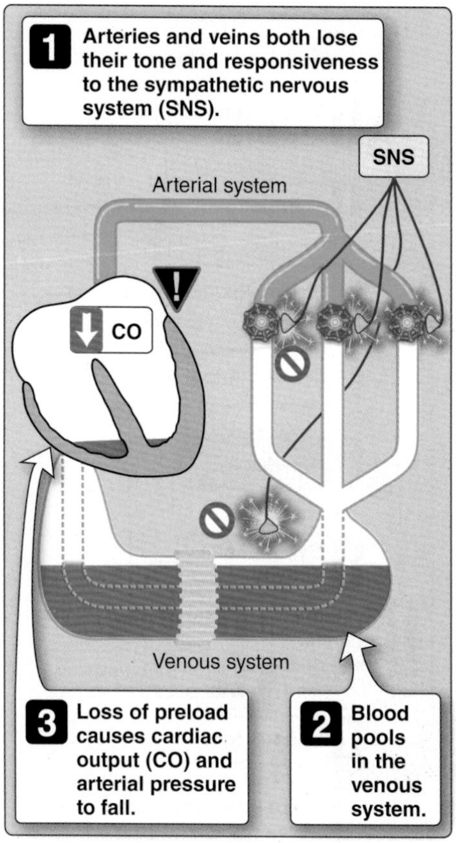

Figure 40.10
Distributive shock.

IV. SHOCK STAGES

The progression of shock can be divided into three stages, beginning with the initial causal event and then progressing in a sequential manner through **preshock**, shock, and **end-organ failure**. The following discussion uses **hemorrhagic shock** as an example to illustrate how the body responds to the initial event and how the systems that attempt to compensate for loss of MAP can create positive feedback spirals that may ultimately hasten failure and lead to death.

A. Preshock

Hemorrhage depletes blood volume and drains the venous reservoir. Hemorrhage depletes veins preferentially because the heart continues transferring blood from the venous compartment to arteries and their dependent capillaries until the venous compartment is depleted. Loss of preload causes MAP to begin to fall, triggering an SNS-mediated baroreceptor reflex (see Figure 20.14 and 20·III). The SNS redirects blood flow away from nonessential organs, increases cardiac inotropy and heart rate (Figure 40.11), and mobilizes blood reservoirs. These pathways are summarized in Figure 40.12.

1. **Redirection of flow:** Flow to nonessential organs is reduced by selective SNS-mediated constriction of resistance vessels. Systemic vascular resistance (SVR) rises as blood flow is directed away from splanchnic, cutaneous, and muscle vascular beds. Reduced flow to the kidney triggers *renin* release from the juxtaglomerular apparatus (JGA) and activates the long-term fluid retention pathways (see below). Two key components (**angiotensin II** and **antidiuretic hormone**) are vasoactive and potentiate SNS-mediated vasoconstriction (see 28·III).
2. **Cardiac efficiency:** SNS stimulation of the myocardium increases HR and contractility to help compensate for loss of preload (see Figure 40.11). Epinephrine release from adrenal glands during SNS activation contributes to tachycardia and increased inotropy during preshock.
3. **Venoconstriction:** SNS stimulation of veins increases their tone and decreases their capacity, forcing blood back to the heart. Venous return (VR) is aided by a steepening of the pressure gradient between capillary beds and the right atrium.

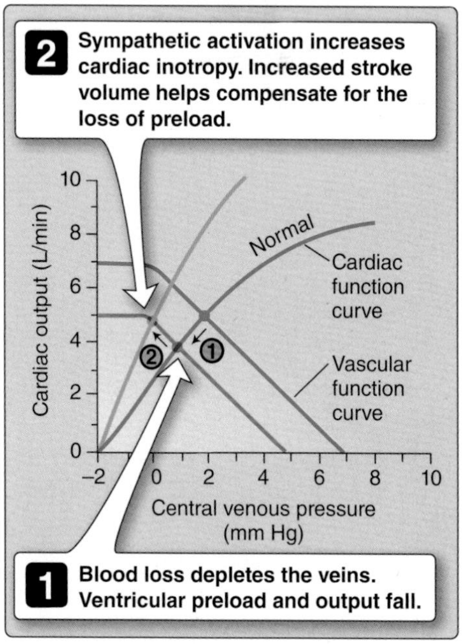

Figure 40.11
Reflex increases in cardiac inotropy during hypovolemia.

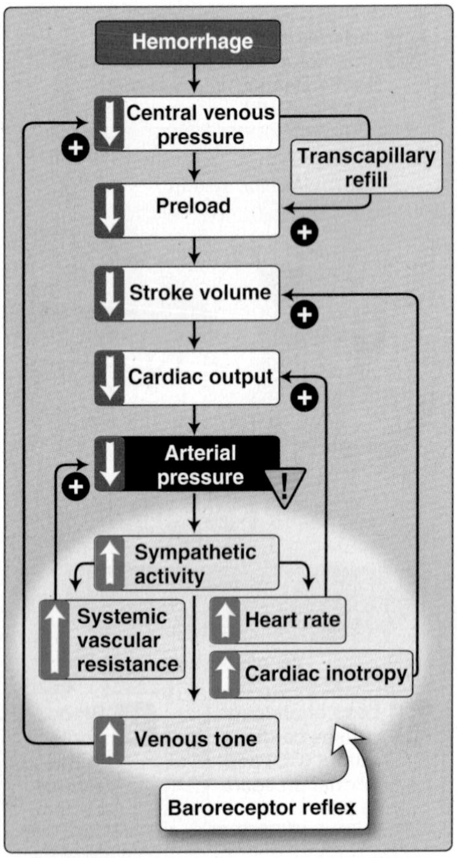

Figure 40.12
Pathways that preserve arterial pressure during preshock.

4. **Transcapillary refill:** The baroreceptor reflex helps preserve flow to critical organs during the first few minutes after a hemorrhage. It also buys time that allows fluid to migrate from the interstitium to the vasculature, a process is known as **transcapillary refill.** Transcapillary refill is a primary survival mechanism that relies on Starling forces to recruit interstitial fluid (see above). Resistance vessel constriction reduces capillary hydrostatic pressure and allows plasma colloid osmotic pressure (π_c) to drive fluid movement from the interstitium into the vasculature (see Figure 19.28). Fluid influx dilutes the plasma proteins and reduces π_c, but transcapillary refill can still replace ~75% of lost blood volume during hemorrhage.

B. Shock

The baroreceptor reflex effectively compensates for decreases in circulating blood volume of ~10%, and, thus, the only early sign of imminent shock may be mild tachycardia. Once blood volume drops by greater than ~10%, however, compensatory mechanisms are no longer adequate to sustain perfusion in critical circulations, and signs of shock become overt. These signs include **hypotension**; cold, clammy skin; decreased urine output; and a rise in plasma lactate levels.

> Most organ systems have **functional reserves** that permit homeostasis even as system capacity is reduced. The efficacy of cardiovascular reserves explains why an individual can donate a unit of blood with little or no detrimental effect on MAP.

1. **Hypotension:** When blood volume drops below ~10%, increases in HR and inotropy alone are unable to compensate for loss of preload, and systolic blood pressure drops to 90 mm Hg or below. Intense SNS-mediated constriction of the vasculature limits blood outflow from the arterial system and keeps diastolic blood pressure high, and, thus, MAP is maintained at a level that allows blood to reach the critical cerebral and coronary circulations. These circulations are regulated primarily through autoregulatory mechanisms (e.g., CO_2, K^+, and lactate release), and, thus, are not directly influenced by SNS activity.

2. **Skin:** Intense SNS activation raises SVR by effectively shutting off flow to the vascular beds that occupy the lowest positions on the circulatory hierarchy, including the splanchnic and cutaneous circulations. The intensity of SNS activation is clearly apparent in the skin, which becomes cold and clammy. Cooling is due to intense cutaneous vasoconstriction, which reduces flow to <6 mL/min. Blood drains from the skin, and its temperature cools accordingly. SNS activation also stimulates sweat glands. Because sweat is a modified blood filtrate, when blood flow is curtailed, output is minimal. The skin becomes slightly clammy to the touch.

3. **Urinary output:** The renal glomerular afferent and efferent arterioles are both resistance vessels. During intense SNS activation, flow through both is restricted severely, and glomerular hydrostatic pressure (P_{GC}) falls dramatically (Figure 40.13; see 25·IV·F). P_{GC}

determines glomerular filtration rate (GFR), so flow through the tubule and urine production slows also to <30 mL of urine output per hour (**oliguria**).

4. **Metabolic acidosis:** Plasma lactate levels are normally 0.5–1.5 mmol/L, but hypoxia forces many tissues to rely on anaerobic metabolism, and, hence, lactate levels rise. A plasma lactate of >4 mmol/L is consistent with shock, although lactate levels can rise under other circumstances also (e.g., ketoacidosis and anaerobic exercise).

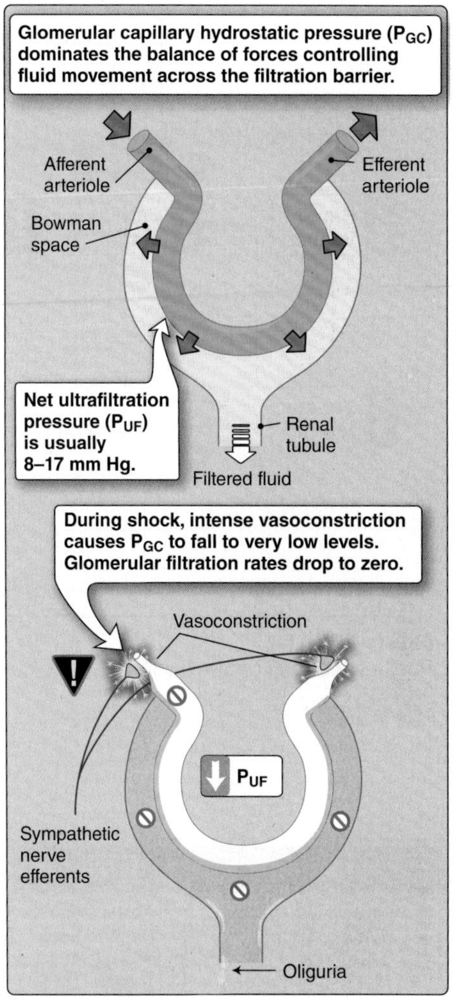

Figure 40.13
Effects of intense sympathetic activation on glomerular blood flow.

C. System failure

The actions described above may be insufficient to ensure patient survival, even though arterial pressure may renormalize for an hour or two. Blood pressure alone may not reliably reflect adequacy of perfusion in early shock because the central nervous system (CNS) cardiovascular control centers have the ability and determination to maintain MAP at levels that ensure continued flow to the cerebral circulation to the very last. In cases of severe hemorrhage, this is accomplished by holding SVR at levels that compromise organs that occupy lower positions on the circulatory hierarchy (see 20·II·F), including the kidneys and gastrointestinal (GI) system. Once the invisible line demarcating reversible from irreversible shock has been crossed, a positive feedback spiral begins that leads inevitably to organ failure and death (Figure 40.14).

> The need to restore circulating blood volume as soon as possible after trauma is a major reason for the widespread use of mobile trauma teams and Medivac helicopters. Rapid-response units allow medical personnel to reach the scene of an accident and administer intravenous fluids to a patient within the critical window before irreversible tissue damage occurs (a period of variable duration often referred to in Emergency Medicine as the **"golden hour"**).

1. **Cardiac depression:** If MAP drops below 60 mm Hg, the myocardium becomes ischemic through inadequacy of coronary perfusion. Ischemia impairs myocardial contractility, and, therefore, pressure falls further. So begins a positive feedback cycle that results in acute heart failure. During hemorrhage, one or more **myocardial-depressant factors** may be released from ischemic tissues that further challenge cardiac function.

2. **Sympathetic escape:** The SNS cannot maintain intense vasoconstriction for prolonged periods, so SVR eventually falls. Resistance vessel dilation (**"sympathetic escape"**) may be due to SNS neurotransmitter depletion, α-adrenergic receptor desensitization, or chronically elevated metabolite concentrations overriding central control. Venoconstrictor influence ultimately fails also, thereby impairing VR and preload (see Figure 40.10).

3. **Acidemia:** Lactic acid and high P_aCO_2 (due to inadequacy of tissue perfusion and pulmonary and renal dysfunction) together cause significant acidemia. Acidemia impairs myocyte function and further

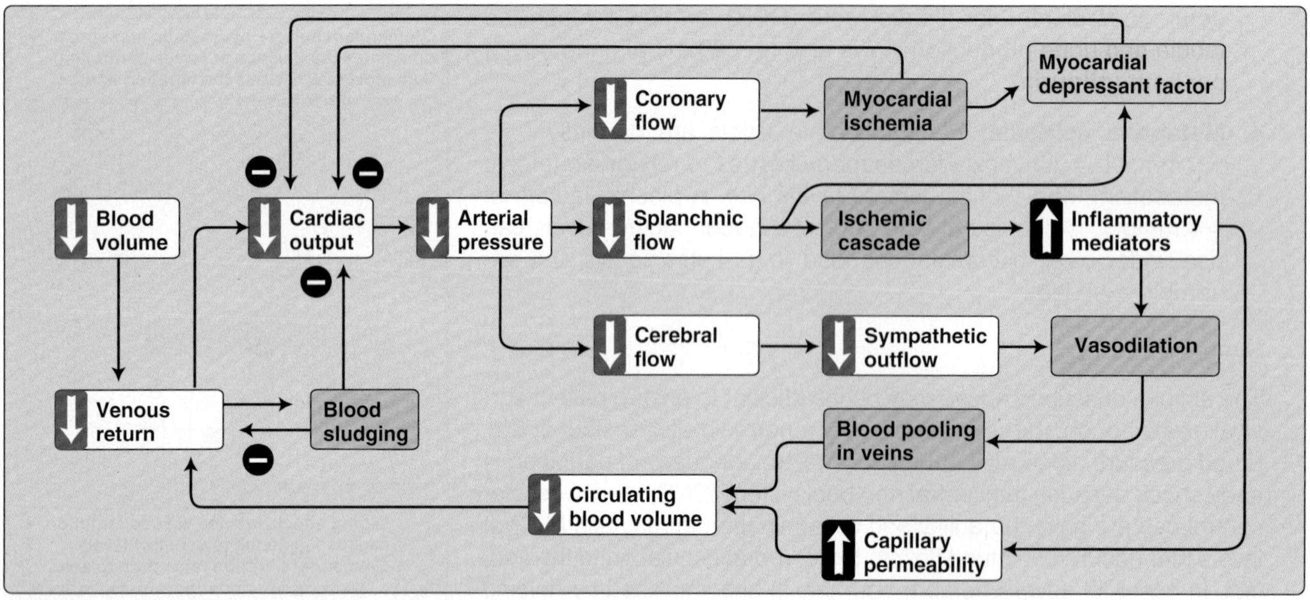

Figure 40.14
Positive feedback pathways that cause cardiovascular system failure.

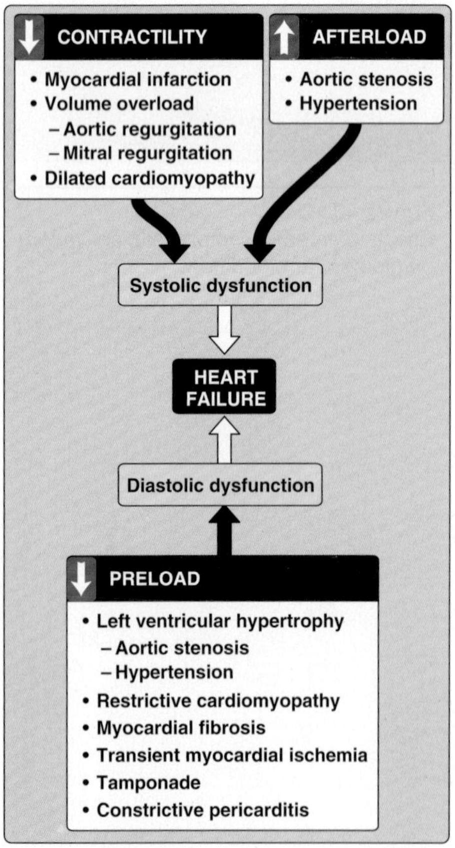

Figure 40.15
Common causes of heart failure.

challenges the ability of the myocardium and vasculature to sustain CO and SVR, respectively. MAP falls yet further as a result.

4. **Increased blood viscosity:** When blood is moving slowly, red blood cells (RBCs), and other blood components adhere to each other, which raises blood viscosity (see 19·III·C). The process is exacerbated by acidemia and involves not only RBCs, but also leukocytes and platelets, causing blood **"sludging."** Sludging increases resistance to flow through the vasculature, and, because MAP cannot rise to compensate, tissue perfusion falls as a consequence. In time, the microvessels (i.e., capillaries, arterioles) become plugged with clots.
5. **Cellular deterioration:** With prolonged hypoxemia, cell integrity breaks down, triggering an inflammatory response. Inflammatory mediators increase vascular permeability and plasma exudes into the interstitial space at the expense of blood volume. Deterioration of the GI epithelial lining breaches the barrier separating the gut contents from the vasculature, allowing microorganisms to gain access to the circulation. The likelihood of septic shock when (and if) circulation is restored is increased greatly as a result.
6. **Cerebral depression:** Prolonged hypoxemia ultimately impacts the brain. Neural activity is depressed, and the cardiovascular and respiratory control centers fail. As sympathetic output wanes, MAP declines. Cerebral perfusion pressure decreases also, and brain death soon follows.

V. HEART FAILURE

Heart failure can be the final common pathway for virtually all forms of cardiac disease, and, therefore, there are many underlying causes (Figure 40.15). Although the time course of failure can vary widely, it may ultimately result in cardiogenic shock.

A. Causes

There can be considerable overlap in the ways in which the various underlying causes of heart failure impact cardiac performance. The right and left ventricles face different challenges and can fail independently of each other, but the left heart is so dependent on the right (and *vice versa*) that failure of either side independently elicits similar compensatory mechanisms.

1. **Right heart:** The right heart is a thin-walled chamber that generates low peak systolic pressures against a low pulmonary vascular resistance (PVR). If flow resistance increases as a result of PE or pulmonary hypertension, for example, it has limited ability to compensate, and, hence, failure develops.

> The most common cause of right heart failure is left heart failure.

2. **Left heart:** The left heart is a thick-walled chamber well adapted to stress associated with generating high peak systolic pressures against a high SVR. Causes of left-heart failure can be grouped according to whether they impair filling (**diastolic heart failure**, also known as **heart failure with preserved ejection fraction**), or ejection (**systolic heart failure**).

   a. **Diastolic:** One common cause of diastolic failure is LV hypertrophy, due either to a chronically increased afterload or a cardiomyopathy. Untreated hypertension and aortic stenosis both impair CO by increasing LV afterload. The myocardium hypertrophies in order to generate the high pressures required to maintain a normal CO (see Clinical Application 18.2). New myofibrils are added in parallel with old myofibrils, causing individual myocytes to increase their girth, thickening the ventricular wall (Figure 40.16). The advantage to a thicker wall is that it helps offset the effects of high intraventricular pressure on wall stress, as described by the law of Laplace (Figure 40.17; see 18·IV). The disadvantages to hypertrophy are twofold. First, myocyte diameter may exceed the diffusional limits for O_2, which increases the likelihood of ischemia (see Figure 40.16) and arrhythmias. Second, the ventricle stiffens and becomes increasingly difficult to fill, requiring higher RV ejection pressures. Ultimately, both ventricles fail under such circumstances.

   b. **Systolic:** Systolic heart failure occurs when the LV fails to maintain adequate output. This can be due to impaired contractility or an excessive afterload, but the most common cause of systolic failure is MI, as discussed below.

B. Myocardial infarction

MI, or a "heart attack," is one of the most common causes of heart failure. An MI typically occurs when an atherosclerotic plaque ruptures and forms a blood clot that occludes a coronary supply vessel. Myocytes that had previously been served by the occluded vessel become ischemic and die, which impairs myocardial contractility.

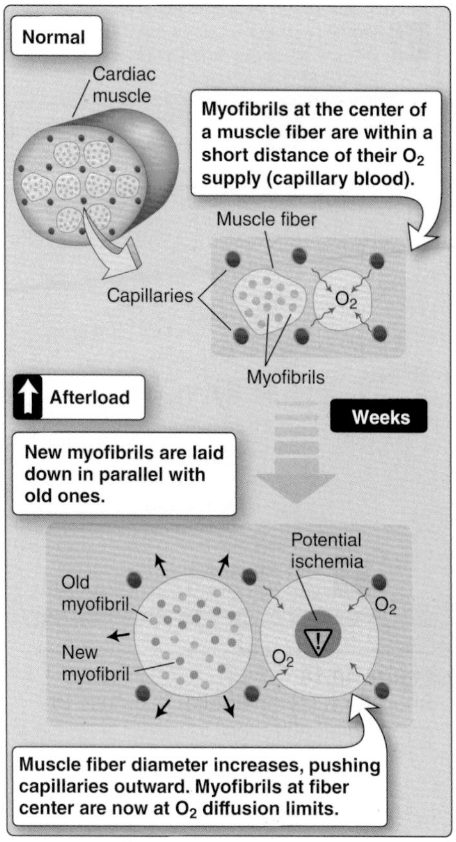

Figure 40.16
Effects of myocardial hypertrophy on O_2 delivery to myofibrils.

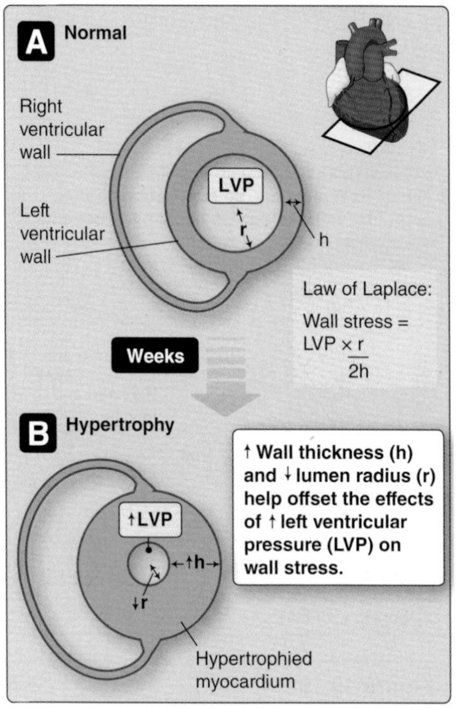

Figure 40.17
Increases in ventricular wall thickness during cardiac hypertrophy.

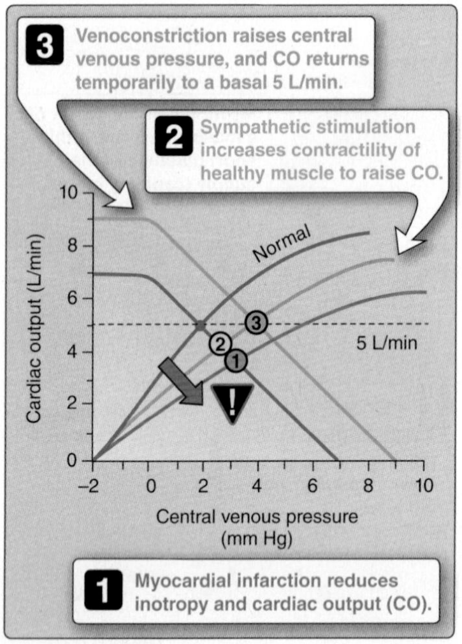

Figure 40.18
Short-term (sympathetic) response to myocardial infarction.

The chances of surviving such an event depend on many factors. If the infarcted area is relatively small, short-term responses may allow the patient to survive the initial insult until long-term compensatory mechanisms become active.

C. Compensation

A small infarct triggers short-term and long-term compensatory mechanisms simultaneously. Short-term events help maintain output until the longer-term pathways have had time to activate fully.

1. **Short term:** The short-term response to myocardial ischemia includes both local and central reflexes.

    a. **Local:** Interruption of blood flow to myocytes causes interstitial metabolite levels (e.g., adenosine, K^+, CO_2, lactate) to rise. All resistance vessels in the immediate vicinity dilate reflexively through local vascular control mechanisms. Collaterals are tonically constricted normally, but they also participate in the vasodilatory response to rising metabolite levels. Blood flow through collaterals may allow areas peripheral to a focal infarct to survive the initial ischemic event (see 21·III·E).

    b. **Central:** Death of myocytes impairs myocardial contractility, which reduces LV stroke volume and CO (Figure 40.18). MAP falls as a result, triggering a baroreceptor reflex that involves all of same effector mechanisms described in Section IV·A above. If the infarct is small, these pathways may be sufficient to restore MAP.

2. **Long term:** A decrease in MAP also activates the *renin–*angiotensin–aldosterone system, regardless of cause (see 20·IV). It takes 24–48 hours for Na^+ and water retention mechanisms to expand ECF volume and support the failing myocardium with increased preload. In the days and weeks following an ischemic event, the body begins repairing some of the tissue damage wrought by infarction. Collateral vessels enlarge, and the myocardium hypertrophies to help compensate for the loss of contractility.

D. Preload penalty

The Frank-Starling mechanism is highly effective in compensating for minor decreases in cardiac inotropy (see 18·III·D). Some individuals may suffer a series of these insults and remain unaware for years until compensation causes symptoms (e.g., dyspnea associated with pulmonary congestion, as discussed below). Dyspnea is only one of several penalties associated with preloading, however. Others include limits to the benefits of length-dependent sarcomeric activation, excessive ventricular wall stress, dysrhythmias, valve incompetency, and edema.

1. **Length-dependent activation limits:** Preloading a healthy heart increases CO through length-dependent activation of sarcomeres (see 13·IV). The end-systolic pressure–volume relationship has a plateau region, however, and, once myocytes have been stretched to a length that optimizes force generation, further increases in preload are ineffective in generating additional force (Figure 40.19).

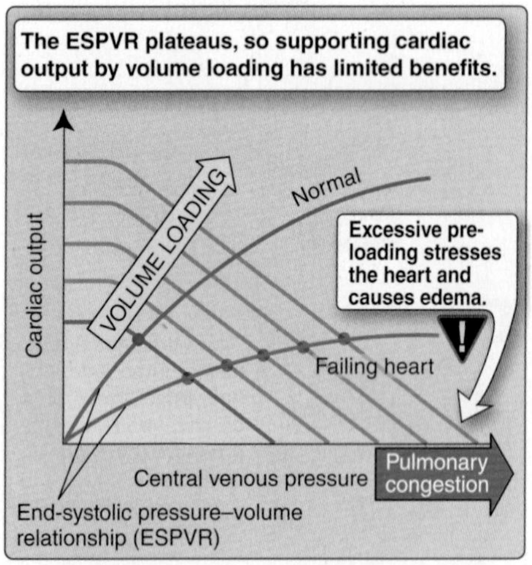

Figure 40.19
Limits to the beneficial effects of preloading.

2. **Wall stress:** Preloading dilates the ventricle and increases wall tension, as predicted by the law of Laplace. Wall tension contributes to afterload, so, although preloading does help support output within normal physiologic ranges, high preloads also increase cardiac workload and reduce its efficacy.

3. **Dysrhythmias:** Excessive preloading stretches the ventricular wall and distorts conduction pathways, which predisposes the myocardium to potentially fatal dysrhythmias and arrhythmias.

4. **Valve incompetency:** Excessive preloading also stretches and distorts the cartilaginous valve rings and unseats the valves. In practice, this means that the valve leaflets no longer come into close apposition upon closure, and blood flows in a retrograde manner. Regurgitation further increases the workload required of a failing heart.

5. **Edema:** Central venous pressure (CVP) elevation raises mean capillary pressure and favors fluid filtration from blood into the interstitium. In the systemic vasculature, interstitial fluid excess manifests as swollen ankles and feet. The lower extremities are particularly prone to edema in an erect individual because vascular pressures in these regions are enhanced by gravity. In the lungs, fluid filters from the pulmonary capillaries and collects in the alveolar sacs, where it interferes with gas exchange (pulmonary congestion) as shown in Figure 40.20. Pulmonary edema may cause **orthopnea** (shortness of breath when lying flat), forcing patients to sleep sitting upright. Gravity helps reduce pulmonary perfusion pressures and, thus, decreases the likelihood of fluid accumulation in the air spaces.

E. System failure

A failing heart becomes locked in a decompensatory spiral in which preload supports output yet ultimately limits efficiency through its effects on wall tension. Unless this cycle is interrupted and managed, it can prove fatal. Thus, the goal of medical intervention is to decrease preload using diuretics while simultaneously supporting the myocardium with inotropes that help it work more efficiently at a lower filling volume.[1] In end-stage heart failure, myocytes continue to die one by one, slowly chipping away at contractility and the ability to sustain MAP. Even mild physical exertion causes severe dyspnea because cardiac reserve has dropped to the point where even minimal muscular activity places demands on output that exceed myocardial capabilities, so patients become bedridden (see 21·III·B). Bedrest exacerbates frailty by causing muscle atrophy and decreased bone density. Excessive volume loading causes pulmonary edema and hypoxic respiratory failure. The liver fails due to passive congestion and restricted O_2 delivery caused by systemic edema. Loss of glomerular pressure precipitates renal failure. Each additional organ loss raises mortality risk by ~20%.

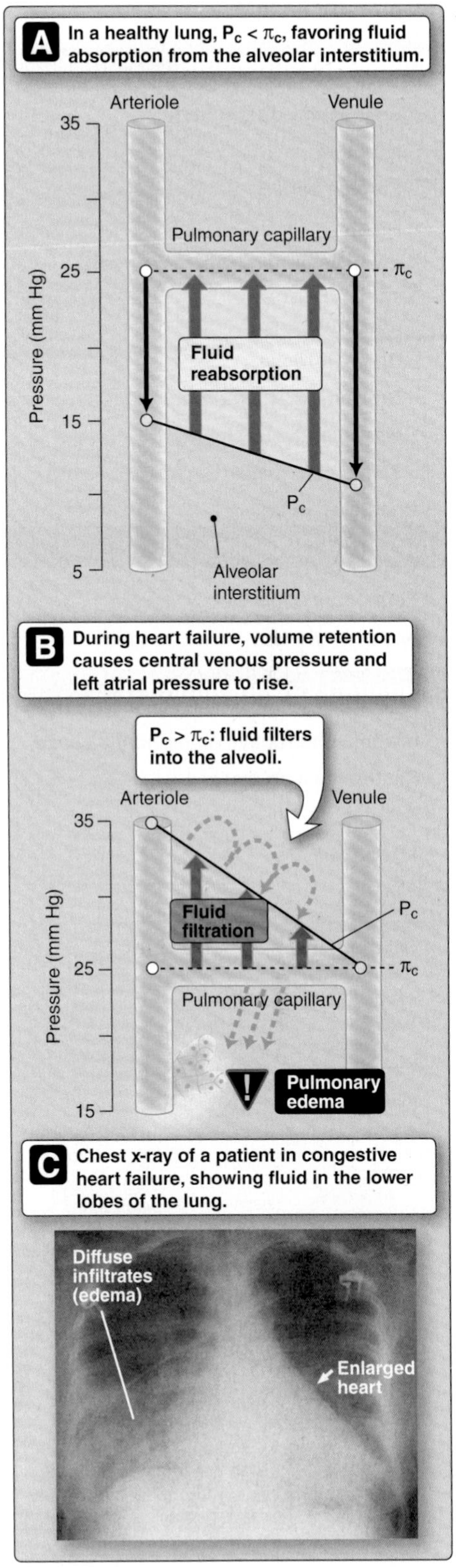

Figure 40.20
Pulmonary edema resulting from heart failure. P_c = capillary hydrostatic pressure; π_c = plasma colloid osmotic pressure.

[1]For more information on pharmaceutical approaches to treatment of heart failure, see *LIR Pharmacology*, 5e, p. 193.

Table 40.2: Common Causes of Respiratory Failure

Impaired Ventilation
• Upper airway obstruction
▫ Infection
▫ Foreign body
▫ Tumor
• Weakness or paralysis of respiratory muscles
▫ Cerebral trauma
▫ Drug overdose
▫ Guillain-Barré syndrome
▫ Muscular dystrophy
▫ Spinal cord injury
• Chest wall injury
Impaired Diffusion
• Pulmonary edema
• Acute respiratory distress syndrome
Impaired $\dot{V}_A/\dot{Q}$ matching
• Chronic obstructive pulmonary disease
• Restrictive lung disease
• Pneumonia
• Atelectasis

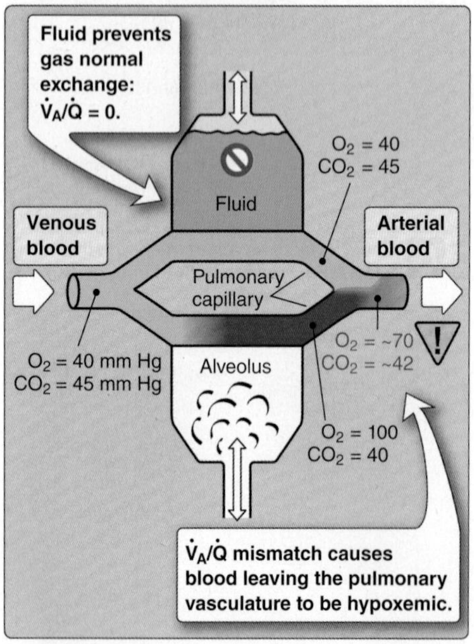

Figure 40.21
A $\dot{V}_A/\dot{Q}$ mismatch, one common cause of hypoxemia. $\dot{V}_A$ = alveolar ventilation; $\dot{Q}$ = alveolar perfusion, all partial-pressure values are given in mm Hg.

VI. RESPIRATORY FAILURE

Respiratory failure occurs when the respiratory system is unable to fulfill one or both of its gas exchange functions, namely O_2 uptake or CO_2 elimination. Clinically, this manifests as **hypoxemic respiratory failure** or **hypercapnic respiratory failure**, respectively. The two failure types represent **syndromes** (sets of related symptoms) rather than the end result of any specific disease.

A. Causes

Respiratory failure may develop chronically or acutely, usually as a result of trauma (see Section D below for a discussion of **acute respiratory distress syndrome** [**ARDS**]). Conditions causing respiratory failure can be grouped according to whether they impair ventilation (air-pump function and control), diffusion (integrity of the blood–gas interface), or ventilation–perfusion ($\dot{V}_A/\dot{Q}$) matching (Table 40.2).

B. Hypoxemic respiratory failure

Hypoxemic respiratory failure is characterized by a P_aO_2 <~60 mm. Hypoxemia can be caused by hypoventilation, but, because attaining a normal P_aO_2 (100 mm Hg) requires that the full area of the blood–gas interface be functional, processes that decrease this area and allow venous blood to pass through the lungs without being arterialized cause some degree of hypoxemia. Thus, hypoxemic respiratory failure usually occurs when air or pulmonary blood is unable to access the interface (i.e., $\dot{V}_A/\dot{Q}$ mismatch).

1. **Ventilation/perfusion mismatch:** Regional $\dot{V}_A/\dot{Q}$ mismatch is common in a healthy lung but has minimal impact on overall respiratory function (see 23·IV·B). In disease states, large numbers of alveoli may collapse and seal (**atelectasis**), or fill with fluid (pulmonary edema), pus (**pneumonia**), or blood (hemorrhage), all of which effectively prohibit O_2 uptake and cause hypoxemia (Figure 40.21).

2. **Compensation:** Hypoxemia is detected primarily by the aortic and carotid chemoreceptors (see 24·III·C). CNS respiratory control centers respond by increasing minute ventilation, and CNS cardiovascular control centers simultaneously increase CO to help maximize the O_2 diffusion gradient across the exchange barrier.

3. **Consequences:** The physiologic consequences of hypoxemia were discussed in relation to the effects of ascent to altitude (see 24·V·A). Mild hypoxemia causes slight impairment of mental function and visual acuity. When P_aO_2 drops below ~40–50 mm Hg, patients become confused and prone to personality changes and irritability. Hypoxemia also initiates a positive feedback spiral in which the pulmonary vasculature constricts reflexively and further reduces O_2 uptake. Vascular constriction also increases right ventricular afterload and induces pulmonary hypertension, which stresses the RV. These symptoms can usually be reversed clinically by administering O_2 to maximize $\dot{V}_A/\dot{Q}$ ratios and, at least temporarily, restore P_aO_2 until the underlying cause of hypoxemia can be evaluated and addressed.

C. Hypercapnic respiratory failure

Hypercapnic respiratory failure is indicated by an acute rise in P_aCO_2 to $>\sim$50 mm Hg. Hypercapnia that develops over a course of months is tolerated well, however, so failure may not occur until P_aCO_2 reaches $\sim$70–90 mm Hg. Unlike hypoxemia, hypercapnia can be corrected relatively easily by adjusting alveolar ventilation. Thus, hypercapnic respiratory failure usually only occurs when ventilatory control is impaired. Hypercapnia is usually associated with varying degrees of hypoxemia.

1. **Ventilation:** Ventilatory failure occurs if the respiratory center or its neural pathways are damaged by stroke, drug overdose, or neuromuscular diseases (e.g., myasthenia gravis), but the most common causes of hypercapnic respiratory failure are impairment of air-pump function (chest wall and respiratory muscles) and chronic airway disorders.

    a. **Chest wall:** Movement of the chest wall can become severely limited by obesity and by abnormal spine curvature. Kyphosis (forward flexion curvature), as shown in Figure 40.22A, and scoliosis (lateral curvature), as shown in Figure 40.22B, are congenital disorders, but the former is also seen in association with arthritis and osteoporosis (see Figure 40.22A). Both can develop into debilitating curvatures that severely limit chest wall excursions and hasten failure in a compromised lung.

    b. **Muscles:** The respiratory muscles (diaphragm and intercostals) increase intrathoracic volume and expand the lungs during inspiration. They are skeletal muscles and, therefore, susceptible to wasting diseases such as muscular dystrophy. They are also subject to **fatigue**, which is a principal concern when addressing problems underlying respiratory failure. Chronic conditions that reduce chest wall or lung compliance (restrictive lung diseases) increase the work of breathing, and fatigue ultimately reduces contractility and causes hypoventilation and hypercapnia.

    c. **Airways:** Chronic obstructive pulmonary disease and asthma increase airway resistance and can reduce alveolar ventilation and raise P_aCO_2.

2. **Compensation:** P_aCO_2 is monitored by central and peripheral chemoreceptors. The respiratory center responds to acute hypercapnia by increasing ventilation rate, even if such an action causes respiratory muscle fatigue and precipitates a respiratory crisis. During chronic hypercapnia, the chemoreceptors adapt to persistent elevation of P_aCO_2, and, thus, ventilation rates remain normal. CO_2 retention decreases plasma pH (respiratory acidosis), but the kidneys compensate by retaining HCO_3^-, allowing pH to remain within a normal range even as P_aCO_2 climbs above $\sim$70–90 mm Hg (see 28·VI·C). Such patients typically have a limited pulmonary reserve, however, and may decompensate quickly if illness creates additional demands on an already compromised system.

3. **Consequences:** The cerebral vasculature is highly sensitive to P_aCO_2. Acute CO_2 retention causes cerebral vasodilation, which causes headaches and intracranial hypertension. The latter may

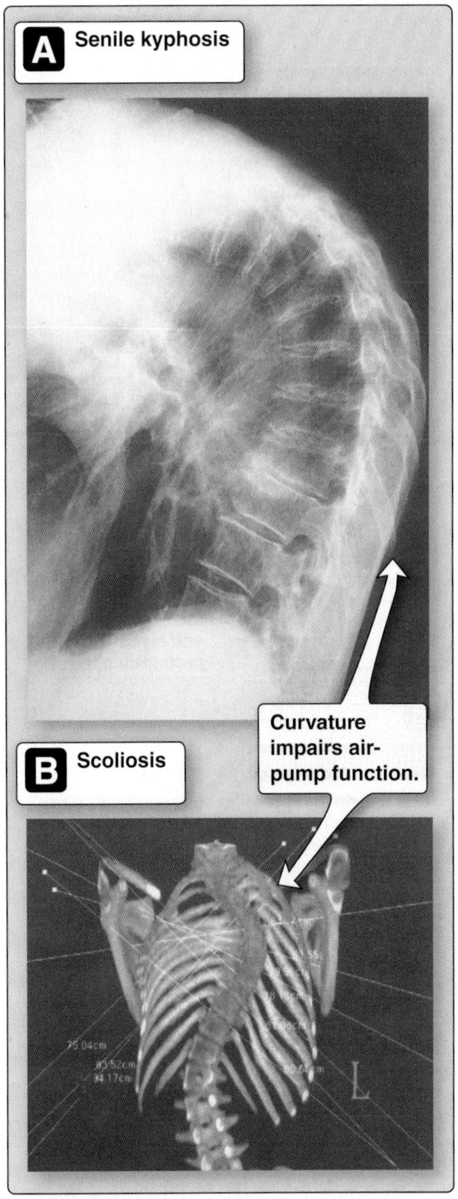

Figure 40.22
Kyphosis and scoliosis.

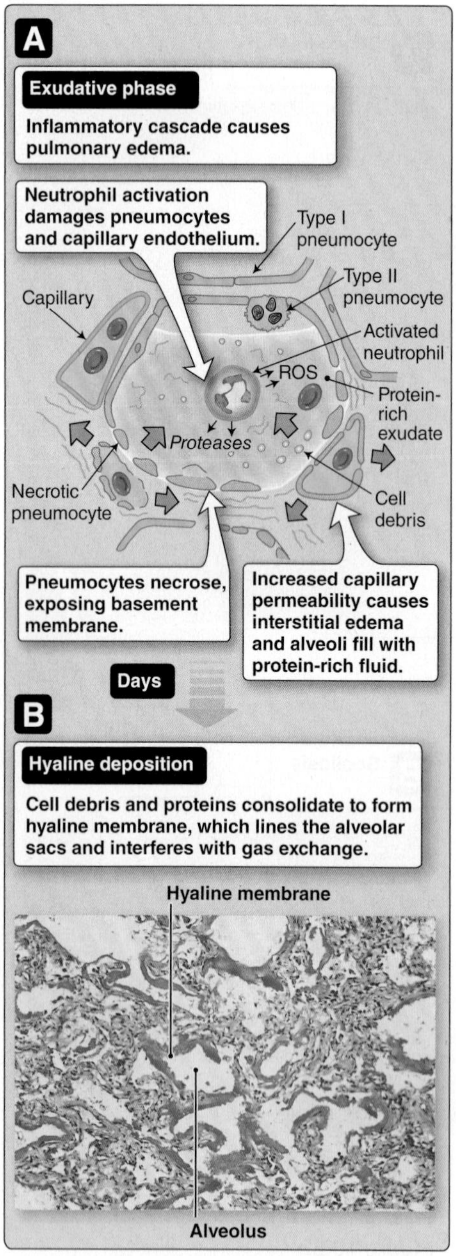
A
Exudative phase
Inflammatory cascade causes pulmonary edema.
Neutrophil activation damages pneumocytes and capillary endothelium.
Type I pneumocyte
Type II pneumocyte
Capillary
Activated neutrophil
ROS
Protein-rich exudate
Proteases
Necrotic pneumocyte
Cell debris
Pneumocytes necrose, exposing basement membrane.
Increased capillary permeability causes interstitial edema and alveoli fill with protein-rich fluid.
Days
B
Hyaline deposition
Cell debris and proteins consolidate to form hyaline membrane, which lines the alveolar sacs and interferes with gas exchange.
Hyaline membrane
Alveolus

Figure 40.23
Acute respiratory distress syndrome. ROS = reactive oxygen species.

manifest as a swelling of the optic disc and can cause blindness. High CO_2 levels also cause dyspnea and neurologic symptoms, such as involuntary wrist movements and hand tremors.

D. Acute respiratory distress syndrome

ARDS is the leading cause of respiratory failure in young adults, with mortality rates as high as ~58%. In the military, ARDS was known originally as "shock lung," reflecting the similarities between the onset of ARDS and septic shock.

1. **Causes:** ARDS may develop in association with a wide range of conditions, the most common being sepsis, aspiration of stomach contents, near drowning, multiple blood transfusions, trauma, bone fractures, and pneumonia.

2. **Stages:** ARDS is precipitated by local or circulating inflammatory mediators (e.g., histamine), which trigger an inflammatory cascade that severely damages the alveolar endothelial cells (pneumocytes) and capillary endothelial cells that comprise the blood–gas interface (Figure 40.23; see 22·II·C). There are three discrete stages to ARDS, characterized by exudate formation, hyaline membrane deposition, and fibrosis.

   a. **Exudates:** The initial stages of ARDS (24–48 hr) are characterized by an inflammatory reaction within the lung parenchyma that damages the alveolar epithelium and causes a profound increase in capillary permeability. The lungs fill with a bloody exudate containing plasma proteins and cellular debris (see Figure 40.23A). Chest x-rays typically reveal diffuse bilateral infiltrates that are reminiscent of pulmonary edema but that occur when CVP and left atrial pressure is normal.

   b. **Hyaline membrane:** In the following 2–8 days, hyaline membranes begin to settle from the exudate and cover the alveolar lining (see Figure 40.23B). Hyaline is a fibrous matrix of plasma proteins and disrupted cell debris.

   c. **Fibrosis:** After ~8 days, the interstitium is infiltrated by fibroblasts, which deposit collagen and other fibrous materials.

3. **Consequences:** The alveolar infiltrates and hyaline membranes prevent gas exchange, causing hypoxemia. The infiltrates also inactivate surfactant and suppress surfactant production by type II pneumocytes, causing alveolar collapse. Surfactant loss and atelectasis makes the lung extremely stiff and difficult to expand, which is one of the hallmarks of ARDS (Figure 40.24). Atelectasis not only impairs ventilation, it also reduces the area of the blood–gas interface and, thereby, exacerbates hypoxemia. Support from a mechanical ventilator is life saving while the underlying process is addressed.

E. System failure

Hypoxemia associated with acute respiratory failure elicits the same SNS responses as shock, but in the context of an intact and functional vasculature. SNS-stimulated increases in CO and SVR can cause

MAP to rise to levels that rupture cerebral blood vessels. Chronic hypoxemia causes a gradual decline in neural function that inhibits the pathways controlling ventilation and blood pressure. The most obvious external sign of hypoxemia is cyanosis, a blue discoloration of the skin and mucous membranes that reflects the color of deoxyhemoglobin (cyanosis occurs when deoxyhemoglobin levels rise to ~5 g/dL). Acute hypercapnia produces respiratory acidosis through CO_2 retention, but the body's response is dominated by reflex responses to the hypoxemia that accompanies hypercapnia. Acute hypercapnia causes anesthetic-like effects on the CNS (**CO_2 narcosis**). Narcosis appears at a P_{CO_2} of ~90 mm Hg, causing confusion and lethargy. High P_{CO_2} depresses respiratory center function and suppresses ventilatory drive, thereby creating a positive feedback cycle that potentiates CO_2 retention and, ultimately, results in coma and death (P_{CO_2} ~130 mm Hg).

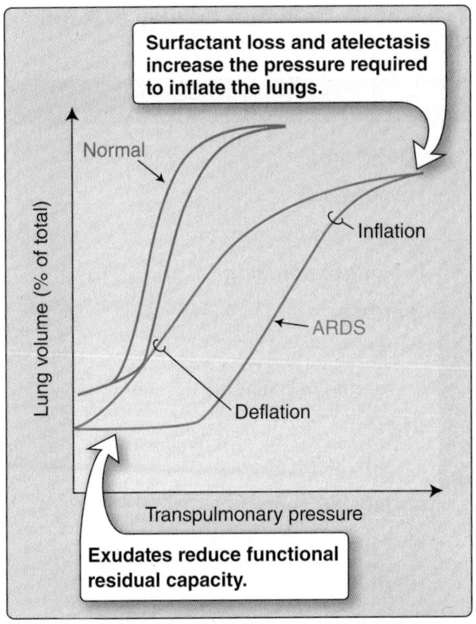

Figure 40.24
Change in lung function accompanying acute respiratory distress syndrome (ARDS).

VII. KIDNEY FAILURE

Two forms of kidney failure are recognized. **Acute kidney injury** (**AKI**) develops abruptly (within 48 hours) but usually can be treated if patients have no other complicating medical issues. Kidney function and risk of failure can be assessed using RIFLE criteria as shown in Table 40.3. **Chronic kidney disease** (**CKD**) develops over the course of many years. CKD is characterized by progressive and irreversible loss of nephrons, but the development of dialysis and transplant technologies means that CKD is not necessarily fatal. The mortality rate of patients on dialysis is very high (dialysis increases lifespan by only 4.5 years in 60–64-year-olds), but death usually occurs from cardiovascular disease, infection, or **cachexia** (a wasting syndrome). AKI is a primary cause of death (~75%) in urgent care facilities, however, where patients may be elderly and have other underlying pathologies.

Table 40.3: Assessing Kidney Injury and Failure

	GFR (Serum Creatinine)*	Urine Output
Risk	↓25% (↑1.5×)	<0.5 mL/kg for 6 hr
Injury	↓50% (↑2×)	<0.5 mL/kg for 12 hr
Failure	↓75% (↑3×)	<0.3 mL/kg for 24 hr or anuria for 12 hr
Loss	Loss of renal function for >4 weeks	
ESRD	End-stage renal disease (ESRD)	

*Deviation from baseline values. GFR = glomerular filtration rate.

A. Causes

AKI can be precipitated by numerous factors, which are usually grouped according to where in the nephron they act: **prerenal**, **intrarenal**, and **postrenal** (Table 40.4).

1. **Prerenal:** Prerenal failure is characterized by a profound drop in GFR due to decreased renal blood flow and glomerular perfusion pressure (see Figure 40.13). Prerenal failure usually occurs secondarily to shock and ischemia.

2. **Intrarenal:** Intrarenal failure occurs when the renal tubule or surrounding interstitium is injured. The most common cause of intrarenal failure is **acute tubular nephrosis** (**ATN**). ATN usually results from ischemia, but the tubule may also be injured by drugs and other toxins.

   a. **Ischemia:** The renal epithelium's transport functions create a high ATP dependency and coincident susceptibility to ischemia. The transport epithelium receives O_2 via the peritubular capillary network, whose flow is governed by glomerular arterioles (see 26·II·C). Tubule ischemia usually occurs during a hypotensive crisis when both arterioles are constricted,

Table 40.4: Common Causes of Acute Renal Injury

Prerenal
• Hypovolemia
▫ Hemorrhage
▫ Dehydration
▫ Prolonged vomiting
▫ Diarrhea
▫ Severe burns
• Peripheral vasodilation
▫ Sepsis
▫ Anaphylactic shock
• Cardiogenic shock
▫ Myocardial infarction
▫ Heart failure
▫ Cardiac tamponade
• Renal vasoconstriction
▫ Vasoactive drugs
Intrarenal
• Vascular occlusion
▫ Renal artery occlusion
▫ Renal vein thrombosis
▫ Vasculitis
• Glomerulonephritis
• Acute tubular necrosis
▫ Ischemia
▫ Nephrotoxic drugs, heavy metals, organic solvents
▫ Intratubular deposits (uric acid casts, muscle proteins)
Postrenal
• Urolithiasis
• Ureterocele
• Prostatic hyperplasia
• Malignancies

filtration has ceased, and peritubular flow no longer meets the epithelium's basal O_2 needs. Ischemic cells may respond by sloughing their apical villi into the tubule lumen, which reduces overall surface area, transporter density, and O_2 requirements.

b. **Toxins:** The renal tubule's ability to concentrate drugs and toxins makes it very vulnerable to nephrotoxicity. Although all tubule regions are at risk, necrosis most commonly occurs in the proximal tubule, which is responsible for secreting many drugs (see 26·IV).

3. **Postrenal:** Postrenal failure is caused by urinary outflow obstruction, which can occur at any point within the tubule, collecting system, ureters, bladder, or urethra. Common causes include kidney stones (**calculi**; see Clinical Application 4.2) and an enlarged prostate (**prostatic hyperplasia**). Obstruction causes pressure in more proximal tubule segments to build to the point where they negate P_{UF}, and glomerular filtration stops. In time, the affected tubule segments may dilate and atrophy.

B. Consequences

Renal failure is considered to have occurred when GFR is reduced to 25% of normal values. Kidneys lose their ability to control water or electrolyte levels when GFR is so low, which manifests as hypervolemia, hyperkalemia, metabolic acidosis, and accumulation of nitrogenous wastes (azotemia).

C. System failure

An inability to excrete nitrogenous wastes and maintain electrolyte balance negatively impacts all organs, but the neurologic effects dominate. Patients become lethargic, drowsy, and delirious and eventually slip into a coma. Death typically occurs due to cardiac arrhythmia caused by hyperkalemia.

VIII. MULTIPLE ORGAN DYSFUNCTION SYNDROME

Emily Dickinson's generation was well acquainted with the signs of organ system failure and impending death. People usually died at home in the care of family and loved ones. The poem that opened this final chapter accurately notes the effects of acute hypoxemia on the CNS (convulsion) and the consequences of intense SNS-mediated stimulation of sweat glands (*The Beads upon the Forehead / By homely Anguish strung*). In more recent times, the final throes occur in medical facilities, witnessed mainly by health care professionals. Patients arriving at emergency departments and intensive care units often already have long-standing diseases and have coped with progressive failure of one or more organs for many months or years. They present when an infection or some other seminal event has precipitated **multiple organ dysfunction syndrome** (a medical crisis involving two or more organ systems), at which point medical intervention is required for continued survival. It is the job of care providers to help restore homeostasis and create an environment in which the body may recover from an acute disease state.

In the final reckoning, however, the physiologic systems that maintain homeostasis and that have been described in the preceding chapters are robust and have remarkable recuperative capabilities. They readily reassert control if given a chance, either on their own or by timely medical intervention as needed. Whether or not they take the chance is the medical mystery that is Life.

Chapter Summary

- **Aging** is accompanied by a progressive reduction in total cell number and organ functionality. Aging ultimately results in organ failure and death.
- The final common pathway for most instances of cell death is **ischemia**. Ischemia initiates a series of events that comprise an **ischemic cascade**. Significant events include acidosis, dissipation of ion gradients, Ca^{2+} activation of *proteases* and other degradative enzymes, and mitochondrial lysis.
- Ischemia usually results from lack of perfusion due to **circulatory shock**. **Hypovolemic shock** results from hemorrhage or reduced extracellular fluid volume. **Cardiogenic shock** is caused by loss of cardiac pump function. **Distributive shock** occurs when the central nervous system loses vascular control, and the resulting systemic vasodilation allows blood to become trapped in capillaries and veins.
- Shock can be divided into three stages: **preshock**, **shock**, and **organ failure**. During preshock, sympathetic nervous system–mediated increases in cardiac and vascular function compensate for falling mean arterial pressure.
- During shock, perfusion of critical circulations (cerebral, coronary) becomes suboptimal, and signs of intense sympathetic nervous system activity become overt (hypotension, decreased urinary output, acidosis).
- **System failure** occurs when shock becomes irreversible. The cardiovascular system becomes locked in a positive feedback spiral that results in loss of myocardial and vascular contractility; acidemia; blood clotting; cellular deterioration; and, ultimately, loss of cerebral perfusion pressure and brain death.
- **Heart failure** is the leading cause of death in the United States and is the final common pathway for many cardiac diseases. The right heart typically fails as a result of increased pulmonary vascular resistance. Causes of **left-heart failure** include filling impairment (**diastolic failure**), loss of contractility, or an excessive afterload that impairs output (**systolic failure**).
- **Myocardial infarction** is a common cause of heart failure. Long-term compensatory mechanisms that support cardiac output though volume retention and invocation of the **Frank-Starling mechanism** ultimately become counterproductive and precipitate failure. Symptoms of congestive heart failure include **edema** and **shortness of breath** upon exertion.
- Respiratory failure occurs when the pulmonary system is unable to take up O_2 or eliminate CO_2 from the body. **Hypoxemic respiratory failure** usually occurs as a result of impaired alveolar ventilation due to alveolar collapse or accumulation of fluid, pus, or blood in alveoli.
- The central respiratory control centers readily adapt to increases in $P_{a}CO_2$, so **hypercapnic respiratory failure** usually reflects impairment of air-pump function (respiratory muscles and chest wall).
- **Acute respiratory distress syndrome** (**ARDS**) is a leading cause of respiratory failure. ARDS is associated with inflammatory reactions that damage the lung parenchyma and increase pulmonary capillary permeability. Lungs fill with infiltrates and become stiff and difficult to expand.
- **Renal failure** can be precipitated by inadequacy of perfusion, kidney tubule deterioration, or obstruction of urinary outflow. The kidney's resulting inability to control water and electrolyte levels allows K^+ levels to rise (**hyperkalemia**), and death usually results from cardiac arrhythmia.

Study Questions

Choose the ONE best answer.

IX.1 A 23-year-old pregnant woman in her third trimester complains to her friend that her feet and ankles are frequently swollen. The swelling is most likely caused by which of the following?

A. High pedal venous pressures
B. Increased left ventricular preload
C. Hypertension (preeclampsia)
D. Decreased blood viscosity
E. Excessive fluid retention

Best answer = A. The gravid uterus compresses veins returning blood from the lower extremities, causing pedal venous pressures to rise (37·IV·D). Pedal capillary hydrostatic pressure rises as a result, promoting fluid filtration and edema. Filtration is potentiated by a coincident fall in plasma colloid osmotic pressure during pregnancy. Maternal cardiac output increases through fluid retention to increase left ventricular preload, but the additional output is needed to supply the placenta and does not contribute significantly to a rise in venous pressure. Blood viscosity changes and hypertension do not affect the Starling forces significantly under physiologic conditions.

IX.2 Healthy pregnant women and athletes engaged in an aerobic exercise training routine both show which of the following changes?

A. Afterload-induced ventricular hypertrophy
B. Resting heart rate increases
C. Resting diastolic pressure increases
D. Minute ventilation increases
E. Hemoglobin concentration decreases

Best answer = D. Both pregnancy and aerobic training increase minute ventilation to maximize tissue O_2 delivery (37·IV·E; 37·VI·D). Hemoglobin (Hb) levels fall during pregnancy, whereas training increases Hb. Ventricular hypertrophy during pregnancy and training occurs in response to a chronically increased preload and stroke volume (SV). Resting heart rate (HR) rises during pregnancy to help meet the increased demands for cardiac output (CO) placed on the maternal cardiovascular system. Resting CO is not changed by training, however, so the SV increase decreases resting HR. Resting diastolic blood pressure falls during pregnancy due to runoff into the low-resistance placental circuit and decreases minimally with aerobic exercise training.

IX.3 The answer choices below compare pairs of cardiovascular system variables. Which of these choices best describes the fetal cardiovascular system?

A. Vascular resistance: systemic > pulmonary
B. Flow: pulmonary vein > descending aorta
C. Atrial pressure: left > right
D. Hemoglobin levels: adult > fetal
E. O_2 saturation: inferior vena cava > aorta

Best answer = E. Fetal blood is oxygenated in the placenta, then flows via the umbilical vein at ~85% saturation into the inferior vena cava (~70%), through the heart, and into the aorta (37·V·C). Saturation has fallen to ~65% by venous admixture during passage. The fetal cardiovascular system is characterized by its high pulmonary vascular resistance compared with systemic vascular resistance. Shunts (foramen ovale and ductus arteriosus) direct blood past the high-resistance pulmonary circuit, and, therefore, pulmonary blood flow is lower than aortic flow. Fetal blood is enriched with hemoglobin to enhance its O_2-carrying capacity.

IX.4 A 52-year-old female with multiple sclerosis has been admitted to the emergency department on three occasions with mild hypothermia (rectal temperature 34°–35°C) during fall camping trips. What is the most likely reason that this individual might be experiencing low temperatures during prolonged cold exposure?

A. Decreased cutaneous warmth sensation
B. Decreased cutaneous pain sensation
C. Increased sweat secretion
D. Preoptic hypothalamic lesions
E. Caudal cerebellar lesions

Best answer = D. Multiple sclerosis causes demyelination and decreased axonal conduction. If a sclerotic lesion occurs in the preoptic area, then central sensation and processing of signals from skin temperature receptors may not be regulated appropriately (38·II·B). This can effectively blunt responses to thermal stress and allow body temperature to fluctuate more with ambient temperatures. Lesions occurring in the cerebellum would affect movement and coordination rather than temperature regulation. Individuals with multiple sclerosis can develop peripheral neuropathy, but skin warmth and pain receptors are not directly involved in cold sensation. Sweating occurs during heat exposure, not cold exposure (38·II·D·2).

IX.5 A 6-month-old female is inadvertently exposed to a cold environment after she and her parents were in a rainstorm on a cold day. What thermogenic tissue has a mitochondrial uncoupling protein that can aid her thermoregulation?

A. White adipose tissue
B. Brown adipose tissue
C. Skeletal muscle
D. Cardiac muscle
E. Smooth muscle

Best answer = B. Nonshivering thermogenesis is a process of increasing metabolic rate to generate heat without shivering (38·II·D·4). Brown adipose tissue, which is proportionally higher in infants than in adults, has a specialized mitochondrial uncoupling protein (thermogenin) that generates heat without producing useful work. White adipose tissue does not have this capability. Muscle, primarily skeletal, can participate in nonshivering thermogenesis but does not contain a thermogenin-like protein. Cardiac and smooth muscle do not participate directly in cold responses.

IX.6 In which of the following conditions would a *cyclooxygenase* inhibitor bring internal temperature back within the 36.5°–37.5°C range?

A. Severe hypothermia
B. Ambient cold stress
C. Ambient heat stress
D. Heat exhaustion
E. Fever

Best answer = E. Nonsteroidal anti-inflammatory medications such as aspirin are *cyclooxygenase* inhibitors that block prostaglandin synthesis. Their actions include inhibiting prostaglandin EP_3-receptor activation in the preoptic hypothalamus, thereby lowering a thermoregulatory set point that has been elevated by pyrogens during fever (38·IV·A). Ambient cold and heat stress challenge thermoregulation, but the internal set point is still within the normal range. Severe hypothermia is defined as an internal temperature $<$28°C, and heat exhaustion is a hyperthermic condition, but the body again attempts to regulate internal temperature to the set point, which is within normal range (38·IV·B).

IX.7 A 25-year-old woman recently underwent surgery for a strained anterior cruciate ligament. Postsurgical exercises included isometric quadriceps exercises that are held until fatigue ($\sim$60 s). Which energy system is primarily used in these exercises?

A. Stored adenosine triphosphate
B. Adenosine triphosphate–creatine phosphate system
C. Lactic acid system
D. Citric acid cycle
E. Oxidative phosphorylation

Best answer = C. The lactic acid system predominates during maximal exercise that fatigues with durations between 30 s to 2.5 min (39·III·A). In this system, glycolysis produces pyruvic acid, which is then shuttled to lactic acid. Adenosine triphosphate (ATP) stores can support exercise for a few seconds and the ATP–creatine phosphate system only extends this time to 8–10 s. Aerobic metabolism (citric acid cycle and oxidative phosphorylation) is the primary energy system used to synthesize ATP during maximal exercise lasting 2.5 min or longer.

IX.8 At the beginning of exercise, a feedforward mechanism increases heart and respiratory rates. What is the best term or receptor responsible for this mechanism?

A. Class III muscle afferents
B. Class IV muscle afferents
C. Arterial baroreceptors
D. Peripheral chemoreceptors
E. Central command

Best answer = E. Central command is the feedforward signal that increases cardiovascular and respiratory system function at the beginning or in the anticipation of exercise (39·IV·B·2). Class III and IV muscle afferents provide feedback regarding stretch, compression, and metabolic status of the muscle. Baroreceptors provide feedback regarding arterial pressures at the aortic arch and carotid arteries. Chemoreceptors provide feedback regarding arterial partial pressure of CO_2 and O_2, and H^+ in similar locations (39·IV·A).

IX.9 An exercise-induced rightward shift in the O_2-hemoglobin dissociation curve is likely responsible for increasing which of the following respiratory parameters during aerobic exercise?

A. Alveolar ventilation
B. Excess postexercise oxygen consumption
C. Work of breathing
D. Arteriovenous O_2 difference
E. O_2-carrying capacity

Best answer = D. The arteriovenous (a–v) O_2 difference widens with aerobic exercise through increased O_2 offloading by hemoglobin ([Hb] 39·VI·B). This manifests in a decrease in venous O_2 content, so that even though arterial levels are unchanged, the a–v difference increases. Increased offloading occurs due to a rightward shift in the Hb-O_2 dissociation curve (23·VI·B). Postexercise O_2 consumption does not impact O_2 usage during exercise. Alveolar ventilation and work of breathing both increase during exercise but are unrelated to the Hb-O_2 dissociation curve. Blood's O_2-carrying capacity is determined primarily by Hb concentration, not by Hb's O_2 affinity.

IX.10 A 21-year-old soldier suffers extensive blood loss from deep wounds inflicted by an improvised explosive device. Which of the following identifies the primary mechanism responsible for maintaining blood volume until fluids can be administered?

A. Venoconstriction
B. Resistance vessel constriction
C. Aldosterone release
D. Decreased renal blood flow
E. Recruitment of interstitial fluid

Best answer = E. Hemorrhage causes central venous pressure to fall, which reduces mean capillary hydrostatic pressure in all circulations (40·IV·A). This causes fluid to move into the vasculature from the interstitium ("transcapillary refill") under the influence of plasma colloid osmotic pressure, thereby helping support blood volume. Constriction of resistance vessels directs blood away from nonessential organs but does not increase blood volume. Venoconstriction forces blood out of veins but does not affect total blood volume. Aldosterone increases fluid retention by the kidney but only after several hours.

IX.11 A 67-year-old woman is brought to the emergency department in shock. Assessment of her cardiovascular function shows that heart rate is high and cardiac output is increased, whereas left atrial pressure, mean arterial pressure, and systemic vascular resistance are all low. What is the likely cause?

A. Septic shock
B. Hypovolemic shock
C. Cardiogenic shock
D. Cardiac tamponade
E. Pulmonary embolism

Best answer = A. Shock triggers an intense sympathetic response in attempts to raise arterial pressure and restore O_2 delivery to the brain (40·IV·B). Such a response includes systemic vasoconstriction to increase systemic vascular resistance (SVR). Inflammatory reactions associated with sepsis damage the vasculature and prevent vasoconstriction, so SVR inevitably falls. Hypovolemic shock reduces cardiac output (CO). In cardiogenic shock (including tamponade), left atrial pressure ([LAP] or preload) would be increased in attempts to support CO. Pulmonary embolism would cause LAP and CO to fall, but SVR would be very high.

IX.12 A 48-year-old man with pneumonia is hospitalized when he develops acute respiratory distress syndrome and requires a mechanical ventilator to support breathing. Why is a mechanical ventilator helpful?

A. It increases cardiac output.
B. Alveolar fluid decreases compliance.
C. Inflammatory exudates impair surfactant.
D. It prevents hyaline membrane formation.
E. It prevents pulmonary fibrosis.

Best answer = C. Acute respiratory distress syndrome (ARDS) patients' lungs are highly noncompliant and require the assistance of a mechanical ventilator to expand, mainly because the inflammatory exudates inactivate surfactant and inhibit its production (40·VI·D). Fibrosis, which may further reduce compliance over time, is not prevented by ventilation. Ventilation has no effect on hyaline membrane formation, which interferes with gas exchange, and may often decrease left ventricular preload and cardiac output. Fluid in the pulmonary interstitium reduces lung compliance, but not within alveoli (fluid-filled lungs are easier to expand than normal because surface tension effects are negated; 22·IV·B).

Figure Credits

Fig. 5.1: Modified from Jennings, H.S. *Behavior of the Lower Organisms.* The Columbia University Press, 1906.

Fig. 5.13: Modified from Moore, K.L. and Dalley, A.F. *Clinical Oriented Anatomy.* Fourth Edition. Lippincott Williams & Wilkins, 1999.

Fig. 12.3: Modified from Seifter, J., Ratner, A., and Sloane, D. *Concepts in Medical Physiology.* Lippincott Williams & Wilkins, 2005.

Fig. 12.2A: Photograph from Cohen, B.J. and Taylor, J.J. *Memmler's the Human Body in Health and Disease.* Eleventh Edition. Lippincott Williams & Wilkins, 2009.

Fig. 14.1: Micrographs from Moore, K.L. and Agur, A. *Essential Clinical Anatomy.* Second Edition. Lippincott Williams & Wilkins, 2002.

Fig. 14.2: Data from Seow, C.Y. and Fredberg, J.J. *J. Appl. Physiol.* 110: 1130–1135, 2011.

Fig. 14.3: Data from Kuo, K.H. and Seow, C.Y. *J. Cell Science.* 117:1503–1511, 2003.

Fig. 15.2: Crystal data from Robinson, R.A. *J. Bone Joint Surg. Am.* 34:389–476, 1952.

Fig. 15.3: Model (lower) from Thurner, P.J. *Nanomed. Nanobiotechnol.* 1:624–629, 2009.

Fig. 21.4: Data from Harper, A.M. *Acta Neurol. Scand.* [Suppl] 14:94, 1965.

Fig. 21.5: Data from Ingvar, D.H. *Brain Res.* 107:181–197, 1976.

Fig. 22.8: From Kahn, G.P. and Lynch, J.P. *Pulmonary Disease Diagnosis and Therapy: A Practical Approach.* Lippincott Williams & Wilkins, 1997.

Fig. 22.16 (panels A, B, and D): Cagle, P.T. *Color Atlas and Text of Pulmonary Pathology.* Lippincott Williams & Wilkins, 2005.

Fig. 23.11: Modified from Daffner, R.H. *Clinical Radiology—The Essentials.* Third Edition. Lippincott Williams & Wilkins, 2007.

Fig. 23.20: From Anderson, S.C. *Anderson's Atlas of Hematology.* Lippincott Williams & Wilkins, 2003.

Fig. 24.15: From The National Oceanic and Atmospheric Administration. Photo credit: Doug Kesling.

Fig. 25.6 (lower two micrographs): From Schrier, R.W. *Diseases of the Kidney and Urinary Tract.* Eighth Edition. Lippincott Williams & Wilkins, 2006.

Fig. 26.1B: Micrograph reprinted with permission from Clapp, W.L., Park, C.H., Madsen, K.M., *et al. Lab. Invest.* 58:549–558, 1988.

Fig. 26.5: Data from Rector, F.C. *Am. J. Physiol.* 244:F461–F471, 1983.

Fig. 27.5: Based on data from Pannabecker, T.L., Dantzler, W.H., Layton, H.E., *et al. Am. J. Physiol.* 295:F1271–F1285, 2008.

Fig. 27.15 (lower): Micrograph reprinted with permission from Clapp, W.L., Madsen, K.M., Verlander, J.W., *et al. Lab. Invest.* 60:219–230, 1989.

Fig. 31.12: Radiograph from Dean, D. and Herbener, T.E. *Cross-Sectional Human Anatomy.* Lippincott Williams & Wilkins, 2000.

Fig. 35.7: Modified from Golan, D.E., Tashjian, A.H., and Armstrong, E.J. *Principles of Pharmacology: The Pathophysiologic Basis of Drug Therapy.* Second Edition. Wolters Kluwer Health, 2008.

Fig. 36.6 (top panel): Modified from Bear, M.F., Connors, B.W., and Paradiso, M.A. *Neuroscience—Exploring the Brain.* Second Edition. Lippincott Williams & Wilkins, 2001.

Fig. 38.9: From Fleisher, G.R. and Ludwig, S. *Textbook of Pediatric Emergency Medicine.* Sixth Edition. Lippincott Williams & Wilkins, 2010.

Fig. 40.20C: Radiograph from Topol, E.J., Califf, R.M., and Isner, J. *Textbook of Cardiovascular Medicine.* Third Edition. Lippincott Williams & Wilkins, 2006.

Fig. 40.22B: Radiograph from Frymoyer, J.W., Wiesel, S.W. *The Adult and Pediatric Spine.* Lippincott Williams & Wilkins, 2004.

Fig. 40.23 (lower): Radiograph from Rubin, R. *Pathology.* Fourth Edition. Lippincott Williams & Wilkins, 2005.

Clinical Application 3.1: Photograph from Eisenberg, R.L. *An Atlas of Differential Diagnosis.* Fourth Edition. Lippincott Williams & Wilkins, 2003.

Clinical Application 4.1: Photograph from Berg, D. and Worzala, K. *Atlas of Adult Physical Diagnosis.* Lippincott Williams & Wilkins, 2006.

Clinical Application 5.2: Photograph from Smeltzer, S.C., Bare, B.G., Hinkle, J.L., *et al. Brunner and Suddarth's Textbook of Medical—Surgical Nursing.* Twelfth Edition. Lippincott Williams & Wilkins, 2009.

Clinical Application 6.2: Modified from Taylor, C., Lillis, C.A., and LeMone, P. *Fundamentals of Nursing.* Second Edition. Lippincott Williams & Wilkins, 2009.

Clinical Application 11.1: Courtesy of Steven R. Nokes, M.D., Little Rock, Arkansas.

Clinical Application 15.1: Courtesy of Tyrone Wei, D.C., D.A.C.B.R., Portland, Oregon.

Clinical Application 15.3 (lower): Photograph from Rubin, R. and Strayer, D.S. *Rubin's Pathology: Clinicopathologic Foundations of Medicine.* Fifth Edition. Lippincott Williams & Wilkins, 2008.

Clinical Application 21.2: Courtesy of Medtronics, Peripheral Division, Santa Rosa, California.

Clinical Application 26.2: Reproduced from Fiechtner, J.J. and Simkin, P.A. *JAMA.* 245:1533–1536, 1981, with permission.

Clinical Application 33.1 (upper): Photograph from Smeltzer, S.C. and Bare, B.G. *Textbook of Medical-Surgical Nursing.* Ninth Edition. Lippincott Williams & Wilkins, 2000.

Clinical Application 33.2: Photograph from Willis, M.C. *A Programmed Learning Approach to the Language of Health Care.* Lippincott Williams & Wilkins, 2002.

Clinical Application 34.2: Photograph from Sadler, T.W. *Langman's Medical Embryology.* Seventh Edition. Lippincott Williams & Wilkins, 1995.

Clinical Application 34.3: Photograph from Rubin, R. *Essential Pathology.* Third Edition. Lippincott Williams & Wilkins, 2000.

Clinical Application 35.1: Photograph from Weber, J. and Kelley, J. *Health Assessment in Nursing.* Second Edition. Lippincott Williams & Wilkins, 2003.

Clinical Application 35.4: Photograph from Becker, K.L., Bilezikian, J.P., Brenner, W.J., *et al. Health Assessment in Nursing. Principles and Practice of Endocrinology and Metabolism.* Third Edition. Lippincott Williams & Wilkins, 2001.

Modified from Bear, M.F., Connors, B.W., and Paradiso, M.A. *Neuroscience—Exploring the Brain.* Third Edition. Lippincott Williams & Wilkins, 2007: Figs. 5.7, 7.4, 7.11, 7.12, 9.3, 9.4, 9.6, 9.7, 9.13, 12.7, 16.12A, and Clinical Application 8.2.

From the Centers for Disease Control and Prevention, Public Health Image Library: Photographs appearing in Clinical Applications 5.1 (photo credit: Dr. Fred Murphy, Sylvia Whitfield, 1975), 12.4, 19.3, and 20.1 (photo credit: Dr. Edwin P. Ewing, Jr., 1972). Data appearing in Tables 37.1 and 40.1.

Modified from Chandar, N. and Viselli, S. *Lippincott's Illustrated Reviews: Cell and Molecular Biology.* Lippincott Williams & Wilkins, 2010: Figs. 1.2, 1.3, 1.4, 1.5, 1.6, 1.12, 1.13, 1.18, 1.21, 1.22, 1.23, 4.5, and 4.10.

Modified from Clarke, M.A., Finkel, R., Rey, J.A., *et al. Lippincott's Illustrated Reviews: Pharmacology.* Fifth Edition. Lippincott Williams & Wilkins, 2012: Figs. 1.17 and 1.20.

From Daffner, R.H. *Clinical Radiology—The Essentials.* Third Edition. Lippincott Williams & Wilkins, 2007: Photographs appearing in Fig. 22.11 (lower panel) and 23.11 and Clinical Applications 4.2 and 29.1 (left).

From Feigenbaum, H., Armstrong, W.F., and Ryan, T. *Feigenbaum's Echocardiography.* Sixth Edition. Lippincott Williams & Wilkins, 2004: Photographs appearing in Clinical Applications 18.1 (upper panel) and 19.2 (upper panel).

From Fleisher, G.R., Ludwig, W., and Baskin M.N. *Atlas of Pediatric Emergency Medicine.* Lippincott Williams & Wilkins, 2004: Photographs appearing Clinical Applications 1.2, 6.1, and 25.2 (lower).

From Goodheart H.P. *Goodheart's Photoguide of Common Skin Disorders.* Second Edition. Lippincott Williams & Wilkins, 2003: Photographs appearing in Fig. 16.2 and Clinical Applications 35.3, 36.1, and 36.2.

From Gorbach, S.L., Bartlett, J.G. , and Blacklow, N.R. *Infectious Diseases.* Lippincott Williams & Wilkins, 2004: Photographs appearing in Clinical Applications 4.4 and 29.1 (right).

Modified from Harvey, R.A. and Ferrier, D.R. *Lippincott's Illustrated Reviews: Biochemistry.* Fifth Edition. Lippincott Williams & Wilkins, 2011: Figs. 23.16, 23.17, 23.19, 23.21, 33.1, 40.6, and Clinical Application 33.1 (lower panel).

Modified from Klabunde, R.E. *Cardiovascular Physiology Concepts.* Lippincott Williams & Wilkins, 2005: Figs. 17.2, 17.14, 18.1, 18.4, 20.5, and 21.14.

From Klossner, N.J. and Hatfield, N. *Introductory Maternity and Pediatric Nursing.* Lippincott Williams & Wilkins, 2005: Fig. 36.9 and Photographs appearing in Fig. 37.12 and Clinical Applications 16.1.

From Koopman, W.J. and Moreland, L.W. *Arthritis and Allied Conditions—A Textbook of Rheumatology.* Fifteenth Edition. Lippincott Williams & Wilkins, 2005: Clinical Applications 15.2 and 30.1.

Modified from Krebs, C., Weinberg, J., and Akesson, E. *Lippincott's Illustrated Review of Neuroscience.* Lippincott Williams & Wilkins, 2012: Figs. 6.5, 6.7, 6.11, 6.12, 7.9, 8.1, 8.9, 8.10, 8.14, 9.2, 9.15, 10.5A and B, 10.6, 11.4, 16.13, and 21.2.

Modified from McCardle, W.D., Katch, F.I., and Katch, V.L. *Exercise Physiology.* Seventh Edition. Lippincott Williams & Wilkins, 2010: Figs. 12.6, 12.11, 23.15, and 24.12.

Modified from McConnell, T.H. *The Nature of Disease Pathology for the Health Professions.* Lippincott Williams & Wilkins, 2007: Clinical Application 16.2 and Photographs appearing in Fig. 23.22 and Clinical Applications 12.1, 18.2, 23.1, 25.1, 30.3, 32.2, and 34.1.

From Mills, S.E. *Histology for Pathologists.* Third Edition. Lippincott Williams & Wilkins, 2007: Photographs appearing in Figs. 1.9, 4.4, 10.5C, 14.1, 15.6, 25.12 (lower panel), 31.2, and 34.1 (lower panel) and Clinical Application 34.3.

Modified from Porth, C.M. *Essentials of Pathophysiology.* Second Edition. Lippincott Williams & Wilkins, 2007: Figs. 15.5 and 25.14.

Modified from Premkumar, K. *The Massage Connection Anatomy and Physiology.* Lippincott Williams & Wilkins, 2004: Figs. 16.3 and 36.7 (lower panel).

Modified from Rhoades, R.A. and Bell, D.R. *Medical Physiology.* Third Edition. Lippincott Williams & Wilkins, 2009: Figs. 12.14, 20.15, 21.9, 22.22, 25.5, 25.9, and 37.5.

From Ross M.H., Kaye G.I., and Pawlina, W. *Histology: A Text and Atlas.* Fourth Edition. Lippincott Williams & Wilkins, 2003: Photographs appearing in Figs. 4.15 (upper panel), 12.4B, and 13.1.

Modified from Rubin, E. and Farber J.L. *Pathology.* Third Edition. Lippincott Williams & Wilkins, 1999: Clinical Application 36.3, and Photographs appearing in Figs. 15.4, 21.10A, 22.16C, and 40.5; and Clinical Applications 13.2, 25.2 (upper panel), and 32.3.

Modified from Siegel, A. and Sapru, H.N. *Essential Neuroscience.* Second Edition. Lippincott Williams & Wilkins, 2011: Figs. 6.6, 6.8, 11.11, 11.12, and 11.13.

From Tasman, W. and Jaeger, E. *The Wills Eye Hospital Atlas of Clinical Ophthalmology.* Second Edition. Lippincott Williams & Wilkins, 2007: Photographs appearing in Fig. 8.4 and Clinical Applications 8.1 and 12.2.

Modified from Taylor, C.R., Lillis C., LeMone, P., *et al. Fundamentals of Nursing, The Art And Science of Nursing Care.* Sixth Edition. Lippincott Williams & Wilkins, 2008: Fig 38.7 and Clinical Application 24.1.

From Uflacker, R. *Feigenbaum's Atlas of Vascular Anatomy: An Angiographic Approach.* Second Edition. Lippincott Williams & Wilkins, 2006: Photographs appearing in Figs. 25.4 and 25.6 (upper panel). Reprinted with permission from Sampaio, F.J.B.

From Yochum, T.R. and Rowe, L.J. *Yochum and Rowe's Essentials of Skeletal Radiology.* Third Edition. Lippincott Williams & Wilkins, 2004: Photographs appearing in Fig. 40.22A and Clinical Application 15.3 (lower).

Modified from West, J.B. *Best and Taylor's Physiological Basis of Medical Practice.* Twelfth Edition. Williams & Wilkins, 1991: Figs. 37.10 and 37.11.

Modified from West, J.B. *Respiratory Physiology—The Essentials.* Seventh Edition. Lippincott Williams & Wilkins, 2005: Figs. 22.1, 22.2, 22.13, 22.18, 22.21, 23.9, 23.10, 24.9, 24.11, 28.12, 28.14, 28.15, 28.16, and 28.17.

Index

Page numbers followed by "*f*" and "*t*" denotes figures and tables, respectively.

A

B

C

D

E

F

G

H

I

J

K

L

T

U

V

W

X

Y

Z

CCS1015